# AP® BIOLOGY 2

## Student Workbook

# AP® Biology 2

## Student Workbook

First edition 2012
Sixth printing (with corrections)

**ISBN 978-1-927173-12-1**


Published by **BIOZONE International Ltd**

Printed by REPLIKA PRESS PVT LTD using paper produced from renewable and waste materials

## About the Writing Team

Tracey Greenwood joined the staff of Biozone at the beginning of 1993. She has a Ph.D in biology, specializing in lake ecology, and taught undergraduate and graduate biology at the University of Waikato for four years.

Lissa Bainbridge-Smith worked in industry in a research and development capacity for eight years before joining Biozone in 2006. Lissa has an M.Sc from Waikato University.

Kent Pryor has a BSc from Massey University majoring in zoology and ecology. He was a secondary school teacher in biology and chemistry for 9 years before joining Biozone as an author in 2009.

Richard Allan has had 11 years experience teaching senior biology at Hillcrest High School in Hamilton, New Zealand. He attained a Masters degree in biology at Waikato University, New Zealand.



Purchases of this workbook may be made direct from the publisher:

www.theBIOZONE.com

**USA & CANADA:**

**BIOZONE** Corporation,
18801 E. Mainstreet
Suite 240, Parker, CO, 80134-3445
United States
Toll FREE phone: 1-855-246-4555
Toll FREE fax: 1-855-935-3555
Email: sales@thebiozone.com
Website: www.theBIOZONE.com

**UNITED KINGDOM & EUROPE:**

**BIOZONE** Learning Media (UK) Ltd.
Unit 5/6, Greenline Business Park,
Wellington Street, Burton-on-Trent,
DE14 2AS, United Kingdom
Telephone: +44 1283 530 366
Fax: +44 1283 530 961
Email: sales@biozone.co.uk
Website: www.BIOZONE.co.uk

**AUSTRALIA:**

**BIOZONE** Learning Media Australia
P.O. Box 2841, Burleigh BC,
QLD 4220,
Australia
Telephone: +61 7 5535 4896
Fax: +61 7 5508 2432
Email: sales@biozone.com.au
Website: www.BIOZONE.com.au

**NEW ZEALAND & REST OF WORLD:**

**BIOZONE** International Ltd.
P.O. Box 5002,
Hamilton 3242,
New Zealand
Telephone: +64 7-856 8104
Fax: +64 7-856 9243
Email: sales@biozone.co.nz
Website: www.theBIOZONE.com

# Contents



**Activity** is marked: ❑ to be done; ✓ when completed

# Contents



## ECOLOGY

### The Nature of Ecosystems



### Energy Flow and Nutrient Cycles



### Populations



### The Diversity and Stability of Ecosystems



**Activity** is marked: ☐• to be done; ☑ when completed

# Getting The Most From This Resource

This workbook is designed as a resource that will help to increase your understanding of the content and skills requirements of **AP Biology**, and reinforce and extend the ideas developed by your teacher. This workbook includes many useful features to help you locate activities and information relating to each of the big ideas and their enduring understandings.

### Constructing New Ideas: The Five Es

| | |
|---|---|
| Engage: | Object, event, or question used to engage students. |
| Explore: | Objects and phenomena are explored. |
| Explain: | Student explains their understanding of concepts and processes. |
| Elaborate: | Student can apply concepts in contexts, and build on or extend their understanding and skills. |
| Evaluate: | Students assess their knowledge, skills and abilities. |

### Features of the Section Concept Map

Chapter panels identify and summarize the essential knowledge covered within each chapter.

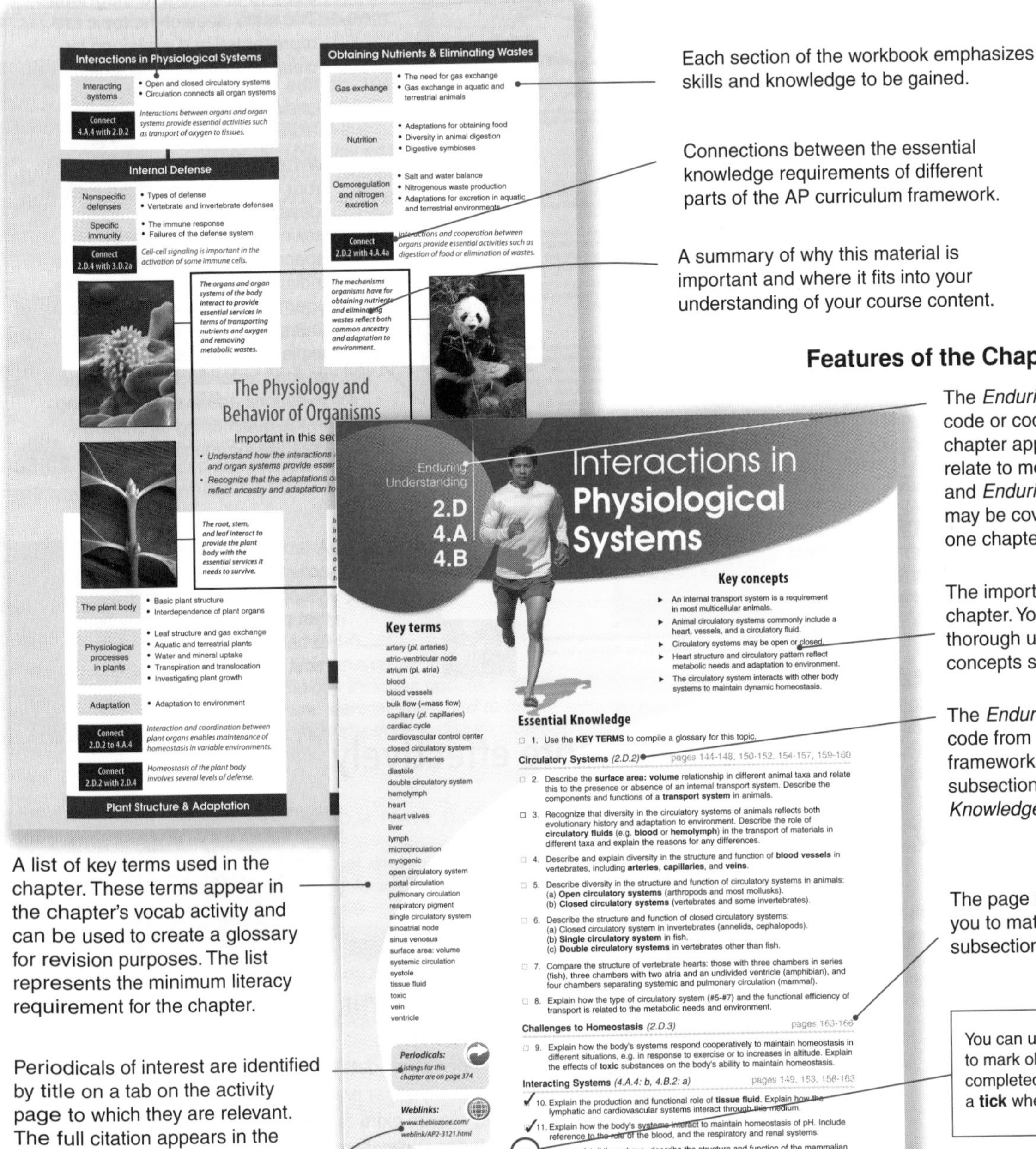

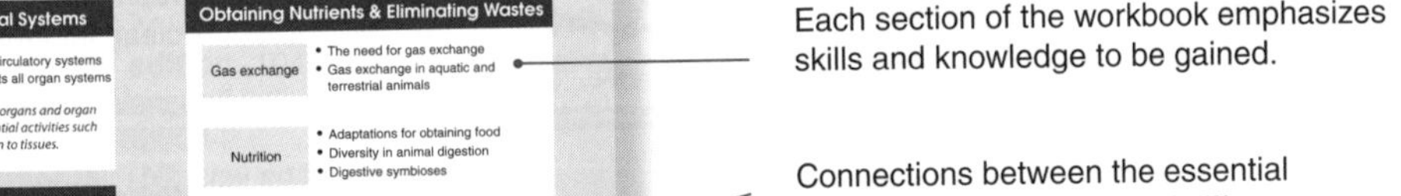


Each section of the workbook emphasizes skills and knowledge to be gained.

Connections between the essential knowledge requirements of different parts of the AP curriculum framework.

A summary of why this material is important and where it fits into your understanding of your course content.

### Features of the Chapter Topic Page

The *Enduring Understanding* code or codes to which this chapter applies. Some chapters relate to more than one big idea and *Enduring Understandings* may be covered over more than one chapter.

The important key ideas in this chapter. You should have a thorough understanding of the concepts summarized here.

The *Enduring Understanding* code from the AP curriculum framework is indicated for each subsection of the *Essential Knowledge* key points.

The page numbers direct you to material related to this subsection of work.

You can use the check boxes to mark objectives to be completed (a **dot** to be done; a **tick** when completed).

Essential knowledge provides a point by point summary of what you should have achieved by the end of the chapter.

A list of key terms used in the chapter. These terms appear in the chapter's vocab activity and can be used to create a glossary for revision purposes. The list represents the minimum literacy requirement for the chapter.

Periodicals of interest are identified by title on a tab on the activity page to which they are relevant. The full citation appears in the **Appendix** on the page indicated.

The Weblinks cited on many of the activity pages can be accessed through the web links page at: *www.thebiozone.com/weblink/AP2-3121.html*

See page 4 for more details.

**Student Review Series** provide color review slides for purchase. Download via the free BIOZONE App, available on the App Store.

## Energy in Living Systems

| | |
|---|---|
| Order in living systems | • Entropy vs order<br>• Energy inputs and outputs |
| ATP | • Energy transformations<br>• Use of ATP in living systems |
| Energy-yielding pathways | • Glycolysis and Krebs cycle<br>• Fermentation |
| Photosynthesis | • Photosynthesis vs chemosynthesis<br>• Capture of energy from light<br>• Production of carbohydrates |
| **Connect 2.A.2 with 4.A.2** | *Mitochondria and chloroplasts provide essential cell processes.* |

## Enzymes and Metabolism

| | |
|---|---|
| Enzyme structure | • Enzymes as globular proteins<br>• Enzyme shape and the active site |
| Enzyme function | • Activation energy and catalysis<br>• Enzyme reaction rates<br>• Substrate specificity |
| Cofactors and inhibitors | • Cofactors and coenzymes<br>• Enzymes inhibition<br>• Allosteric interactions |
| **Connect 4.B.1 with 3.D.3** | *Change in the structure of a molecular system may result in change in function.* |

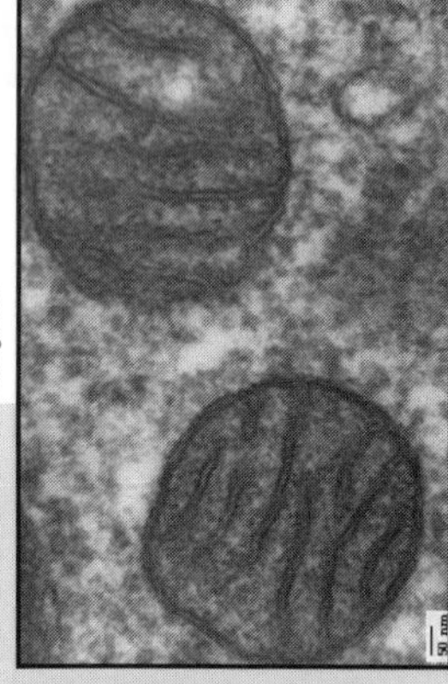
Loisa Howard: Dartmouth College PD

*Organisms capture and store free energy for use in biological processes, such as photosynthesis, that maintain or increase order.*

*The interactions between molecules, e.g. between enzyme, substrate, and required cofactors affect their structure and function.*

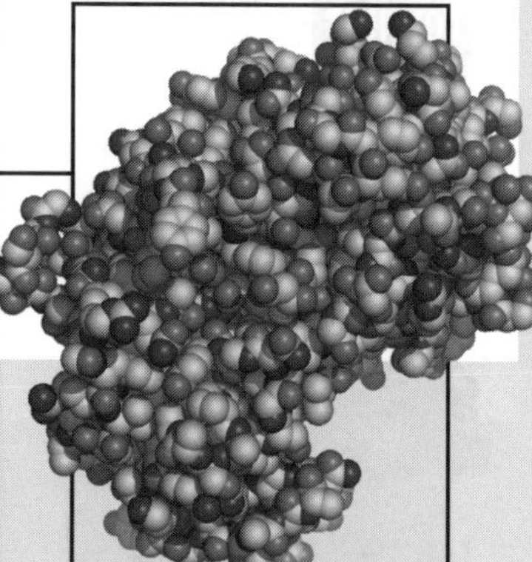

# Metabolism

### Important in this section ...

- *Understand how organisms allocate free energy.*
- *Understand how negative feedback mechanisms maintain dynamic homeostasis.*
- *Recognize the role of positive feedback in escalating specific biological responses.*

*Organisms use free energy to maintain dynamic homeostasis, and to grow and reproduce. Feedback mechanisms regulate these processes. How individuals invest in reproduction and self-maintenance defines their life history strategy.*

| | |
|---|---|
| Use of free energy | • Regulation of body temperature<br>• Reproductive strategies<br>• Size and metabolic rate<br>• Energy storage and growth |
| Feedback mechanisms | • Negative feedback: operons, thermoregulation, blood glucose<br>• Positive feedback: lactation, labor, fruit ripening<br>• Disorders in feedback regulation: diabetes, ADH disorders |
| **Connect 2.A.1 with 2.C.1** | *Feedback mechanisms are used to regulate growth and reproduction and maintain dynamic homeostasis.* |

## Homeostasis and Energy Allocation

Enduring Understanding

2.A
4.A

# Energy in Living Systems

## Key concepts

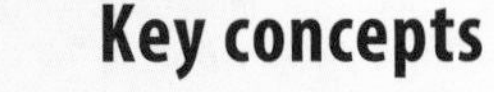

- ATP is the universal energy currency in cells.
- Cellular respiration and photosynthesis are important energy transformation processes.
- Both cellular respiration and photosynthesis involve the use of ATP and electron carriers.
- Cellular respiration involves the stepwise oxidation of glucose in the mitochondria.
- Photosynthesis uses light energy to fix carbon in organic compounds. It occurs in the chloroplasts.

## Key terms

absorption spectrum
accessory pigment
acetyl coA
action spectrum
anaerobic metabolism
ATP
autotroph
Calvin cycle
cellular respiration
chemiosmosis
chlorophyll
chloroplast
consumer
cyclic photophosphorylation
electron transport chain
entropy
exergonic reaction
fermentation
glycolysis
$H^+$ acceptor
heterotroph
Krebs cycle
light dependent reactions
light independent reactions
mitochondrion
NAD/NADP
non-cyclic photophosphorylation
oxidative phosphorylation
photolysis
photosynthesis
respiratory substrate
ribulose bisphosphate
substrate level phosphorylation
thylakoid discs

**Periodicals:**
*Listings for this chapter are on page 374*

**Weblinks:**
*www.thebiozone.com/weblink/AP2-3121.html*

*BIOZONE APP: Student Review Series Cellular Energetics*

## Essential Knowledge

☐ 1. Use the **KEY TERMS** to compile a glossary for this topic.

### Energy in Living Systems *(2.A.1: a-c, 2.A.2: a-b)* pages 10-17, 20, 30

☐ 2. Explain how order in biological systems is maintained by constant input of free energy. Explain what happens when there is loss of order or free energy flow. Distinguish between **autotrophs** and **heterotrophs** with respect to their source of free energy and carbon.

☐ 3. Using examples, explain how **exergonic reactions** are coupled with energetically unfavorable reactions to offset **entropy** in biological systems.

☐ 4. Using examples, describe how energy-related pathways in biological systems are sequential and may be entered at multiple points in the pathway.

### Energy Yielding Pathways *(2.A.2: b-c, f-g, 4.A.2: d)* pages 13-22

☐ 5. Explain the role of **ATP** in metabolism. Describe the synthesis of ATP and explain how it stores and releases its energy. Compare cellular respiration and photosynthesis as energy transformation processes.

☐ 6. Describe the structure and function of a mitochondrion. Identify the location of each step in glucose catabolism: **glycolysis**, **link reaction**, **Krebs cycle**, and **electron transport chain**.

☐ 7. Describe glycolysis and recognize it as the major anaerobic pathway in cells. State the net yield of ATP and $NADH_2$ from glycolysis.

☐ 8. Describe the complete oxidation of glucose to $CO_2$, including reference to:
- The conversion of pyruvate to acetyl-coenzyme A.
- The stepwise oxidation of intermediates in the Krebs cycle.
- Generation of ATP by **chemiosmosis** in the electron transport chain.
- The role of oxygen as the terminal electron acceptor.

☐ 9. Describe **fermentation** in mammalian muscle and in yeast, identifying the **$H^+$ acceptor** to each case. Compare and explain the differences in the yields of ATP from aerobic respiration (#8) and from fermentation.

### Photosynthesis *(2.A.2: a, c-e, 4.A.2: g)* pages 23-28

☐ 10. Describe the structure and role of **chloroplasts**. Explain the role of **chlorophyll** and **accessory pigments** in light capture by green plants. Explain what is meant by the **absorption spectrum** and **action spectrum** of pigments.

☐ 11. Describe and explain **photosynthesis** in a $C_3$ plant, including reference to:
- The generation of ATP and $NADPH_2$ in the **light dependent phase**.
- The **Calvin cycle** and the fixation of $CO_2$ using ATP and $NADPH_2$ in the **light independent phase**. Include reference to the reduction of GP and the regeneration of **ribulose bisphosphate**.

☐ 12. Describe and explain factors affecting **photosynthetic rate** and yield.

# Entropy and Order

Thermodynamics lays out the fundamental laws of energy that govern the universe. Firstly, the energy in an isolated system is constant. Secondly, disorder (**entropy**) increases over time. In other words, to maintain order within a system, there must be an input of energy. Living organisms follow these laws by using energy in the form of light (in plants), or food (in animals) to drive the chemical reactions that maintain order in their bodies. Without the input of energy, order is quickly lost and organisms die. The disorder in their constituent parts then continues to increase until it reaches an equilibrium with the environment.

## Entropy

**Entropy** is a measure of the disorder in a system, or the amount of energy not available to do work, and increases with time. The greater the entropy, the greater the disorder in the system.

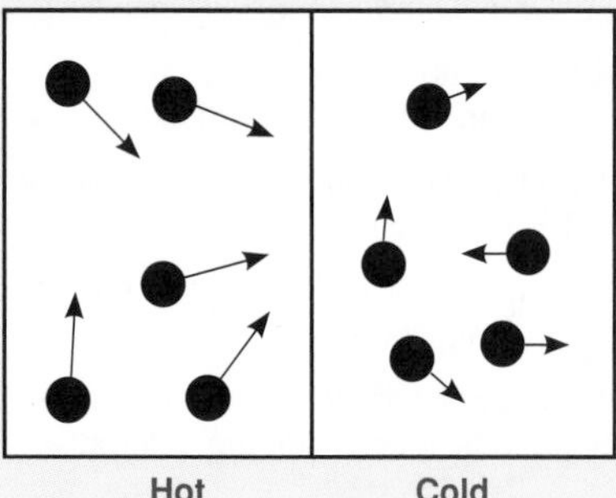



The system shown above has order. Hot, fast-moving molecules (high energy) are separated from cold, slow moving molecules (low energy).

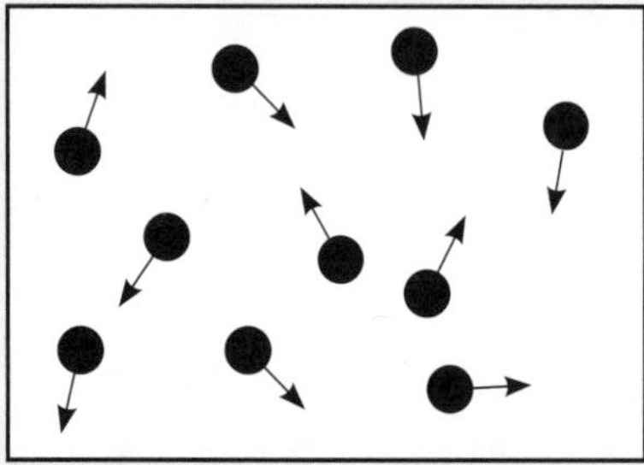

If the separation in the system is removed, the high energy molecules move into the area of low energy molecules. Collisions of the high energy molecules with the low energy molecules decrease the energy of the high energy molecules and increase the energy of the low energy molecules. Energy is also lost as heat. Eventually all molecules will have the same energy (equilibrium is reached). Entropy in the system has increased to its maximum level.

## Order and the Cell

Cells maintain order by using energy, either directly or indirectly. ATP can be used to directly drive a chemical reaction, or it can used to produce a gradient which is then used to move molecules into or out of the cell.

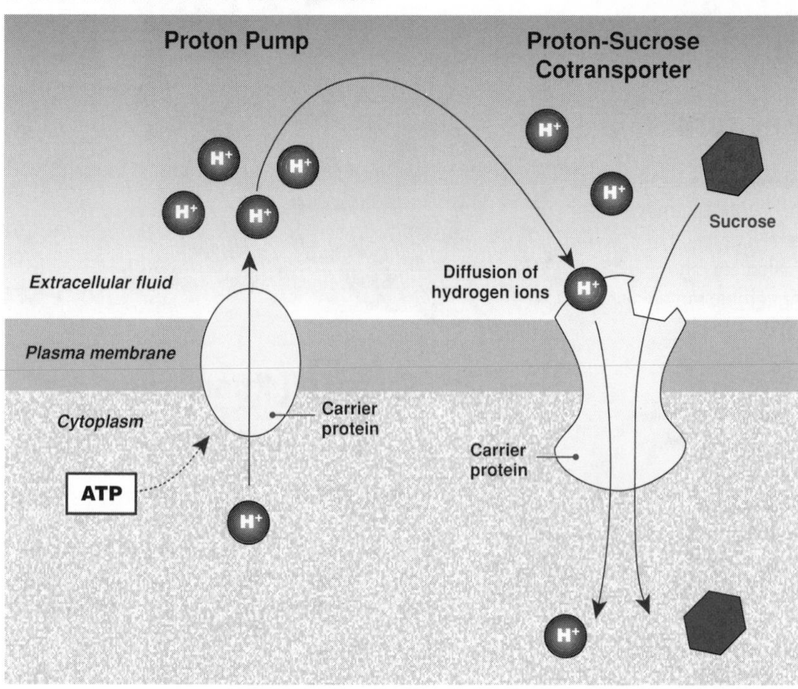



The proton-sucrose co-transporter shown above is common in the plasma membranes of plant cells. Energy in ATP is used to pump protons out of the cell, decreasing entropy with respect to the protons and producing a gradient. Sucrose is coupled to the flow of protons back down the gradient (increasing entropy with respect to the protons). Thus as the entropy with respect to the protons increases, the entropy with respect to the sucrose decreases.

## The Arrow of Time

Entropy provides an answer as to why time moves forward (never backwards) and why we age. The universe is (presumably) an isolated system, and its entropy can never decrease. Thus time is a result of the entropy of the universe moving towards its maximum state. Aging can be viewed as an increase in the body's entropy. Death is effectively the highest and therefore most thermodynamically favored level of entropy. So favorable, in fact, that no amount of energy input can prevent it.

1. (a) Define **entropy**: ___

   (b) Why has the entropy in the hot-cold system in the blue panel increased in the second image? ___

2. (a) Explain how living organisms maintain order in their cells: ___

   (b) How does the proton-sucrose co-transporter decrease entropy with respect to the sucrose inside and outside a cell.




ISBN: 978-1-927173-12-1

# Energy Inputs and Outputs

The way living things obtain their energy can be classified into two broad categories. The group upon which all others depend are called **autotrophs** (or **producers**). These organisms make their food from simple inorganic substances using the free energy in sunlight (**photoautotrophs**) or inorganic molecules such as sulfur (**chemoautotrophs**). **Heterotrophs** (consumers) feed on autotrophs or other heterotrophs to obtain their energy. The energy flow into and out of each trophic level in a food chain can be identified and represented diagrammatically using arrows of different sizes. The sizes of the arrows (see the diagrams below and on the next page) represent relative amounts of energy lost from different trophic levels.

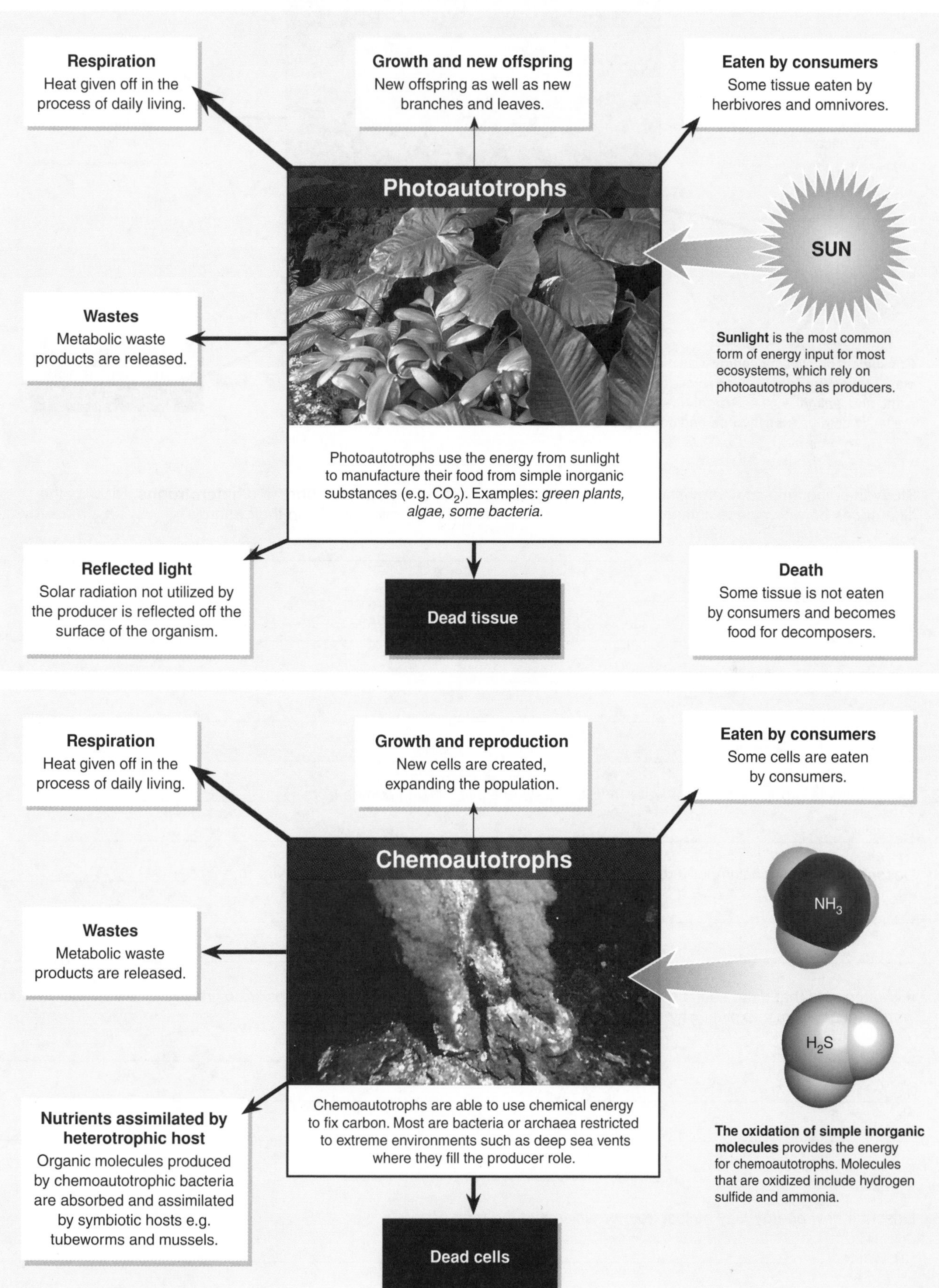



ISBN: 978-1-927173-12-1

**Respiration**
Heat given off in the process of daily living.

**Growth and new offspring**
New offspring as well as growth and weight gain.

**Eaten by carnivores**
Some tissue eaten by carnivores and omnivores.

**Heterotrophs**

**Death**
Some tissue is not eaten by other consumers and becomes food for detritivores and decomposers.

**Wastes**
Metabolic waste products are released (e.g. as urine, feces, carbon dioxide).

Heterotrophs rely on other living organisms or organic particulate matter for their energy. Heterotrophs include herbivores, carnivores, detritivores, and decomposers. Examples: *animals, some protists, some bacteria*

**Food**
Consumers obtain their energy from a variety of sources: plant tissues (**herbivores**), animal tissues (**carnivores**), plant and animal tissues (**omnivores**), dead organic matter or detritus (**detritivores** and **decomposers**).

**Dead tissue**

1. Study the diagrams on energy flow relating to **photoautotrphs**, **chemoautotrophs**, and **heterotrophs**. Discuss the differences between these categories of organisms with respect to how they obtain their energy:

2. Many animals group together for warmth. Explain where this warmth comes from:

3. Describe the ecological importance of chemoautotrophic organisms in deep sea environments:

4. It is often said that deep sea hydrothermal vent communities have no requirement for or no link to the rest of the outside world's ecosystems. Explain why this is not the case:

5. Describe how energy may be lost from organisms in the form of:

(a) Wastes:

(b) Respiration:

ISBN: 978-1-927173-12-1

# The Role of ATP in Cells

All organisms require energy for their metabolism. The universal energy carrier for the cell is the molecule ATP (**adenosine triphosphate**). ATP transports chemical energy within the cell for use in metabolic processes such as biosynthesis, cell division, cell signaling, thermoregulation, cell motility, and active transport.

The hydrolysis of ATP is catalyzed by the enzyme ATPase. Once hydrolyzed, ATP becomes ADP (adenosine diphosphate), which has less stored energy than ATP. ATP can be reformed from ADP and Pi using energy from the controlled breakdown of respiratory substrates (commonly glucose) in cellular respiration.

## Adenosine Triphosphate (ATP)

The ATP molecule consists of three components; a purine base (**adenine**), a pentose sugar (**ribose**), and **three phosphate groups** which attach to the 5' carbon of the pentose sugar. The three dimensional structure of ATP is described below.

The bonds between the phosphate groups contain electrons in a high energy state which store a large amount of energy. The energy is released during ATP hydrolysis. Typically, hydrolysis is coupled to another cellular reaction to which the energy is transferred. The end products of the reaction are adenosine diphosphate (ADP) and an inorganic phosphate (Pi).

Note that energy is released during the formation of bonds during the hydrolysis reaction, not the breaking of bonds between the phosphates (which requires energy input).

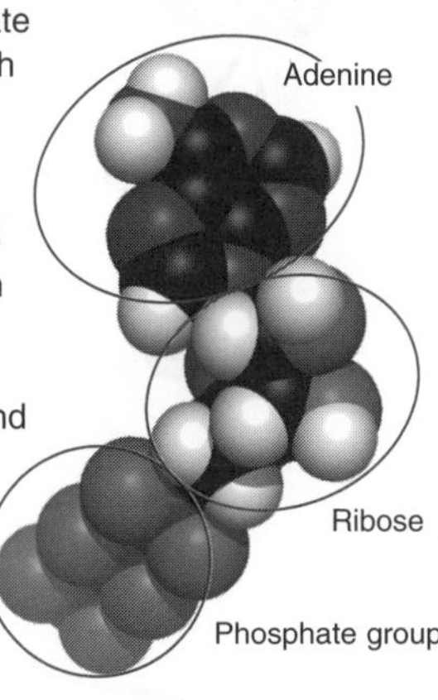


## The Mitochondrion

Cellular respiration and ATP production occur in the mitochondria. A mitochondrion is bound by a double membrane. The inner and outer membranes are separated by an intermembrane space, compartmentalizing the regions where the different reactions of cellular respiration take place.

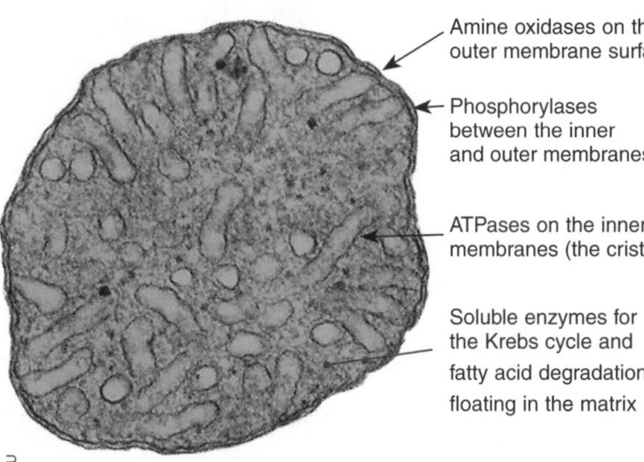


## ATP Powers Metabolism

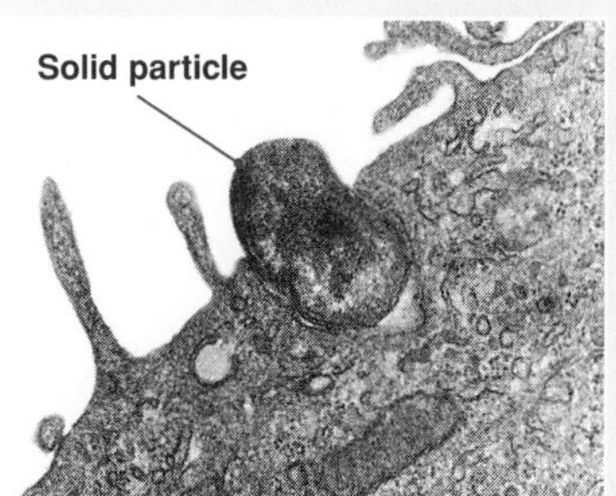


The energy released from the hydrolysis of ATP is used to actively transport molecules and substances across the cellular membrane. **Phagocytosis** (left), which involves the engulfment of solid particles, is one such example.

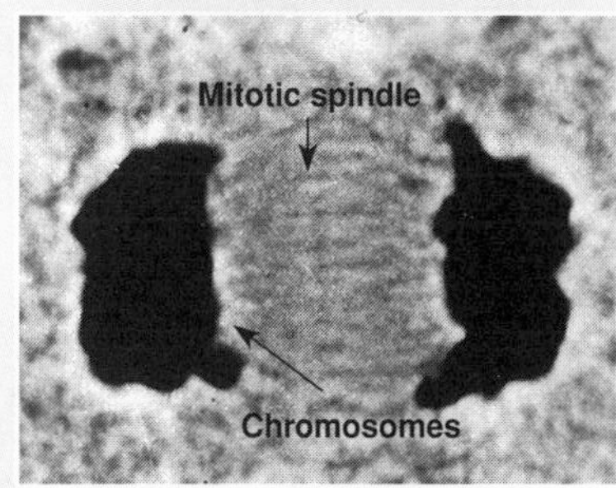


Cell division (mitosis), as observed in this onion cell, requires ATP to proceed. Formation of the mitotic spindle and chromosome separation are two aspects of cell division which require energy from ATP hydrolysis to occur.

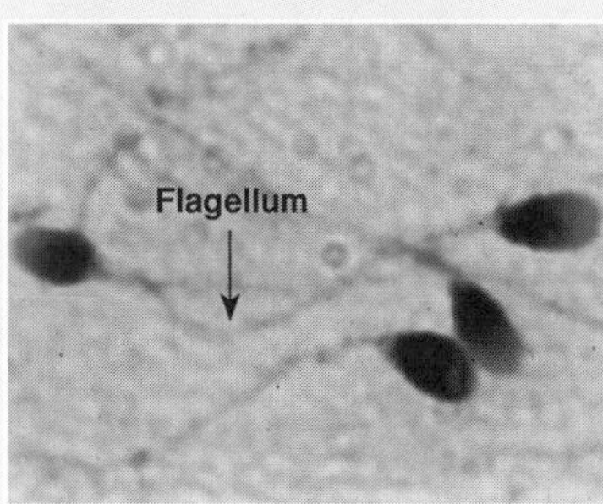


The hydrolysis of ATP provides the energy for motile cells to achieve movement via a tail-like structure called a flagellum. For example, mammalian sperm must be able to move to the ovum to fertilize it.

The maintenance of body temperature requires energy. To maintain body heat, muscular activity increases (e.g. shivering, erection of body hairs). Cooling requires expenditure of energy too. For example, sweating is an energy requiring process involving secretion from glands in the skin.

1. In which organelle is ATP produced in the cell? ______

2. Which enzyme catalyzes the hydrolysis of ATP? ______

3. (a) Explain how ATP is involved in thermoregulation: ______

   (b) Explain how ATP is involved in cell motility: ______

4. (a) How does ATP supply energy to power metabolism? ______

## How does ATP provide energy?

ATP releases its energy during hydrolysis. Water is split and added to the terminal phosphate group resulting in ADP and Pi. For every mole of ATP hydrolyzed, **30.7 kJ** of energy is released. Note that energy is released during the formation of chemical bonds not from the breaking of chemical bonds.



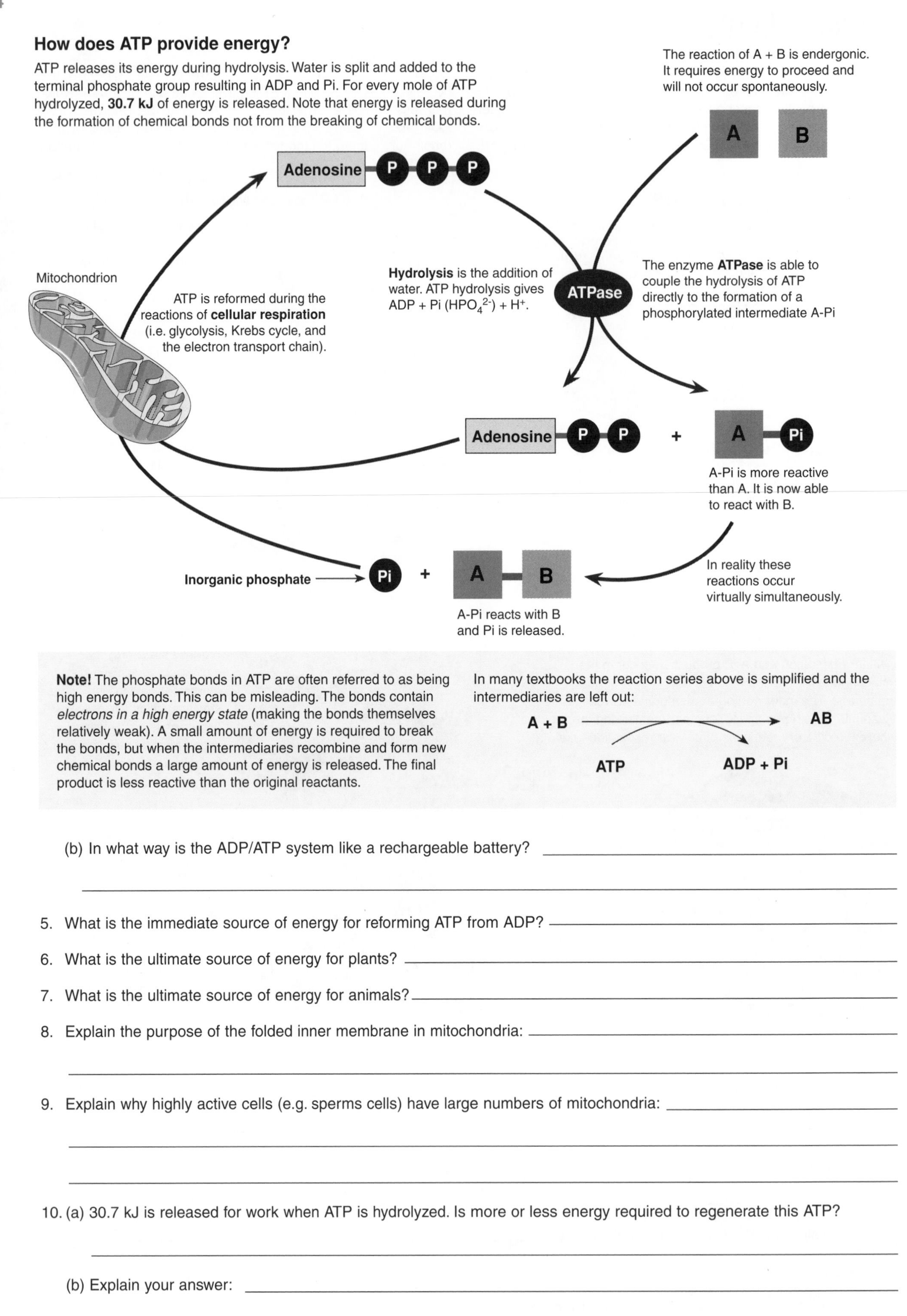

**Note!** The phosphate bonds in ATP are often referred to as being high energy bonds. This can be misleading. The bonds contain *electrons in a high energy state* (making the bonds themselves relatively weak). A small amount of energy is required to break the bonds, but when the intermediaries recombine and form new chemical bonds a large amount of energy is released. The final product is less reactive than the original reactants.

In many textbooks the reaction series above is simplified and the intermediaries are left out:



(b) In what way is the ADP/ATP system like a rechargeable battery? ______

5. What is the immediate source of energy for reforming ATP from ADP? ______

6. What is the ultimate source of energy for plants? ______

7. What is the ultimate source of energy for animals? ______

8. Explain the purpose of the folded inner membrane in mitochondria: ______

9. Explain why highly active cells (e.g. sperms cells) have large numbers of mitochondria: ______

10. (a) 30.7 kJ is released for work when ATP is hydrolyzed. Is more or less energy required to regenerate this ATP?

______

(b) Explain your answer: ______

ISBN: 978-1-927173-12-1

# Energy Transformations in Cells

A summary of the flow of energy within a plant cell is illustrated below. Animal cells have a similar flow except the glucose is supplied by ingestion rather than by photosynthesis. The energy not immediately stored in chemical bonds is lost as heat. Note the role of ATP; it is made in cellular respiration and provides the energy for metabolic reactions, including photosynthesis.

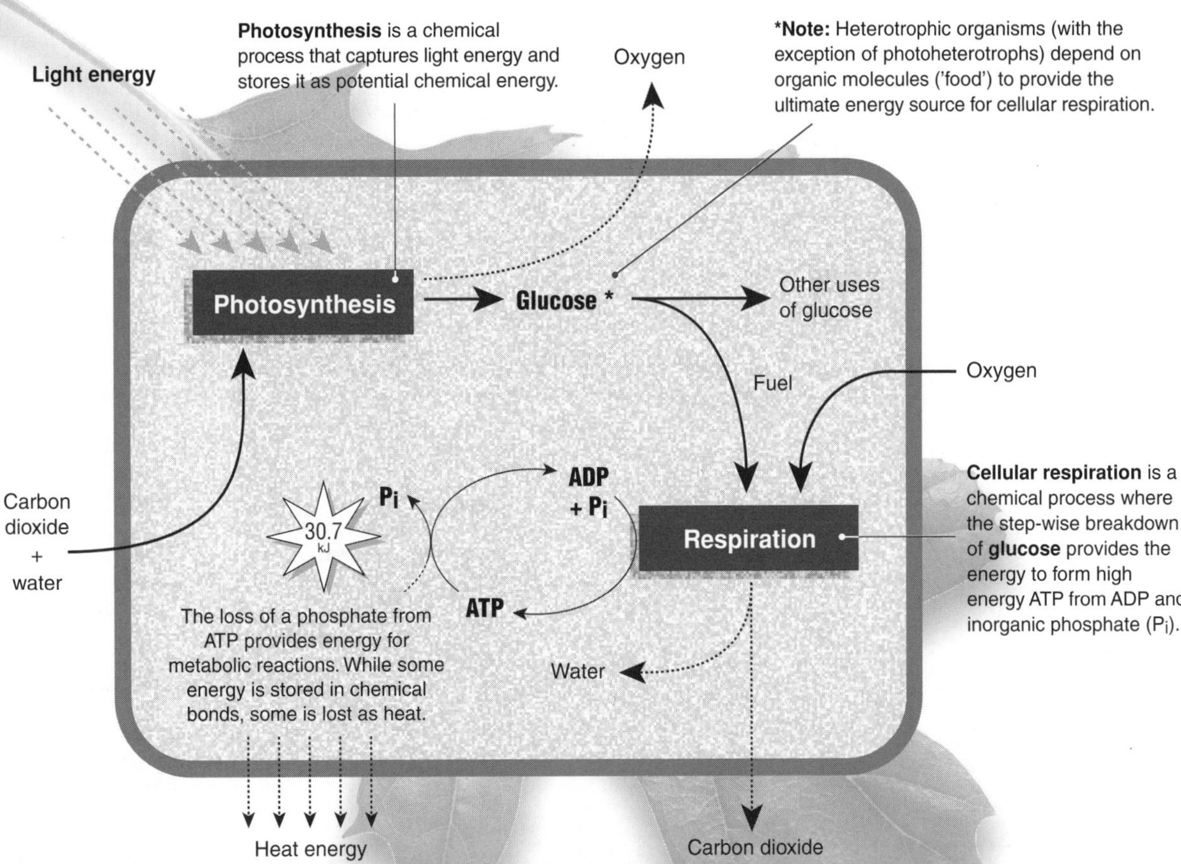




1. Discuss the differences between **heterotrophs, photosynthetic autotrophs**, and **chemosynthetic autotrophs** with respect to how these organisms derive their source of energy for metabolism:

2. In 1977, scientists working near the Galapagos Islands in the equatorial eastern Pacific found warm water spewing from cracks in the mid-oceanic ridges 2600 metres below the surface. Clustered around these hydrothermal vents were strange and beautiful creatures new to science. The entire community depends on sulfur-oxidizing bacteria that use hydrogen sulfide dissolved in the venting water as an energy source to manufacture carbohydrates. This process is similar to photosynthesis, but does not rely on sunlight to provide the energy for generating ATP and fixing carbon:

(a) Explain why a community based on photosynthetic organisms is not found at this site:

(b) Name the ultimate energy source for the bacteria:

(c) This same chemical that provides the bacteria with energy is also toxic to the process of cellular respiration; a problem that the animals living in the habitat have resolved by evolving various adaptations. Explain what would happen if these animals did not possess adaptations to reduce the toxic effect on cellular respiration:

(d) Name the energy source classification for these sulfur-oxidizing bacteria:

ISBN: 978-1-927173-12-1


*Related activities:* *The Role of ATP in Cells*

# ATP Production in Cells

Glycolysis and cellular respiration are the processes by which organisms break down energy rich molecules (e.g. glucose) to release the energy in a useable form (ATP). All living cells respire in order to exist, although the substrates they use may vary. **Aerobic respiration** requires oxygen. Some plants and animals can generate ATP using anaerobic pathways for short periods of time but the ATP yields from these pathways are low. Other organisms (bacteria) use only **anaerobic respiration** and live in oxygen-free environments. These organisms use some other final electron acceptor rather than oxygen (e.g. nitrate or $Fe^{2+}$).

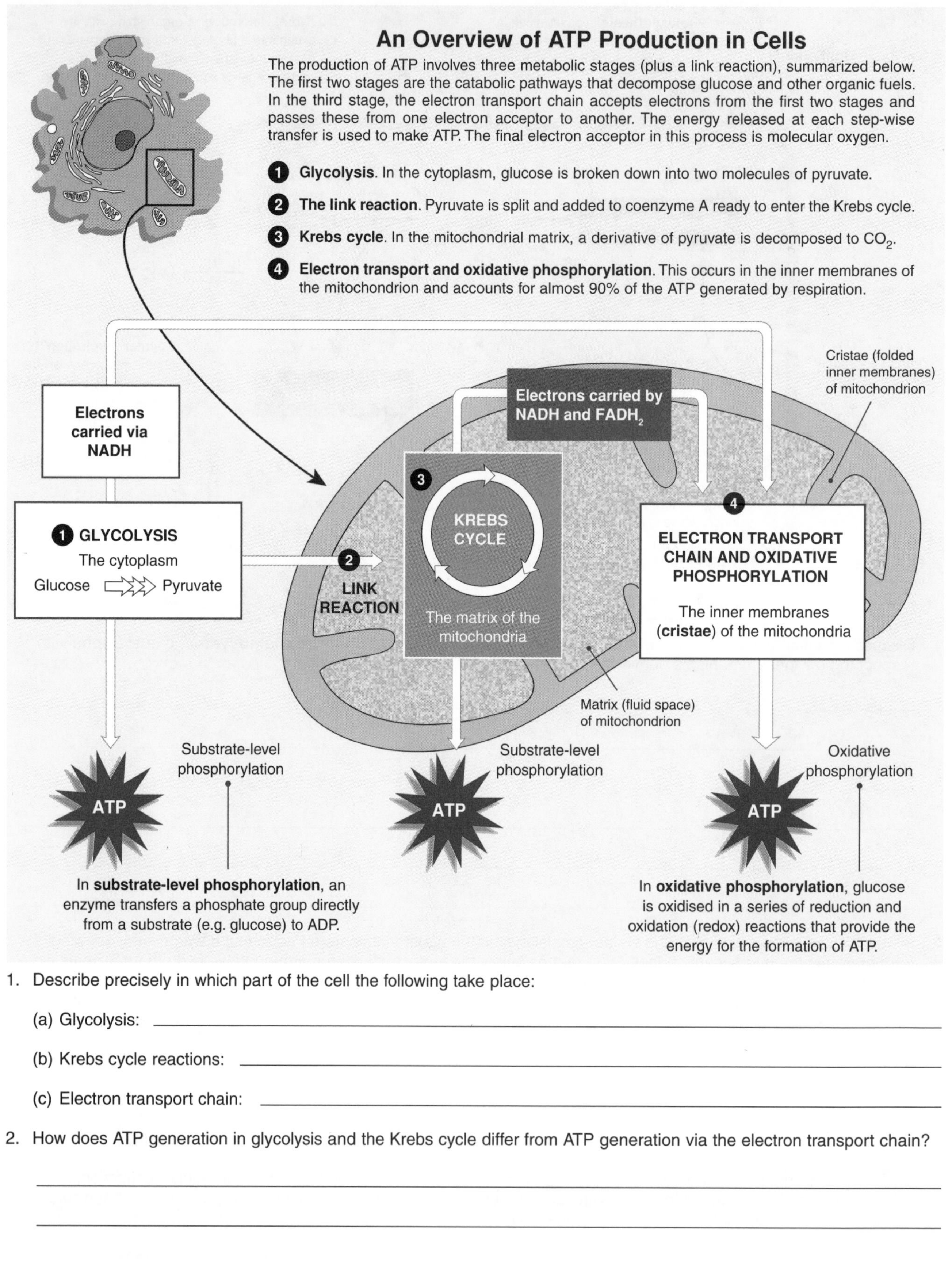


1. Describe precisely in which part of the cell the following take place:

   (a) Glycolysis: ______________________

   (b) Krebs cycle reactions: ______________________

   (c) Electron transport chain: ______________________

2. How does ATP generation in glycolysis and the Krebs cycle differ from ATP generation via the electron transport chain?

______________________

______________________

______________________

______________________



**Related activities:** *Energy Transformations in Cells*

***Periodicals:***
*Fuelled for life*


**ISBN: 978-1-927173-12-1**

# The Biochemistry of Respiration

**Glycolysis** and **cellular respiration** are catabolic, energy yielding pathways. The breakdown of glucose and other organic fuels (such as fats and proteins) to simpler molecules releases energy for the synthesis of ATP. All living cells respire in order to exist, although the substrates they use may vary. Glycolysis is the first stage of glucose catabolism and is anaerobic. The reactions of cellular respiration oxidize the end product of glycolysis (**pyruvate**) in multiple steps to produce ATP. ATP generation in respiration is coupled to the movement of hydrogen ions down their electrochemical gradient via the membrane-bound enzyme ATP synthase. The final electron acceptor in this electron transport pathway is oxygen.

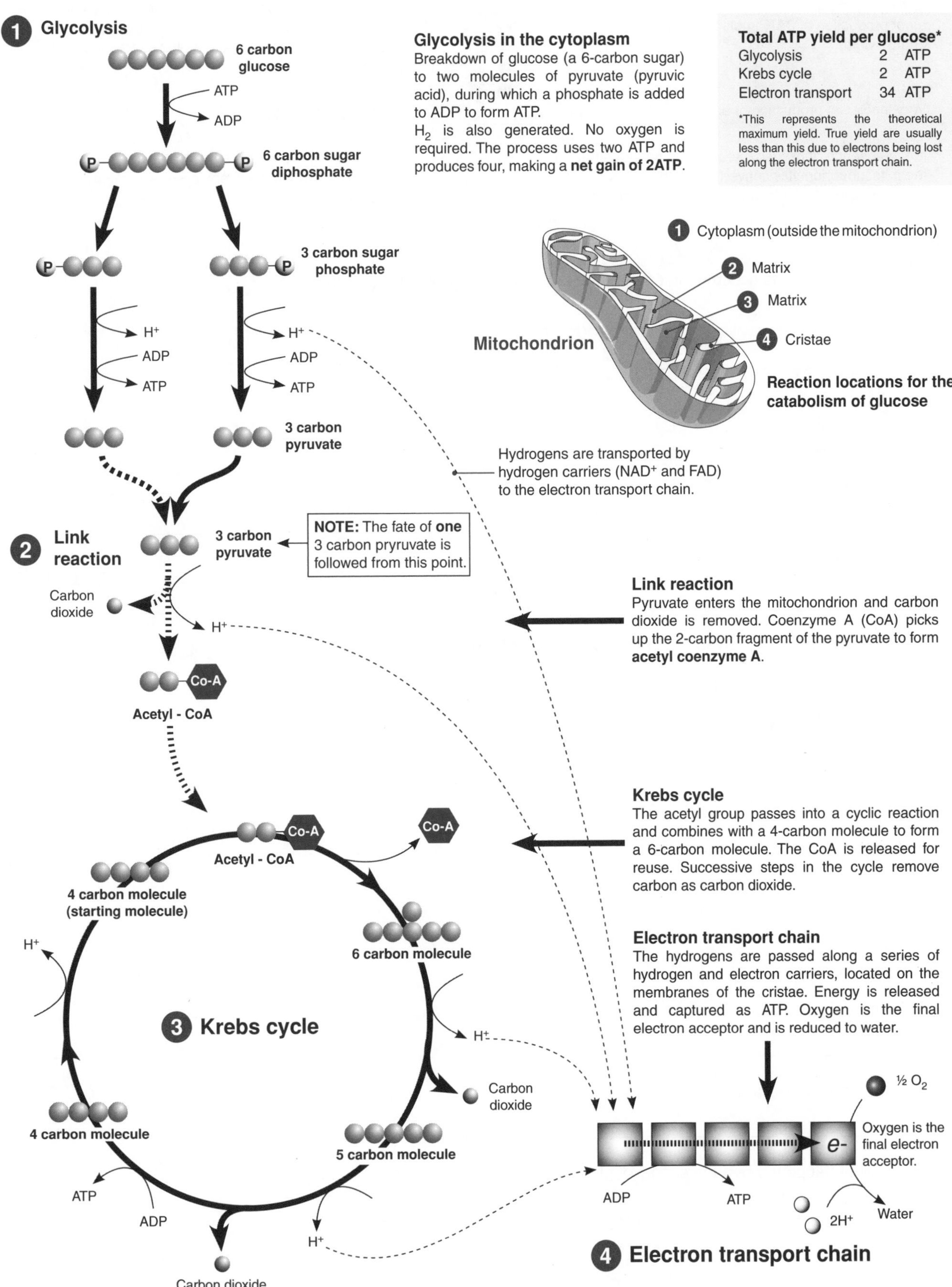





ISBN: 978-1-927173-12-1

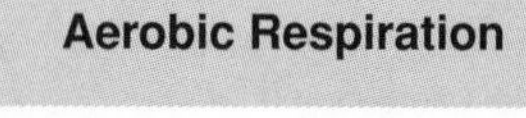

| Glycolysis and Fermentation | Aerobic Respiration |
|---|---|
| Occurs in the cytoplasm of the cell. | Occurs in the mitochondria of eukaryotic cells. |
| Oxygen is not required. | Oxygen is required. |
| Glycolysis: glucose is converted to pyruvic acid with a net production of 2ATP molecules (a very low energy yield). | Pyruvic acid is converted to carbon dioxide, water, and a further 36 ATP molecules from the Krebs cycle and electron transport chain (a high energy yield per glucose molecule). |
| Very inefficient production of energy. | An energy efficient process. |
| In the absence of oxygen, pyruvic acid cannot enter the mitochondrion. It is converted to ethanol and carbon dioxide in plants, and lactic acid in animals. | When oxygen is present, pyruvic acid can be oxidized further via the Krebs cycle and electron transport chain. |

1. Summarize the events occurring in each of the following stages of cellular respiration:

   (a) Glycolysis: ______________________________

   (b) Link reaction: ______________________________

   (c) Krebs cycle: ______________________________

   (d) Electron transport chain: ______________________________

2. Determine how many ATP molecules **per molecule of glucose** are generated during the following stages of respiration:

   (a) Glycolysis: ________ (b) Krebs cycle: ________ (c) Electron transport chain: ________ (d) Total: ________

3. Explain what happens to the carbon atoms lost during respiration: ______________________________

4. Describe the role of each of the following in cellular respiration:

   (a) Hydrogen atoms: ______________________________

   (b) NAD and FAD: ______________________________

   (c) Oxygen: ______________________________

   (d) Acetyl coenzyme A: ______________________________

5. Explain what happens when the supply of glucose for cellular respiration is limited: ______________________________


ISBN: 978-1-927173-12-1

# Chemiosmosis

**Chemiosmosis** is the process in which electron transport is coupled to ATP synthesis. It takes place in the membranes of the mitochondria of all eukaryotic cells, the chloroplasts of plants, and across the plasma membrane of bacteria. Chemiosmosis involves the establishment of a proton (hydrogen) gradient across biological membranes (shown below for cellular respiration). The concentration gradient is a form of potential energy, which in chemiosmosis is used to generate ATP. Chemiosmosis has two key components: an **electron transport chain** (ETC) sets up a proton gradient as electrons pass along it to a final electron acceptor, and an enzyme called **ATP synthase** which uses the proton gradient to catalyze ATP synthesis In cellular respiration, electron transport carriers on the inner membrane of the mitochondrion oxidize NADH + $H^+$ and $FADH_2$. Energy from this process forces protons to move, against their concentration gradient, from the mitochondrial matrix into the space between the two membranes. The protons then flow back into the matrix via ATP synthase molecules in the membrane. As the protons flow down their concentration gradient, energy is released and ATP is synthesized. In the chloroplasts of green plants, the process is similar as ATP is produced when protons pass from the thylakoid lumen to the chloroplast stroma via ATP synthase.



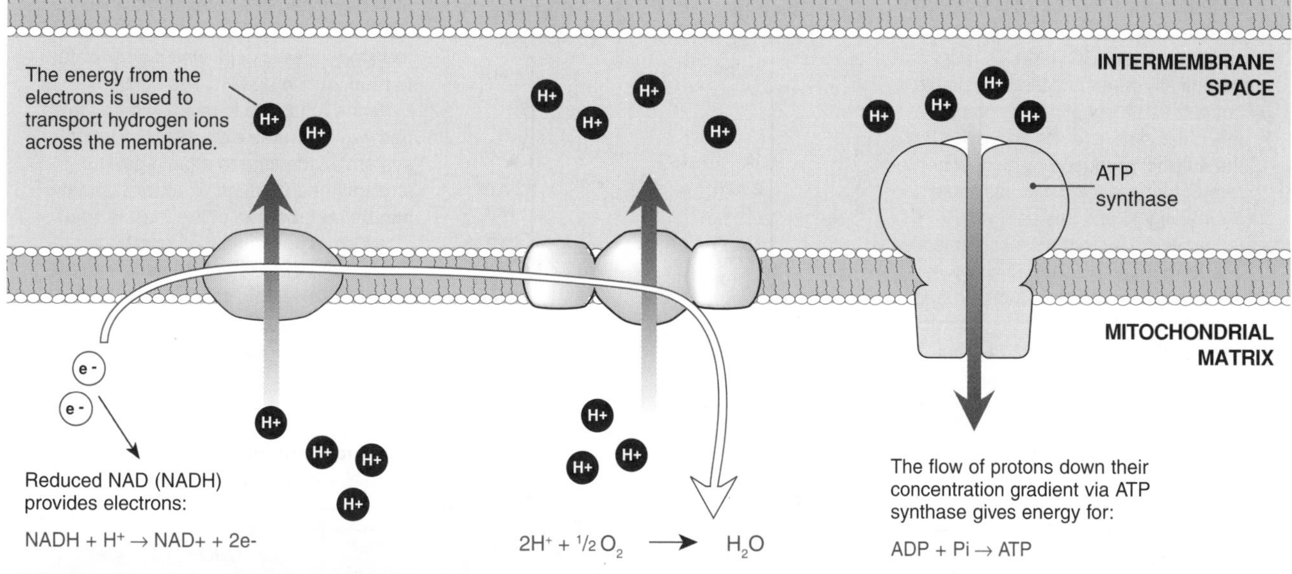


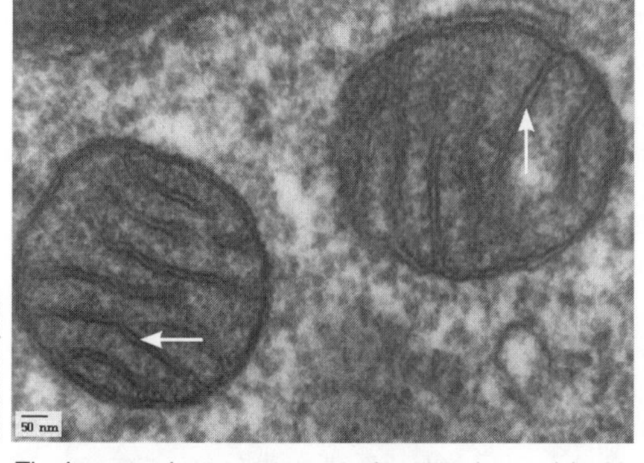

*The intermembrane spaces can be seen (arrows) in this transverse section of miotchondria.*

### The Evidence for Chemiosmosis

The British biochemist Peter Mitchell proposed the chemiosmotic hypothesis in 1961. He proposed that, because living cells have membrane potential, electrochemical gradients could be used to do work, i.e. provide the energy for ATP synthesis. Scientists at the time were skeptical, but the evidence for chemiosmosis was extensive and came from studies of isolated mitochondria and chloroplasts. Evidence included:

- The outer membranes of mitochondria were removed leaving the inner membranes intact. Adding protons to the treated mitochondria increased ATP synthesis.
- When isolated chloroplasts were illuminated, the medium in which they were suspended became alkaline.
- Isolated chloroplasts were kept in the dark and transferred first to a low pH medium (to acidify the thylakoid interior) and then to an alkaline medium (low protons). They then spontaneously synthesized ATP (no light was needed).

1. Briefly summarize the process of chemiosmosis: ______

2. Explain why adding protons to the treated mitochondria increased ATP synthesis: ______

3. Explain why the suspension of isolated chloroplasts became alkaline when illuminated? ______

4. (a) Explain the purpose of transferring the chloroplasts first to an acid then to an alkaline medium: ______

   (b) Explain why ATP synthesis occurred spontaneously in these treated chloroplasts: ______

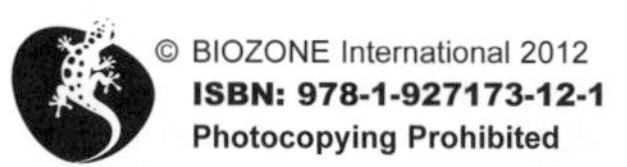

**ISBN: 978-1-927173-12-1**

# Anaerobic Pathways for ATP Production

All organisms can metabolize glucose anaerobically (without oxygen) using glycolysis in the cytoplasm, but the energy yield from this process is low and few organisms can obtain sufficient energy this way. In the absence of oxygen, glycolysis soon stops unless there is an alternative acceptor for the electrons produced from the glycolytic pathway. In yeasts and the root cells of higher plants this acceptor is ethanal, and the pathway is called alcoholic fermentation. In the skeletal muscle of mammals, the acceptor is pyruvate itself and the end product is lactic acid. In the case of lactic acid fermentation, lactate can concert back to pyruvate to provide a fuel for intense activity. Although fermentation is often used synonymously with anaerobic respiration, they are not the same. Respiration always involves hydrogen ions passing down a chain of carriers to a terminal acceptor, and this does not occur in fermentation. In anaerobic respiration, the terminal $H^+$ acceptor is a molecule other than oxygen, e.g. $Fe^{2+}$ or nitrate.

## Alcoholic Fermentation

In alcoholic fermentation, the $H^+$ acceptor is ethanal which is reduced to ethanol with the release of carbon dioxide ($CO_2$). Yeasts respire aerobically when oxygen is available but can use alcoholic fermentation when it is not. At levels above 12-15%, the ethanol produced by alcoholic fermentation is toxic and this limits their ability to use this pathway indefinitely. The root cells of plants also use fermentation as a pathway when oxygen is unavailable but the ethanol must be converted back to respiratory intermediates and respired aerobically.

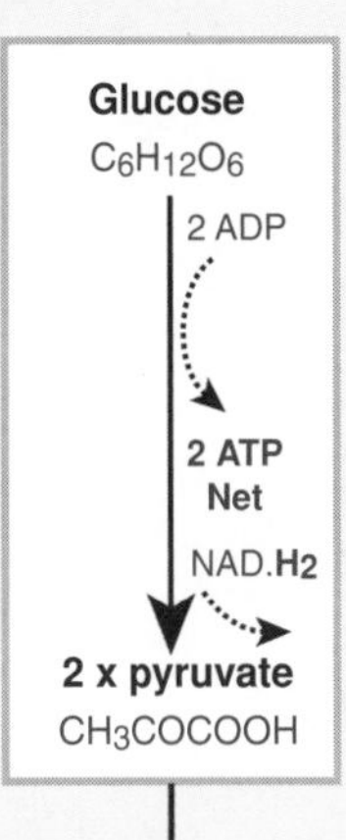


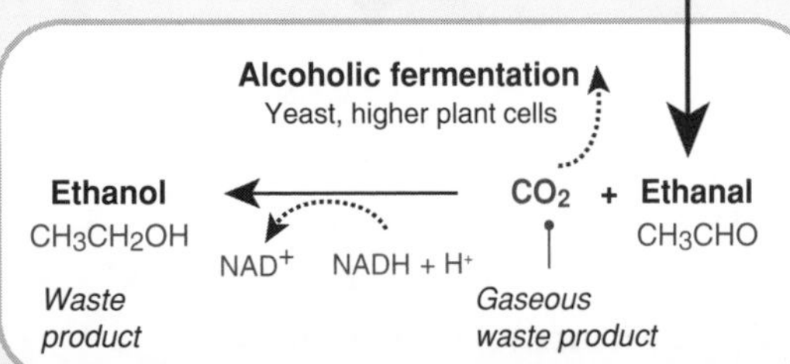


## Lactic Acid Fermentation

Skeletal muscles produce ATP in the absence of oxygen using lactic acid fermentation. In this pathway, pyruvate is reduced to lactic acid, which dissociates to form lactate and $H^+$. The conversion of pyruvate to lactate is reversible and this pathway operates alongside the aerobic system all the time to enable greater intensity and duration of activity. Lactate can be metabolized in the muscle itself or it can enter the circulation and be taken up by the liver to replenish carbohydrate stores. This 'lactate shuttle' is an important mechanism for balancing the distribution of substrates and waste products.

Glucose
$C_6H_{12}O_6$
2 ADP
2 ATP Net
NAD.H2
2 x pyruvate
$CH_3COCOOH$

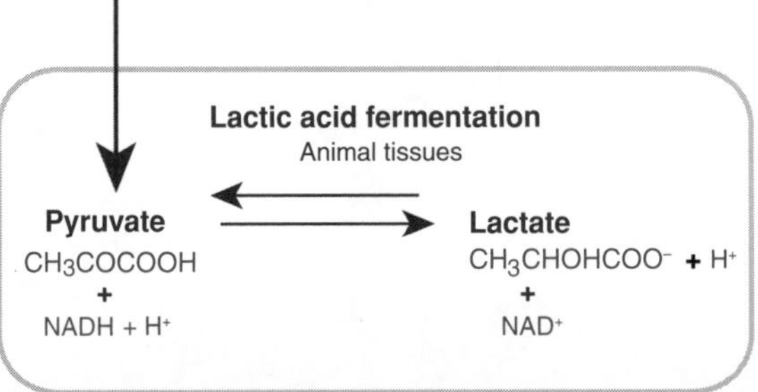


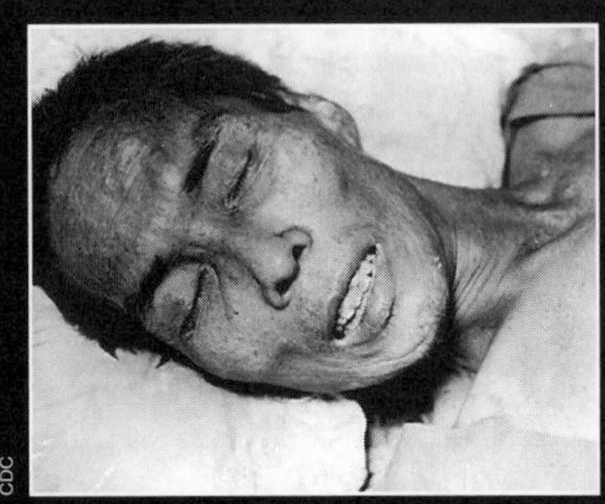

Some organisms respire only in the absence of oxygen and are known as obligate anaerobes. Many of these organisms are bacterial pathogens and cause diseases such as tetanus (above), gangrene, and botulism.

The lactate shuttle in vertebrate skeletal muscle works alongside the aerobic system to enable maximal muscle activity. Lactate moves from its site of production to regions within and outside the muscle where it can be respired aerobically.

The products of alcoholic fermentation have been utilized by humans for centuries. The alcohol and carbon dioxide produced from this process form the basis of the brewing and baking industries.

1. Describe the key difference between aerobic respiration and fermentation: ____________________

____________________

2. (a) Refer to the previous activity and determine the efficiency of fermentation compared to aerobic respiration: ______ %

   (b) Why is the efficiency of these anaerobic pathways so low? ____________________

   ____________________

3. Why can't alcoholic fermentation go on indefinitely? ____________________

____________________

____________________



*Related activities:* *The Biochemistry of Respiration*
*Weblinks:* *Lactate and Alcoholic Fermentation*


ISBN: 978-1-927173-12-1

# Investigating Yeast Fermentation

Any practical investigation requires you to critically evaluate your results in the light of your own hypothesis and your biological knowledge. A critical evaluation of any study involves analyzing, presenting, and discussing the results, as well as accounting for any deficiencies in your procedures and erroneous results. This activity describes an experiment comparing different carbohydrates for their effectiveness as substrates for fermentation. Brewer's yeast is a **facultative anaerobe** (meaning it can respire aerobically or use fermentation). It will preferentially use alcoholic fermentation when sugars are in excess. One would expect glucose to be the preferred substrate, as it is the starting molecule in cellular respiration, but yeast are capable of utilizing a variety of sugars, including disaccharides that can be broken down into single units. Completing this activity, which involves a critical evaluation of the second-hand data provided, will help to prepare you for your own evaluative task.



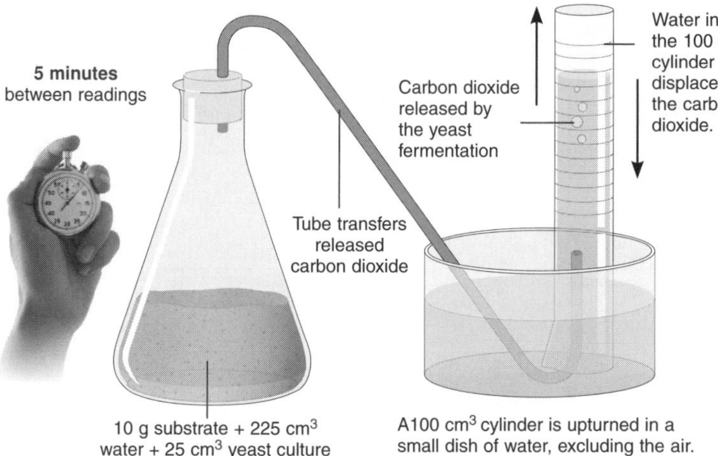


### The Apparatus

In this experiment, all substrates tested used the same source culture of 30 g active yeast dissolved in 150 cm$^3$ of room temperature (24°C) tap water. For each substrate, 25 g of the substrate to be tested was added to 225 cm$^3$ room temperature (24°C) tap water buffered to pH 4.5. Then 25 cm$^3$ of source culture was added to the test solution. The control contained yeast solution but no substrate:

| Substrate / Time (min) | **Group 1**: Volume of carbon dioxide collected (cm$^3$) | | | | |
|---|---|---|---|---|---|
| | **None** | **Glucose** | **Maltose** | **Sucrose** | **Lactose** |
| 0 | 0 | 0 | 0 | 0 | 0 |
| 5 | 0 | 0 | 0.8 | 0 | 0 |
| 10 | 0 | 0 | 0.8 | 0 | 0 |
| 15 | 0 | 0 | 0.8 | 0.1 | 0 |
| 20 | 0 | 0.5 | 2.0 | 0.8 | 0 |
| 25 | 0 | 1.2 | 3.0 | 1.8 | 0 |
| 30 | 0 | 2.8 | 3.6 | 3.0 | 0.5 |
| 35 | 0 | 4.2 | 5.4 | 4.8 | 0.5 |
| 40 | 0 | 4.6 | 5.6 | 4.8 | 0.5 |
| 45 | 0 | 7.4 | 8.0 | 7.2 | 1.0 |
| 50 | 0 | 10.8 | 8.9 | 7.6 | 1.3 |
| 55 | 0 | 13.6 | 9.6 | 7.7 | 1.3 |
| 60 | 0 | 16.1 | 10.4 | 9.6 | 1.3 |
| 65 | 0 | 22.0 | 12.1 | 10.2 | 1.8 |
| 70 | 0 | 23.8 | 14.4 | 12.0 | 1.8 |
| 75 | 0 | 26.7 | 15.2 | 12.6 | 2.0 |
| 80 | 0 | 32.5 | 17.3 | 14.3 | 2.1 |
| 85 | 0 | 37.0 | 18.7 | 14.9 | 2.4 |
| 90 | 0 | 39.9 | 21.6 | 17.2 | 2.6 |

| Substrate / Time (min) | **Group 2**: Volume of carbon dioxide collected (cm$^3$) | | | | |
|---|---|---|---|---|---|
| | **None** | **Glucose** | **Maltose** | **Sucrose** | **Lactose** |
| 90 | 0 | 24.4 | 19.0 | 17.5 | 0 |

### The Aim

To investigate the suitability of different mono- and disaccharide sugars as substrates for alcoholic fermentation in yeast.

### Background

The rate at which brewer's or baker's yeast (*Saccharomyces cerevisiae*) metabolizes carbohydrate substrates is influenced by factors such as temperature, solution pH, and type of carbohydrate available. The literature describes yeast metabolism as optimal in warm, slightly acid environments. High levels of sugars suppress aerobic respiration in yeast, so yeast will preferentially use the fermentation pathway in the presence of excess substrate.

**Substrates**: Glucose is a monosaccharide, maltose (glucose-glucose), sucrose (glucose-fructose), and lactose (glucose-galactose) are disaccharides.

1. Write the equation for the fermentation of glucose by yeast:

2. Calculate the rate of carbon dioxide production per minute for each substrate in group 1's results:

   (a) None:

   (b) Glucose:

   (c) Maltose:

   (d) Sucrose:

   (e) Lactose:

3. A second group of students performed the same experiment. Their results are summarized, below left. Calculate the rate of carbon dioxide production per minute for each substrate in group 2's results:

   (a) None:

   (b) Glucose:

   (c) Maltose:

   (d) Sucrose:

   (e) Lactose:

*Experimental design and results adapted from Tom Schuster, Rosalie Van Zyl, & Harold Coller , California State University Northridge 2005*


ISBN: 978-1-927173-12-1


***Related activities:*** *Anaerobic Pathways for Metabolism*
***Weblinks:*** *Biomolecules: Carbohydrates*

4. What assumptions are being made in this experimental design and do you think they were reasonable?

5. Use the tabulated data to plot an appropriate graph of group 1's results on the grid provided:

6. (a) Summarize the results of group 1's fermentation experiment:

(b) Explain the findings based on your understanding of cellular respiration and carbohydrate chemistry:

7. (a) Plot a column chart to compare the results of the two groups in the volume of $CO_2$ collected after 90 minutes for each substrate (axes have been completed):

(b) Compare the results of the two groups:



(c) Provide a probable explanation for any differences in the results:

(d) Describe one improvement you could make to the experiment in order to generate more reliable data:


ISBN: 978-1-927173-12-1

# Chloroplasts

**Chloroplasts** are specialized plastids where photosynthesis occurs. A mesophyll leaf cell will contain between 50-100 chloroplasts. The chloroplasts are generally aligned so that their broad surface runs parallel to the cell wall to maximize the surface area available for light absorption. Chloroplasts have an internal structure characterized by a system of membranous structures called **thylakoids** arranged into stacks called **grana**. Special pigments, called **chlorophylls** and **carotenoids**, are bound to the membranes as part of light-capturing photosystems. They absorb light of specific wavelengths (mostly reds and blues) and thereby capture the light energy. Chlorophylls give leaves their green color (green light is reflected and not absorbed).



## The Structure of a Chloroplast

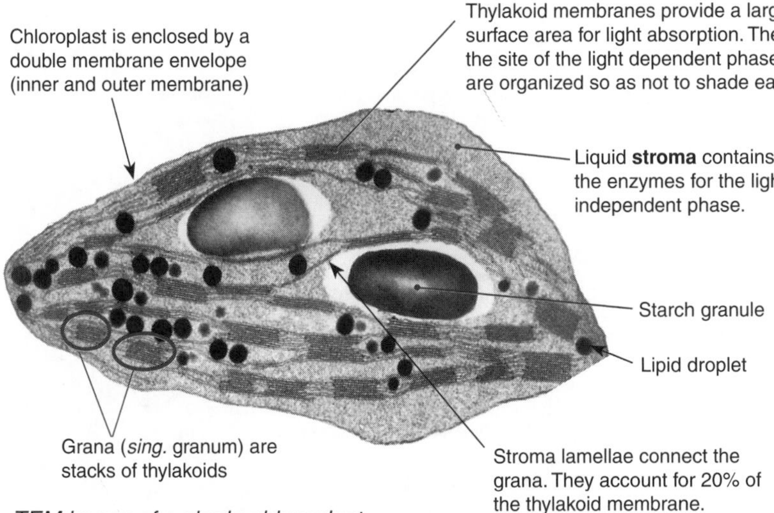


*TEM image of a single chloroplast*

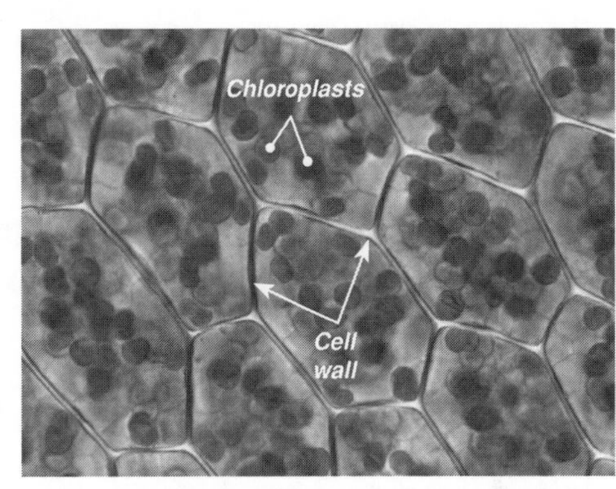


*Chloroplasts visible in plant cells*

1. Label the transmission electron microscope image of a chloroplast below:

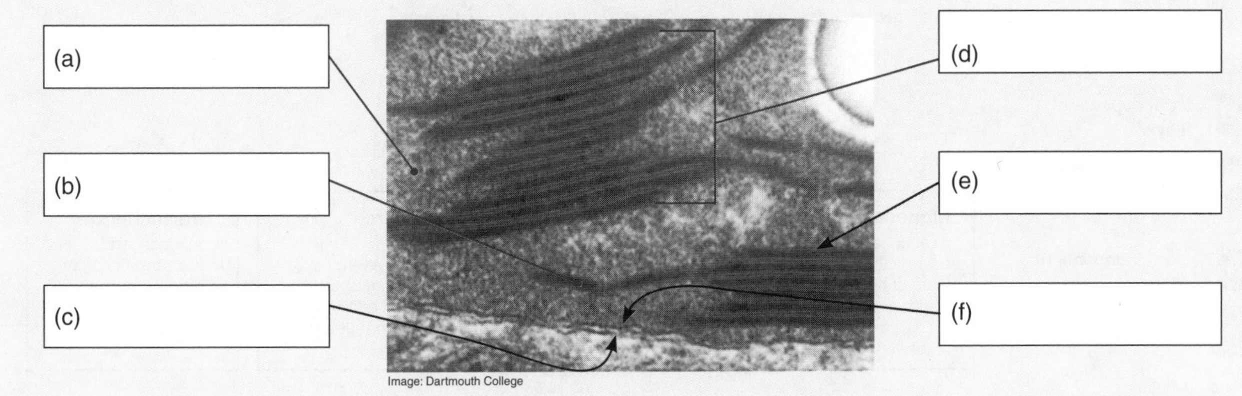


2. (a) Describe where chlorophyll is found in a chloroplast: ______

   (b) Explain why chlorophyll is found there: ______

3. Explain how the internal structure of chloroplasts helps absorb the maximum amount of light: ______

4. Explain why plant leaves appear green: ______


ISBN: 978-1-927173-12-1

# Photosynthesis

**Photosynthesis** is of fundamental importance to living things because it transforms sunlight energy into chemical energy stored in molecules, releases free oxygen gas, and absorbs carbon dioxide (a waste product of cellular metabolism). Photosynthetic organisms use special pigments, called **chlorophylls**, to absorb light of specific wavelengths and thereby capture the light energy.

Visible light is a small fraction of the total **electromagnetic radiation** reaching Earth from the sun. Of the visible spectrum, only certain wavelengths (red and blue) are absorbed for photosynthesis. Other wavelengths, particularly green, are reflected or transmitted. Photosynthesis is summarized in the chemical equation and diagram (below).

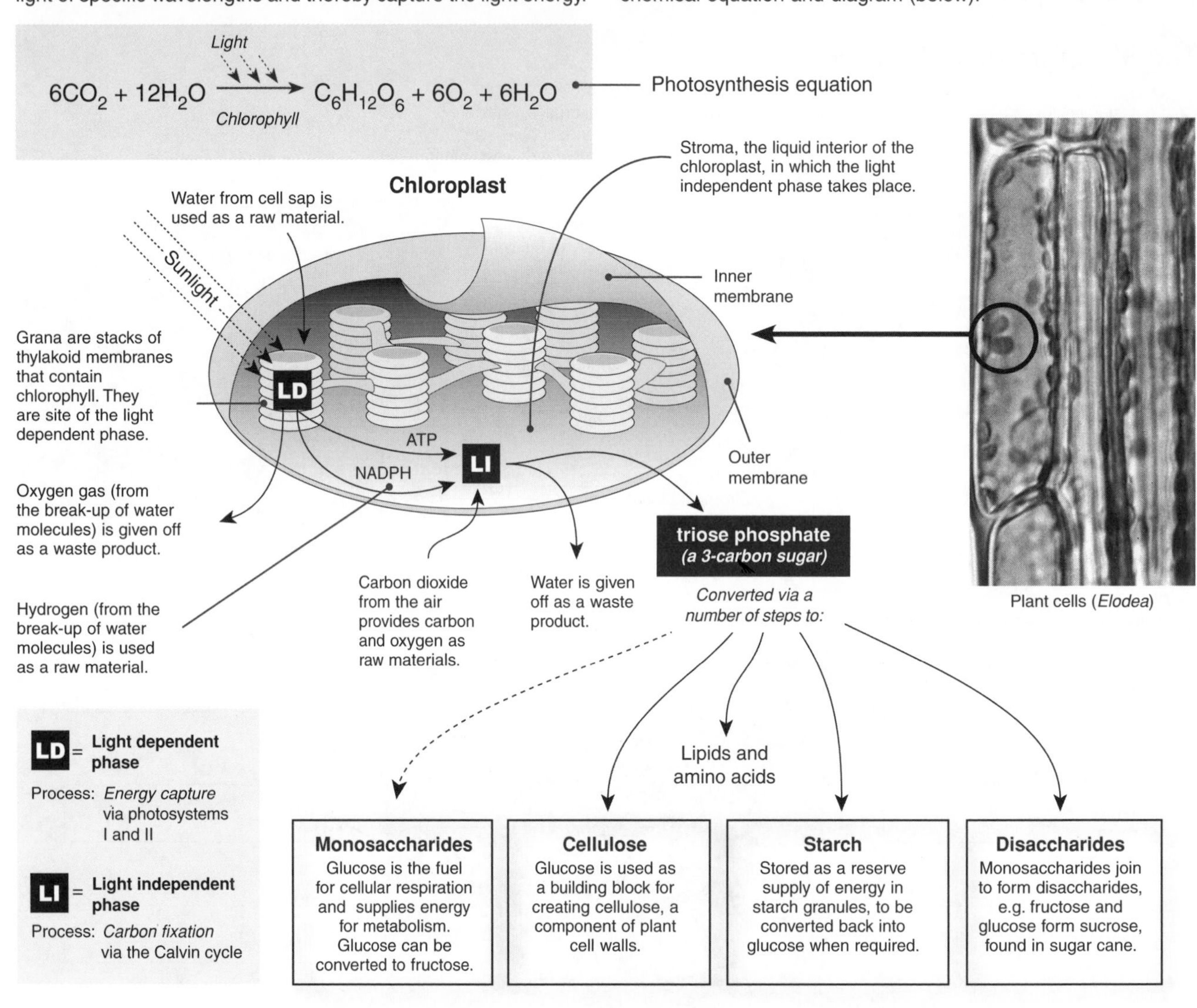


1. Describe the three things of fundamental biological importance provided by photosynthesis:

   (a) ____________________

   (b) ____________________

   (c) ____________________

2. Describe the role of the following in photosynthesis:

   (a) The carrier molecule NADP: ____________________

   (b) ATP: ____________________

   (c) Chlorophyll molecules: ____________________

   (d) Light: ____________________




**ISBN: 978-1-927173-12-1**

# Pigments and Light Absorption

As light meets matter, it may be reflected, transmitted, or absorbed. Substances that absorb visible light are called **pigments**, and different pigments absorb light of different wavelengths. The ability of a pigment to absorb particular wavelengths of light can be measured with a spectrophotometer. The light absorption vs the wavelength is called the **absorption spectrum** of that pigment. The absorption spectrum of different photosynthetic pigments provides clues to their role in photosynthesis, since light can only perform work if it is absorbed. An **action spectrum** profiles the effectiveness of different wavelength light in fuelling photosynthesis. It is obtained by plotting wavelength against some measure of photosynthetic rate (e.g. $CO_2$ production). Some features of photosynthetic pigments and their light absorbing properties are outlined below.



## The Electromagnetic Spectrum

Light is a form of energy known as electromagnetic radiation. The segment of the electromagnetic spectrum most important to life is the narrow band between about 380 nm and 750 nm. This radiation is known as visible light because it is detected as colors by the human eye (although some other animals, such as insects, can see in the UV range). It is the visible light that drives photosynthesis.

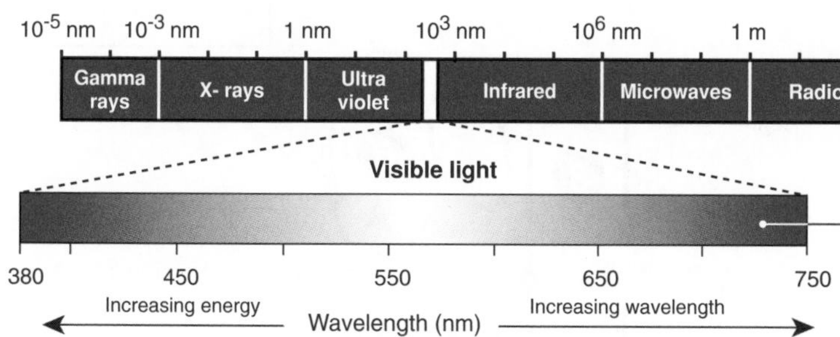


Electromagnetic radiation (EMR) travels in waves, where wavelength provides a guide to the energy of the photons; the greater the wavelength of EMR, the lower the energy of the photons in that radiation.

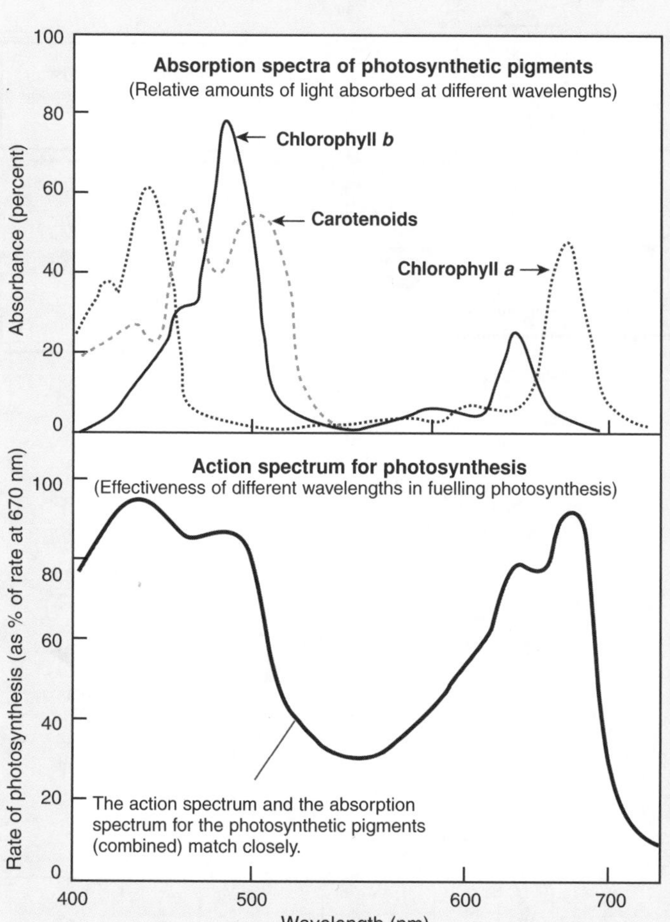


### The photosynthetic pigments of plants

The photosynthetic pigments of plants fall into two categories: chlorophylls (which absorb red and blue-violet light) and carotenoids (which absorb strongly in the blue-violet and appear orange, yellow, or red). The pigments are located on the chloroplast membranes (the thylakoids) and are associated with membrane transport systems.

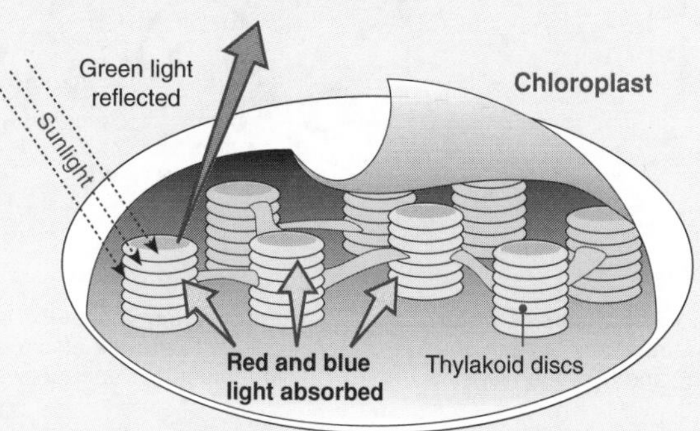


The pigments of chloroplasts in higher plants (above) absorb blue and red light, and the leaves therefore appear green (which is reflected). Each photosynthetic pigment has its own characteristic absorption spectrum (left, top graph). Although only chlorophyll a can participate directly in the light reactions of photosynthesis, the accessory pigments (chlorophyll *b* and carotenoids) can absorb wavelengths of light that chlorophyll *a* cannot. The accessory pigments pass the energy (photons) to chlorophyll a, thus broadening the spectrum that can effectively drive photosynthesis.

**Left**: Graphs comparing absorption spectra of photosynthetic pigments compared with the action spectrum for photosynthesis.

1. What is meant by the absorption spectrum of a pigment? ____________________

____________________

2. Why doesn't the **action spectrum** for photosynthesis exactly match the absorption spectrum of chlorophyll *a*?

____________________

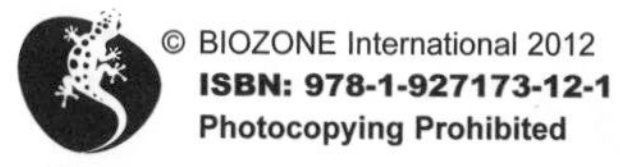


**ISBN: 978-1-927173-12-1**

# Light Dependent Reactions

Like cellular respiration, photosynthesis is a redox process, but in photosynthesis, water is split, and electrons and hydrogen ions, are transferred from water to $CO_2$, reducing it to sugar. The electrons increase in potential energy as they move from water to sugar. The energy to do this is provided by light. Photosynthesis has two phases. In the **light dependent reactions**, light energy is converted to chemical energy (ATP and NADPH). In the **light independent reactions**, the chemical energy is used to synthesize carbohydrate. The light dependent reactions most commonly involve **non-cyclic phosphorylation**, which produces ATP and NADPH in roughly equal quantities. The electrons lost are replaced from water. In **cyclic phosphorylation**, the electrons lost from photosystem II are replaced by those from photosystem I. ATP is generated, but not NADPH.

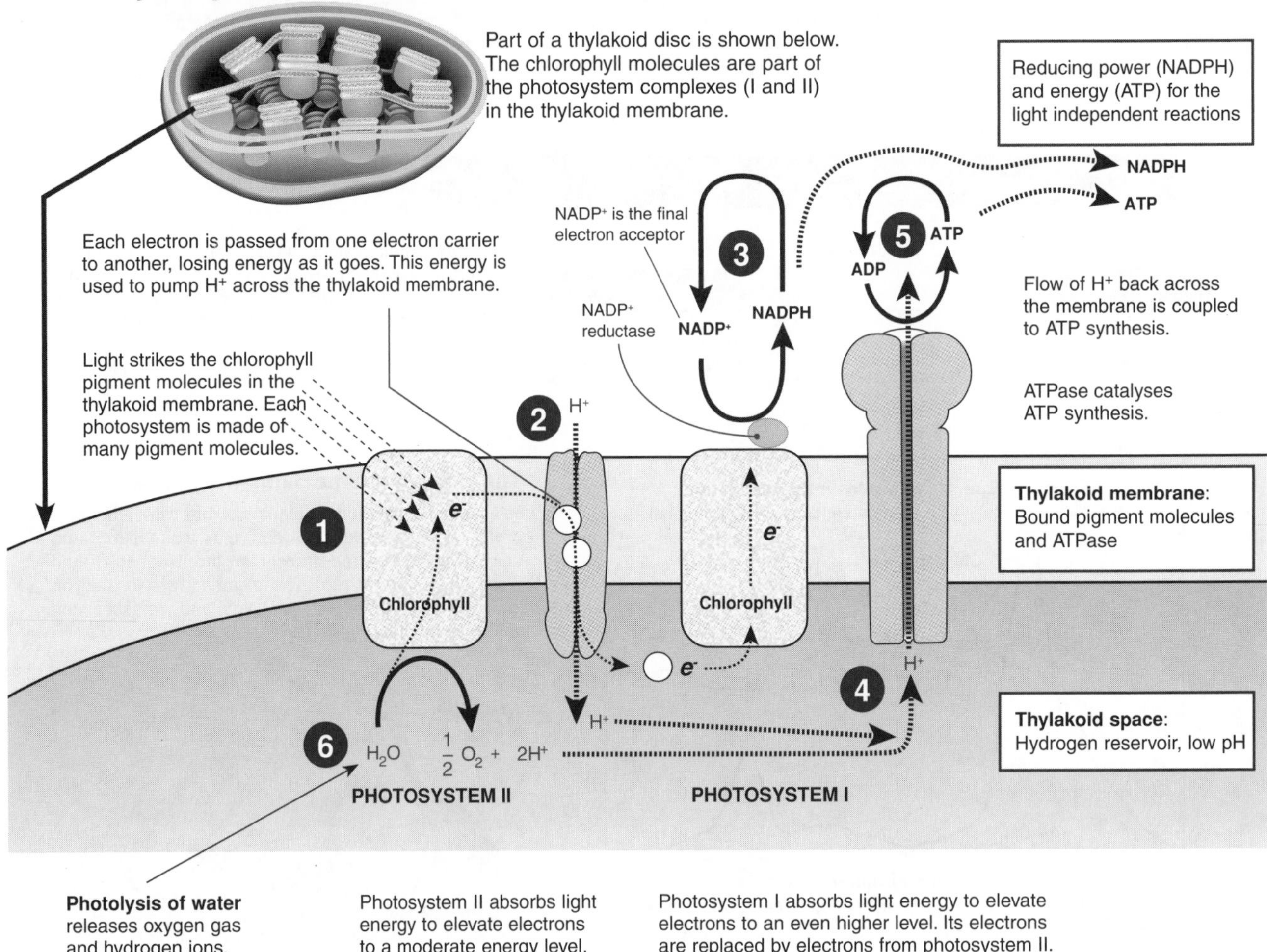

## Non-cyclic phosphorylation

Part of a thylakoid disc is shown below. The chlorophyll molecules are part of the photosystem complexes (I and II) in the thylakoid membrane.

Reducing power (NADPH) and energy (ATP) for the light independent reactions

NADPH

ATP

NADP$^+$ is the final electron acceptor

Each electron is passed from one electron carrier to another, losing energy as it goes. This energy is used to pump H$^+$ across the thylakoid membrane.

3

5

ADP

ATP

NADP$^+$ reductase

NADP$^+$

NADPH

Flow of H$^+$ back across the membrane is coupled to ATP synthesis.

Light strikes the chlorophyll pigment molecules in the thylakoid membrane. Each photosystem is made of many pigment molecules.

2

H$^+$

ATPase catalyses ATP synthesis.

**Thylakoid membrane**: Bound pigment molecules and ATPase

1

$e^-$

$e^-$

Chlorophyll

Chlorophyll

$e^-$

H$^+$

4

**Thylakoid space**: Hydrogen reservoir, low pH

H$^+$

6

$H_2O \quad \frac{1}{2} O_2 + 2H^+$

PHOTOSYSTEM II

PHOTOSYSTEM I

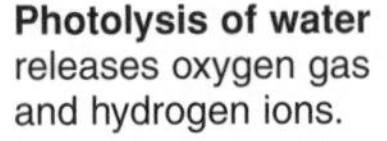

**Photolysis of water** releases oxygen gas and hydrogen ions.

Photosystem II absorbs light energy to elevate electrons to a moderate energy level.

Photosystem I absorbs light energy to elevate electrons to an even higher level. Its electrons are replaced by electrons from photosystem II.

## Cyclic phosphorylation

Cyclic phosphorylation involves only photosystem I and NADPH is not generated. Electrons from photosystem I are shunted back to the electron carriers in the membrane. This pathway produces ATP only. The Calvin cycle uses more ATP than NADPH, so cyclic phosphorylation makes up the difference. It is activated when NADPH levels build up, and remains active until enough ATP is made to meet demand.

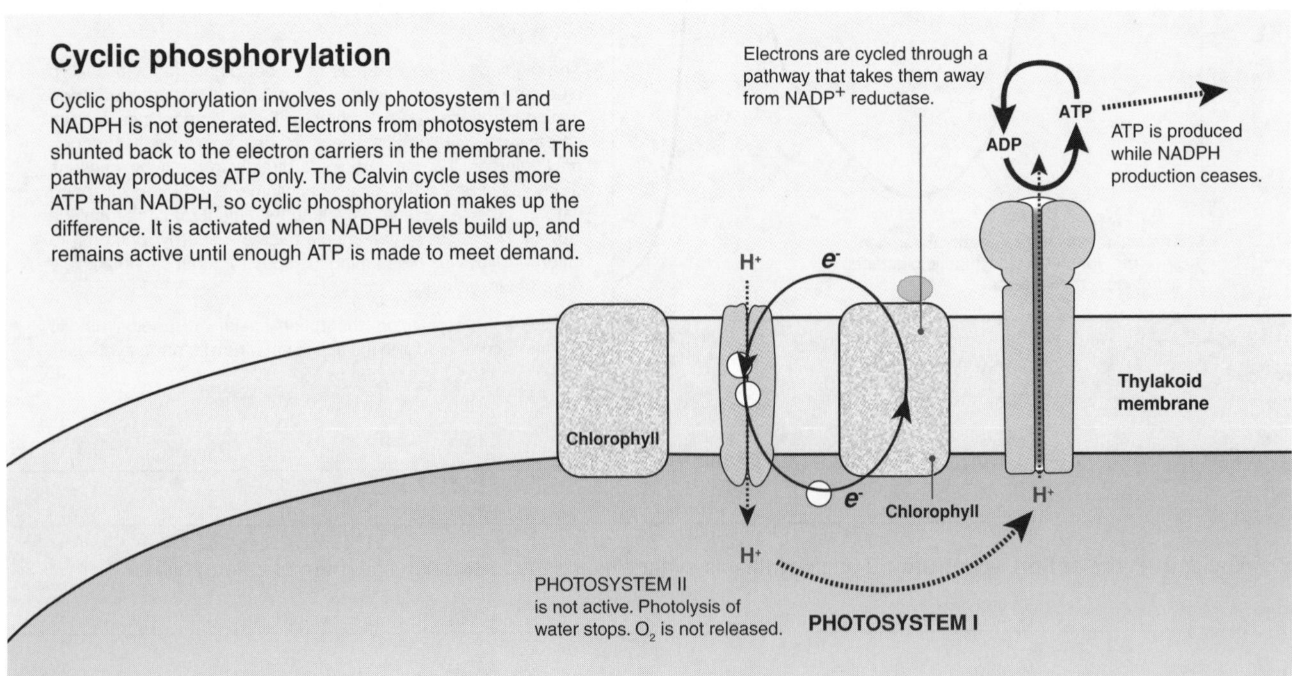




***Related activities:** Chloroplasts, Light Independent Reactions*

***Weblinks:** Photosynthesis Light Reactions, Photosystem II*


**ISBN: 978-1-927173-12-1**

1. Describe the role of the carrier molecule **NADP** in photosynthesis:

2. Explain the role of chlorophyll molecules in photosynthesis:

3. Summarize the events of the light dependent reactions:

4. Describe how ATP is produced as a result of light striking chlorophyll molecules during the light dependent phase:

5. (a) Explain what you understand by the term **non-cyclic phosphorylation**:

   (b) Suggest why this process is also known as non-cyclic **photo**phosphorylation:

6. (a) Describe how **cyclic photophosphorylation** differs from non-cyclic photophosphorylation:

   (b) Both cyclic and noncyclic pathways operate to varying degrees during photosynthesis. Since the non-cyclic pathway produces both ATP and NAPH, explain the purpose of the cyclic pathway of electron flow:

7. Explain how the independence of photosystem I gives a mechanism for evolution of the photosynthetic pathway:



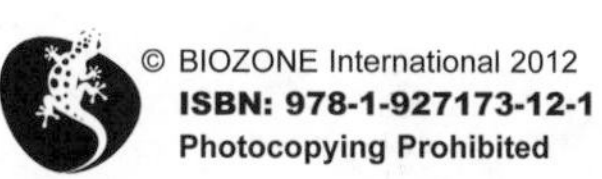


**ISBN: 978-1-927173-12-1**

# Light Independent Reactions

The **light independent reactions** of photosynthesis (the **Calvin cycle**) take place in the stroma of the chloroplast, and do not require light to proceed. Here, hydrogen ($H^+$) is added to $CO_2$ and a 5C intermediate to make carbohydrate. The $H^+$ and ATP are supplied by the light dependent reactions. The Calvin cycle uses more ATP than NADPH, but the cell uses cyclic phosphorylation (which does not produce NADPH) when it runs low on ATP to make up the difference.

KEY:
Carbon atom
P Phosphate group

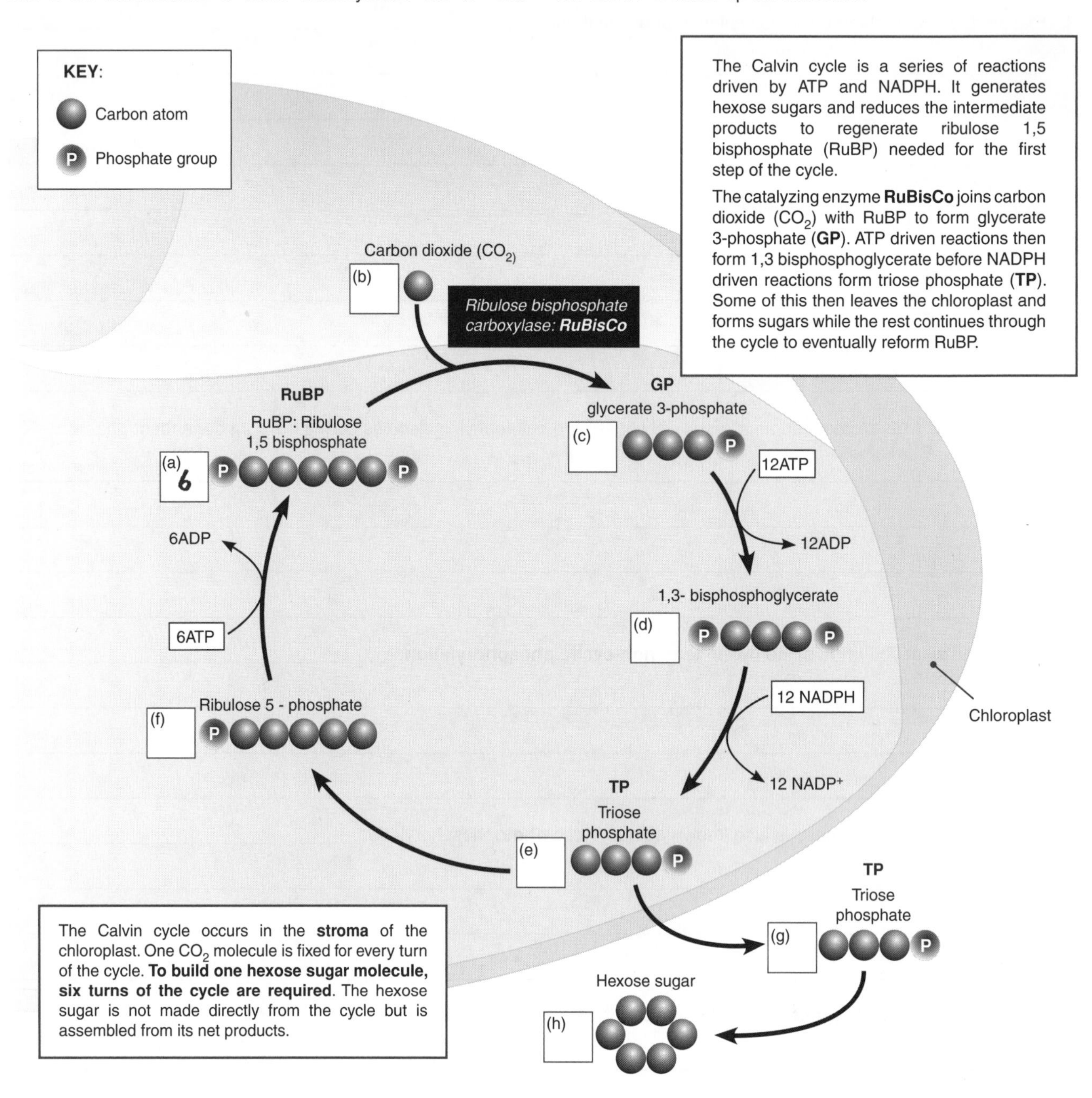

The Calvin cycle is a series of reactions driven by ATP and NADPH. It generates hexose sugars and reduces the intermediate products to regenerate ribulose 1,5 bisphosphate (RuBP) needed for the first step of the cycle.

The catalyzing enzyme **RuBisCo** joins carbon dioxide ($CO_2$) with RuBP to form glycerate 3-phosphate (**GP**). ATP driven reactions then form 1,3 bisphosphoglycerate before NADPH driven reactions form triose phosphate (**TP**). Some of this then leaves the chloroplast and forms sugars while the rest continues through the cycle to eventually reform RuBP.



The Calvin cycle occurs in the **stroma** of the chloroplast. One $CO_2$ molecule is fixed for every turn of the cycle. **To build one hexose sugar molecule, six turns of the cycle are required**. The hexose sugar is not made directly from the cycle but is assembled from its net products.

1. In the boxes on the diagram above, write the number of molecules formed at each step during the formation of **one hexose sugar molecule**. The first one has been done for you:

2. Explain the importance of RuBisCo in the Calvin cycle: ______________________

______________________

3. Identify the actual end product on the Calvin cycle: ______________________

4. Write the equation for the production of one hexose sugar molecule from carbon dioxide: ______________________

______________________

5. Explain why the Calvin cycle is likely to cease in the dark for most plants, even though it is independent of light:

______________________



***Related activities:*** *Light Dependent Reactions*
***Weblinks:*** *Photosynthesis Biochemistry*

***Periodicals:***
*Rubisco*

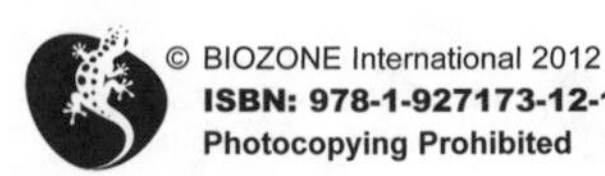


**ISBN: 978-1-927173-12-1**

# KEY TERMS: Crossword

Complete the crossword below, which will test your understanding of key terms in this chapter and their meanings:



### Clues Across

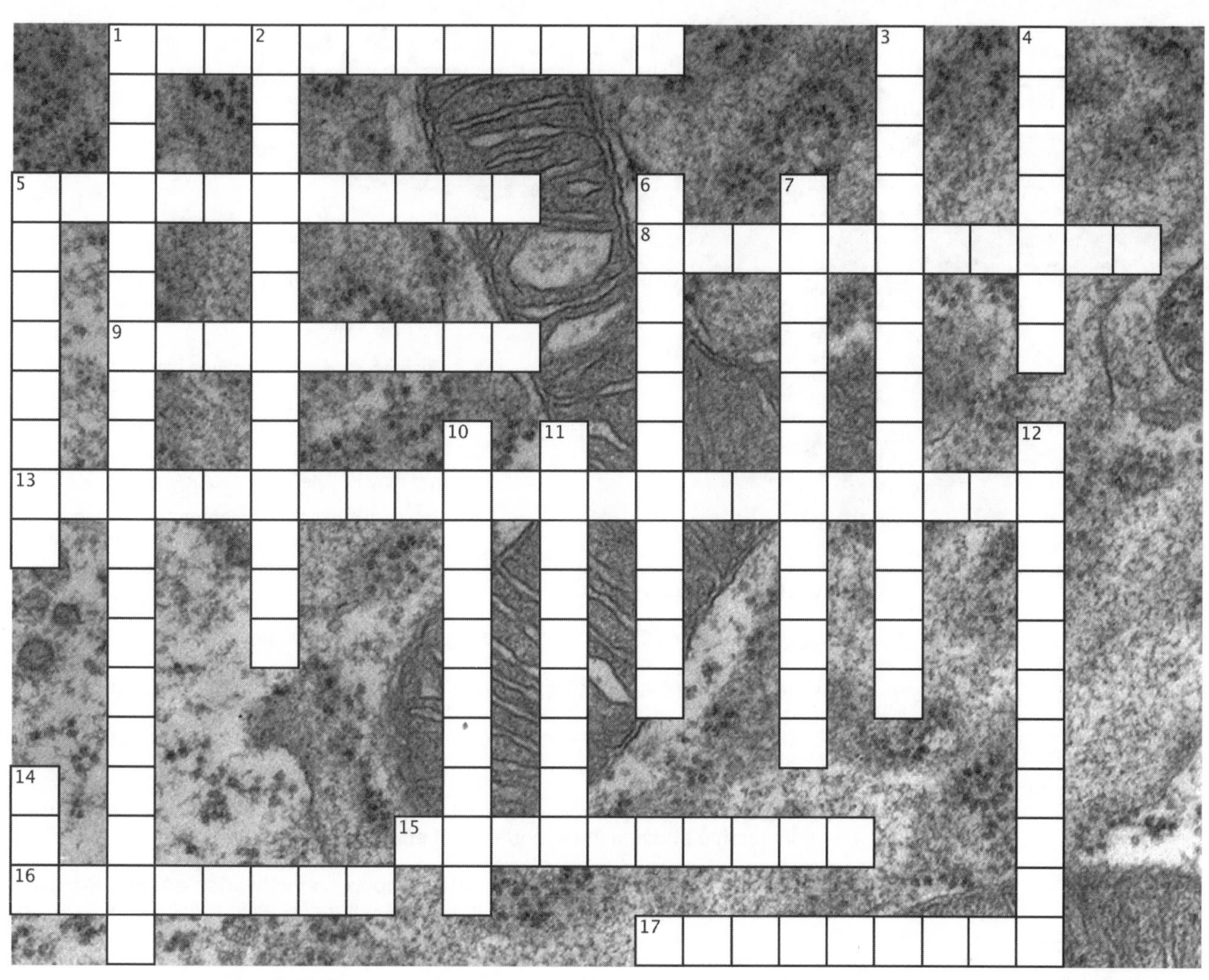

1. The process where the synthesis of ATP is coupled to electron transport and the movement of protons.
5. The organelle found in the cells of all green plants and where photosynthesis takes place. It contains the green pigment chlorophyll and other pigments involved with photosynthesis.
8. Obtaining energy from other living organisms or their dead remains.
9. Organisms that manufacture their own food from simple inorganic substances.
13. The chain of enzyme based redox reactions which pass electrons from high to low redox potentials. The energy released is used to pump protons across a membrane and produce ATP. (2 words: 8, 9,5).
15. A series of (ten) reactions that convert glucose into pyruvate. The energy released is used to produce ATP.
16. An organism that produces its own food, usually gaining the required energy via photosynthesis or chemosynthesis.
17. The reactions in photosynthesis which use energy from light to initiate photochemical reactions which result in the production of ATP and NADPH are called the light _ _ _ _ _ _ _ _ _ reactions.

### Clues Down

1. The catabolic process in which the chemical energy in complex organic molecules is coupled to ATP production. (2 words: 8, 11).
2. Organelle responsible for producing the cell's ATP. It appears oval in shape with an outer double membrane and a convoluted interior membrane. Contains its own circular DNA.
3. The biochemical process that uses light energy to convert carbon dioxide and water into glucose molecules and oxygen.
4. A measure of the disorder in an isolated system.
5. An organism that obtains its carbon and energy from other organisms.
6. A photosynthetic pigment which strongly absorbs red and blue-violet light and appears green in color.
7. Process of deriving energy from oxidation of organic compounds using an endogenous acceptor rather than an exogenous one (such as oxygen).
10. Also known as the citric acid cycle. Part of a metabolic pathway involved in the chemical conversion of carbohydrates, fats and proteins to $CO_2$ and water to generate a form of usable energy (ATP). (2 words 5, 5).
11. Metabolic process that yields energy without the need for molecular oxygen.
12. The reactions in photosynthesis that take place in the stroma of chloroplasts, in which inorganic carbon is incorporated into organic molecules are called the light _ _ _ _ _ _ _ _ _ _ _ _ reactions.
14. A nucleotide comprising a purine base, a pentose sugar, and three phosphate groups, which acts as the cell's energy carrier.

ISBN: 978-1-927173-12-1

Enduring Understanding

2.D
4.B

# Enzymes and Metabolism

## Key terms

activation energy
active site
allosteric interaction
anabolism
biofilm
catabolism
catalyst
coenzyme
cofactor
denaturation
end-product inhibition
endergonic reaction
enzyme
enzyme inhibition
exergonic reaction
induced fit model
lock and key model
metabolic pathway
optimum
regional specialization

**Periodicals:**
*Listings for this chapter are on page 374*

**Weblinks:**
*www.thebiozone.com/weblink/AP2-3121.html*

*BIOZONE APP: Student Review Series The Molecules of Life*

## Key concepts

- Enzymes are biological catalysts whose specificity is dictated by the shape and charge of the active site.
- Enzyme-catalyzed reactions can be followed by measuring the rate of product formation or substrate use.
- Enzyme activity may be dependent on cofactors and may be altered by the presence of inhibitors.
- Regional specialization promotes efficiency in multicellular organisms but populations of unicellular organisms can behave similarly.

## Essential Knowledge

☐ 1. Use the **KEY TERMS** to compile a glossary for this topic.

### Enzyme Structure and Function *(4.B.1)* — pages 31-38

☐ 2. Using **enzymes** as an illustrative example, explain how change in the structure of a molecular system may change its function.

☐ 3. Describe the general role of enzymes as biological **catalysts**. Include reference to the **active site** and the importance of **specificity**.

☐ 4. With reference to **enzyme-substrate complex** and **activation energy**, explain how enzymes work as catalysts to bring about reactions in cells.

☐ 5. Explain the **induced fit model** of enzyme function. Compare and contrast it with the older **lock and key model**. Recall the difference between **endergonic** and **exergonic reactions** and outline the role of enzymes in **anabolism** and **catabolism**.

☐ 6. Describe ways in which the time course of an enzyme-catalyzed reaction can be followed by measuring the rate of product formation (e.g. catalase) or by measuring the rate of substrate use (e.g. starch breakdown by amylase).

☐ 7. Describe the effect of substrate concentration, enzyme concentration, pH, and temperature on enzyme activity. Explain the term **optimum** with respect to enzyme activity. Recognize that enzymes (as proteins) can be denatured.

☐ 8. Distinguish between **cofactors** and **coenzymes**. Using examples, explain the role of cofactors in enzyme activity.

☐ 9. Describe **enzyme inhibition**. Distinguish **reversible inhibition** from **irreversible inhibition**. Describe the effects of **competitive inhibitors** and **non-competitive inhibitors** on enzyme activity. Interpret graphs of enzyme activity showing competitive and non-competitive inhibition.

☐ 10. Explain the role of **allosteric interactions** in the control of **metabolic pathways** by **end-product inhibition**.

### Achieving Metabolic Efficiencies *(4.B.2, 2.D.1)* — pages 39-42

☐ 11. Explain how compartmentalization within cells and organisms contributes to functional efficiency. Describe levels of organization in multicellular organisms and explain how **regional specialization** contributes to efficient functioning of the whole organism.

☐ 12. Explain how interactions between cells in a population of unicellular organisms (e.g. in a **biofilm**) can lead to increased efficiency and use of energy and matter. Comment on the similarity of such populations to a multicellular organism.

# Enzymes

Most enzymes are proteins. They are capable of catalyzing (speeding up) biochemical reactions and are therefore called biological **catalysts**. Enzymes act on one or more compounds (called the **substrate**). They may break down a single substrate molecule into simpler substances, or join two or more substrate molecules together. The enzyme itself is unchanged in the reaction. Its presence merely allows the reaction to take place more rapidly. The part of the enzyme into which the substrate binds and undergoes reaction is the **active site**. It is a function of the polypeptide's complex tertiary structure.

## Enzyme Structure

The model on the right illustrates the enzyme *Ribonuclease S*, which breaks up RNA molecules. It is a typical enzyme, being a globular protein and composed of up to several hundred amino acids. The darkly shaded areas are part of the **active site** and make up the **cleft**; the region into which the substrate molecule(s) are drawn.

The correct positioning of these sites is critical for the catalytic reaction to occur. The substrate (RNA in this case) is drawn into the cleft by the active sites. By doing so, it puts the substrate molecule under stress, causing the reaction to proceed more readily.

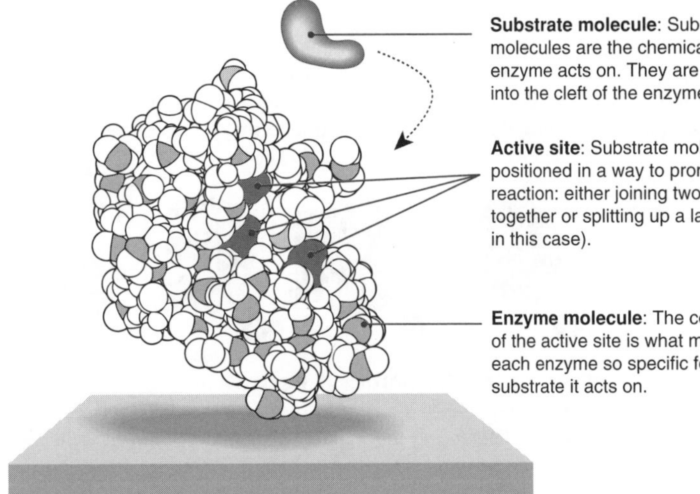


Source: After *Biochemistry*, (1981) by Lubert Stryer

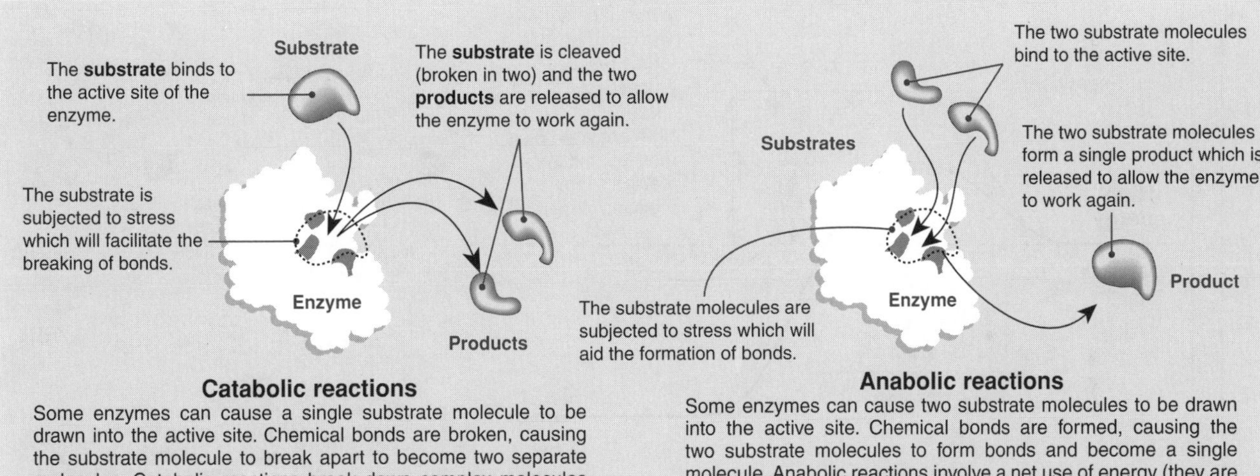


### Catabolic reactions

Some enzymes can cause a single substrate molecule to be drawn into the active site. Chemical bonds are broken, causing the substrate molecule to break apart to become two separate molecules. Catabolic reactions break down complex molecules into simpler ones and involve a net release of energy, so they are called exergonic. **Examples**: *hydrolysis, cellular respiration.*

### Anabolic reactions

Some enzymes can cause two substrate molecules to be drawn into the active site. Chemical bonds are formed, causing the two substrate molecules to form bonds and become a single molecule. Anabolic reactions involve a net use of energy (they are endergonic) and build more complex molecules and structures from simpler ones. **Examples:** *protein synthesis, photosynthesis.*

1. Explain what is meant by the **active site** of an enzyme and relate it to the enzyme's tertiary structure:

2. What might happen to an enzyme's activity if the gene encoding its production was altered by a mutation?

3. Distinguish between **catabolism** and **anabolism**, giving an example of each and identifying each reaction as **endergonic** or **exergonic**:


**ISBN: 978-1-927173-12-1**


***Related Activities***: *How Enzymes Work*

# How Enzymes Work

Chemical reactions in cells are accompanied by energy changes. Any reaction, even an exergonic reaction, needs to raise the energy of the substrate to an unstable **transition state** before the reaction will proceed (below left). The amount of energy required to do this is the activation energy ($E_a$). Enzymes work by lowering the $E_a$ for any given reaction. They do this by orienting the substrate, or by adding charges or otherwise inducing strain in the substrate so that bonds are destabilized and the substrate is more reactive. The current 'induced-fit' model of enzyme function is supported by studies of enzyme inhibitors, which show that enzymes are flexible and change shape when interacting with the substrate.

### How Enzymes Work

The **lock and key** model proposed earlier last century suggested that the (perfectly fitting) substrate was simply drawn into a matching cleft on the enzyme molecule (below). This model was supported by early X-ray crystallography but has since been modified to recognize the flexibility of enzymes (the **induced fit** model, described right).

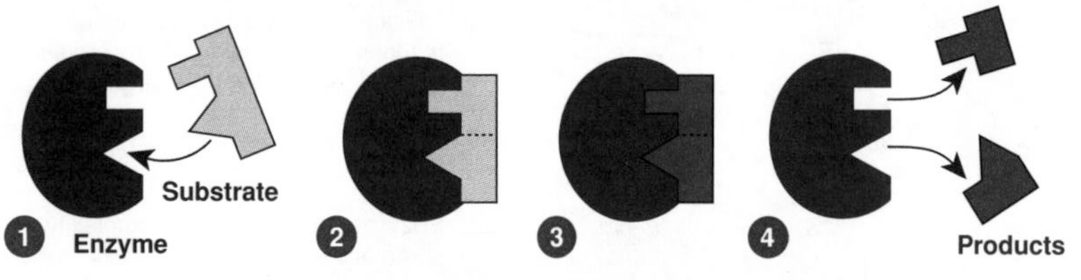


### Lowering the Activation Energy

The presence of an enzyme simply makes it easier for a reaction to take place. All catalysts speed up reactions by influencing the stability of bonds in the reactants. They may also provide an alternative reaction pathway, thus lowering the activation energy ($E_a$) needed for a reaction to take place (see the graph below).

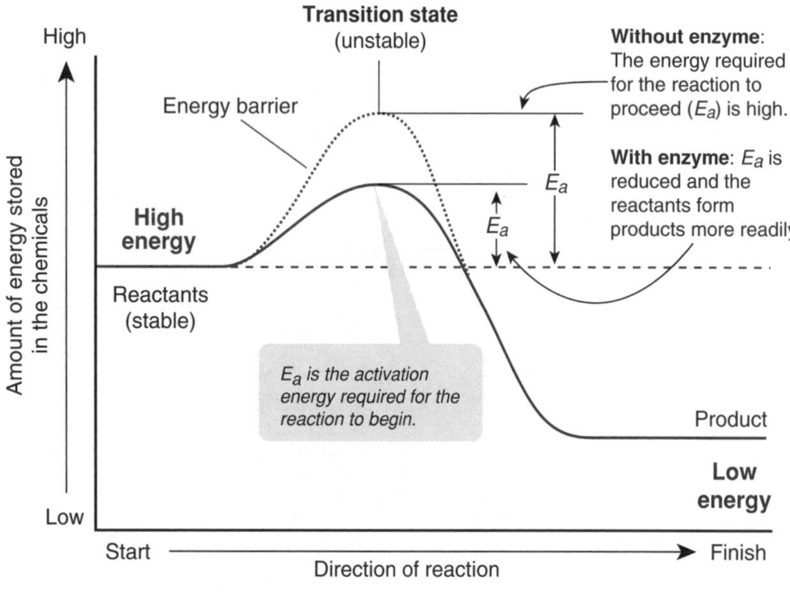


### The Current Model: Induced Fit

An enzyme's interaction with its substrate is best regarded as an induced fit (below). The shape of the enzyme changes when the substrate fits into the cleft. The reactants become bound to the enzyme by weak chemical bonds. This binding can weaken bonds within the reactants themselves, allowing the reaction to proceed more readily.

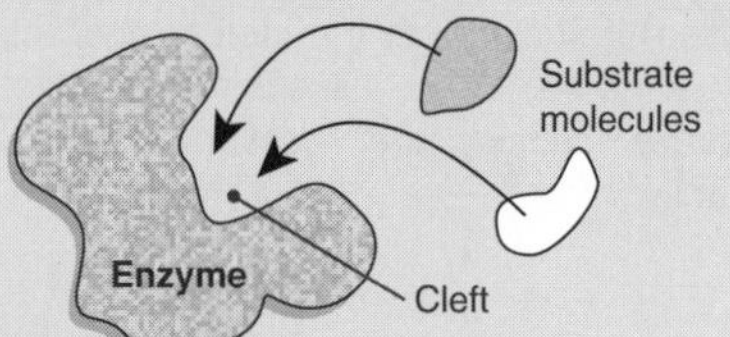


1. Two substrate molecules are drawn into the cleft of the enzyme.

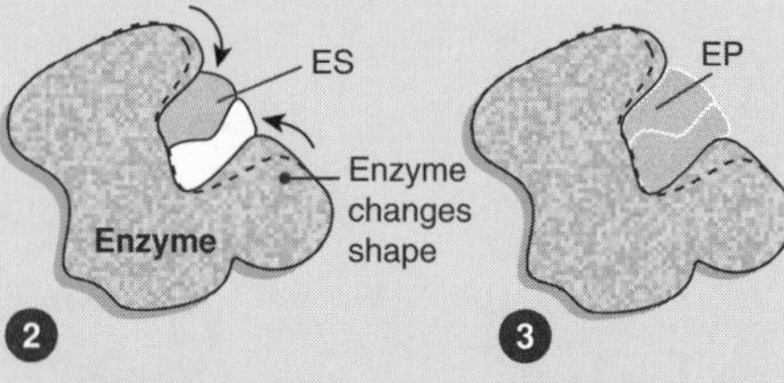


The enzyme changes shape as the substrate molecules bind in an enzyme-substrate complex (ES). The enzyme-substrate interaction results in an intermediate enzyme product (EP) complex.

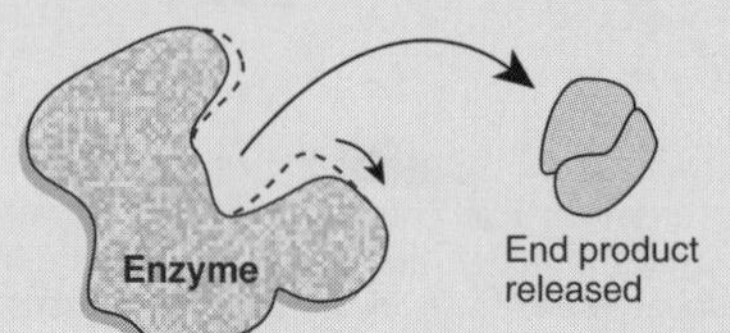


4. The end product is released and the enzyme returns to its previous shape.

1. Explain how enzymes act as **biological catalysts**: ____________________

2. Describe the key features of the '**lock and key**' model of enzyme action and explain its deficiencies as a working model:

3. Describe the current '**induced fit**' model of enzyme action, explaining how it differs from the lock and key model:




**ISBN: 978-1-927173-12-1**

# Enzyme Reaction Rates

Enzymes are sensitive molecules. They often have a narrow range of conditions under which they operate properly. For most of the enzymes associated with plant and animal metabolism, there is little activity at low temperatures. As the temperature increases, so too does the enzyme activity, until the point is reached where the temperature is high enough to damage the enzyme's structure. At this point, the enzyme ceases to function: a phenomenon called **denaturation**. Extremes in acidity and alkalinity (pH) can also cause the protein structure of enzymes to denature. Poisons often work by denaturing enzymes or occupying the enzyme's active site so that it does not function. In some cases, enzymes will not function without cofactors, such as vitamins or trace elements. In the four graphs below, the *rate of reaction* or *degree of enzyme activity* is plotted against each of four factors that affect enzyme performance. Answer the questions relating to each graph:

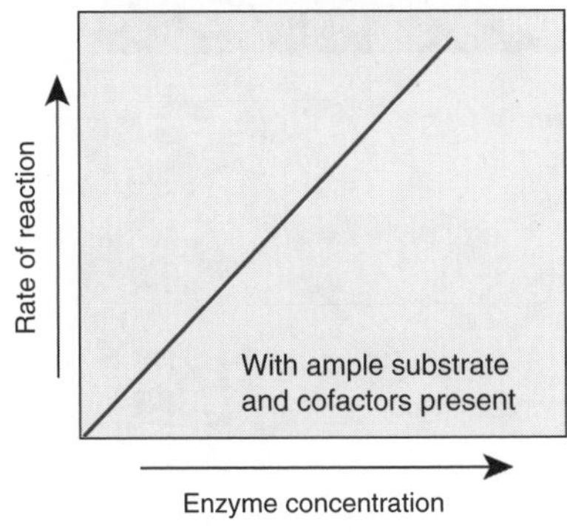


1. **Enzyme concentration**

   (a) Describe the change in the rate of reaction when the enzyme concentration is increased (assuming there is plenty of the substrate present):

   (b) Suggest how a cell may vary the amount of enzyme present in a cell:

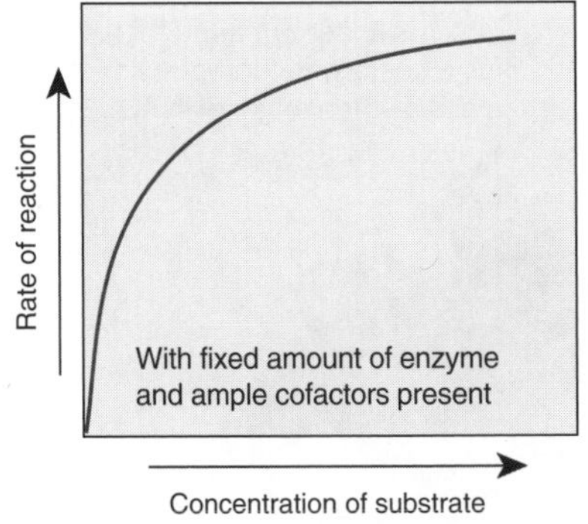


2. **Substrate concentration**

   (a) Describe the change in the rate of reaction when the substrate concentration is **increased** (assuming a fixed amount of enzyme and ample cofactors):

   (b) Explain why the rate changes the way it does:

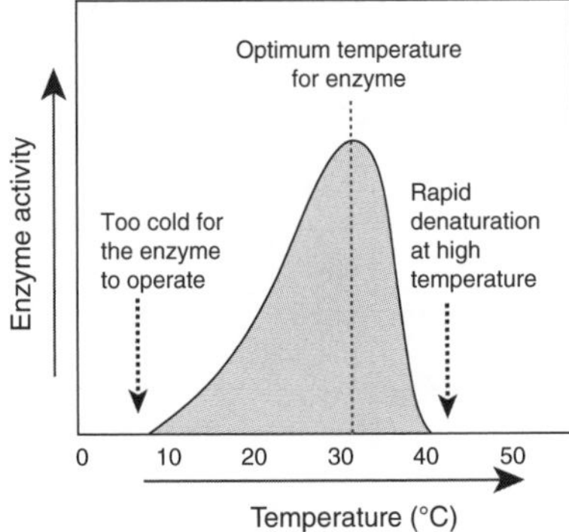


3. **Temperature**

   Higher temperatures speed up all reactions, but few enzymes can tolerate temperatures higher than 50–60°C. The rate at which enzymes are **denatured** (change their shape and become inactive) increases with higher temperatures.

   (a) Describe what is meant by an *optimum temperature* for enzyme activity:

   (b) Explain why most enzymes perform poorly at low temperatures:

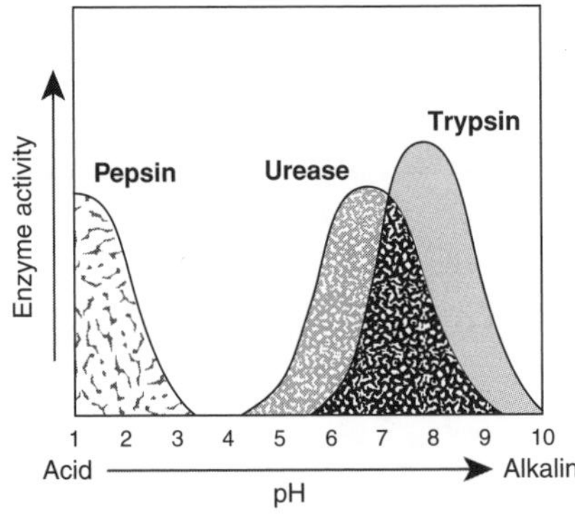


4. **Acidity or alkalinity (pH)**

   Like all proteins, enzymes are **denatured** by *extremes* of **pH** (very acid or alkaline). Within these extremes, most enzymes are still influenced by pH. Each enzyme has a preferred pH range for optimum activity.

   (a) State the optimum pH for each of the enzymes:

   Pepsin: ____________ Trypsin: ____________ Urease: ____________

   (b) Pepsin acts on proteins in the stomach. Explain how its optimum pH is suited to its working environment:

**ISBN: 978-1-927173-12-1**


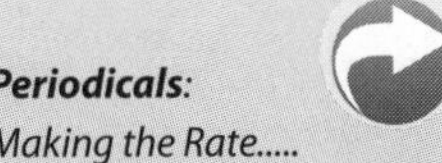

# Enzyme Cofactors

Nearly all enzymes are made of protein, although RNA has been demonstrated to have enzymatic properties. Some enzymes (e.g. pepsin) consist of only protein. Other enzymes require the addition of extra non-protein components to be functional. In these cases, the protein portion is called the **apoenzyme**, and the additional chemical component is called a **cofactor**. Neither the apoenzyme nor the cofactor has catalytic activity on its own. Cofactors may be organic molecules (e.g. vitamin C and the coenzymes in the respiratory chain) or inorganic ions (e.g. $Ca^{2+}$, $Zn^{2+}$). They also may be tightly or loosely bound to the enzyme. Permanently bound cofactors are called **prosthetic groups**, whereas temporarily attached molecules, which detach after a reaction are called **coenzymes**. Some cofactors include both an organic and a non-organic component. Examples include the heme prosthetic groups, which consist of an iron atom in the center of a porphyrin ring.

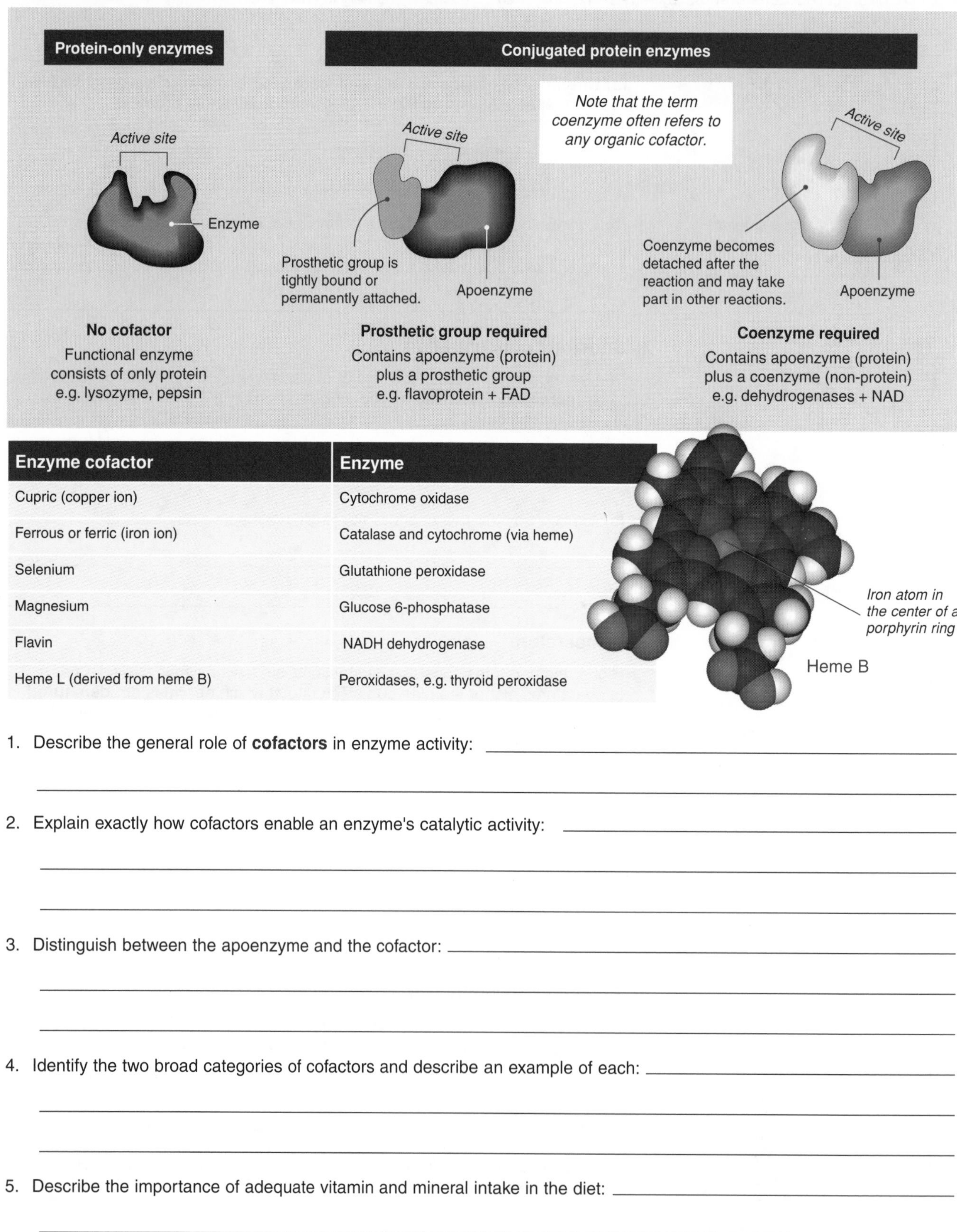


| Enzyme cofactor | Enzyme |
|---|---|
| Cupric (copper ion) | Cytochrome oxidase |
| Ferrous or ferric (iron ion) | Catalase and cytochrome (via heme) |
| Selenium | Glutathione peroxidase |
| Magnesium | Glucose 6-phosphatase |
| Flavin | NADH dehydrogenase |
| Heme L (derived from heme B) | Peroxidases, e.g. thyroid peroxidase |

1. Describe the general role of **cofactors** in enzyme activity:

2. Explain exactly how cofactors enable an enzyme's catalytic activity:

3. Distinguish between the apoenzyme and the cofactor:

4. Identify the two broad categories of cofactors and describe an example of each:

5. Describe the importance of adequate vitamin and mineral intake in the diet:




**ISBN: 978-1-927173-12-1**

# Enzyme Inhibitors

Enzymes may be deactivated, temporarily or permanently, by chemicals called enzyme inhibitors. **Irreversible inhibitors** bind tightly to the enzyme, either at the active site or remotely from it, and are not easily displaced. **Reversible inhibitors** can be displaced from the enzyme and have a role as enzyme regulators in metabolic pathways. **Competitive inhibitors** compete directly with the substrate for the active site, and their effect can be overcome by increasing the concentration of available substrate. A **non-competitive inhibitor** does not occupy the active site, but distorts it so that the substrate and enzyme can no longer interact. Both competitive and non-competitive inhibition may be irreversible, in which case the inhibitors involved act as poisons.

## Allosteric Enzyme Regulation

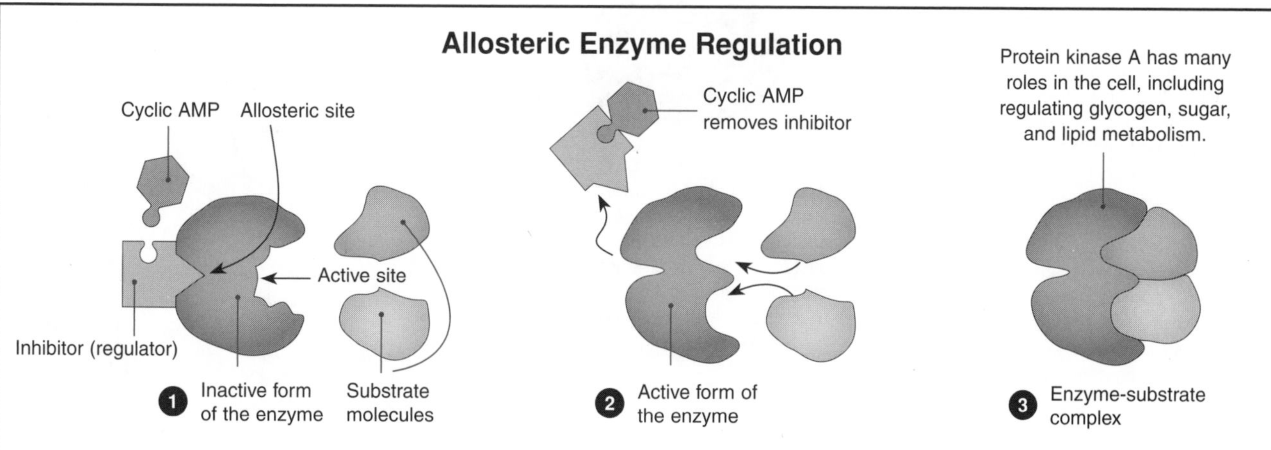


Allosteric regulators have a receptor site, called the **allosteric site**, on a part of the enzyme other than the active site. When a substance binds to the allosteric site, it regulates the activity of the enzyme. Often the action is inhibitory (as shown above for protein kinase A), but allosteric regulators can also switch an enzyme from its inactive to its active form. Thus, they can serve as regulators of metabolic pathways. The activity of the enzyme **protein kinase A** is regulated by the level of **cyclic AMP** in the cell. When a regulatory inhibitor protein binds reversibly to its allosteric site, the enzyme is inactive. Cyclic AMP removes the allosteric inhibitor and activates the enzyme.

## Competitive Inhibition

Competitive inhibitors compete with the normal substrate for the enzyme's active site.

If a competitive inhibitor occupies the active site only temporarily then the inhibition will be reversible.

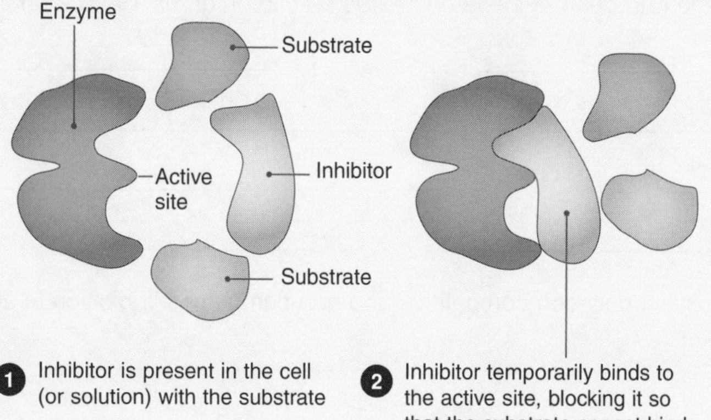


1 Inhibitor is present in the cell (or solution) with the substrate

2 Inhibitor temporarily binds to the active site, blocking it so that the substrate cannot bind

Figure 1 Effect of competitive inhibition on enzyme reaction rate at different substrate concentration

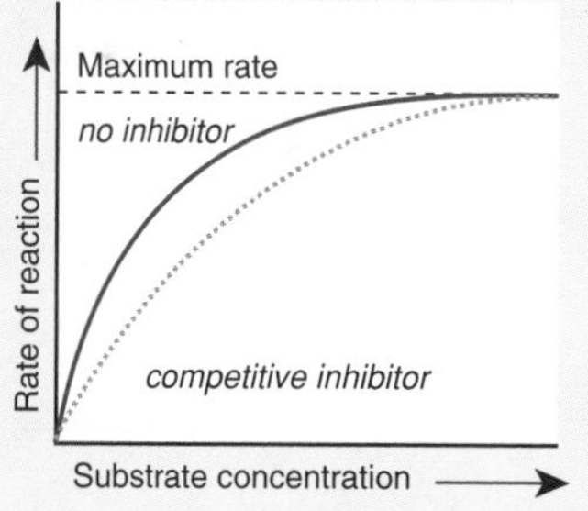


## Non-competitive Inhibition

Non-competitive inhibitors bind with the enzyme at a site other than the active site. They inactivate the enzyme by altering its shape.

Substrate

Inhibitor

Enzyme

Substrate

Active site cannot bind the substrates

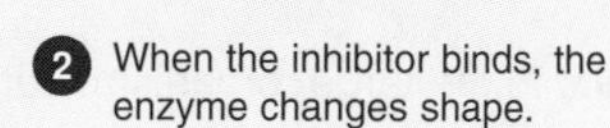

1 Without the inhibitor bound, the enzyme can bind the substrate.

2 When the inhibitor binds, the enzyme changes shape.

Figure 2 Effect of non-competitive inhibition on enzyme reaction rate at different substrate concentration

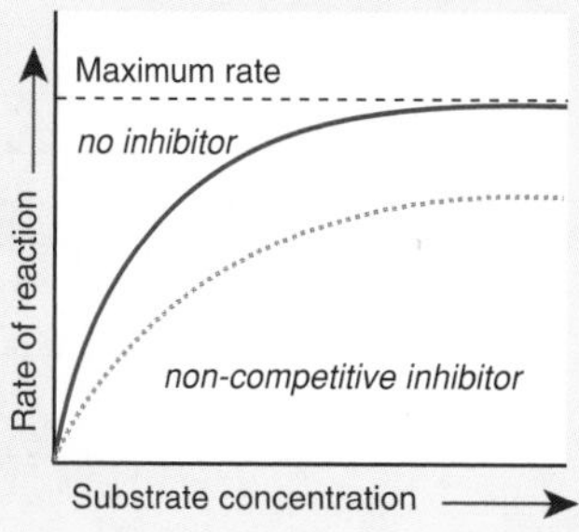

ISBN: 978-1-927173-12-1


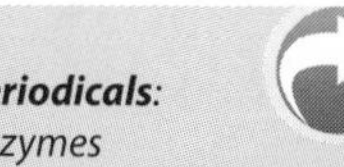

## Poisons are Irreversible Inhibitors

Some enzyme inhibitors are poisons because the enzyme-inhibitor binding is irreversible. Irreversible inhibitors form strong covalent bonds with an enzyme. These inhibitors may act at, near, or remotely from the active site and modify the enzyme's structure to such an extent that it ceases to work. For example, the poison **cyanide** is an irreversible enzyme inhibitor that combines with the copper and iron in the active site of **cytochrome *c* oxidase** and blocks cellular respiration.

Since many enzymes contain sulfhydryl (-SH), alcohol, or acidic groups as part of their active sites, any chemical that can react with them may act as an irreversible inhibitor. Heavy metals, $Ag^{+}$, $Hg^{2+}$, or $Pb^{2+}$, have strong affinities for -SH groups and destroy catalytic activity. Most heavy metals are non-competitive inhibitors.

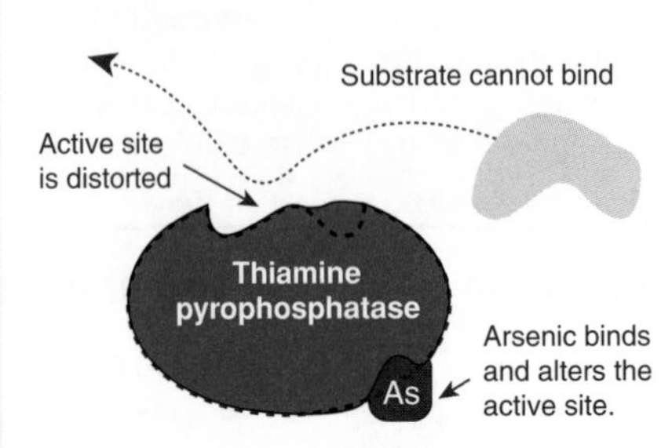


Arsenic and phosphorus share some structural similarities so arsenic will often substitute for phosphorus in biological systems. It therefore targets a wide variety of enzyme reactions. Arsenic can act as either a competitive or a non-competitive inhibitor (as above) depending on the enzyme.

### Drugs

Many drugs work by irreversible inhibition of a pathogen's enzymes. Penicillin and related antibiotics inhibit transpeptidase, a bacterial enzyme which forms some of the linkages in the bacterial cell wall. Susceptible bacteria cannot complete cell wall synthesis and cannot divide. Human cells are unaffected by the drug.

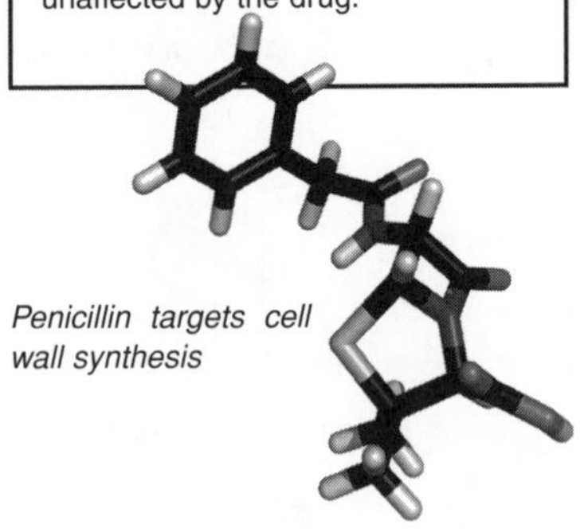

*Penicillin targets cell wall synthesis*

1. Distinguish between **competitive** and **non-competitive** inhibition:

2. (a) Compare and contrast the effect of competitive and non-competitive inhibition on the relationship between the substrate concentration and the rate of an enzyme controlled reaction (figures 1 and 2 on the previous page):

   (b) Suggest how you could distinguish between competitive and non-competitive inhibition in an isolated system:

3. Describe how an **allosteric regulator** can regulate enzyme activity:

4. Explain why heavy metals, such as lead and arsenic, are poisonous:

5. (a) Using an example, explain how enzyme inhibition is exploited to control human diseases:

   (b) Explain why the drug is poisonous to the target organism, but not to humans:


ISBN: 978-1-927173-12-1

# Catalase Activity in Germinating Seeds

Once you have completed the practical part of an experiment, the next task is to evaluate your results in the light of your own hypothesis and your current biological knowledge. A critical evaluation of any study involves analyzing, presenting, and discussing the results, as well as accounting for any deficiencies in your procedures and erroneous results. This activity describes an experiment in which germinating seeds of different ages were tested for their level of catalase activity using hydrogen peroxide solution as the substrate and a simple apparatus to measure oxygen production (see background). Completing this activity, which involves a critical evaluation of the second-hand data provided, will help to prepare you for your own investigation.

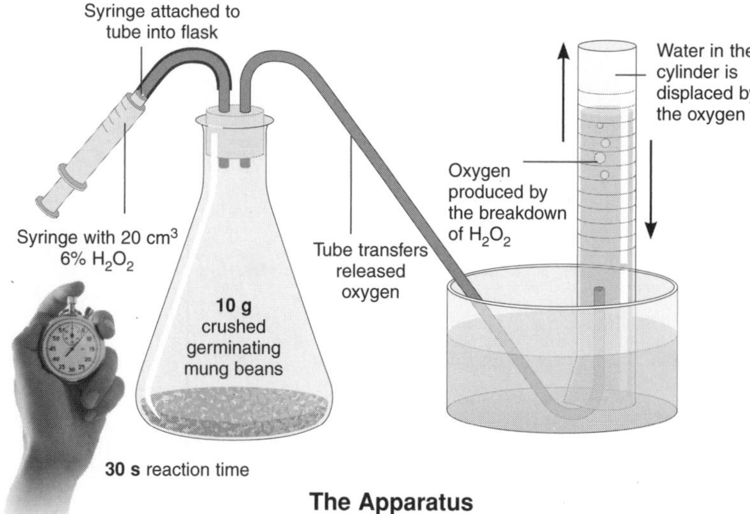


**The Apparatus**

In this experiment, 10 g germinating mung bean seeds (0.5, 2, 4, 6, or 10 days old) were ground by hand with a mortar and pestle and placed in a conical flask as above. There were six trials at each of the five seedling ages. With each trial, 20 cm$^3$ of 6% $H_2O_2$ was added to the flask at time 0 and the reaction was allowed to run for 30 seconds. The oxygen released by the decomposition of the $H_2O_2$ by catalase in the seedlings was collected via a tube into an inverted measuring cylinder. The volume of oxygen produced is measured by the amount of water displaced from the cylinder. The results from all trials are tabulated below:

### The Aim

To investigate the effect of germination age on the level of catalase activity in mung beans.

### Background

Germinating seeds are metabolically very active and this metabolism inevitably produces reactive oxygen species, including hydrogen peroxide ($H_2O_2$). $H_2O_2$ helps germination by breaking dormancy, but it is also toxic. To counter the toxic effects of $H_2O_2$ and prevent cellular damage, germinating seeds also produce **catalase**, an enzyme that catalyses the breakdown of $H_2O_2$ to water and oxygen.

A class was divided into six groups with each group testing the seedlings of each age. Each group's set of results (for 0.5, 2, 4, 6, and 10 days) therefore represents one trial.

| Stage of germination (days) \ Trial # | Volume of oxygen collected after 30s (cm$^3$) | | | | | | | | Mean rate (cm$^3$ s$^{-1}$ g$^{-1}$) |
|---|---|---|---|---|---|---|---|---|---|
| | 1 | 2 | 3 | 4 | 5 | 6 | Mean | Standard deviation | |
| 0.5 | 9.5 | 10 | 10.7 | 9.5 | 10.2 | 10.5 | | | |
| 2 | 36.2 | 30 | 31.5 | 37.5 | 34 | 40 | | | |
| 4 | 59 | 66 | 69 | 60.5 | 66.5 | 72 | | | |
| 6 | 39 | 31.5 | 32.5 | 41 | 40.3 | 36 | | | |
| 10 | 20 | 18.6 | 24.3 | 23.2 | 23.5 | 25.5 | | | |

1. Write the equation for the catalase reaction with hydrogen peroxide: ______

2. Complete the table above to summarize the data from the six trials:

   (a) Calculate the mean volume of oxygen for each stage of germination and enter the values in the table.

   (b) Calculate the standard deviation for each mean and enter the values in the table (you may use a spreadsheet).

   (c) Calculate the mean rate of oxygen production in cm$^3$ per second per gram. For the purposes of this exercise, assume that the weight of germinating seed in every case was 10.0 g.

3. In another scenario, group (trial) #2 obtained the following measurements for volume of oxygen produced: 0.5 d: 4.8 cm$^3$, 2 d: 29.0 cm$^3$, 4 d: 70 cm$^3$, 6 d: 30.0 cm$^3$, 10 d: 8.8 cm$^3$ (pencil these values in beside the other group 2 data set).

   (a) Describe how group 2's new data compares with the measurements obtained from the other groups: ______

   (b) Describe how you would approach a reanalysis of the data set incorporating group 2's new data: ______

ISBN: 978-1-927173-12-1

(c) Explain the rationale for your approach ______

4. Use the tabulated data to plot an appropriate graph of the results on the grid provided:

5. (a) Describe the trend in the data: ______

(b) Explain the relationship between stage of germination and catalase activity shown in the data: ______

6. Describe any potential sources of errors in the apparatus or the procedure: ______

7. Describe two things that might affect the validity of findings in this experimental design: ______

8. Describe one improvement you could make to the experiment in order to generate more reliable data: ______

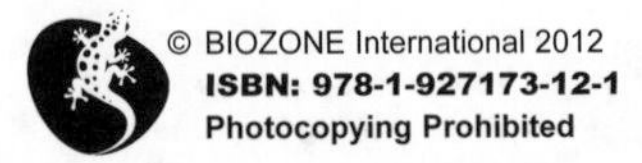

ISBN: 978-1-927173-12-1

# Levels of Organization

Organization and the emergence of novel properties in complex systems are two of the defining features of living organisms. Organisms are organized according to a hierarchy of structural levels (below), each level building on the one before it. At each level, novel properties emerge that were not present at the simpler level. Hierarchical organization allows specialized cells to group together into tissues and organs to perform a particular function. This improves efficiency of function in the organism.

In the spaces provided for each question below, assign each of the examples listed to one of the levels of organization as indicated.

1. **Animals**: *epinephrine, blood, bone, brain, cardiac muscle, cartilage, collagen, DNA, heart, leukocyte, lysosome, mast cell, nervous system, neuron, phospholipid, reproductive system, ribosomes, Schwann cell, spleen, squamous epithelium.*

   (a) Molecular level: ____________________

   ____________________

   (b) Organelles: ____________________

   ____________________

   (c) Cells: ____________________

   ____________________

   (d) Tissues: ____________________

   ____________________

   (e) Organs: ____________________

   ____________________

   (f) Organ system: ____________________

   ____________________

   ____________________

2. **Plants**: *cellulose, chloroplasts, collenchyma, companion cells, DNA, epidermal cell, fibers, flowers, leaf, mesophyll, parenchyma, pectin, phloem, phospholipid, ribosomes, roots, sclerenchyma, tracheid.*

   (a) Molecular level: ____________________

   ____________________

   (b) Organelles: ____________________

   ____________________

   (c) Cells: ____________________

   ____________________

   (d) Tissues: ____________________

   ____________________

   (e) Organs: ____________________

   ____________________

   ____________________

**MOLECULAR LEVEL**

Atoms and molecules form the most basic level of organization. This level includes all the chemicals essential for maintaining life, e.g. water, ions, fats, carbohydrates, amino acids, proteins, and nucleic acids.

**ORGANELLE LEVEL**

Many diverse molecules may associate together to form complex, specialized cellular organelles, where metabolic reactions may be compartmentalized, e.g. mitochondria, Golgi apparatus, endoplasmic reticulum, chloroplasts.

**CELLULAR LEVEL**

Cells are the basic structural and functional units of an organism. Each specialized cell type has a different structure and role as a result of cellular differentiation during development.
***Animal examples*** *include: epithelial cells, osteoblasts, muscle fibers.*
***Plant examples*** *include: sclereids, xylem vessels, sieve tubes.*

**TISSUE LEVEL**

Tissues are collections of specialized cells of the same origin that together carry out a specific function.
***Animal examples*** *include: epithelial tissue, bone, muscle.*
***Plant examples*** *include: phloem, chlorenchyma, endodermis, xylem.*

**ORGAN LEVEL**

Organs are formed by the functional grouping together of multiple tissues. They have a definite form and structure.
***Animal examples*** *include: stomach, heart, lungs, brain, kidney.*
***Plant examples*** *include: leaves, roots, storage organs, ovary.*

**ORGAN SYSTEM LEVEL**

In animals, organs form parts of larger units called **organ systems**. An organ system is an association of organs with a common function, e.g. digestive system, cardiovascular system, urinary system. In all, eleven organ systems make up a mammalian **organism.**

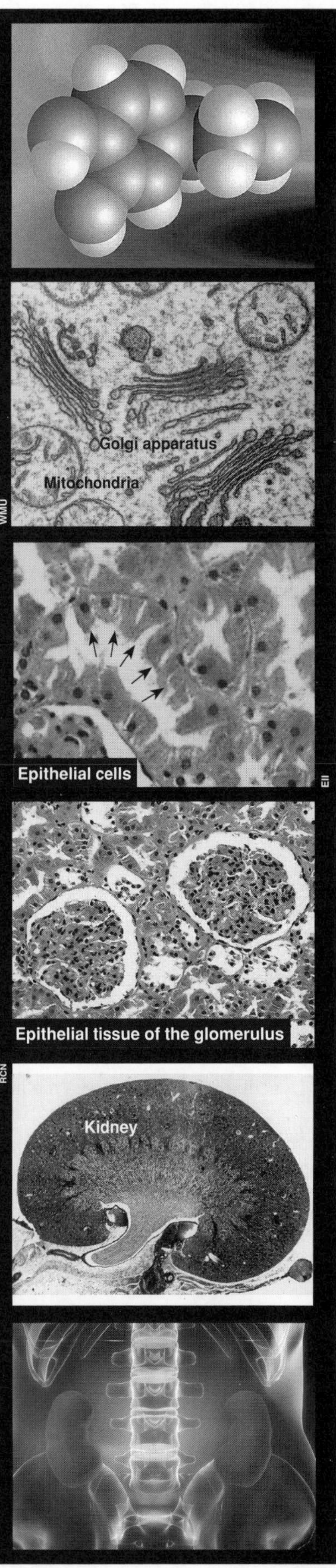




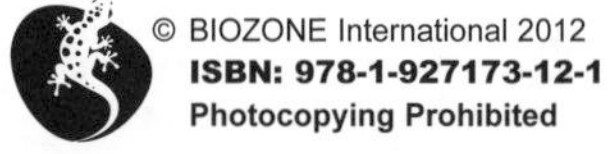


ISBN: 978-1-927173-12-1


***Related activities**: Regional Specialization and Functional Efficiency*

# Achieving Metabolic Efficiency

**Metabolic pathways** are linked biochemical reactions that occur within living organisms to maintain life. Each step in a metabolic pathway relies on the completion of the previous step to progress through to the end. Each step is controlled by specific **enzymes**. The end product of one enzyme-controlled step provides the substrate for the next step, so failure of one step causes failure of all subsequent steps. Metabolic pathways are tightly controlled to prevent unnecessary energy being wasted. This energy conservation is termed **metabolic efficiency**. Metabolic reactions are often localized within regions of a cell, or within specific organelles so that all the necessary components of a metabolic pathway are kept together.

## Achieving Efficiency by Compartmentalization

Membranes separate regions of the cell into compartments. This increases metabolic efficiency because certain metabolic reactions are restricted to regions where all the necessary metabolic components are located. Compartmentalization stops interference between different reaction pathways and enables different reaction environments to be provided within different organelles.

**Example: Cellular Respiration in the Mitochondrion**

The membrane system of the mitochondrion divides it into several regions. Glycolysis takes place outside of the mitochondrion, in the cell's cytoplasm, but the remaining steps take place in different specialized regions of the mitochondrion. This helps to regulate movement of substrates and end-products and therefore reaction rates, increasing efficiency of the process (below).

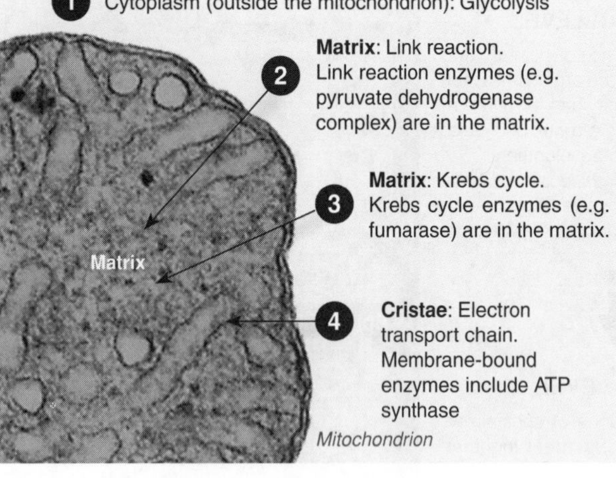
1 Cytoplasm (outside the mitochondrion): Glycolysis
2 **Matrix**: Link reaction. Link reaction enzymes (e.g. pyruvate dehydrogenase complex) are in the matrix.
3 **Matrix**: Krebs cycle. Krebs cycle enzymes (e.g. fumarase) are in the matrix.
Matrix
4 **Cristae**: Electron transport chain. Membrane-bound enzymes include ATP synthase
Mitochondrion

## Achieving Efficiency by Inhibition

Many metabolic pathways are controlled by **feedback inhibition** (negative feedback loop). The pathway is stopped when a build-up of end product (or certain intermediate products) occurs. The build-up stops the enzymes in the pathway from working and allows cells to shut down a pathway when it is not needed. This conserves the cell's energy, so it is not manufacturing products it does not need. Both linear pathways (e.g. glycolysis), and cyclic pathways (e.g. the Krebs cycle) and can be regulated this way (below).

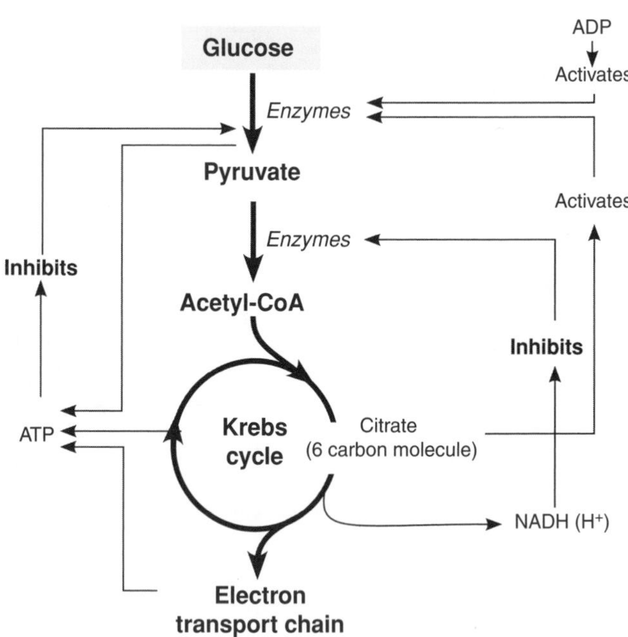
Glucose
Enzymes
Pyruvate
Enzymes
Acetyl-CoA
Krebs cycle
Citrate
(6 carbon molecule)
Electron transport chain
ADP
Activates
Activates
Inhibits
Inhibits
ATP
NADH ($H^+$)

1. What does metabolic efficiency mean?

2. Describe how cells achieve metabolic efficiency through:

   (a) **Compartmentailization**:

   (b) **Feedback inhibition**:

3. What would happen if cells could not regulate their metabolic pathways?

***Related activities**: Levels of Organization, The Biochemistry of Respiration*

**ISBN: 978-1-927173-12-1**

# Regional Specialization and Functional Efficiency

Functional efficiency in multicellular organisms is enhanced by having specialized organs and groups of organs (**organ systems**) that perform a specific function. Examples include the digestive system and the immune system. Organs that perform a specialized role can work more efficiently than if they were required to carry out a number of different tasks. Within an organ system, further specialization occurs. Specific organs or regions perform certain tasks, and this increases the functional efficiency of the system as a whole. The example below illustrates how efficiency is achieved in the digestive system of a mammal.

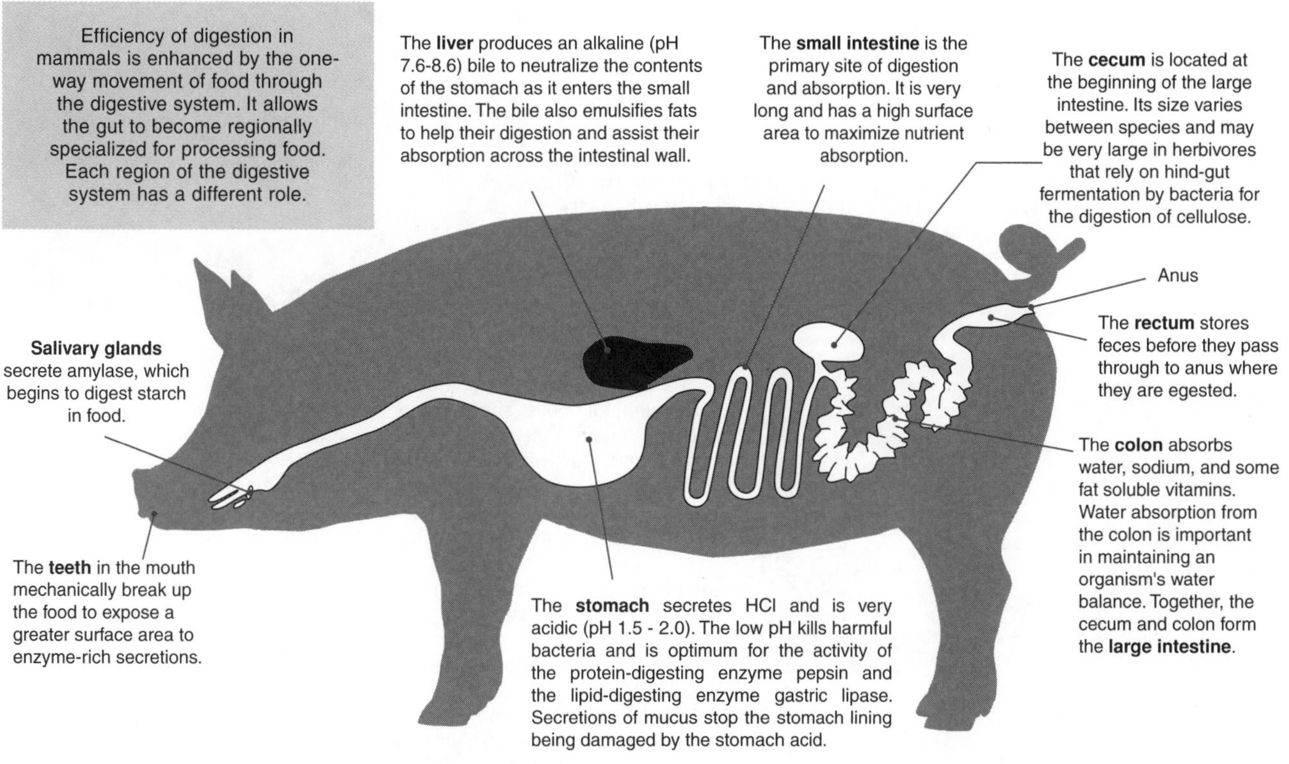


### Organ Systems Work Together

Although multicellular organisms have specialized organ systems that perform specific roles, the various organ systems must interact and work together to contribute to the overall functioning of the organism. For example:

- The skeletal and muscular systems protect the heart and lungs from damage and enable ventilation of the lungs.
- The respiratory system provides the environment for the exchange of $O_2$ and $CO_2$ between the blood and the air.
- The circulatory system transports $O_2$ from the gas exchange surfaces to the tissues and transports $CO_2$ in the opposite direction.

### Specialization of the Internal Environment

The chemical environment within each region of the mammalian digestive tract varies. For example, the stomach is a very acidic environment, whereas the small intestine has a more neutral pH. Each region of the gut provides the optimal chemical environment for the enzymes operating there. This helps to increase digestive efficiency, reducing the time it takes for food to pass through the gut and enabling the animal to maximize food intake.

*Pepsin*

For example, the protein-digesting enzymes pepsin (left) and trypsin both degrade proteins to peptides. Pepsin is found in the stomach, operates at pH 2.0, and is inactive in the neutral pH of the small intestine. Trypsin works optimally in the more neutral pH of the small intestine. Having several different enzymes performing a similar task maximizes the digestion of food.

1. (a) What is the advantage of organ specialization? ______________________________

______________________________

______________________________

(b) How do multicellular organisms achieve functional efficiency? ______________________________

______________________________

______________________________

2. Explain how regional functionality of the mammalian digestive system is achieved: ______________________________

______________________________

______________________________

______________________________

______________________________

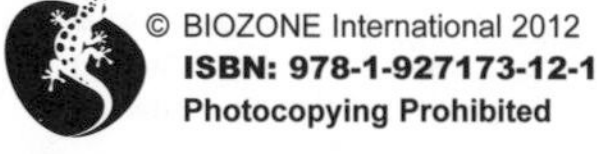

**ISBN: 978-1-927173-12-1**


***Related activities:*** *Diversity in Tube Guts, Mammalian Guts*

# Metabolism in Bacterial Communities

Until recent years, bacteria were viewed as unicellular organisms that showed individual behaviors. There are now many examples of bacterial communities showing behavior that is closer to multicellularity than unicellularity. By behaving somewhat like a multicellular organism, bacterial communities can access nutrient and energy sources that would otherwise be unavailable to them.

Under certain conditions microbial communities can attach to surfaces, forming organized structures called **biofilms**. During the formation of biofilms, cells undergo numerous changes in gene expression, which lead to changes in the metabolism of the community as a whole. The changes offer new properties to the biofilm including enhanced production of extracellular polymeric substances (for anchoring the colony), antimicrobial resistance, quorum sensing, and gene transfer. The biofilm also offers protection against dehydration, osmotic stress, and nutrient limitations. Biofilms allow individual cells to interact in a way that optimizes use of available resources. Biofilms are important in both medicine and industry. In medicine they are responsible for producing plaques (e.g. dental plaque) and infections that are extremely difficult to eliminate. In industry, biofilms can cause numerous problems, including reducing fuel efficiency in ships, fouling heat exchangers, and contaminating food processing machinery.

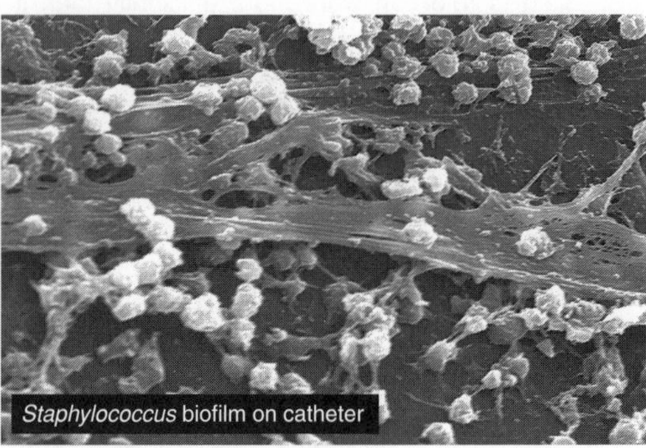

*Staphylococcus* biofilm on catheter

CDC

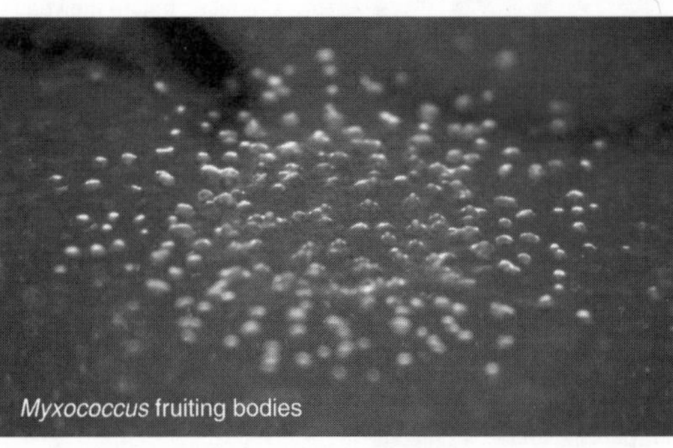

*Myxococcus* fruiting bodies

Trance Gemini GFDL

*Myxococcus xanthus* is well known for the cooperative ability of individual cells. When starved, cells come together to form fruiting bodies, but when food is plentiful, cells show cooperative predatory behavior. When in an aqueous environment with cyanobacteria, *M. xanthus* will secrete lytic enzymes and feed off the cyanobacteria. However, the lytic enzymes are diluted in water. *M. xanthus* deals with this by forming spherical colonies with cyanobacteria trapped inside. The colony is then able to feed effectively on mass, thus utilizing the resource by multicellular behavior.

Heterocysts

*Anabaena*

B Hicks

Cyanobacteria (e.g. *Anabaena*) are a phylum of bacteria that are able to photosynthesize. The bacteria often form long filaments of individual cells joined together. Under low-nitrogen conditions, some of these cells will differentiate into **heterocysts**. These cells are able to fix nitrogen from molecular $N_2$. They show quite different gene expression from neighboring undifferentiated cells, most importantly in the production of the enzyme **nitrogenase** and their inability to photosynthesize. Moreover, the heterocysts share the nitrogen they fix with neighboring cells, while receiving other nutrients from them.

In marine bacterial communities, the bacterium *Thiovulum* forms meshes of up to a million cells per $cm^2$. These are formed when *Thiovulum* cells swarm to a transition zone between oxygen rich water above and sulfide rich water below (e.g. seeping from a vent). The population produces slime threads that hold the cells together. The mesh separates the two layers of water and enables *Thiovulum* to regulate the flow of each. By swimming as a group, the mesh community is able to move the mesh up or down to maintain its position on the boundary and control the amount of oxygen or sulfide they receive.

1. (a) Explain how **biofilms** are formed: ______________________________

   (b) How does a bacterial community benefit by forming a biofilm? ______________________________

   (c) Why is understanding biofilm formation and metabolism important for medicine and industry? ______________________________

2. Using an example from above, explain how unicellular organisms benefit by forming communities that behave like multicellular organisms:

______________________________




**ISBN: 978-1-927173-12-1**

# KEY TERMS: Crossword

Complete the crossword below, which will test your understanding of key terms in this chapter and their meanings

**Clues Across**

2. A chemical that lowers the activation energy of a reaction but is not itself used up during the reaction.
6. A metabolic process in which complex molecules are broken down into simpler ones.
7. A chain of enzyme-catalyzed biochemical reactions in living cells (2 words: 9, 7).
9. A type of enzyme regulation that occurs when a substance binds to a site on an enzyme other than the active site (2 words:10, 10).
10. A type of enzyme inhibition where the substance and inhibitor compete to bind in the active site.
12. A process where an enzyme is deactivated, either permanently or temporarily.
13. A reaction that absorbs or requires energy in order to proceed is described as this.
14. The name of a reaction that releases energy.
15. A microbially-derived sessile community attached to a surface with the help of extracellular polymeric substances.
16. A globular protein which acts as catalyst to speed up a specific biological reaction.

**Clues Down**

1. The term that describes the conditions at which an enzyme is most active.
2. An additional non-protein substance essential for the operation of some enzymes.
3. The loss of the three-dimensional structure of proteins, often caused by pH or heat.
4. Energy required to initiate a chemical reaction (2 words; 10, 6).
5. A metabolic process that builds more complex molecules from simple ones.
8. The region of an enzyme responsible for substrate binding and reaction catalysis (2 words; 6, 4).
11. An organic molecule that acts as a cofactor in an enzyme reaction but is only loosely bound to the enzyme.



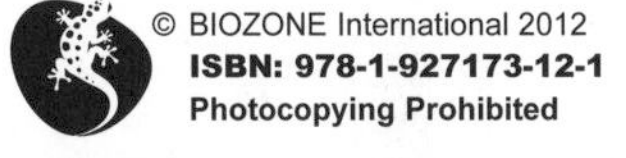


ISBN: 978-1-927173-12-1

Enduring Understanding

2.A

2.C

# Homeostasis and Energy Allocation

## Key concepts

- Free energy is required to maintain organization, and for growth and reproduction in organisms.
- An organism's reproductive strategy may vary depending on the availability of free energy.
- Homeostasis is maintained using hormonal and nervous mechanisms via negative feedback.
- Negative feedback stabilizes systems against excessive change. Positive feedback leads to instability but has some specific physiological roles.
- Disruptions to feedback controls may cause disease.

## Key terms

blood clot
cold stratification
countercurrent exchange
diabetes mellitus
diapause
dormancy
ectothermic
endothermic
fibrin
glucagon
hemostasis
homeostasis
homeotherm
hyperthermia
hypothalamus
hypothermia
insulin
life history strategy
metabolism
negative feedback
poikilotherm
positive feedback
reproductive strategy
thermoregulation
vernalization

## Essential Knowledge

☐ 1. Use the **KEY TERMS** to compile a glossary for this topic.

### Maintaining Organization *(2.A.1: d)*

pages 13,46-50, 55-59, 64

☐ 2. Use examples to illustrate that organisms must use free energy to maintain organization of their systems, and to grow and reproduce.

☐ 3. Describe adaptations in animals for **thermoregulation** *(also 2.C.2)*. Compare and contrast mechanisms of thermoregulation in **ectotherms** and **endotherms** (i.e. endothermic homeotherms). Explain the role of **countercurrent exchange mechanisms** in thermoregulation in some animals *(also 2.D.2)*.

☐ 4. Describe mechanisms to maintain elevated floral temperatures in some plants.

☐ 5. Appreciate that reproduction requires free energy beyond that used for maintenance and growth. Describe how organisms allocate energy to reproduction, recognizing that different **reproductive strategies** reflect evolutionary history and adaptation to environment. Describe how organisms change their allocation to reproduction based on the free energy available.

☐ 6. Describe the relationship between metabolic rate per unit body mass and organism size. Discuss possible reasons for the relationship.

☐ 7. Describe and explain the result of excess and insufficient acquired free energy.

### Feedback Mechanisms *(2.C.1)*

pages 51-63

☐ 8. Explain how **negative feedback** mechanisms maintain dynamic **homeostasis** by stabilizing systems against excessive change. Use examples (e.g. thermoregulation in animals) to explain how negative feedback operates in biological systems.

☐ 9. Describe examples of how **positive feedback** operates in biological systems (e.g. in lactation, labor (parturition), and blood clotting). Appreciate that positive feedback is unstable because it amplifies disturbance, but it has a role in certain physiological functions.

☐ 10. Explain situations where disruptions to feedback mechanisms result in disease, e.g. **diabetes mellitus** in response to decreased **insulin**, diabetes insipidus in response to decreased secretion of ADH, and blood clotting disorders.

*Periodicals:*
*Listings for this chapter are on page 374*

*Weblinks:*
*www.thebiozone.com/weblink/AP2-3121.html*

*BIOZONE APP:*
*Student Review Series*
*Integument & Homeostasis*

# Metabolism and Body Size

During a deep dive, the largest whales are able to stay submerged for almost two hours. However, small diving mammals such as the water shrew, can only stay submerged for thirty seconds. Part of the reason for these variations is the obvious difference in the size of their lungs and their body's ability to store oxygen. The most important factor is the difference in the rate at which their body's consume oxygen. The water shrew uses oxygen at a rate of 7.4 L $kg^{-1}$ $h^{-1}$. The blue whale uses just 0.02 L $kg^{-1}$ $h^{-1}$. The rate of oxygen consumption is a measure of the metabolic rate of an organism (since oxygen is required to produce the ATP used for metabolic activities). In general, the larger an organism is the lower its specific metabolic rate.

| Animal | Body mass (kg) | $O_2$ consumption (L $kg^{-1}$ $h^{-1}$) |
|---|---|---|
| Shrew | 0.005 | 7.40 |
| Harvest mouse | 0.009 | 2.50 |
| Kangaroo mouse | 0.015 | 1.80 |
| Mouse | 0.025 | 1.65 |
| Ground squirrel | 0.096 | 1.03 |
| Rat | 0.290 | 0.87 |
| Cat | 2.5 | 0.68 |
| Dog | 11.7 | 0.33 |
| Sheep | 42.7 | 0.22 |
| Human | 70 | 0.21 |
| Horse | 650 | 0.11 |
| Elephant | 3830 | 0.07 |

Data: Schmidt-Nielsen

**Body Mass vs Oxygen Consumption Rate in Mammals**

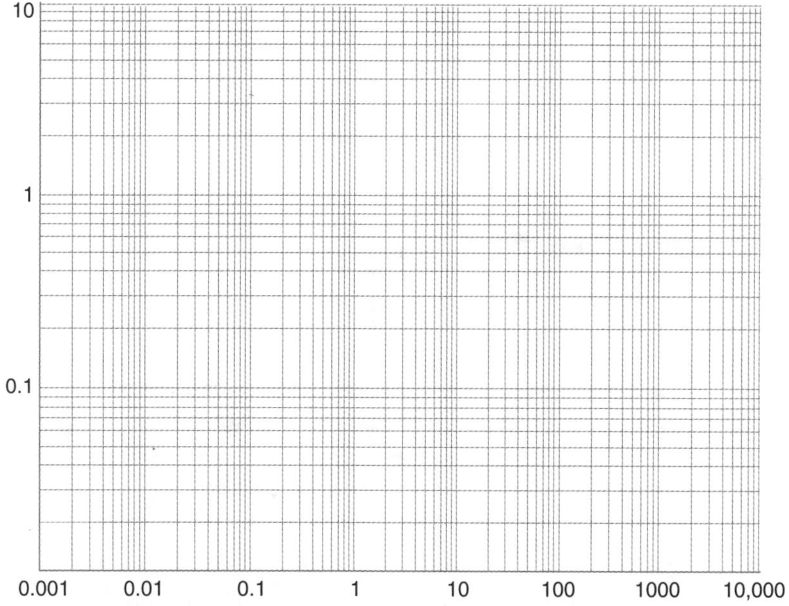

The relationship between specific metabolic rate and body mass for mammals is expressed as $B \sim 70\ M^{-0.25}$. That is, the specific metabolic rate (B) of any mammal is proportional to its mass (M (kg)) to the power of -0.25. Thus a 2 kg cat has a specific metabolic rate 3.1 times less than a 0.02 kg mouse. However, there is debate over the accuracy of this law, as there appear to be many exceptions.

The relationship of specific metabolic rate and size also appears in organisms other than mammals. Birds follow a relationship very close to that of mammals. Invertebrates and unicellular organisms all follow similar patterns in that the bigger they are, the slower their metabolic rate. Trees follow a relationship close to 1 (i.e. a tree half the size of another will have a metabolic rate twice as great).

Exactly why metabolic rates are slower in large animals is uncertain. Certainly the surface area to volume ratio of organisms appears to be involved (small animals radiate more heat per volume than large ones). But this does not explain why poikilothermic animals follow a similar relationship to mammals. It may be due to smaller animals having more structural mass per volume, which metabolically costs more to maintain.

1. (a) Use the logarithmic grid above to plot the rate of oxygen consumption for the animals listed in the table:

   (b) What is the shape of the graph? ____________________

2. Explain why a small mammal (e.g. a shrew) needs to proportionally eat more than a large mammal (e.g. an elephant):

   ____________________

   ____________________

3. Calculate the following:

   (a) How many times greater the specific metabolic rate of a shrew is than a horse: ____________________

   (b) How many times less the specific metabolic rate of a human is than a rat: ____________________

4. Explain how specific metabolic rate may affect the diving times of mammals: ____________________

   ____________________

Homeostasis and Energy Allocation


ISBN: 978-1-927173-12-1

# How Organisms Allocate Energy

The energy available to any organism is limited. Even plants, which use the virtually unlimited solar energy to produce glucose, can only produce it during daylight and, in many places, only during spring and summer. The result of this limited energy supply is that organisms must make trade-offs and compromises with respect to how they allocate energy. Energy used for locomotion cannot be used for growing. Organisms may travel great distances to find the energy source they need. Provided they expend less energy travelling than they gain once they arrive, there will be a net energy gain.

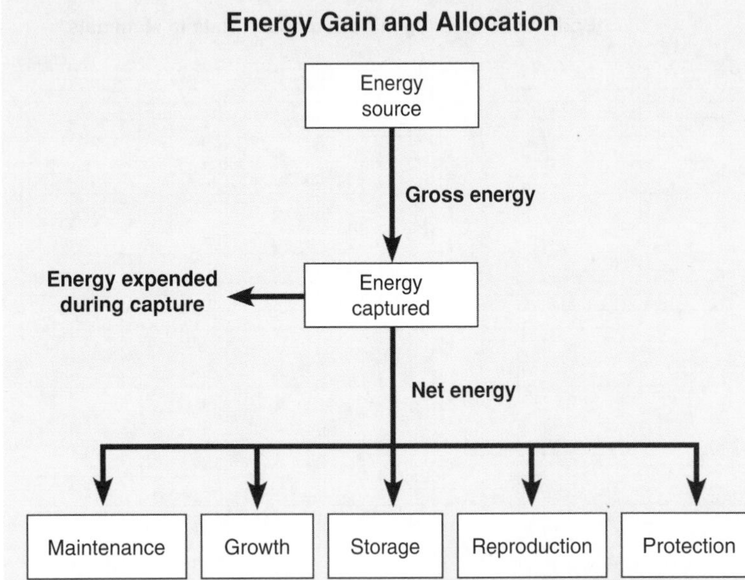


The net energy gain from an energy source depends on the size of the energy source and the energy expended to obtain it. In the case of a predator, such as a cheetah, the energy expended in the chase must be less than the energy gained from the eating the prey. Enough energy must be left over to maintain the body, grow, or reproduce.

A cheetah's energy expenditure during a hunt is very precise. If it doesn't make a kill in under a minute, it stops the chase.

A large amount of energy is expended during migration. The payoff is that the energy gained from abundant food sources at the destination is often enough to allow the animal to breed, rear its young, and carry it through the return migration and winter.

Magne Flåten cc 3.0

Hibernation saves energy. The low metabolic rate of a hibernating animal reduces its energy expenditure, allowing it to survive through the winter when food is scarce. However, large amounts of energy must be stored during summer to carry it through the hibernation period.

Deciduous trees expend energy growing new leaves every spring. They produce all the energy needed to grow and reproduce before losing the leaves in autumn. Losing leaves reduces frost or storm damage to the tree during winter months.

1. Why is it important for organisms to accurately assess energy gains against expenditure when selecting food sources:

2. In terms of energy allocation and expenditure, explain the following:

   (a) Why many animals migrate every year despite the large energy cost:

   (b) Why hibernation is a strategy for animals living in regions with extreme seasonal variations:




**ISBN: 978-1-927173-12-1**

### Energy Requirements for Infants Aged 0-12 Months

| Age (months) | Total energy expenditure (MJ day$^{-1}$) | | Energy deposition (MJ day$^{-1}$) | |
|---|---|---|---|---|
| | Boys | Girls | Boys | Girls |
| 0-1 | 1.3 | 1.2 | 0.88 | 0.75 |
| 1-2 | 1.6 | 1.5 | 0.76 | 0.67 |
| 2-3 | 1.9 | 1.7 | 0.58 | 0.56 |
| 3-4 | 2.1 | 2.0 | 0.22 | 0.29 |
| 4-5 | 2.4 | 2.1 | 0.19 | 0.24 |
| 5-6 | 2.5 | 2.3 | 0.15 | 0.20 |
| 6-7 | 2.6 | 2.4 | 0.069 | 0.083 |
| 7-8 | 2.8 | 2.5 | 0.065 | 0.069 |
| 8-9 | 2.9 | 2.6 | 0.057 | 0.063 |
| 9-10 | 3.0 | 2.7 | 0.089 | 0.074 |
| 10-11 | 3.1 | 2.8 | 0.087 | 0.063 |
| 11-12 | 3.2 | 2.9 | 0.093 | 0.060 |

Data: FAO

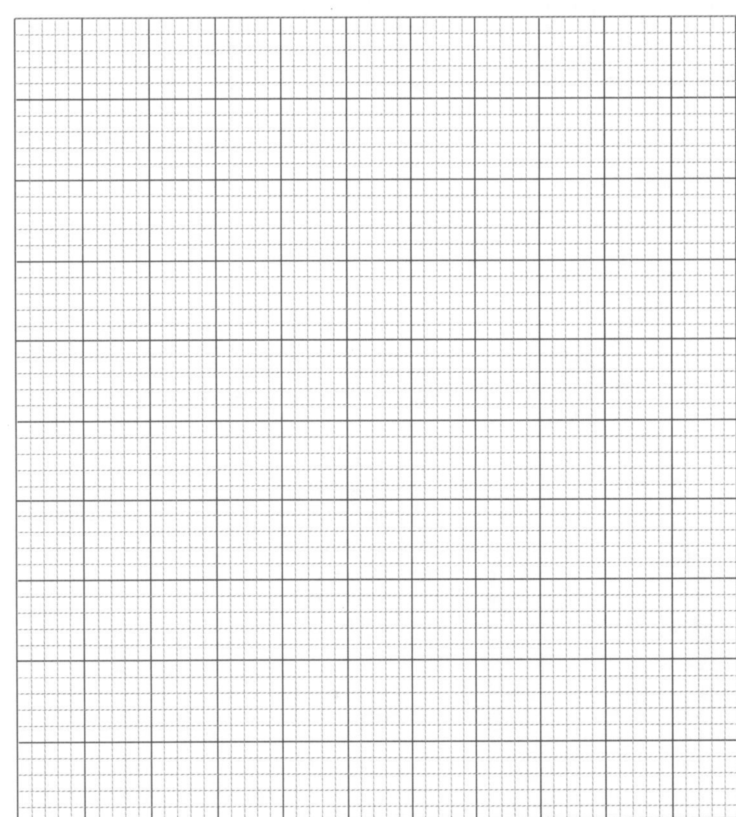

In order to grow, an organism must obtain more energy than it expends in activity and maintenance. As the body grows, new tissue also requires maintaining, and so the daily energy needs of a growing organism increase. As growth slows, the daily energy requirements above those needed for maintenance are reduced. The table above lists the daily total energy expenditure and the daily energy deposition (used for growth) for boys and girls from birth to 12 months. Adding the two values gives the total daily energy requirement for boys or girls.

When people obtain more energy than they need for daily activities or growth, the extra energy is stored as fat. This is a growing problem many countries where high energy food is easily obtainable.

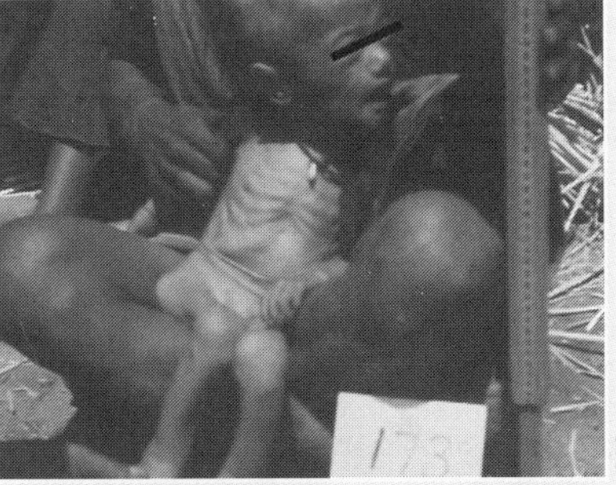

CDC

When the body does not get enough energy to carry out its daily activities, it starts to obtain energy by breaking down tissues in the body. This begins with fat tissue, then proteins (normally muscle).

Balancing energy input and energy output is important for managing weight or being able to carry out high energy physical tasks like building muscle or playing sport. High energy tasks require greater energy input.

3. (a) Plot a graph of the total energy expenditure and energy deposition for boys and girls aged 0 to 12 months.

   (b) What happens to the total energy expenditure for infants as they grow? ______

   (c) What happens to the energy deposition for infants as they grow? ______

   (d) How does the ratio of energy deposition and total energy requirement change for infants as they grow? ______

4. For someone with a sedentary (inactive) lifestyle, how does the body deal with an excess energy intake? ______



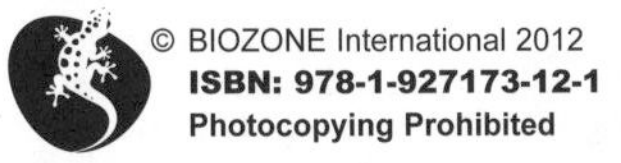


ISBN: 978-1-927173-12-1

# Energy and Seasonal Breeding

Reproduction is an energetically expensive process. Energy is required to find a mate, produce gametes, and raise, feed, and protect the young. The breeding cycles of plants and animals are normally timed to maximize the use of available energy (sunlight or food). Animals and plants most often reproduce during the spring months, although some animals will begin the reproductive process in autumn, so that spring and summer can be entirely devoted to feeding and raising young.

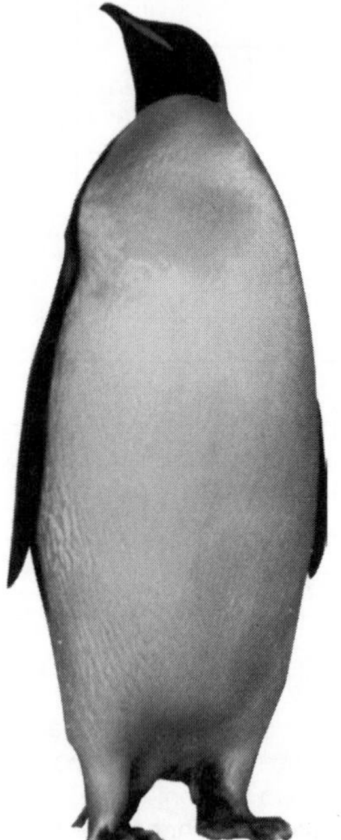

The reproductive cycle for seasonally reproductive animals often begins in late autumn, with gestation being over the winter months, so that the young are born in spring. This maximizes the time available for growth and raising young before the next winter and synchronizes it with the time of maximum food availability.

Emperor penguins begin their reproductive cycle in March, at the beginning of the Antarctic winter. Having spent the summer feeding, they proceed to breeding colonies. Chicks are raised through the winter and are independent by November-December, the start of the Antarctic summer. Raising chicks through the harsh Antarctic winter allows them to be independent by summer and so be better able to survive the following winter.

Emperor penguins must also have enough time after raising their chick to feed before their annual molt, during which they can not enter the sea and so must again rely on their fat stores.

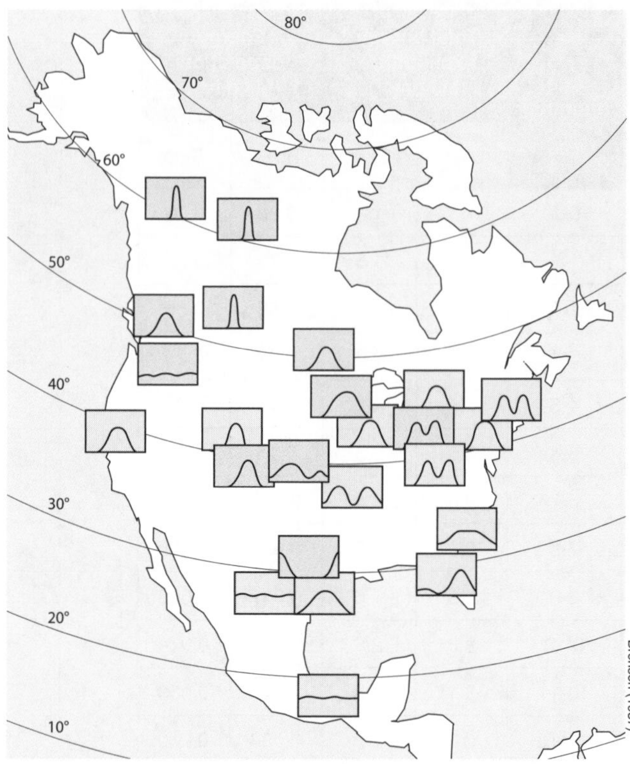


The diagram above shows the relative proportions of female deer mice (*Peromyscus*) pregnant at any one time of the year (January on the left side of the graphs, December on the right). Note that at higher latitudes, females breed during summer months only, whereas at lower latitudes they breed throughout the year (the blue boxes show one single species).

Red deer stags

In many animals, male-male competition develops where the benefits (access to females or resources) outweigh the energy costs (fighting or territory defense). The amount of energy expended by males in attracting or monopolizing several females needs to be weighed against the assistance the females might need to raise the young and how far they range. It is energetically uneconomic for males to defend a large number of females if they require male assistance to rear the young or travel a great distance to find food. Generally, the most energy expensive reproductive activities for males involve courtship and mating. In females, it is in pregnancy and rearing young.

Red deer does

1. Explain why many animals and plants have annual reproductive cycles: ______________________

2. (a) In which part of the reproductive process do males tend to spend the most energy? ______________________

   (b) In which part of the reproductive process do females tend to spend the most energy? ______________________

3. Explain why raising young is often timed to coincide with Spring: ______________________



**Related activities:** *How Organisms Allocate Energy*


**ISBN: 978-1-927173-12-1**

# Reproductive Allocation and Parental Care

Organisms allocate a certain amount of energy to reproduction (the **reproductive effort**). How the reproductive effort is apportioned is termed the reproductive strategy and it is highly variable between and even within species. Generally, species can divide their reproductive effort into many small units (offspring), or into a smaller number of larger units. A reproductive strategy involving the release of millions of eggs into the environment involves little risk to the parent, but few of the offspring will survive. Reproductive strategies in which more energy is invested in the young are usually associated with a higher degree of parental care before and often after the young are born or hatched. The more energy a parent invests in each young, the greater (in general) the degree of parental care. Parental care is energy expensive and, in many species, the parent will not have the resources to care for all the young.

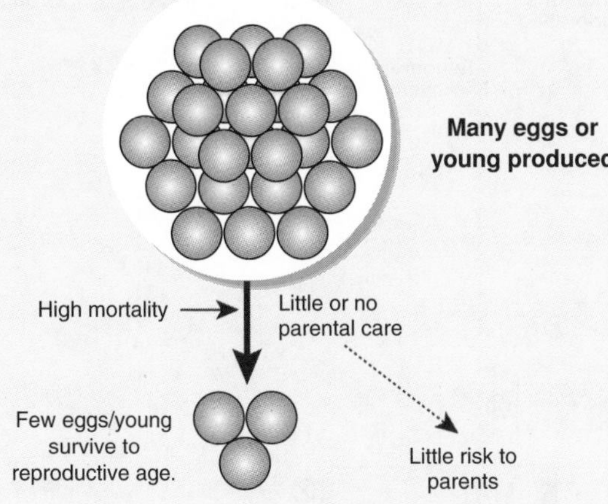


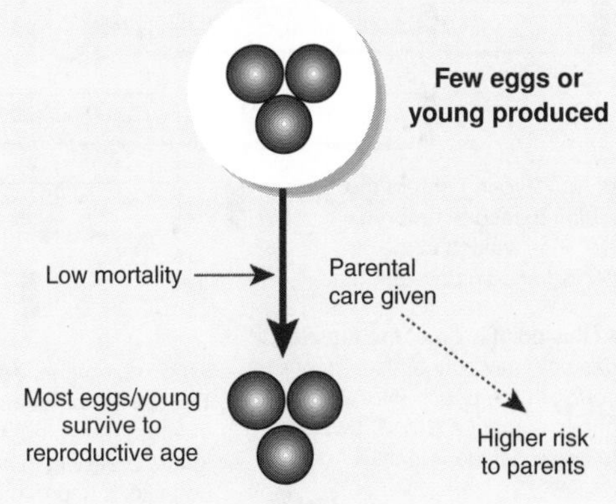


### Little or No Parental Care

- Large number of offspring produced.
- The reproductive effort per offspring is low.
- Little or no parental care of the offspring.
- Reproductive effort is directed at producing the offspring and not at parental care after birth.

### Parental Care

- Few offspring produced.
- The reproductive effort per offspring is high.
- Moderate/substantial care of young after birth.
- Greater reproductive effort is directed at parental care after birth.

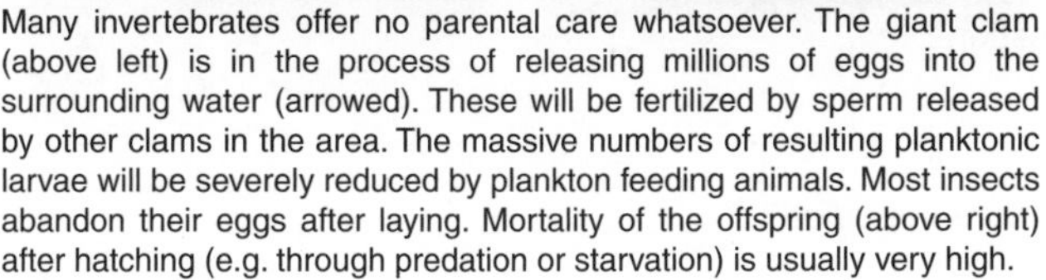

Many invertebrates offer no parental care whatsoever. The giant clam (above left) is in the process of releasing millions of eggs into the surrounding water (arrowed). These will be fertilized by sperm released by other clams in the area. The massive numbers of resulting planktonic larvae will be severely reduced by plankton feeding animals. Most insects abandon their eggs after laying. Mortality of the offspring (above right) after hatching (e.g. through predation or starvation) is usually very high.

Both mammals and birds are well known for their high levels of parental care. Other vertebrates, such as some amphibians, fish, and reptiles also provide care until the offspring are capable of fending for themselves. Bird parents are required to incubate their eggs in a nest and then feed the chicks until they are independent. Although most mammals give birth to well developed offspring, they are dependent on their mother for nourishment via suckling milk, as well as learning valuable behaviors.

1. What factors might be important in determining if an animal reproduces: ______________________

2. Discuss patterns of parental investment in reproduction, explaining how these relate to the extent of parental care:




ISBN: 978-1-927173-12-1


***Periodicals:*** *Evolution of parental care*

***Related activities:*** *Diapause as a Reproductive Strategy*

# Diapause as a Reproductive Strategy

**Diapause** describes a strategy in which growth and development is suspended in response to unfavorable conditions. It is a survival mechanism most often associated with insects. However, some of Australia's marsupials incorporate embryonic diapause into their reproductive strategy as an adaptation to a highly seasonal environment subject to frequent droughts. The **red kangaroo** is one such marsupial. In good conditions, a female may have a joey at heel, one in the pouch, and a dormant embryo ready to replace the pouch offspring as soon as it leaves. Conversely, in prolonged periods of drought and low forage, breeding may cease, as illustrated below. The numbers in the diagram indicate offspring number, showing how the young develop through time.

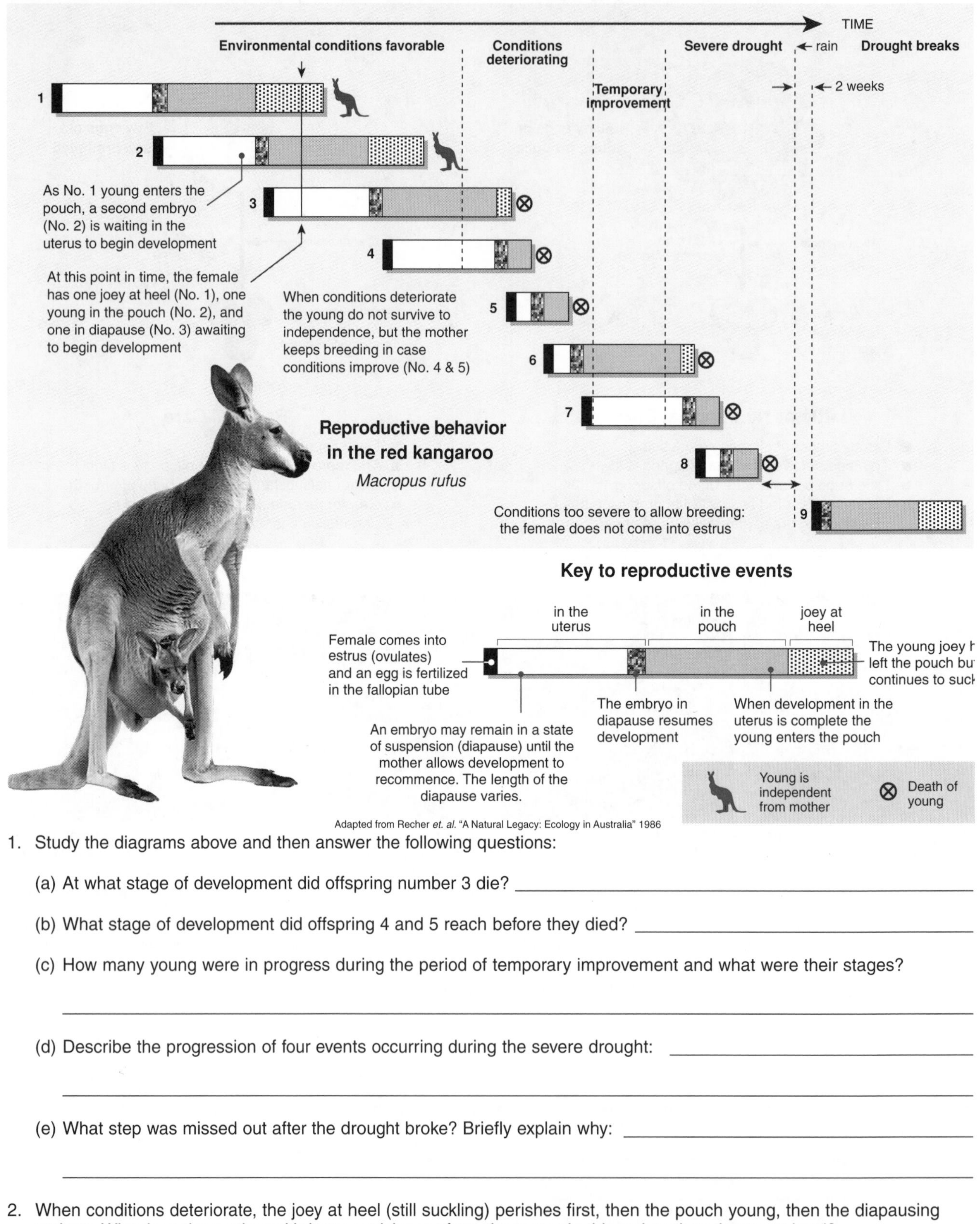


Adapted from Recher *et. al.* "A Natural Legacy: Ecology in Australia" 1986

1. Study the diagrams above and then answer the following questions:

   (a) At what stage of development did offspring number 3 die? ________________

   (b) What stage of development did offspring 4 and 5 reach before they died? ________________

   (c) How many young were in progress during the period of temporary improvement and what were their stages?

   (d) Describe the progression of four events occurring during the severe drought: ________________

   (e) What step was missed out after the drought broke? Briefly explain why: ________________

2. When conditions deteriorate, the joey at heel (still suckling) perishes first, then the pouch young, then the diapausing embryo. Why does the mother withdraw nourishment from the young in this order when times are hard?


**ISBN: 978-1-927173-12-1**

# Negative Feedback

**Homeostasis** refers to the relative physiological constancy of the body, despite external fluctuations. Homeostasis of the internal environment is an essential feature of complex animals and it is the job of the body's **organ systems** to maintain it, even as they make necessary exchanges with the environment. Homeostatic control systems have three functional components: a receptor to detect change, a control center, and an effector to direct an appropriate response. In **negative feedback** systems, movement away from a steady state triggers a mechanism to counteract further change in that direction. Using negative feedback systems, the body counteracts disturbances and restores the steady state. **Positive feedback** is also used in physiological systems, but to a lesser extent since positive feedback leads to the response escalating in the same direction.

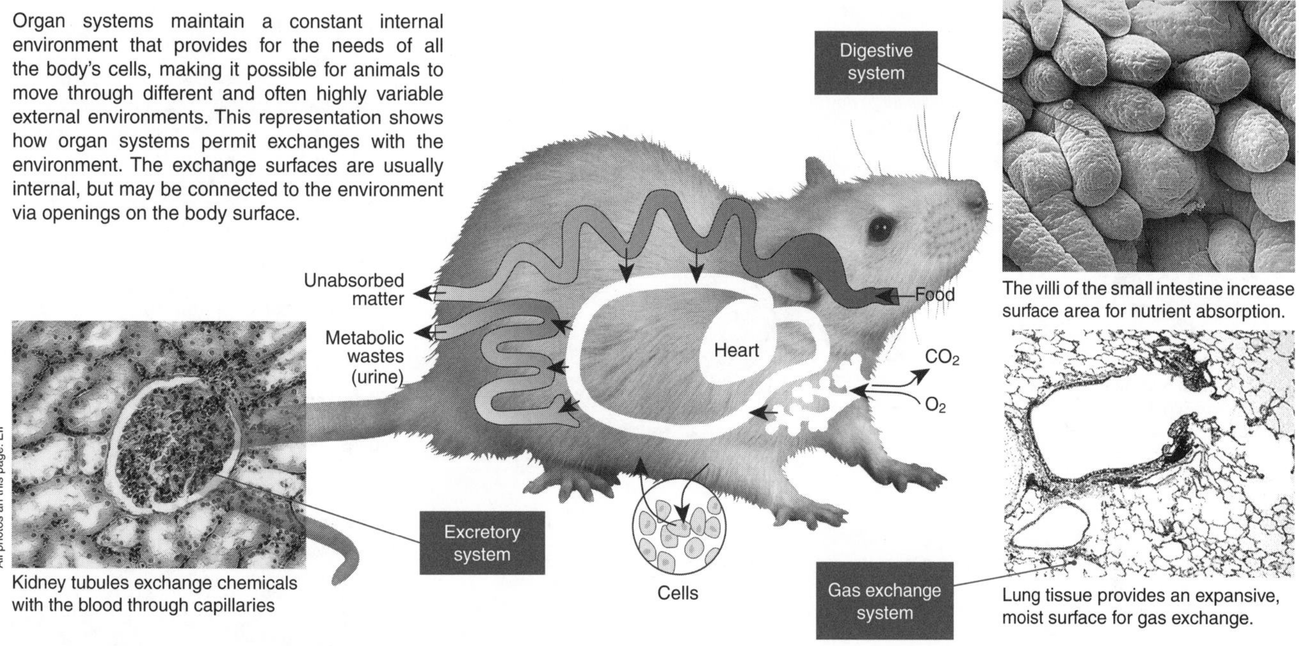


Kidney tubules exchange chemicals with the blood through capillaries

The villi of the small intestine increase surface area for nutrient absorption.

Lung tissue provides an expansive, moist surface for gas exchange.

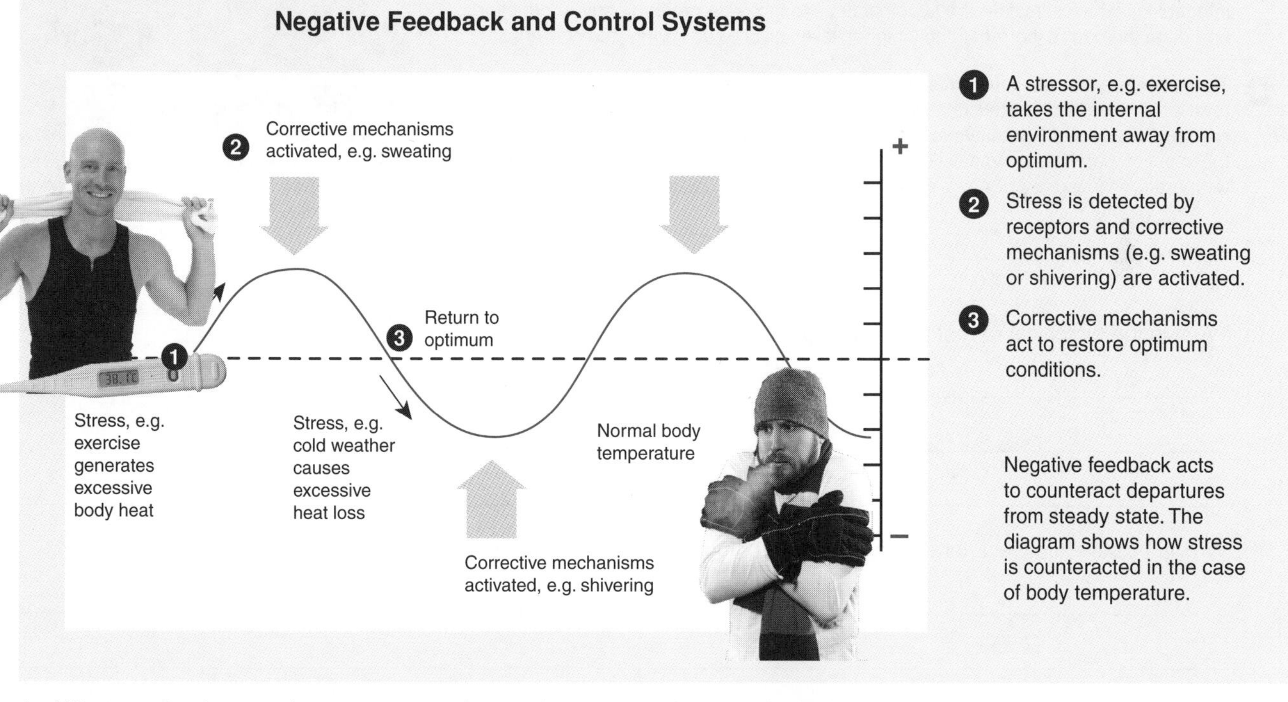


Homeostasis and Energy Allocation

1. What are the three main components of a regulatory control system in the human body and what is the role of each?

2. How do negative feedback mechanisms maintain homeostasis in a variable environment?



**ISBN: 978-1-927173-12-1**

# Positive Feedback

**Positive feedback** mechanisms amplify a physiological response in order to achieve a particular result. Labor, lactation, fever, blood clotting, and fruit ripening all involve positive feedback mechanisms. Normally, a positive feedback loop is ended when the natural resolution is reached (e.g. baby is born, pathogen is destroyed, fruit falls off). Few homeostatic mechanisms are controlled by positive feedback because such mechanisms are unstable. If left unchecked, they can be dangerous or even fatal.

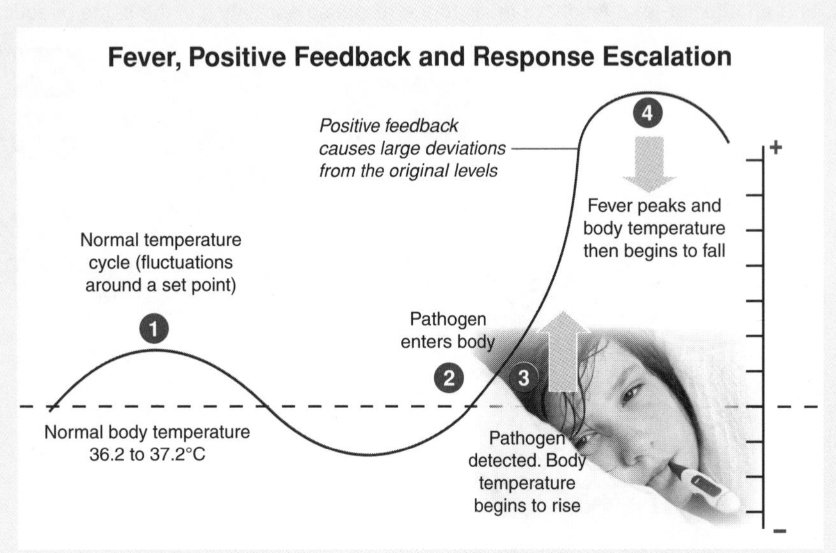


1. Body temperature fluctuates on a normal, regular basis around a narrow set point.
2. Pathogen enters the body.
3. The body detects the pathogen and macrophages attack it. Macrophages release interleukins which stimulate the hypothalamus to increase prostaglandin production and reset the body's thermostat to a higher 'fever' level by shivering (the chill phase).
4. The fever breaks when the infection subsides. Levels of circulating interleukins (and other fever-associated chemicals) fall, and the body's thermostat is reset to normal. This ends the positive feedback escalation and normal controls resume. If the infection persists, the escalation may continue, and the fever may intensify. Body temperatures in excess of 43°C are often fatal or result in brain damage.

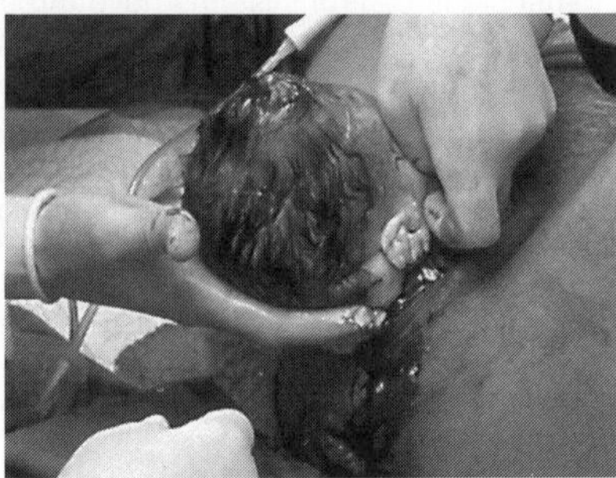

**Labor and lactation**: During childbirth (above), the release of oxytocin intensifies the contractions of the uterus so that labor proceeds to its conclusion. The birth itself restores the system by removing the initiating stimulus. After birth, levels of the milk-production hormone prolactin increase. Suckling maintains prolactin secretion and causes the release of oxytocin, resulting in milk release. The more an infant suckles, the more these hormones are produced.

Ethylene is a gaseous plant hormone involved in fruit ripening. It accelerates the ripening of fruit in its vicinity so nearby fruit also ripens, releasing more ethylene. Over-exposure to ethylene causes fruit to over-ripen (rot).

1. (a) What is the biological role of positive feedback loops? Describe an example: ____________________

(b) Why is positive feedback inherently unstable (compare with negative feedback)? ____________________

(c) How is a positive feedback loop normally stopped? ____________________

(d) Describe a situation in which this might not happen. What would be the result? ____________________



*Related activities:* Negative Feedback, Maintaining Homeostasis
*Weblinks:* Control, Regulation and Feedback


ISBN: 978-1-927173-12-1

# Maintaining Homeostasis

The various organ systems of the body act to maintain homeostasis through a combination of hormonal and nervous mechanisms. In everyday life, the body must regulate respiratory gases, protect itself against agents of disease (pathogens), maintain fluid and salt balance, regulate energy and nutrient supply, and maintain a constant body temperature. All these must be coordinated and appropriate responses made to incoming stimuli. In addition, the body must be able to repair itself when injured and be capable of reproducing (leaving offspring).

## Regulating Respiratory Gases

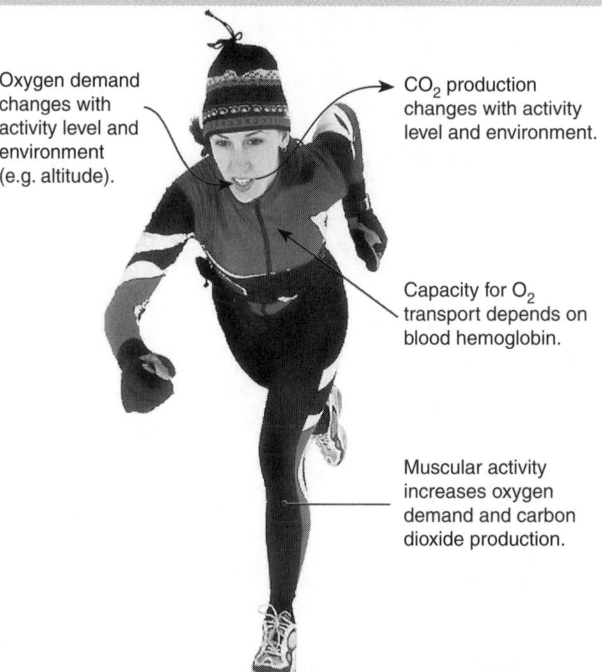


Oxygen must be delivered to all cells and carbon dioxide (a waste product of cellular respiration) must be removed. **Breathing** brings in oxygen and expels $CO_2$, and the cardiovascular and lymphatic systems circulate these respiratory gases (the oxygen mostly bound to hemoglobin). The rate of breathing is varied according to oxygen demands (as detected by $CO_2$ levels in the blood).

## Coping with Pathogens

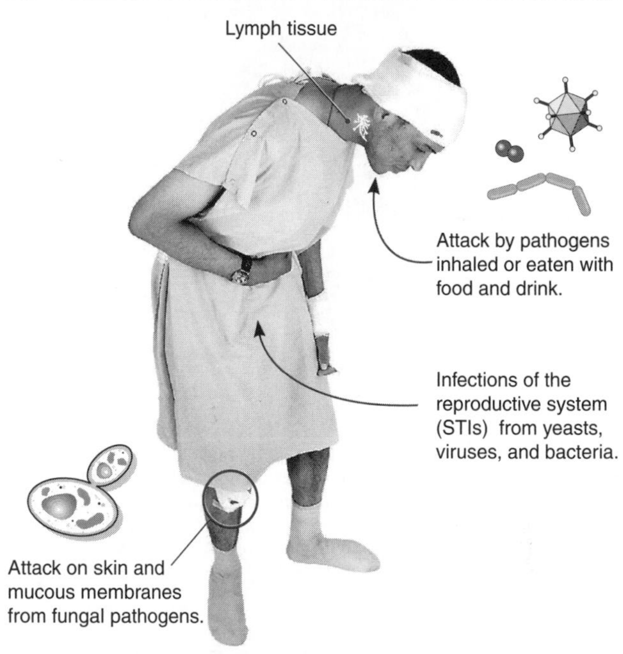


All of us are under constant attack from pathogens (disease causing organisms). The body has a number of mechanisms that help to prevent the entry of pathogens and limit the damage they cause if they do enter the body. The skin, the digestive system, and the immune system are all involved in the body's defense, while the cardiovascular and lymphatic systems circulate the cells and antimicrobial substances involved.

## Maintaining Nutrient Supply

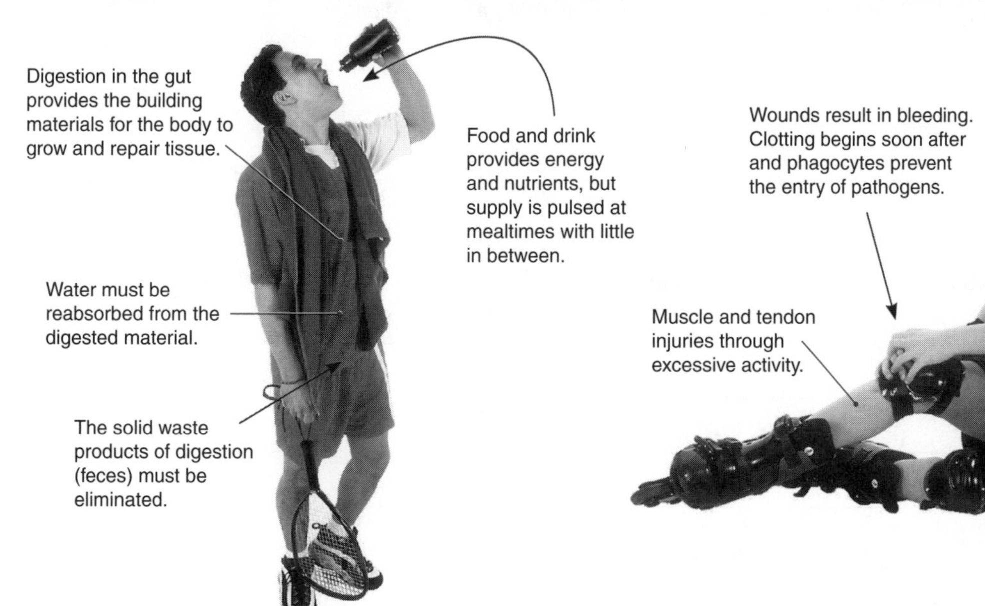


Food and drink must be taken in to maintain the body's energy supplies. The digestive system makes these nutrients available, and the cardiovascular system distributes them throughout the body. Food intake is regulated largely through nervous mechanisms while hormones control the regulation of cellular uptake of glucose.

## Repairing Injuries

Damage to body tissues triggers the **inflammatory response** and white blood cells move to the injury site. The inflammatory response is started (and ended) by chemical signals (e.g. from histamine and prostaglandins) released when tissue is damaged. The cardiovascular and lymphatic systems distribute the cells and molecules involved.

**ISBN: 978-1-927173-12-1**


***Related activities:*** *The Body's Defenses, Nervous Regulatory Systems, Exercise & Blood Flow*

## Maintaining Fluid and Ion Balance

Water loss through breathing.

Water and ions taken in with food and drink.

Loss of water and ions via sweat.

Loss of water and ions via urine and feces.

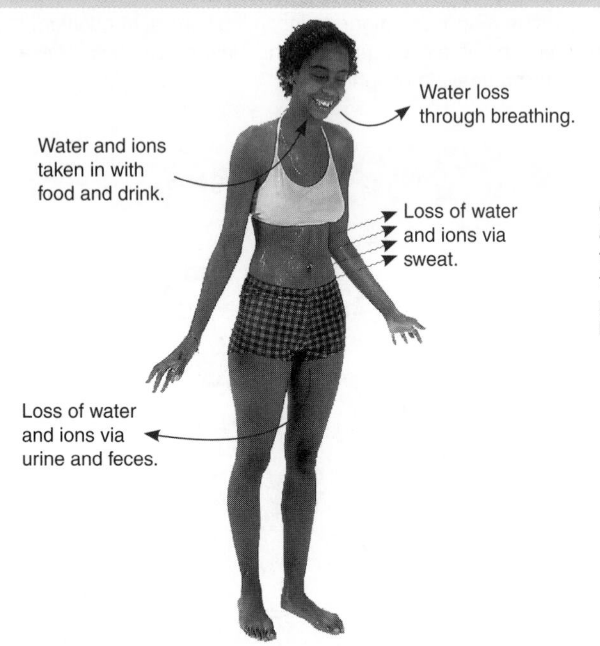

Fluid and electrolyte balance in the body is maintained by the kidneys (although the skin is also important). Osmoreceptors monitor blood volume and bring about the release of regulatory hormones; the kidneys regulate reabsorption of water and sodium from blood in response to levels of the hormones ADH and aldosterone. The cardiovascular and lymphatic system distribute fluids around the body.

## Coordinating Responses

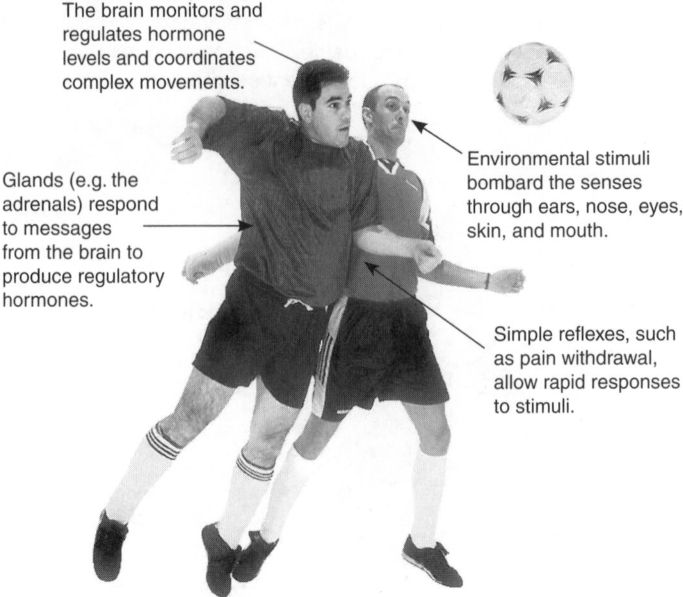


The body is constantly bombarded by stimuli from the environment. The brain sorts these stimuli into those that require a response and those that do not. Responses are coordinated via nervous or hormonal controls. Simple nervous responses (reflexes) act quickly. Hormones, which are distributed by the cardiovascular and lymphatic systems, take longer to produce a response and the response is more prolonged.

1. Describe two mechanisms that operate to restore homeostasis after infection by a pathogen:

   (a) ___

   (b) ___

2. Describe two mechanisms by which responses to stimuli are brought about and coordinated:

   (a) ___

   (b) ___

3. Explain two ways in which water and ion balance are maintained. Name the organ(s) and any hormones involved:

   (a) ___

   (b) ___

4. Explain two ways in which the body regulates its respiratory gases during exercise:

   (a) ___

   (b) ___

ISBN: 978-1-927173-12-1

# Mechanisms of Thermoregulation

The process of controlling body temperature is called **thermoregulation**. For many years, animals were classified as either homeotherms (= constant body temperature) or poikilotherms (= variable body temperature). Unfortunately, these terms are not particularly accurate for many animals; for example, some mammals (typical homeotherms), may have unstable body temperatures. A more recent, thermal classification of animals is based on the source of the body heat: whether it is largely from the environment (**ectothermic**) or from metabolic activity (**endothermic**). This classification can be more accurately applied to most animals but, in reality, many animals still fall somewhere between the two extremes.

## How Body Temperature Varies

Aquatic invertebrates like jellyfish are true poikilotherms: their temperature is the same as the environment.

Tuna and some of the larger sharks can maintain body temperatures up to 14°C above the water temperature.

Hibernating rodents and bats let their body temperature drop to well below what is typical for most mammals.

Most birds and mammals maintain a body temperature that varies less than 2°C: they are true homeotherms.

*Increasingly homeothermic* →

### Poikilothermic

Body temperature varies with the environmental temperature. Traditionally includes all animals other than birds and mammals, but many reptiles, some large insects and some large fish are not true poikilotherms because they may maintain body temperatures that are different from the surrounding environment.

### Homeothermic

Body temperature remains almost constant despite environmental fluctuations. Traditionally includes birds and mammals, which typically maintain body temperatures close to 37-38°C. Many reptiles are partially homeothermic and achieve often quite constant body temperatures through behavioral mechanisms.

## Source of Body Heat

With a few exceptions, most fish are fully ectothermic. Unlike many reptiles they do not usually thermoregulate.

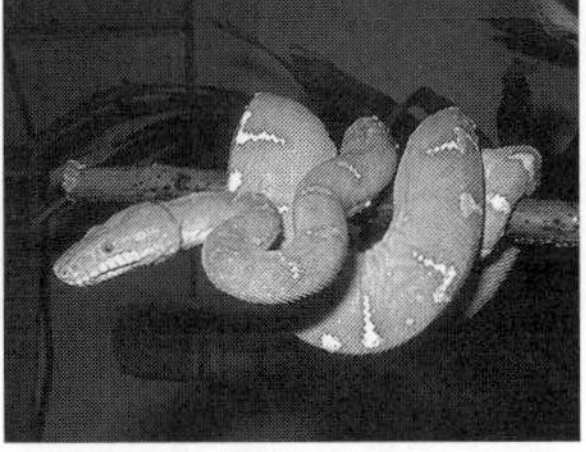

Snakes use heat energy from the environment to increase their body temperature for activity.

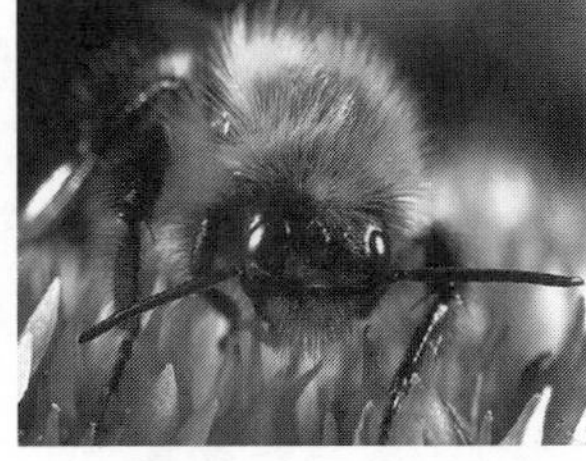

Some large insects like bumblebees may raise their temperature for short periods through muscular activity.

Mammals (and birds) achieve high body temperatures through metabolic activity and reduction of heat losses.

*Increasingly endothermic* →

### Ectothermic

Ectotherms depend on the environment for their heat energy. The term ectotherm is often equated with poikilotherm, although they are not the same. Poikilotherms are also ectotherms but many ectotherms may regulate body temperature (often within narrow limits) by changing their behavior (e.g. snakes and lizards).

### Endothermic

Endotherms rely largely on metabolic activity for their heat energy. Since they usually maintain a constant body temperature, most endotherms are also homeotherms. As well as birds and mammals, some fast swimming fish, like tuna, and some large insects may also use muscular activity to maintain a high body temperature.

### Daily temperature variations in ectotherms and endotherms

**Ectotherm:** Diurnal lizard (top right)
Body temperature is regulated by behavior so that it does not rise above 40°C. Basking increases heat uptake from the sun. Activity occurs when body temperature is high. Underground burrows are used for retreat.

**Endotherm:** Human (bottom right)
Body temperature fluctuates within narrow limits over a 24 hour period. Exercise and eating increase body temperature for a short time. Body temperature falls during rest and is partly controlled by an internal rhythm.

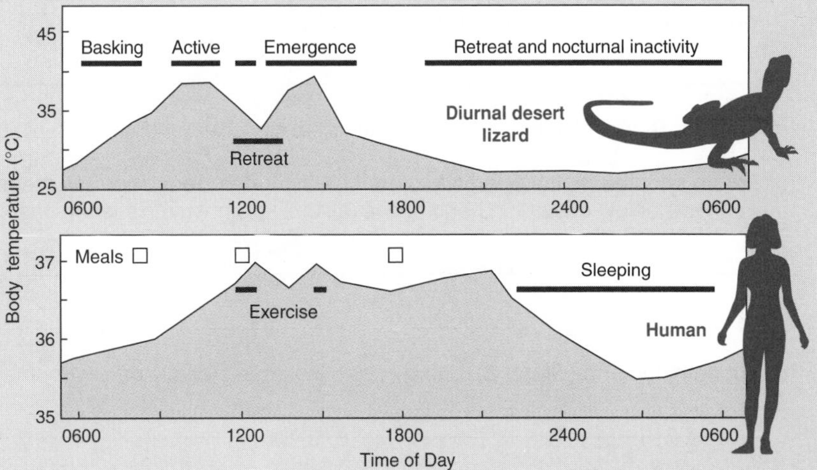

ISBN: 978-1-927173-12-1


*Periodicals:* *Temperature regulation*

*Related activities:* *Thermoregulation in Mammals*
*Weblinks:* *Countercurrent Heat Exchange, Frozen Alive!*

1. (a) Explain what is meant by a **homeothermic endotherm**:

   (b) Explain why the term **poikilotherm** is not necessarily a good term for classifying many terrestrial lizards and snakes:

2. **Ectotherms** will often maintain high, relatively constant body temperatures for periods in spite of environmental fluctuations, yet they also tolerate marked declines in body temperature to levels lower than are tolerated by endotherms.
   (a) Describe the advantages of letting body temperature fluctuate with the environment (particularly at low temperature):

   (b) Suggest why ectothermy is regarded as an adaptation to low or variable food supplies:

3. Some **endotherms** do not always maintain a high body temperature. Some, such as small rodents, allow their body temperatures to fall during **hibernation**. Explain the advantage of this behavior:

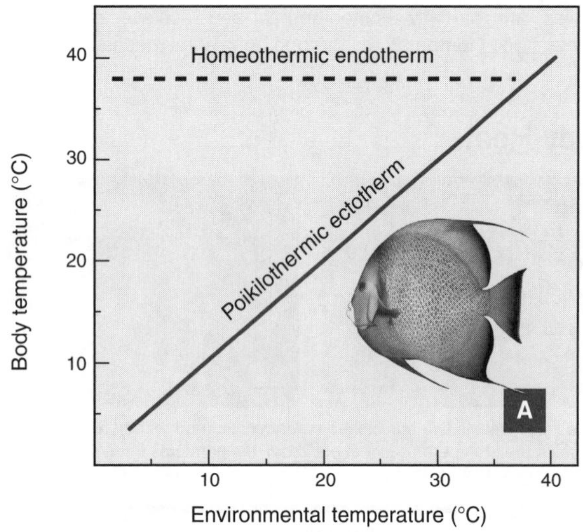

Homeothermic endotherm
Poikilothermic ectotherm
Body temperature (°C)
Environmental temperature (°C)
A

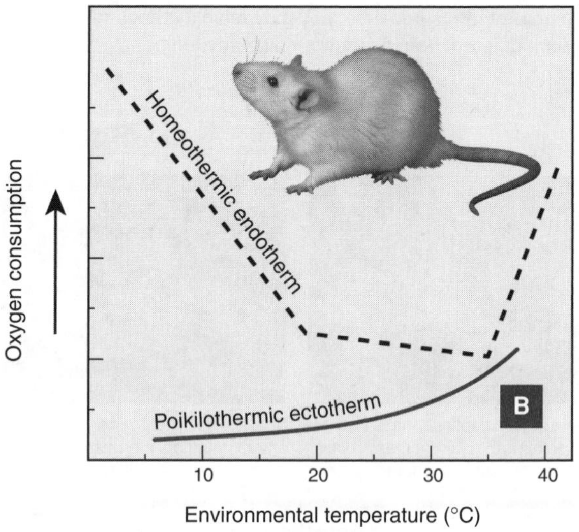

Homeothermic endotherm
Poikilothermic ectotherm
Oxygen consumption
Environmental temperature (°C)
B

4. The two graphs above illustrate the differences in temperature regulation between a homeothermic endotherm and a poikilothermic ectotherm (such as a fish). Graph A shows change in body temperature with environmental temperature. Graph B shows change in oxygen consumption with environmental temperature. Use the graphs to answer the following:
   (a) Explain how ectotherms and endotherms differ in their response to changes in environmental temperature (graph **A**):

   (b) Explain why a poikilothermic ectotherm (no behavioral regulation of temperature) would be limited to environments where temperatures were below about 40°C:

   (c) In graph **B**, state the optimum temperature range for an endotherm:

   (d) For an endotherm, the energetic costs of temperature regulation (as measured by oxygen consumption) increase markedly below about 15°C and above 35°C. Explain why this is the case:

   (e) For an ectotherm (Graph B), energy costs increase steadily as environmental temperature increases. Explain why:


**ISBN: 978-1-927173-12-1**

# Thermoregulation in Animals

For a body to maintain a constant temperature heat losses must equal heat gains. Heat exchanges with the environment occur via three mechanisms: **conduction** (direct heat transfer), **radiation** (indirect heat transfer), and **evaporation**. The importance of each of these depends on environment. For example, there is no evaporative loss in water, but such losses in air can be very high. Coverage of temperature regulation in humans, including the role of the skin, is covered elsewhere in this chapter.

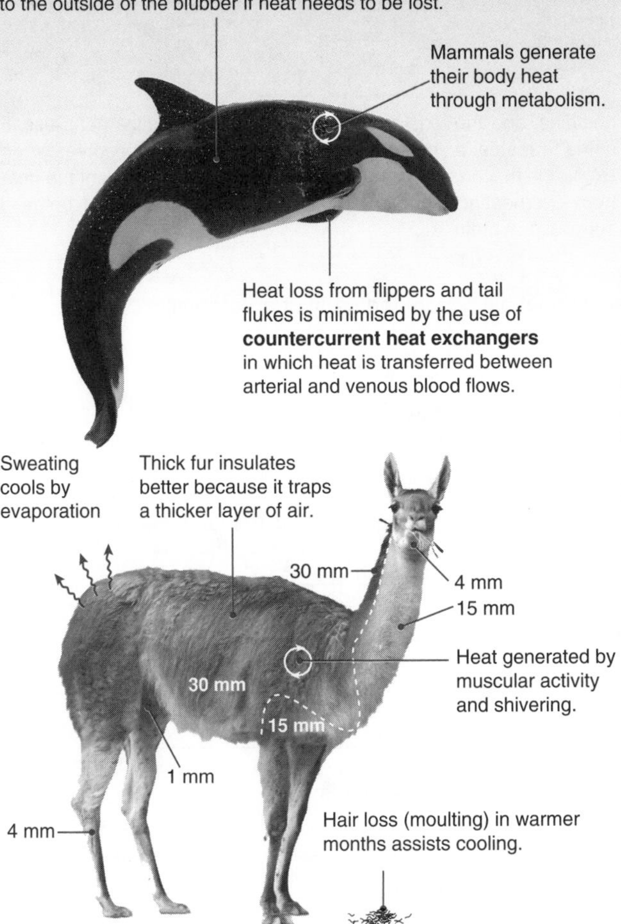


### Temperature regulation mechanisms in water

- Heat generation from metabolic activity
- Insulation layer of blubber
- Changes in circulation patterns when swimming
- Large body size
- Heat exchange systems in limbs or high activity muscle

In fast swimming fish, such as tuna, heat exchangers are used to maintain muscle temperatures up to 14°C above the water temperature.


### Temperature regulation mechanisms in air

- Behaviour or habitat choice
- Heat generation from metabolic activity, including shivering.
- Insulation (fat, fur, feathers)
- Circulatory changes including constriction and dilation of blood vessels
- Large body size
- Sweating and panting
- Tolerance of fluctuation in body temperature

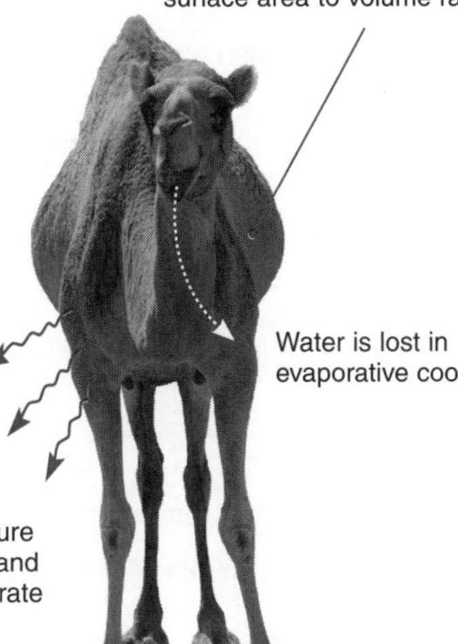


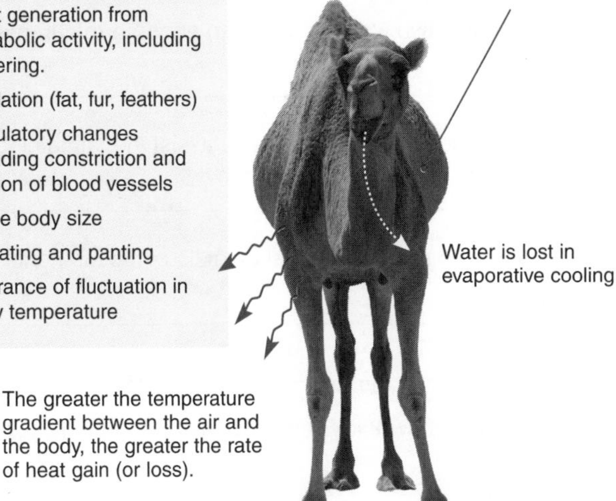

For most mammals, the thickness of the fur or hair varies around the body (as indicated above). Thermoregulation is assisted by adopting body positions that expose or cover areas of thin fur (the figures above are for the llama-like guanaco).

Animals adapted to temperature extremes (hot or cold) often tolerate large fluctuations in their body temperature. In well watered camels, body temperature fluctuates less than 2°C, but when they are deprived of water, the body temperature may fluctuate up to 7°C (34°C to 41°C) over a 24 hour period. By allowing their body temperature to rise, heat gain is reduced and the animal conserves water and energy.

Panting to lose accumulated heat is important in dogs, which have sweat glands only on the pads of their feet.

Circulation changes slow heat loss in water and speed heat gain when basking on land in marine iguanas.

Thick blubber and large body size in seals and other marine mammals provide an effective insulation.

Mammals and birds in cold climates, like the musk oxen above, cluster together to retain body heat.

Behaviours to reduce heat uptake via conduction, e.g. standing on two legs, are important in desert lizards.

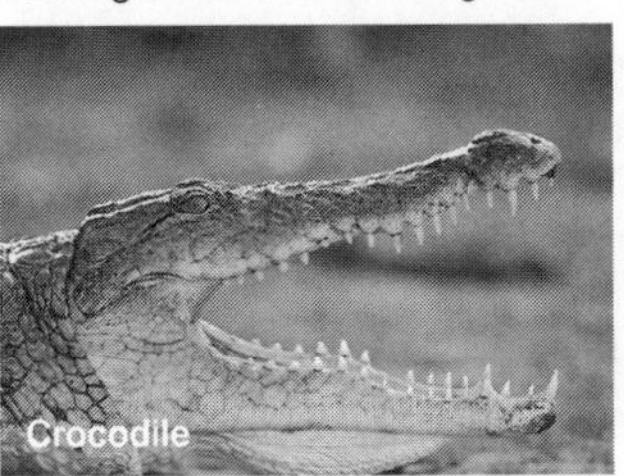

Gaping is a behavioural mechanism in large ectotherms to protect the brain from overheating.

Hair, fur, wool, or feathers trap air next to the skin. The air layer slows the loss and gain of heat.

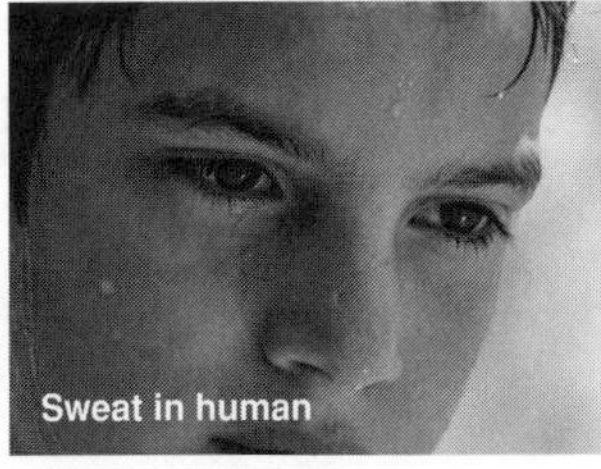

Most mammals can sweat to cool down. Heat is lost when sweat evaporates from the body surface.

Homeostasis and Energy Allocation

**ISBN: 978-1-927173-12-1**


***Related activities:*** *Thermoregulation in Humans, Hypothermia*
***Weblinks:*** *Blood Flow and Thermoregulation*

## Countercurrent Heat Exchange Systems

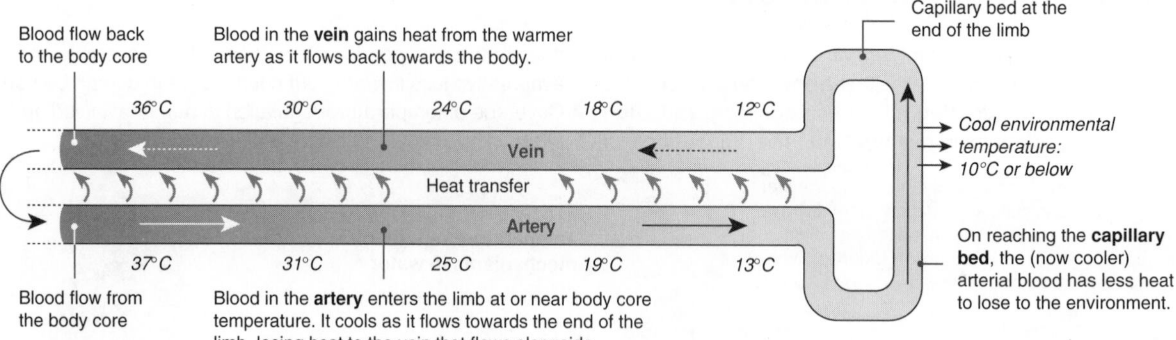


Countercurrent heat exchange systems occur in both aquatic and terrestrial animals as an adaptation to maintaining a stable core temperature. The diagram illustrates the general principle of countercurrent heat exchangers: heat is exchanged between incoming and outgoing blood. In the flippers and fins of whales and dolphins, and the legs of aquatic birds, they minimize heat loss. In some terrestrial animals adapted to hot climates, the countercurrent exchange mechanism works in the opposite way to prevent the head from overheating: venous blood cools the arterial blood before it supplies the brain.

1. (a) How does water differ from air in the way in which it transmits heat away from the body of an organism?

   (b) Identify two ways in which an endotherm can maintain its internal temperature in water:

   (c) How does a large body size assist in maintaining body temperature in both aquatic and terrestrial species?

   (d) How do small terrestrial animals compensate for more rapid heat loss from a high surface area?

2. (a) Explain how thick hair or fur assists in the regulation of body temperature in mammals:

   (b) Explain why fur/hair thickness varies over different regions of a mammal's body:

   (c) Explain how you would expect fur thickness to vary between related mammal species at high and low altitude:

   (d) Explain how marine mammals compensate for lack of thick hair or fur:

3. Giving an example, explain how countercurrent heat exchange systems assist in temperature regulation in mammals:

4. (a) Describe the role that group behavior plays in temperature regulation in some mammals:

   (b) Describe another behavior, not reliant on a group, that is important in thermoregulation. For the behavior, suggest when and where it would occur, and comment on its adaptive value:


**ISBN: 978-1-927173-12-1**

# Thermoregulation in Humans

In humans and other placental mammals, the temperature regulation center of the body is in the **hypothalamus**. In humans, it has a '**set point**' temperature of 36.7°C. The hypothalamus responds directly to changes in core temperature and to nerve impulses from receptors in the skin. It then coordinates appropriate nervous and hormonal responses to counteract the changes and restore normal body temperature. Like a thermostat, the hypothalamus detects a return to normal temperature and the corrective mechanisms are switched off (**negative feedback**). Toxins produced by pathogens, or substances released from some white blood cells, cause the set point to be set to a higher temperature. This results in **fever** and is an important defense mechanism in the case of infection.

## Counteracting Heat Loss

**Heat promoting center*** in the hypothalamus monitors fall in skin or core temperature below 35.8°C and coordinates responses that generate and conserve heat. These responses are mediated primarily through the **sympathetic nerves** of the autonomic nervous system.

**Thyroxine** (together with epinephrine) **increases metabolic rate**.

Under conditions of *extreme* cold, epinephrine and thyroxine increase the energy releasing activity of the liver. Under normal conditions, the liver is thermally neutral.

**Muscular activity** (including *shivering*) produces internal heat.

Erector muscles of hairs contract to raise hairs and increase insulating layer of air. Blood flow to skin decreases (**vasoconstriction**).

### Factors causing heat loss

- Wind chill factor accelerates heat loss through conduction.
- Heat loss due to temperature difference between the body and the environment.
- The rate of heat loss from the body is increased by being wet, by inactivity, dehydration, inadequate clothing, or shock.

### Factors causing heat gain

- Gain of heat directly from the environment through radiation and conduction.
- Excessive fat deposits make it harder to lose the heat that is generated through activity.
- Heavy exercise, especially with excessive clothing.

*NOTE: The heat promoting center is also called the cold center and the heat losing center is also called the hot center. We have used the terminology descriptive of the activities promoted by the center in each case.*

## Counteracting Heat Gain

**Heat losing center*** in the hypothalamus monitors any rise in skin or core temperature above 37.5°C and coordinates responses that increase heat loss. These responses are mediated primarily through the **parasympathetic nerves** of the autonomic nervous system.

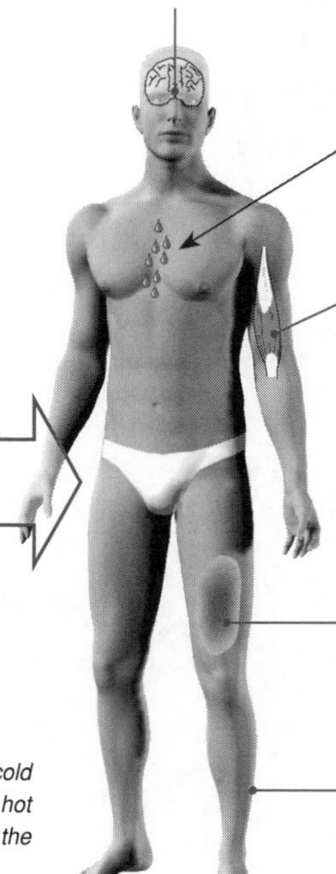

**Sweating increases.** Sweat cools by evaporation.

**Muscle tone** and **metabolic rate** are decreased. These mechanisms reduce the body's heat output.

Blood flow to skin (**vasodilation**) increases. This increases heat loss.

Erector muscles of hairs relax to flatten hairs and decrease insulating air layer.

### The Skin and Thermoregulation

**Thermoreceptors** in the dermis (probably free nerve endings) detect changes in skin temperature outside the normal range and send nerve impulses to the hypothalamus, which mediates a response. Thermoreceptors are of two types: **hot thermoreceptors** detect a rise in skin temperature above 37.5°C while the **cold thermoreceptors** detect a fall below 35.8°C. Temperature regulation by the skin involves **negative feedback** because the output is fed back to the skin receptors and becomes part of a new stimulus-response cycle.

Note that the thermoreceptors detect the temperature change, but the hair erector muscles and blood vessels are the **effectors** for mediating a response.

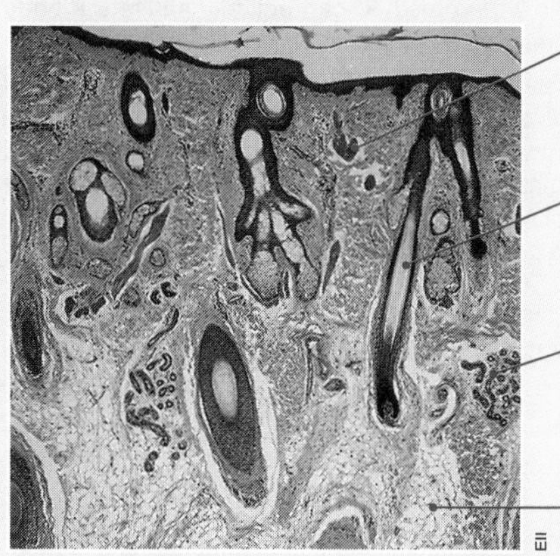


*Cross section through the skin of the scalp.*

**Blood vessels** in the dermis dilate (vasodilation) or constrict (vasoconstriction) to respectively promote or restrict heat loss.

**Hairs** raised or lowered to increase or decrease the thickness of the insulating air layer between the skin and the environment.

**Sweat glands** produce sweat in response to parasympathetic stimulation from the hypothalamus. Sweat cools through evaporation.

**Fat** in the subdermal layers insulates the organs against heat loss.

1. State two mechanisms by which body temperature could be reduced after intensive activity (e.g. hard exercise):

   (a) ______________________ (b) ______________________

2. Briefly state the role of the following in regulating internal body temperature:

   (a) The hypothalamus: ______________________

   (b) The skin: ______________________

   (c) Nervous input to effectors: ______________________

   (d) Hormones: ______________________


ISBN: 978-1-927173-12-1


Related activities: Hypothermia

# Hypothermia

**Hypothermia** is a condition experienced when the core body temperature drops below 35ºC. Hypothermia is caused by exposure to low temperatures, and results from the body's inability to replace the heat being lost to the environment. The condition ranges from mild to severe depending how low the body temperature has dropped. Severe hypothermia results in severe mental confusion, including inability to speak and amnesia, organ and heart failure, and death.

Maintaining a normal body temperature of around 37ºC allows the body's metabolism to function optimally. At temperatures below 35ºC, metabolic reactions begin to slow, resulting in a loss of coordination, difficulty in moving, and mental fatigue. Hypothermia can result from exposure to very low temperatures for a short time or to moderately low temperatures for a long time. Exposure to cold water (even just slightly cold) will produce symptoms of hypothermia far more quickly than exposure to the same temperature of air. This is because water is much more effective than air at conducting heat away from the body.

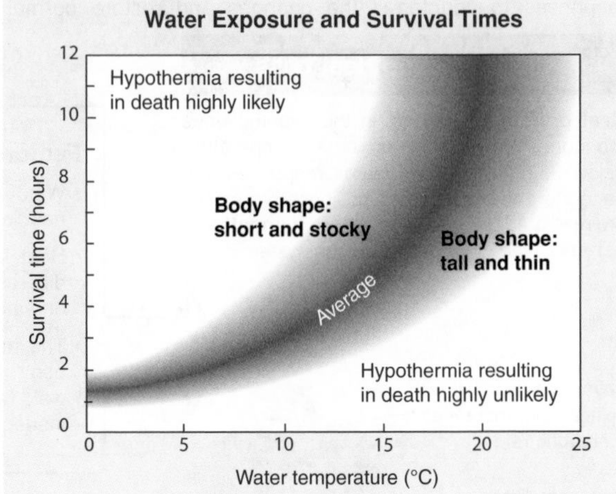


Normal body temperature → 37°

39° 37° 35° 33° 31° 29° 27° °C

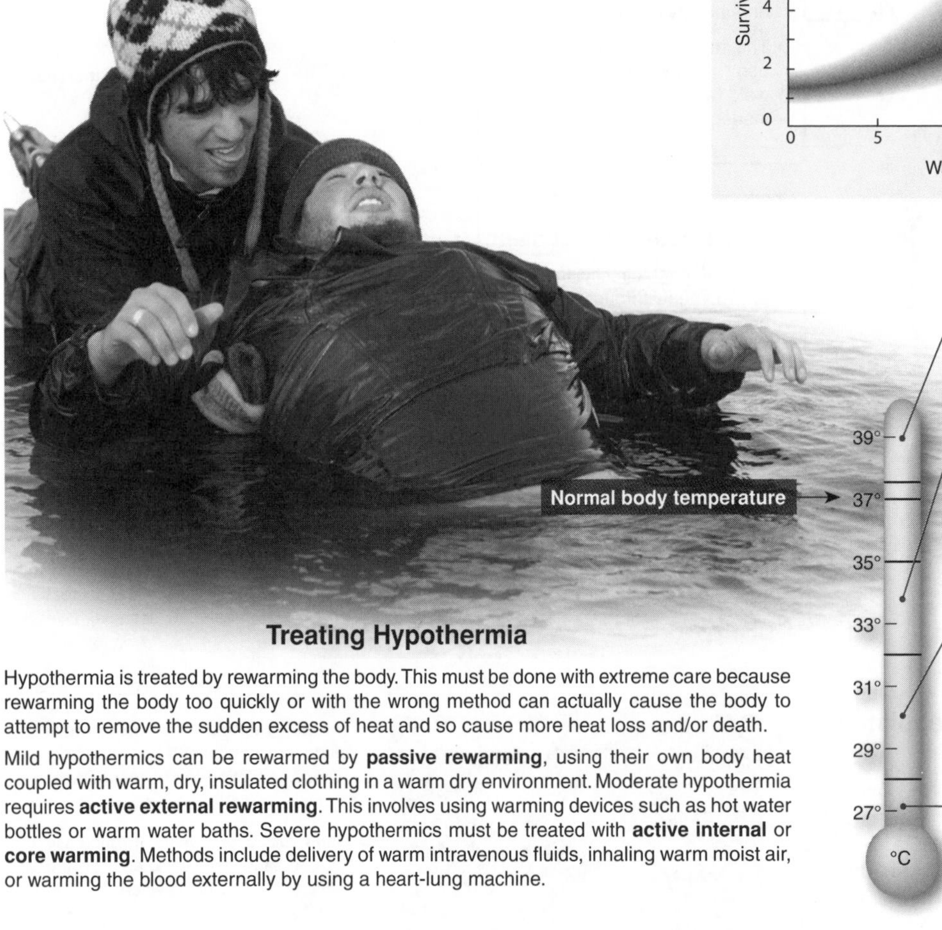

**Hyperthermia**: Body temperatures above normal cause metabolic problems that can lead to death.

**Mild hypothermia:** Shivering. Vasoconstriction reduces blood flow to the extremities. Hypertension and cold diuresis (increased urine production due to the cold).

**Moderate hypothermia:** Muscle coordination becomes difficult. Movements slow or laboured. Blood vessels in ears, nose, fingers, and toes constrict further resulting in these turning a blue color. Mental confusion sets in.

**Severe hypothermia:** Speech fails. Mental processes become irrational, victim may enter a stupor. Organs and heart eventually fail resulting in death.

## Treating Hypothermia

Hypothermia is treated by rewarming the body. This must be done with extreme care because rewarming the body too quickly or with the wrong method can actually cause the body to attempt to remove the sudden excess of heat and so cause more heat loss and/or death.

Mild hypothermics can be rewarmed by **passive rewarming**, using their own body heat coupled with warm, dry, insulated clothing in a warm dry environment. Moderate hypothermia requires **active external rewarming**. This involves using warming devices such as hot water bottles or warm water baths. Severe hypothermics must be treated with **active internal** or **core warming**. Methods include delivery of warm intravenous fluids, inhaling warm moist air, or warming the blood externally by using a heart-lung machine.

1. Describe the conditions that may cause a person to become hypothermic: ____________________

2. (a) With reference to the graph (above), identify which body shape has best survival at 15°C: ____________________

   (b) Explain your choice: ____________________

3. Describe the methods used to rewarm hypothermics and the importance of using the correct methods.




**ISBN: 978-1-927173-12-1**

# Control of Blood Glucose

The endocrine portion of the **pancreas** (the α and β cells of the **islets of Langerhans**) produces two hormones, **insulin** and **glucagon**, which maintain blood glucose at a steady state through **negative feedback**. Insulin promotes a decrease in blood glucose by promoting cellular uptake of glucose and synthesizing glycogen. **Glucagon** promotes an increase in blood glucose through the breakdown of glycogen and the synthesis of glucose from amino acids. When normal blood glucose levels are restored, negative feedback stops hormone secretion. Regulating blood glucose to within narrow limits allows energy to be available to cells as needed. Extra energy is stored as glycogen or fat, and is mobilized to meet energy needs as required. The liver is pivotal in these carbohydrate conversions. One of the consequences of a disruption to this system is the disease **diabetes mellitus**. In type 1 diabetes, the insulin-producing β cells are destroyed as a result of autoimmune activity and insulin is not produced. In type 2 diabetes, the pancreatic cells produce insulin, but the body's cells become increasingly resistant to it.

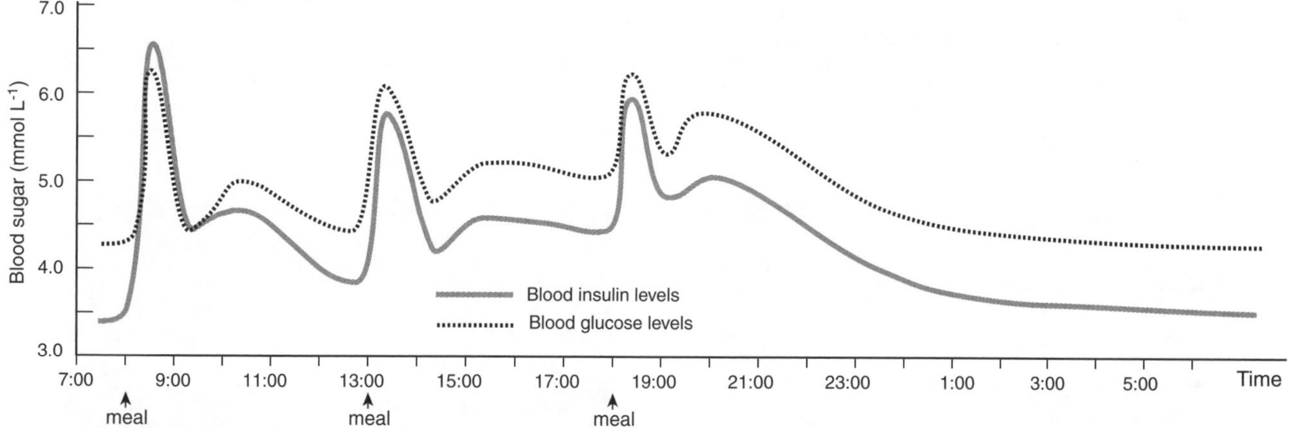


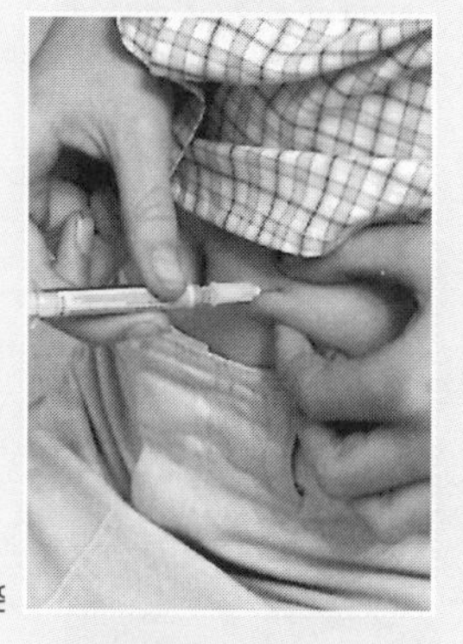


In type 1 diabetes mellitus, the β cells of the pancreas are destroyed and insulin must be delivered to the bloodstream by injection. Type 2 diabetics produce insulin, but their cells do not respond to it.

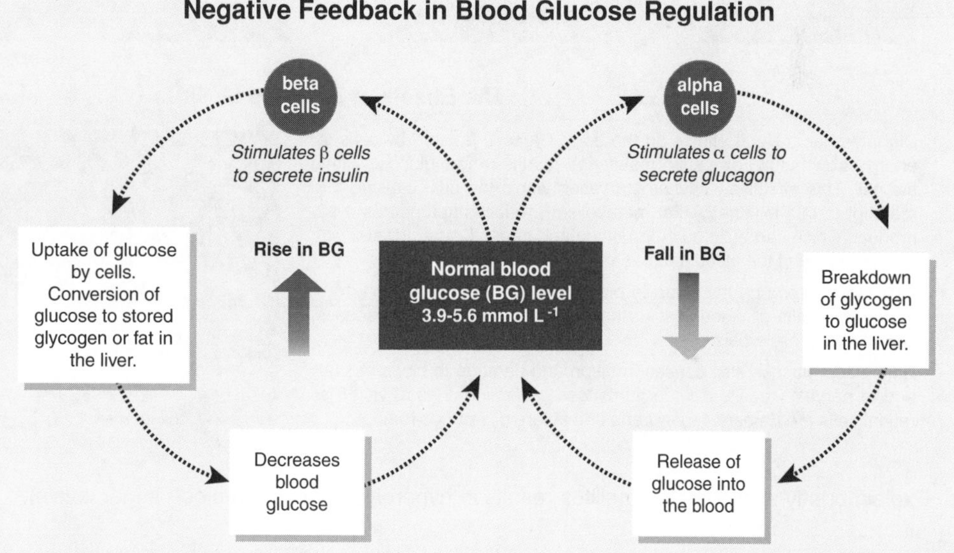


1. (a) Identify the stimulus for the release of insulin: ________________

   (b) Identify the stimulus for the release of glucagon: ________________

   (c) Explain how glucagon brings about an increase in blood glucose level: ________________

   (d) Explain how insulin brings about a decrease in blood glucose level: ________________

2. Explain the pattern of fluctuations in blood glucose and blood insulin levels in the graph above:

3. Identify the mechanism regulating insulin and glucagon secretion: ________________

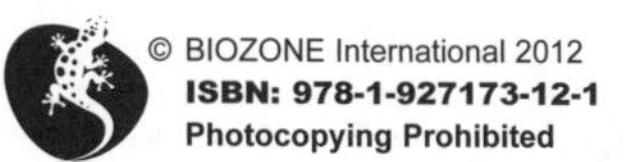


ISBN: 978-1-927173-12-1


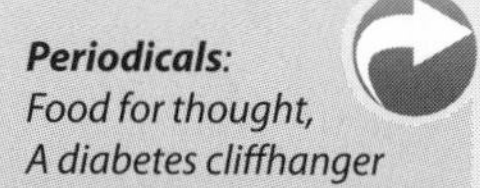

*Periodicals:*
*Food for thought,*
*A diabetes cliffhanger*

***Related activities:*** *Negative Feedback*
***Weblinks:*** *Insulin and Glucose Regulation*

## Type 1 Diabetes mellitus (insulin dependent)

### Deficient Insulin Production

**Age at onset**: Early in life; often in childhood (type 1 diabetes mellitus is often called juvenile onset diabetes).

**Cause**: Absolute deficiency of insulin due to lack of insulin production (pancreatic $\beta$ cells are destroyed in an autoimmune reaction). There is a genetic component but usually a childhood viral infection triggers the development of the disease. Mumps, coxsackie, and rubella are implicated.

**Treatment**: Blood glucose is monitored regularly and insulin injections combined with dietary management have been used to keep blood sugar levels stable. New therapies involving transplants of insulin-producing islet cells have been used with varying degrees of success. In the future, the option of gene therapy, where the gene for insulin production is inserted into the patient's cells, may be possible.

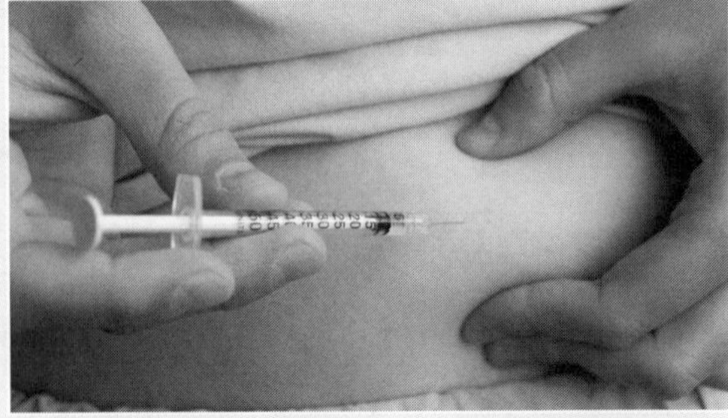

## Type 2 Diabetes mellitus (insulin resistant)

### Deficient Insulin Response

**Age at onset**: Historically, type 2 diabetes mellitus has usually occurred in adults over the age of 40, but its incidence is increasing in younger adults and obese children.

**Cause**: Type 2 diabetes may have a genetic component (in susceptibility), but it occurs most commonly as a result of lifestyle factors. Obesity (BMI > 27), a sedentary lifestyle, hypertension, high blood lipids, and a poor diet all contribute to make a person more susceptible to developing type 2 diabetes. Ethnicity may also be a contributing factor.

**Treatment**: Increasing general physical activity, losing weight (especially abdominal fat), and improving diet may be sufficient to control type 2 diabetes in many cases. The use of prescribed anti-diabetic drugs and insulin therapy (injections) may be required if lifestyle changes are insufficient on their own.

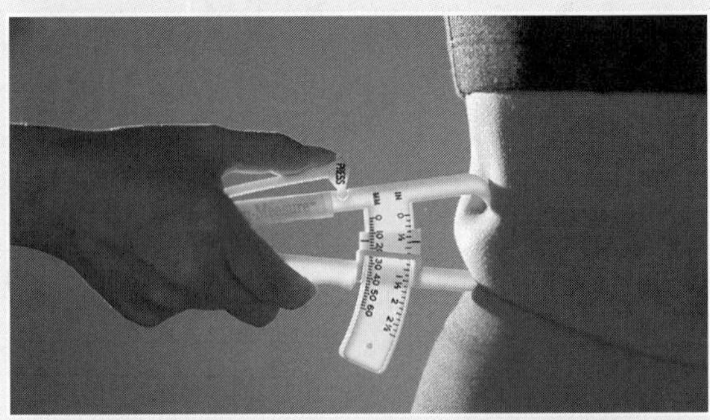

### The Effects of Diabetes Mellitus

Diabetes has several effects on the body. Glucose is used by cells for energy. Low insulin or insulin resistance results in low glucose within the cell. This effectively causes starvation within the cell, causing it to metabolize fats for energy. Fat metabolism produces ketones as a by-product, which can build up in the blood and cause **ketoacidosis**, with lowers the pH of the blood and can result in death.

Fats moving through the blood to places of metabolism can stick to or irritate the walls of blood vessels and cause cardiovascular diseases, such as atherosclerosis.

High blood glucose also causes irritation and damage to blood vessels and to nerves. Results include numbness or tingling, loss of vision as retinal cells are damaged, gangrene and failure or wounds to heal.

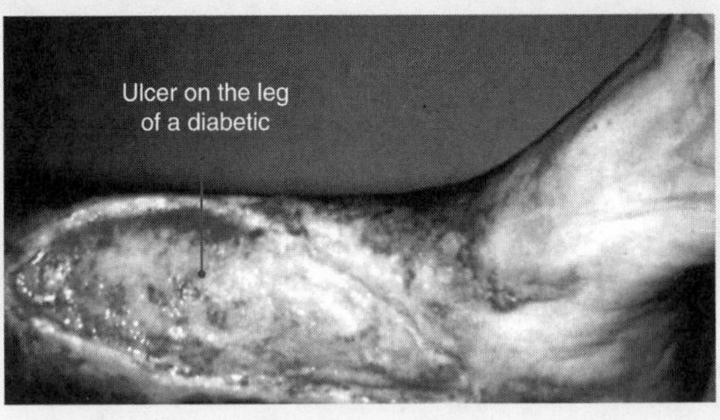


4. Explain briefly why diabetes mellitus results in hyperglycaemia (high blood sugar levels):

5. Discuss the differences between type 1 and type 2 diabetes, including causes and treatments:

6. Explain why the increase in type 2 diabetes is considered epidemic in the developed world:

7. Describe some effects of high glucose levels in the blood:


ISBN: 978-1-927173-12-1

# Blood Clotting

Apart from its transport role, **blood** has a role in the body's defense against infection and **hemostasis** (the prevention of bleeding and maintenance of blood volume). The tearing or puncturing of a blood vessel initiates **clotting**. Clotting is a rapid process that seals off the tear, preventing blood loss and the invasion of bacteria into the site. Clotting is the result of a **positive feedback loop** initiated by the release of chemicals from injured tissue. The chemicals activate platelets, which release chemicals to activate more platelets, causing a rapid cascade that ends with the formation of the clot.

When tissue is wounded, the blood quickly coagulates to prevent further blood loss and maintain the integrity of the circulatory system. For external wounds, clotting also prevents the entry of pathogens. Blood clotting involves a cascade of reactions involving at least twelve clotting factors in the blood. The end result is the formation of an insoluble network of fibers, which traps red blood cells and seals the wound.

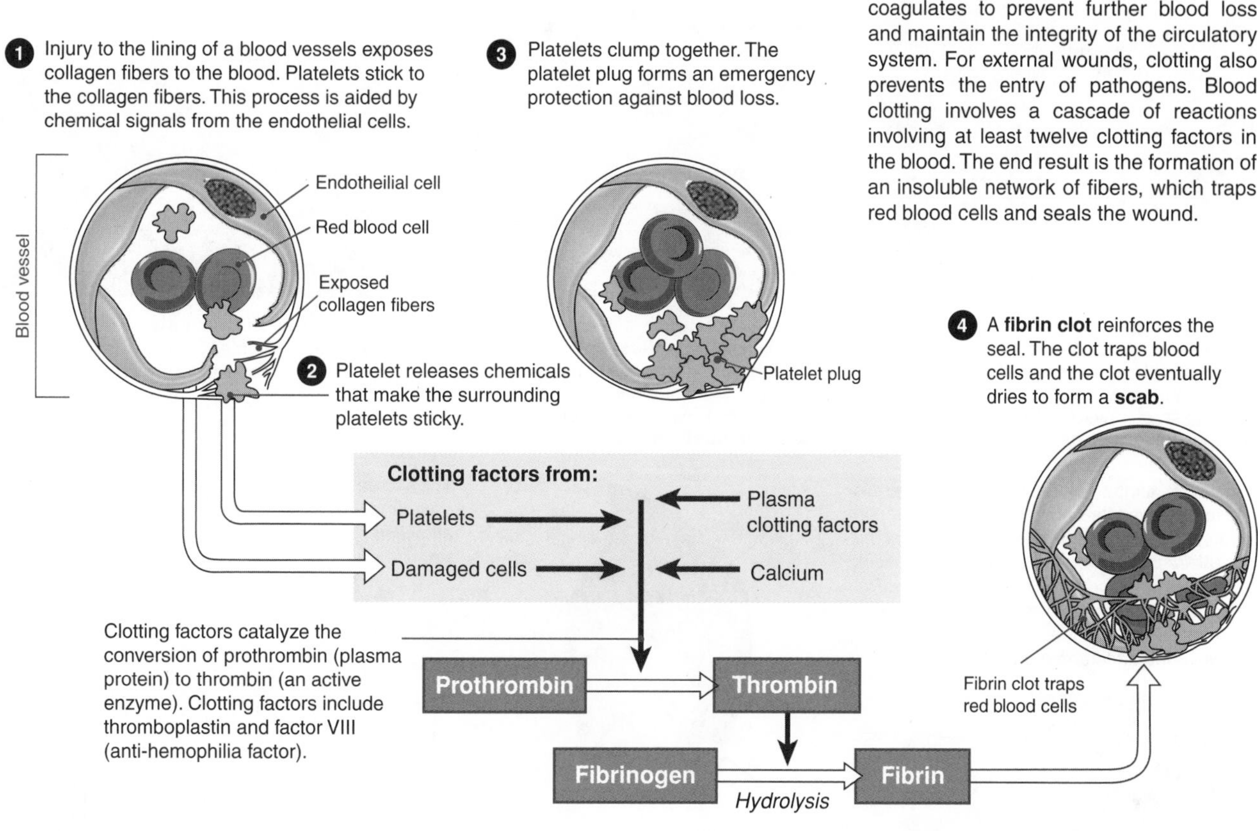


1. Explain two roles of the blood clotting system in internal defense and haemostasis:

   (a) ______

   (b) ______

2. Explain the role of each of the following in the sequence of events leading to a blood clot:

   (a) Injury: ______

   (b) Release of chemicals from platelets: ______

   (c) Clumping of platelets at the wound site: ______

   (d) Formation of a fibrin clot: ______

3. (a) What is the role of clotting factors in the blood in formation of the clot? ______

   (b) Why are these clotting factors not normally present in the plasma? ______

4. (a) Name one inherited disease caused by the absence of a clotting factor: ______

   (b) Name the clotting factor involved: ______

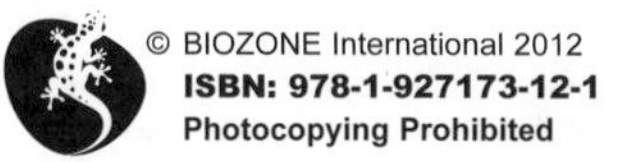


ISBN: 978-1-927173-12-1


*Periodicals:* *Code red*

***Related activities:*** *Positive Feedback*
***Weblinks:*** *Hemostasis*

# Temperature Regulation in Plants

Plants are capable of marked physiological responses to a wide range of environmental variables, including temperature, light, and water availability. Gardeners are well aware of the seasonal effects of temperature on plant growth and development. Woody plants are able to survive freezing temperatures because of metabolic changes that occur in the plant between summer and winter. This process of **acclimation** involves the production of thicker cell walls and accumulation of growth inhibitors in the plant tissues. Some plants, especially those in the Arum family, are also able to regulate the temperature of their flowers.

Alternation of periods of growth with periods of **dormancy** allows the plant to survive water shortages and extremes of hot or cold. When dormant, growth will not resume until the right combination of environmental cues are met e.g. exposure to cold, and suitable photoperiod.

Low temperature stimulation of seed germination (**stratification**) is common in many species. The seeds of many cold-climate plants will not germinate until exposed to a period of wet, cold (5°C) conditions. This is called **cold stratification**.

Bud burst and flowering follow exposure to a cold period in many plants, including bulbs and many perennials. This process is called **vernalization.**

Although a number of plants are endothermic (able to produce internal warmth) only a few can thermoregulate. Two of these are the sacred lotus (*Nelumbo nucifera*) and the eastern skunk cabbage (*Symplocarpus foetidus*) (below). Both of these are able to maintain steady floral temperatures well above the air temperature even when the air temperature changes.

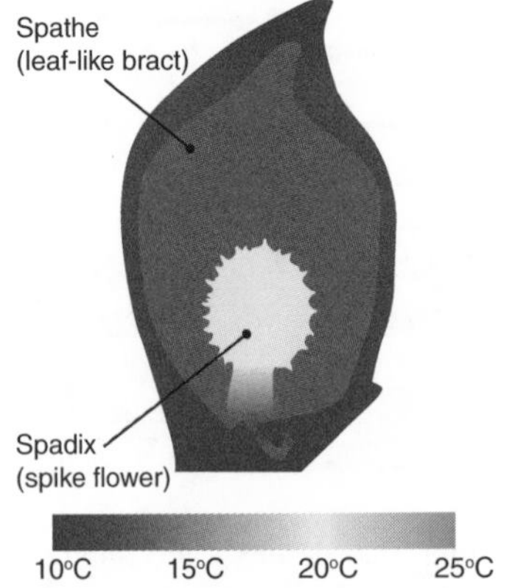


High floral temperatures are maintained by using an alternative electron transport chain in which the energy is used to produce heat instead of ATP. The skunk cabbage (above and left) produces enough heat to melt the snow around it.

The sacred lotus is able to maintain a floral temperature of around 30°C. This may help in dispersing scent and attracting pollinating insects to the warmth.

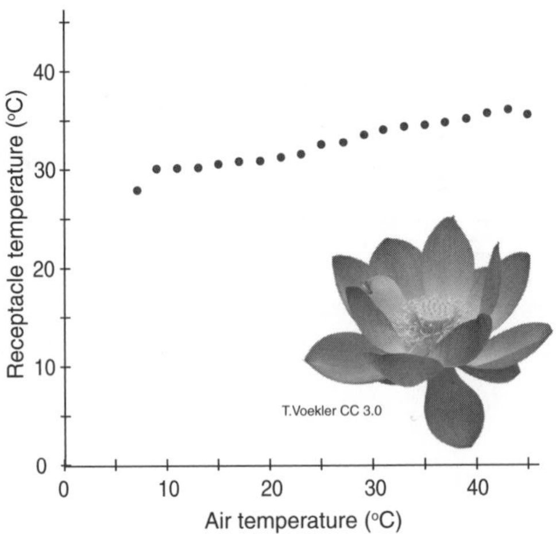


1. Describe three physiological responses of plants to changes in temperature:

   (a) ____________________

   (b) ____________________

   (c) ____________________

2. What is the adaptive value of **cold stratification** in seeds? ____________________

3. (a) Describe how some plants are able to produce heat internally: ____________________

   (b) Give two reasons why regulating floral temperature could be an advantage to plants: ____________________




ISBN: 978-1-927173-12-1

# KEY TERMS: Word Find

Use the clues below to find the relevant key terms in the WORD FIND grid

I G Y K M J H E N D O T H E R M I C P T S N P Y X

L L Z D R B R Y X M D Y A N F O E F C X H O A B P

I U P N I V Z U P W E I R T S N K C O S E X E H X

F C S E N A D H O O M K R F O K C P U T M B C Y P

E A T G S N B E O R T W K I O Y Y X N R O L T P O

H G N A U N I E E M O H T R C M Z B T A S O O E I

I O V T L S G H T S E A A N E G P C E T T O T R K

S N H I I B T R A E Z O A L J U L R R I A D H T I

T R D V N O F J D I S M T F A V K P C F S C E H L

O D O E P L P E L D R M O H I M S J U I I L R E O

R J K Y S K N A I O V Y E C E B U F R C S O M R T

Y U H L C L N Y D N P S T L I R R S R A O T I M H

A J W C O R P P W Z N J E H L H M I E T M E C I E

U Y T H E R M O R E G U L A T I O N N I Z V K A R

M K C V Y C D P O S I T I V E L T Y T O Y T O B M

T E L K O H O M E O S T A S I S M U G N U H G S R

R Y J B S H P M E T A B O L I S M M S A S X C B M

The final product of the blood coagulation step in hemostasis, essential to help prevent bleeding after an injury.

Process in which seeds are subjected to cold, moist conditions to break dormancy and encourage germination is called cold _ _ _ _ _ _ _ _ _ _ _ _ _ _ .

The thermoregulation system that occurs in certain animals, where heat is exchanged between incoming and outgoing blood, for example in the flipper and fins of whales and dolphins is called _ _ _ _ _ _ _ _ _ _ _ _ _ _ heat exchange.

A disease caused by the body's inability to produce or react to insulin.

A condition of resting with minimal metabolism and general cessation of growth. Usually used with respect to plants.

An animal dependent on the environment for its heat energy.

An animal that relies on metabolic activity for its heat energy.

A fibrous protein involved in blood clotting and made from fibrinogen.

The hormone that elevates blood glucose levels if they become too low.

The prevention of bleeding and maintenance of blood volume.

The relative physiological constancy of the body despite external fluctuations.

An animal that maintains a constant body temperature despite the environment.

The region of the brain which coordinates the nervous and endocrine systems via the pituitary gland.

A lowered body temperature (in a warm blooded animal), to such an extent that it cannot be rapidly restored to normal without an external source of heat.

The hormone that induces cellular uptake of glucose.

The allocation of an organism's energy throughout its lifetime among the three competing goals of maintenance, growth, and reproduction is called its _ _ _ _ _ _ _ _ _ _ _ strategy.

All of the chemical processes occurring within a living organism in order to maintain life.

The mechanism by which receptors and effectors stabilize body systems against excessive fluctuations is called _ _ _ _ _ _ _ _ feedback.

An animal that has a body temperature which varies with the environmental temperature.

The mechanism by which the body can speed up or escalate a physiological response is called _ _ _ _ _ _ _ _ feedback.

The process by which animals regulate body temperature.

Promotion of flowering by the application of a cold period.



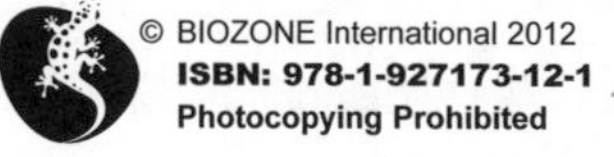


ISBN: 978-1-927173-12-1

## Interactions in Physiological Systems

| | |
|---|---|
| Interacting systems | • Open and closed circulatory systems<br>• Circulation connects all organ systems |
| Connect 4.A.4 with 2.D.2 | *Interactions between organs and organ systems provide essential activities such as transport of oxygen to tissues.* |

## Defense Mechanisms

| | |
|---|---|
| Nonspecific defenses | • Types of defense<br>• Vertebrate and invertebrate defenses |
| Specific immunity | • The immune response<br>• Failures of the defense system |
| Connect 2.D.4 with 3.D.2a | *Cell-cell signaling is important in the activation of some immune cells.* |

## Obtaining Nutrients & Eliminating Wastes

| | |
|---|---|
| Gas exchange | • The need for gas exchange<br>• Gas exchange in aquatic and terrestrial animals |
| Nutrition | • Adaptations for obtaining food<br>• Diversity in animal digestion<br>• Digestive symbioses |
| Osmoregulation and nitrogen excretion | • Salt and water balance<br>• Nitrogenous waste production<br>• Adaptations for excretion in aquatic and terrestrial environments |
| Connect 2.D.2 with 4.A.4a | *Interactions and cooperation between organs provide essential activities such as digestion of food or elimination of wastes.* |

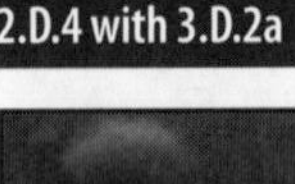

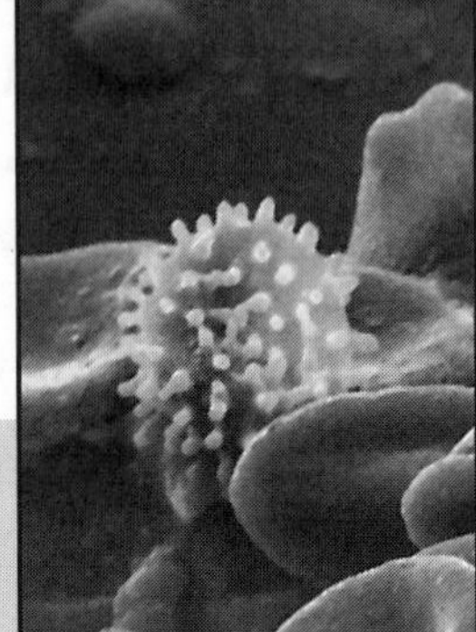

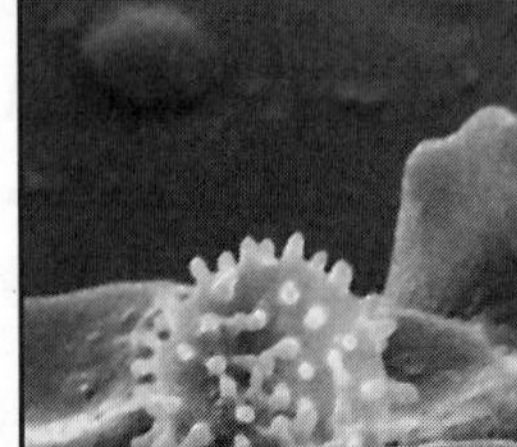

*The organs and organ systems of the body interact to provide essential services in terms of transporting nutrients and oxygen and removing metabolic wastes.*

*The mechanisms organisms have for obtaining nutrients and eliminating wastes reflect both common ancestry and adaptation to environment.*

# The Physiology and Behavior of Organisms

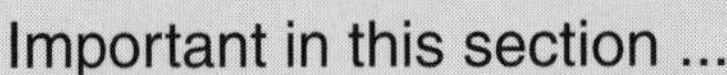

### Important in this section ...

- *Understand how the interactions between organs and organ systems provide essential activities.*
- *Recognize that the adaptations of body systems reflect ancestry and adaptation to environment.*

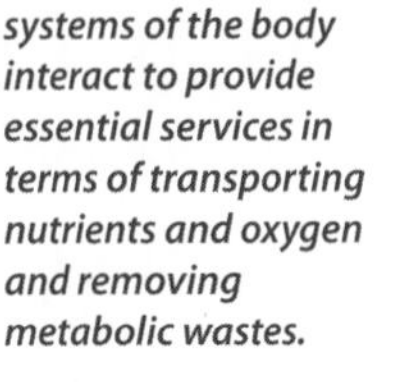

*The root, stem, and leaf interact to provide the plant body with the essential services it needs to survive.*

*Individuals act on information about their environment and can communicate it to others. Sociality and cooperation contribute to population survival.*

| | |
|---|---|
| The plant body | • Basic plant structure<br>• Interdependence of plant organs |
| Physiological processes in plants | • Leaf structure and gas exchange<br>• Aquatic and terrestrial plants<br>• Water and mineral uptake<br>• Transpiration and translocation<br>• Investigating plant growth |
| Adaptation | • Adaptation to environment |
| Connect 2.D.2 to 4.A.4 | *Interaction and coordination between plant organs enables maintenance of homeostasis in variable environments.* |
| Connect 2.D.2 with 2.D.4 | *Homeostasis of the plant body involves several levels of defense.* |

## Plant Structure & Adaptation

| | |
|---|---|
| Nervous systems | • Receiving and responding to stimuli<br>• Signal transduction<br>• Integrating information |
| Connect 3.E.2 with 4.A.4 | *Nervous and muscular systems interact to respond appropriately to stimuli.* |

## Nervous Systems & Responses

| | |
|---|---|
| Regulation of timing | • Tropisms and photoperiodism<br>• Circadian and seasonal rhythms<br>• Regulation of microbial physiology |
| Social behavior | • Innate and learned behavior<br>• Communication and sociality |
| Connect 2.E.2-3 with 2.C.2 | *Individuals respond to changes in their external environments, act on information, and communicate it to others.* |

## Timing, Coordination, & Social Behavior

Enduring Understanding

2.D
4.A

# Plant Structure and Adaptation

## Key concepts

- Plant diversity reflects adaptation environment.
- Plant roots are specialized to provide a high surface area for water and mineral uptake.
- Gases enter and leave the plant mainly through stomata, but the consequence of this is water loss.
- Plants have adaptations to reduce water loss.
- Transpiration drives water uptake in plants.
- Translocation moves carbohydrate around the plant from sources to sinks.

## Key terms

active transport
apoplast (apoplastic route)
bulk (=mass) flow
capillary action
Casparian strip
companion cell
cortex
endodermis
flaccid
gas exchange
guard cells
hydrophyte
leaf
lenticels
mesophyll
mineral
osmosis
phloem
pressure-flow hypothesis
root
root hair
root pressure
sieve tube cell
sink
source
stem
stomata
symplast (symplastic route)
tension-cohesion hypothesis
tracheid
translocation
transpiration
transpiration pull
transpiration rate
turgid
xerophyte
xylem

**Periodicals:**
*Listings for this chapter are on page 374*

**Weblinks:**
*www.thebiozone.com/weblink/AP2-3121.html*

## Essential Knowledge

☐ 1. Use the **KEY TERMS** to compile a glossary for this topic.

### Maintaining the Plant Body *(2.D.2: b)* pages 68-69, 74, 78

☐ 2. Outline the fundamental differences between plants and animals in terms of their exchanges with the environment, including the way they exchange gases, obtain nutrients, and eliminate wastes. Recognize that all green algae and land plants form a single evolutionary lineage with a history back to the Paleozoic era.

### Plant Structure in Relation to Function *(4.A.4, 4.B.2)* pages 69-84

☐ 3. Describe the basic morphology of the plant body, explaining the relationship between the **roots**, **stems**, and **leaves** and the interdependence of the support and transport systems.

☐ 4. Use explanatory examples (as below) to explain how interaction and coordination between plant systems provides essential biological activities, such as gas exchange, transport of water, and transport of sucrose.

☐ 5. Explain the relationship between the distribution of tissues (e.g. **phloem**, **xylem**) in a leaf and the functions of these tissues. Include reference to light absorption, gas exchange, support, water conservation, and transport. Describe the adaptations of leaves to maximize photosynthesis in different environments.

☐ 6. Describe the mechanism and pathways for water uptake in plant roots. Describe passive and active uptake of **minerals** in plant roots. Identify the source and role of some of the mineral ions important to plants.

☐ 7. Describe and explain **transpiration** in a flowering plant. Include reference to the roles of xylem, **tension-cohesion**, **transpiration pull** and evaporation, and **root pressure**. Investigate and explain the effect of humidity, light, air movement, temperature, and water availability on **transpiration rate**.

☐ 8. Describe the role of **stomata** and explain the movement of gases into and out of the spongy **mesophyll** of the leaf. Recognize transpiration as a consequence of gas exchange in plants. Explain how **guard cells** regulate water losses by opening and closing the stomata. Describe the role of **lenticels** in gas exchange in woody plants. Compare gas exchange in aquatic and terrestrial plants.

☐ 9. Describe and explain **translocation** in the **phloem**, identifying **sources** and **sinks** in sucrose transport. Evaluate the evidence for and against the **mass flow** (pressure-flow) **hypothesis** for the mechanism of translocation.

### Plant Growth *(2.A.1: d)* pages 85-87

☐ 10. Describe how plants allocate energy to growth. Explain how this is affected by the availability of nutrients required for the manufacture of essential compounds.

### Adaptation to Environment *(2.D.1)* pages 88-90

☐ 11. Describe adaptations in **xerophytes** and/or **hydrophytes** to the abiotic environment. In each case, explain how the adaptation enhances survival.

# Plant Structure in Relation to Function

The basic morphology of plants reflect their evolutionary history. As terrestrial organisms they have developed two interdependent systems to take advantage of and to solve the problems of living on land. The **shoot system**, consisting of **stems**, **leaves** and **reproductive structures**, has evolved to collect carbon dioxide, oxygen and light, and to disperse pollen and seeds. The **root system** has evolved to collect water and nutrients from the soil and to provide anchorage to the ground or substrate.

These systems also combine to form the support and transport systems, which are closely linked. If a plant is to grow to any size, it must have ways to hold itself up against gravity and to move materials around its body. Vascular tissues (xylem and phloem) link all plant parts. Water and minerals are transported in the xylem, while manufactured food is transported in the phloem. All plants rely on fluid pressure within their cells (turgor) to give some support to their structure.

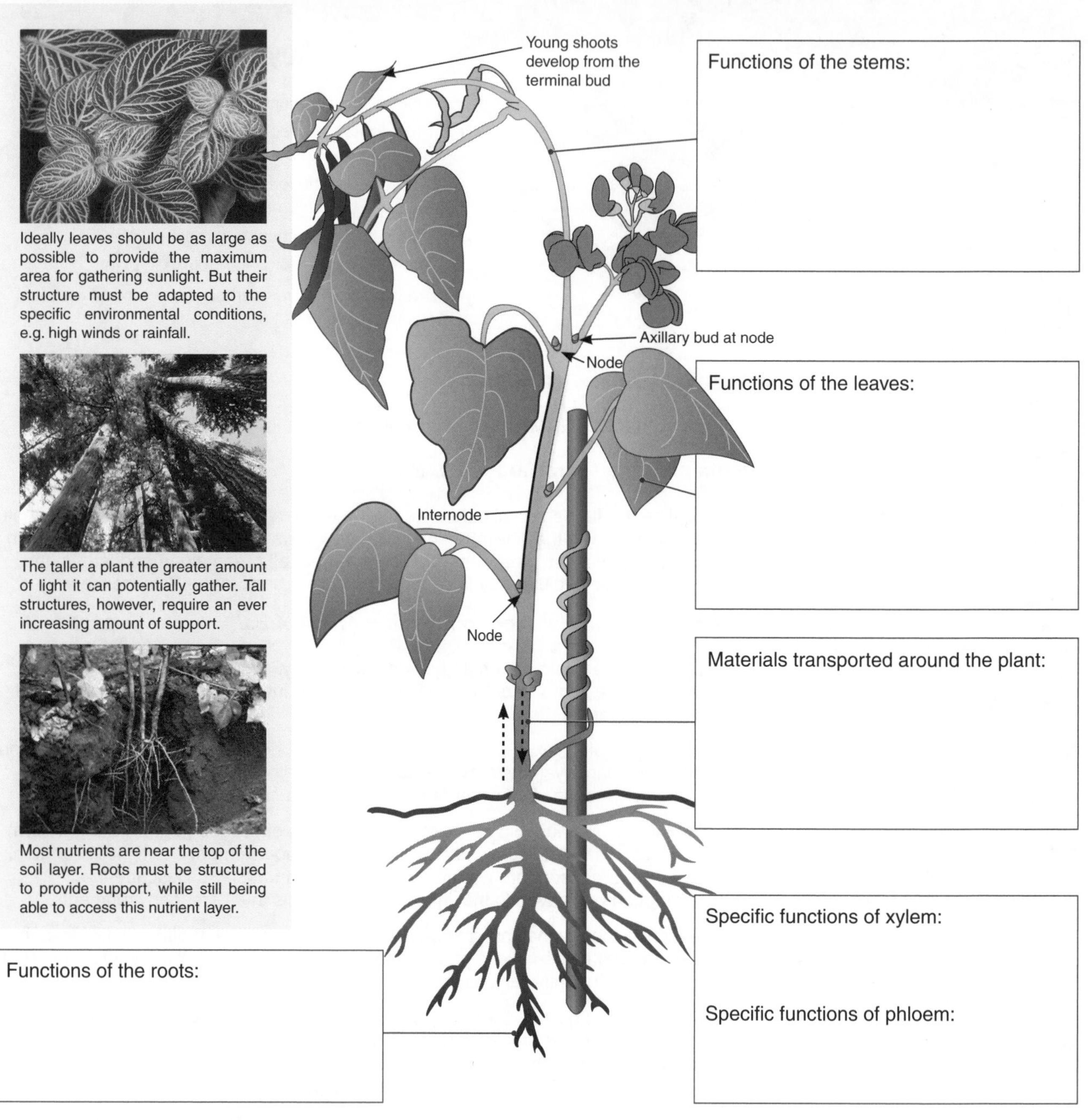


Ideally leaves should be as large as possible to provide the maximum area for gathering sunlight. But their structure must be adapted to the specific environmental conditions, e.g. high winds or rainfall.

The taller a plant the greater amount of light it can potentially gather. Tall structures, however, require an ever increasing amount of support.

Most nutrients are near the top of the soil layer. Roots must be structured to provide support, while still being able to access this nutrient layer.

Functions of the roots:

1. In the boxes provided in the diagram above:
   (a) Describe the main functions of the leaves, roots and stems (remember that the leaves themselves have leaf veins).
   (b) List the materials that are transported around the plant body.
   (c) Describe the functions of the transport tissues: xylem and phloem.

2. Name the solvent for all the materials that are transported around the plant: ______________________

3. What factors are involved in determining how tall a plant could potentially grow? ______________________

______________________________________________________________________

______________________________________________________________________



***Related activities:*** *Xylem and Phloem*


**ISBN: 978-1-927173-12-1**

# Gas Exchange in Plants

Respiring tissues require oxygen, and the photosynthetic tissues of plants also require carbon dioxide in order to produce the sugars needed for their growth and maintenance. The principal gas exchange organs in plants are the leaves, and sometimes the stems. In most plants, the exchange of gases directly across the leaf surface is prevented by the waterproof, waxy cuticle layer. Instead, access to the respiring cells is by means of **stomata**, which are tiny pores in the leaf surface. The plant has to balance its need for carbon dioxide (keeping stomata open) against its need to reduce water loss (stomata closed).

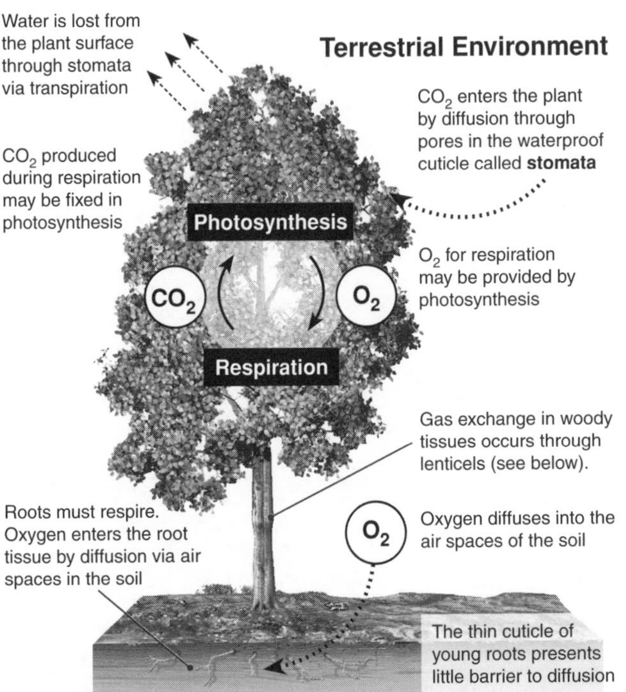


Most gas exchange in plants occurs through the leaves, but some also occurs through the stems and the roots. The shape and structure of leaves (very thin with a high surface area) assists gas exchange by diffusion.

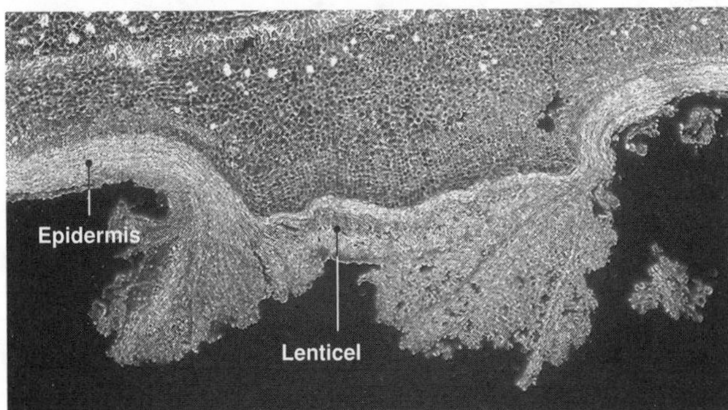


In woody plants, the wood prevents gas exchange. A lenticel is a small area in the bark where the loosely arranged cells allow entry and exit of gases into the stem tissue underneath.

## Aquatic Environment

The aquatic environment presents special problems for plants. Water loss is not a problem, but $CO_2$ availability is often very limited because most of the dissolved $CO_2$ is present in the form of bicarbonate ions, which is not directly available to plants. Maximizing uptake of gaseous $CO_2$ by reducing barriers to diffusion is therefore important.

Absorption of $CO_2$ by direct diffusion

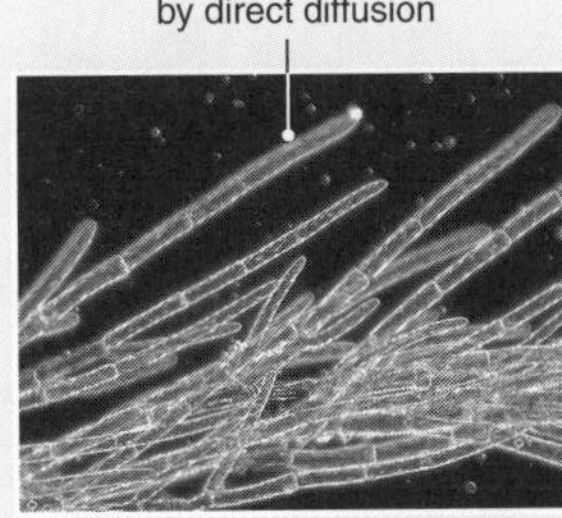

Algae lack stomata but achieve adequate gas exchange through simple diffusion into the cells.

Gas exchange through stomata on the upper surface

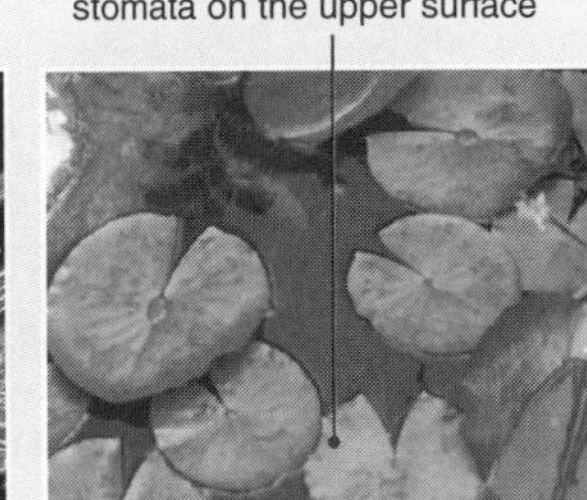

Floating leaves, such as the water lilies above, generally lack stomata on their lower surface.

With the exception of liverworts, all terrestrial plants and most aquatic plants have stomata to provide for gas exchange. $CO_2$ uptake is aided in submerged plants because they have little or no cuticle to form a barrier to diffusion of gases. The few submerged aquatics that lack stomata altogether rely only on diffusion through the epidermis. Most aquatic plants also have air spaces in their spongy tissues (which also assist buoyancy).

## Transitional Environment

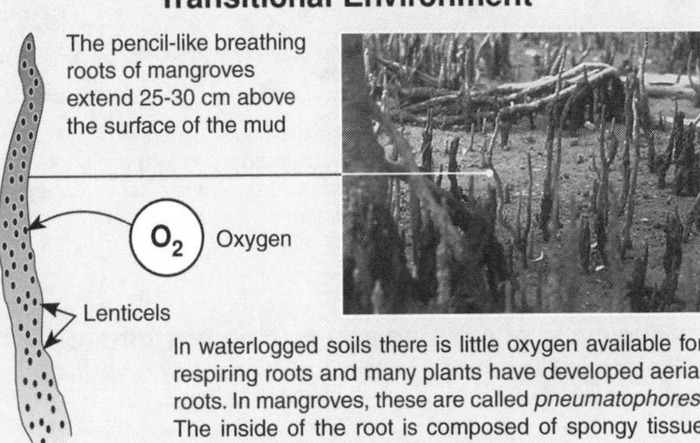


In waterlogged soils there is little oxygen available for respiring roots and many plants have developed aerial roots. In mangroves, these are called *pneumatophores*. The inside of the root is composed of spongy tissue filled with air from lenticels in the bark.

1. Name the gas produced by cellular respiration that is also a raw material for photosynthesis: ______

2. Describe the role of lenticels in plant gas exchange: ______

3. Identify two properties of leaves that assist gas exchange: ______

4. With respect to gas exchange and water balance, describe the most important considerations for:

   (a) Terrestrial plants: ______

   (b) Aquatic plants: ______

5. Describe an adaptation for gas exchange in the following plants:

   (a) A submerged aquatic angiosperm: ______

   (b) A mangrove in a salty mudflat: ______

ISBN: 978-1-927173-12-1

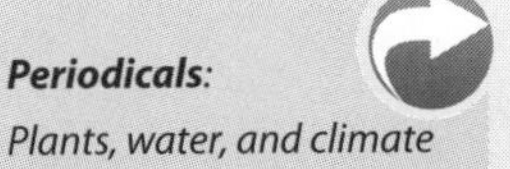

*Periodicals:*
*Plants, water, and climate*

***Related activities:*** *Gas Exchange and Stomata*

# Gas Exchange and Stomata

The leaf epidermis of angiosperms is covered with tiny pores, called **stomata**. Angiosperms have many air spaces between the cells of the stems, leaves, and roots. These air spaces are continuous and gases are able to move freely through them and into the plant's cells via the stomata. Each stoma is bounded by two **guard cells**, which together regulate the entry and exit of gases and water vapor. Although stomata permit gas exchange between the air and the photosynthetic cells inside the leaf, they are also the major routes for water loss through transpiration.

## Gas Exchanges and the Function of Stomata

Gases enter and leave the leaf by way of stomata. Inside the leaf (as illustrated by a dicot, right), the large air spaces and loose arrangement of the spongy mesophyll facilitate the diffusion of gases and provide a large surface area for gas exchanges.

Respiring plant cells use oxygen ($O_2$) and produce carbon dioxide ($CO_2$). These gases move in and out of the plant and through the air spaces by diffusion.

When the plant is photosynthesizing, the situation is more complex. Overall there is a net consumption of $CO_2$ and a net production of oxygen. The fixation of $CO_2$ maintains a gradient in $CO_2$ concentration between the inside of the leaf and the atmosphere. Oxygen is produced in excess of respiratory needs and diffuses out of the leaf. These **net** exchanges are indicated by the arrows on the diagram.

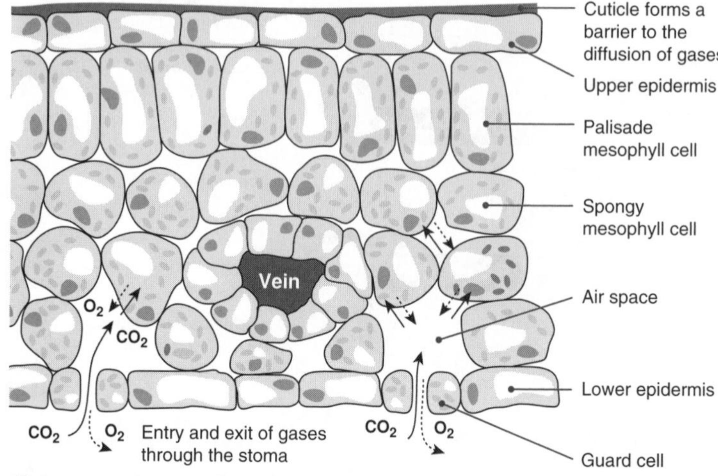


**Net gas exchanges** in a photosynthesizing dicot leaf

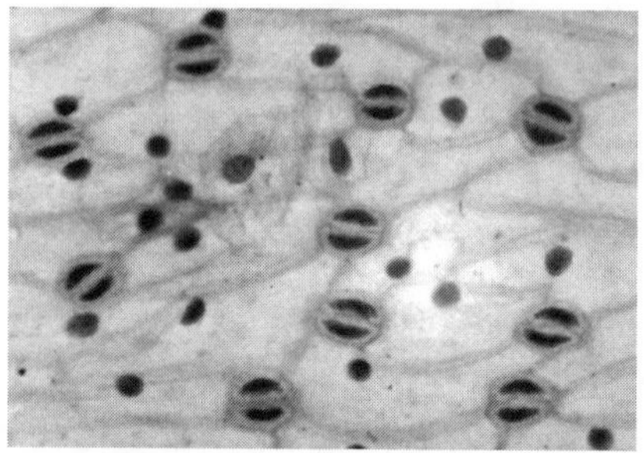

RCN

A surface view of the leaf epidermis of a dicot (above) illustrating the density and scattered arrangement of stomata. In dicots, stomata are usually present only on the lower leaf surface.

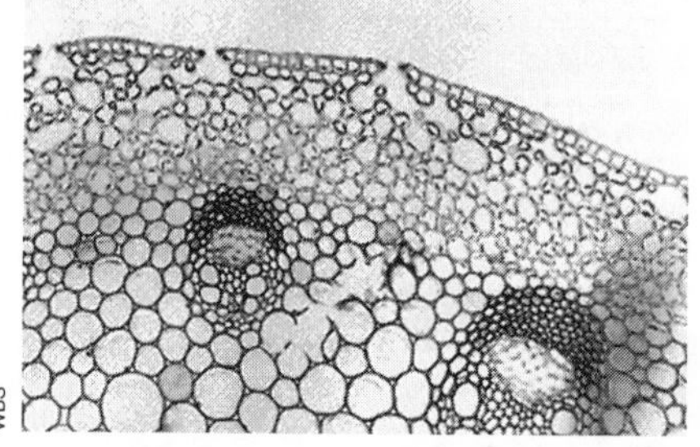

WBS

The stems of some plants (e.g. the buttercup above) are photosynthetic. Gas exchange between the stem tissues and the environment occurs through stomata in the outer epidermis.

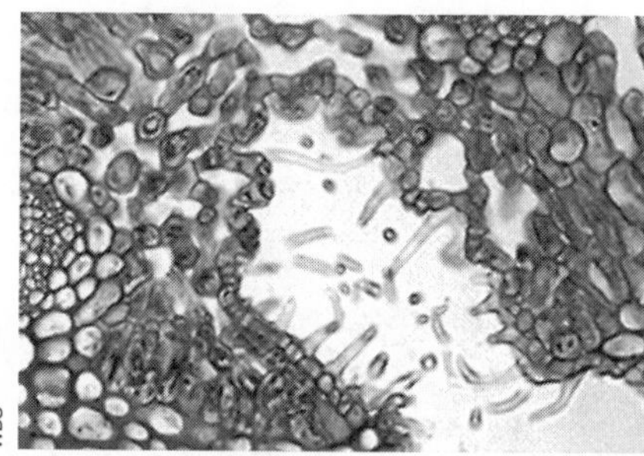

WBS

Oleander (above) is a xerophyte with many water conserving features. The stomata are in pits on the leaf underside. The pits restrict water loss to a greater extent than they reduce $CO_2$ uptake.

### The cycle of opening and closing of stomata

The opening and closing of stomata shows a daily cycle that is largely determined by the hours of light and dark.

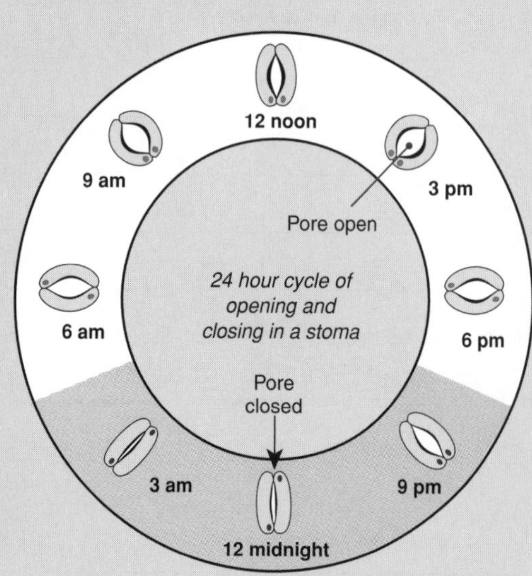


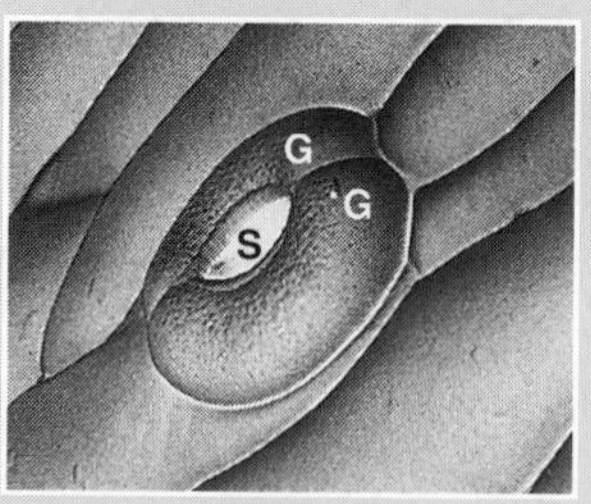


EII

The image left shows a scanning electron micrograph (SEM) of a single stoma from the leaf epidermis of a dicot.

Note the guard cells (G), which are swollen tight and open the pore (S) to allow gas exchange between the leaf tissue and the environment.

### Factors influencing stomatal opening

| Stomata | Guard cells | Daylight | $CO_2$ | Soil water |
|---|---|---|---|---|
| **Open** | Turgid | Light | Low | High |
| **Closed** | Flaccid | Dark | High | Low |

The opening and closing of stomata depends on environmental factors, the most important being light, carbon dioxide concentration in the leaf tissue, and water supply. Stomata tend to open during daylight in response to light, and close at night (left and above). Low $CO_2$ levels also promote stomatal opening. Conditions that induce water stress cause the stomata to close, regardless of light or $CO_2$ level.



***Related activities:*** *Gas Exchange in Plants*
***Weblinks:*** *Structure and Working of Stomata*

**ISBN: 978-1-927173-12-1**

The regulation of stomatal size by the guard cells is the primary way in which plants can balance their need for carbon dioxide against their need to limit water loss. The guard cells that lie each side of a stoma control the diameter of the pore by changing shape. When the guard cells take up water (by osmosis) they swell and become turgid, making the pore wider. When the guard cells lose water, they become flaccid, and the pore closes up. By opening and closing the stomata a plant can control the amount of gas entering, or water leaving, the plant.

**Stomatal Pore Open**

K+ enters the guard cells from the epidermal cells (active transport coupled to a proton pump).

Water follows K+ by osmosis.

Guard cell swells and becomes turgid.

Thickened ventral wall

Pore opens

Nucleus of guard cell

$H_2O$ $K^+$

**Water enters the guard cells**

Stomata open when the guard cells actively take up $K^+$ from the neighboring epidermal cells. The ion uptake results in a lower concentration of water molecules in the guard cells. As a consequence, water is taken up by the cells and they swell and become turgid. The walls of the guard cells are thickened more on the inside surface (the ventral wall) than the outside wall, so that when the cells swell they buckle outward, opening the pore.

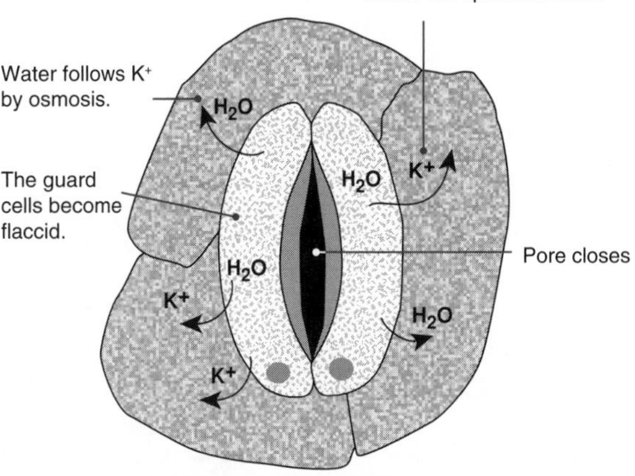


**Water leaves the guard cells**

Stomata close when $K^+$ leaves the guard cells. The loss of these ions increases the concentration of water molecules in the guard cells relative to the epidermal cells. As a consequence, water is lost by osmosis and the guard cells sag together and close the pore. The $K^+$ movements in and out of the guard cells are thought to be triggered by blue-light receptors in the plasma membrane, which activate the active transport mechanisms involved.

1. With respect to a mesophytic, terrestrial flowering plant:

   (a) Describe the **net** gas exchanges between the air and the cells of the mesophyll in the dark (no photosynthesis):

   (b) Explain how this situation changes when a plant is photosynthesizing:

2. Identify two ways in which the continuous air spaces through the plant facilitate gas exchange:

   (a)

   (b)

3. Briefly outline the role of stomata in gas exchange in an angiosperm:

4. Summarize the mechanism by which the guard cells bring about:

   (a) Stomatal opening:

   (b) Stomatal closure:


ISBN: 978-1-927173-12-1

# Leaf and Stem Adaptations

In order to photosynthesize, plants must obtain a regular supply of carbon dioxide ($CO_2$) gas; the raw material for the production of carbohydrate. In green plants, the systems for gas exchange and photosynthesis are linked; without a regular supply of $CO_2$, photosynthesis ceases. The leaf, as the primary photosynthetic organ, is adapted to maximize light capture and facilitate the entry of $CO_2$, while minimizing water loss. There are various ways in which plant leaves are adapted to do this. The ultimate structure of the leaf reflects the environment of the leaf (sun or shade, terrestrial or aquatic), its resistance to water loss, and the importance of the leaf relative to other parts of the plant that may be photosynthetic, such as the stem.

## Sun plant

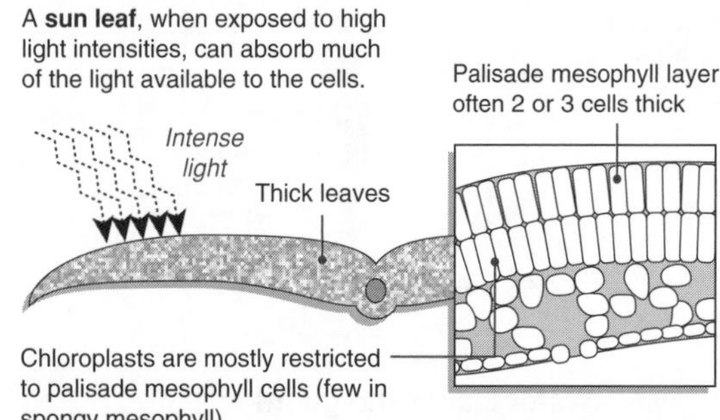

**Sun plants** are adapted for growth in full sunlight. They have higher levels of respiration but can produce sugars at rates high enough to compensate for this. Sun plants include many weed species found on open exposed grassland. They expend more energy on the construction and maintenance of thicker leaves than do shade plants. The benefit of this investment is that they can absorb the higher light intensities available and grow rapidly.

## Shade plant

A **shade leaf** can absorb the light available at lower light intensities. If exposed to high light, most would pass through.

Low light intensity

Thin leaves

Palisade mesophyll layer only 1 cell thick

Chloroplasts occur throughout the mesophyll (as many in the spongy as palisade mesophyll).

**Shade plants** typically grow in forested areas, partly shaded by the canopy of larger trees. They have lower rates of respiration than sun plants, mainly because they build thinner leaves. The fewer number of cells need less energy for their production and maintenance. In competition with sun plants, they are disadvantaged by lower rates of sugar production, but in low light environments this is offset by their lower respiration rates.

1. (a) Identify the structures in leaves that facilitate gas exchange: ______

   (b) Explain their critical role in plant nutrition: ______

2. (a) State which type of plant (sun or shade adapted) has the highest level of respiration: ______

   (b) Explain how the plant compensates for the higher level of respiration: ______

3. Discuss the adaptations of leaves in **sun** and **shade plants**: ______




**ISBN: 978-1-927173-12-1**

## Adaptations for Photosynthesis and Gas Exchange in Plants

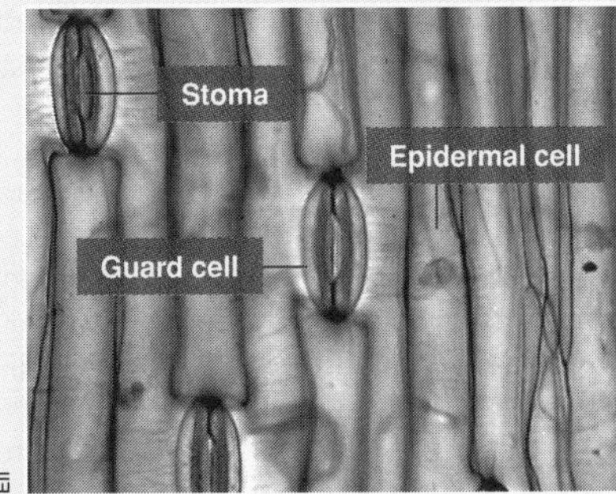


Surface view of stomata on a monocot leaf (grass). The parallel arrangement of stomata is a typical feature of monocot leaves. Grass leaves show properties of **xerophytes**, with several water conserving features (see right).

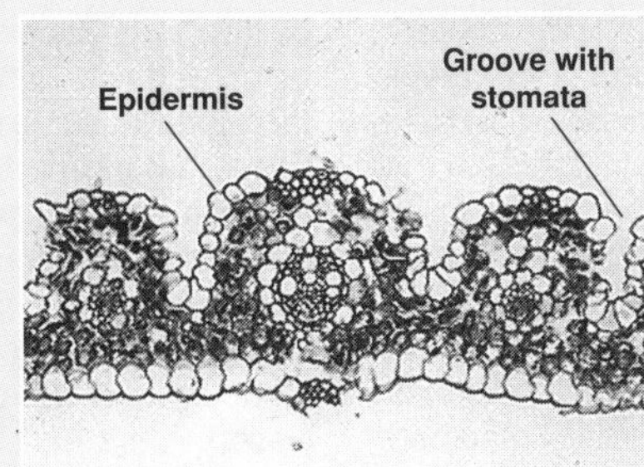


Cross section through a grass leaf showing the stomata housed in grooves. When the leaf begins to dehydrate, it may fold up, closing the grooves and thus preventing or reducing water loss through the stomata.

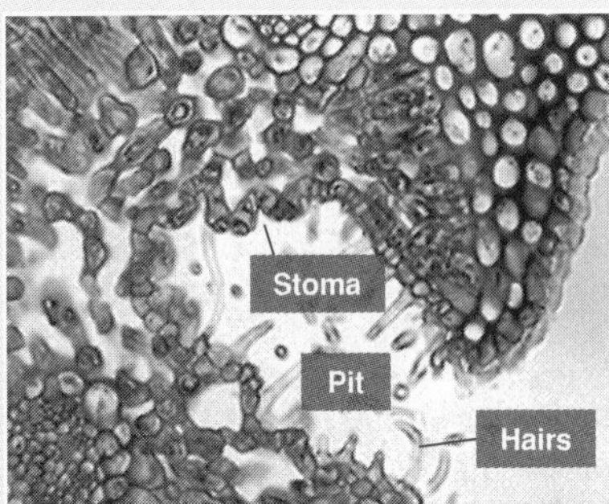


Oleander (above) is a xerophyte that displays many water conserving features. The stomata are found at the bottom of pits on the underside of the leaf. The pits restrict water loss to a greater extent than they reduce $CO_2$ uptake.

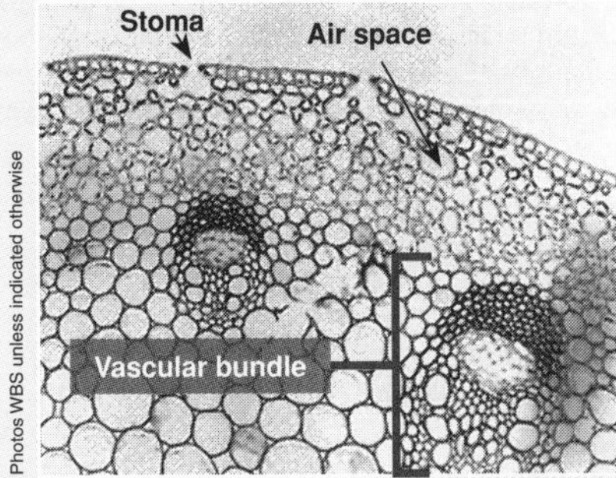


Some plants (e.g. buttercup above) have photosynthetic stems, and $CO_2$ enters freely into the stem tissue through stomata in the epidermis. The air spaces in the cortex are more typical of leaf mesophyll than stem cortex.

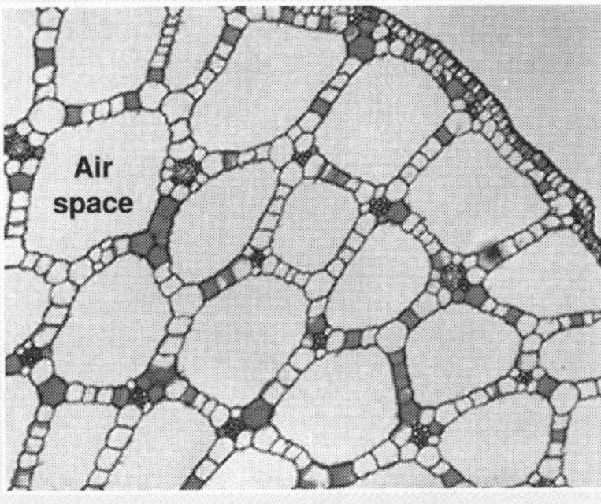


**Hydrophytes**, such as *Potamogeton*, above, have stems with massive air spaces. The presence of air in the stem means that they remain floating in the zone of light availability and photosynthesis is not compromised.

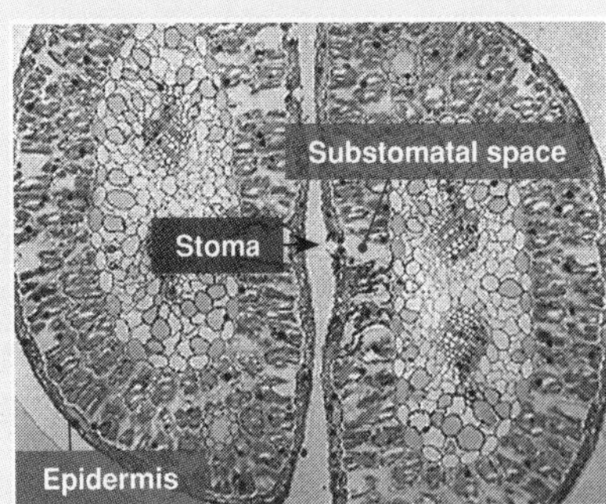


This transverse view of the twin leaves of a two-needle pine shows the sunken stomata and substomatal spaces. This adaptation for arid conditions reduces water loss by creating a region of high humidity around the stoma.

4. Describe two adaptations in plants for reducing water loss while maintaining entry of gas into the leaf:

   (a) ______________________________

   (b) ______________________________

5. Describe two adaptations of photosynthetic stems that are not present in non-photosynthetic stems, and explain the reasons for these:

   (a) ______________________________

   (b) ______________________________

6. The example of a photosynthetic stem above is from a buttercup, a plant in which the leaves are still the primary organs of photosynthesis.

   (a) Identify an example of the plant where the stem is the **only** photosynthetic organ: ______________

   (b) Describe the structure of the leaves in your example and suggest a reason for their particular structure:

   ______________________________

7. Describe one role of the air spaces in the stems of *Potamogeton* related to maintaining photosynthesis: ______________

   ______________________________


ISBN: 978-1-927173-12-1

# Excretion in Plants

All organisms must dispose of the waste products of metabolism. Of these, nitrogenous wastes from the breakdown of proteins are among the most difficult to detoxify and eliminate. There are important differences between plant and animal metabolism that make excretion in plants of less significance than excretion in animals. Firstly, plants are **producers** and synthesize all their organic requirements. They manufacture, from raw material (e.g. $NH_4^+$), only the protein they require to meet immediate needs. There is rarely excess protein and therefore very little excretion of nitrogenous wastes. Three of the waste products of metabolic processes in plants (oxygen, carbon dioxide, and water) are the raw materials for other reactions, and so excesses are used up in this way. Secondly, plants have very **low metabolic rates** compared with animals of the same size. Metabolic wastes therefore accumulate only slowly. Thirdly, plant structure is based around **carbohydrate** rather than protein, and structural proteins are far less important in plants than they are in animals. Many of the substances that accumulate in plants and were once regarded as wastes are now known to have specific defense functions, either as **deterrents** to grazers, as **defenses** against pathogens, or as **allelopathic chemicals** to inhibit competition. Nevertheless, such accumulated substances are important in the structure and seasonal activities of plants. They are considered, along with more usual excretory products, below.

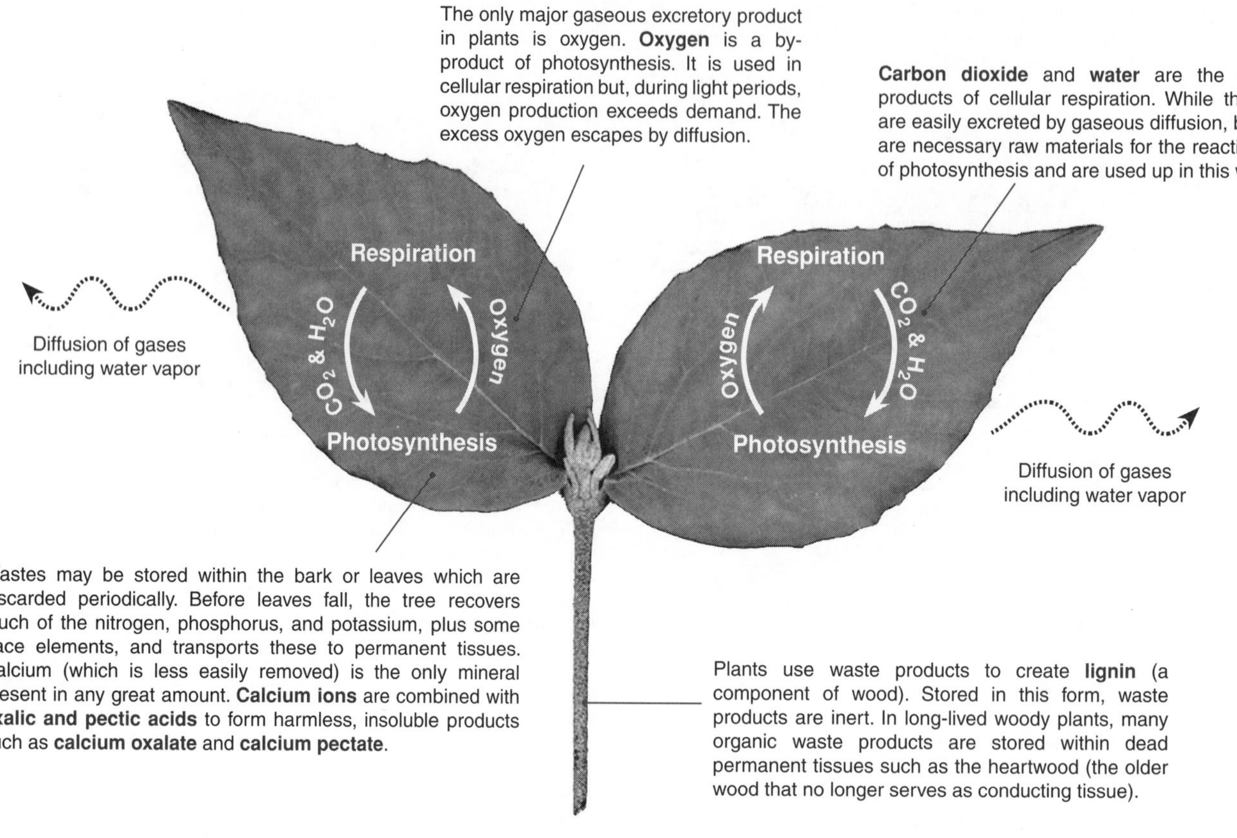


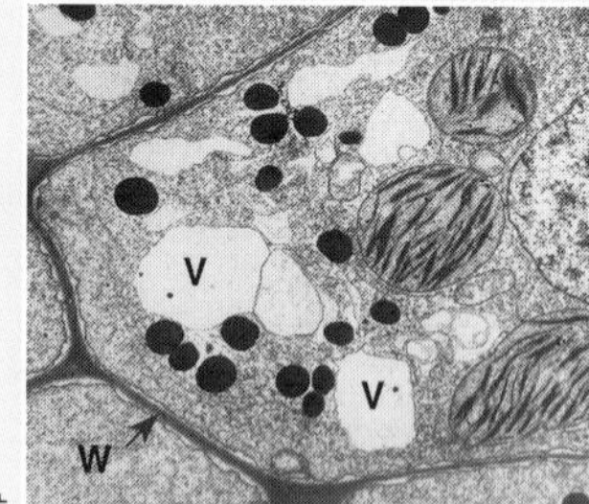

Non-woody plants store wastes in the vacuoles (V) and walls (W) of their cells until they die back.

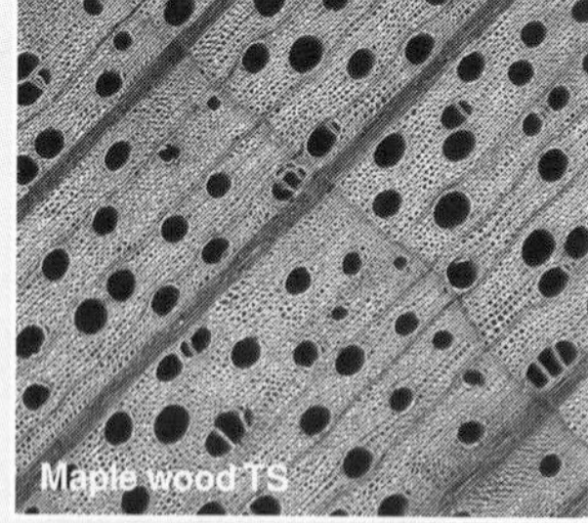

Compounds stored in wood, e.g. **oils**, **gums**, **resins**, and **tannins**, make the wood colored and aromatic.

The color of leaves in fall is partly due to the breakdown of chlorophyll, which normally masks other pigments.

In some plants, e.g. eucalypts (above), toxins deposited in the cell walls make the old leaves tough and unpalatable.

1. (a) Name the three primary waste products of plants: ____________

   (b) How are these excreted from the plant? ____________

2. Describe two purposes for the wastes that are stored in plant tissues:

   (a) ____________

   (b) ____________

3. Name three regions where wastes in plants can be stored: ____________



Weblinks: *Photographic Atlas of Plant Anatomy*

**ISBN: 978-1-927173-12-1**

# Xylem and Phloem

The two main kinds of supporting tissues in plants are **xylem** and **phloem**. As in animals, tissues in plants are groupings of different cell types that work together for a common function. Xylem and phloem are complex tissues composed of a number of cell types. They are specialized for the transport of water and dissolved sugars respectively. Most of xylem tissue is composed of large vessels, which have thickened and strengthened walls and conduct water. Xylem also contains packing cells and fibers, which provide support to the tissue. When mature, xylem is dead. Phloem comprises packing cells and supporting fiber cells, and two special cell types: **sieve tubes** and **companion cells**. Unlike xylem, phloem tissue is alive when mature.

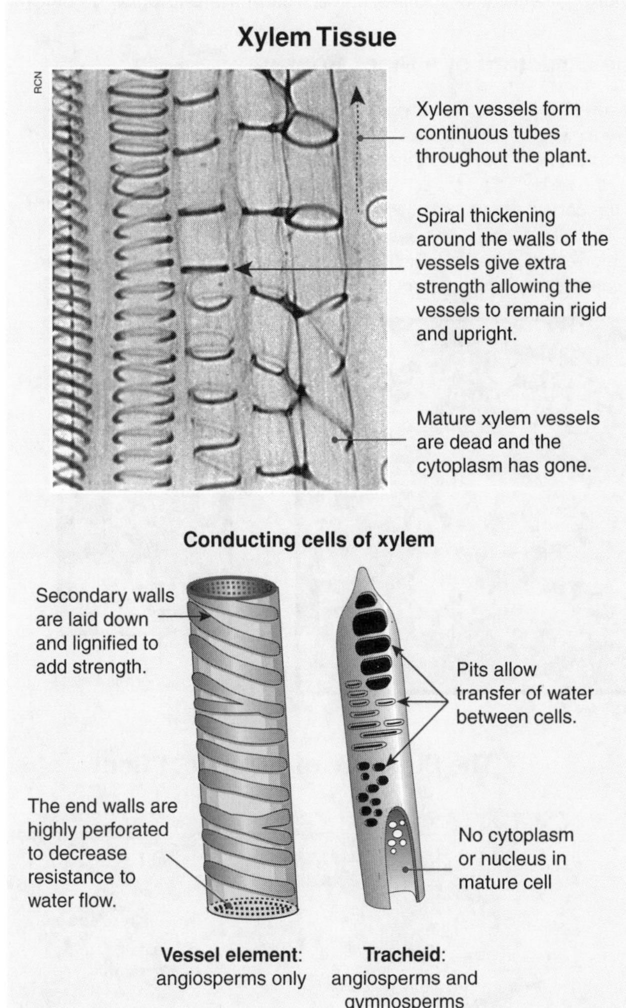


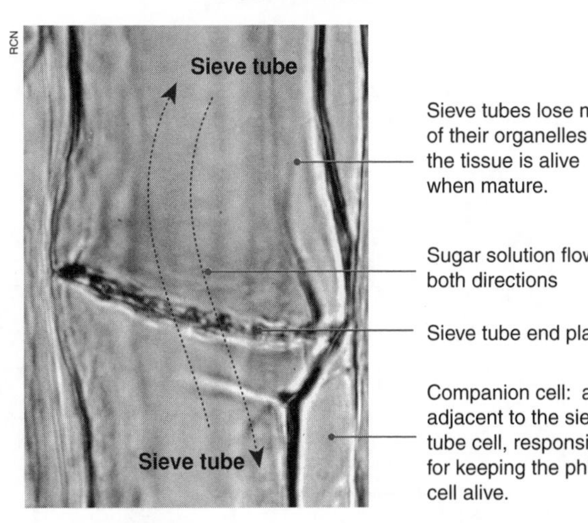


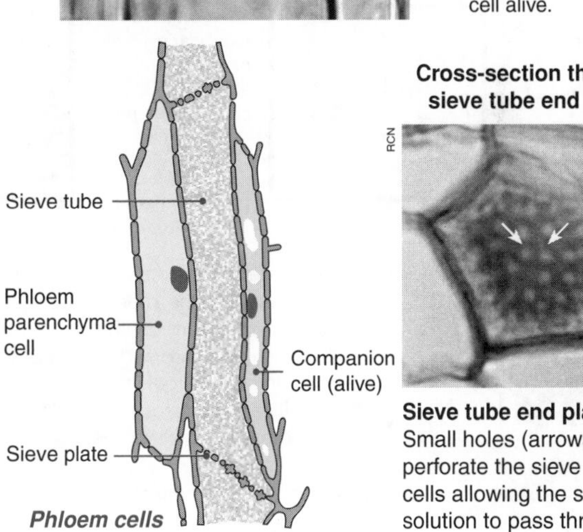


**Sieve tube end plate**: Small holes (arrows) perforate the sieve tube cells allowing the sugar solution to pass through.

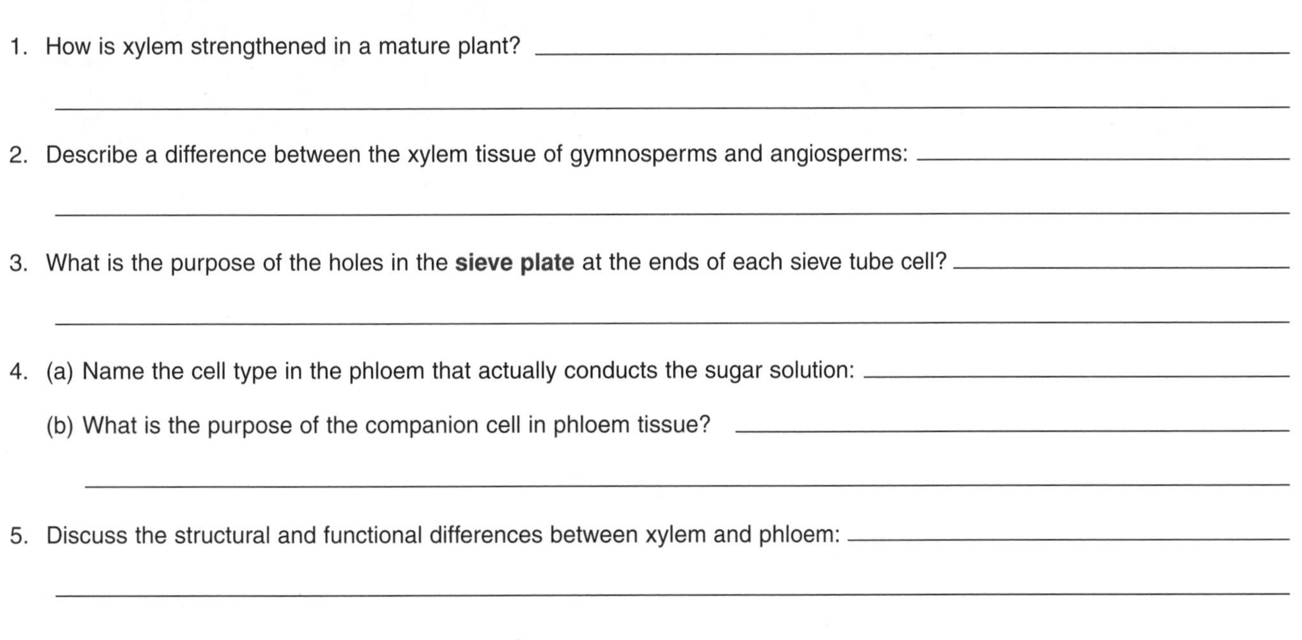

1. How is xylem strengthened in a mature plant? ______

2. Describe a difference between the xylem tissue of gymnosperms and angiosperms: ______

3. What is the purpose of the holes in the **sieve plate** at the ends of each sieve tube cell? ______

4. (a) Name the cell type in the phloem that actually conducts the sugar solution: ______

   (b) What is the purpose of the companion cell in phloem tissue? ______

5. Discuss the structural and functional differences between xylem and phloem: ______


ISBN: 978-1-927173-12-1


*Related activities*: *Specialization in Plant Cells, Translocation*
*Weblinks*: *Photographic Atlas of Plant Anatomy*

# Angiosperm Root Structure

Roots are essential plant organs. They anchor the plant in the ground, absorb water and minerals from the soil, and transport these materials to other parts of the plant body. Roots may also act as storage organs, storing excess carbohydrate reserves until they are required by the plant. Roots are covered in an epidermis but, unlike the epidermis of leaves, the root epidermis has only a thin cuticle that presents no barrier to water entry. Young roots are also covered with **root hairs** (see below). Much of a root comprises a cortex of parenchyma cells. The air spaces between the cells are essential for aeration of the root tissue. Minerals and water must move from the soil into the xylem before they can be transported around the plant. Compared with stems, roots are relatively simple and uniform in structure. The structure of monocot and dicot roots is compared in the photographs below.

## The Structure of a Dicot Root

These photographs (left and below) show cross sections through a young dicot root (i.e. primary tissues). In the photograph to the left, note the large area of the root occupied by the cortex. The parenchyma (packing) cells of the cortex store starch and other substances. The air spaces between the cells are essential for aeration of the root tissue, which is non-photosynthetic.

The vascular tissue, xylem (X) and phloem (P) forms a central cylinder through the root and is surrounded by the pericycle, a ring of cells from which lateral roots arise. The primary xylem of dicot roots forms a star shape in the center of the vascular cylinder with usually 3 or 4 points. Unlike monocots, there is no central pith of parenchyma cells.

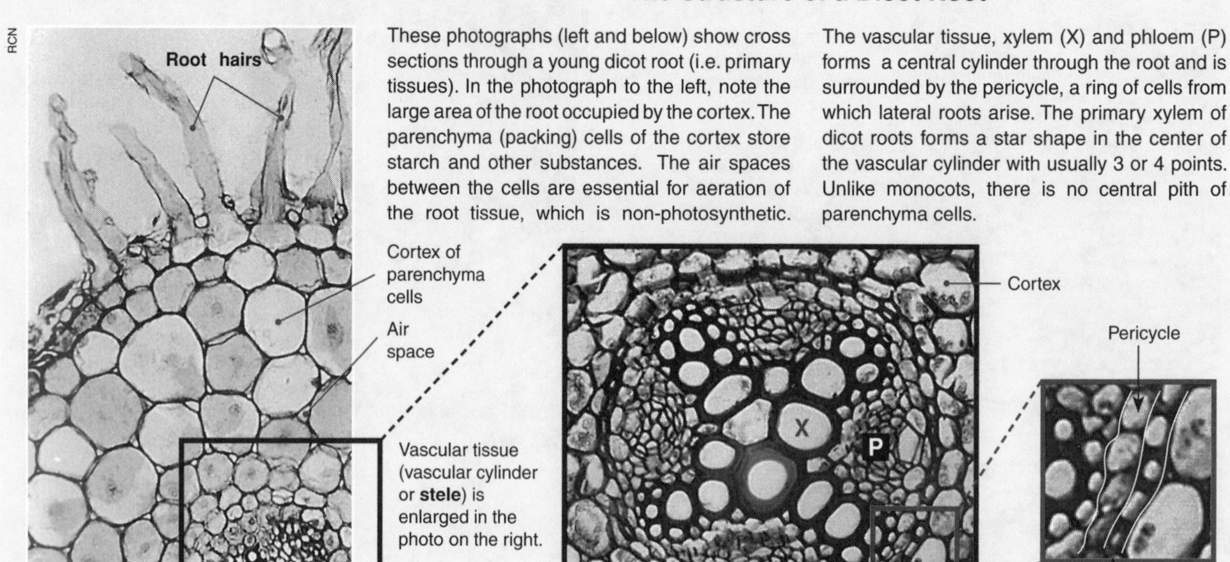

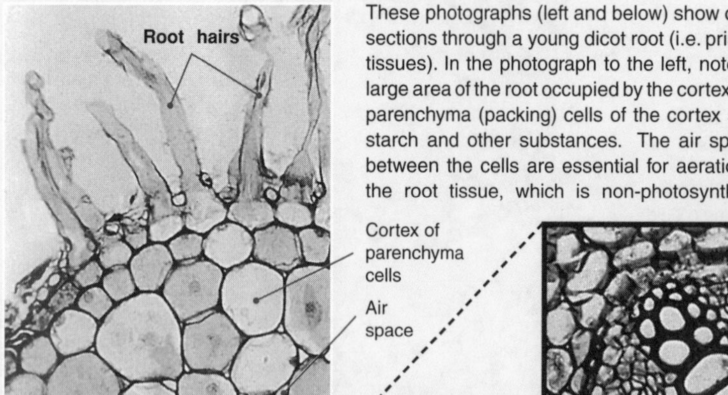


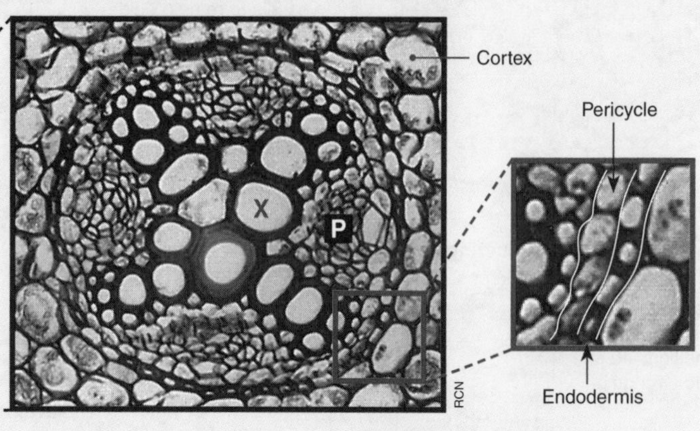


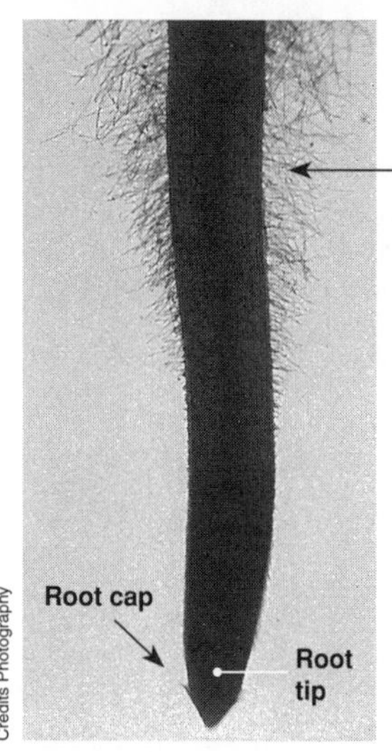


**Root hairs** are located behind the region of cell elongation in the root tip. They are single celled extensions of the epidermal cells that increase the surface area for absorption. Individual root hairs are short lived, but they are produced continually. The root tip is covered by a slimy **root cap**. This protects the dividing cells of the tip and aids the root's movement through the soil.

## The Structure of a Monocot Root

Monocot roots (right) vary from dicot roots in several ways. As in dicots, there is a large cortex but the **endodermis** is very prominent and heavily thickened. The stele (ring of vascular tissue) is large compared with the size of the root and there are many xylem points. There is a central pith inside the vascular tissue that is absent in dicot roots.

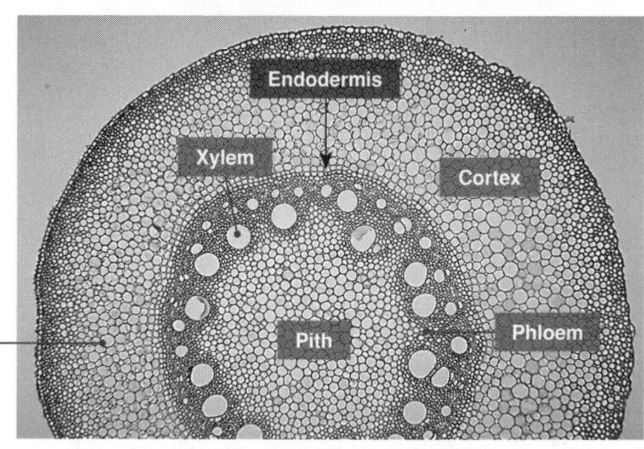


*Cross section through an old root of corn (Zea mays), a typical monocot.*

1. Explain the purpose of the root hairs: ______

2. Explain why the root tip is covered by a cap of cells: ______

3. Describe two features of internal anatomy that distinguish monocot and dicot roots:

   (a) ______

   (b) ______

4. Describe one feature that monocot and dicot roots have in common: ______

5. Describe the role of the parenchyma cells of the cortex: ______




**ISBN: 978-1-927173-12-1**

# Uptake in the Root

Plants need to take up water and minerals constantly to compensate for water losses and obtain the materials they need to manufacture food. The uptake of water and minerals is mostly restricted to the younger cells of the roots and root hairs. Water moves by osmosis because of the gradient in solute concentration from soil to leaves. Some travels through the plasmodesmata of the cells (the **symplastic route**), but most passes through the free spaces outside the plasma membranes (the **apoplast**).

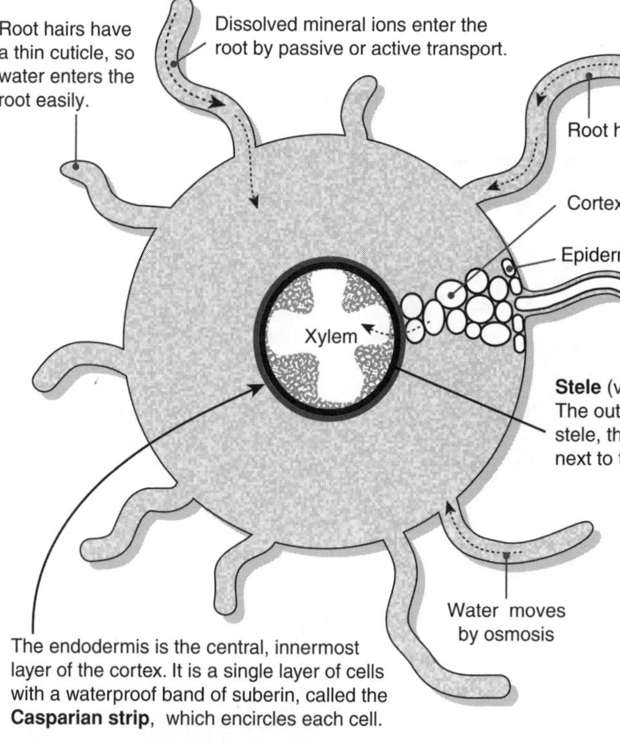


*Schematic cross-section through a dicot root*

Root hairs are extensions of the root epidermal cells and provide a large surface area for absorbing water and nutrients.

## Mineral Uptake

Many minerals are absorbed passively along with the water they are dissolved in. However, some minerals are in such low concentration in the soil that they must be taken up actively by the root. This **active transport** requires energy and is therefore a cost for the plant.

## Water Uptake

The uptake of water through root tissue occurs by **osmosis**, the diffusion of water molecules from a lower to higher solute concentration. There are three pathways of water movement:

**Apoplast pathway:** About 90% of water moves through the non-living spaces of the plant: within the cellulose cell walls and the water-filled spaces of dead cells, such as xylem vessels.

**Symplast pathway:** Some water also moves through the living contents of cells. Water enters the cytoplasm and moves between cells through the cytoplasmic connections called **plasmodesmata** (*sing.* plasmodesma) that cross the cell walls through pits.

**Vacuolar pathway:** A small amount of water passes into the cell vacuoles by osmosis. This vacuolar pathway is the route by which individual cells absorb water. It is not shown on the diagram.

### Enlargement of root cells

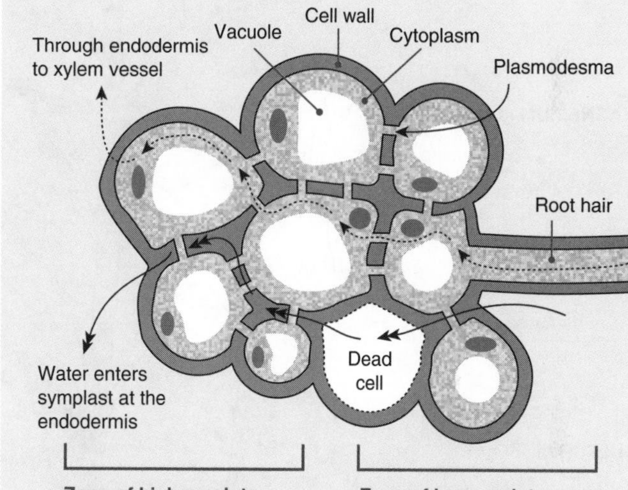


**Zone of higher solute concentration [lower water]**
May be due to: less turgid cells, lower wall pressure or higher concentration of dissolved substances

**Zone of lower solute concentration [higher water]**
May be due to: fully turgid cells, higher wall pressure or lower concentration of dissolved substances

Apoplast pathway (cell wall)
Symplast pathway (cytoplasm)

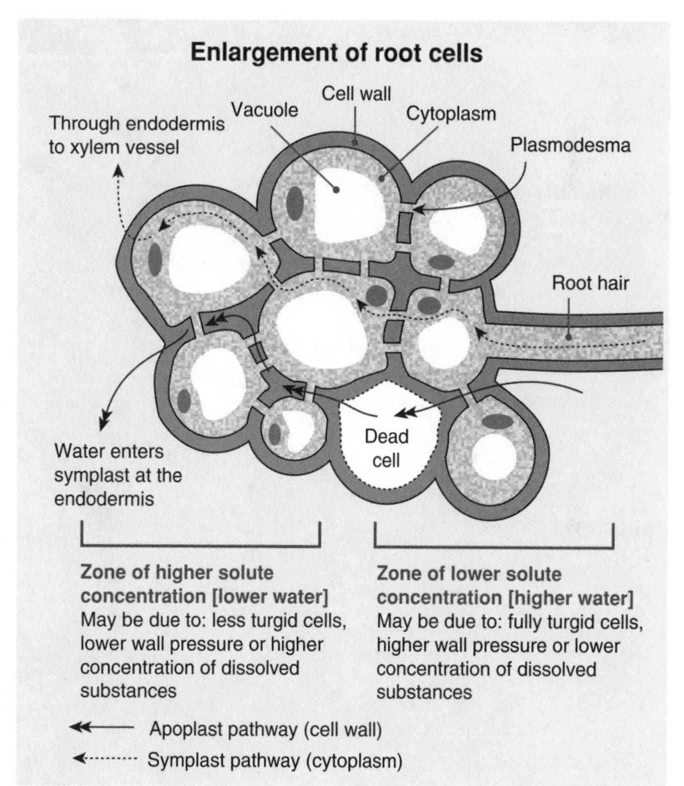

1. (a) Identify the two mechanisms that plants use to absorb nutrients: ______

   (b) Describe the two principal pathways by which water moves through a plant: ______

2. Plants take up water constantly to compensate for transpirational loss. Describe a benefit of a large water uptake: ______

3. (a) Describe the consequence of the **Casparian strip** to the route water takes into the stele: ______

   (b) Suggest why this feature might be advantageous in terms of selective mineral uptake: ______

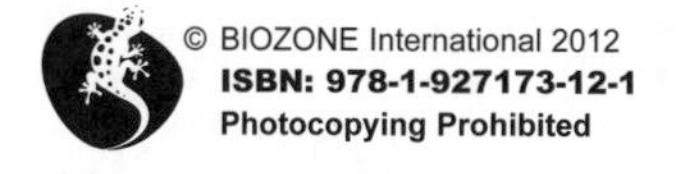


**ISBN: 978-1-927173-12-1**

# Plant Mineral Requirements

Plants normally obtain minerals from the soil. The availability of mineral ions to plant roots depends on soil texture, since this affects the permeability of the soil to air and water. Mineral ions may be available to the plant in the soil water, adsorbed on to clay particles, or via release from humus and soil weathering. **Macronutrients** (e.g. nitrogen, sulfur, phosphorus) are required in large amounts for building basic constituents such as proteins. **Trace elements** (e.g. manganese, copper, and zinc) are required in small amounts. Many are components of, or activators for, enzymes. After being absorbed, mineral ions diffuse into the endodermis and may diffuse, or be actively transported, to the xylem for transport to other regions of the plant.

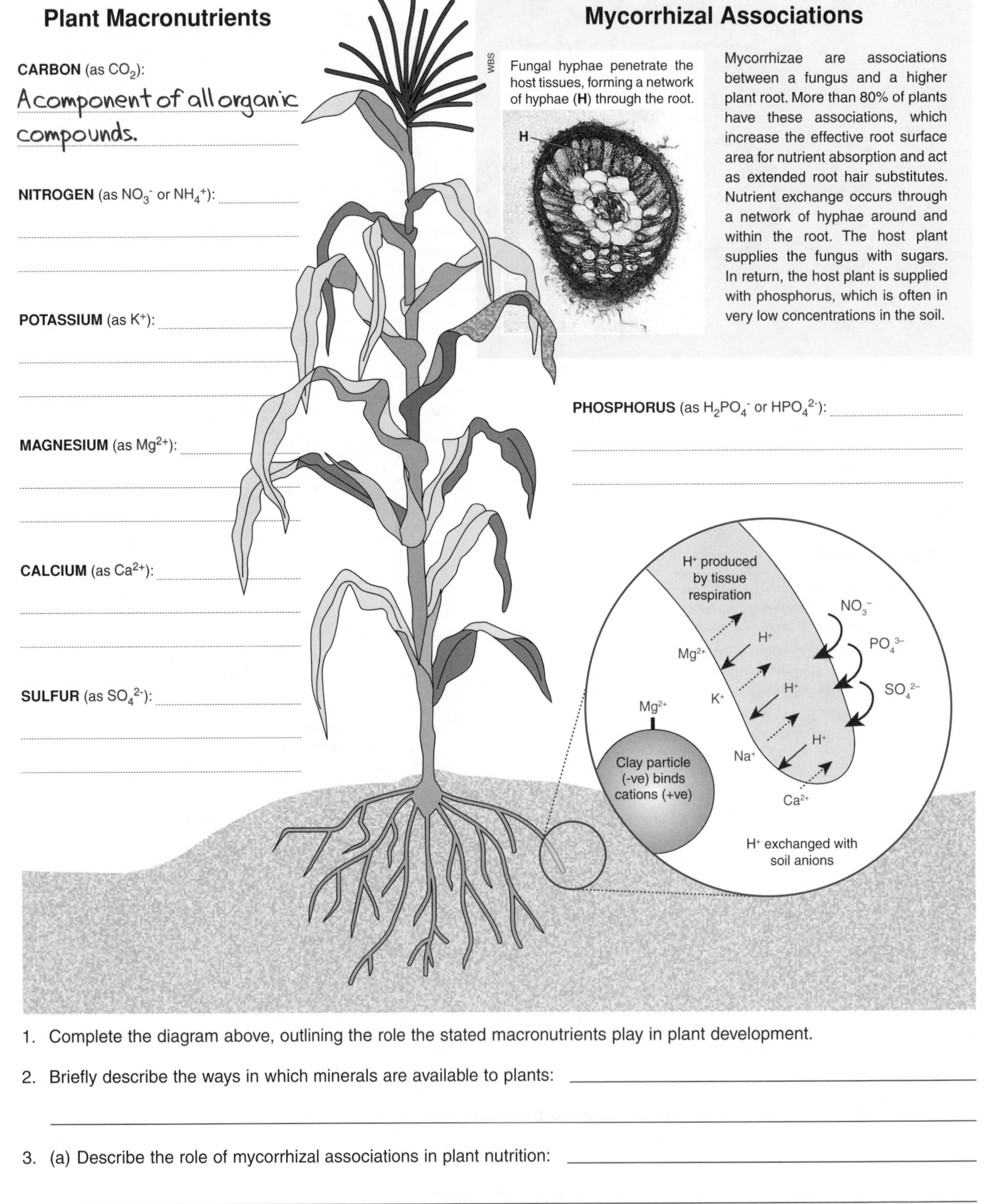


1. Complete the diagram above, outlining the role the stated macronutrients play in plant development.

2. Briefly describe the ways in which minerals are available to plants:

3. (a) Describe the role of mycorrhizal associations in plant nutrition:

   (b) Suggest why plants without mycorrhizal associations might show poor growth:



***Related activities**: Uptake in the Root*
***Weblinks**: Mineral Uptake in Plants*


**ISBN: 978-1-927173-12-1**

# Transpiration in Plants

Plants lose water all the time, despite the adaptations they have to help prevent it (e.g. waxy leaf cuticle). Approximately 99% of the water a plant absorbs from the soil is lost by evaporation from the leaves and stem. This loss, mostly through stomata, is called **transpiration** and the flow of water through the plant is called the **transpiration stream**. Plants rely on a gradient in solute concentration from the roots to the air to move water through their cells. Water flows passively from soil to air along a gradient of increasing solute (decreasing water) concentration. This gradient is the driving force in the ascent of water up a plant. A number of processes contribute to water movement up the plant: transpiration pull, cohesion, and root pressure. Transpiration may seem wasteful, but it has benefits; evaporative water loss cools the plant and the transpiration stream helps the plant to maintain an adequate mineral uptake, as many essential minerals occur in low concentrations in the soil.

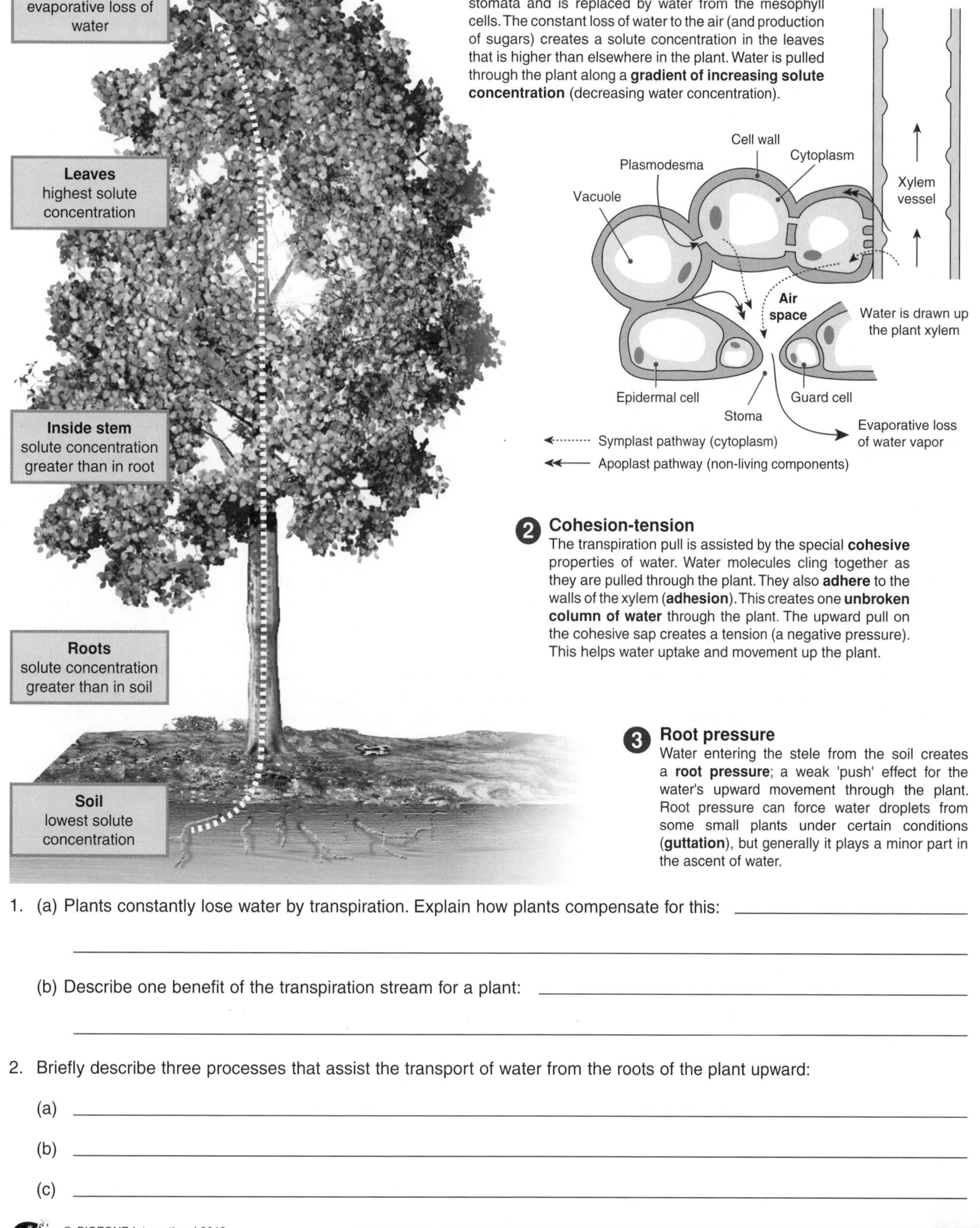


**1 Transpiration pull**
Water is lost from the air spaces by evaporation through stomata and is replaced by water from the mesophyll cells. The constant loss of water to the air (and production of sugars) creates a solute concentration in the leaves that is higher than elsewhere in the plant. Water is pulled through the plant along a **gradient of increasing solute concentration** (decreasing water concentration).

**2 Cohesion-tension**
The transpiration pull is assisted by the special **cohesive** properties of water. Water molecules cling together as they are pulled through the plant. They also **adhere** to the walls of the xylem (**adhesion**). This creates one **unbroken column of water** through the plant. The upward pull on the cohesive sap creates a tension (a negative pressure). This helps water uptake and movement up the plant.

**3 Root pressure**
Water entering the stele from the soil creates a **root pressure**; a weak 'push' effect for the water's upward movement through the plant. Root pressure can force water droplets from some small plants under certain conditions (**guttation**), but generally it plays a minor part in the ascent of water.

1. (a) Plants constantly lose water by transpiration. Explain how plants compensate for this: ______

______

(b) Describe one benefit of the transpiration stream for a plant: ______

______

2. Briefly describe three processes that assist the transport of water from the roots of the plant upward:

(a) ______

(b) ______

(c) ______


ISBN: 978-1-927173-12-1

## The Potometer

A potometer is a simple instrument for investigating transpiration rate (water loss per unit time). The equipment is simple and easy to obtain. A basic potometer, such as the one shown right, can easily be moved around so that transpiration rate can be measured under different environmental conditions.

Some of the physical conditions investigated are:

- Humidity or vapor pressure (high or low)
- Temperature (high or low)
- Air movement (still or windy)
- Light level (high or low)
- Water supply

It is also possible to compare the transpiration rates of plants with different adaptations e.g. comparing transpiration rates in plants with rolled leaves vs rates in plants with broad leaves. If possible, experiments like these should be conducted simultaneously using replicate equipment. If conducted sequentially, care should be taken to keep the environmental conditions the same for all plants used.



3. Describe three environmental conditions that increase the rate of transpiration in plants, and explain how they operate:

(a) ______

(b) ______

(c) ______

4. The **potometer** (above) is an instrument used to measure transpiration rate. Briefly explain how it works:

______

______

5. An experiment was conducted on transpiration from a hydrangea shoot in a potometer. The experiment was set up and the plant left to stabilize (environmental conditions: still air, light shade, 20°C). The plant was then subjected to different environmental conditions and the water loss was measured each hour. Finally, the plant was returned to original conditions, allowed to stabilize and transpiration rate measured again. The data are presented below:

| **Experimental conditions** | **Temperature** (°C) | **Humidity** (%) | **Transpiration** (g $h^{-1}$) |
|---|---|---|---|
| (a) Still air, light shade | 20 | 70 | 1.20 |
| (b) Moving air, light shade | 20 | 70 | 1.60 |
| (c) Still air, bright sunlight | 23 | 70 | 3.75 |
| (d) Still air and dark, moist chamber | 19.5 | 100 | 0.05 |

(a) Name the control in this experiment: ______

______

(b) Identify the factors that increased transpiration rate, explaining how each has its effect: ______

______

______

(c) Suggest a possible reason why the plant had such a low transpiration rate in humid, dark conditions:

______

______

ISBN: 978-1-927173-12-1

# Investigating Plant Transpiration

Once you have your experimental results (data), it is often helpful to tabulate and graph the information. Graphs and tables display data in a way that makes it easy to see trends or relationships between different variables. Presenting graphs properly requires attention to a few basic details, including correct orientation and labelling of the axes, and accurate plotting of points. This activity describes a plant transpiration experiment and provides guidelines for drawing line graphs. Put these guidelines into practice by graphing the second hand data provided as part of this activity.

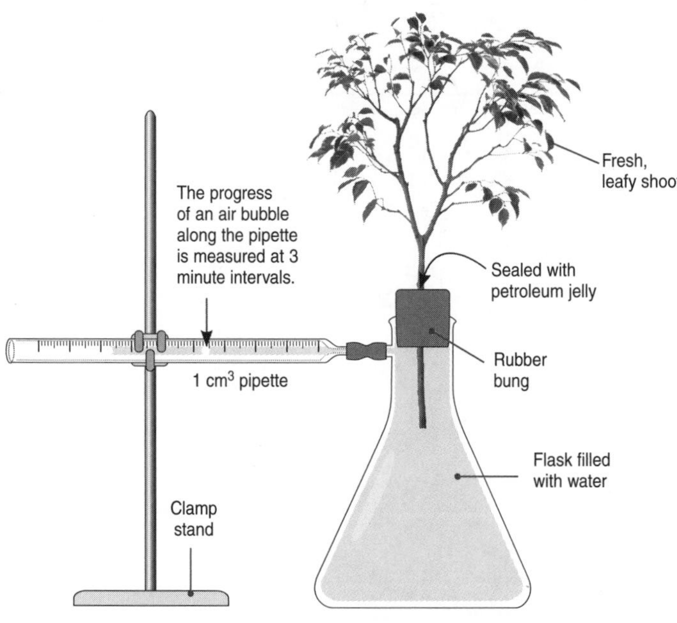

The progress of an air bubble along the pipette is measured at 3 minute intervals.
Fresh, leafy shoot
Sealed with petroleum jelly
Rubber bung
$1\ cm^3$ pipette
Flask filled with water
Clamp stand

### The Apparatus

This experiment investigated the influence of environmental conditions on plant transpiration rate. Four conditions were studied: room conditions (ambient), wind, bright light, and high humidity. After setting up the potometer, the apparatus was equilibrated for 10 minutes, and the position of the air bubble in the pipette was recorded. This is the time 0 reading. The plant was then exposed to one of the environmental conditions. Students recorded the location of the air bubble every three minutes over a 30 minute period. The potometer readings for each environmental condition are presented in Table 1 (next page).

### The Aim

To investigate the effect of environmental conditions on the transpiration rate of plants.

### Background

Plants lose water all the time by evaporation from the leaves and stem. This loss, mostly through pores in the leaf surfaces, is called **transpiration**. Despite the adaptations plants have to help prevent water loss (e.g. waxy leaf cuticle), 99% of the water a plant absorbs from the soil is lost by evaporation. Environmental conditions can affect transpiration rate.

A class was divided into four groups to study how four different environmental conditions (ambient, wind, bright light, and high humidity) affected transpiration rate. A **potometer** was used to measure transpiration rate (water loss per unit time). A basic potometer, such as the one shown left, can easily be moved around so that transpiration rate can be measured under different environmental conditions.

## Guidelines for Drawing Line Graphs

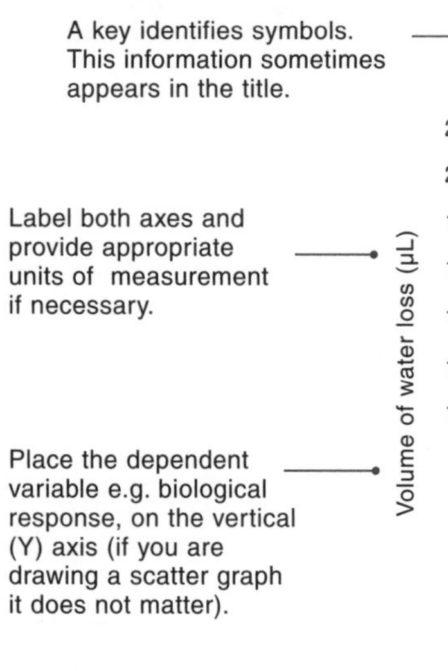

Line graphs are used when one variable (the independent variable) affects another, the dependent variable.
A key identifies symbols. This information sometimes appears in the title.
Label both axes and provide appropriate units of measurement if necessary.
Place the dependent variable e.g. biological response, on the vertical (Y) axis (if you are drawing a scatter graph it does not matter).

Fig. 1: Cumulative water loss in μL from a geranium shoot in still and moving air.

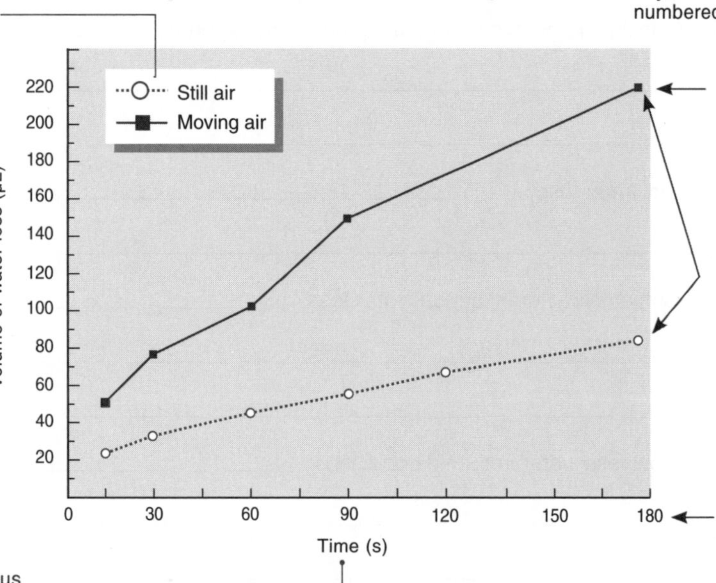

Still air
Moving air
Volume of water loss (μL)
220
200
180
160
140
120
100
80
60
40
20
0
30
60
90
120
150
180
Time (s)

Graphs (called figures) should have a concise, explanatory title. If several graphs appear in your report they should be numbered consecutively.

Plot points accurately. Different responses can be distinguished using different symbols, lines or bar colors.

Two or more sets of results can be plotted on the same figure and distinguished by a key. For a time series, it is appropriate to join the plotted points with a line.

Each axis should have an appropriate scale. Decide on the scale by finding the maximum and minimum values for each variable.

NOTE: The data must be continuous for both variables.

Place the independent variable e.g. time or treatment, on the horizontal (X) axis

ISBN: 978-1-927173-12-1

**Table 1. Potometer readings**

| Treatment \ Time (min) | 0 | 3 | 6 | 9 | 12 | 15 | 18 | 21 | 24 | 27 | 30 |
|---|---|---|---|---|---|---|---|---|---|---|---|
| Ambient | 0 | 0.002 | 0.005 | 0.008 | 0.012 | 0.017 | 0.022 | 0.028 | 0.032 | 0.036 | 0.042 |
| Wind | 0 | 0.025 | 0.054 | 0.088 | 0.112 | 0.142 | 0.175 | 0.208 | 0.246 | 0.283 | 0.325 |
| High humidity | 0 | 0.002 | 0.004 | 0.006 | 0.008 | 0.011 | 0.014 | 0.018 | 0.019 | 0.021 | 0.024 |
| Bright light | 0 | 0.021 | 0.042 | 0.070 | 0.091 | 0.112 | 0.141 | 0.158 | 0.183 | 0.218 | 0.239 |

1. (a) Plot the potometer data from Table 1 on the grid provided. Use the guidelines for drawing line graphs on the previous page as a reference if you need help:

   (b) Identify the independent variable: ______________________

2. (a) Identify the control: ______________________

   (b) Explain the purpose of including an experimental control in an experiment: ______________________

   (c) Which factors increased water loss? ______________________

   (d) How does each environmental factor influence water loss? ______________________

   (e) Explain why the plant lost less water in humid conditions: ______________________


ISBN: 978-1-927173-12-1

# Translocation

Phloem transports the organic products of photosynthesis (sugars) through the plant in a process called **translocation**. In angiosperms, the sugar moves through the sieve elements, which are arranged end-to-end and perforated with sieve plates. Apart from water, phloem sap comprises mainly sucrose (up to 30%). It may also contain minerals, hormones, and amino acids, in transit around the plant. Movement of sap in the phloem is from a **source** (a plant organ where sugar is made or mobilized) to a **sink** (a plant organ where sugar is stored or used). Loading sucrose into the phloem at a source involves energy expenditure; it is slowed or stopped by high temperatures or respiratory inhibitors. In some plants, unloading the sucrose at the sinks also requires energy, although in others, diffusion alone is sufficient to move sucrose from the phloem into the cells of the sink organ.

Phloem sap moves from source to sink at rates as great as 100 m $h^{-1}$: too fast to be accounted for by cytoplasmic streaming. The most acceptable model for phloem movement is the **pressure-flow** (bulk flow) hypothesis. Phloem sap moves by bulk flow, which creates a pressure (hence the term "pressure-flow"). The key elements in this model are outlined below and in steps 1-4 at right. Note that, for simplicity, the cells that lie between the source (and sink) cells and the phloem sieve-tube have been omitted.

1. Loading sugar into the phloem increases the solute concentration inside the sieve-tube cells. This causes the sieve-tubes to take up water by osmosis.
2. The water uptake creates a hydrostatic pressure that forces the sap to move along the tube, just as pressure pushes water through a hose.
3. The pressure gradient in the sieve tube is reinforced by the active unloading of sugar and consequent loss of water by osmosis at the sink (e.g. root cell).
4. Xylem recycles the water from sink to source.

### Measuring Phloem Flow

Experiments investigating flow of phloem often use aphids. Aphids feed on phloem sap (left) and act as natural **phloem probes**. When the mouthparts (stylet) of an aphid penetrate a sieve-tube cell, the pressure in the sieve-tube force-feeds the aphid. While the aphid feeds, it can be severed from its stylet, which remains in place in the phloem. The stylet serves as a tiny tap that exudes sap. Using different aphids, the rate of flow of this sap can be measured at different locations on the plant.

### Phloem Transport

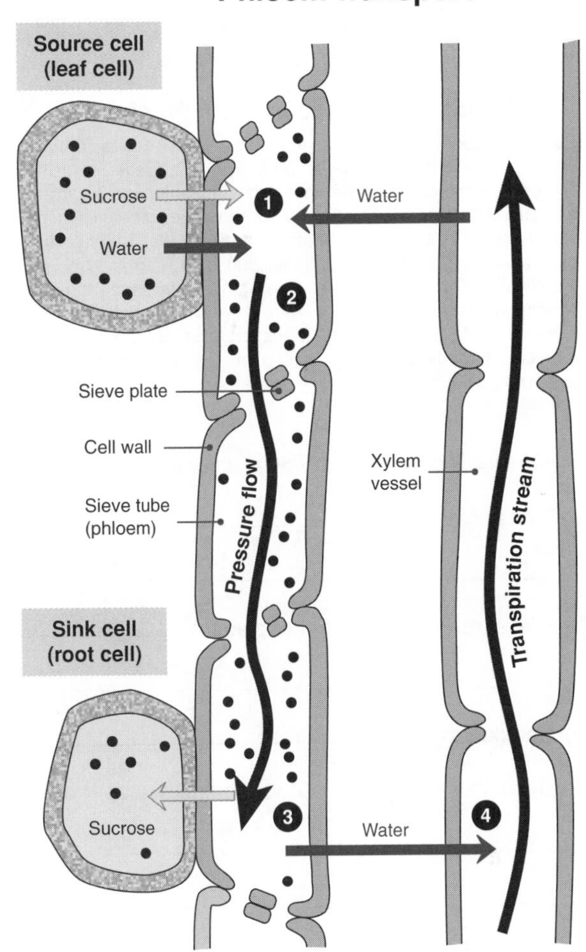


Source: Modified after Campbell *Biology* 1993

1. (a) Explain what is meant by '**source to sink**' flow in phloem transport:

   (b) Name the usual **source** and **sink** in a growing plant:

   Source: Sink:

   (c) Name another possible **source** region in the plant and state when it might be important:

   (d) Name another possible **sink** region in the plant and state when it might be important:

2. Explain why energy is required for translocation and where it is used:

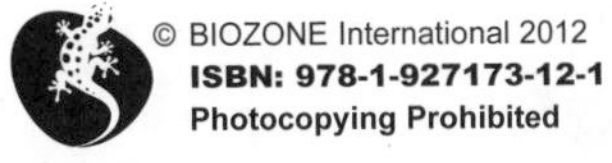


**ISBN: 978-1-927173-12-1**


***Related activities:*** *Xylem and Phloem*
***Weblinks:*** *Sucrose Transport, Sucrose Transport in the Phloem*

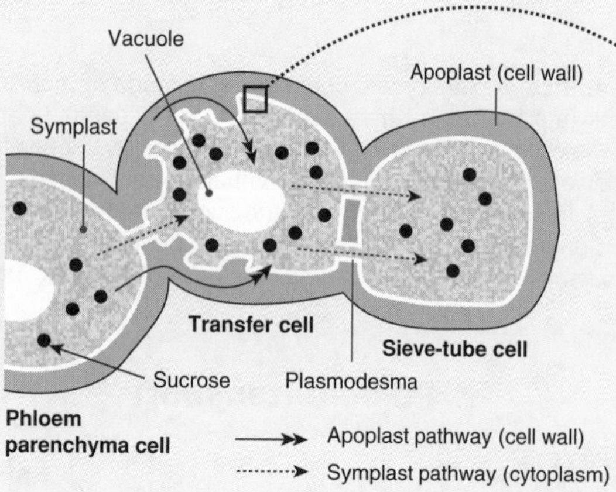

## Coupled Transport of Sucrose

## Loading Sucrose into the Phloem

Sugar (sucrose) can travel to the phloem sieve-tubes through both apoplastic and symplastic pathways. It is loaded into the phloem sieve-tube cells via modified companion cells, called **transfer cells** (above). Loading sucrose into the phloem requires active transport. Using a **coupled transport** (secondary pump) mechanism (right), transfer cells expend energy to accumulate the sucrose. The sucrose then passes into the sieve tube through plasmodesmata. The transfer cells have wall ingrowths that increase surface area for the transport of solutes. Using this mechanism, some plants can accumulate sucrose in the phloem to 2-3 times the concentration in the mesophyll.

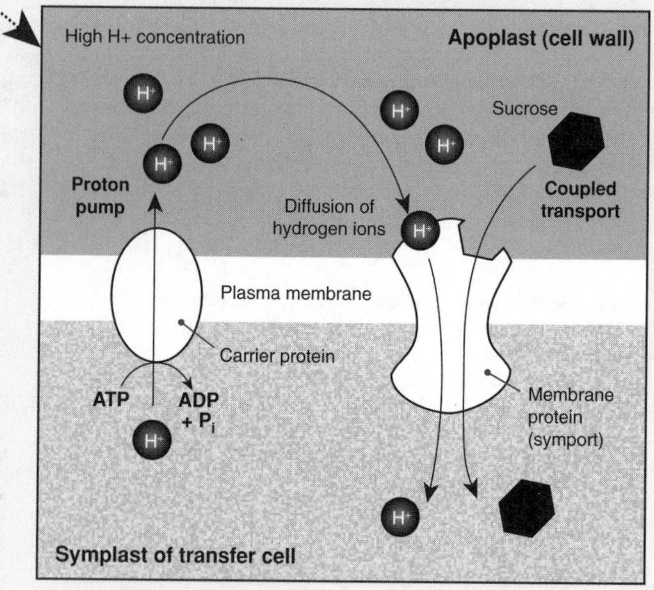

Above: Proton pumps generate a hydrogen ion gradient across the membrane of the transfer cell. This process requires expenditure of energy. The gradient is then used to drive the transport of sucrose, by coupling the sucrose transport to the diffusion of $H^+$ back into the cell.

3. In your own words, describe what is meant by the following:

   (a) Translocation:

   (b) Pressure-flow movement of phloem:

   (c) Coupled transport of sucrose:

4. Briefly explain why water follows the sucrose as the sucrose is loaded into the phloem sieve-tube cell:

5. Explain the role of the companion (transfer) cell in the loading of sucrose into the phloem:

6. Contrast the composition of phloem sap and xylem sap:

7. Explain why it is necessary for phloem to be alive to be functional, whereas xylem can function as a dead tissue:

8. The sieve plate represents a significant barrier to effective mass flow of phloem sap. Suggest why the presence of the sieve plate is often cited as evidence against the pressure-flow model for phloem transport:


**ISBN: 978-1-927173-12-1**

# Investigating Plant Growth

Recording data from an experiment is an important skill. Using a table is the preferred way to record your results **systematically**, both during the course of your experiment and in presenting your results. A table can also show calculated values, such as rates or means. An example of a table for recording results is shown at the bottom of the page. It relates to a student investigation that followed the observation that plants in a paddock fertilized with a nitrogen fertilizer grew more vigorously than plants in a non-fertilized paddock. The table's first column shows the range of the independent variable. There are spaces for multiple samples, and calculated mean values. The students tested which concentration of a soluble nitrogen fertilizer produced optimal growth.

Radishes

### The Aim

To investigate the effect of a nitrogen fertilizer on the growth of plants.

### Background

Inorganic fertilizers revolutionized crop farming when they were introduced during the late 19th and early 20th century. Crop yields soared and today it is estimated around 50% of crop yield is attributable to the use of fertilizer. Nitrogen is a very important element for plant growth and several types of purely nitrogen fertilizer are manufactured to supply it, e.g. urea.

### Experimental Method

This experiment was designed to test the effect of nitrogen fertilizer on plant growth. Radish seeds were planted in separate identical pots (5 cm x 5 cm wide x 10 cm deep) and grown together in normal room conditions. The radishes were watered every day at 10 am and 3 pm with 1.25 L per treatment. Water soluble fertilizer was mixed and added with the first watering on the 1st, 11th and 21st days. The fertilizer concentrations used were: 0.00, 0.06, 0.12, 0.18, 0.24, and 0.30 g $L^{-1}$ with each treatment receiving a different concentration. The plants were grown for 30 days before being removed, washed, and the root (radish) weighed. Results were tabulated below:

To investigate the effect of nitrogen on plant growth, a group of students set up an experiment using different concentrations of nitrogen fertilizer. Radish seeds were planted into a standard soil mixture and divided into six groups each, with five sample plants (30 plants in total).

Tables should have an accurate, descriptive title. Number tables consecutively through the report.

Heading and subheadings identify each set of data and show units of measurement.

Independent variable in the left column.

**Table 1:** Mass (g) of radish plant roots under six different fertilizer concentrations (data given to 1dp).

| Fertilizer concentration (g $L^{-1}$) | Mass of radish root (g)† | | | | | | |
|---|---|---|---|---|---|---|---|
| | Sample (*n*) | | | | | Total mass | Mean mass |
| | 1 | 2 | 3 | 4 | 5 | | |
| **0** | 80.1 | 83.2 | 82.0 | 79.1 | 84.1 | 408.5 | 81.7 |
| **0.06** | 109.2 | 110.3 | 108.2 | 107.9 | 110.7 | | |
| **0.12** | 117.9 | 118.9 | 118.3 | 119.1 | 117.2 | | |
| **0.18** | 128.3 | 127.3 | 127.7 | 126.8 | DNG* | | |
| **0.24** | 23.6 | 140.3 | 139.6 | 137.9 | 141.1 | | |
| **0.30** | 122.3 | 121.1 | 122.6 | 121.3 | 123.1 | | |

* DNG: Did not germinate

Control values (if present) should be placed at the beginning of the table.

Values should be shown only to the level of significance allowable by your measuring technique.

Organize the columns so that each category of like numbers or attributes is listed vertically.

Each row should show a different experimental treatment, organism, sampling site etc.

† Based on data from M S Jilani, *et al* Journal Agricultural Research

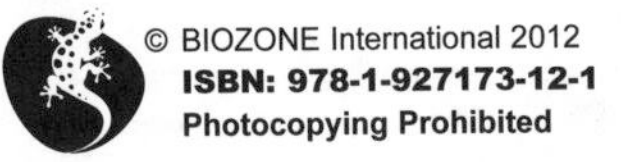


ISBN: 978-1-927173-12-1

1. Identify the independent variable for the experiment and its range: ____________________

____________________

2. What is the sample size for each concentration of fertilizer? ____________________

3. One of the radishes recorded in Table 1 did not grow as expected and produced an extreme value. Record the **outlying value** here and decide whether or not you should include it in future calculations:

____________________

____________________

4. Complete the table on the previous page by calculating the **total mass** and **mean mass** of the radish roots:

5. Use the grid below to draw a **line graph** of the experimental results. Use your calculated means and remember to include a title and correctly labelled axes.

6. The students recorded the wet mass of the root (the root still containing water) in their table. What mass should they have actually recorded to get a better representation of the effect of the fertilizer on root mass?

____________________

____________________

7. Why would measuring just root mass not be a totally accurate way of measuring the effect of fertilizer on radish growth?

____________________

____________________

8. Describe some other measurements the students could have taken to make their experiment more complete:

____________________

____________________

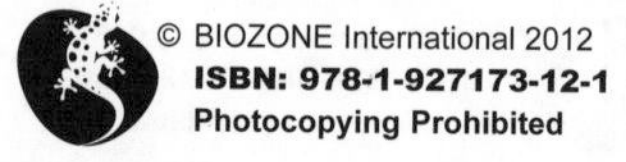

**ISBN: 978-1-927173-12-1**

## Calculating Simple Statistics for a Data Set

| Statistic | Definition and use | Method of calculation |
|---|---|---|
| **Mean** | • The average of all data entries.<br>• Measure of central tendency for normally distributed data. | • Add up all the data entries.<br>• Divide by the total number of data entries. |
| **Median** | • The middle value when data entries are placed in rank order.<br>• A good measure of central tendency for skewed distributions. | • Arrange the data in increasing rank order.<br>• Identify the middle value.<br>• For an even number of entries, find the mid point of the two middle values. |
| **Mode** | • The most common data value.<br>• Suitable for bimodal distributions and qualitative data. | • Identify the category with the highest number of data entries using a tally chart or a bar graph. |
| **Range** | • The difference between the smallest and largest data values.<br>• Provides a crude indication of data spread. | • Identify the smallest and largest values and find the difference between them. |

Data can be simply summarized using **descriptive statistics**. Descriptive statistics, such as mean, median, and mode, can highlight trends or patterns in the data. The mean can be used to compare different groups. You can use more complex statistics to determine if the means of different groups are significantly different.

### When NOT to calculate a mean:

In certain situations, calculation of a simple arithmetic mean is inappropriate.

**Remember:**

- *DO NOT* calculate a mean from values that are already means (averages) themselves.
- *DO NOT* calculate a mean of ratios (e.g. percentages) for several groups of different sizes; go back to the raw values and recalculate.
- *DO NOT* calculate a mean when the measurement scale is not linear, e.g. pH units are not measured on a linear scale.

The students decided to further their experiment by recording the number of leaves on each radish plant:

**Table 2:** Number of leaves on radish plant under six different fertilizer concentrations.

| Fertilizer concentration (g L$^{-1}$) | Number of leaves: Sample ($n$) 1 | 2 | 3 | 4 | 5 | Mean | Median | Mode |
|---|---|---|---|---|---|---|---|---|
| **0** | 9 | 9 | 10 | 8 | 7 | | | |
| **0.06** | 15 | 16 | 15 | 16 | 16 | | | |
| **0.12** | 16 | 17 | 17 | 17 | 16 | | | |
| **0.18** | 18 | 18 | 19 | 18 | DNG* | | | |
| **0.24** | 6 | 19 | 19 | 18 | 18 | | | |
| **0.30** | 18 | 17 | 18 | 19 | 19 | | | |

* DNG: Did not germinate

9. Complete Table 2 by calculating the mean, median and mode for each concentration of fertilizer:

10. Which concentration of fertilizer appeared to produce the best growth results? ___________

11. Describe some sources of error for the experiment: ___________

___________

12. Write a brief conclusion for the experiment. Include a reference to the aim and results: ___________

___________

___________

___________

13. The students decided to replicate the experiment (carry it out again). How might this improve the experiment's results?

___________

___________

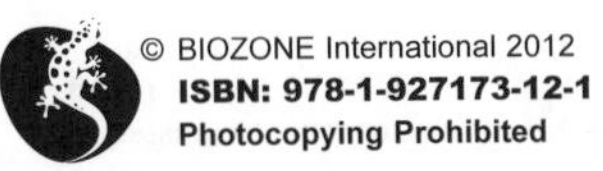


ISBN: 978-1-927173-12-1

# Adaptations of Hydrophytes

**Hydrophytes** are plants that have adapted to living partially or fully submerged in water. Survival in water poses different problems to those faced by terrestrial plants. Hydrophytes have a reduced root system. This is related to the relatively high concentration of nutrients in the sediment and the plant's ability to remove nitrogen and phosphorus directly from the water. The leaves of submerged plants are thin to increase the surface area of photosynthetic tissue and reduce internal shading. Hydrophytes typically have no cuticle (waterproof covering) or the cuticle is very thin. This allows the plant to absorb minerals and gases directly from the water. In addition, they are supported by the water, so they need very little structural support tissue.

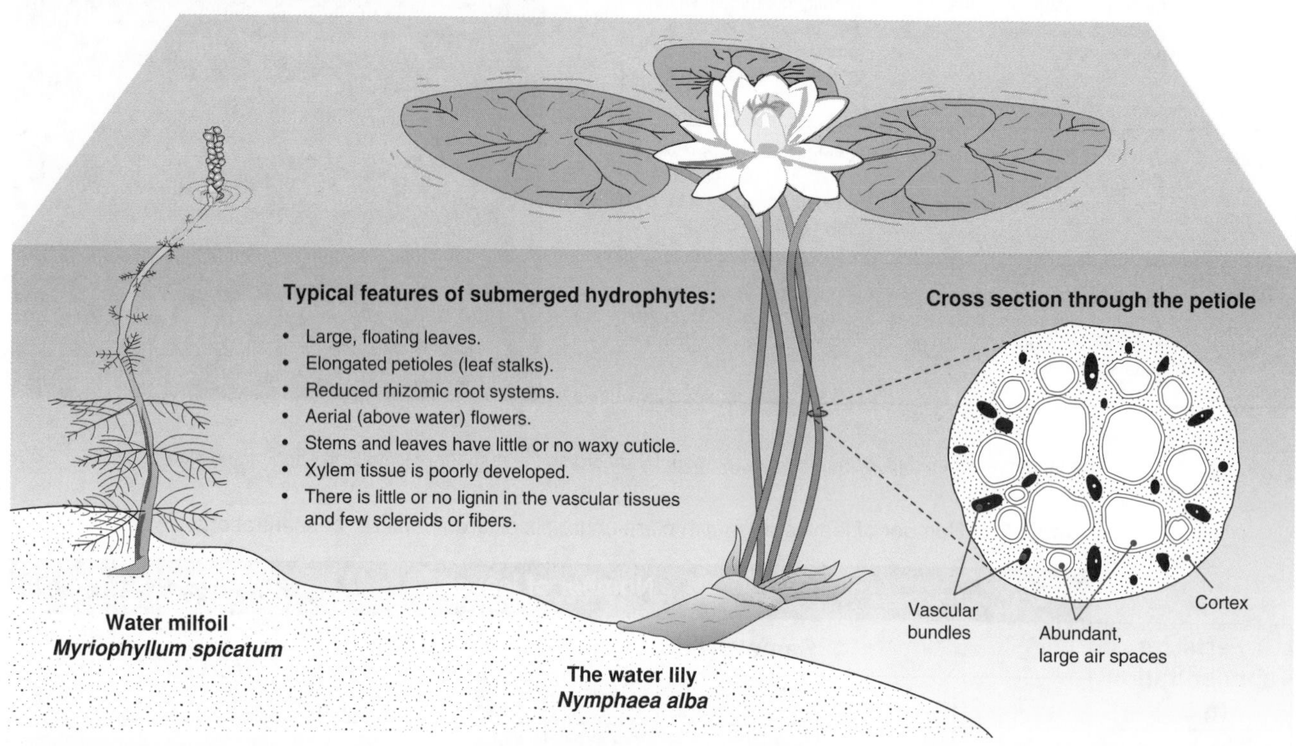

*Myriophyllum*'s submerged leaves are well spaced and taper towards the surface to assist with gas exchange and distribution of sunlight.

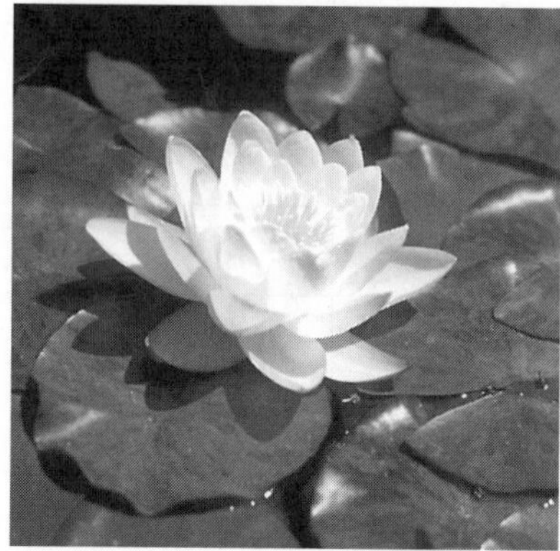
The floating leaves of water lilies (*Nymphaea*) have a high density of stomata on the upper leaf surface so they are not blocked by water.

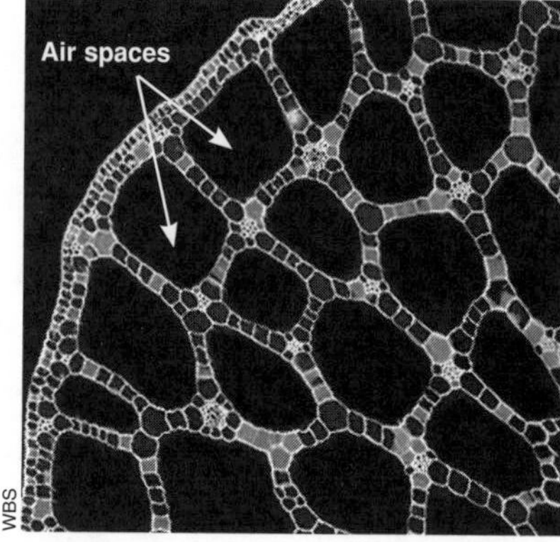

Cross section through *Potamogeton*, showing large air spaces which assist with flotation and gas exchange.

1. Explain how the following adaptations assist hydrophytes to survive in an aquatic environment:

   (a) Large air spaces within the plants tissues: ______________________

   ______________________

   (b) Thin cuticle: ______________________

   ______________________

   (c) High stomatal densities on the upper leaf surface: ______________________

   ______________________

2. Explain why water loss through transpiration is not a problem for hydrophytes: ______________________

______________________



***Related activities**: Leaf and Stem Adaptations*
***Weblinks**: Some Adaptations to Habitat*


**ISBN: 978-1-927173-12-1**

# Adaptations of Xerophytes

Plants adapted to dry conditions are called **xerophytes** and they show structural (xeromorphic) and physiological adaptations for water conservation. These typically include small, hard leaves, and epidermis with a thick cuticle, sunken stomata, succulence, and permanent or temporary absence of leaves. Xerophytes may live in humid environments, provided that their roots are in dry microenvironments (e.g. the roots of epiphytic plants that grow on tree trunks or branches). The nature of the growing environment is important in many other situations too. **Halophytes** (salt tolerant plants) and alpine species may also show xeromorphic features in response to the scarcity of obtainable water and high transpirational losses in these environments.

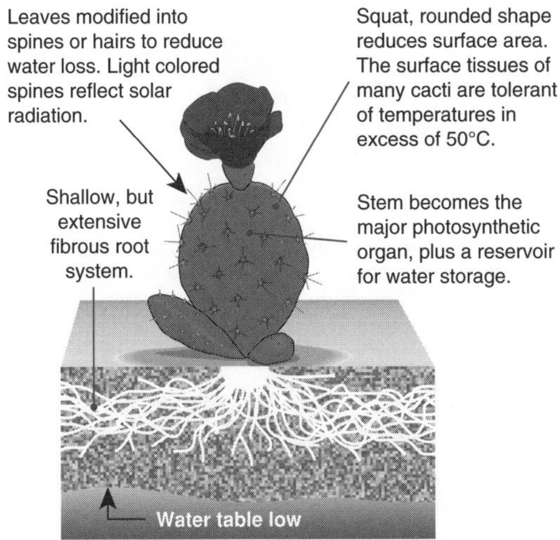


## Dry Desert Plant

Desert plants, such as cacti, must cope with low or sporadic rainfall and high transpiration rates. A number of structural adaptations (diagram left) reduce water losses, and enable them to access and store available water. Adaptations such as waxy leaves also reduce water loss and, in many desert plants, germination is triggered only by a certain quantity of rainfall.

Acacia trees have **deep root systems**, allowing them to draw water from lower water table systems.

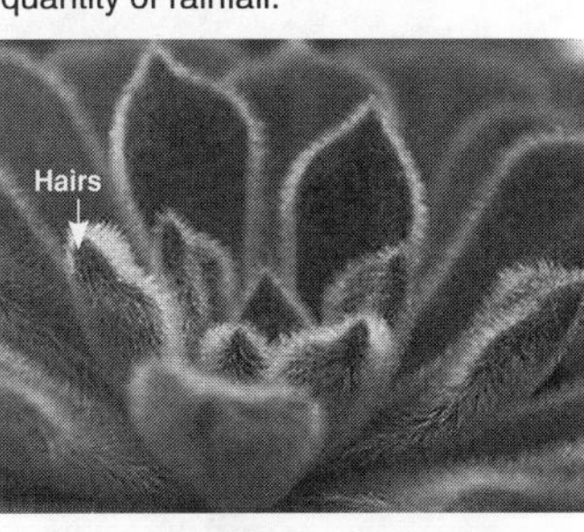


The outer surface of many succulents are coated in fine hairs, which traps air close to the surface reducing transpiration rate.

Seaweeds, which are protists, not plants, tolerate drying between tides even though they have no xeromorphic features.

A waxy coating of **suberin** on mangrove roots excludes 97% of salt from the water.

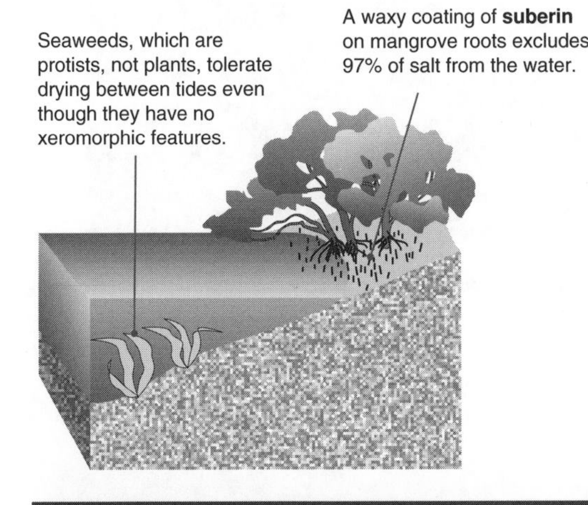

## Ocean Margin Plant

Land plants that colonize the shoreline must have adaptations to obtain water from their saline environment while maintaining their osmotic balance. In addition, the shoreline is often a windy environment, so they frequently show xeromorphic adaptations that enable them to reduce transpirational water losses.

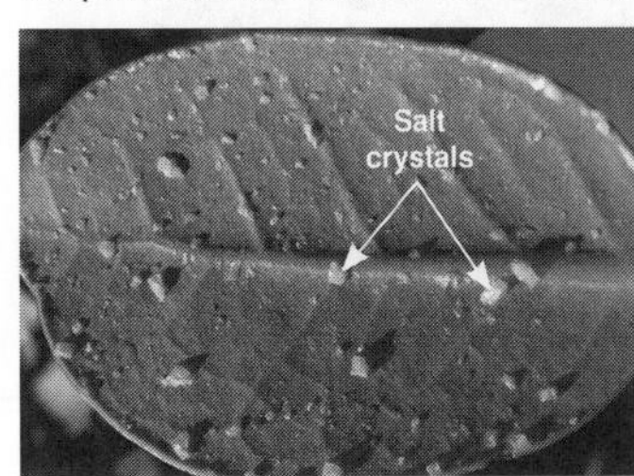


To maintain osmotic balance, mangroves can secrete absorbed salt as salt crystals (above), or accumulate salt in old leaves which are subsequently shed.

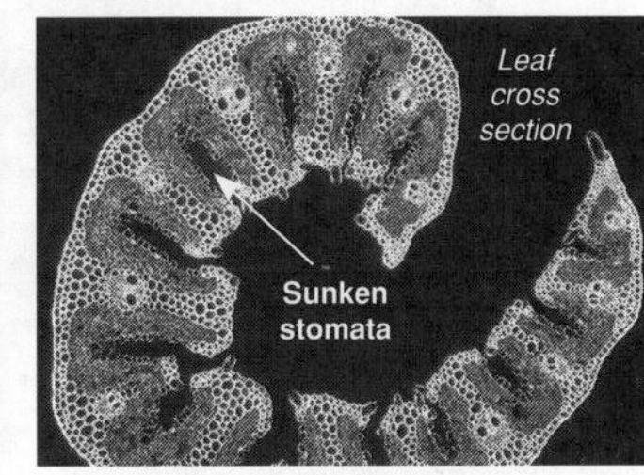


Grasses found on shoreline coasts (where it is often windy), curl their leaves and have sunken stomata to reduce water loss by transpiration.

### Methods of water conservation in various plant species

| Adaptation for water conservation | Effect of adaptation | Example |
|---|---|---|
| Thick, waxy cuticle to stems and leaves | Reduces water loss through the cuticle. | *Pinus* sp. ivy (*Hedera*), sea holly (*Eryngium*), prickly pear (*Opuntia*). |
| Reduced number of stomata | Reduces the number of pores through which water loss can occur. | Prickly pear (*Opuntia*), *Nerium* sp. |
| Stomata sunken in pits, grooves, or depressions<br>Leaf surface covered with fine hairs<br>Massing of leaves into a rosette at ground level | Moist air is trapped close to the area of water loss, reducing the diffusion gradient and therefore the rate of water loss. | **Sunken stomata:** *Pinus* sp., *Hakea* sp. Hairy leaves: lamb's ear. **Leaf rosettes:** dandelion (*Taraxacum*), daisy. |
| Stomata closed during the light, open at night | CAM metabolism: $CO_2$ is fixed during the night, water loss in the day is minimized. | **CAM plants,** e.g. American aloe, pineapple, *Kalanchoe*, *Yucca*. |
| Leaves reduced to scales, stem photosynthetic<br>Leaves curled, rolled, or folded when flaccid | Reduction in surface area from which transpiration can occur. | **Leaf scales:** broom (*Cytisus*). Rolled leaf: marram grass (*Ammophila*), *Erica* sp. |
| Fleshy or succulent stems<br>Fleshy or succulent leaves | When readily available, water is stored in the tissues for times of low availability. | **Fleshy stems: Opuntia**, candle plant (*Kleinia*). **Fleshy leaves:** *Bryophyllum*. |
| Deep root system below the water table | Roots tap into the lower water table. | Acacias, oleander. |
| Shallow root system absorbing surface moisture | Roots absorb overnight condensation. | Most cacti |


**ISBN: 978-1-927173-12-1**


***Periodicals:*** *Cacti*

***Related activities:*** *Leaf and Stem Adaptations*
***Weblinks:*** *Some Adaptations to Habitat*

# Adaptations in halophytes and drought tolerant plants

**Ice plant** (*Carpobrotus*): The leaves of many desert and beach dwelling plants are fleshy or succulent. The leaves are triangular in cross section and crammed with water storage cells. The water is stored after rain for use in dry periods. The shallow root system is able to take up water from the soil surface, taking advantage of any overnight condensation.

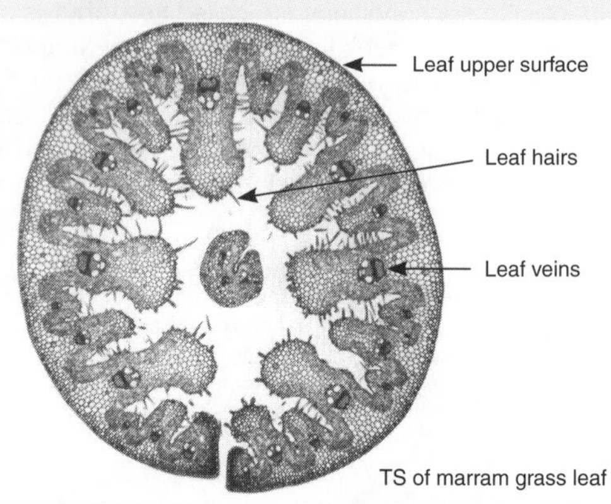


**Marram grass** (*Ammophila*): The long, wiry leaf blades of this beach grass are curled downwards with the stomata on the inside. This protects them against drying out by providing a moist microclimate around the stomata. Plants adapted to high altitude often have similar adaptations.

**Ball cactus** (*Echinocactus grusonii*): In cacti, the leaves are modified into long, thin spines which project outward from the thick fleshy stem (see close-up above right). This reduces the surface area over which water loss can occur. The stem takes over the role of producing the food for the plant and also stores water during rainy periods for use during drought. As in succulents like ice plant, the root system in cacti is shallow to take advantage of surface water appearing as a result of overnight condensation.

1. Explain the purpose of **xeromorphic** adaptations: ____________________

2. Describe three xeromorphic adaptations of plants:

   (a) ____________________

   (b) ____________________

   (c) ____________________

3. Describe a physiological mechanism by which plants can reduce water loss during the daylight hours:

4. Explain why creating a moist microenvironment around the areas of water loss reduces transpiration rate:

5. Explain why seashore plants (halophytes) exhibit many desert-dwelling adaptations: ____________________


**ISBN: 978-1-927173-12-1**

# KEY TERMS: Mix and Match

*INSTRUCTIONS: Test your vocabulary by matching each term to its definition, as identified by its preceding letter code.*

active transport ..........

bulk (=mass) flow ..........

capillary action ..........

Casparian strip ..........

companion cells ..........

cortex ..........

endodermis ..........

flaccid ..........

gas exchange ..........

guard cells ..........

hydrophyte ..........

mineral ..........

osmosis ..........

phloem ..........

pressure-flow hypothesis ..........

root hair ..........

root pressure ..........

sieve tube cells ..........

stomata ..........

tracheid ..........

translocation ..........

transpiration ..........

transpiration pull ..........

tension-cohesion hypothesis ..........

turgid ..........

vessel ..........

xerophyte ..........

xylem ..........

**A** Without turgor, limp.

**B** Water conducting cells in the xylem of angiosperms but absent from most gymnosperms.

**C** The passive movement of water molecules across a partially permeable membrane down a water potential gradient.

**D** The outer layer of a plant stem or root, bounded on the outside by the epidermis and on the inside by the endodermis.

**E** One proposed mechanism for the movement of sugars from source to sink in the phloem.

**F** A plant that has adaptations to survive in arid conditions.

**G** Complex plant tissue specialized for the transport of water and dissolved mineral ions.

**H** Elongated cells in phloem for transporting carbohydrate (sugar).

**I** Living cells in close association with the sieve-tube members in phloem.

**J** Specialized cells, which occur in pairs and which regulate movement of gases and water vapor through the stomata.

**K** Pores in the leaf surface through which gases can move.

**L** The tendency of fluids in narrow tubes to move upwards, against the pull of gravity.

**M** The hypothesis for the movement of water through the plant based on transpiration pull and the cohesive properties of water.

**N** The exchange of oxygen and carbon dioxide across the respiratory membrane.

**O** Movement of water and solutes together as a single mass due to a pressure gradient.

**P** A thin layer of parenchyma tissue found just outside the vascular cylinder in roots, which helps to regulate the passive movements of water and ions.

**Q** Vascular tissue that conducts sugars through the plant. Characterized by the presence of sieve tubes.

**R** The transport of materials within a plant by the phloem.

**S** Elongated cells in the xylem that transport water and mineral salts.

**T** One of the many chemical elements required by living organisms, other than the four elements carbon, hydrogen, nitrogen, and oxygen present in common organic molecules.

**U** The loss of water vapor by plants, mainly from leaves via the stomata.

**V** A term meaning swollen or tight.

**W** Osmotic pressure within the cells of a root system that aids the movement of water through the plant. It occurs when the soil moisture level is high during the night or when transpiration is low during the day.

**X** A tubular outgrowth of an epidermal cell on a plant root that increases the surface area for water absorption.

**Y** A band of waterproof material in the radial and transverse walls of the endodermis, which blocks the passive movement of materials, such as water and solutes into the stele.

**Z** The loss of water vapor by plants, mainly from leaves via the stomata.

**AA** The energy-requiring movement of substances across plasma membranes into cells.

**BB** A plant which has adapted to living either partially or fully submerged in water.


**ISBN: 978-1-927173-12-1**

Enduring Understanding

2.D
4.A

# Obtaining Nutrients and **Eliminating Wastes**

## Key terms

alveoli
antidiuretic hormone (ADH)
dentition
digestive enzymes
excretion
extracellular digestion
gastric ceca
gastrovascular cavity
gills
gut (=digestive tract)
heterotroph
holozoic
intestinal villi
intracellular digestion
kidney
loop of Henle
lungs
malpighian tubules
nephridia
nitrogenous waste
osmoregulation
parasite
predator
protonephridia
respiratory gas
respiratory membrane
spiracles
spiral valve
tracheae (tracheal tubes)

***Periodicals:***
*Listings for this chapter are on page 374*

***Weblinks:***
*www.thebiozone.com/weblink/AP2-3121.html*

*BIOZONE APP: Student Review Series Integument & Homeostasis*

## Key concepts

- Animals exchange matter with their environment.
- Gas exchange surfaces maximize exchange rates.
- Structural and functional diversity in animal guts is related to the volume and type of food ingested.
- Maintenance of fluid and ion balance ensures optimum conditions for metabolism.
- Diversity in excretory systems reflects the range of excretory products and the environments in which animals live.
- Urine production in the kidney is the result of ultrafiltration, secretion, and reabsorption.

## Essential Knowledge

□ 1. Use the **KEY TERMS** to compile a glossary for this topic.

### Maintaining the Animal Body *(2.D.2: b, 4.A.4)* pages 94, 105, 109-110, 134

□ 2. Describe the diversity in the systems by which animals exchange matter with the environment, including the way they exchange gases, obtain nutrients, and eliminate wastes. Recognize that this diversity reflects both evolutionary history and adaptation to environment.

### Gas Exchange *(2.D.2: b, 4.A.4)* pages 93-104

□ 3. Describe structural and functional diversity in animal gas exchange systems, relating specific features to their suitability in the environment in each case. Examples could include the gas exchange systems of insects (**tracheae**), fish (**gills**), or air-breathing vertebrates (**lungs**). Explain the close relationship between the gas exchange and circulatory systems in mammals.

### Digestive Mechanisms *(2.D.2: b)* pages 105-127

□ 4. Describe the three principal nutritional modes in **heterotrophs**: **parasites**, saprophytes, and **holozoic animals**. Describe diversity in the ways animals feed.

□ 5. Compare and contrast **intracellular digestion** and **extracellular digestion**. Describe diversity in the digestive mechanisms of animals, including reference to **food vacuoles** and **gastrovascular cavities**, and one-way tube guts.

□ 6. Compare and contrast the basic structure of the gut in insects, fish, and mammals. Describe and explain specific adaptations of the gut for **digestion** and nutrient **absorption** in chosen examples, e.g. carnivorous mammals, carnivorous fish, ruminant herbivores, omnivores, wood-eating insects, nectar feeders.

### Excretion and Osmoregulation *(2.D.2: b-c)* pages 128-141

□ 7. Distinguish between **excretion** and **osmoregulation** and explain the link between these processes. Describe the problems associated with maintaining water and salt (ion) balance in marine and freshwater. Describe diversity in mechanisms and strategies for salt and water balance in representative taxa.

□ 8. Identify the form of the **nitrogenous waste** excreted by different animal taxa and relate the energy costs of production and excretion to life history and environment.

□ 9. Describe diversity in the structure and function of systems for excretion and osmoregulation in representative taxa. Examples could include the excretory systems of flatworms (**protonephridia**), annelids (**nephridia**), insects (**malpighian tubules**), or vertebrates (**kidneys**).

□ 10. Explain the basic processes common to urine formation in vertebrate nephrons: **ultrafiltration**, **secretion**, and **reabsorption**.

□ 11. Describe the role of **ADH** in regulating urine output. Explain how ADH secretion is controlled via **negative feedback** in response to blood osmolarity.

# The Need for Gas Exchange

Living cells require energy for the activities of life. Energy is released in cells by the breakdown of sugars and other substances in the metabolic process called **cellular respiration**. As a consequence of this process, gases need to be exchanged (by **diffusion**) between the respiring cells and the environment. In most organisms these gases are carbon dioxide and oxygen.

The diagram below illustrates this for an animal. Plant cells also respire, but their gas exchange budget is different because they consume $CO_2$ and produce $O_2$ in photosynthesis. Effective gas exchange surfaces are thin so that the barrier they present to diffusion is minimized. Diffusion gradients are maintained by transport of gases away from the gas exchange surface.

## The Need for Gas Exchange

**Gas exchange** is the process by which oxygen is acquired and carbon dioxide is removed. Cellular respiration creates a constant demand for oxygen ($O_2$) and a need to eliminate carbon dioxide gas ($CO_2$).

**Gas exchange** surfaces provide a means for gases to enter and leave the body. Some organisms use the body surface as the sole gas exchange surface, but many have specialized gas exchange structures (e.g. lungs, gills, or stomata). Amphibians use the body surface and simple lungs to provide for their gas exchange requirements.

Carbon dioxide gas

Oxygen gas

**Cellular respiration** takes place in every cell of an organism's body

Water ($H_2O$)

Carbon dioxide ($CO_2$)

Glucose ($C_6H_{12}O_6$)

Energy

Oxygen ($O_2$)

**Mitochondria** are the main site where glucose is broken down to release energy. In the process, oxygen is used to make water and carbon dioxide is released as a waste product.

**Fick's Law**

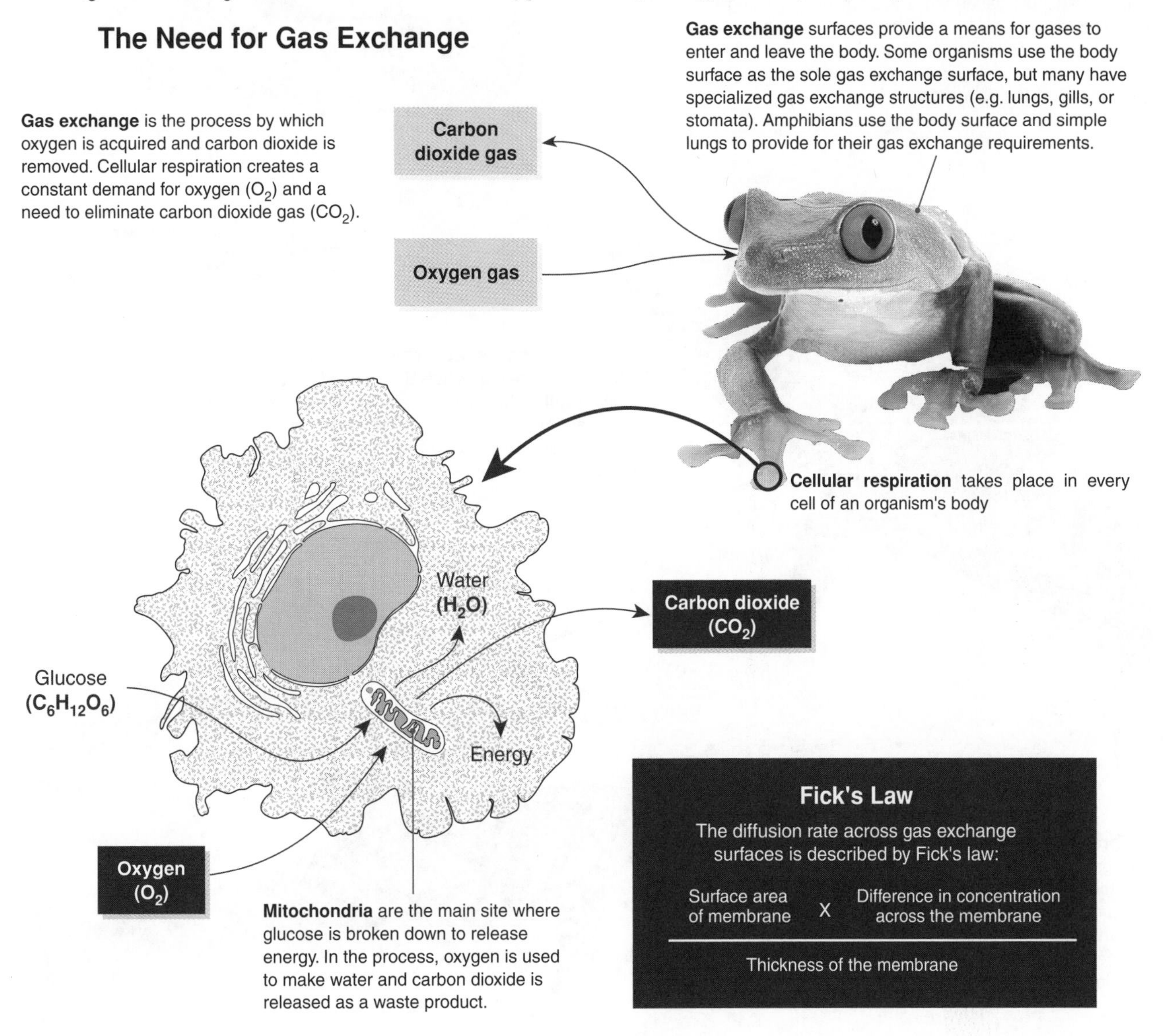

The diffusion rate across gas exchange surfaces is described by Fick's law:

$$\frac{\text{Surface area of membrane} \times \text{Difference in concentration across the membrane}}{\text{Thickness of the membrane}}$$

1. Distinguish between cellular respiration and gas exchange: ______

2. (a) What gases are involved in cellular respiration? ______

   (b) By which transport process do these gases move? ______

3. What is the main function of a gas exchange surface? ______

4. Describe the three properties that all gas exchange surfaces have in common and state the significance of each:

   (a) ______

   (b) ______

   (c) ______

ISBN: 978-1-927173-12-1




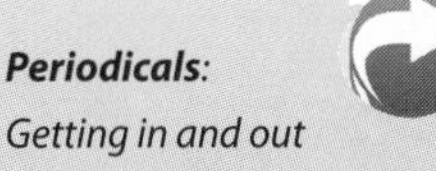

Obtaining Nutrients & Eliminating Wastes

# Gas Exchange in Animals

The way an animal exchanges gases with its environment is influenced by the animal's general body form and by the environment in which the animal lives. Small, aquatic organisms, such as sponges and flatworms, require no specialized respiratory structures, but larger animals need more complex exchange systems to support their metabolic activities. The complexity of these is related to the efficiency of gas exchange required. This is determined by the oxygen demands of the organism.

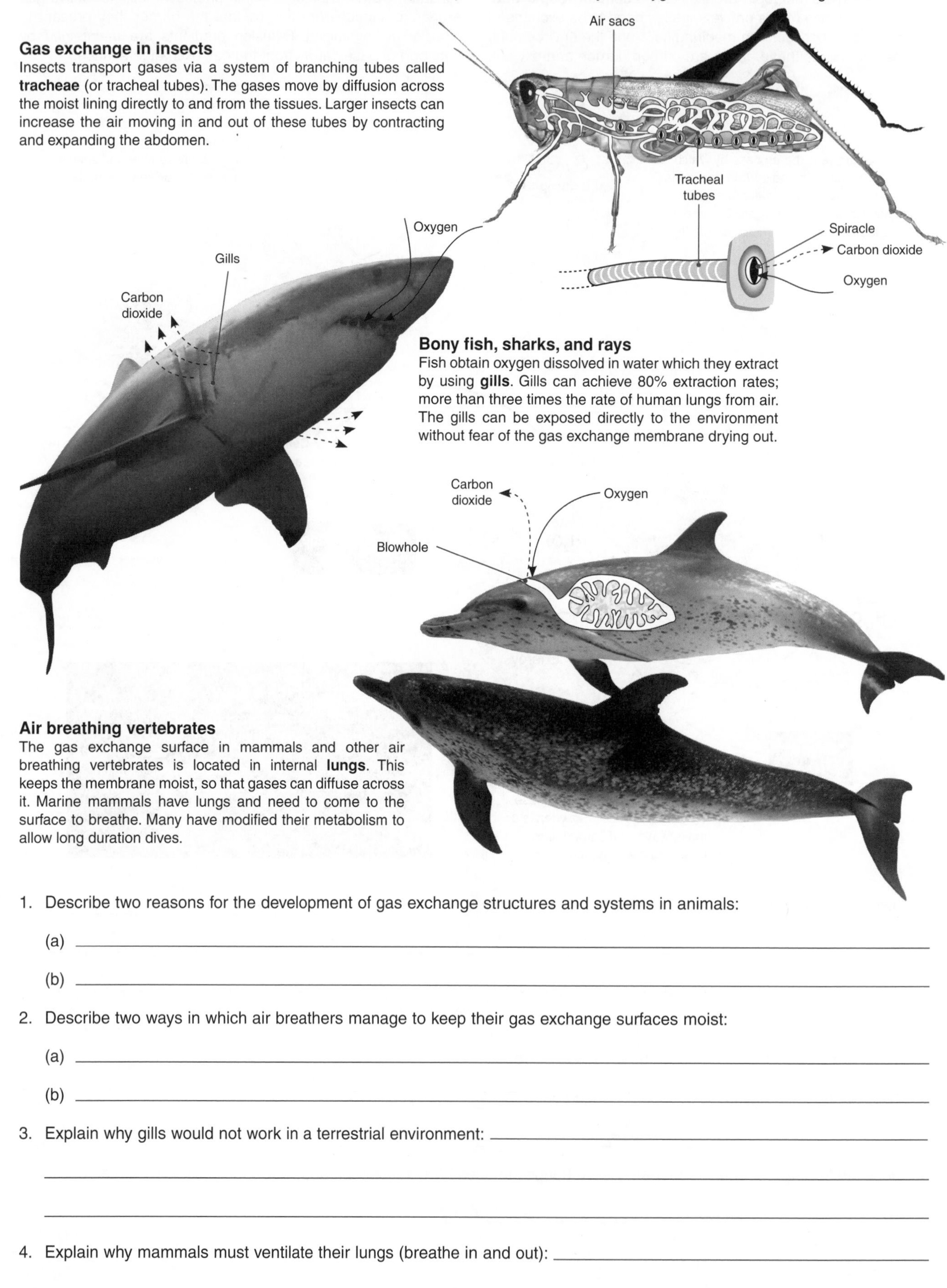

### Gas exchange in insects

Insects transport gases via a system of branching tubes called **tracheae** (or tracheal tubes). The gases move by diffusion across the moist lining directly to and from the tissues. Larger insects can increase the air moving in and out of these tubes by contracting and expanding the abdomen.

### Bony fish, sharks, and rays

Fish obtain oxygen dissolved in water which they extract by using **gills**. Gills can achieve 80% extraction rates; more than three times the rate of human lungs from air. The gills can be exposed directly to the environment without fear of the gas exchange membrane drying out.

### Air breathing vertebrates

The gas exchange surface in mammals and other air breathing vertebrates is located in internal **lungs**. This keeps the membrane moist, so that gases can diffuse across it. Marine mammals have lungs and need to come to the surface to breathe. Many have modified their metabolism to allow long duration dives.

1. Describe two reasons for the development of gas exchange structures and systems in animals:

   (a) ______

   (b) ______

2. Describe two ways in which air breathers manage to keep their gas exchange surfaces moist:

   (a) ______

   (b) ______

3. Explain why gills would not work in a terrestrial environment: ______

4. Explain why mammals must ventilate their lungs (breathe in and out): ______



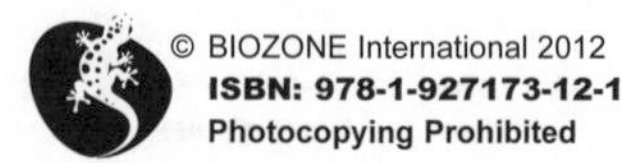


**ISBN: 978-1-927173-12-1**

# Gas Exchange in Insects

Oxygen is proportionately more abundant in air than in water. However, breathing air also presents problems. Water can be easily lost through any exposed surface that is moist, thin, permeable, and vascular enough to serve as a gas exchange membrane. Most insects are small terrestrial animals with a large surface area to volume ratio. Although they are highly susceptible to drying out, they are covered by a hard **exoskeleton** with a waxy outer layer that minimizes water loss. **Tracheal systems** are the most common gas exchange organs of terrestrial arthropods, including insects. Most body segments have paired openings called **spiracles** in the body wall. Filtering devices prevent small particles from clogging the system and valves control the degree to which the spiracles are open. In small insects, diffusion is the only mechanism needed to exchange gases, because it occurs so rapidly through the air-filled tubules. Larger, more active insects, such as locusts (below) have a tracheal system that includes air sacs, which can be compressed and expanded to assist in moving air through the tubules.

### Insect Tracheal Tubes

Insects, and some spiders, transport gases via a system of branching tubes called tracheae or tracheal tubes. Respiratory gases move by diffusion across the moist lining directly to and from the tissues. The end of each tube contains a small amount of fluid in which the respiratory gases are dissolved. The fluid is drawn into the muscle tissues during their contraction, and is released back into the tracheole when the muscle rests. Insects ventilate their tracheal system by making rhythmic body movements to help move the air in and out of the tracheae.

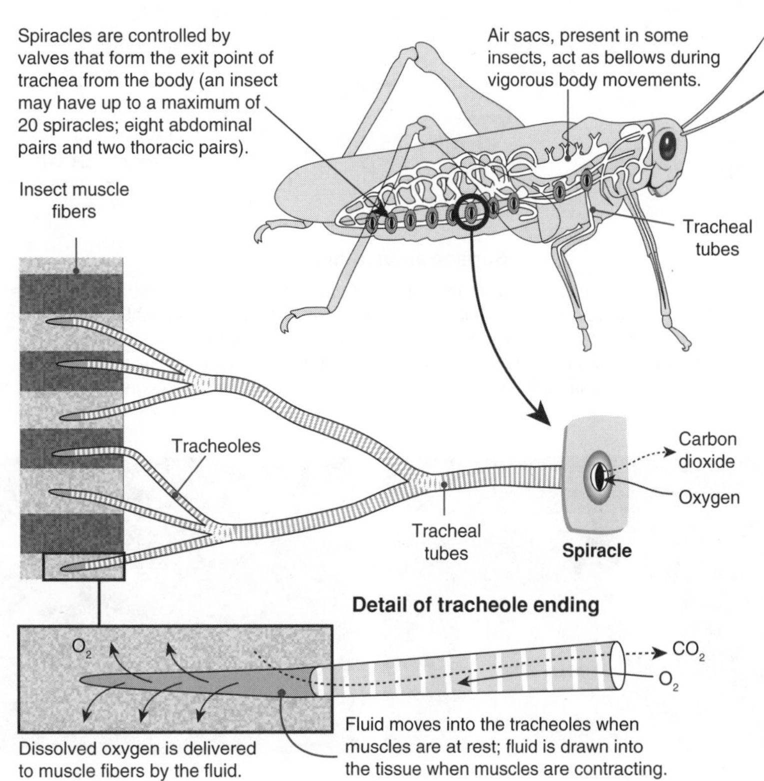


1. How are oxygen and carbon dioxide exchanged between the air and tissues at the end of insect tracheoles?

2. Valves in the spiracles can regulate the amount of air entering the tracheal system. Suggest a reason for this adaptation:

3. How is ventilation achieved in a terrestrial insect?

4. Even though most insects are small, they have evolved an efficient and highly developed gas exchange system that is independent of diffusion across the body surface. Why do you think this is the case?

Obtaining Nutrients & Eliminating Wastes


**ISBN: 978-1-927173-12-1**


***Related activities:*** *Gas Exchange in Animals*
***Weblinks:*** *Insect Respiratory System, Bubble Breathing*

Many insects live in freshwater. Aquatic insects, like terrestrial insects, exchange gases via the system of air-filled **tracheae**. What varies is the method by which oxygen enters this system. Aquatic insect larvae rely on **diffusion** across the body surface, with or without gills. Adult insects carry air with them when submerged. The air may be carried as a bubble, or stay trapped by regions of unwettable (hydrofuge) hairs. A thin film of air trapped by hairs is called a **plastron**. It provides a source of oxygen and acts as a diffusion gill, into which oxygen can diffuse from the water.

## Gas Exchange in Aquatic Invertebrates

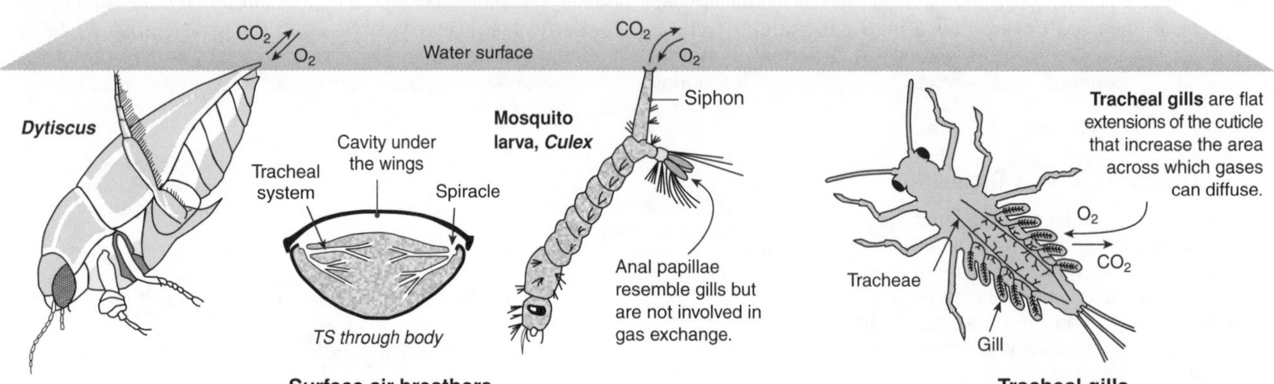


### Surface air breathers

The **diving beetle**, *Dytiscus*, traps air from the surface beneath its wings where it forms a compressible gill. The spiracles open into the air space and lead to the tracheal tubes. As the submerged insect respires, the oxygen is gradually used up and the bubble decreases in size. A **mosquito larva** penetrates the water surface with a siphon extending from a spiracle at the tip of the abdomen. The larva hangs at the surface while gas exchange occurs by diffusion.

### Tracheal gills

In the larvae of many aquatic insects, gas exchange occurs by diffusion across the body surface. This is enhanced by the presence of **tracheal gills** which may account for 20-70% of $O_2$ uptake depending on their surface area.

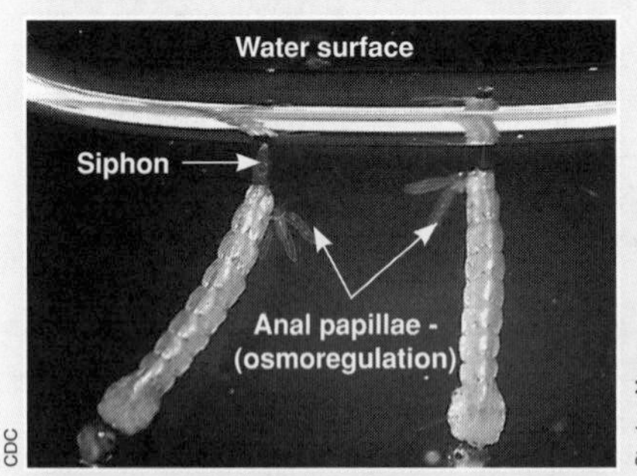


Gas exchange in mosquito larvae occurs with the air and is independent of $O_2$ content of the water.

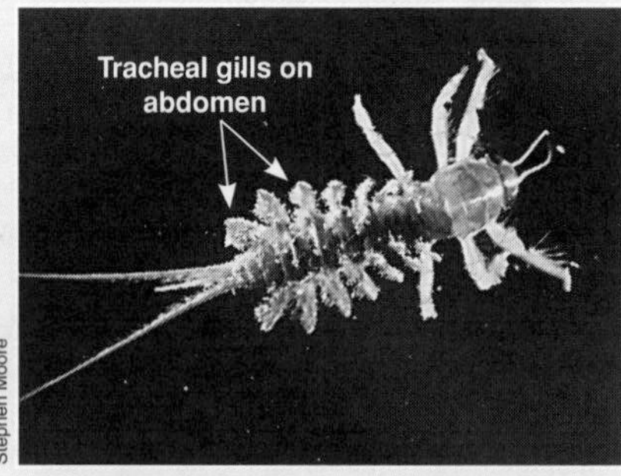


The tracheal gills of this spiny gilled mayfly (a very active species), are located on the abdomen.

Hydrophilid beetles use hydrofuge hairs to trap a film of air against the spiracles (a plastron).

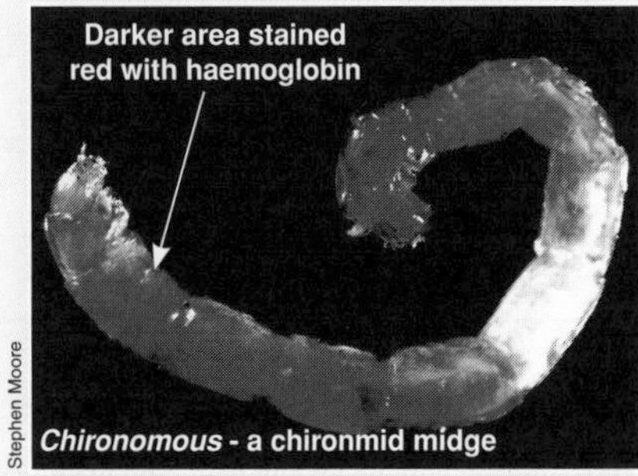


The blood of a few insect larvae, e.g. *Chironomus*, contains the $O_2$-carrying pigment **hemoglobin**, which allows them to survive when $O_2$ levels fall.

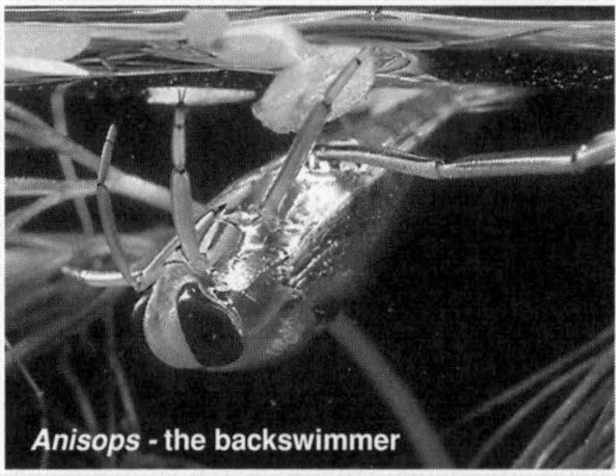


*Anisops* carries only a small air mass when diving but can exploit oxygen-poor waters because it has large hemoglobin-filled cells in its abdomen.

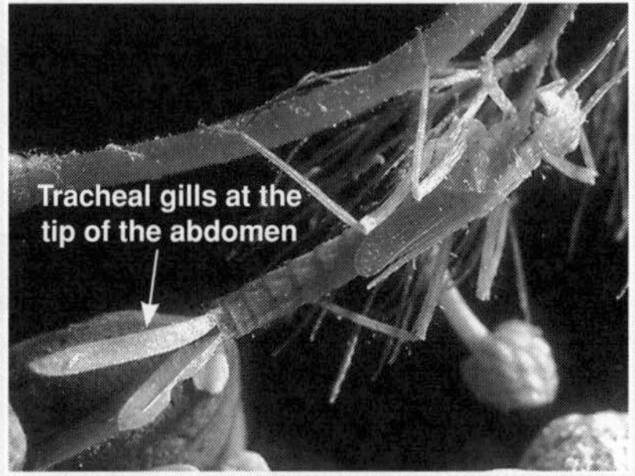


The tracheal gills of this damselfly larva are located at the tip of the abdomen (arrows). Like other insects with gills they are intolerant of low oxygen.

5. Giving an example for each, briefly describe three structural adaptations of freshwater insects for gas exchange:

   (a) ______________________________

   (b) ______________________________

   (c) ______________________________

6. Describe one physiological adaptation of freshwater insects for gas exchange: ______________________________


ISBN: 978-1-927173-12-1

# Gas Exchange in Fish

Fish obtain the oxygen they need from the water by means of gills: membranous structures supported by cartilaginous or bony struts. Gill surfaces are very large and respiratory gases are exchanged between the blood and the water as the water flows over the gill surface. The percentage of dissolved oxygen in a volume of water is much less than in the same volume of air; air is 21% oxygen while in water, dissolved oxygen is about 1% by volume. Efficient uptake of oxygen from the water (as achieved by gills) is therefore necessary for active organisms in an aquatic environment. Fish facilitate gas exchange by ventilating the gill surfaces, either by actively pumping water across the gill or by swimming continuously with the mouth open.

Bony fish have four pairs of gills, each supported by a bony arch. The operculum (gill cover) is involved in ventilating the gills.

Cartilaginous fish have five or six pairs of gills. Water enters via the mouth and spiracle and exits through gill slits (there is no operculum).

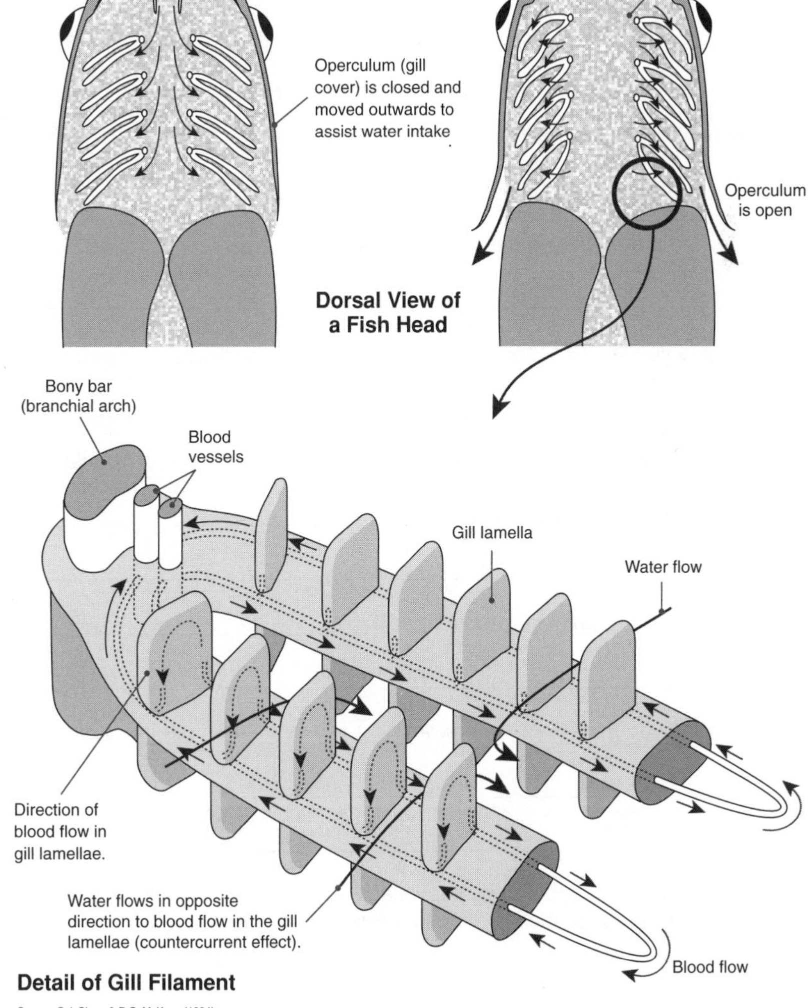

Fish have a closed circulatory system where the blood is entirely contained within vessels. In fish, the blood is pumped from the heart to the gills (the gas exchange surface) then directly to the body and back to the heart, flowing only once through the heart in each circulation of the body. The blood loses pressure when passing through the gills and, on leaving them, flows at low pressure around the body before returning to the heart.

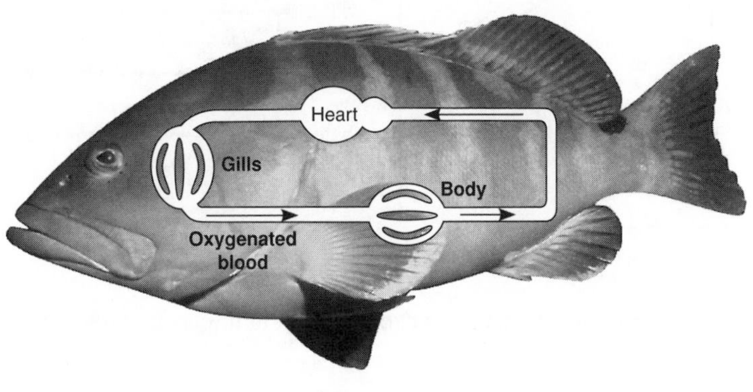


### Fish Gills

The gills of fish have a great many folds, which are supported and kept apart from each other by the water. This gives them a high surface area for gas exchange. The outer surface of the gill is in contact with the water, and blood flows in vessels inside the gill. Gas exchange occurs by diffusion between the water and blood across the gill membrane and capillaries. The operculum (gill cover) permits exit of water and acts as a pump, drawing water past the gill filaments. Fish gills are very efficient and can achieve an 80% extraction rate of oxygen from water; over three times the rate of human lungs from air.

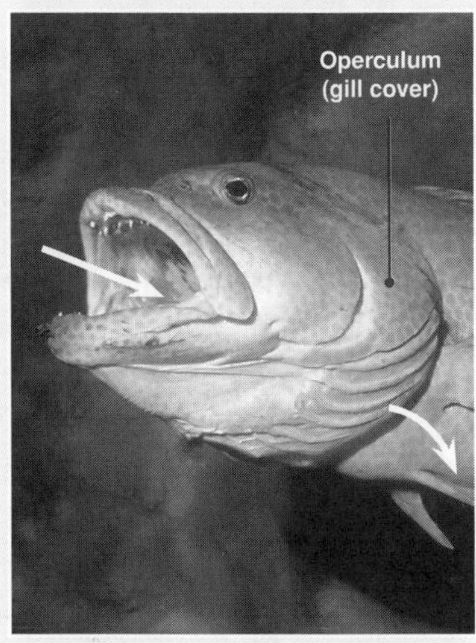


**INSPIRATION** (mouth open)

Oral valve opens

Mouth cavity expands, taking in water through the open mouth.

Operculum (gill cover) is closed and moved outwards to assist water intake.

**EXPIRATION** (mouth closed)

Oral valve shuts

Mouth cavity contracts forcing water back across the gills

Operculum is open

**Dorsal View of a Fish Head**

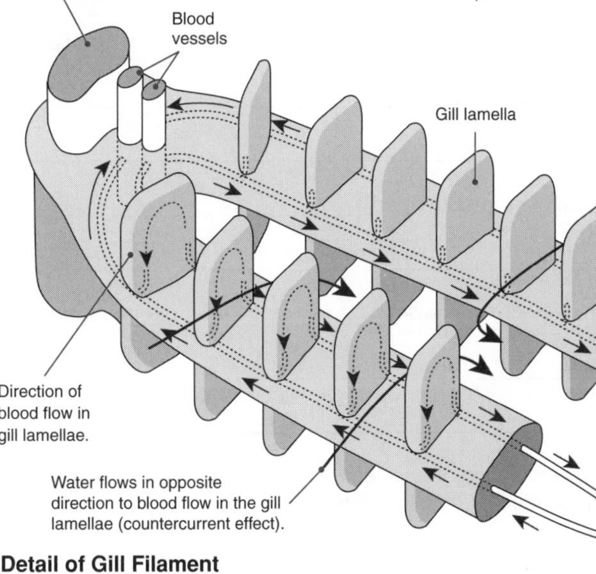


**Detail of Gill Filament**

Source: C.J. Clegg & D.G. McKean (1994)

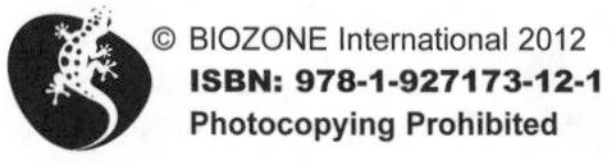


ISBN: 978-1-927173-12-1


*Periodicals:* Breathless

*Related activities:* Closed Circulatory Systems

The structure of fish gills and their physical arrangement in relation to the blood flow maximizes gas exchange rates. A constant stream of oxygen-rich water flows over the gill filaments in the **opposite** direction to the direction of blood flow through the gills (below, left). This is called **countercurrent flow** and it is an adaptation for maximizing the amount of $O_2$ removed from the water. Blood flowing through the gill capillaries encounters water of increasing oxygen content. Therefore, the concentration gradient (for oxygen uptake) across the gill is maintained across the entire distance of the gill lamella and oxygen diffuses into the bloodstream ($CO_2$ diffuses out at the same time) A parallel current flow would not achieve the same oxygen extraction rates because the concentrations across the gill would quickly equalize (below, right).

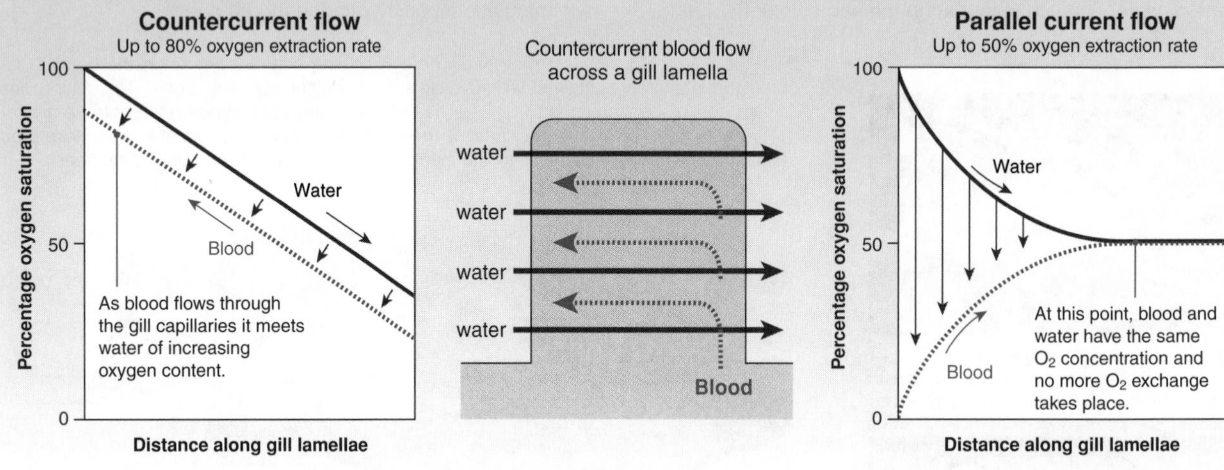


1. Describe three features of a fish gas exchange system (gills and related structures) that facilitate gas exchange:

   (a) ______

   (b) ______

   (c) ______

2. (a) How does the countercurrent system in a fish gill increase the efficiency of oxygen extraction from the water?

   ______

   (b) Why wouldn't parallel flow achieve adequate rates of gas exchange? ______

3. (a) What is meant by ventilation of the gills? ______

   (b) Why is ventilation necessary? ______

   (c) Describe the two ways in which bony fish achieve adequate ventilation of the gills:

   Pumping (mouth and operculum): ______

   Continuous swimming (mouth open): ______

   (d) Why do you think large, fast swimming fish (e.g. tuna) die in aquaria that restrict continuous swimming movement?

   ______

4. In terms of the amount of oxygen available in the water, explain why fish are very sensitive to increases in water temperature or suspended organic material in the water:

   ______


ISBN: 978-1-927173-12-1

# Gas Exchange in Mammals

Lungs are internal sac-like organs found in most amphibians, and all reptiles, birds, and mammals. The **alveolar lungs** of mammals occupy about 6% of the body volume, irrespective of body mass or lifestyle (figure below). They are connected to the outside air by way of a system of tubular passageways, ending in the respiratory bronchioles from which arise 2-11 alveolar ducts and many **alveoli**. These provide a large surface area (70 $m^2$) for gas exchange by diffusion between the alveoli and the blood in the capillaries. Ciliated, mucus secreting epithelium lines the tubules, trapping and removing dust and pathogens before they reach the gas exchange surface. The concentration gradient for diffusion of respiratory gases is maintained by breathing.

Nasal passages warm and moisten the air entering through the nostrils. Each nostril has a border of hairs to trap particles and filter them out of the system.

Air entering the body through the mouth enters the pharynx and mixes with air from the nasal passages.

The **trachea** lies in front of the esophagus and extends into the thorax. It is strengthened with C-shaped bands of cartilage and lined with ciliated epithelium.

The trachea splits into two **bronchi** (*sing.* bronchus) These are also supported by cartilage bands.

**Bronchioles** branch off the bronchi and divide into progressively smaller branches. The cartilage is gradually lost as the bronchioles decrease in diameter.

Lung

Lung

**Detail of a teminal bronchiole and its branches**

Air movement

Lymph vessels

The smallest respiratory bronchioles lack cartilage. They subdivide into the alveolar ducts, from which arise the alveoli.

Alveoli

EII

Bronchiole

Alveolar duct

Alveolus

Photo right: Bronchiole and alveolar duct leading to alveoli. Note the very thin alveolar walls. The alveoli tend to recoil inward (deflate) after each breath out. A phospholipid **surfactant** helps to prevent collapse of the alveoli by decreasing surface tension.

In mammals, lung size is a regular function of body size as this graph shows. Small mammals, with high metabolic rates, obtain sufficient oxygen with lungs of the same relative size as in large mammals.

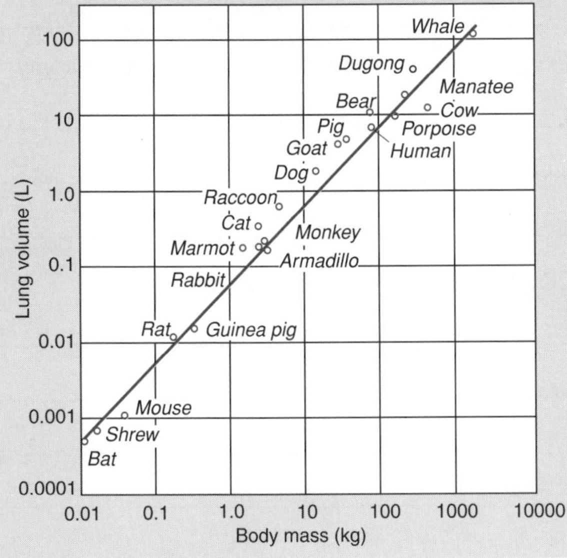

Air pressure at high altitude is far less than at sea level and less oxygen is available for gas exchange. Above 2000 metres, breathing in humans becomes labored. Above 5000 metres, oxygen equipment is needed. Llamas, vicunas, and Bactrian camels are well suited to high altitude life. Vicunas and llamas, which live in the Andes, have high blood cell counts and their red blood cells live almost twice as long as those in humans. Their hemoglobin also picks up and off-loads oxygen more efficiently than the hemoglobin of most mammals.

**ISBN: 978-1-927173-12-1**


***Related activities:*** *Breathing, Gas Transport in Mammals*

***Weblinks:*** *Respiratory Basics Learning Activity, Interactive Lungs*

## An Alveolus

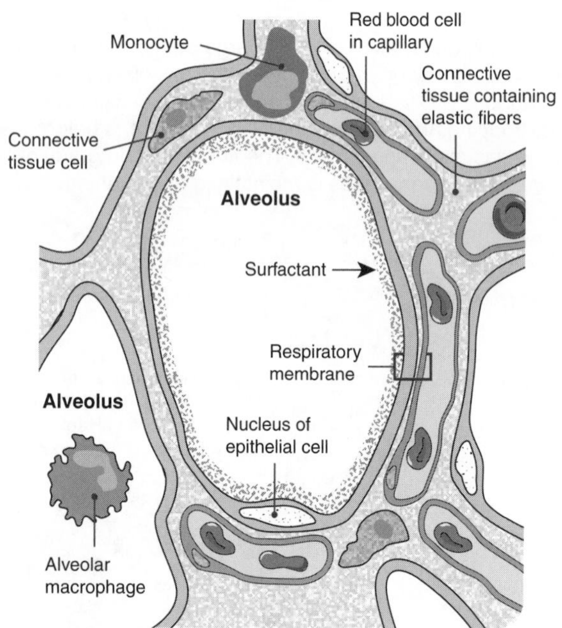


The diagram above illustrates the physical arrangement of the alveoli to the capillaries through which the blood moves. Phagocytic monocytes and macrophages are also present to protect the lung tissue. Elastic connective tissue gives the alveoli their ability to expand and recoil.

## The Respiratory Membrane

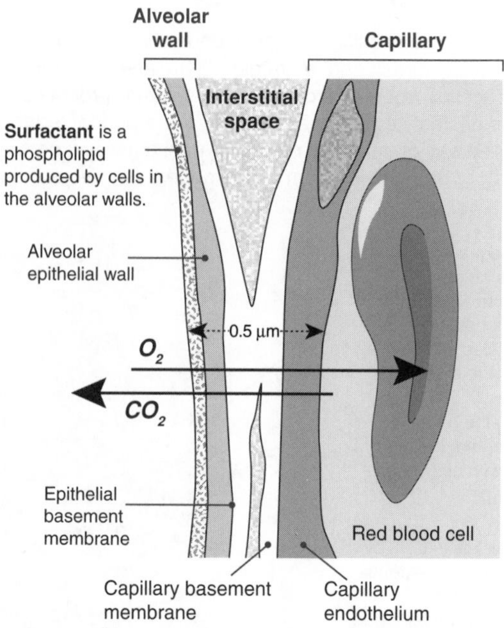


The **respiratory membrane** is the term for the layered junction between the alveolar epithelial cells, the endothelial cells of the capillary, and their associated basement membranes (thin, collagenous layers that underlie the epithelial tissues). Gases move freely across this membrane.

1. Explain how the structure of alveolar lungs provides a large area for gas exchange: ____________________

2. Describe the main features and functional role of the following parts of the gas exchange system in mammals,:

   (a) Trachea: ____________________

   (b) Bronchioles: ____________________

   (c) Alveoli: ____________________

3. The table (right) gives the approximate percentages of respiratory gases in the lungs, and in inhaled and exhaled air. Study the table and then answer the following questions.

| GAS | INHALED AIR | AIR IN LUNGS | EXHALED AIR |
|---|---|---|---|
| Oxygen | 21% | 15% | 16% |
| Carbon dioxide | 0.04% | 5.5% | 3.6% |

   (a) Calculate the difference in $CO_2$ between inhaled and exhaled air and explain where this 'extra' $CO_2$ comes from:

   (b) Calculate the difference in oxygen between inhaled and exhaled air: ____________________

   (c) Explain why exhaled air has slightly more oxygen and less $CO_2$ than air in the lungs: ____________________

4. Surfactants are found in the lungs of all vertebrates. Explain their role: ____________________

**ISBN: 978-1-927173-12-1**

# Breathing in Humans

In mammals, the mechanism of breathing (ventilation) provides a continual supply of fresh air to the lungs and helps to maintain a large diffusion gradient for respiratory gases across the gas exchange surface. Oxygen must be delivered regularly to supply the needs of respiring cells. Similarly, carbon dioxide, which is produced as a result of cellular metabolism, must be quickly eliminated from the body. Adequate lung ventilation is essential to these exchanges. The cardiovascular system participates by transporting respiratory gases to and from the cells of the body. The basic rhythm of breathing is controlled by the **respiratory center**, a cluster of neurons located in the medulla oblongata. This rhythm is adjusted in response to the physical and chemical changes that occur when we carry out different activities (e.g. exercise produces more $CO_2$, which result in an increased rate of breathing).

### Inspiration (inhalation or breathing in)

During quiet breathing, inspiration is achieved by increasing the space (therefore decreasing the pressure) inside the lungs. Air then flows into the lungs in response to the decreased pressure inside the lung. Inspiration is always an active process involving muscle contraction.

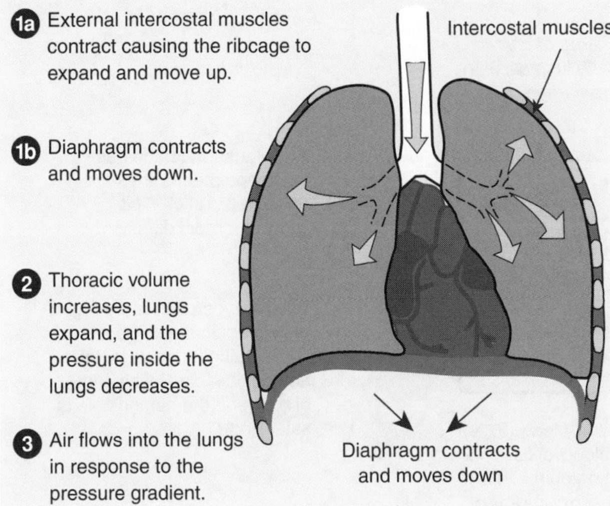


### Expiration (exhalation or breathing out)

During quiet breathing, expiration is achieved passively by decreasing the space (thus increasing the pressure) inside the lungs. Air then flows passively out of the lungs to equalise with the air pressure. In active breathing, muscle contraction is involved in bringing about both inspiration and expiration.

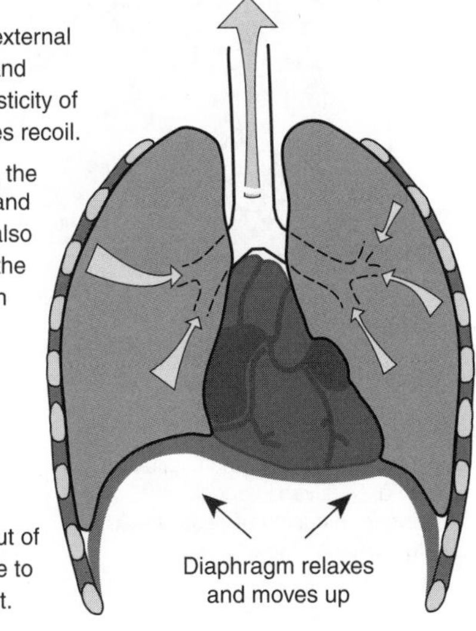

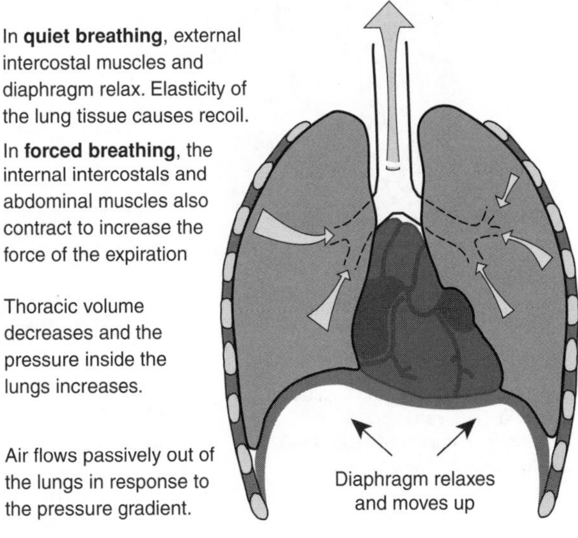


1. Explain the purpose of breathing: ____________________

2. (a) Describe the sequence of events involved in quiet breathing: ____________________

   (b) Explain the essential difference between this and the situation during heavy exercise or forced breathing: ____________________

3. Identify what other gas is lost from the body in addition to carbon dioxide: ____________________

4. Explain the role of the elasticity of the lung tissue in normal, quiet breathing: ____________________

5. Breathing rate is regulated through the medullary respiratory center in response to demand for oxygen. The trigger for increased breathing rate is a drop in blood pH. Suggest why this is an appropriate trigger to increase breathing rate: ____________________

Obtaining Nutrients & Eliminating Wastes


ISBN: 978-1-927173-12-1


*Related activities: Measuring Lung Function, Gas Transport in Humans*
*Weblinks: Respiratory Basics Learning Activity*

# Gas Transport in Mammals

The transport of respiratory gases around the body is the role of the blood and its respiratory pigments. Oxygen is transported throughout the body chemically bound to the respiratory pigment **hemoglobin** inside the red blood cells. In the muscles, oxygen from hemoglobin is transferred to and retained by **myoglobin**, a molecule that is chemically similar to hemoglobin except that it consists of only one heme-globin unit. Myoglobin has a greater affinity for oxygen than hemoglobin and acts as an oxygen store within muscles, releasing the oxygen during periods of prolonged or extreme muscular activity. If the myoglobin store is exhausted, the muscles are forced into oxygen debt and must respire anaerobically. The waste product of this, lactic acid, accumulates in the muscle and is transported (as lactate) to the liver where it is metabolized under aerobic conditions.

## Gas Exchange and Transport

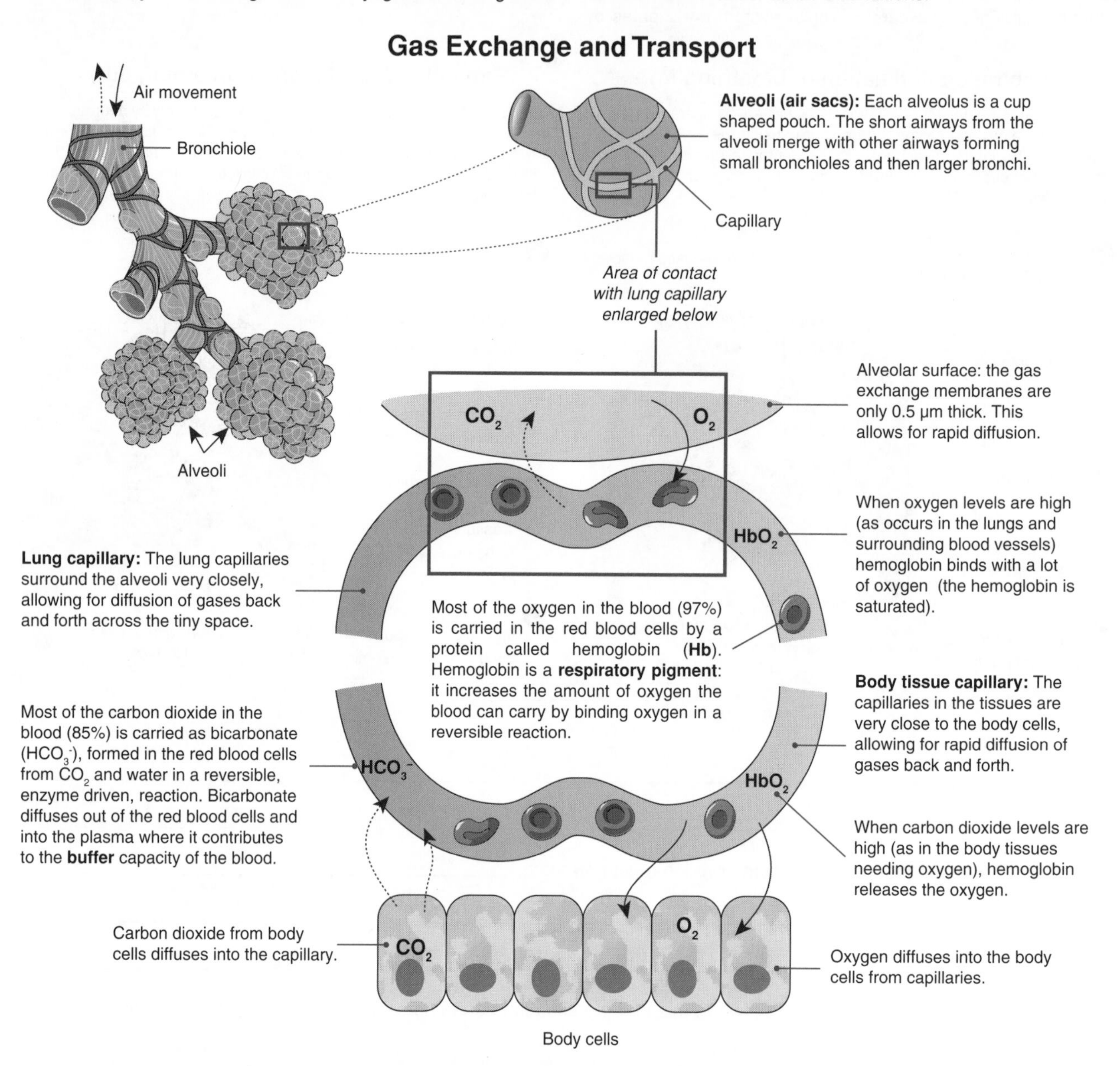

Air movement

Bronchiole

Alveoli

**Alveoli (air sacs):** Each alveolus is a cup shaped pouch. The short airways from the alveoli merge with other airways forming small bronchioles and then larger bronchi.

Capillary

*Area of contact with lung capillary enlarged below*

$CO_2$ $O_2$

Alveolar surface: the gas exchange membranes are only 0.5 μm thick. This allows for rapid diffusion.

$HbO_2$

When oxygen levels are high (as occurs in the lungs and surrounding blood vessels) hemoglobin binds with a lot of oxygen (the hemoglobin is saturated).

**Lung capillary:** The lung capillaries surround the alveoli very closely, allowing for diffusion of gases back and forth across the tiny space.

Most of the oxygen in the blood (97%) is carried in the red blood cells by a protein called hemoglobin (**Hb**). Hemoglobin is a **respiratory pigment**: it increases the amount of oxygen the blood can carry by binding oxygen in a reversible reaction.

**Body tissue capillary:** The capillaries in the tissues are very close to the body cells, allowing for rapid diffusion of gases back and forth.

Most of the carbon dioxide in the blood (85%) is carried as bicarbonate ($HCO_3^-$), formed in the red blood cells from $CO_2$ and water in a reversible, enzyme driven, reaction. Bicarbonate diffuses out of the red blood cells and into the plasma where it contributes to the **buffer** capacity of the blood.

$HCO_3^-$

$HbO_2$

When carbon dioxide levels are high (as in the body tissues needing oxygen), hemoglobin releases the oxygen.

Carbon dioxide from body cells diffuses into the capillary.

$CO_2$ $O_2$

Oxygen diffuses into the body cells from capillaries.

Body cells

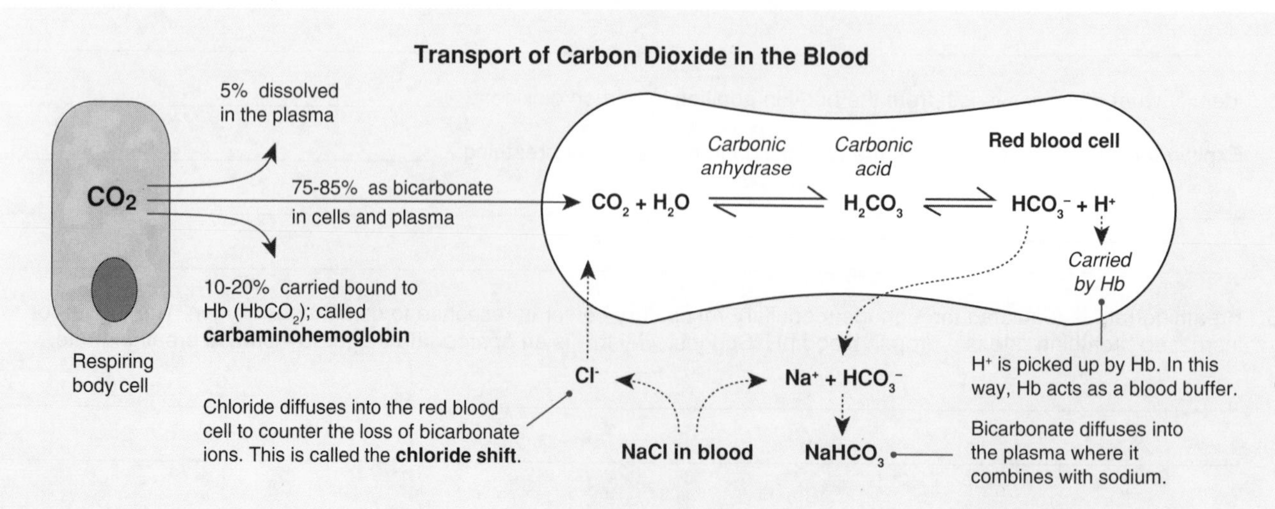




**Related activities:** *Adaptations of Vertebrate Blood, Gas Exchange in Mammals*


ISBN: 978-1-927173-12-1

Oxygen does not dissolve easily in blood, but is carried in chemical combination with hemoglobin (Hb) in red blood cells. The most important factor determining how much oxygen is carried by Hb is the level of oxygen in the blood. The greater the oxygen tension, the more oxygen will combine with Hb. This relationship can be illustrated with an oxygen-hemoglobin dissociation curve as shown below (Fig. 1). In the lung capillaries, (high $O_2$), a lot of oxygen is picked up and bound by Hb. In the tissues, (low $O_2$), oxygen is released. In skeletal muscle, myoglobin picks up oxygen from hemoglobin and therefore serves as an oxygen store when oxygen tensions begin to fall. The release of oxygen is enhanced by the **Bohr effect** (Fig. 2).

## Respiratory Pigments and the Transport of Oxygen

*Fig. 1: Dissociation curves for hemoglobin and myoglobin at normal body temperature for fetal and adult human blood.*

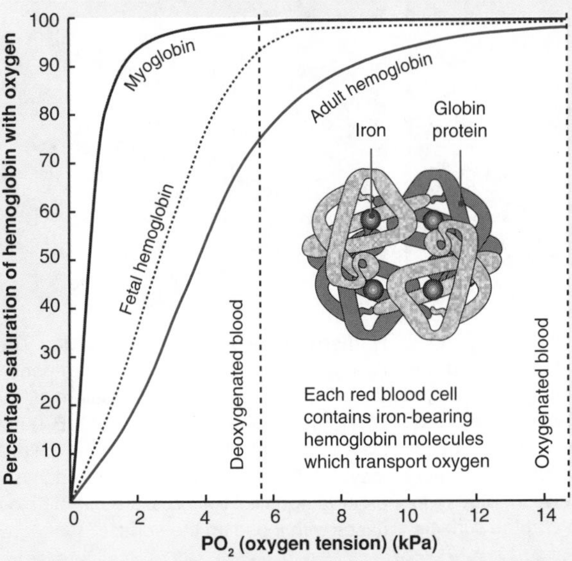


As oxygen level increases, more oxygen combines with hemoglobin (Hb). Hb saturation remains high, even at low oxygen tensions. Fetal Hb has a high affinity for oxygen and carries 20-30% more than maternal Hb. Myoglobin in skeletal muscle has a very high affinity for oxygen and will take up oxygen from hemoglobin in the blood.

*Fig. 2: Oxygen-hemoglobin dissociation curves for human blood at normal body temperature at different blood pH.*

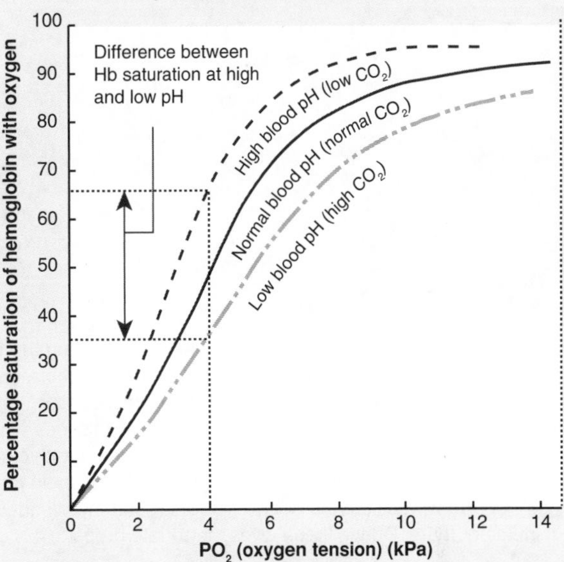


As pH increases (lower $CO_2$), more oxygen combines with Hb. As the blood pH decreases (higher $CO_2$), Hb binds less oxygen and releases more to the tissues (**the Bohr effect**). The difference between Hb saturation at high and low pH represents the amount of oxygen released to the tissues.

1. (a) Identify two regions in the body where oxygen levels are very high: __________

   (b) Identify two regions where carbon dioxide levels are very high: __________

2. Explain the significance of the **reversible binding** reaction of hemoglobin (Hb) to oxygen: __________

3. (a) Hemoglobin saturation is affected by the oxygen level in the blood. Describe the nature of this relationship:

   (b) Comment on the significance of this relationship to oxygen delivery to the tissues: __________

4. (a) Describe how fetal Hb is different to adult Hb: __________

   (b) Explain the significance of this difference to oxygen delivery to the fetus: __________

5. At low blood pH, less oxygen is bound by hemoglobin and more is released to the tissues:

   (a) Name this effect: __________

   (b) Comment on its significance to oxygen delivery to respiring tissue: __________

6. Explain the significance of the very high affinity of myoglobin for oxygen: __________

7. Identify the two main contributors to the buffer capacity of the blood: __________

Obtaining Nutrients & Eliminating Wastes

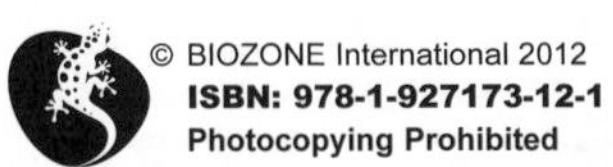

ISBN: 978-1-927173-12-1

# Adaptations of Diving Mammals

An air breathing animal with an aquatic lifestyle involving long dives requires special **adaptations**. All air breathing animals that dive must cope with the problem of oxygen supply to the tissues. This is particularly a problem for mammals because of their high metabolic rate and high oxygen demand. In addition, resurfacing from deep dives of 20 meters or more brings the risk of decompression sickness (**the bends**). Primates (including humans) are one of the few orders of mammals without diving representatives. The problems that humans encounter when diving while breathing compressed air (e.g. the bends) are the result of continuing to breathe during the dive (animals adapted for diving do not do this).

**Diving mammals** (e.g. dolphins, whales, seals) are among the most well adapted divers. Before diving, they store oxygen in the muscles and exhale any air left in the lungs. At depth, the lungs are compressed and only the trachea contains air. This stops nitrogen entering the blood (nitrogen causes problems in the tissues when surfacing from dives). During dives, heart rate slows and blood flow is redistributed to supply only critical organs. Diving mammals have high levels of myoglobin and their muscles work well without oxygen. Sperm whales are the deepest divers (3000 m). Weddell seals dive to 1000 m for 40 minutes or more. During these dives, heart rate drops to 4% of the rate at the surface.

**Humans** are poorly adapted for diving. They lack adequate body fat for long periods underwater and they inhale before diving. Divers using compressed air (SCUBA) may stay submerged for much longer. However, they must take care when ascending in order to equalise the pressure of their blood gases with those on the surface.

1. (a) Describe an advantage gained from breathing out before diving: ____________________

   (b) Explain how this behavior is different from a human diving (unaided by equipment): ____________________

   (c) Describe the adaptive advantage of reducing heart rate during a dive: ____________________

2. Remaining submerged for long periods of time requires an ability to maintain oxygen supply to the tissues. This depends on oxygen stores. This table compares the oxygen in different regions of the body during a dive in a small seal and a human (not on scuba). In the spaces, calculate the amount of oxygen (in mL) per kilogram of body weight for both.

| Location of oxygen in the body | Seal (30 kg) | | Human (70 kg) | |
|---|---|---|---|---|
| | Amount of oxygen (mL) | Oxygen (mL $kg^{-1}$) | Amount of oxygen (mL) | Oxygen (mL $kg^{-1}$) |
| Alveolar air | 55 | | 720 | |
| Blood | 1125 | | 1000 | |
| Muscle | 270 | | 240 | |
| Tissue water | 100 | | 200 | |
| Total | 1550 | 51.67 | 2160 | 30.86 |

3. (a) Describe the most striking difference between a seal and a human in terms of the oxygen stores during a dive: ____________________

   (b) With respect to diving adaptations, suggest why this is the case: ____________________



*Related activities:* Gas Exchange in Mammals
*Weblinks:* How Low Can You Go?

*Periodicals:* Deep science


ISBN: 978-1-927173-12-1

# Obtaining Food

All animals are **heterotrophs**, meaning they feed on other things, either dead or alive. However, they display a wide range of methods for obtaining the food they need. Animals may feed on solid or fluid food and may suck, bite, lap, or swallow it whole. The **adaptations** of mouthparts and other feeding appendages reflects both the diet and the way in which they obtain their food. Various behavioral adaptations also contribute to the way animals obtain food. Some animals are predators and hunt other animals. Others are herbivorous and graze or browse continually. Omnivores feed off both meat and vegetable material.

## Bulk Feeding

Many animals feed on large food masses which are ingested whole or in pieces. Carnivorous mammals and fish eat large pieces of food, and have teeth specialized for cutting flesh (below).

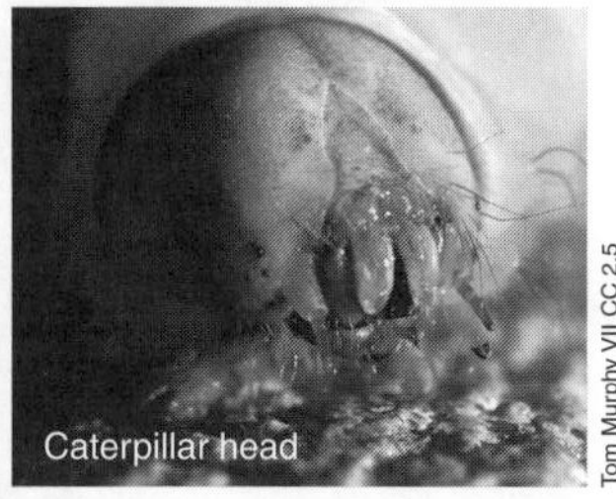


Grazing or cropping herbivores (e.g. many insects and mammalian herbivores), cut or tear vegetation and have mouthparts or teeth to chew the plant material into pieces.

## Filter Feeding

Filter feeders remove food that is suspended in the water using specialized filtering structures. Humpback and other baleen whales use comb-like plates suspended from the upper jaw to sieve shrimps and fish from large volumes of water. Food is trapped against the plates when the mouth closes and water is forced out. Some fish are also filter feeders and use sieve-like modifications of the gill rakers to strain plankton from the water. Some aquatic insects, e.g. some caddisfly larvae are also filter feeders.

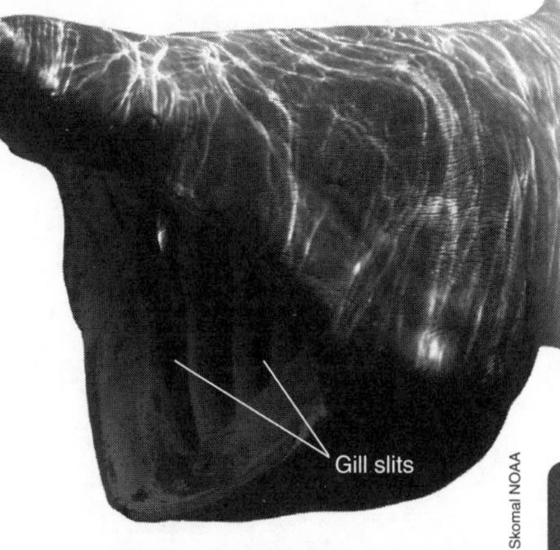


The basking shark is a passive filter feeder, swimming with its mouth open and trapping plankton against the sieve-like lining of its gills.

## Fluid Feeding

Many insects (e.g. flies, moths, butterflies, aphids), and some fish and mammals feed on fluids such as blood, plant sap, or nectar. Fluid feeders have mouthparts and guts that enable them to obtain and process a liquid diet. Insect fluid feeders have piercing or tubular mouthparts to obtain fluids directly. Mammalian and fish fluid feeders have small sharp teeth designed to pierce or even remove the skin of prey and lap the blood.

## Deposit Feeding

Deposit feeders feed by sifting the substrate to find food. Many invertebrates, but relatively few insects, are deposit feeders. Insects that feed while burrowing through detritus or dung are deposit feeders as are some **benthic** fish. For example, mullet, catfish, and **carp** suck up mud and ingest the plant and animal material in it.

1. Describe one **structural** adaptation for obtaining food in each of the following:

   (a) A blood sucking mammal: ______

   (b) A filter feeding shark: ______

   (c) A mammalian predatory carnivore: ______

   (d) A leaf chewing caterpillar: ______

   (e) A deposit feeding fish: ______

2. Use lines to match each of the following modes of feeding with its correct description and example:

| | | |
|---|---|---|
| (a) Filter feeding | Obtaining nutrients by ingesting prey and its surrounding fluid | Mosquito |
| (b) Deposit feeding | Obtaining nutrients by eating whole organisms | Stoat |
| (c) Suction feeding | Obtaining nutrients from particles suspended in water | Dung beetle |
| (d) Bulk feeding | Obtaining nutrients by ingesting only the fluids of another organism | Zooplanktivorous fish |
| (e) Fluid feeding | Obtaining nutrients by sifting through or ingesting sediment or detritus | Baleen whale |

3. Common carp is an introduced species in the US, and responsible for the deterioration of water quality in some lakes and rivers. Based on their method of feeding, suggest why carp have such a negative effect on the environment:

______

______


ISBN: 978-1-927173-12-1


*Related activities: Insect Mouthparts, Dentition in Mammals, The Teeth of Fish*
*Weblinks: Suction Feeding, Deposit Feeding in Carp*

# Adaptations of Predators

A predator is any animal that hunts, captures, and consumes prey. In most cases, the predator and prey are different species, although **cannibalism** is not uncommon in many animal species. The adaptations of predators for prey capture and consumption include those associated with sensing and locating prey (e.g. vibration sense, vision), capturing and subduing it (e.g. speed, camouflage, claws), and then consuming and digesting the flesh or fluids (e.g. teeth, low stomach pH).

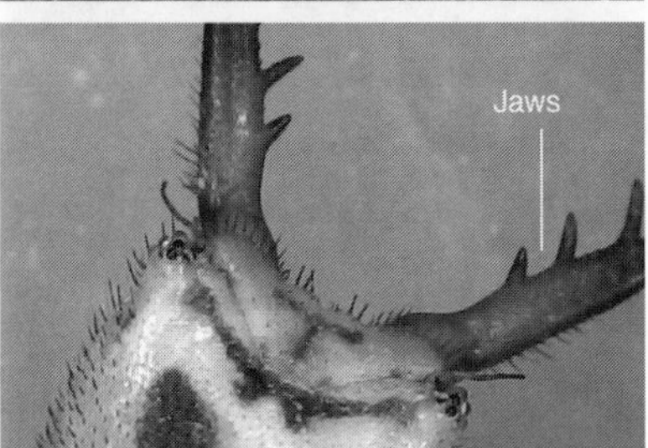


**Traps and lures** allow predators to wait in ambush for their prey. **Antlion larvae** build pits in sand as insect traps. The larva waits at the bottom of the pit (photo, top), buried in the soil with only the jaws projecting above the surface, often open wide (above). The loose sand of the pit sides makes it difficult for insects to escape once they have fallen in, and they are quickly captured.

Anglerfish also use lures to attract prey. The lure is a fleshy growth adapted from the dorsal fin and extending over the fish's head. It can be moved independently of the fish and a reflex closes the jaw on the prey as soon as the lure is touched.

**Echolocation** ('sonar') enables bats to hunt their insect prey while flying at night. Sonar allows the flying bat to locate the exact position of its prey even when both prey and predator are in flight. Some bats (above) also hunt on the ground amongst the leaf litter, but only use sonar for location of prey in the air.

**Concealment and ambush** are common strategies amongst insect predators. Mantids are well camouflaged against the vegetation in their environment. Unsuspecting prey are ambushed and captured with a rapid movement of the highly specialized, elongated forelegs.

**Speed, agility, and cooperation** enable dolphins to hunt together to exploit large schools of fish. The dolphin group will encircle a large school of fish, driving them together into a dense mass or bait ball. The dolphins then take turns to charge through the school to feed. Dolphins use echolocation to explore their environment, navigate, and to locate prey.

1. (a) Describe the role of camouflage in the predatory strategy of ambush predators such as mantids:

   (b) Describe some other adaptations commonly associated with prey capture in ambush predators:

2. Describe the adaptations of antlion larvae for prey capture and identify them as structural, behavioral, or physiological:

3. Describe the adaptations of dolphins for prey capture and identify them as structural, behavioral, or physiological:



***Related activities:*** *Insect Mouthparts, Dentition in Mammals, The Teeth of Fish*
***Weblinks:*** *Antlion Predation*


**ISBN: 978-1-927173-12-1**

# Food Vacuoles and Simple Guts

The simplest form of digestion occurs inside cells (**intracellularly**) within food vacuoles. This process is relatively slow and digestion is exclusively intracellular only in protozoa and sponges. In animals with simple, sac-like guts, digestion begins **extracellularly** (with secretion of enzymes to the outside or into the digestive cavity) and is completed intracellularly.

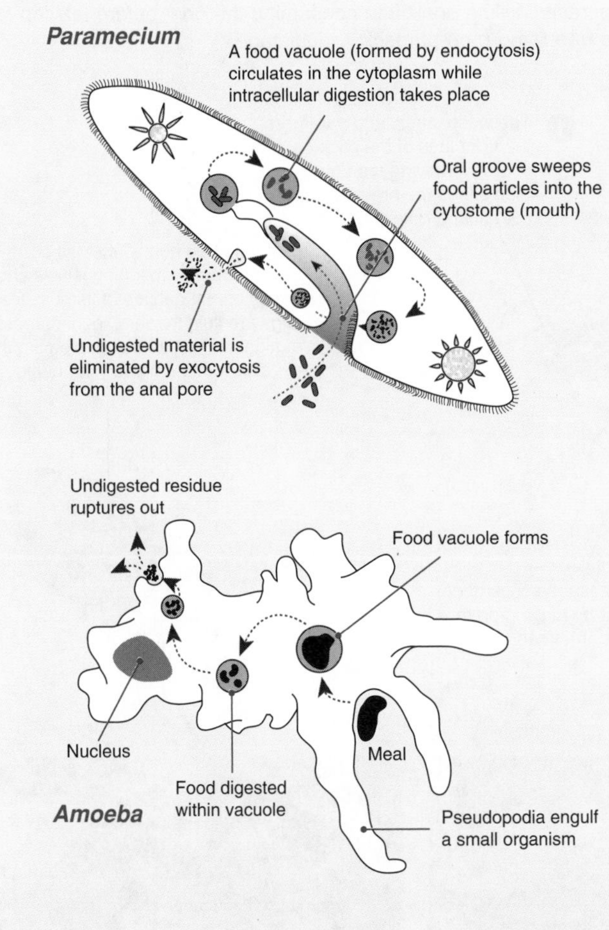


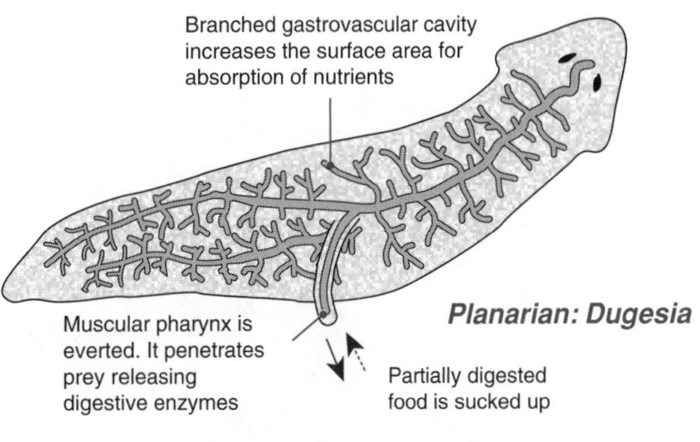


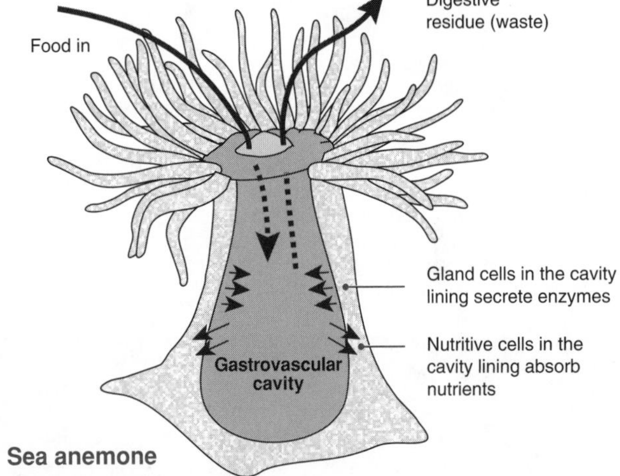


### Intracellular Digestion In Food Vacuoles

EXAMPLES: *Protozoans (above), sponges*

The simplest digestive compartments are food vacuoles: organelles where a single cell can digest its food without the digestive enzymes mixing with the cell's own cytoplasm. Sponges and protozoans (e.g. *Paramecium* and *Amoeba*) digest food in this way. *Paramecium* sweeps food into a food groove, from where vacuoles form. *Amoeba* engulf food using cytoplasmic extensions called pseudopodia. Digestion is intracellular, occurring within the cell itself.

### Digestion In A Gastrovascular Cavity

EXAMPLES: *Cnidarians, flatworms (above)*

Some of the simplest animals have a digestive sac or gastrovascular cavity with a single opening through which food enters and digested waste passes out. In organisms with this system, digestion is both extra- and intracellular. Digestion begins (using secreted enzymes) either in the cavity (in cnidarians) or outside it (flatworms). In both these groups, the digestion process is completed intracellularly within the **vacuoles** in cells.

1. Describe two ways in which simple saclike gastrovascular cavities differ from tubelike guts:

   (a) ______________________________

   (b) ______________________________

2. (a) Distinguish between intracellular and extracellular digestion: ______________________________

   ______________________________

   (b) Explain why intracellular digestion is not suitable as the only means of digestion for most animals: ______________________________

   ______________________________

   ______________________________

3. State the main difference between extracellular digestion in sea anemones and *Dugesia*: ______________________________

   ______________________________

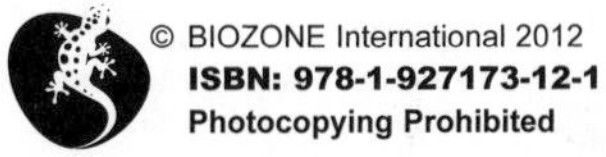

ISBN: 978-1-927173-12-1


**Related activities:** Diversity in Tube Guts

# Parasitism

Parasitism is the most common type of **symbiosis** and a specialized way to obtain nutrients. Parasites always harm their host, but do not usually kill it. The parasite benefits by obtaining nutrition, but there are often other advantages, such as protection. Many animal groups have **parasitic** representatives, although parasites are more common in some taxa than in others. For example, there are many parasitic insects, but relatively few parasitic fish. Animal parasites are highly specialized carnivores, feeding off the body fluids or skin of **host** species. **Parasitoids** are animals adapted to spend only part of their life cycle as a parasite, killing and often consuming the host before leaving for a free-living (non-parasitic) adult stage.

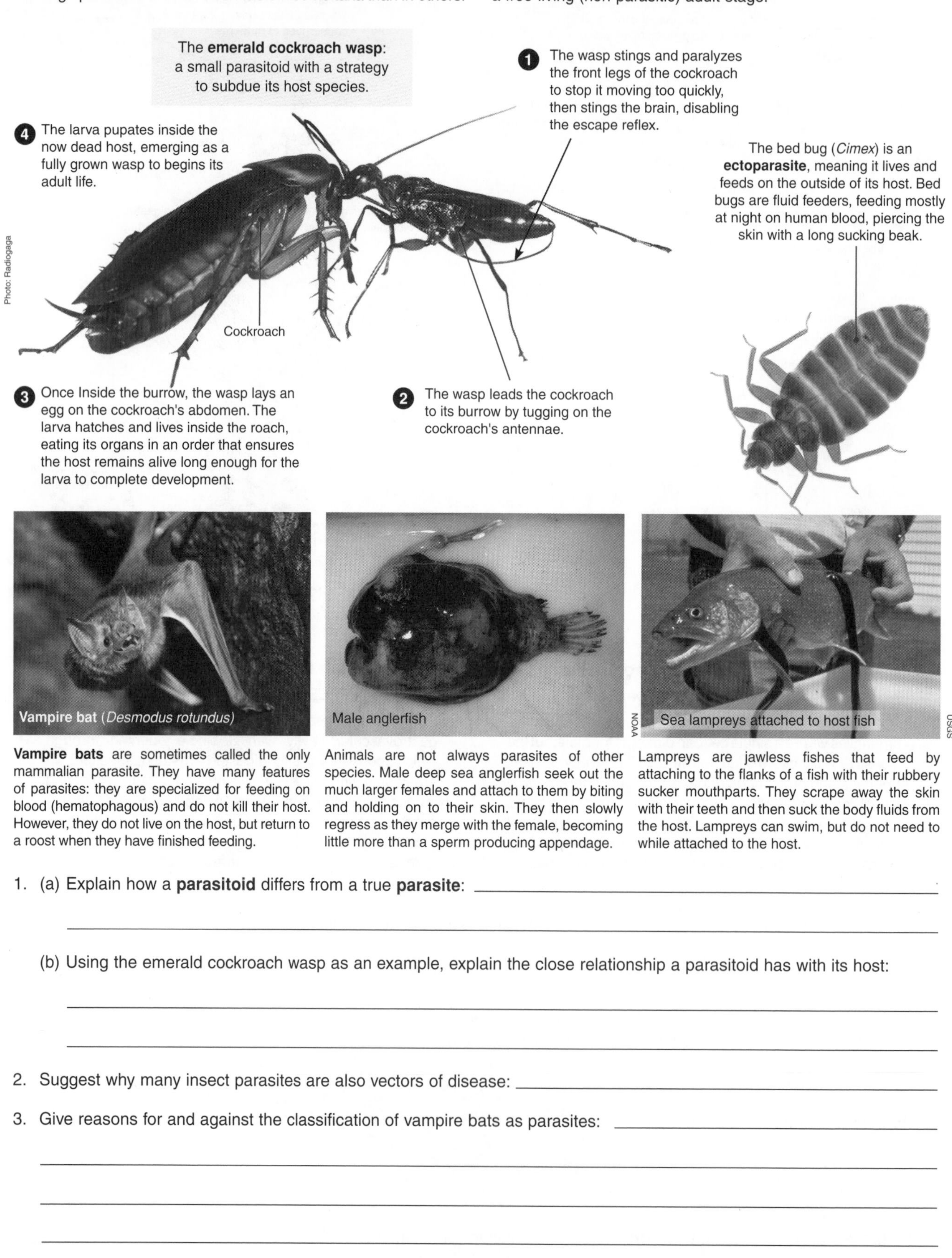


**Vampire bats** are sometimes called the only mammalian parasite. They have many features of parasites: they are specialized for feeding on blood (hematophagous) and do not kill their host. However, they do not live on the host, but return to a roost when they have finished feeding.

Animals are not always parasites of other species. Male deep sea anglerfish seek out the much larger females and attach to them by biting and holding on to their skin. They then slowly regress as they merge with the female, becoming little more than a sperm producing appendage.

Lampreys are jawless fishes that feed by attaching to the flanks of a fish with their rubbery sucker mouthparts. They scrape away the skin with their teeth and then suck the body fluids from the host. Lampreys can swim, but do not need to while attached to the host.

1. (a) Explain how a **parasitoid** differs from a true **parasite**: __________

   (b) Using the emerald cockroach wasp as an example, explain the close relationship a parasitoid has with its host:

2. Suggest why many insect parasites are also vectors of disease: __________

3. Give reasons for and against the classification of vampire bats as parasites: __________




**ISBN: 978-1-927173-12-1**

# Diversity in Tube Guts

Most animals have digestive tubes running between two openings, a **mouth** and an **anus**. Two exceptions to this are cnidarians and flatworms, which possess sac-like cavities. One-way movement of food allows the gut to become regionally specialized for processing food. Tube guts are quite uniform in general structure, with regions for storing, digesting, absorbing, and eliminating the food. Particular specializations, e.g. expansion of some areas, are related to the food type and how it is ingested. Usually, food **ingested** at the mouth and pharynx passes through an esophagus to a crop, gizzard or stomach. In the intestine, digestive enzymes break down the food molecules and nutrients are absorbed across the epithelium of the gut wall. Undigested wastes are **egested** through an anus or cloaca. Some typical tube guts are described below.

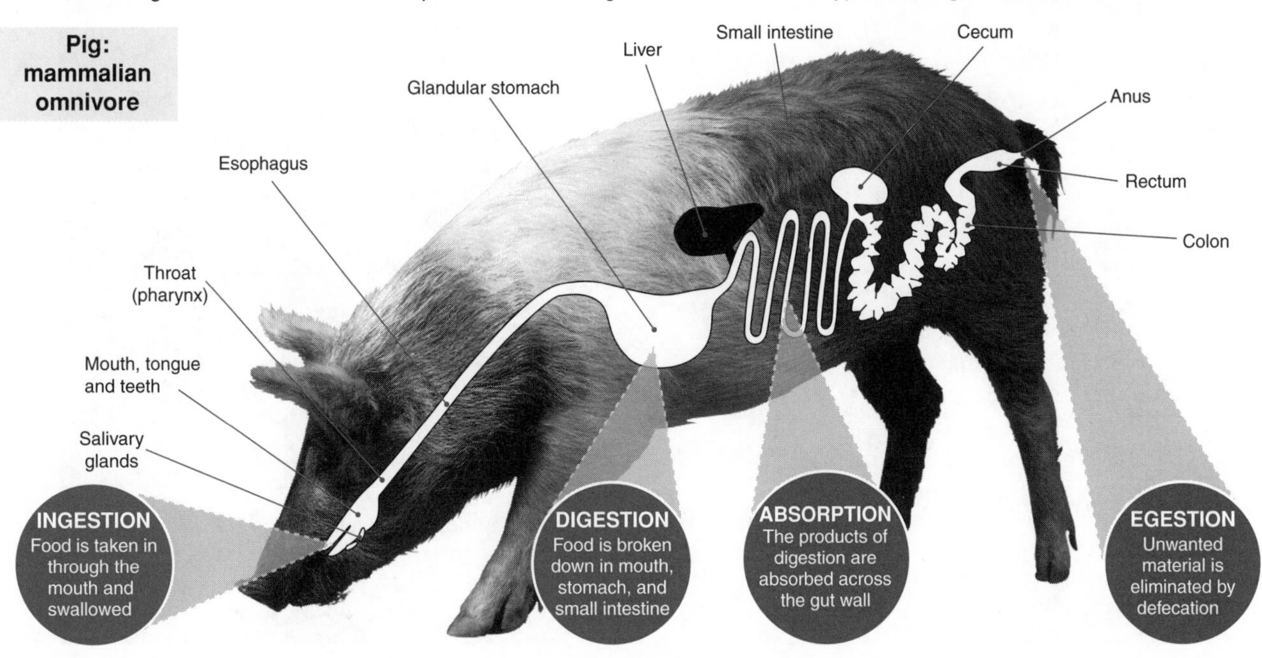


## Peristalsis

Solid food in a tube gut is usually formed into small lumps (each is called a **bolus**). These are moved through the gut by waves of muscular contraction called **peristalsis**. The gut wall has two layers of muscle. The inner circular muscles contract to narrow the tube, and the outer longitudinal muscles contract to widen and shorten the tube. When one set of muscles contracts, the other relaxes.

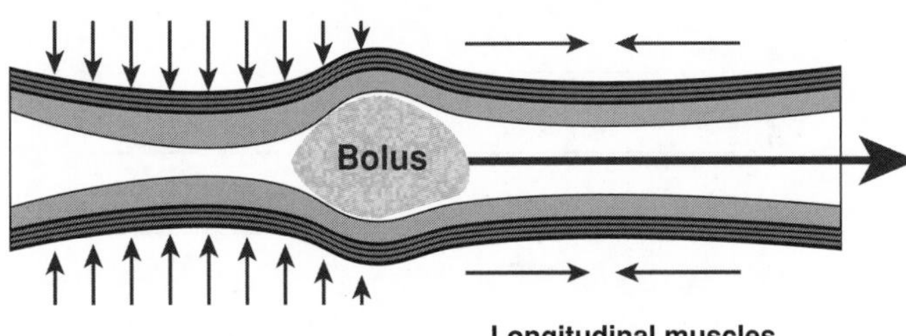


**Circular muscles** contract behind the plug of food (the bolus)

**Longitudinal muscles** contract ahead of the food, causing the tube to shorten and widen to receive the bolus

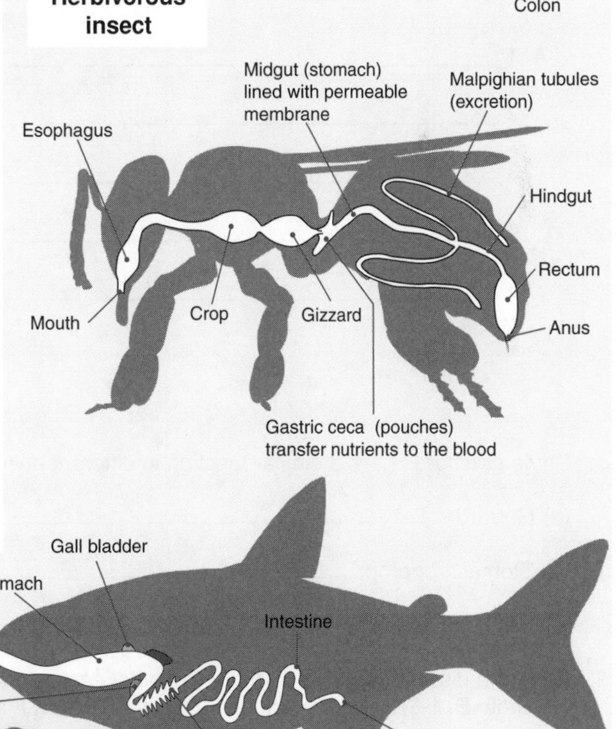


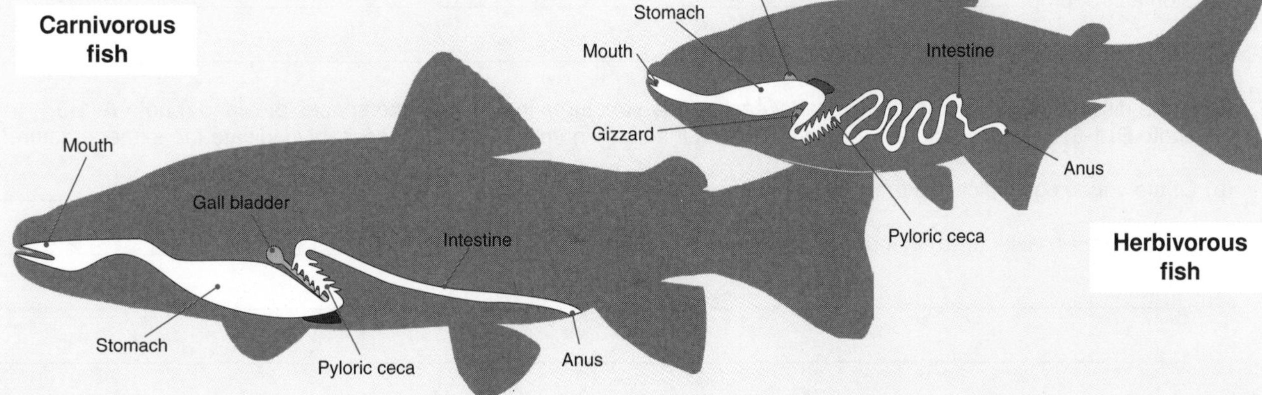



ISBN: 978-1-927173-12-1


***Related activities:*** *Mammalian Guts*
***Weblinks:*** *Digestion Animation, Peristalsis and Gastric Emptying, Peristalsis Animation*

## Recognizing Digestive Organs in a Dissection of a Rat

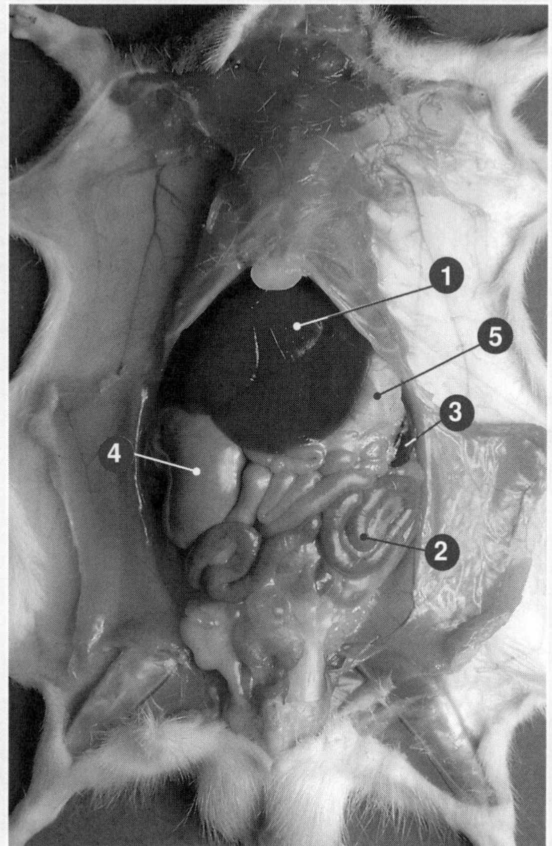


A: Undisturbed organs in the abdominal cavity

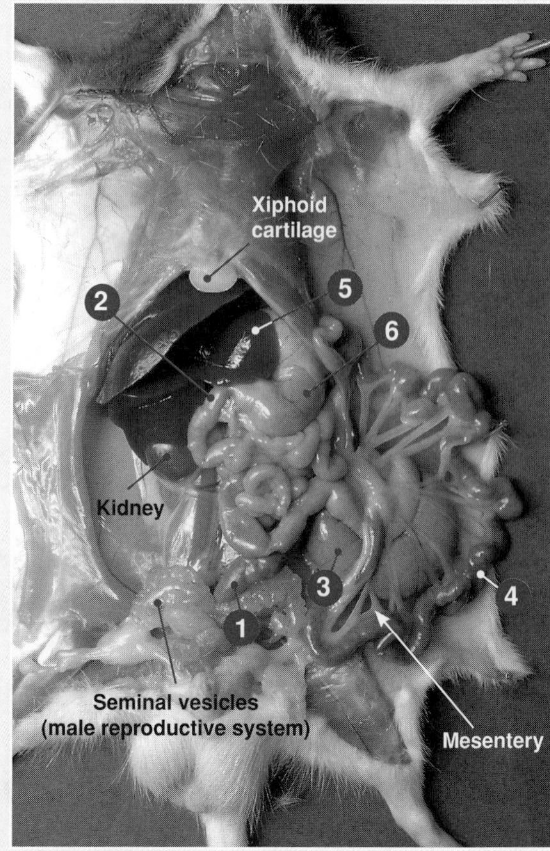


B: Abdominal organs partially dissected

Dissections are often required when studying animal systems. The photographs above show two stages of a rat dissection Use the information provided on the previous page to help you identify and label the structures indicated.

A. 1 ______________________

2 ______________________

3 ______________________

4 ______________________

5 ______________________

B. 1 ______________________

2 ______________________

3 ______________________

4 ______________________

5 ______________________

6 ______________________

1. Some structures have a similar function in different animals. What is the general function of the following gut structures?

   (a) Gizzard: ______________________

   (b) Stomach or crop: ______________________

   (c) Intestine (midgut in insects): ______________________

2. (a) In the dissections of the rat (above), label each of the structures indicated in the spaces provided (photo **A**: **1**-**5**, photo **B**: **1**-**6**). Some structures are the same, but the same numbers do not necessarily indicate the same structure.

   (b) Of the various guts pictured previously, which one does the rat gut most closely resemble? ______________________

   (c) Explain your answer to (b) in terms of the structures present and absent: ______________________

   ______________________

   ______________________

   (d) Give one reason why you might expect this similarity: ______________________

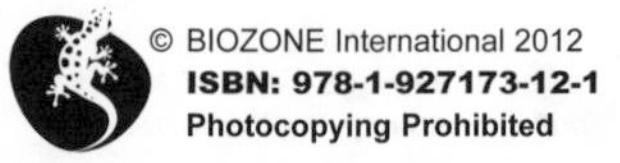


**ISBN: 978-1-927173-12-1**

# Insect Mouthparts

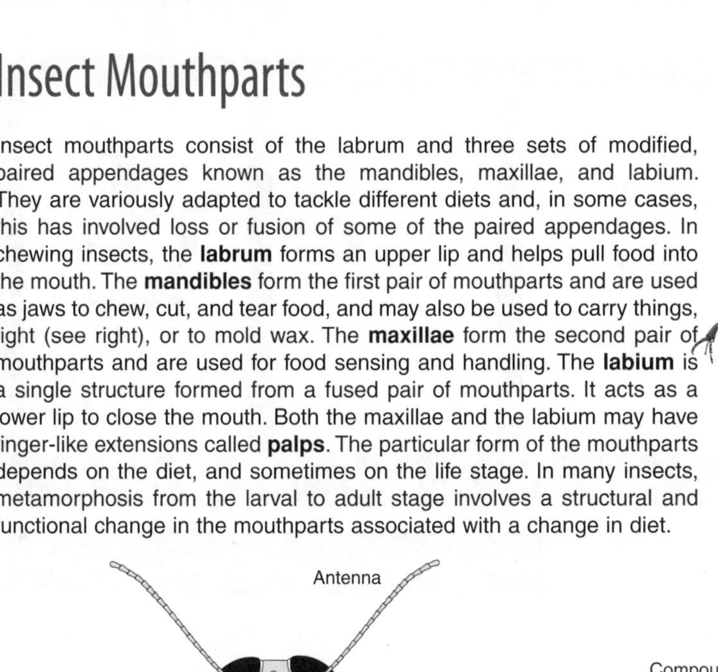

Insect mouthparts consist of the labrum and three sets of modified, paired appendages known as the mandibles, maxillae, and labium. They are variously adapted to tackle different diets and, in some cases, this has involved loss or fusion of some of the paired appendages. In chewing insects, the **labrum** forms an upper lip and helps pull food into the mouth. The **mandibles** form the first pair of mouthparts and are used as jaws to chew, cut, and tear food, and may also be used to carry things, fight (see right), or to mold wax. The **maxillae** form the second pair of mouthparts and are used for food sensing and handling. The **labium** is a single structure formed from a fused pair of mouthparts. It acts as a lower lip to close the mouth. Both the maxillae and the labium may have finger-like extensions called **palps**. The particular form of the mouthparts depends on the diet, and sometimes on the life stage. In many insects, metamorphosis from the larval to adult stage involves a structural and functional change in the mouthparts associated with a change in diet.

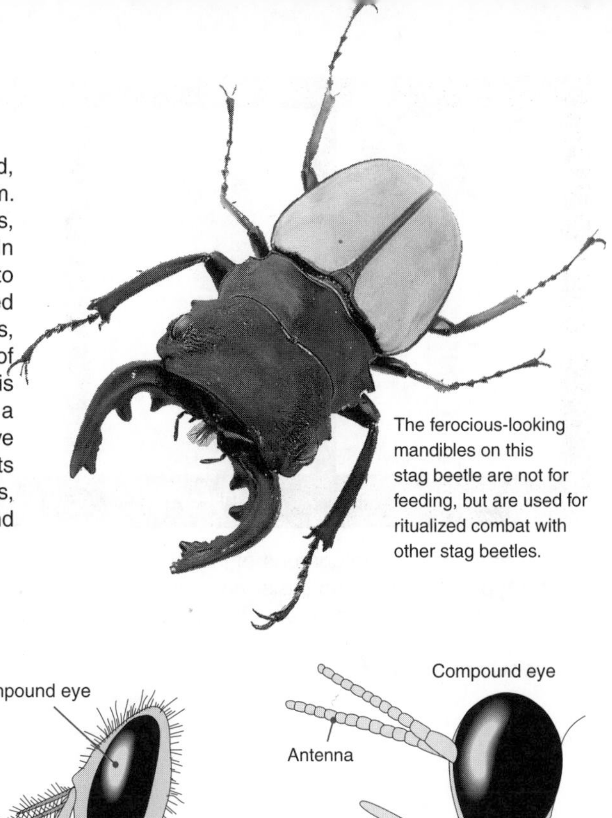

The ferocious-looking mandibles on this stag beetle are not for feeding, but are used for ritualized combat with other stag beetles.

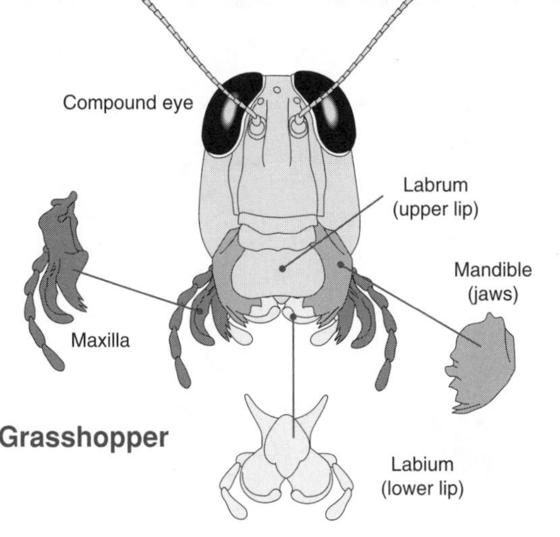


**Grasshopper**

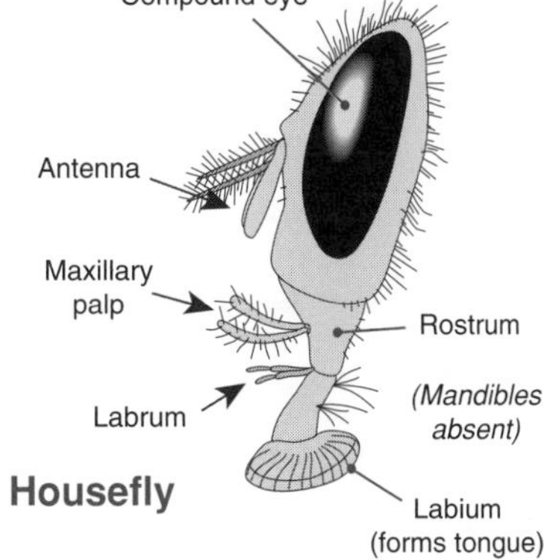


**Housefly**

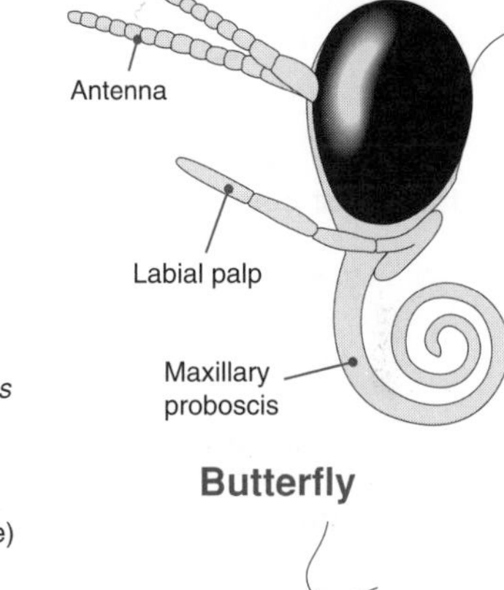


**Butterfly**

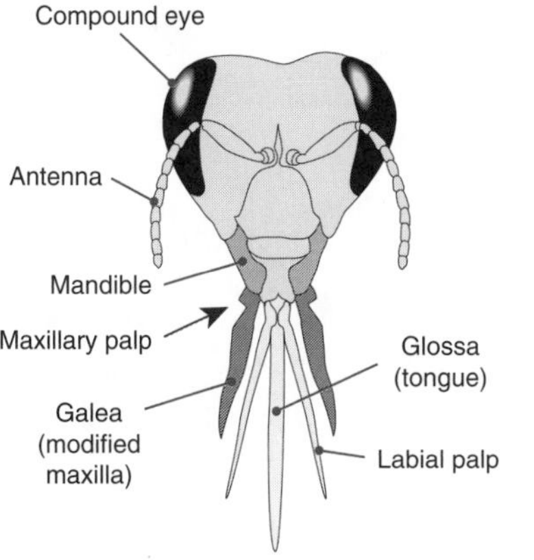


**Honey bee (worker)**

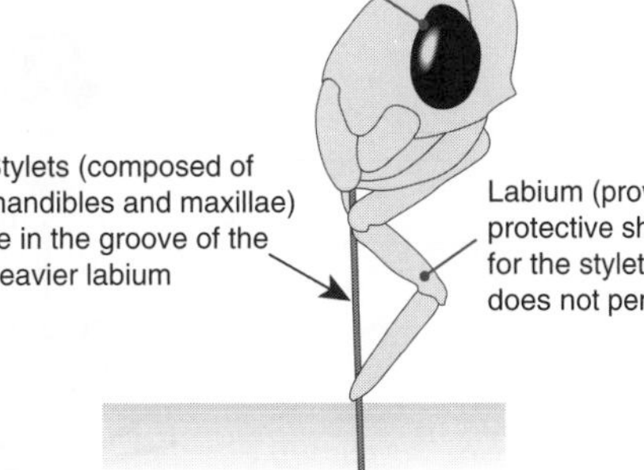


**Shield bug**

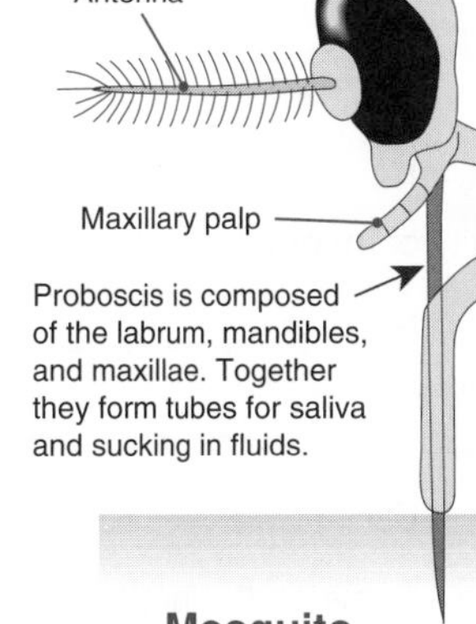


**Mosquito**

1. Describe the components of an insect's mouthparts: ______________________________

______________________________

2. Name the four main modes of feeding carried out by the various insect groups: ______________________________

______________________________

3. Match the following list of insects with their correct mode of feeding: *locust, bee, biting fly, moth, mosquito, beetle, cicada, housefly, flea, dragonfly, aphid, maggot*. There may be more than one example for each feeding mode.

(a) Chewing: ______________________________

(b) Sponging: ______________________________

(c) Chewing/lapping: ______________________________

(d) Seizing/chewing: ______________________________

(e) Sucking: ______________________________

(f) Piercing/sponging: ______________________________

(g) Piercing/sucking: ______________________________

(h) Sucking with mouth hooks: ______________________________


ISBN: 978-1-927173-12-1


***Related activities**: Digestion in Insects*
***Weblinks**: Digital Zoology: Insect Mouthparts*

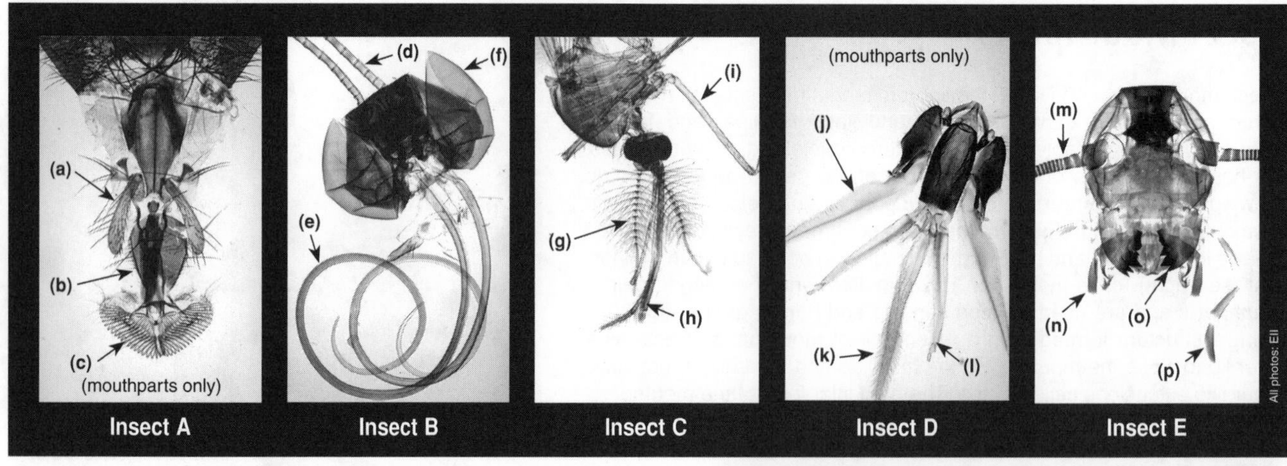


4. For each of the photographs of insects above (**A** - **E**), identify the **type of insect** and the structures labeled (a)-(q). Note that some of the labeled structures are not mouthparts:

Identity of **insect A**: ______________________

(a) ______________ (c) ______________

(b) ______________

Identity of **insect B**: ______________________

(d) ______________ (f) ______________

(e) ______________

Identity of **insect C**: ______________________

(g) ______________ (i) ______________

(h) ______________

Identity of **insect D**: ______________________

(j) ______________ (l) ______________

(k) ______________

Identity of **insect E**: ______________________

(m) ______________ (p) ______________

(n) ______________ (q) ______________

(o) ______________

5. The diagrams on the right illustrate the arrangement of the mouthparts for various insects. Use highlighter pens to create a color key and color in each type of mouthpart.

6. Many insects undergo metamorphosis at certain stages in their life cycle. Butterflies start their active life as caterpillars, after which they pass through a pupal stage, to finally emerge as butterflies. Comment on the diets and changes to the mouthparts of caterpillars and their adult forms (butterflies):

(a) Caterpillar diet: ______________________

Mouthparts: ______________________

______________________

(b) Butterfly diet: ______________________

Mouthparts: ______________________

______________________

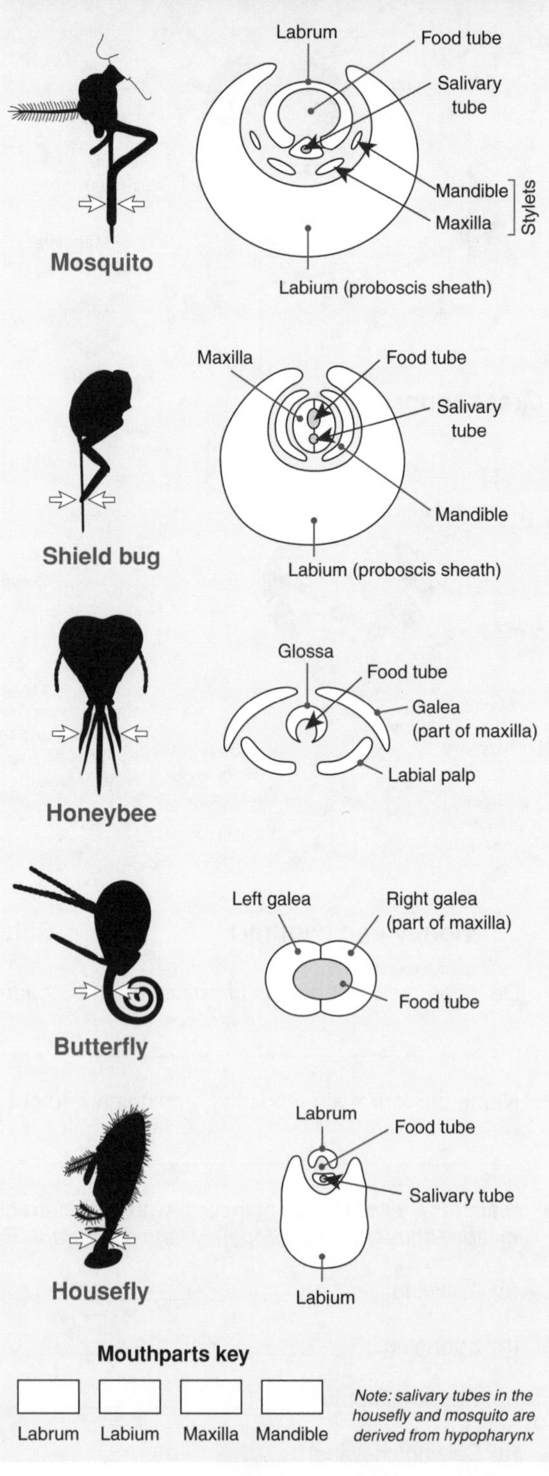



ISBN: 978-1-927173-12-1

# The Teeth of Fish

Fish are **homodonts**, meaning that all their teeth are identical. As a group though, fish show an extremely varied array of teeth types depending on their diet. The teeth are not always found in just the jaw. Fish may have teeth or teeth-like structures on the roof of the mouth, the tongue, and the gill arches. In most cases the teeth are curved backwards to help grip prey.

The teeth of predatory **sharks** are triangular and serrated at the edges to slice through flesh.

Unlike other fish, shark teeth are not attached to the skeleton and are continually lost and replaced.

A shark's upper jaw is not directly connected to the skull. This allows it to push forward as the lower jaw swings open, giving a wider bite than would otherwise be possible.

The premaxillary, maxillary and mandible of many **boney fish** are connected so that as the fish opens its mouth, a large cavity is formed that sucks in prey.

Maxillary

Premaxillary

Mandible

Butterfly fish use narrow pincer like jaws to pluck polyps from coral.

Parrotfish have a horny bill, which they use to bite off chunks of coral. It then grinds these up with hard, flat teeth.

Photo: Drow Male

A

Nare

Mouth

Gills

B

C

D

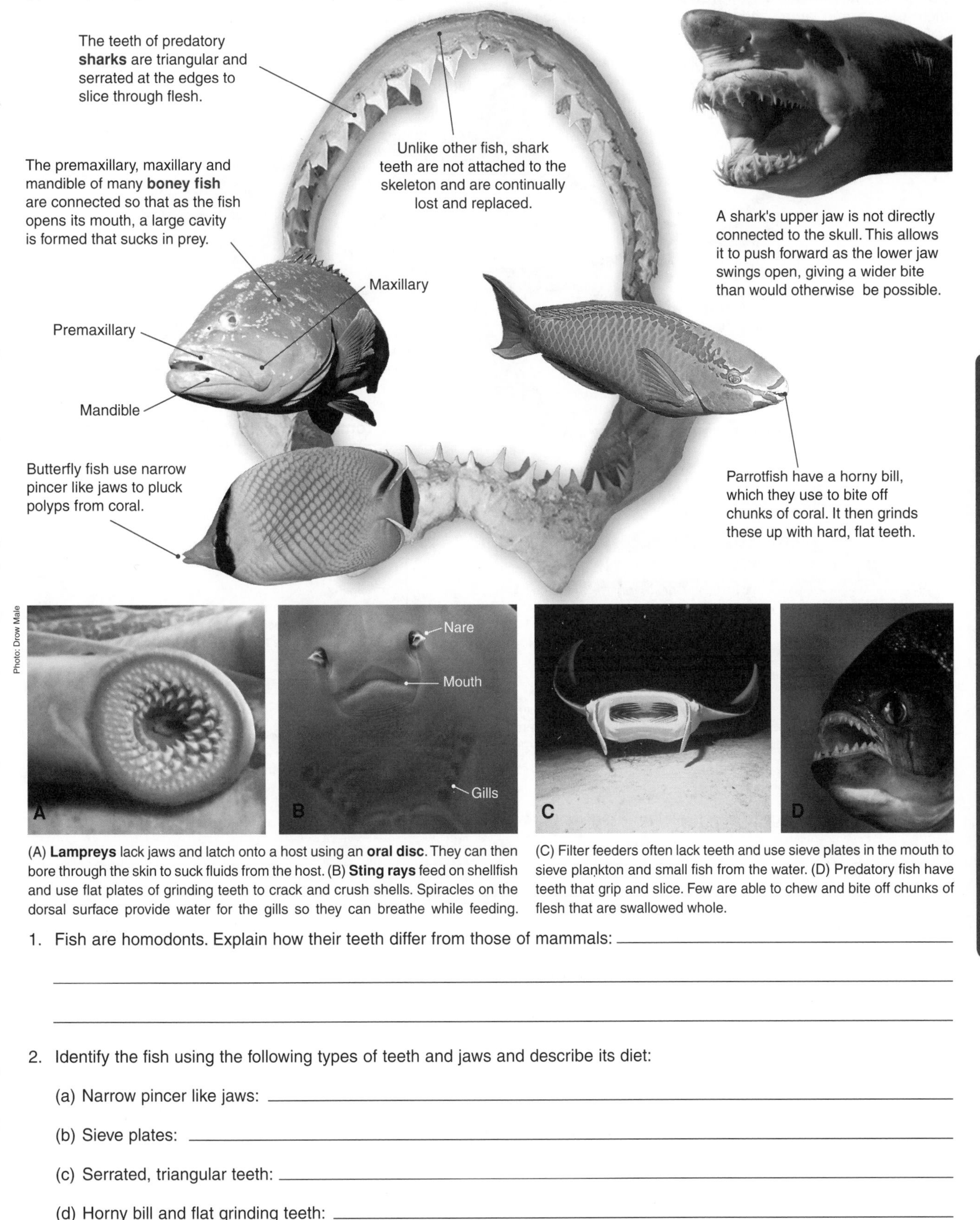

(A) **Lampreys** lack jaws and latch onto a host using an **oral disc**. They can then bore through the skin to suck fluids from the host. (B) **Sting rays** feed on shellfish and use flat plates of grinding teeth to crack and crush shells. Spiracles on the dorsal surface provide water for the gills so they can breathe while feeding. (C) Filter feeders often lack teeth and use sieve plates in the mouth to sieve plankton and small fish from the water. (D) Predatory fish have teeth that grip and slice. Few are able to chew and bite off chunks of flesh that are swallowed whole.

1. Fish are homodonts. Explain how their teeth differ from those of mammals: ______________________

______________________________________________________________

______________________________________________________________

2. Identify the fish using the following types of teeth and jaws and describe its diet:

   (a) Narrow pincer like jaws: ______________________

   (b) Sieve plates: ______________________

   (c) Serrated, triangular teeth: ______________________

   (d) Horny bill and flat grinding teeth: ______________________

   (e) No jaws, teeth-like structure on the tongue: ______________________

3. Explain why sharks need to continually replace their teeth: ______________________

______________________________________________________________



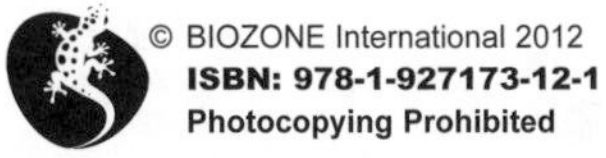


ISBN: 978-1-927173-12-1

# Digestion in Fish

Fish are generally **ectothermic**, meaning that their body temperature is dependent on external sources of energy, not metabolism. For most fish, the temperature of the body and the environment are the same. However, some large predatory fish, such as tuna and large sharks, can maintain at least some of the body at temperatures above those of the seawater (**regional endothermy**). Increasing body temperature enables them to maintain high levels of activity and process greater quantities of food more rapidly. They maintain these higher temperatures using a **countercurrent network** of blood capillaries.

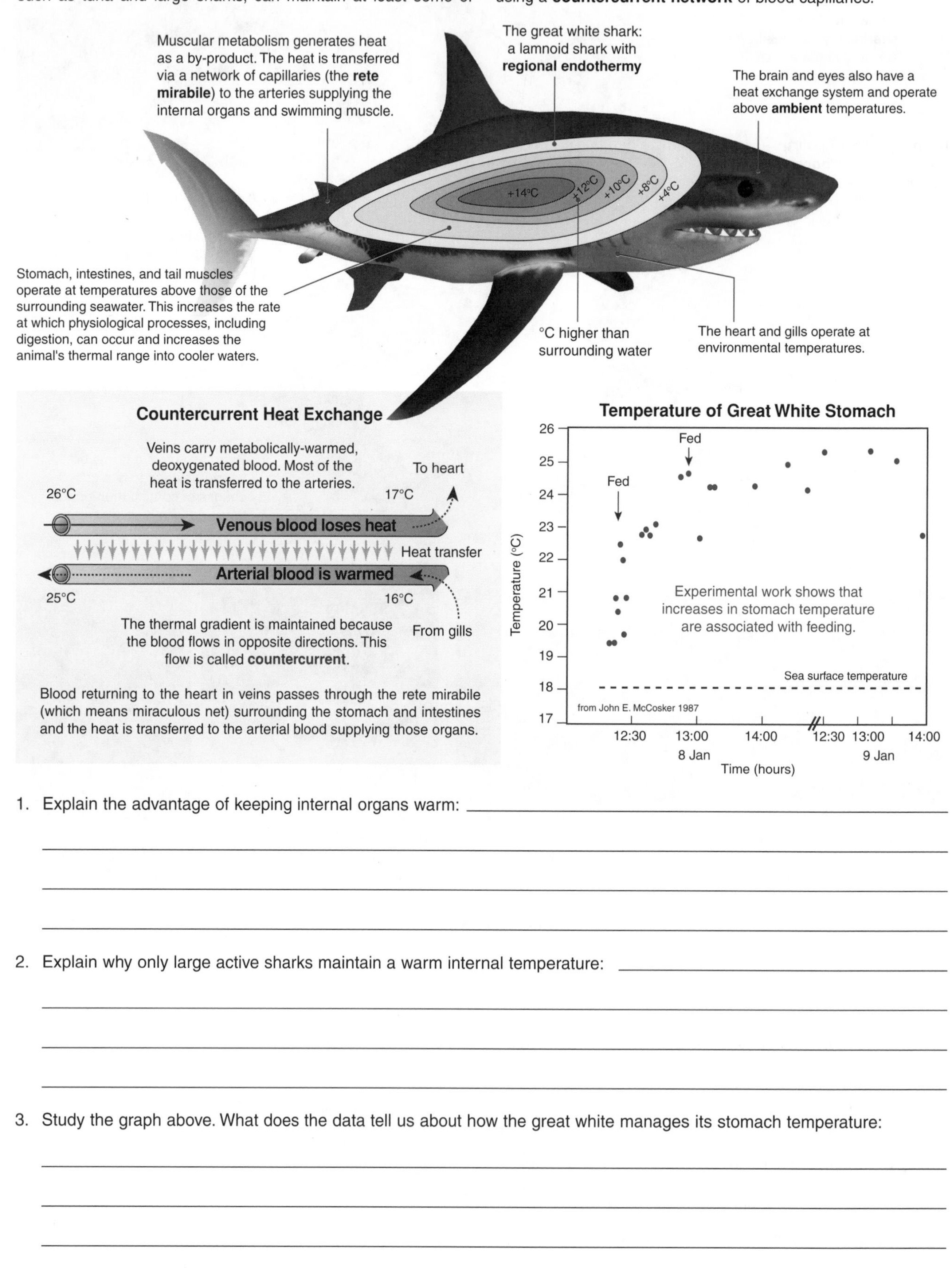


1. Explain the advantage of keeping internal organs warm:

2. Explain why only large active sharks maintain a warm internal temperature:

3. Study the graph above. What does the data tell us about how the great white manages its stomach temperature:



***Related activities:*** *Diversity in Tube Guts*


**ISBN: 978-1-927173-12-1**

# Dentition in Mammals

Many of an organism's adaptations are related to the successful procurement of food. Even within the class Mammalia, there is great diversity in the adaptations for different diets. These adaptations are most obvious in the dentition and jaw musculature, as well as in the length and organization of the gut. The diversity of dentition amongst the mammals reflects the wide range of foods and feeding modes within the class; some mammals have relatively generalized dentition, while others are highly specialized, even to the extent of losing teeth entirely. This activity explores some mammalian dietary specializations by examining the dentition of different mammalian representatives.

Using the examples provided, identify the skulls and describe the diet associated with each. Describe the dental adaptations of each, looking for differences in tooth size and arrangement, and whether teeth are absent or modified.

For further help, visit *Will's skull page* web site, which provides information on the structure of mammalian skulls and their dental formulae.

**Animals included in this exercise**:
*Lion, rabbit, mountain sheep, pig, giant anteater, black and white ruffed lemur, dolphin, gray whale, tree shrew.*

**Diets included in this exercise**:
*Rough vegetation and grasses; herbs and grasses; crustaceans; omnivorous; fish; insects and worms; leaves, fruit and flowers; meat; termites and ants.*

1. Animal: ______________________

   Diet: ______________________

   Dental adaptations to diet: ______________________

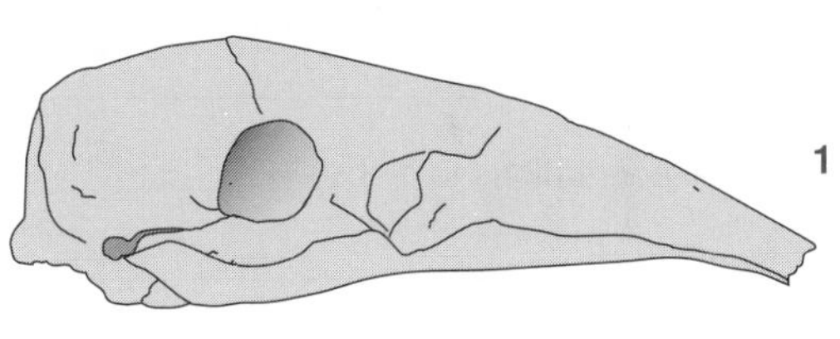


2. Animal: ______________________

   Diet: ______________________

   Dental adaptations to diet: ______________________

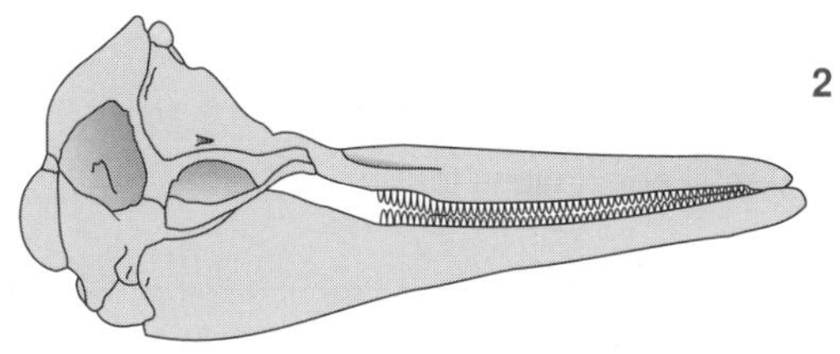


3. Animal: ______________________

   Diet: ______________________

   Dental adaptations to diet: ______________________

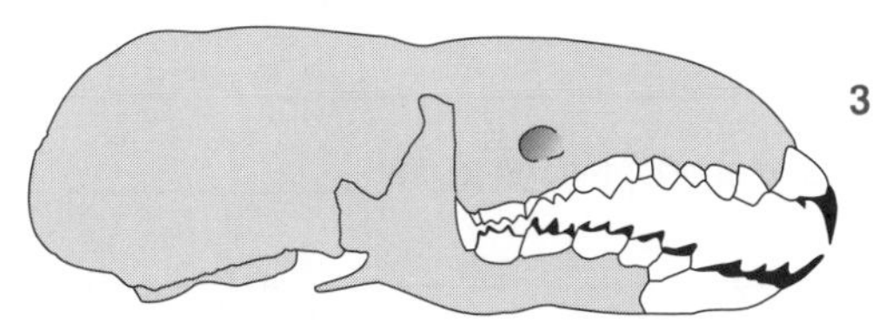



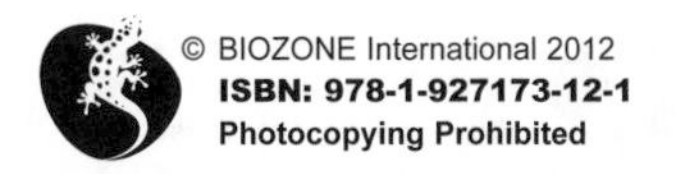

ISBN: 978-1-927173-12-1


***Related activities**: Obtaining Food, Diversity in Tube Guts*
***Weblinks**: Will's Skull Page*

4. Animal: ______________________________

   Diet: ______________________________

   Dental adaptations to diet: ______________________________

4

5. Animal: ______________________________

   Diet: ______________________________

   Dental adaptations to diet: ______________________________

5

6. Animal: ______________________________

   Diet: ______________________________

   Dental adaptations to diet: ______________________________

6

7. Animal: ______________________________

   Diet: ______________________________

   Dental adaptations to diet: ______________________________

7

8. Animal: ______________________________

   Diet: ______________________________

   Dental adaptations to diet: ______________________________

8

9. Animal: ______________________________

   Diet: ______________________________

   Dental adaptations to diet: ______________________________

9

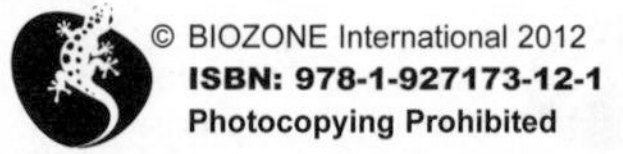


ISBN: 978-1-927173-12-1

# Mammalian Guts

Among animals, bulky, high fiber diets are harder to digest than diets containing very little plant material. Herbivores therefore tend to have longer guts with larger chambers than carnivores. Grazing mammals are dependent on symbiotic microorganisms to digest plant cellulose for them. This microbial activity may take place in the stomach (foregut fermentation) or the colon and cecum (hindgut fermentation). Some grazers are ruminants; regurgitating and rechewing the partially digested food which is then reswallowed. The kangaroo is a ruminant equivalent within the marsupials. The diagrams below compare gut structure in representative mammals. Further comparisons of the guts of carnivores and ruminants are made on the next page.



### Omnivore

*Human: Homo sapiens*

Omnivorous diets can vary enormously and the regional specializations of the gut vary according to the balance of animal and plant based material eaten. In all mammals (and all vertebrates) the gastric phase of digestion in the stomach always involves secretion of acid and the protein-digesting enzyme pepsin. The pepsin requires an acid (low pH) environment to function. Further enzymic digestion and absorption of nutrients occurs in the small intestine.

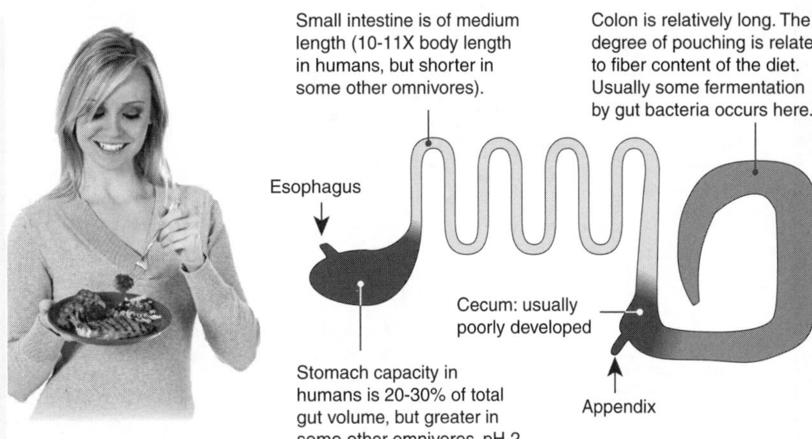


### Carnivore

*Dog: Canis familiaris*

The guts of carnivores are adapted for processing animal flesh. The viscera (gut and internal organs) of killed or scavenged animals are eaten as well as the muscle, and provide valuable nutrients. Regions for microbial fermentation are poorly developed or absent. Some animals evolved as carnivores, but have since become secondarily adapted to a more omnivorous diet (bears) or a highly specialized herbivorous diet (pandas). Their guts retain the basic features of a carnivore's gut.

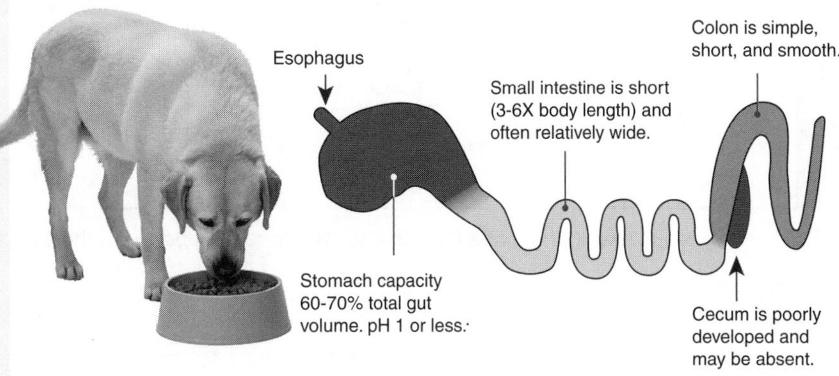


### Herbivore: foregut digestion

*Red kangaroo: Macropus rufus*

Kangaroos and ruminants are both foregut fermenters. A large part of the enlarged stomach acts as fermentation chamber, where bacteria and protozoan ciliates break down plant cellulose. Volatile fatty acids released by the microbes provide energy, and digestion of the microbes themselves provides protein. The kangaroo foregut is smaller and more tubular than a ruminant foregut (described later in this chapter). This reduces the weight of a full stomach and increases the rate at which food passes through the gut.

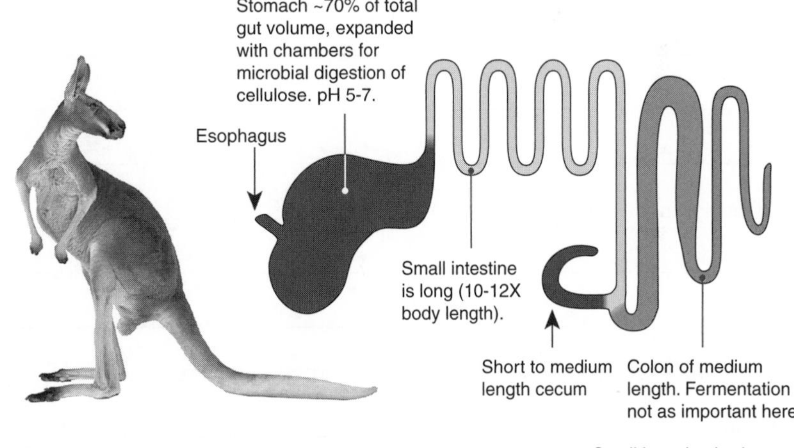


### Herbivore: hindgut digestion

*European rabbit: Oryctolagus cuniculus*

Rabbits are specialized herbivores. The cecum is expanded into a very large chamber for digestion of cellulose. At the junction between the ileum and the colon, indigestible fiber is pushed into the colon where it forms hard feces. Digestible matter passes into the cecum where anaerobic bacteria ferment the material and more absorption takes place. Vitamins and microbial proteins from this fermentation are included in specially formed soft faecal pellets, which pass out of the gut and are eaten directly from the anus (an action called cecotrophy).

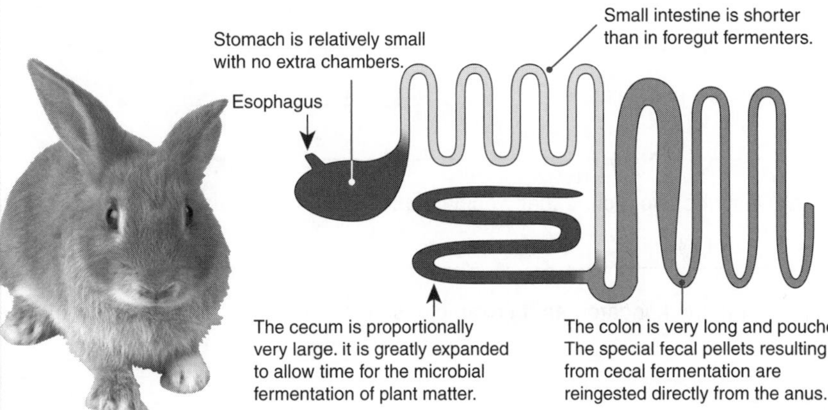

**ISBN: 978-1-927173-12-1**

## Ruminant herbivore: Cattle (*Bos taurus*)

Ruminants are specialized herbivores with teeth adapted for chewing and grinding. Their nutrition is dependent on their mutualistic relationship with their microbial gut flora (bacteria and ciliates), which digest plant material and provide the ruminant with energy and protein. In return, the rumen provides the microbes with a warm, oxygen free, nutrient rich environment.

## Carnivore: Lion (*Panthera leo*)

The teeth and guts of carnivores are superbly adapted for eating animal flesh. The canine and incisor teeth are specialized to bite down and cut, while the carnassials are enlarged, lengthened, and positioned to act as shears to slice through flesh. As meat is easier to digest than cellulose, the guts of carnivores are comparatively more uniform and shorter than those of herbivores.

Dental adaptations

Small temporalis muscle

Horny pad

Incisors

Canine

Masseter muscle is very large to assist in chewing.

Premolars

Diastema (toothless space)

Molars are large for grinding

Flow of food in the stomach

Passage of food

Regurgitation, rechewing, and reswallowing

**Omasum:** removes water

**Abomasum:** true stomach secretes gastric juices.

**Rumen:** contains bacteria for the digestion of cellulose and ciliates to digest starch.

**Reticulum:** forms the cud which is returned to the mouth.

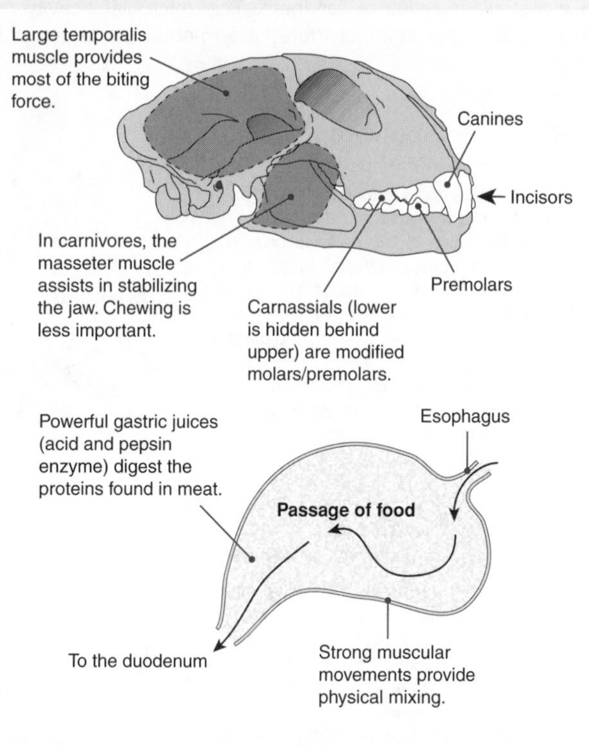


1. For each of the following, describe the **structural** differences between the guts of a carnivore and a named herbivore:

   (a) Size of stomach (relative to body size): ______________________

   (b) Relative length of small intestine: ______________________

   (c) Development of hind gut (cecum and colon): ______________________

2. Identify and explain a structural difference between carnivores and ruminant herbivores with respect to:

   (a) The teeth: ______________________

   (b) The jaw musculature: ______________________

3. (a) Describe two factors that would influence the time it takes for food to pass through the gut (the **gut transit time**):

   ______________________

   (b) Using the kangaroo as an example, suggest how gut transit time might be adapted to a mammal's lifestyle:

   ______________________

**ISBN: 978-1-927173-12-1**

# Microbes and Digestion in Mammals

Mammals cannot directly digest the cellulose in plants. They produce all the enzymes necessary to digest protein, but lack those required to break the bonds joining the glucose molecules in cellulose. Herbivorous mammals are able to digest cellulose in two main ways and both involve a **mutualistic relationship** with microorganisms in their gut. Ruminants, such as cattle, rely on microbes in the **rumen** to break down the cellulose, regurgitating, rechewing, and then reswallowing the partially digested material. **Hindgut fermenters**, such as horses and rodents, house their cellulose-digesting microbes in the hindgut. Some, including rabbits, have a further strategy, called **cecotrophy**, in which some feces are reingested to be processed for a second time. These strategies help to increase the efficiency of digestion, maximizing the energetic gain from the food that is ingested.

## Ruminants and their microbes

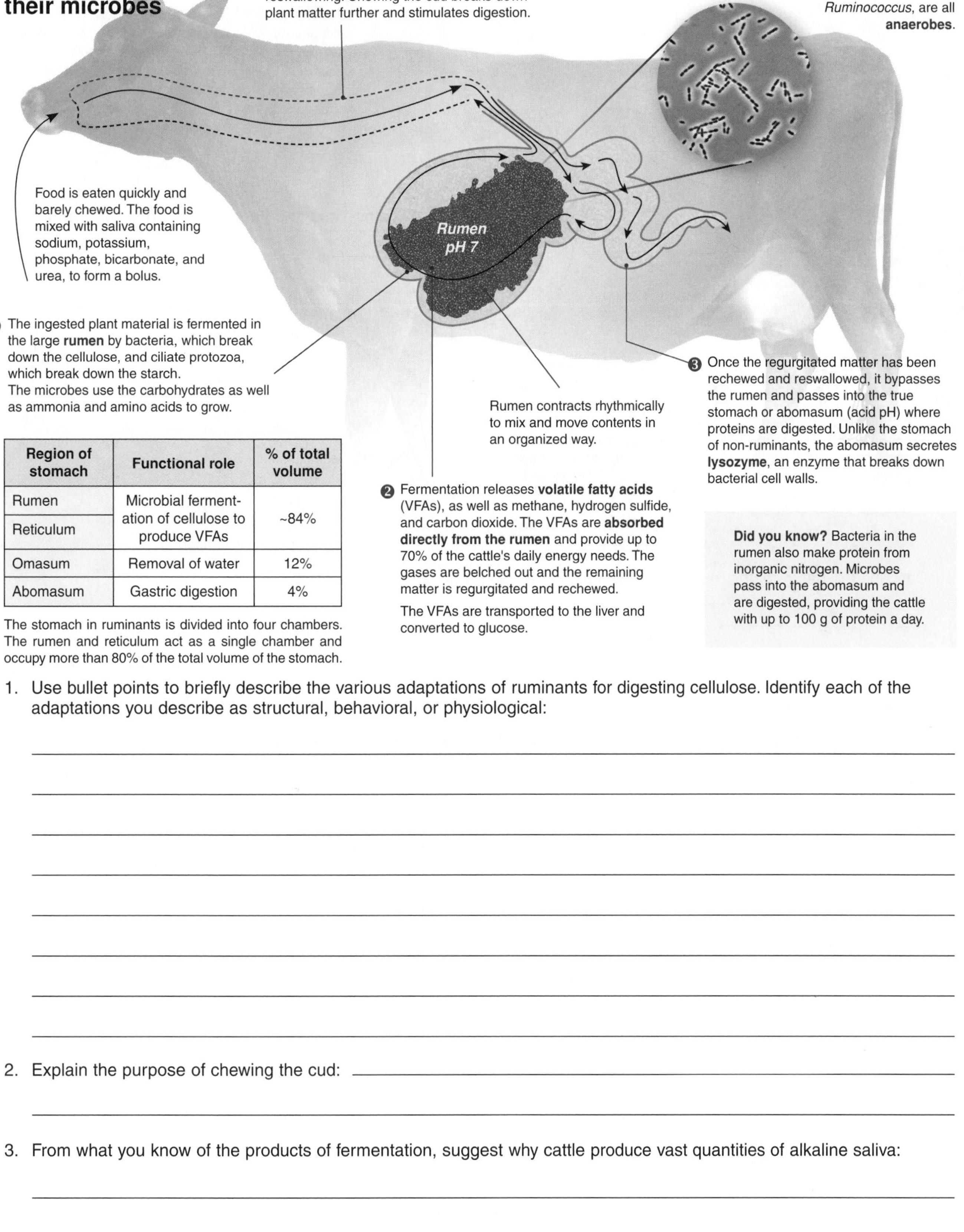


| Region of stomach | Functional role | % of total volume |
|---|---|---|
| Rumen | Microbial fermentation of cellulose to produce VFAs | ~84% |
| Reticulum | | |
| Omasum | Removal of water | 12% |
| Abomasum | Gastric digestion | 4% |

The stomach in ruminants is divided into four chambers. The rumen and reticulum act as a single chamber and occupy more than 80% of the total volume of the stomach.

**Did you know?** Bacteria in the rumen also make protein from inorganic nitrogen. Microbes pass into the abomasum and are digested, providing the cattle with up to 100 g of protein a day.

1. Use bullet points to briefly describe the various adaptations of ruminants for digesting cellulose. Identify each of the adaptations you describe as structural, behavioral, or physiological:

2. Explain the purpose of chewing the cud:

3. From what you know of the products of fermentation, suggest why cattle produce vast quantities of alkaline saliva:


ISBN: 978-1-927173-12-1


***Related activities:*** *Mammalian Guts, Digestion in Insects*
***Weblinks:*** *Ruminant Digestion, A Cow's Digestive System*

# Cecotrophy: An Adaptation for Better Nutrition

Rabbits and hares are small active mammals. Like all hindgut fermenters, they require a high throughput of material, so they eat frequently.

Like ruminants, rabbits derive nutrients from the products of microbial metabolism in the gut. However, in rabbits, the bacteria reside in a pouch-like region of the hindgut called the **cecum**. The bacteria in the cecum break down the cellulose-based plant material to release VFAs, which are absorbed through the lining of the cecum and provide up to 40% of the animal's energy requirements.

Rabbits produce two types of faecal pellets: soft pellets, which are eaten and 'normal' hard pellets. Production of the soft and hard feces follows a daily rhythm as shown below for rabbits with free access to food under a 12 hour light: 12 hour dark light regime.

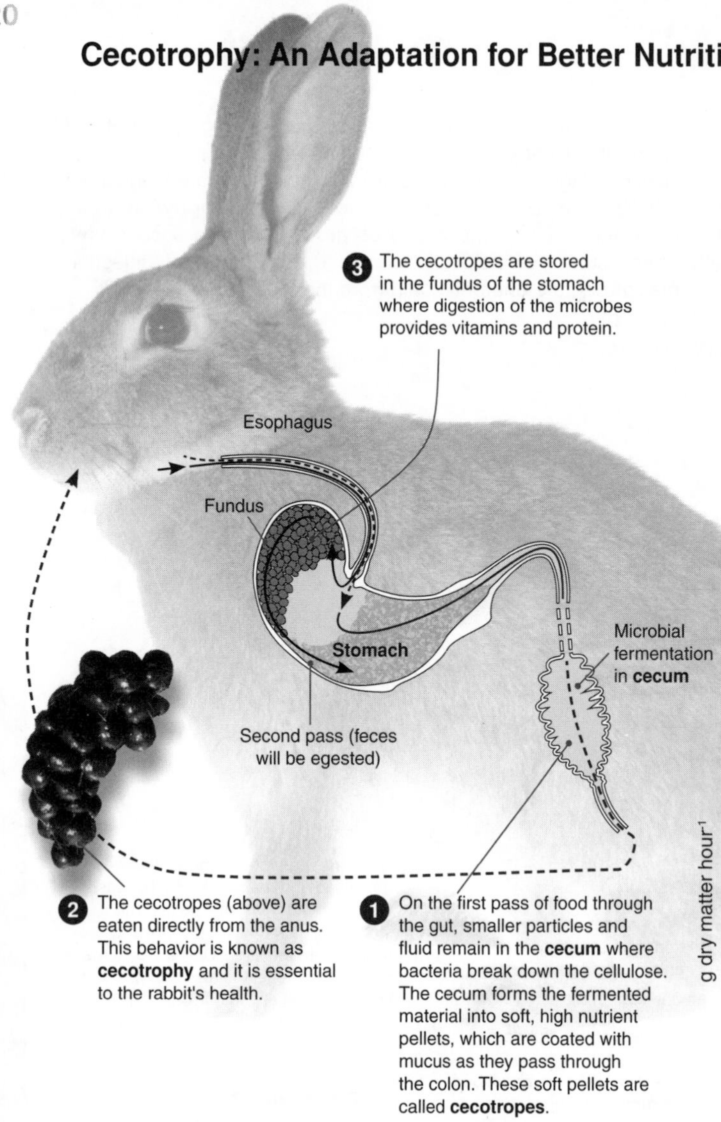

3 The cecotropes are stored in the fundus of the stomach where digestion of the microbes provides vitamins and protein.

Esophagus

Fundus

Stomach

Microbial fermentation in **cecum**

Second pass (feces will be egested)

2 The cecotropes (above) are eaten directly from the anus. This behavior is known as **cecotrophy** and it is essential to the rabbit's health.

1 On the first pass of food through the gut, smaller particles and fluid remain in the **cecum** where bacteria break down the cellulose. The cecum forms the fermented material into soft, high nutrient pellets, which are coated with mucus as they pass through the colon. These soft pellets are called **cecotropes**.

**Feeding Time and Feces Production in Rabbits**

Cecotrophy occurs in the morning when the rabbits are in the burrow

Hard feces are produced during the evening, when rabbits are outside the burrow grazing

Soft feces
Feed intake
Hard feces

g dry matter $hour^{-1}$
0 2 4 6 8 10
10 12 14 16 18 20 22 24 2 4 6 8
Time (hour)

Carabaño and Merino, 1996

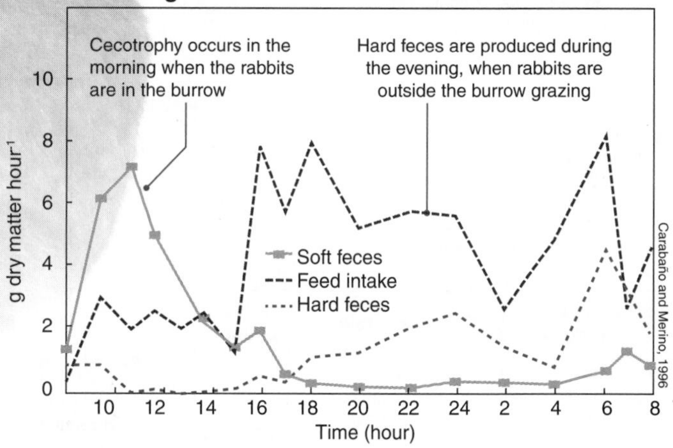

4. Describe the nature of the mutualistic relationship between ruminants and their gut microbes:

5. Explain the adaptive value of a daily rhythm for hard and soft feces production in rabbits:

6. Compare and contrast the adaptations for digesting cellulose in ruminants and rabbits:

**ISBN: 978-1-927173-12-1**

# Digestion in Insects

During digestion, food is changed by physical and chemical means from its original state until its constituents are released as small, simple molecules that can be **absorbed** and **assimilated** (taken up by cells). Animal diets are highly variable and so are the adaptations for digesting these diets. Some foods, such as egg and honey, are pure, concentrated nutriment. Some diets are nutritionally rich but contain large volumes of water (e.g. blood), while others are bulky and difficult to digest (e.g. vegetation).

## Feeding on Fluid

Insects that feed on fluids must be able to store large volumes during feeding and remove the water from the fluid to concentrate it before it is digested.

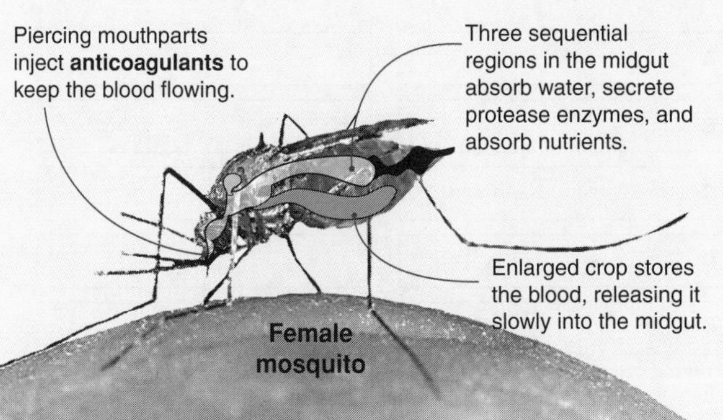


**Blood: A high protein fluid**

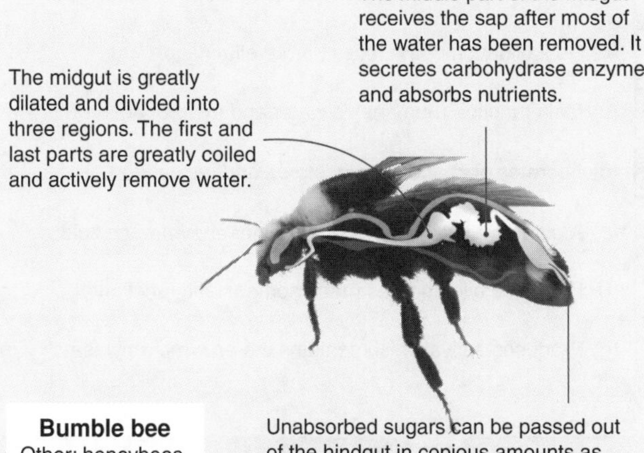


**Plant sap: A high volume, low protein fluid**

## Feeding on Wood

Insects that eat wood rely on microbial populations to break the chemical bonds between the glucose molecules and provide them with molecules they can use.

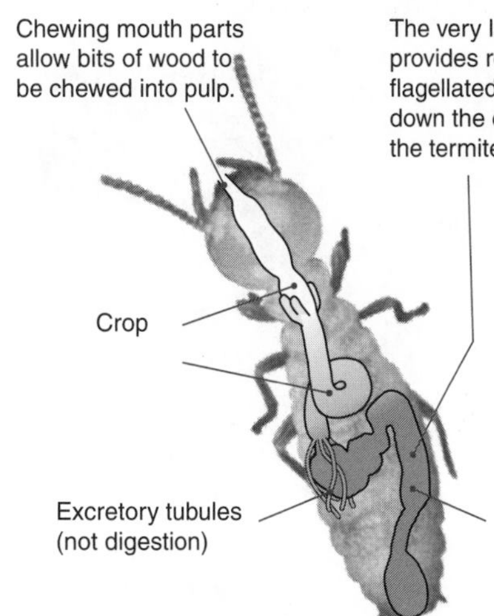


**It's complicated! A symbiont within a symbiont**

| Termites | Protozoa | Bacteria |
|---|---|---|
| Chew and ingest woody material | Engulf wood fibers using phagocytosis | Break down cellulose using cellulase enzymes |
| Absorb sugars released from protozoan symbionts | Use some sugars and release some to the termite symbiont | Use some sugars and release some to the protozoan symbiont |

The flagellated protozoans in the termite gut are very mobile. They are more easily kept as gut residents than free bacteria, which would be lost.

1. Describe one common adaptation of the guts of fluid feeders and its role: ______________________

2. Use brief notes to compare and contrast the digestion of cellulose in termites and ruminants: ______________________




ISBN: 978-1-927173-12-1

# The Human Digestive Tract

An adult consumes an estimated metric tonne of food a year. Food provides the source of the energy required to maintain **metabolism**. The human digestive tract, like most tube-like guts, is regionally specialized to maximize the efficiency of physical and chemical breakdown (**digestion**), **absorption**, and **elimination**. The gut is essentially a hollow, open-ended, muscular tube, and the food within it is essentially outside the body, having contact only with the cells lining the tract. Food is physically moved through the gut tube by waves of muscular contraction, and subjected to chemical breakdown by enzymes contained within digestive secretions. The products of this breakdown are then absorbed across the gut wall. A number of organs are associated with the gut along its length and contribute, through their secretions, to the digestive process at various stages.

Feeding and satiety centers in the hypothalamus regulate eating

A B C D E F G H I J K L

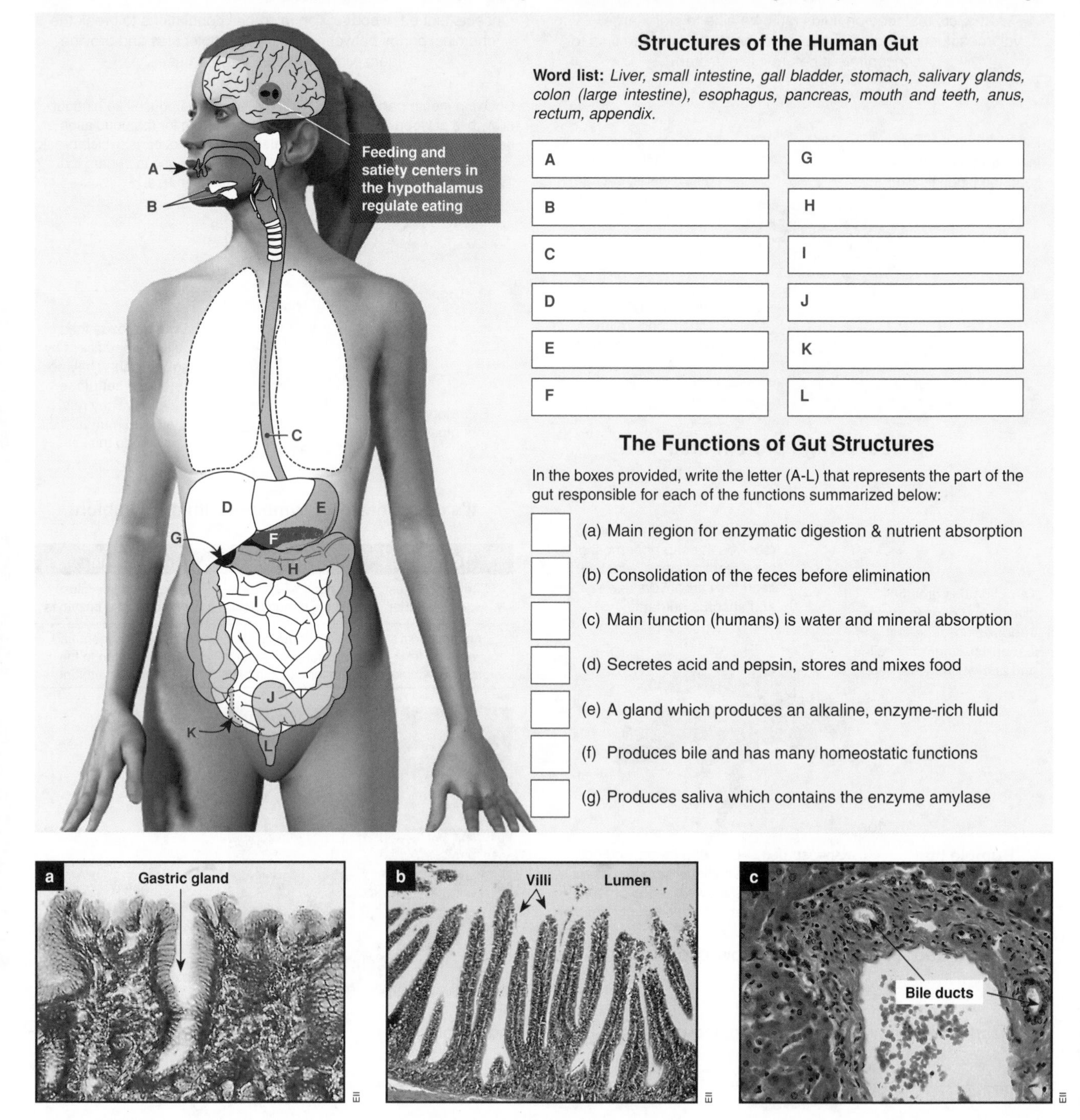

## Structures of the Human Gut

**Word list:** *Liver, small intestine, gall bladder, stomach, salivary glands, colon (large intestine), esophagus, pancreas, mouth and teeth, anus, rectum, appendix.*

| | |
|---|---|
| A | G |
| B | H |
| C | I |
| D | J |
| E | K |
| F | L |

## The Functions of Gut Structures

In the boxes provided, write the letter (A-L) that represents the part of the gut responsible for each of the functions summarized below:

☐ (a) Main region for enzymatic digestion & nutrient absorption

☐ (b) Consolidation of the feces before elimination

☐ (c) Main function (humans) is water and mineral absorption

☐ (d) Secretes acid and pepsin, stores and mixes food

☐ (e) A gland which produces an alkaline, enzyme-rich fluid

☐ (f) Produces bile and has many homeostatic functions

☐ (g) Produces saliva which contains the enzyme amylase

a Gastric gland

b Villi Lumen

c Bile ducts

1. In the spaces provided on the diagram above, identify the parts labeled **A-L** (choose from the word list provided). Match each of the **functions** described (a)-(g) with the letter representing the corresponding structure on the diagram.

2. On the same diagram, mark with lines and labels: anal sphincter (**AS**), pyloric sphincter (**PS**), cardiac sphincter (**CS**).

3. Identify the region of the gut illustrated by the photographs (**a**)-(**c**) above. For each one, explain the identifying features:

(a) ______

(b) ______

(c) ______



*Related activities:* Adaptations for Absorption
*Weblinks:* Digestion Animation

*Periodicals:*
*The anatomy of digestion*

# The Stomach, Duodenum, and Pancreas

Esophagus

Cardiac sphincter

The gall bladder stores bile, which is produced by the liver cells.

Bile from liver

Pyloric sphincter

Duodenum

Pancreatic duct

Three layered muscular wall mixes the stomach contents to produce a soupy mixture called chyme. Stretching of the stomach wall is a stimulus for gastric secretion.

Folds (rugae) in the stomach wall allow the stomach to expand to 1L.

In the pancreas, the acinar cells secrete an alkaline fluid into the pancreatic duct.

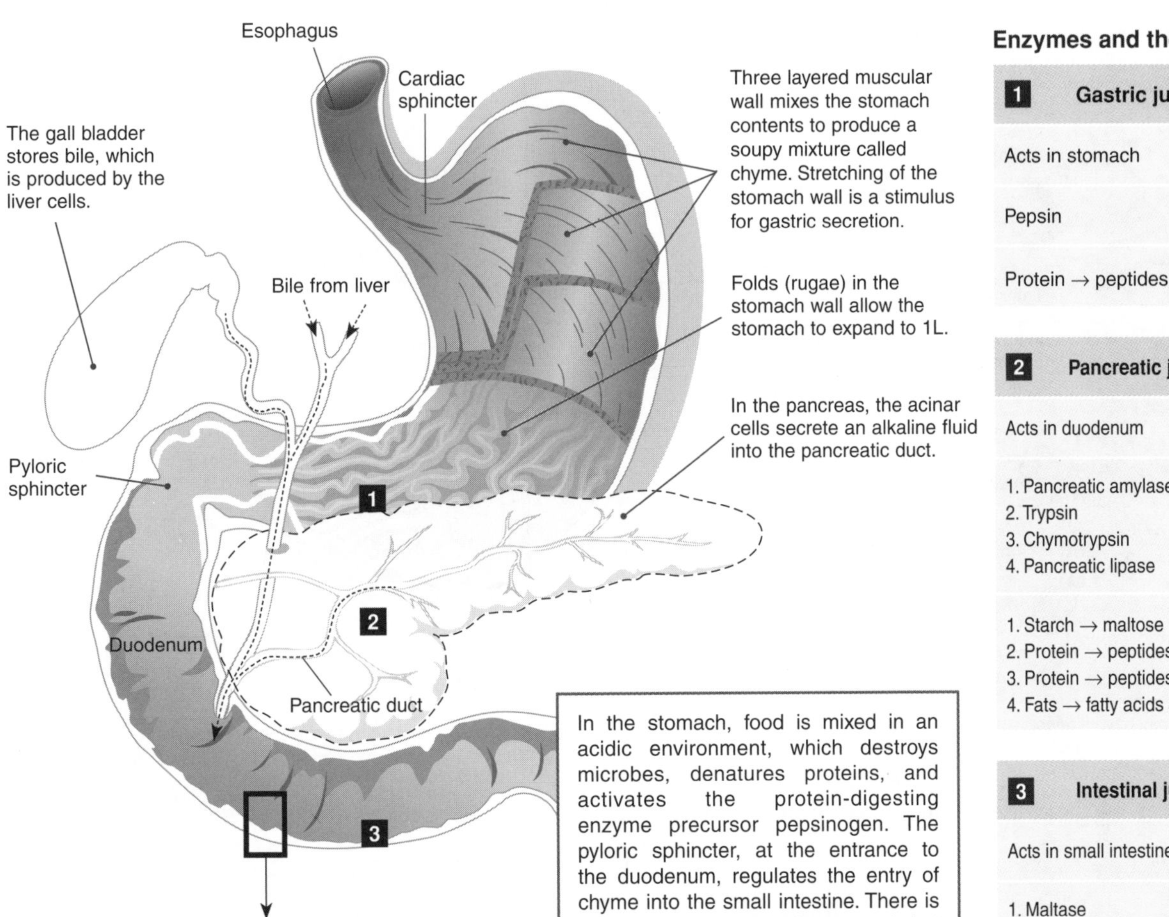

## Enzymes and their actions

| | |
|---|---|
| **1 Gastric juice** | |
| Acts in stomach | |
| Pepsin | Protein → peptides |
| **2 Pancreatic juice** | |
| Acts in duodenum | |
| 1. Pancreatic amylase | 1. Starch → maltose |
| 2. Trypsin | 2. Protein → peptides |
| 3. Chymotrypsin | 3. Protein → peptides |
| 4. Pancreatic lipase | 4. Fats → fatty acids & glycerol |
| **3 Intestinal juice** | |
| Acts in small intestine | |
| 1. Maltase | 1. Maltose → glucose |
| 2. Peptidases | 2. Polypeptides → amino acids |

In the stomach, food is mixed in an acidic environment, which destroys microbes, denatures proteins, and activates the protein-digesting enzyme precursor pepsinogen. The pyloric sphincter, at the entrance to the duodenum, regulates the entry of chyme into the small intestine. There is very little absorption in the stomach, but very small molecules (glucose, aspirin, alcohol) are absorbed directly across the stomach wall into the gastric blood vessels surrounding the stomach.

## Detail of a Villus (Small Intestine)

The **intestinal villi** project into the gut lumen and provide an immense surface area for nutrient absorption. The villi are lined with **epithelial cells** and each has a brush border of many **microvilli** which further increase the surface area.

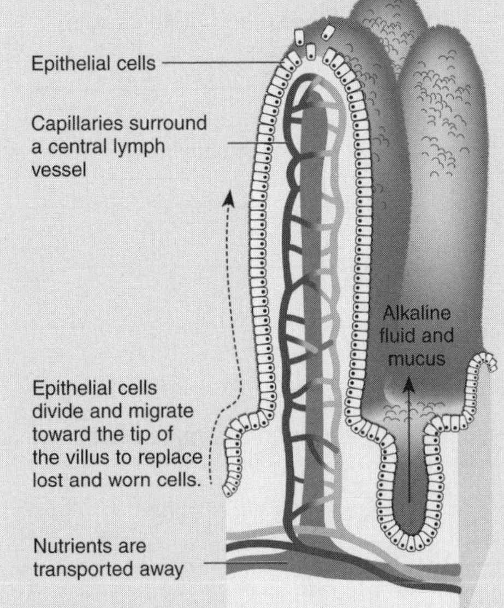


Enzymes bound to the surfaces of the epithelial cells break down peptides and carbohydrate molecules. The breakdown products are then absorbed into the underlying blood and lymph vessels. Tubular exocrine glands and goblet cells secrete alkaline fluid and mucus into the lumen.

## Detail of a Gastric Gland (Stomach Wall)

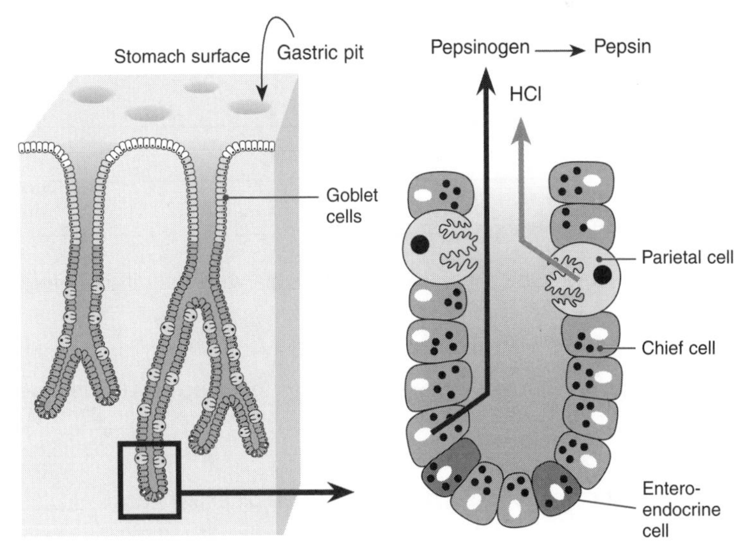


Gastric secretions are produced by **gastric glands**, which pit the lining of the stomach. Chief cells in the gland secrete pepsinogen, a precursor of the enzyme pepsin. Parietal cells produce hydrochloric acid, which activates the pepsinogen. Goblet cells at the neck of the gastric gland secrete mucus to protect the stomach mucosa from the acid. Enteroendocrine cells in the gastric gland secrete the hormone gastrin which acts on the stomach to increase gastric secretion.




ISBN: 978-1-927173-12-1

# The Large Intestine

After most of the nutrients have been absorbed in the small intestine, the remaining fluid contents pass into the large intestine (appendix, cecum, and colon). The fluid comprises undigested or undigestible food, bacteria, dead cells, mucus, bile, ions, and water. In humans and other omnivores, the large intestine's main role is to reabsorb water and electrolytes.

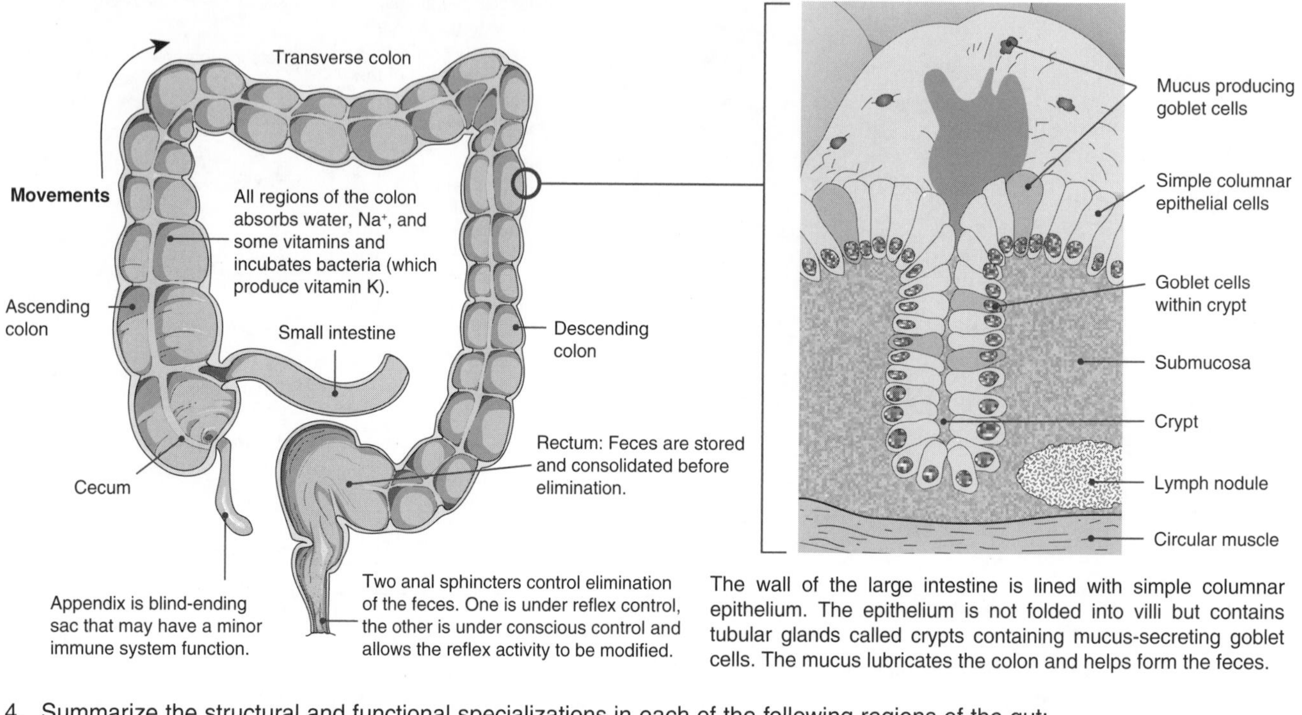


The wall of the large intestine is lined with simple columnar epithelium. The epithelium is not folded into villi but contains tubular glands called crypts containing mucus-secreting goblet cells. The mucus lubricates the colon and helps form the feces.

4. Summarize the structural and functional specializations in each of the following regions of the gut:

   (a) Stomach: ______________________

   (b) Small intestine: ______________________

   (c) Large intestine: ______________________

5. Identify two sites for enzyme secretion in the gut, give an example of an enzyme produced there, and state its role:

   (a) Site: ______________ Enzyme: ______________

   Enzyme's role: ______________________

   (b) Site: ______________ Enzyme: ______________

   Enzyme's role: ______________________

6. (a) Suggest why the pH of the gut secretions varies at different regions in the gut: ______________________

   (b) Explain why protein-digesting enzymes (e.g. pepsin) are secreted in an inactive form and then activated after release:

   ______________________

7. (a) Describe how food is moved through the digestive tract: ______________________

   (b) Explain how the passage of food through the tract is regulated: ______________________

8. (a) Predict the consequence of food moving too rapidly through the gut: ______________________

   (b) Predict the consequence of food moving too slowly through the gut: ______________________

ISBN: 978-1-927173-12-1

# Control of Digestion

The majority of digestive juices are secreted only when there is food in the gut and both nervous and hormonal mechanisms are involved in coordinating and regulating this activity appropriately. The digestive system is innervated by branches of the **autonomic nervous system**. Hormonal regulation is achieved through the activity of several hormones, which are released into the bloodstream in response to nervous or chemical stimuli and influence the activity of gut and associated organs.

**Feeding center**:
The feeding center in the hypothalamus is constantly active. It monitors metabolites in the blood and stimulates hunger when these metabolites reach low levels. After a meal, the neighboring satiety center suppresses the activity of the feeding center for a period of time.

**Salivation**:
Entirely under nervous control. Some saliva is secreted continuously. Food in the mouth stimulates the salivary glands to increase their secretions.

Parasympathetic stimulation of the stomach and pancreas via the vagus nerve increases their secretion. Sympathetic stimulation has the opposite effect. These are simple **reflexes** in response to the sight, smell, or taste of food.

**Pancreatic secretions and bile**:
Cholecystokinin (CCK) stimulates secretion of enzyme-rich fluid from the pancreas and release of bile from the gall bladder. Secretin stimulates the pancreas to increase its secretion of alkaline fluid and the production of bile from the liver cells.

Vagus nerve

Gastrin

**Gastric secretion**:
Physical distension and the presence of food in the stomach causes release of the hormone gastrin from cells in the gastric mucosa. Gastrin in the blood increases gastric secretion and motility.

CCK and secretin

**Intestinal secretion of hormones**:
The entry of **chyme** (especially fat and gastric acid) into the small intestine stimulates the intestinal mucosa to secrete the hormones cholecystokinin (CCK) and secretin.

**Summary of Hormones Acting in the Gut**

| Hormone | Organ | Effect |
|---|---|---|
| Secretin | Pancreas | Increases secretion of alkaline fluid |
| Secretin | Liver | Increases bile production |
| CCK | Pancreas | Increases enzyme secretion |
| CCK | Liver | Stimulates release of bile |
| Gastrin | Stomach | Increases stomach motility and secretion |

1. Describe the role of each of the following stimuli in the control of digestion, identifying both the response and its effect:

   (a) Presence of food in the mouth: __________

   (b) Presence of fat and acid in the small intestine: __________

   (c) Stretching of the stomach by the presence of food: __________

2. Outline the role of the vagus nerve in regulating digestive activity: __________

3. Discuss the role of nerves and hormones in controlling digestion: __________




**ISBN: 978-1-927173-12-1**


***Related activities:*** *The Human Digestive Tract*

# Adaptations for Absorption

The components of food (simple sugars, amino acids, and fatty acids) must be **absorbed** across the gut wall before they can be **assimilated** (taken up by the body's cells). The rate of nutrient transport across a membrane depends partly on the membrane's surface area. As you might expect, adaptations for increasing the surface area of the gut are widespread because they increase rates of nutrient absorption so that more food can be processed more quickly. Gut pouches and folds increase gut surface area. Nutrients are absorbed by diffusion or active transport and transported away in the blood or hemolymph.

### Insect Gastric Ceca

**Examples**: Grasshopper, crickets, cockroaches

In grasshoppers and their relatives, the **gastric ceca** are midgut pouches just behind the gizzard. The ceca improve absorption by transferring nutrients into the hemolymph ('blood'). The membrane lining the midgut is continually worn away and surrounds the feces when they are passed out. The membrane is replaced by the underlying epithelial cells (much like our skin).

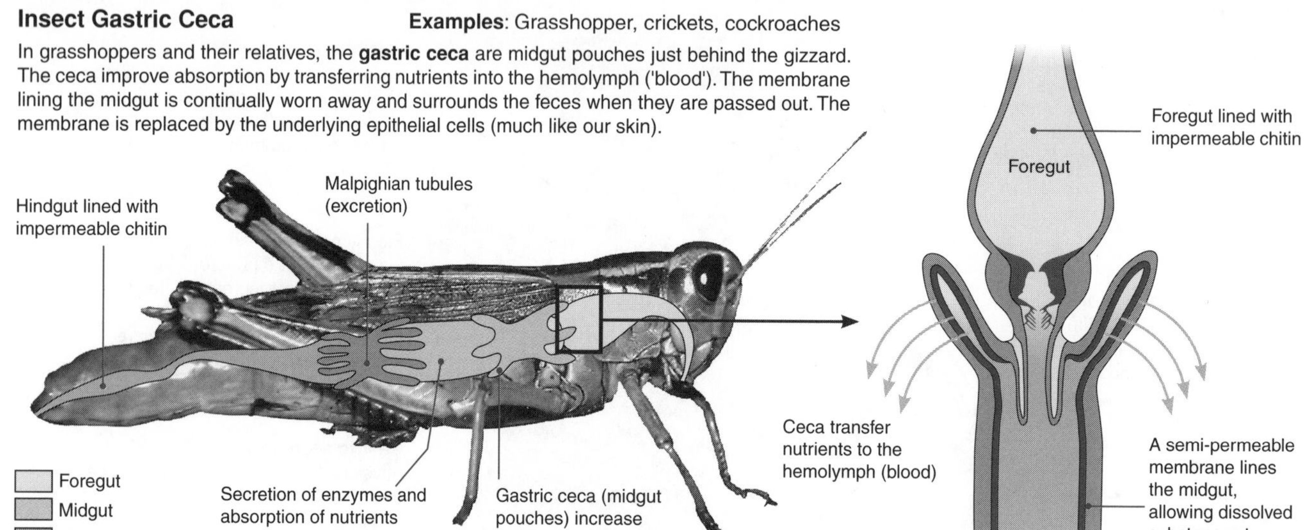


### Intestinal Spiral Valve

**Examples**: Sharks and primitive bony fish

The streamlined shape of sharks (and most other fish) makes it difficult to fit a long intestine into the body. Cartilaginous fish, such as sharks, as well as primitive bony fish, solve this problem by having a short intestine with a spiralling fold in the inner wall. This **spiral valve** increases the length of time digested material remains in the intestine by effectively increasing its length. The spiral valve also prevents large masses of indigestible material passing into the intestines.

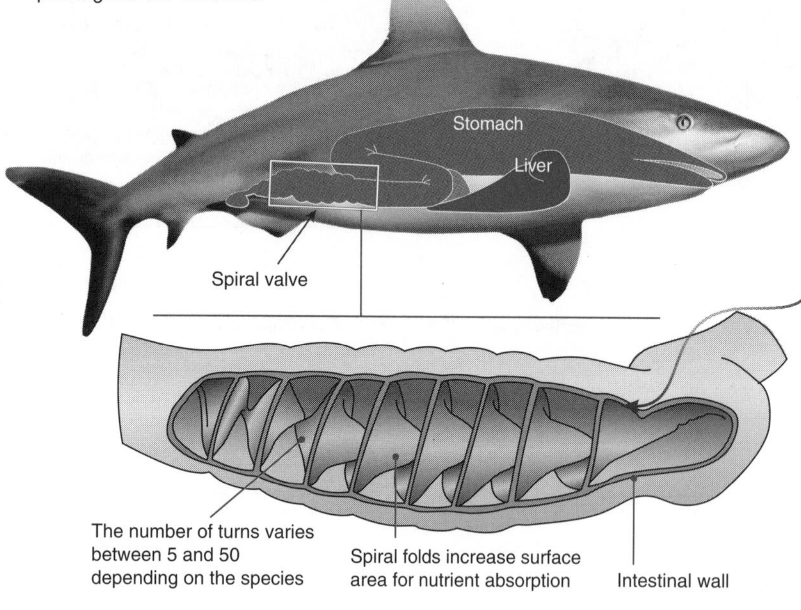


### Pyloric Ceca

**Example**: Most bony fish

In most bony fish, the elongated intestines are folded back on themselves in coils. The intestinal surface area is increased further by tube-like outgrowths called **pyloric ceca** at the junction of the stomach and intestine. Most bony fish have several hundred pyloric ceca and they are the main areas for digesting food and absorbing nutrients.

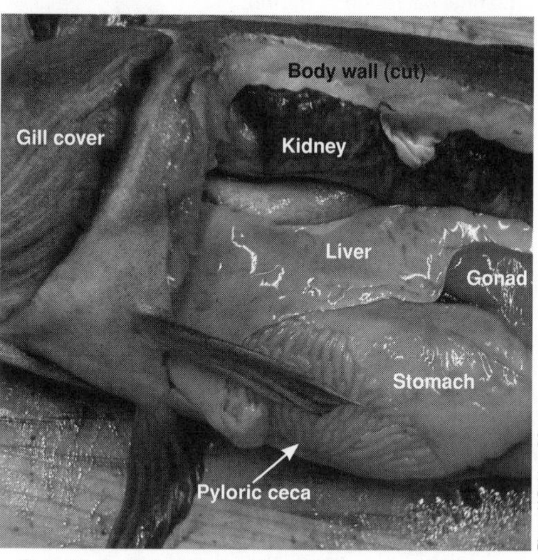


*Gut of bony fish showing the pyloric ceca, which open into the duodenum, the first part of the small intestine*

1. (a) Explain the adaptive value of increasing the surface area available for nutrient absorption: __________

(b) What factors do you think could influence the extent to which surface area is increased? __________



**Related activities:** *Diversity in Tube Guts*

**Periodicals:** *The anatomy of digestion*


**ISBN: 978-1-927173-12-1**

## Mammalian Intestine EXAMPLE: Pig

The gut of a mammal is divided into the stomach, and the intestines and associated glands. Mechanical and enzymic breakdown of food begins in the mouth, continues in the stomach, and is completed in the small intestine. The **pancreas** contributes an alkaline, enzyme-rich fluid to the intestine to break down fats, carbohydrates, and proteins. The **liver** contributes an alkaline fluid to **emulsify*** fats. The simple molecules that result from this digestion are absorbed across the intestinal wall, which is folded into microscopic finger-like structures called **villi**. These increase the surface area for absorption of nutrients into the blood.

*Emulsify means to make a suspension from two fluids that do not normally mix.

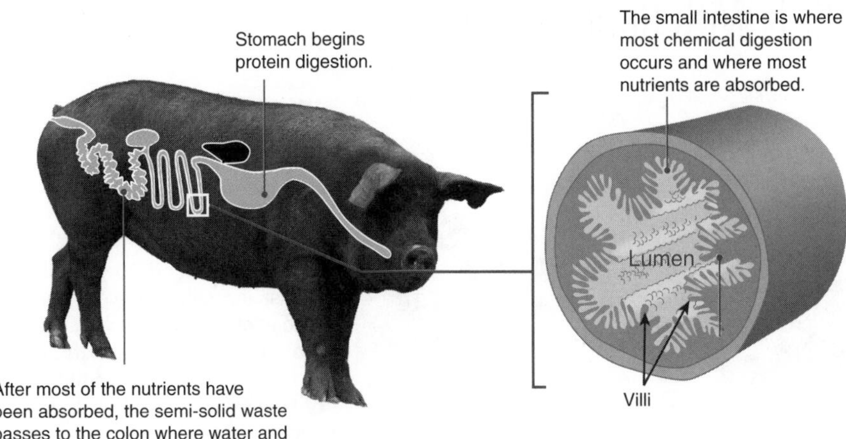


Cross section through a villus, showing how the products of digestion are absorbed across the intestinal epithelium into the capillaries or (for fats) into the lymphatic vessels. These vessels deliver these nutrients to the liver for processing.

### Nutrient Absorption by Intestinal Villi

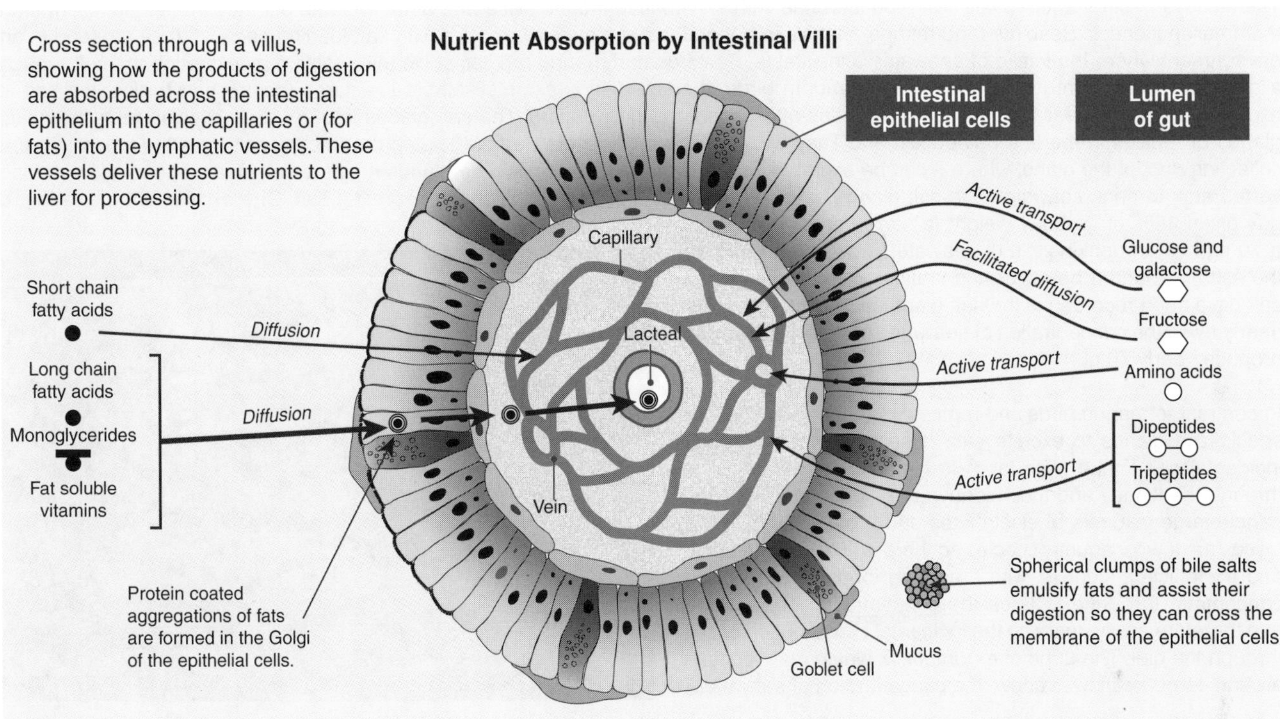


2. Compare and contrast the adaptations for increasing surface area for nutrient absorption in insects, fish, and mammals:

3. Describe how each of the following nutrients is absorbed by the intestinal villi in mammals:

(a) Glucose and galactose:

(b) Fructose:

(c) Amino acids:

(d) Dipeptides:

(e) Tripeptides:

(f) Short chain fatty acids:

(g) Monoglycerides:

(h) Fat soluble vitamins:


**ISBN: 978-1-927173-12-1**

# Birds With Runny Noses

"*Water, water, everywhere and not a drop to drink*".... The line from the 'Rime of the Ancient Mariner' clearly states the need to drink, but in an ocean, none can be found. However it may have been a lot easier for the albatross that accompanied the ship on its voyage. It, and other marine adapted animals can, literally, drink seawater.

There are around 35 g of dissolved salts (mostly sodium chloride with some potassium salts) in every liter of seawater. Human kidneys have an excretory efficiency such that it would take 1.5 L of water from the body to rid the body of the salt in 1 L of seawater. Consequently, drinking seawater causes humans to become more dehydrated than not drinking at all. In addition, in the period of time it takes for the kidneys to excrete that volume of water, the excess salt accumulates in the blood and creates potentially fatal ion imbalances.

The kidneys of birds and reptiles can produce urine with salt concentrations twice that of their blood, but they are less efficient than mammalian kidneys. Seabirds (and marine animals in general) have a special problem with salt loading because they eat food in an environment where the intake of seawater is inevitable. Seabirds and marine reptiles solve this problem by excreting the salt though a different mechanism; salt glands in the head. In birds, the salt glands excrete salt via the nares (nostrils); in turtles, the salt is excreted by glands near the eyes, so that turtles often appear to be weeping. The salt glands in birds likely evolved from the nasal glands of reptiles in the late Palaeozoic era. They work by active transport via Na-K pump that moves salt from the blood into the collecting duct of the gland, where it can be excreted as a concentrated solution. The glands regulate salt balance, and allow marine vertebrates to drink seawater. The salt glands have an extraordinarily high capacity for excreting salt. One experiment found that a gull given 10% of its body weight in seawater (equivalent to a 70 kg human drinking 7 L of seawater) was able to excrete the salt in around 3 hours. Concentrations of up to 64 $gL^{-1}$ of salt have been recorded in the salt gland secretions of petrels; nearly twice the concentration of seawater. Marine iguanas can produce nearly 70 $gL^{-1}$.

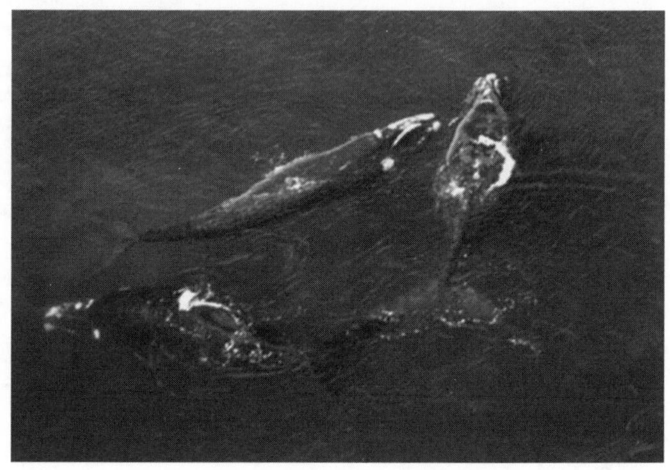

In contrast to marine birds and reptiles, marine mammals lack specialized glands to excrete salt, despite the fact that their incidental salt intake (with food) is high. All salt excretion is through the kidney and their kidneys are highly specialized to handle large volumes of electrolytes and protein. Evaporative losses are low for aquatic species so there is little need to drink, and the efficient kidneys, with their long loops of Henle, can concentrate the urine to twice the concentration of seawater and therefore easily deal with the incidental consumption of salt through the diet. The urine of a humpback whale may contain around 44 $gL^{-1}$ salt; well above the concentration of seawater.

1. Explain why humans can not drink seawater to quench their thirst: ______________________________

2. Describe and explain the role of the salt glands in marine birds and reptiles: ______________________________

3. Discuss the efficiency of the salt glands in birds and reptiles and the kidneys in marine mammals:



**Related activities:** *Nitrogenous Wastes in Animals*


**ISBN: 978-1-927173-12-1**

# Osmoregulation in Water

Animals living in aquatic environments have quite different osmoregulatory problems to those on land. Animals in freshwater tend to gain water osmotically and must prevent dilution of body fluids, so all are **osmoregulators**. Marine species have fewer problems and are either **osmoconformers**, or counter osmotic losses by drinking seawater and excreting the excess ions.

## Osmoregulators vs Osmoconformers

The body fluids of most marine invertebrates, e.g. sea anemones (left), fluctuate with the environment. These animals are **osmoconformers** and cannot regulate salt and water balance. Some intertidal species can tolerate frequent dilutions of normal seawater and some actively take up ions across the gills to compensate for ion loss via the urine.

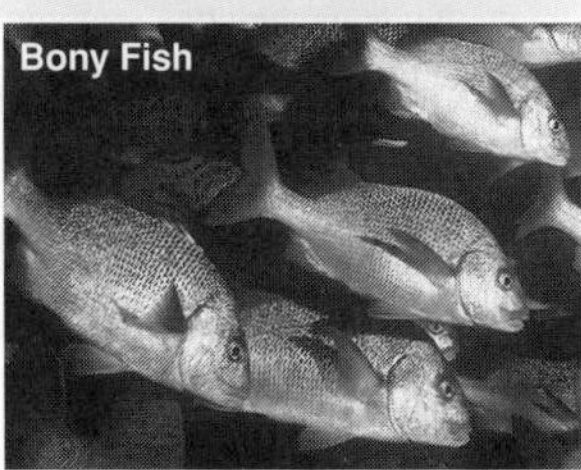

Bony Fish

Animals that regulate their salt and water fluxes, such as fish and marine mammals, are termed **osmoregulators**. Bony fish lose water osmotically and counter the loss by drinking salt water and then excreting the excess salt across the gill surfaces.

Ray

Marine sharks and rays generate osmotic concentrations in their body fluids similar to seawater by tolerating high urea. Excess salt from the diet is excreted via a salt gland in the rectum. Marine mammals produce a urine that is high in both salt and urea.

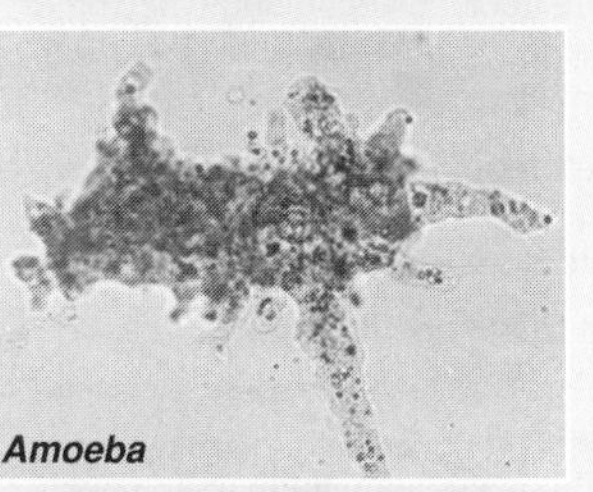

*Amoeba*

**Freshwater animals** have body fluids that are osmotically more concentrated than the water they live in and are all osmoregulators. Water tends to enter their tissues by osmosis and must be expelled to avoid flooding the body. Simple protozoans (*Amoeba*, left) use contractile vacuoles to collect the excess water and expel it.

Damselfly larva

Other invertebrates expel water and nitrogenous wastes using simple nephridial organs. Bony fish and aquatic arthropods produce dilute urine (containing ammonia) and actively take up salts across their gill surfaces.

## Response to Seawater Dilution in Rock Crabs

Species of intertidal crabs vary widely in their ability to regulate their salt and water levels in the face of environmental fluctuations. As shown in the graph below, intertidal crabs face an osmotic influx of water when placed into dilute salinities. The excess water can be excreted in the urine (via the antennal glands), but this is accompanied by a loss of valuable ions. In many species, after a period of adjustment, this loss is met by active ion uptake across the gills.

In an experiment, a student investigated the effect of increasing seawater dilution on the cumulative weight gain of a common rock crab. Six crabs in total were used in the experiment. Three were placed in seawater dilution of 75:25 (75% seawater) and three were placed in a seawater dilution of 50:50 (50% seawater). Cumulative weight gain in each of the six crabs was measured at regular intervals over a period of 30 minutes. The results are plotted below and a line of best fit has been drawn for each set of data.

*Weight gain in crabs at two seawater dilutions*

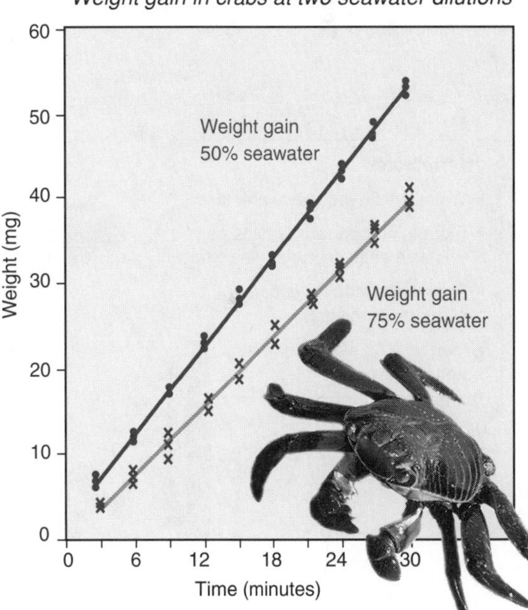


Obtaining Nutrients & Eliminating Wastes

1. Describe one way in which freshwater animals can compensate for ions lost during excretion of excess water:

2. Explain the difference between an **osmoregulator** and an **osmoconformer**:

3. (a) Explain the difference in the two lines plotted on the graph of crab weight gain (above right):

   (b) Explain what this experiment suggests about the osmoregulatory ability of this crab species:


**ISBN: 978-1-927173-12-1**

# Managing Fluid Balance on Land

Many aspects of metabolism, including enzyme activity, membrane transport, and nerve conduction, are dependent on particular concentrations of ions and metabolites. To achieve this ion balance, the salt and fluid content of the internal body fluids must be regulated; a process called **osmoregulation**. **Terrestrial** animals show specific adaptations for obtaining and conserving water in an environment where (to varying degrees) water loss is a constant problem. These adaptations are most well developed in desert animals. Despite the lack of available water in deserts, a large a number of animal taxa have desert-dwelling representatives. The most arid tolerant are even capable of surviving without drinking.

## Obtaining Water

Most animals obtain the majority of their water by **drinking.** Some, such as camels, can retain relatively large volumes of water in the gut, but most will need to regularly visit a water supply.

Obtaining water from **food** is important in dry environments where free standing water is limited. Many large predators obtain a large amount of their water in this way.

US Fish and Wildlife

Some desert animals, such as the kangaroo rat, do not need to drink. 90% of their water comes from metabolism (oxidation of food). The rest comes from the small amount of water present in the food.

Ian Smith

Amphibians can take up water directly through their skin. The skin is water permeable, so they can osmotically acquire water when they need it when submerged or resting on a damp surface.

## Conserving and Losing Water

*Water losses through airways*

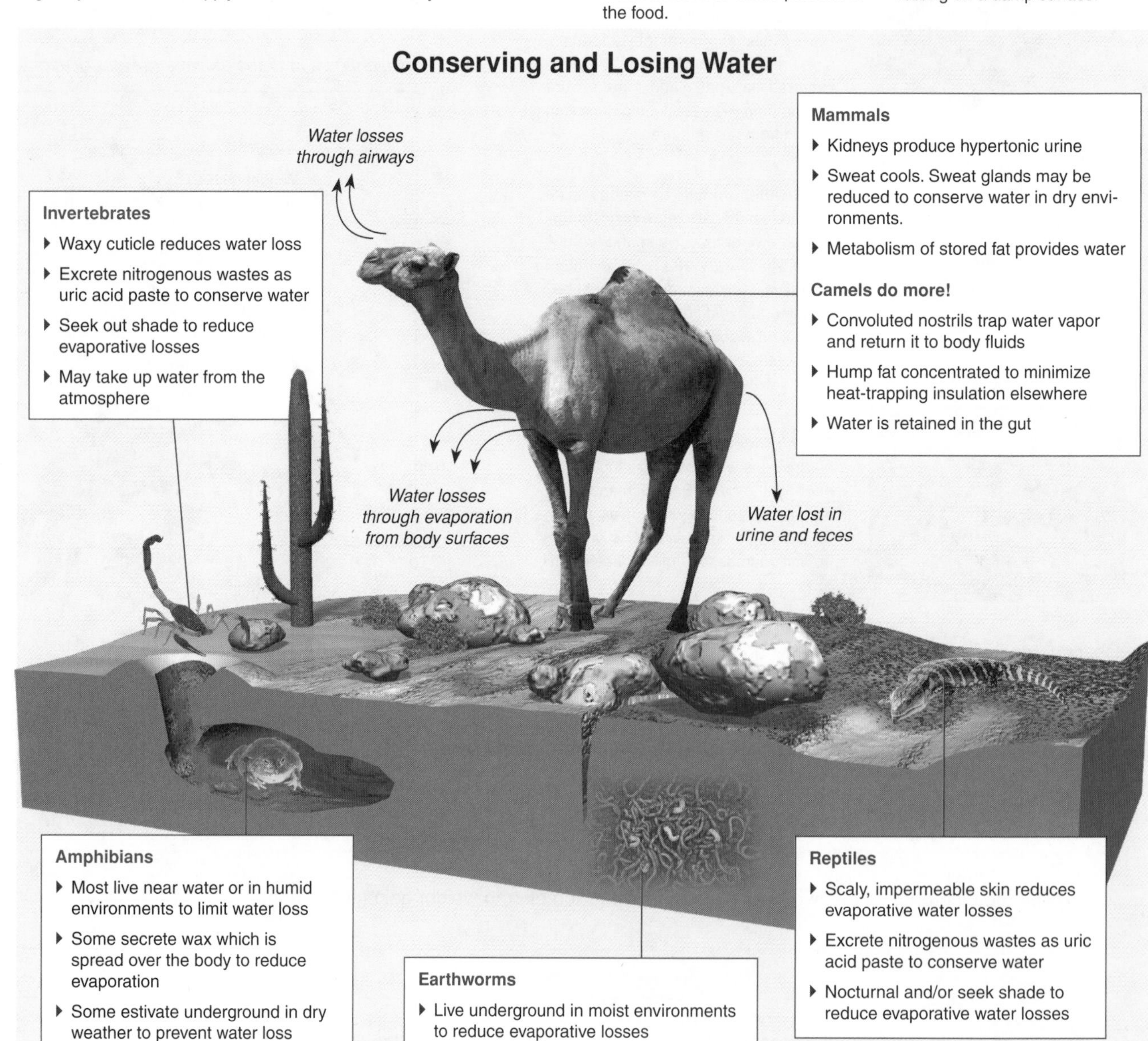

**Invertebrates**

- Waxy cuticle reduces water loss
- Excrete nitrogenous wastes as uric acid paste to conserve water
- Seek out shade to reduce evaporative losses
- May take up water from the atmosphere

**Mammals**

- Kidneys produce hypertonic urine
- Sweat cools. Sweat glands may be reduced to conserve water in dry environments.
- Metabolism of stored fat provides water

**Camels do more!**

- Convoluted nostrils trap water vapor and return it to body fluids
- Hump fat concentrated to minimize heat-trapping insulation elsewhere
- Water is retained in the gut

*Water losses through evaporation from body surfaces*

*Water lost in urine and feces*

**Amphibians**

- Most live near water or in humid environments to limit water loss
- Some secrete wax which is spread over the body to reduce evaporation
- Some estivate underground in dry weather to prevent water loss

**Earthworms**

- Live underground in moist environments to reduce evaporative losses

**Reptiles**

- Scaly, impermeable skin reduces evaporative water losses
- Excrete nitrogenous wastes as uric acid paste to conserve water
- Nocturnal and/or seek shade to reduce evaporative water losses



***Related activities**: Water Balances in Desert Mammals, Water Budget in Humans*


**ISBN: 978-1-927173-12-1**

# The Loop of Henle and Water Conservation

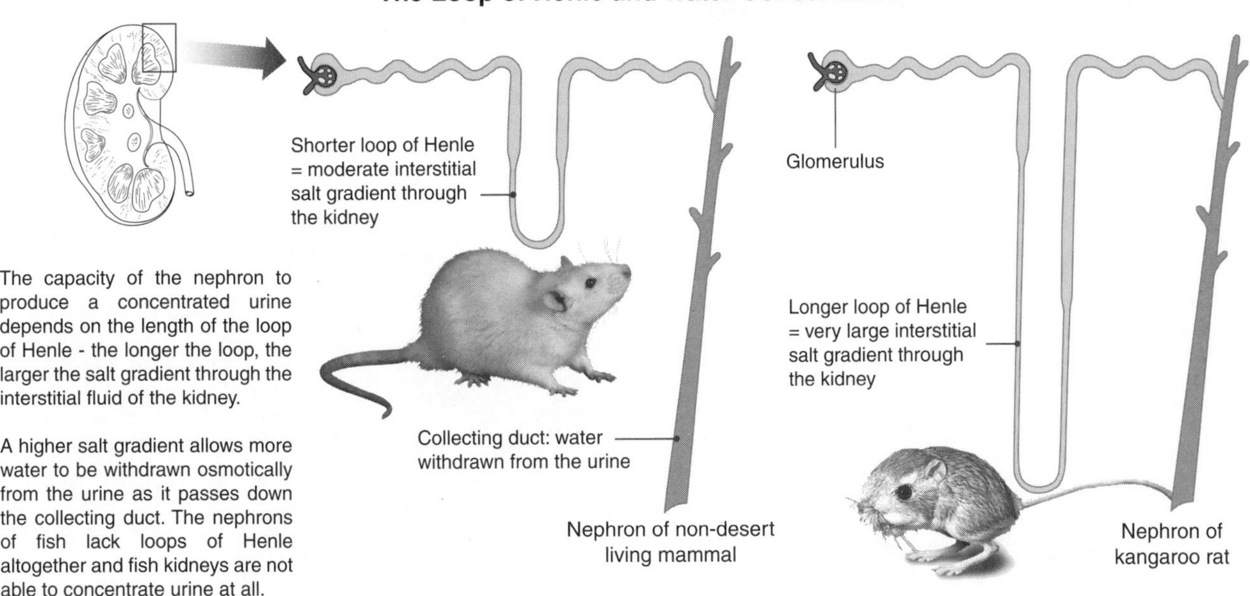


The capacity of the nephron to produce a concentrated urine depends on the length of the loop of Henle - the longer the loop, the larger the salt gradient through the interstitial fluid of the kidney.

A higher salt gradient allows more water to be withdrawn osmotically from the urine as it passes down the collecting duct. The nephrons of fish lack loops of Henle altogether and fish kidneys are not able to concentrate urine at all.

1. Describe four ways in which water can be obtained: ______

2. Describe three ways in which each of the following animals conserves water:

   (a) Mammal: ______

   (b) Reptile: ______

   (c) Arthropod: ______

   (d) Amphibian: ______

   (e) Annelid: ______

3. Animals use structural, physiological, and behavioral adaptations to reduce their water losses. Explain the difference between structural, physiological, and behavioral adaptations using examples to illustrate your answer:

4. Describe three ways in which animals lose water to the environment: ______

5. (a) Explain why only mammals and birds are able to produce urine that is more concentrated than their bodily fluids:

   (b) Explain how kangaroo rats have been able to develop this ability to the extent that they do not need to drink:




ISBN: 978-1-927173-12-1

# Water Balances in Desert Mammals

Water loss is a major problem for most mammals. The behavioral, physiological, and structural adaptations of mammals adapted to desert or arid regions enables them to minimize their water losses and reduce the amount of water they need to drink. Arid-adapted species typically produce very concentrated urine (their kidney nephrons have long loops of Henle), water losses through this route are minimal. In addition, the metabolic breakdown of food contributes a large proportion of daily water needs. Some, like kangaroo rats (below, left) do not need to drink at all, and obtain most of their water this way.

## Adaptations of Arid Adapted Rodents

Most desert-dwelling mammals are adapted to tolerate a low water intake. Arid adapted rodents, such as jerboas and kangaroo rats, conserve water by reducing losses to the environment and obtain the balance of their water needs from the oxidation of dry foods (respiratory metabolism). The table below shows the water balance in a kangaroo rat after eating 100 g of dry pearl barley. Note the high urine to plasma concentration ratio (17) which is more than four times that of a human (4).

### Water balance in a kangaroo rat
(*Dipodomys spectabilis*)

| Water gains | | Water losses | |
|---|---|---|---|
| Absorbed from food | 6.0 ml | Breathing | 43.9 ml |
| From metabolism | 54.0 ml | Urination | 13.5 ml |
| | | Defecation | 2.6 ml |

Urine/plasma concentration ratio = 17

### Adaptations of kangaroo rats

Kangaroo rats, and other arid-adapted rodents, tolerate long periods without drinking, meeting their water requirements from the metabolism of dry foods. They dispose of nitrogenous wastes with very little output of water and they neither sweat nor pant to keep cool.

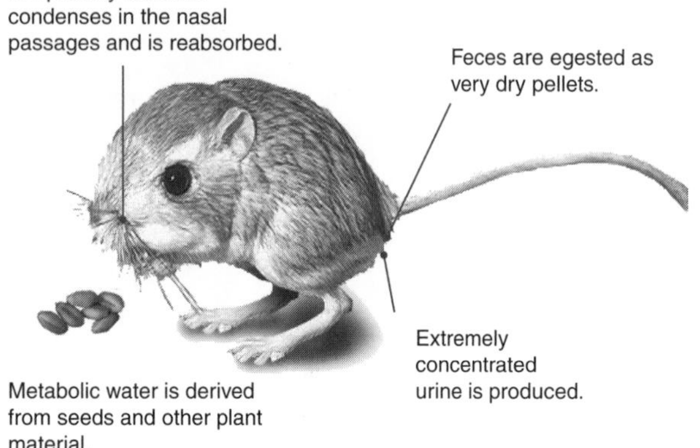


## Large Desert Mammals

Camels are well adapted to desert life. Their nasal passages are highly convoluted to trap moisture when exhaling, and sweating only begins when the body temperature exceeds 41°C. The urine is highly concentrated and syrupy and the feces are dry enough for humans to burn as fuel. Metabolism of the fat in the hump also produces water. Camels can also withstand large losses of water without circulatory disturbance because their oval-shaped blood cells can flow even when they are dehydrated.

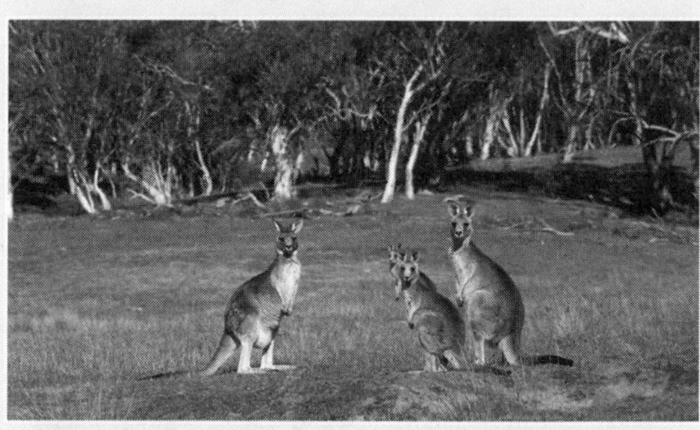

A suite of adaptations enable kangaroos to exploit the vast arid environment of Australia's semi-deserts. Water is taken in with food but is also derived from metabolism. An elongated large intestine maximizes the removal of water from undigested material, producing very dry feces. The urine is also concentrated, reducing water loss. The hopping mode of locomotion is also very energy efficient, and enables them to cover large expanses of country in search of food and water.

1. Explain why most mammals need to drink regularly: ______________________

______________________

2. Use the tabulated data for the kangaroo rat (above) to graph the water gains and losses in the spaces provided below.

3. Describe three physiological adaptations of desert adapted rodents to low water availability:

   (a) ______________________

   (b) ______________________

   (c) ______________________

4. If kangaroo rats neither pant nor sweat, suggest how they might keep cool during the heat of the day

______________________

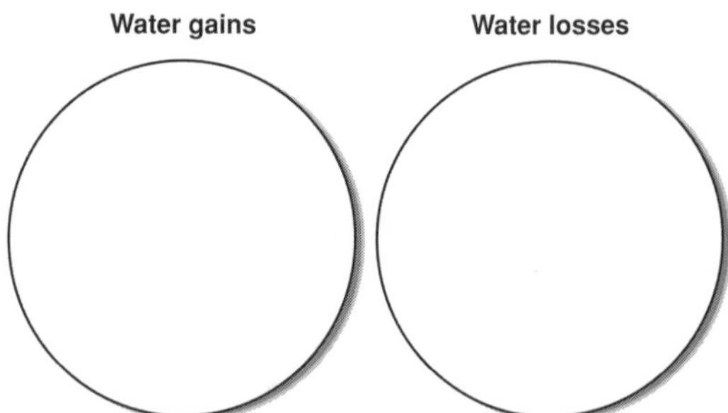




**Related activities:** : *Fluid Regulation on Land, Osmoregulation in Water*
**Weblinks:** *Water Conservation in Kangaroo Rats*

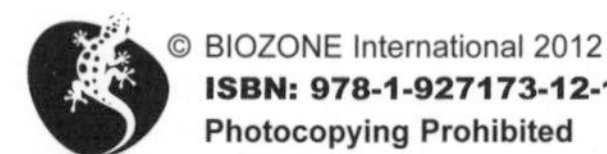

ISBN: 978-1-927173-12-1

# Water Budget in Humans

We cannot live without water for more than about 100 hours and adequate water is a requirement for physiological function and health. Body water content varies between individuals and through life, from above about 90% of total weight as a fetus to 74% as an infant, 60% as a child, and around 50-59% in adults, depending on gender and age. Gender differences (males usually have a higher water content than females) are the result of differing fat levels. Water intake and output are highly variable but closely matched to less than 0.1% over an extended period. Typical values for water gains and losses, as well as daily water transfers are given below. Men need more water than women due to their higher (on average) fat-free mass and energy expenditure. Infants and young children need more water in proportion to their body weight as they cannot concentrate their urine as efficiently as adults. They also have a greater surface area relative to weight, so water losses from the skin are greater.

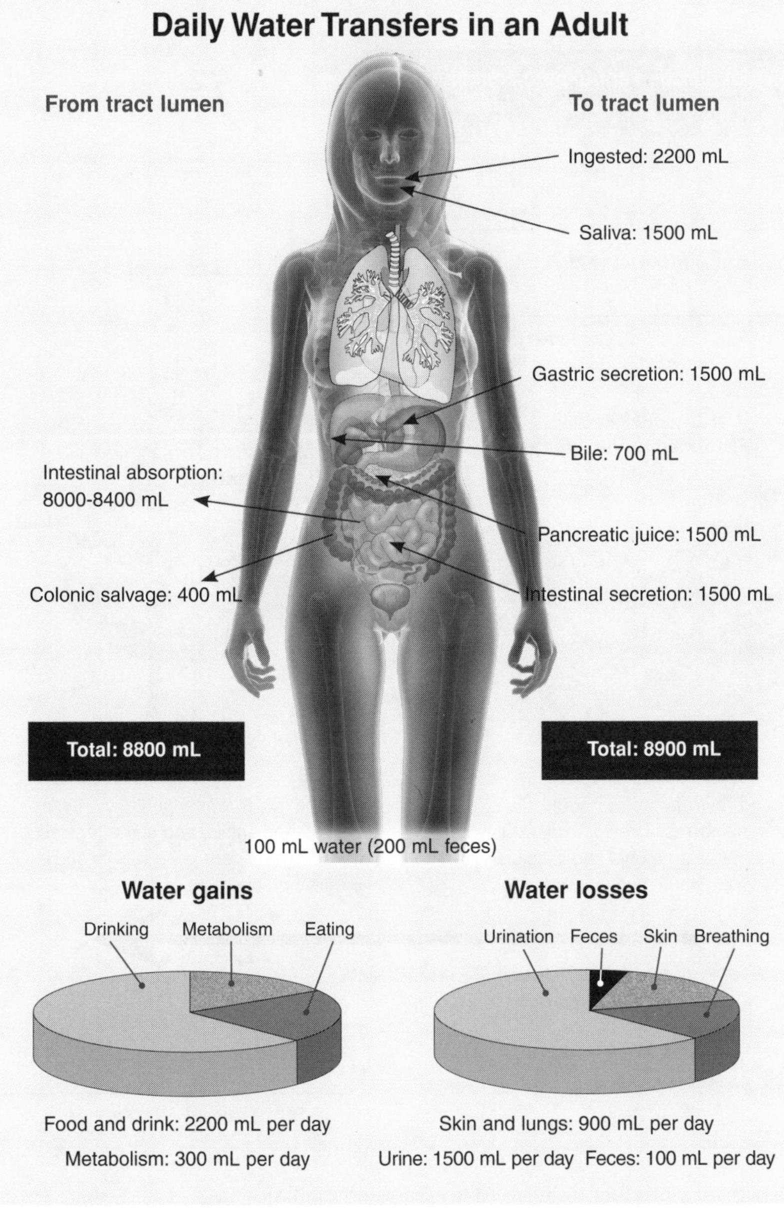

About 63% of our daily requirement for water is met through drinking fluids, 25% is obtained from food, and the remaining 12% comes from metabolism (the oxidation of glucose to ATP, $CO_2$, and water).

Typically, we lose 60% of body water through urination, 36% through the skin and lungs, and 4% in feces. Losses through the skin and from the lungs (breathing) average about 900 mL per day or more during heavy exercise. These are called **insensible losses**.

1. Explain how metabolism provides water for the body's activities: ____________

2. Describe four common causes of physiological dehydration:

   (a) ____________ (c) ____________

   (b) ____________ (d) ____________

3. Some recent sports events have received media coverage because athletes have collapsed after excessive water intakes. This condition, called **hyponatremia** or water intoxication, causes nausea, confusion, diminished reflex activity, stupor, and eventually coma. From what you know of fluid and electrolyte balances in the body, explain these symptoms:

TG




ISBN: 978-1-927173-12-1

# Nitrogenous Wastes in Animals

Waste materials are generated by the metabolic activity of cells. If allowed to accumulate, they would reach toxic concentrations and so must be continually removed. Excretion is the process of removing waste products and other toxins from the body. Waste products include carbon dioxide and water, and the nitrogenous (nitrogen containing) wastes that result from the breakdown of amino acids and nucleic acids. The simplest breakdown product of nitrogen containing compounds is ammonia, a small molecule that cannot be retained for long in the body because of its high toxicity. Most aquatic animals excrete ammonia immediately into the water where it is washed away. Other animals convert the ammonia to a less toxic form that can remain in the body for a short time before being excreted via special excretory organs. The form of the excretory product in terrestrial animals (urea or uric acid) depends on the type of organism and its life history. Terrestrial animals that lay eggs produce uric acid rather than urea, because it is non-toxic and very insoluble. It remains as an inert solid mass in the egg until hatching.

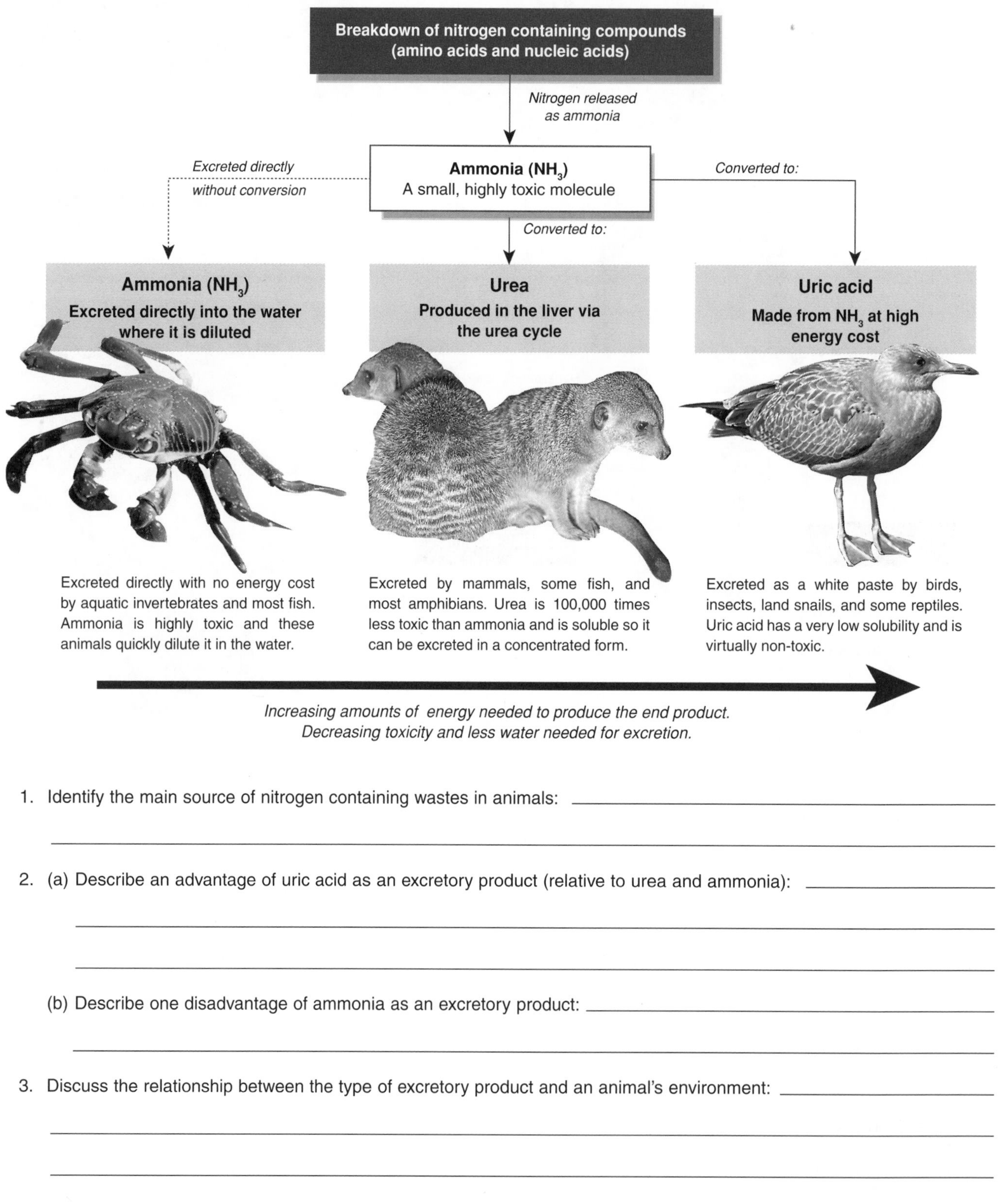


1. Identify the main source of nitrogen containing wastes in animals: ______________________

2. (a) Describe an advantage of uric acid as an excretory product (relative to urea and ammonia): ______________________

   (b) Describe one disadvantage of ammonia as an excretory product: ______________________

3. Discuss the relationship between the type of excretory product and an animal's environment: ______________________



**Related activities:** *The Role of the Liver*


**ISBN: 978-1-927173-12-1**

# Invertebrate Excretory Systems

Metabolism produces toxic by-products. The most troublesome of these to eliminate from the body is nitrogenous waste from the metabolism of proteins and nucleic acids. The simplest and most common type of excretory organs, widely distributed in invertebrates, are simple tubes (**protonephridia** and **nephridia**) opening to the outside through a pore. The **malpighian tubules** of insects are highly efficient, removing nitrogenous wastes from the blood, and also functioning in **osmoregulation**. Note that all three forms of nitrogenous waste are represented here: ammonia (flatworms, annelids), urea (annelids), and uric acid (insects).

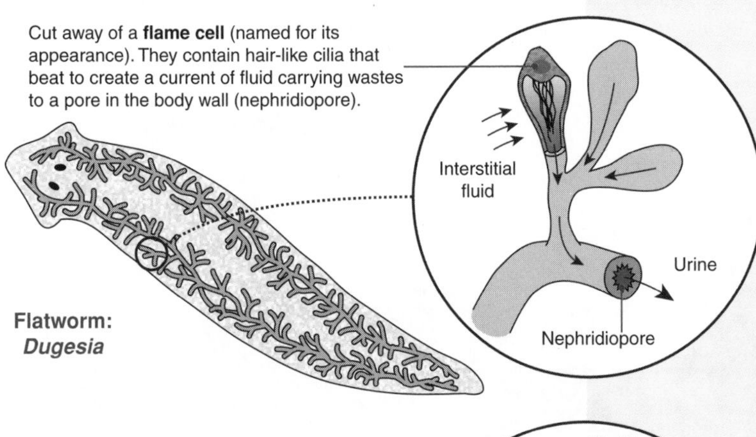


## Platyhelminthes (flatworms)

Excretory system: **protonephridia**

**Protonephridia** are very simple excretory structures. Each protonephridium comprises a branched tubule ending in a number of blind capillaries called **flame cells**. **Ammonia** is excreted directly into the moist environment. Flatworms do not have a circulatory system or fluid-filled inner body spaces. They use their branching network of flame cells to regulate the composition of the fluid bathing the cells (interstitial fluid). Interstitial fluid enters the flame cell and is propelled along the tubule, away from the blind end, by beating cilia. Tubules merge into ducts that expel the urine through **nephridiopores**.

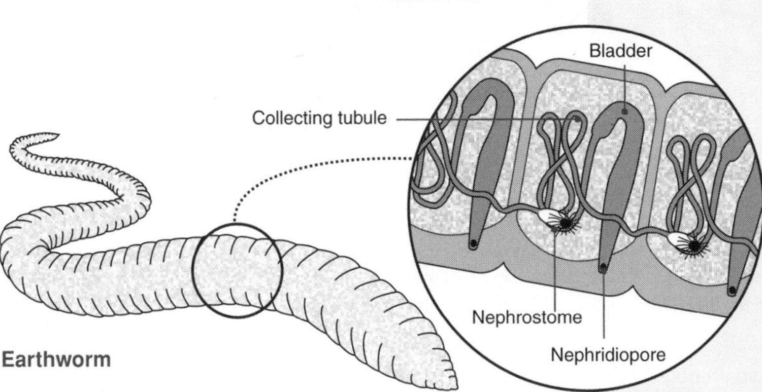


## Annelids (segmented worms)

Excretory system: **nephridia**

In earthworms, each segment has a pair of excretory organs called **nephridia**, which drain the next segment in front. Fluid enters the nephrostome and passes through the collecting tubule. These tubules are surrounded by a capillary network of blood vessels (not shown here) which recover valuable salts from the developing urine. The collecting tubule empties into a storage bladder which expels the dilute urine (a mix of **ammonia** and **urea**) to the outside through the nephridiopore.

## Insects

Excretory system: **malpighian tubules**

Insects have two to several hundred **malpighian tubules** projecting from the junction of the midgut and hindgut. They bathe in the clear fluid (hemolymph) of the insect's body cavity where they actively pump $K^+$ and $Na^+$ into the tubule. Water, uric acid salts, and several other substances follow by passive transport. Water and some ions are reabsorbed in the hindgut, while **uric acid** precipitates out as a paste and is passed out of the anus along with the fecal material. The ability to conserve water by excreting solid uric acid has enabled insects to colonize very arid environments.

1. For each of the following, name the organs for excreting nitrogenous waste and state the form of the waste product:

   (a) Flatworm: ______________________ Waste: ______________

   (b) Insect: ______________________ Waste: ______________

   (c) Earthworm: ______________________ Waste: ______________

2. Explain briefly how insects concentrate their nitrogenous waste into a paste: ______________________

3. For one of the above animals, relate the form of the excretory product to the environment in which the animal lives:

**ISBN: 978-1-927173-12-1**


***Related activities:*** *Nitrogenous Wastes in Animals*
***Weblinks:*** *Ultrafiltration, Transport & Resorption in a Protonephridium*

# Vertebrate Excretory Systems

In vertebrates, the excretory units (**nephrons**) are collected into discrete organs called **kidneys**. The kidneys of most vertebrates are similar in that they produce urine (excretory fluid or paste) by **filtering** the body fluids and then modifying the filtrate by **reabsorption** and **secretion** of ions. The kidneys of a few bony fish differ from this general pattern by lacking a filtration mechanism and producing urine solely through the secretion of ions. Whilst all vertebrates have kidneys and can produce urine, only birds and mammals can produce a urine that is more concentrated than the body fluids. The kidneys of bony fish have few nephrons and can only create a urine that is the same concentration, or less concentrated, than the blood. Nearly all their nitrogenous waste is excreted by the gills, which also have an important role in salt balance. Mammals and birds have highly efficient kidneys that can produce a concentrated urine, excreting nitrogenous wastes whilst conserving water and valuable ions.

## Freshwater Bony Fish

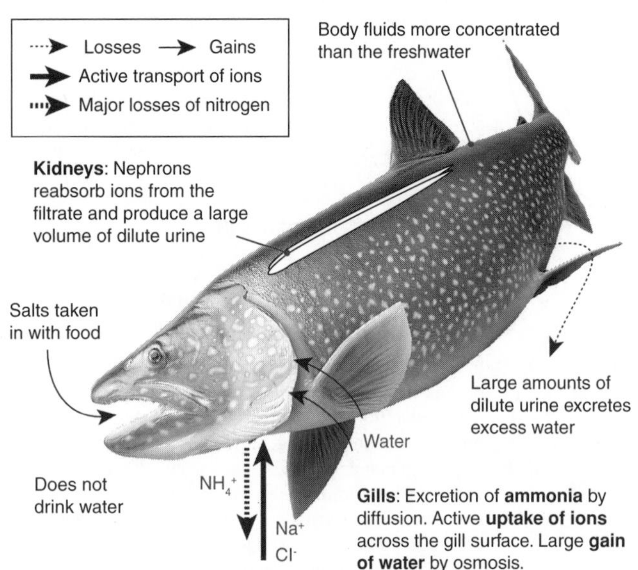


Fish in freshwater must excrete **excess water** gained through osmosis and they must excrete **nitrogenous waste**. Their kidneys excrete large amounts of dilute urine; valuable ions are lost because of the large urine volumes produced. The kidneys reabsorb salts from the filtrate through active transport mechanisms, and the gills take up ions from the water.

## Marine Bony Fish

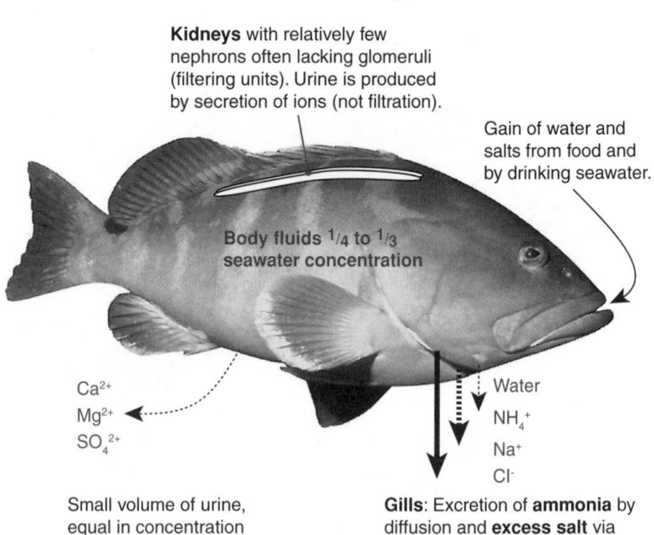


Fish in seawater must excrete **excess salt** gained through diet as well as **nitrogenous waste**. The urine is isotonic, and excess salts are actively excreted across the gill surface into the water (against a concentration gradient). Note that, unlike bony fish, sharks and rays tolerate high urea levels in their tissues and excrete excess salts via a salt gland in the rectum.

## Terrestrial Mammal

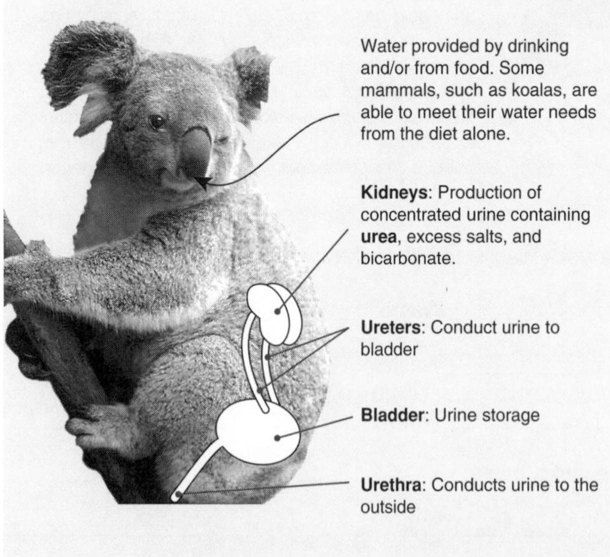


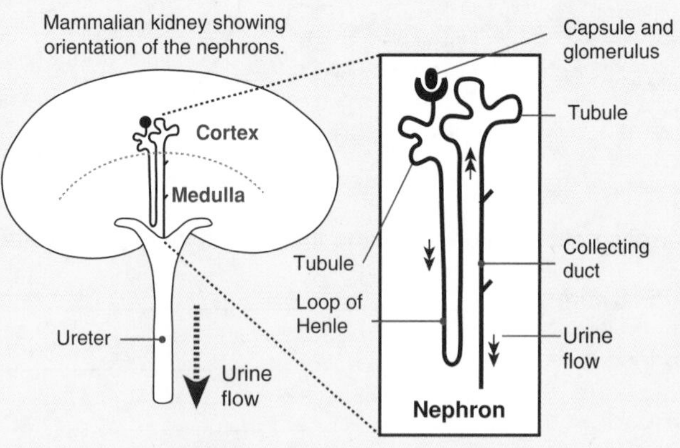


Water loss is a major problem for most mammals. The degree to which urine can be concentrated (and water conserved) depends on the number of nephrons present and the length of the loop of Henle. The highest urine concentrations are found in desert-adapted mammals.

Mammalian **kidneys** each contain more than one million **nephrons**: the selective filter elements which regulate the composition of the blood and excrete wastes. In the nephron, the initial urine is formed by **filtration** in the glomerulus and Bowman's capsule. The filtrate is modified by **secretion** and **reabsorption** of ions and water. These processes create a salt gradient in the fluid around the nephron, which allows water to be withdrawn from the urine in the collecting duct.



***Related activities:*** *Fluid Regulation on Land, Osmoregulation in Water*


**ISBN: 978-1-927173-12-1**

1. Name the primary organ for nitrogen excretion in fish: ______

2. (a) Discuss the problems of excretion and osmoregulation for bony fish in fresh and salt water environments:

(b) Explain how fish in these two environments overcome these difficulties: ______

3. (a) Name the functional excretory unit of the mammalian kidney: ______

(b) Explain how mammals are able to produce a concentrated urine: ______

4. Describe the two factors that determine the degree to which mammalian urine can be concentrated and explain why they are important:

(a) ______

(b) ______

5. Discuss the functional and structural differences between mammalian kidneys and the kidneys of fish:

6. The graph below shows the volume of urine collected from a subject after drinking 1000 mL (1 liter) of distilled water. The subject's urine was collected at 25 minute intervals over a number of hours.

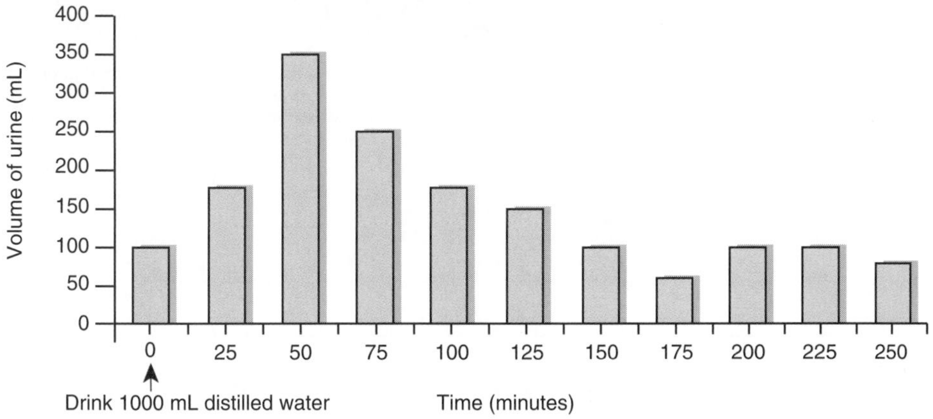

Volume of urine (mL)
400
350
300
250
200
150
100
50
0
0 25 50 75 100 125 150 175 200 225 250
Drink 1000 mL distilled water
Time (minutes)

(a) Describe the changes in urine output during the experiment: ______

(b) Explain the difference in the volume of urine collected at 25 minutes and 50 minutes: ______

**ISBN: 978-1-927173-12-1**

# The Urinary System

The mammalian urinary system consists of the kidneys and bladder, and their associated blood vessels and ducts. The kidneys have a plentiful blood supply from the renal artery. The blood plasma is filtered by the kidneys to form urine. Urine is produced continuously, passing along the ureters to the bladder, a hollow muscular organ lined with smooth muscle and stretchable epithelium. Each day the kidneys filter about 180 liters of plasma. Most of this is reabsorbed, leaving a daily urine output of about 1 liter. By adjusting the composition of the fluid excreted, the kidneys help to maintain the body's internal chemical balance. All vertebrates have kidneys, but their efficiency in producing a concentrated urine varies considerably. Mammalian kidneys are very efficient, producing a urine that is concentrated to varying degrees depending on requirements.

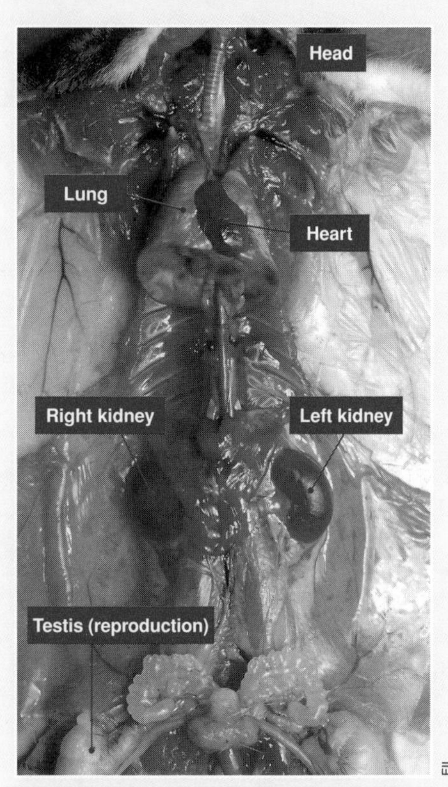


The kidneys of **rats** (above), humans, and other mammals are distinctive, bean shaped organs that lie at the back of the abdominal cavity to either side of the spine. The kidneys lie outside the peritoneum of the abdominal cavity and are partly protected by the lower ribs. Each kidney is surrounded by three layers of tissue. The innermost renal capsule is a smooth fibrous membrane that acts as a barrier against trauma and infection. The two outer layers comprise fatty tissue and fibrous connective tissue. These act to protect the kidney and anchor it firmly in place.

**The Human Urinary System**

**Vena cava** returns blood to the heart.

**Dorsal aorta** supplies oxygenated blood to the body.

**Adrenal glands** are associated with, but not part of, the urinary system.

**Renal vein** returns the blood from the kidney to the venous circulation.

**Renal artery** carries blood from the aorta into the kidney.

**Kidney** produces urine (blood filtration, the removal of waste products, and the regulation of blood volume).

**Ureter** carries urine to the bladder.

**Bladder** (sectioned) stores the urine before it passes out of the body. It can expand to hold about 80% of the daily urine output.

**Urethra** conducts urine from the bladder to the outside. The urethra is regulated by a voluntary sphincter muscle.

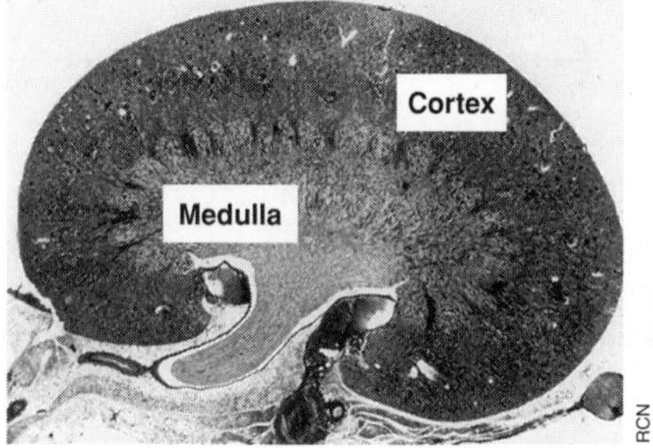


The very precise alignment of the nephrons (the filtering elements of the kidney) and their associated blood vessels gives the kidney tissue a striated appearance, as seen in this cross section.

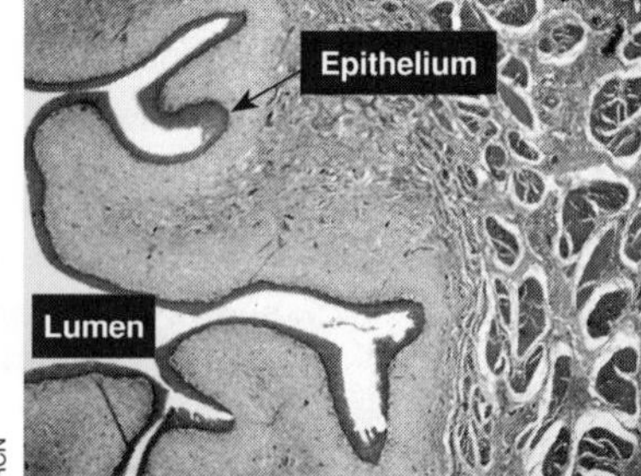


**Transitional epithelium** lines the bladder. This type of epithelium is layered, or stratified, and can be stretched without the outer cells breaking apart from each other.

1. Identify the components of the urinary system and describe their functions: ______________________

______________________

______________________

______________________

2. Calculate the percentage of the plasma reabsorbed by the kidneys: ______________________

3. The kidney receives blood at a higher pressure than other organs. Explain why this is the case: ______________________

______________________

______________________

4. Suggest why the kidneys are surrounded by fatty connective tissue: ______________________

______________________



***Related activities:*** *The Physiology of the Kidney*


**ISBN: 978-1-927173-12-1**

# The Physiology of the Kidney

The functional unit of the kidney, the **nephron**, is a selective filter element, comprising a renal corpuscle and its associated tubules and ducts. Filtration, i.e. forcing fluid and dissolved substances through a membrane by pressure, occurs in the first part of the nephron, across the membranes of the capillaries and the glomerular capsule. The passage of water and solutes into the nephron and the formation of the glomerular filtrate depends on the pressure of the blood entering the afferent arteriole (below). If it increases, filtration rate increases; when it falls, glomerular filtration rate also falls. This process is so precisely regulated that, in spite of fluctuations in arteriolar pressure, glomerular filtration rate per day stays constant. After formation of the initial filtrate, the **urine** is modified through secretion and tubular reabsorption according to physiological needs at the time.

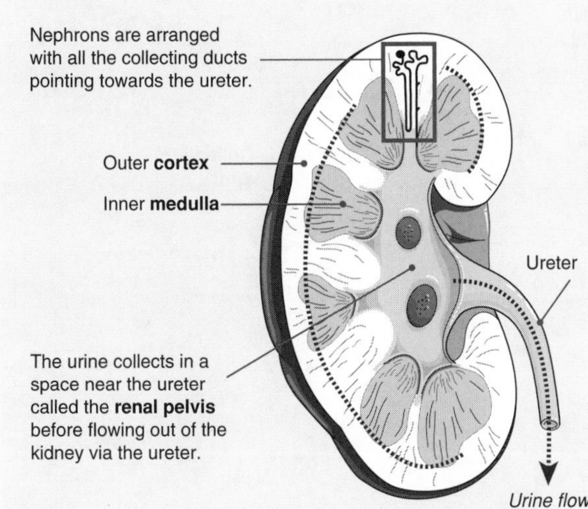


## Internal Structure of the Human Kidney

Human kidneys are about 100-120 mm long and 25 mm thick. The functional unit of the kidney is the **nephron**. The other parts of the urinary system are primarily passageways and storage areas. The inner tissue of the kidney appears striated (striped), because of the orientation and alignment of the nephrons and blood vessels in the kidney tissue. It is this precise arrangement that makes it possible to fit in all the filtering units required.

Each kidney contains more than 1 million nephrons. By selectively filtering the blood plasma, the nephrons regulate blood composition and pH, and excrete wastes (e.g. urea) and toxins. The initial fluid is formed by **filtration** through the specialized epithelium of the glomerulus. This filtrate is then modified as it passes through the tubules of the nephron. The resulting **urine**, passes out the ureter.

## Nephron Structure and Function

About 90% of the filtrate is **reabsorbed** in the **proximal convoluted tubule**. Glucose, $Na^+$, $K^+$, $Mg^{2+}$, and $Ca^{2+}$ are reabsorbed by active transport. Chloride ions and water follow by passive transport.

Efferent arteriole (leaves glomerulus)

Fluid is forced through the capillaries of the **glomerulus**, forming a **filtrate** similar to blood. The filtrate contains water, salts, glucose, **urea**, and small proteins but lacks cells and larger proteins.

Renal corpuscle = glomerulus + Bowman's capsule

Bowman's capsule

Afferent arteriole (enters glomerulus)

Venule

In the **distal convoluted tubule** $Na^+$, $K^+$, and $Ca^{2+}$ are **reabsorbed** by active transport. Water follows by osmosis. $H^+$ and $K^+$ are **secreted** by active transport.

In the **collecting duct tubule**, the urine is concentrated as water leaves the tubule by osmosis. The collecting duct drains to the renal pelvis.

Increasing extracellular salt gradient in the kidney

In the ascending limb of the loop of Henle, salt is actively transported from the filtrate.

Blood vessels surround the loop of Henle

The loop of Henle has a variable permeability to salt and water. The transport of salts and the passive movement of water establish a salt gradient through the kidney. This salt gradient is used for the osmotic withdrawal of water from the collecting duct.

In the descending limb of the loop, water leaves the filtrate by osmosis.

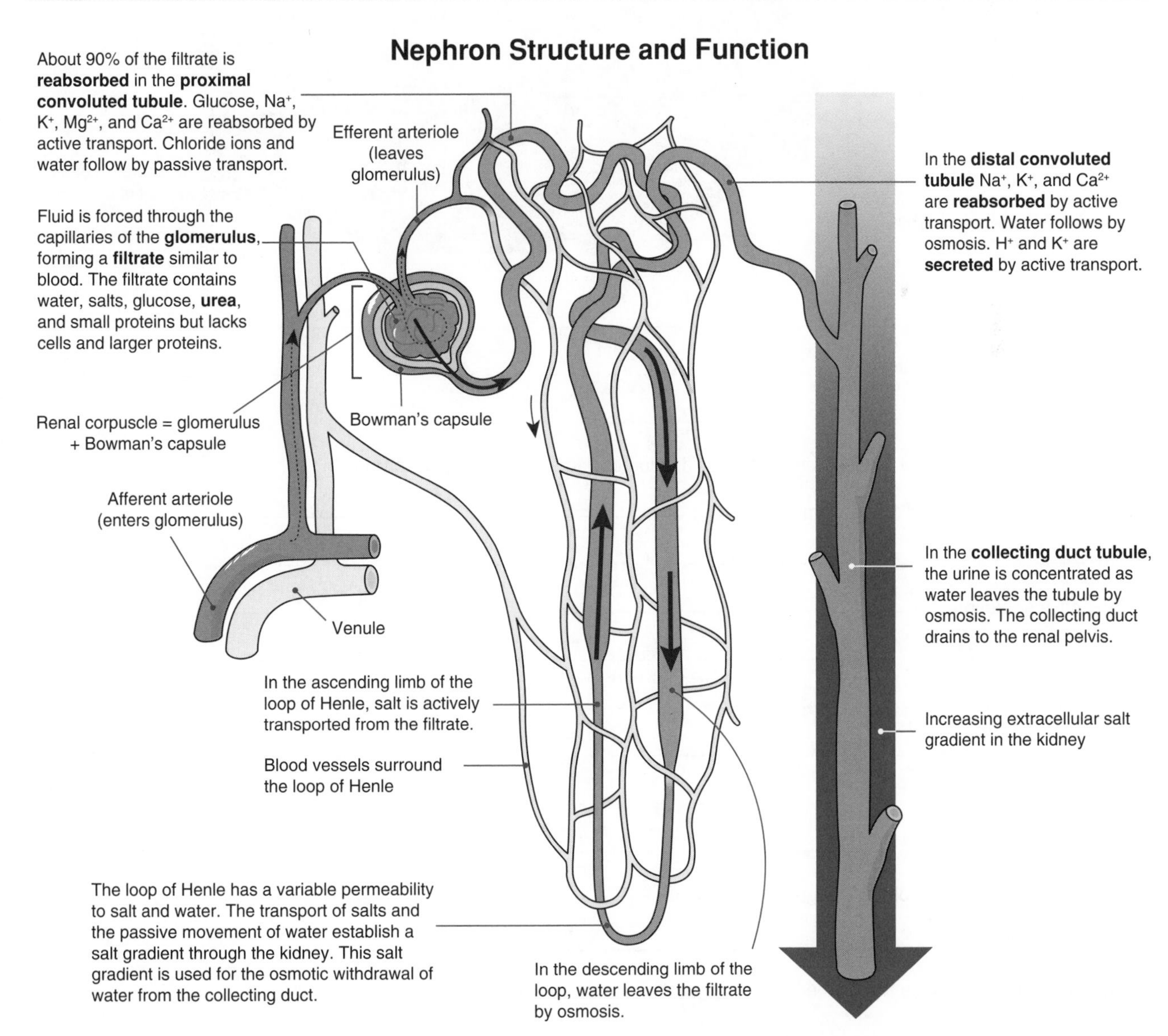


ISBN: 978-1-927173-12-1


***Related activities:*** *The Urinary System* ***Weblinks:*** *Interactive Kidney Quiz, Kidney Vascular System, The Juxtaglomerular Apparatus*

# Summary of Activities in the Kidney Nephron

Urine formation begins by **ultrafiltration** of the blood, as fluid is forced through the capillaries of the glomerulus, forming a filtrate similar to blood but lacking cells and proteins. The filtrate is then modified by secretion and **reabsorption** to add or remove substances (e.g. ions). The processes involved in urine formation are summarized below for each region of the nephron: glomerulus, proximal convoluted tubule, loop of Henle, distal convoluted tubule, and collecting duct.

**Filtrate**
- $H_2O$
- Salts (e.g. NaCl)
- $HCO_3^-$ (bicarbonate)
- $H^+$
- Urea
- Glucose, amino acids
- Some drugs

**Reabsorption**
- Active transport
- Passive transport
- Secretion (active transport)

**Renal corpuscle**

Blood

**Proximal tubule**

NaCl $HCO_3^-$ $H_2O$ Glucose and amino acids

Some drugs and poisons $H^+$

**Distal tubule**

NaCl $HCO_3^-$

$K^+$ and some drugs $H^+$

**Collecting duct**

Cortex

Medulla

**Loop of Henle**

$H_2O$

NaCl

Direction of filtrate flow

NaCl

$H_2O$

NaCl

Urea

$H_2O$

Urine (to renal pelvis)

The loop of Henle has varying permeability to salt and water. The transport of salts establishes and maintains the salt gradient across the medulla needed to concentrate the urine in the collecting duct. Water follows the salt out of the filtrate but it is transported away by the capillaries so that the high interstitial salt gradient is maintained.

Reabsorption of a small amount of urea from the urine helps to maintain the osmotic gradient for the removal of water.

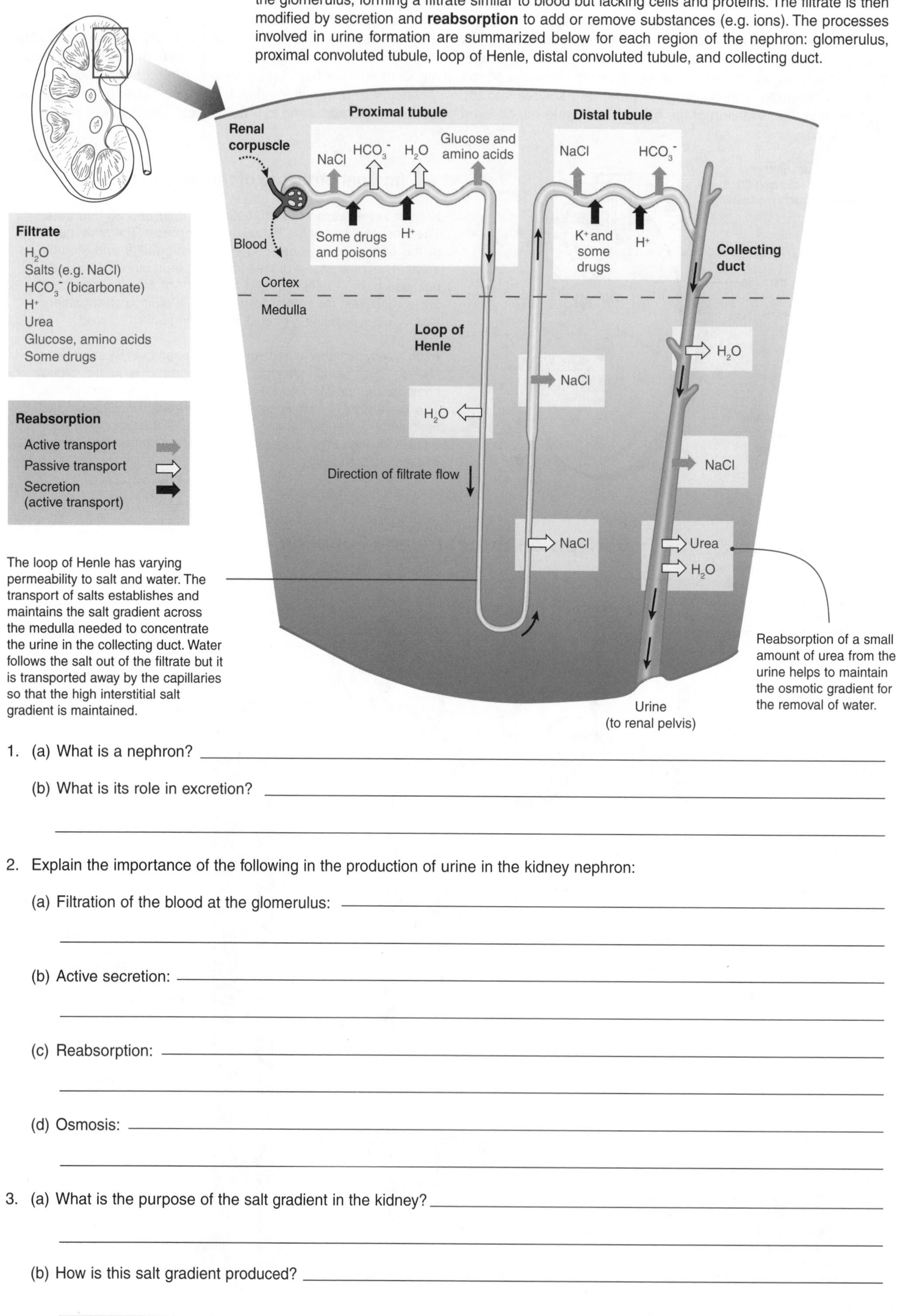

1. (a) What is a nephron? ______________________

   (b) What is its role in excretion? ______________________

2. Explain the importance of the following in the production of urine in the kidney nephron:

   (a) Filtration of the blood at the glomerulus: ______________________

   (b) Active secretion: ______________________

   (c) Reabsorption: ______________________

   (d) Osmosis: ______________________

3. (a) What is the purpose of the salt gradient in the kidney? ______________________

   (b) How is this salt gradient produced? ______________________

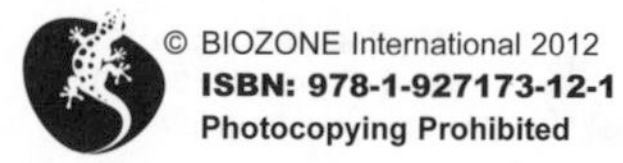

ISBN: 978-1-927173-12-1

# Control of Urine Output

Variations in salt and water intake, and in the environmental conditions to which we are exposed, contribute to fluctuations in blood volume and composition. The primary role of the kidneys is to regulate blood volume and composition (including the removal of nitrogenous wastes), so that homeostasis is maintained. This is achieved through varying the volume and composition of the urine. Two hormones, **antidiuretic hormone** (ADH) and **aldosterone**, are involved in the process.

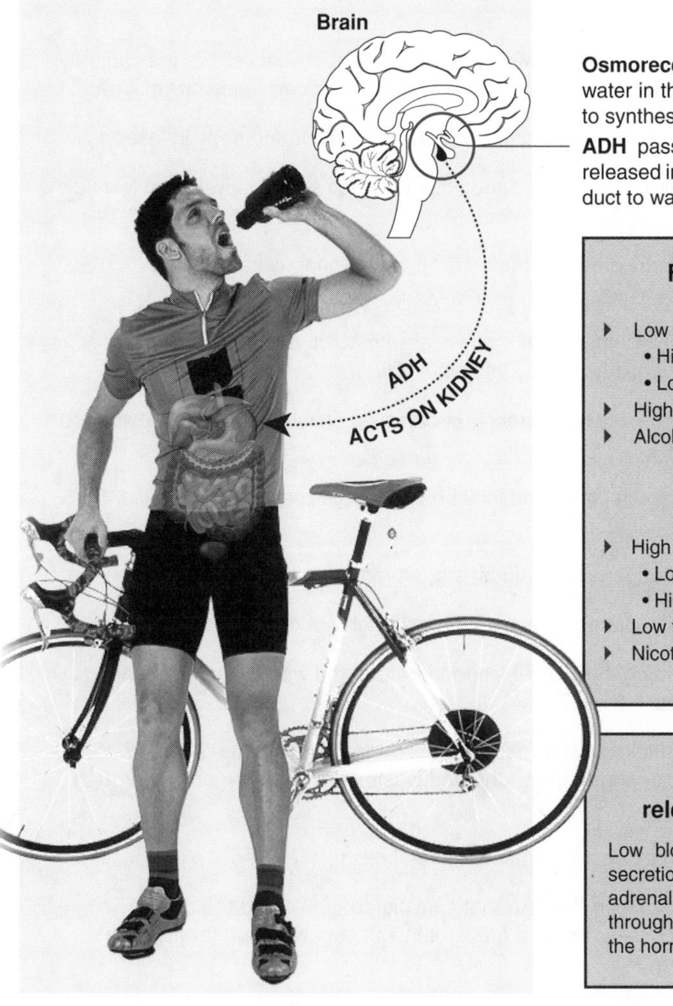


## Control of Urine Output

**Osmoreceptors** in the **hypothalamus** detect a fall in the concentration of water in the blood. They stimulate **neurosecretory cells** in the hypothalamus to synthesize and secrete the hormone ADH (antidiuretic hormone).

**ADH** passes from the hypothalamus to the posterior pituitary where it is released into the blood. ADH increases the permeability of the kidney collecting duct to water so that more water is reabsorbed and urine volume decreases.

**Factors inhibiting ADH release**

- Low solute concentration
  - High blood volume
  - Low blood sodium levels
- High fluid intake
- Alcohol consumption

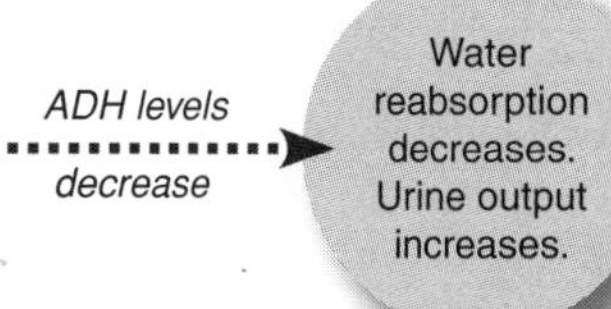

*ADH levels decrease* → Water reabsorption decreases. Urine output increases.

**Factors causing ADH release**

- High solute concentration
  - Low blood volume
  - High blood sodium levels
- Low fluid intake
- Nicotine and morphine

*ADH levels increase* → Water reabsorption increases. Urine output decreases.

**Factors causing release of aldosterone**

Low blood volumes also stimulate secretion of aldosterone from the adrenal cortex. This is mediated through a complex pathway involving the hormone renin from the kidney.

*Aldosterone* → Sodium reabsorption increases, water follows, blood volume restored.

1. (a) **Diabetes insipidus** is a disease caused by a lack of ADH. Based on what you know of the role of ADH in regulating urine volumes, describe the symptoms of this disease:

   (b) Suggest how this disorder might be treated:

2. Explain why alcohol consumption (especially to excess) causes dehydration and thirst:

3. Explain how negative feedback mechanisms operate to regulate blood volume and urine output:

4. **Diuretics** are drugs that increase urine volume. Many work by inhibiting the active transport of sodium and chloride in the nephron. Explain how this would lead to an increase in urine volume:



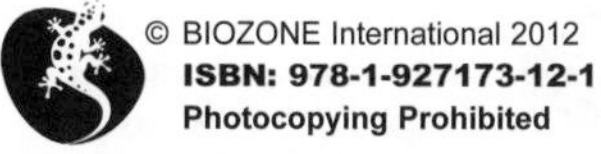


**ISBN: 978-1-927173-12-1**


***Related activities:*** *Water Budget in a Human, The Physiology of the Kidney*
***Weblinks:*** *Adrenaline and ADH*

# KEY TERMS: Mix and Match

*INSTRUCTIONS: Test your vocabulary by matching each term to its definition, as identified by its preceding letter code.*

alveoli

antidiuretic hormone (ADH)

dentition

digestive enzymes

excretion

extracellular digestion

gastric ceca

gastrovascular cavity

gills

gut (=digestive tract)

intestinal villi

intracellular digestion

kidney

loop of Henle

lungs

malpighian tubules

nephridia

nitrogenous waste

osmoregulation

parasite

predator

protonephridia

respiratory gas

respiratory membrane

spiral valve

tracheae (tracheal tubes)

**A** Tube-like structure with two openings in bilateral animals. Food enters through one opening (the mouth) and is digested and nutrients absorbed. Waste is egested at the opposite end (the anus).

**B** The term used for the layered junction between the alveolar epithelial cells of the capillary and their associated basement membranes across which gases can freely move.

**C** Series of tube-like structures which allow gas to be conducted into the body in insects.

**D** The hormone released in response to low blood volumes, high sodium levels and low fluid intake.

**E** The active regulation of osmotic pressure in an organism (through water and ion regulation).

**F** Enzymes in the digestive system that breaks down large macromolecules into their structural units/monomers for absorption by intestinal villi.

**G** Digestion occurring outside the cell by secretion of digestive enzymes to break down food material which is then absorbed, into the cell.

**H** Expansions of the midgut (insects) found directly behind the gizzard which increase the area available for nutrient absorption.

**I** Excretory waste based on nitrogen compounds, e.g, ammonia, urea and uric acid.

**J** The gas exchange organs of most aquatic animals (although not aquatic mammals).

**K** Simple gut with a single opening through which food enters and wastes leave. Found in Cnidaria and Platyhelminthes.

**L** Excretory organs used by annelids, arthropods, and mollusks. They comprise of ciliated tubules which pump water carrying surplus ions and wastes out of the organism through openings.

**M** Elimination (by an organism) of waste products of metabolism.

**N** Part of the kidney nephron between the proximal convoluted tubule and the distal convoluted tubule. Its function is to create a gradient in salt concentration through the medullary region of the kidney.

**O** Microscopic structures in the lungs of most vertebrates that form the terminus of the bronchioles. The site of gas exchange.

**P** Tubules that make up part of the excretory organs of arthropods; they collect wastes from the body fluids and discharge them into the hind gut.

**Q** Digestion occurring within the cell.

**R** Bean shaped organ in vertebrates used to remove and concentrate metabolic wastes from the blood.

**S** Animal that obtains it food (or living space) from a host, with deleterious but usually not fatal effects on the host.

**T** Fingerlike projections lining the surface of the intestine which increase the surface area for absorption.

**U** Feature of the intestine of sharks. The intestine is internally modified into a spiral to increase the surface area of the intestine.

**V** An organism that captures and kills prey for consumption.

**W** The excretory organs of platyhelminthes comprising of blind-ending tubules containing ciliated flame cells.

**X** Any gas that takes part in the respiratory process; oxygen or carbon dioxide.

**Y** The kind, number and arrangement of the teeth in the tooth rows.

**Z** Internal gas exchange structures found in vertebrates.

ISBN: 978-1-927173-12-1

Enduring Understanding

**2.D**
**4.A**
**4.B**

# Interactions in Physiological Systems

## Key concepts

- An internal transport system is a requirement in most multicellular animals.
- Animal circulatory systems commonly include a heart, vessels, and a circulatory fluid.
- Circulatory systems may be open or closed.
- Heart structure and circulatory pattern reflect metabolic needs and adaptation to environment.
- The circulatory system interacts with other body systems to maintain dynamic homeostasis.

## Key terms

artery (*pl.* arteries)
atrio-ventricular node
atrium (pl. atria)
blood
blood vessels
bulk flow (=mass flow)
capillary (*pl.* capillaries)
cardiac cycle
cardiovascular control center
closed circulatory system
coronary arteries
diastole
double circulatory system
hemolymph
heart
heart valves
liver
lymph
microcirculation
myogenic
open circulatory system
portal circulation
pulmonary circulation
respiratory pigment
single circulatory system
sinoatrial node
sinus venosus
surface area: volume
systemic circulation
systole
tissue fluid
toxic
vein
ventricle

*Periodicals:*
*Listings for this chapter are on page 374*

*Weblinks:*
*www.thebiozone.com/weblink/AP2-3121.html*

*BIOZONE APP: Student Review Series The Cardiovascular System*

## Essential Knowledge

☐ 1. Use the **KEY TERMS** to compile a glossary for this topic.

### Circulatory Systems *(2.D.2)* pages 144-148, 150-152, 154-157, 159-160

☐ 2. Describe the **surface area: volume** relationship in different animal taxa and relate this to the presence or absence of an internal transport system. Describe the components and functions of a **transport system** in animals.

☐ 3. Recognize that diversity in the circulatory systems of animals reflects both evolutionary history and adaptation to environment. Describe the role of **circulatory fluids** (e.g. **blood** or **hemolymph**) in the transport of materials in different taxa and explain the reasons for any differences.

☐ 4. Describe and explain diversity in the structure and function of **blood vessels** in vertebrates, including **arteries**, **capillaries**, and **veins**.

☐ 5. Describe diversity in the structure and function of circulatory systems in animals:
(a) **Open circulatory systems** (arthropods and most mollusks).
(b) **Closed circulatory systems** (vertebrates and some invertebrates).

☐ 6. Describe the structure and function of closed circulatory systems:
(a) Closed circulatory system in invertebrates (annelids, cephalopods).
(b) **Single circulatory system** in fish.
(c) **Double circulatory systems** in vertebrates other than fish.

☐ 7. Compare the structure of vertebrate hearts: those with three chambers in series (fish), three chambers with two atria and an undivided ventricle (amphibian), and four chambers separating systemic and pulmonary circulation (mammal).

☐ 8. Explain how the type of circulatory system (#5-#7) and the functional efficiency of transport is related to the metabolic needs and environment.

### Challenges to Homeostasis *(2.D.3)* pages 163-166

☐ 9. Explain how the body's systems respond cooperatively to maintain homeostasis in different situations, e.g. in response to exercise or to increases in altitude. Explain the effects of **toxic** substances on the body's ability to maintain homeostasis.

### Interacting Systems *(4.A.4: b, 4.B.2: a)* pages 149, 153, 158-163

☐ 10. Explain the production and functional role of **tissue fluid**. Explain how the lymphatic and cardiovascular systems interact through this medium.

☐ 11. Explain how the body's systems interact to maintain homeostasis of pH. Include reference to the role of the blood, and the respiratory and renal systems.

☐ 12. In more detail than above, describe the structure and function of the mammalian heart, including its relationship to the pulmonary system. Explain the reasons for the heart's asymmetry and relate this to the pressure of blood flow through the lungs and the kidneys.

☐ 13. Describe the central role of the **liver** in metabolism, including its close functional relationship with the circulatory and excretory systems.

# Transport and Exchange Systems

Living cells require a constant supply of nutrients and oxygen, and continuous removal of wastes. Simple, small organisms can achieve this through **diffusion** across moist body surfaces without requiring specialized transport or exchange systems. Larger, more complex organisms require systems to facilitate exchanges as their surface area to volume ratio decreases. **Mass transport** (also known as mass flow or **bulk flow**) describes the movement of materials at equal rates or as a single mass. Mass transport accounts for the long distance transport of fluids in living organisms. It includes the movement of blood in the circulatory systems of animals and the transport of water and solutes in the xylem and phloem of plants. In the diagram below, exchanges by diffusion are compared with mass transport to specific exchange sites.

## Exchanges Across a Body Surface

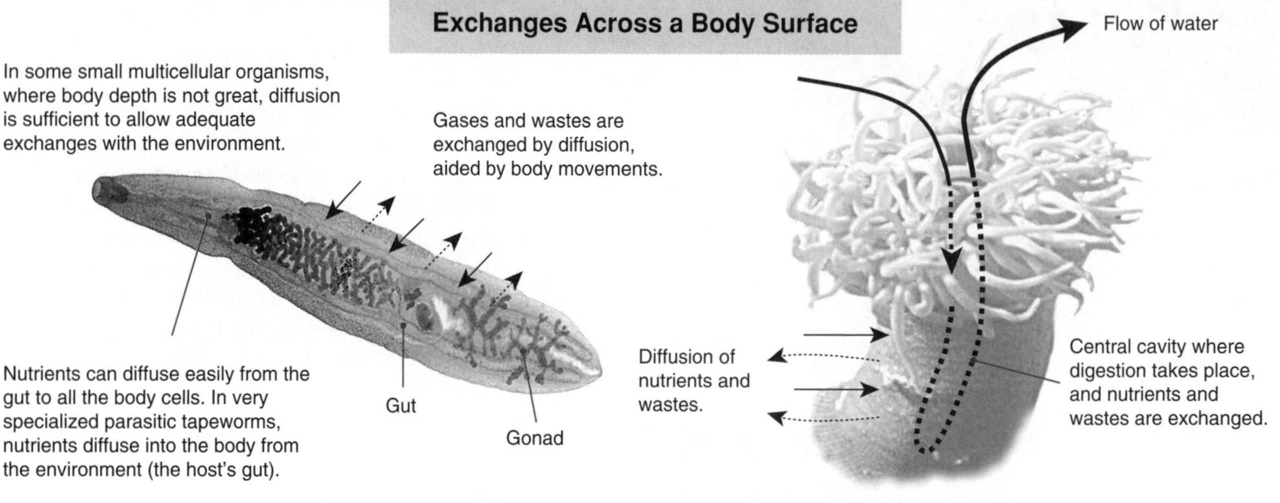


**Platyhelminthes (liver fluke)** **Cnidarians (sea anemone)**

## Systems for Exchange and Transport

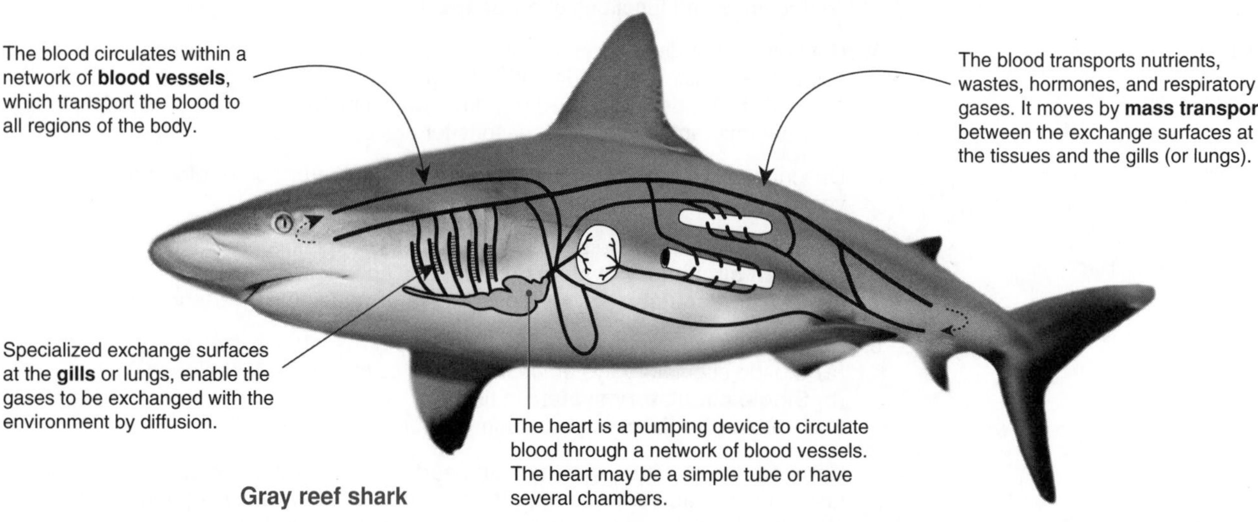


**Gray reef shark**

1. Explain why animals above a certain size or level of complexity require specialized systems for transport and exchange:

_______________

_______________

_______________

2. (a) Describe how materials move within the circulatory system of a vertebrate: _______________

_______________

(b) Contrast this with how materials are transported in a flatworm or single celled eukaryote:

_______________

_______________

(c) Identify two exchange sites in a vertebrate: _______________



***Related activities:*** *Open Circulatory Systems, Closed Circulatory Systems*


**ISBN: 978-1-927173-12-1**

# Circulatory Fluids in Invertebrates

The internal transport system of most animals includes a circulating fluid. In animals with closed systems, the fluid in the blood vessels is distinct from the tissue fluid outside the vessels and is called **blood**. Blood can have many different appearances, depending on the animal group, but it usually consists of cells and cell fragments suspended in a watery fluid. In animals with open systems, there is no difference between the fluid in the vessels and that in the sinuses **(hemocoel)** so the circulating fluid is called **hemolymph.** In insects, the hemolymph carries nutrients but not respiratory gases.

Hemolymph is a blood-like substance found in all invertebrates with open circulatory systems. The hemolymph fills the hemocoel and surrounds all cells.

About 90% of insect hemolymph is plasma, a watery fluid, which is usually clear. Compared to vertebrate blood, it contains relatively high concentrations of amino acids, proteins, sugars, and inorganic ions.

The remaining 10% of hemolymph volume is made up of various cell types (hemocytes). These are involved in clotting and internal defence. Unlike vertebrate blood, insect hemolymph lacks red blood cells and (with a few exceptions) lacks respiratory pigment, because oxygen is delivered directly to tissues by the tracheal system.

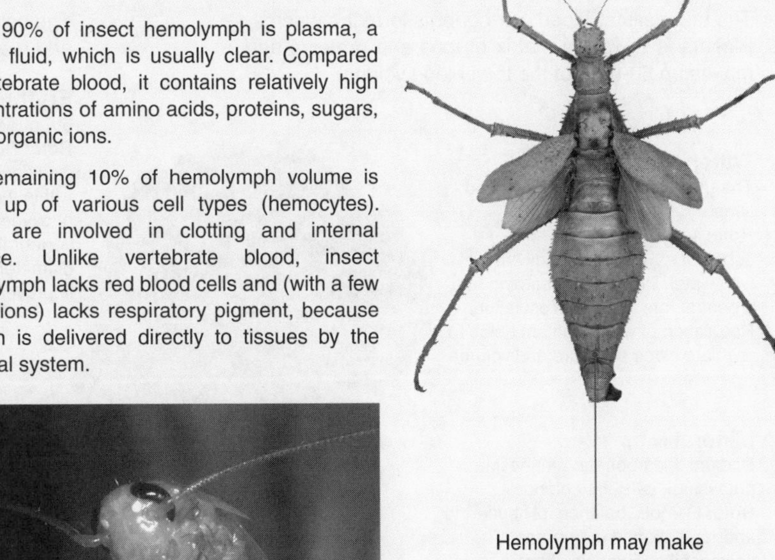

Hemolymph may make up between 11% and 40% of the total body mass of an insect

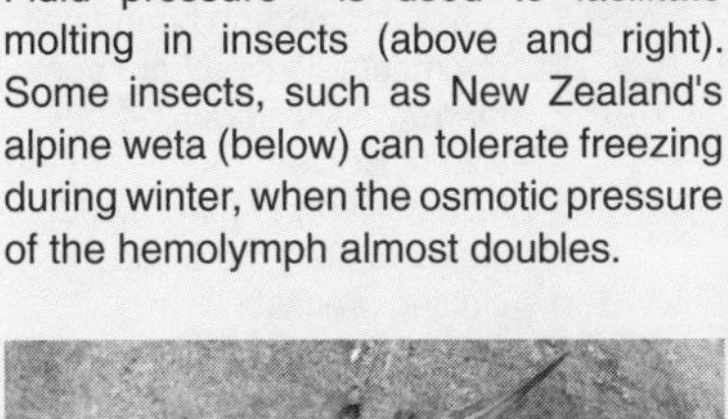

Fluid pressure is used to facilitate molting in insects (above and right). Some insects, such as New Zealand's alpine weta (below) can tolerate freezing during winter, when the osmotic pressure of the hemolymph almost doubles.

New Zealand's alpine weta can freeze solid. Their hemolymph contains the disaccharide trehalose, which acts as a cryoprotectant.

Pressure of the hemolymph enables arthropods to expand the soft cuticle of the body segments before they harden (sclerotize). This enables them to grow.

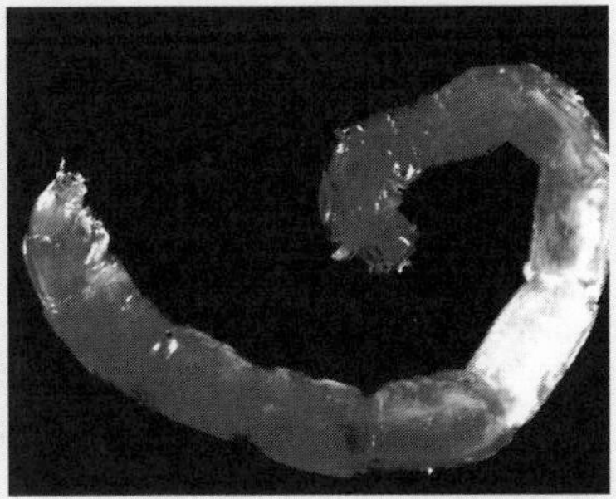

A few insects (but not many), like this midge larva, possess hemoglobin as an adaptation to living in low-oxygen substrates.

1. Describe two common functions of mammalian blood and insect hemolymph:

   (a) ______________________________

   (b) ______________________________

2. Describe one function of mammalian blood not commonly performed by insect hemolymph: ______________________________

   ______________________________

3. Describe one function of insect hemolymph not performed by mammalian blood: ______________________________

   ______________________________

4. Contrast the proportions of cellular and non-cellular components in blood and hemolymph: ______________________________

   ______________________________

   ______________________________


ISBN: 978-1-927173-12-1

# Vertebrate Blood

Blood is a complex liquid tissue comprising cellular components suspended in **plasma**. Blood performs many functions: it transports nutrients, respiratory gases, hormones, and wastes; it has a role in thermoregulation through the distribution of heat; it defends against infection; and its ability to clot protects against blood loss. With the exception of most insects, the blood of animals contains oxygen-carrying **respiratory pigments**, which increase the capacity of the blood to carry oxygen. In vertebrates, the respiratory pigment is carried inside specialized cells, rather than dissolved in the blood itself. This was an important step in vertebrate evolution because it enabled more oxygen to be carried without increasing the blood's viscosity.

## Non-Cellular Blood Components

The non-cellular blood components form the plasma. Plasma is a watery matrix of ions and proteins and makes up 50-60% of the total blood volume.

### Water

The main constituent of blood and lymph.
**Role:** Transports dissolved substances. Provides body cells with water. Distributes heat and has a central role in thermoregulation. Regulation of water content helps to regulate blood pressure and volume.

### Mineral ions

Sodium, bicarbonate, magnesium, potassium, calcium, chloride.
**Role:** Osmotic balance, pH buffering, and regulation of membrane permeability. They also have a variety of other functions, e.g. $Ca^{2+}$ is involved in blood clotting.

### Plasma proteins

7-9% of the plasma volume.
**Serum albumin**
**Role:** Osmotic balance and pH buffering, $Ca^{2+}$ transport.
**Fibrinogen and prothrombin**
**Role:** Take part in blood clotting.
**Immunoglobulins**
**Role:** Antibodies involved in the immune response.
**$\alpha$-globulins**
**Role:** Bind/transport hormones, lipids, fat soluble vitamins.
**$\beta$-globulins**
**Role:** Bind/transport iron, cholesterol, fat soluble vitamins.
**Enzymes**
**Role:** Take part in and regulate metabolic activities.

### Substances transported by non-cellular components

**Products of digestion**
Examples: sugars, fatty acids, glycerol, and amino acids.
**Excretory products**
Example: urea
**Hormones and vitamins**
Examples: insulin, sex hormones, vitamins A and $B_{12}$.
**Importance:** These substances occur at varying levels in the blood. They are transported to and from the cells dissolved in the plasma or bound to plasma proteins.

## Cellular Blood Components

The cellular components of the blood float in the plasma and make up 40-50% of the total blood volume.

### Erythrocytes (red blood cells or RBCs)

5-6 million per $mm^3$ blood and 38-48% of total blood volume.
**Role:** RBCs transport oxygen and a small amount of carbon dioxide. The oxygen is carried bound to hemoglobin (Hb) within the cells. Each Hb molecule can bind four molecules of oxygen. Although the size of RBCs varies widely among vertebrates, the cell width is about 25% larger than capillary diameter. It has been hypothesized that this size differential improves the oxygen transfer from RBCs to the tissues.

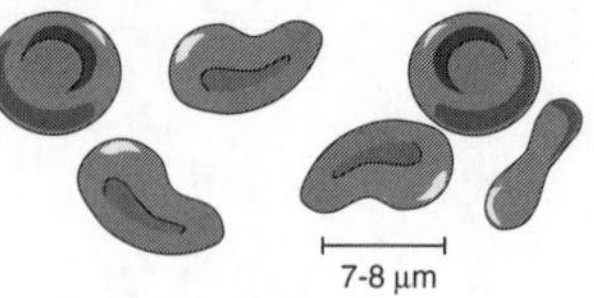


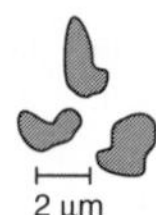


### Platelets

Small, membrane bound cell fragments derived from bone marrow cells; about 1/4 the size of RBCs.
0.25 million per $mm^3$ blood.
**Role:** To start the blood clotting process.

### Leukocytes (white blood cells)

5-10 000 per $mm^3$ blood
2-3% of total blood volume.
**Role:** Involved in internal defense. There are several types of white blood cells (see below)..

**Lymphocytes**
T and B cells.
24% of the white cell count.
**Role:** Antibody production and cell mediated immunity.

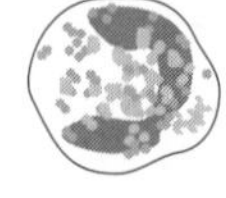

**Neutrophils**
Phagocytes.
70% of the white cell count.
**Role:** Engulf foreign material.

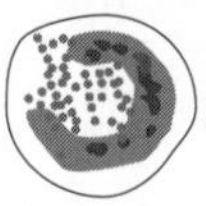

**Eosinophils**
Rare leukocytes; normally 1.5% of the white cell count.
**Role:** Mediate allergic responses such as hayfever and asthma.

**Basophils**
Rare leukocytes; normally 0.5% of the white cell count.
**Role:** Produce heparin (an anti-clotting protein), and histamine. Involved in inflammation.



***Related activities**: Gas Transport in Humans, The Body's Defenses, Hemostasis, Thermoregulation in Humans*

**ISBN: 978-1-927173-12-1**

## The Examination of Blood

The cellular components of blood are normally present in particular specified ratios. Therefore a change in the morphology, type, or proportion of different blood cells can be used to indicate a specific disorder or infection. A SEM (right) shows the detailed external morphology of the blood cells. A fixed smear of a blood sample viewed with a light microscope (far right) can be used to identify the different blood cell types present, and their relative ratios. Determining the types and proportions of different white blood cells in blood is called a **differential white blood cell count.** Elevated counts of particular cell types indicate allergy or infection.

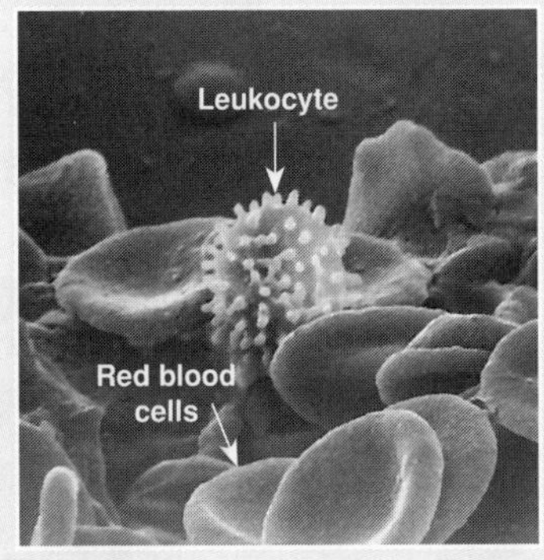


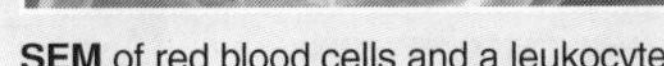

**SEM** of red blood cells and a leukocyte.

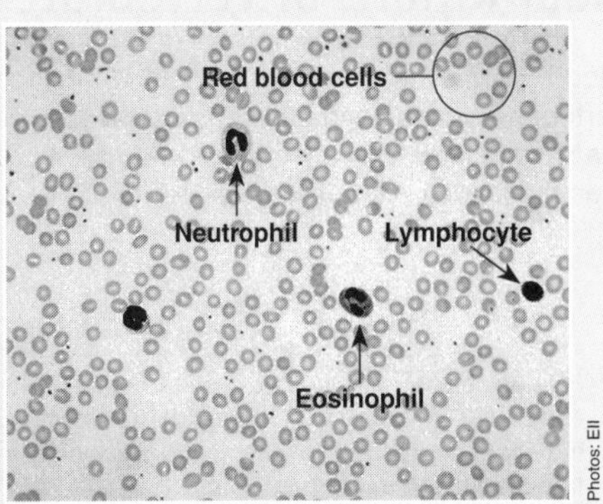


**Light microscope** view of a fixed blood smear.

1. For each of the following blood functions, identify the component (or components) of the blood responsible and state how the function is carried out (the mode of action). The first one is done for you:

   (a) **Temperature regulation**. *Blood component*: Water component of the plasma

   *Mode of action*: Water absorbs heat and dissipates it from sites of production (e.g. organs)

   (b) **Protection against disease.** *Blood component*: ___

   *Mode of action*: ___

   (c) **Communication between cells, tissues, and organs**. *Blood component*: ___

   *Mode of action:* ___

   (d) **Oxygen transport.** *Blood component*: ___

   *Mode of action*: ___

   (e) **$CO_2$ transport.** *Blood components*: ___

   *Mode of action*: ___

   (f) **Buffer against pH changes**. *Blood components*: ___

   *Mode of action*: ___

   (g) **Nutrient supply**. *Blood component*: ___

   *Mode of action*: ___

   (h) **Tissue repair**. *Blood components*: ___

   *Mode of action*: ___

   (i) **Transport of hormones, lipids, and fat soluble vitamins**. *Blood component*: ___

   *Mode of action*: ___

2. Identify a feature that distinguishes red and white blood cells: ___

3. Explain two physiological advantages of red blood cell structure (lacking nucleus and mitochondria):

   (a) ___

   (b) ___

4. Suggest what each of the following results from a differential white blood cell count would suggest:

   (a) Elevated levels of eosinophils (above the normal range): ___

   (b) Elevated levels of neutrophils (above the normal range): ___

   (c) Elevated levels of basophils (above the normal range): ___

   (d) Elevated levels of lymphocytes (above the normal range): ___


ISBN: 978-1-927173-12-1

# Adaptations of Vertebrate Blood

Blood varies in composition and function throughout the vertebrate orders. Mammalian blood is highly complex, containing many different types of cells. A major difference between mammalian blood and the blood of other vertebrates is that mammalian red blood cells have no nucleus. The blood of some animals in Antarctic environments may even contain non-cellular components such as anti-freeze sugars and proteins. These enable them to function effectively at sub-zero temperatures.

## Oxygen Capacity and Hemoglobin

Hemoglobin is a respiratory pigment (a colored protein) capable of combining reversibly with oxygen, hence increasing the amount of oxygen that can be carried by the blood. All vertebrates use hemoglobin to transport blood through the body.

| Taxon | Oxygen capacity ($cm^3$ $O_2$ per 100 $cm^3$ blood) | Pigment |
|---|---|---|
| Fishes | 2 - 4 | Hemoglobin |
| Reptiles | 7 - 12 | Hemoglobin |
| Birds | 20 - 25 | Hemoglobin |
| Mammals | 15 - 30 | Hemoglobin |

The oxygen carrying capacity of hemoglobin is not the same for all vertebrates. Variations in its structure change not only its capacity to carry oxygen, but also its ability to take up and off load oxygen. In general, the oxygen capacity of blood in vertebrates increases from fish, to reptiles, to birds and mammals and may be related metabolic activity. The fishes have a variety of hemoglobin structures. This allows them to move between different environments while maintaining oxygen capacity.

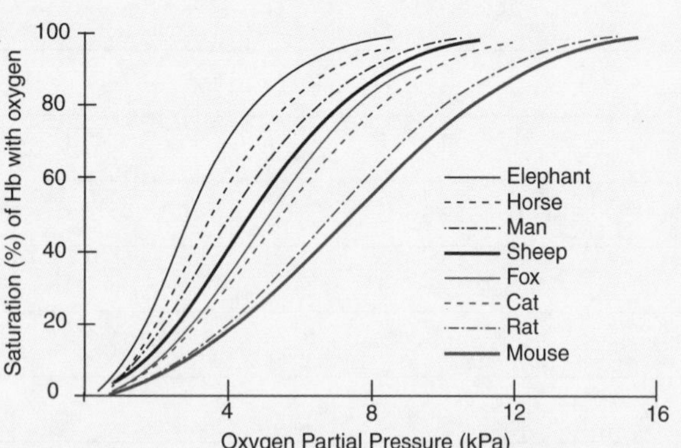


The hemoglobin of small mammals releases oxygen more readily than larger mammals because it has a lower affinity for oxygen. This may be related to the fact that small mammals consume oxygen at a higher rate than larger mammals and so require a faster delivery of oxygen.

Photo: Professor Dr. habil. Uwe Kils CC3.0

Crocodile icefish larva

The Antarctic icefish (Channichthyidae) are the only known vertebrates without hemoglobin. They survive because the extremely cold water (as low as -2°C) contains high concentrations of $O_2$. This diffuses directly across their scaleless skin into the blood plasma and is transported throughout the body. Many icefish contain antifreeze glycoproteins in their blood allowing the blood to remain fluid at freezing temperatures.

Kenneth Catania, Vanderbilt University NSF

Adaptations of hemoglobin have enabled some mammals to live in places with low $O_2$ content. Llamas and their relatives live at high altitudes with low $O_2$ pressure. Moles live underground where $CO_2$ levels can be high and $O_2$ levels low. The hemoglobin in some moles has lost the ability to bind the molecule DPG (which inhibits $O_2$ binding in deoxygenated tissues). This increases the amount of $O_2$ and $CO_2$ that can be transported in the blood.

1. For a human weighing 70 kg, calculate their mass (and therefore approximate volume) of blood: ____________________

2. Describe two adaptations in the blood of icefish to their cold environment: ____________________

3. Explain why the ability to transport greater amounts of both $CO_2$ and $O_2$ is an advantage to moles: ____________________

4. Explain why many fish have a number of different hemoglobin structures in their blood: ____________________

5. In order to achieve 100% oxygen saturation in the blood of a mouse, the partial pressure of oxygen must **higher/lower** (delete one) than that applied to elephant blood.



***Related activities:*** *Gas Transport in Mammals, Blood Vessels*


**ISBN: 978-1-927173-12-1**

# Acid–Base Balance

The pH of the body's fluids must be maintained within a very narrow range (pH 7.35-7.45). The products of metabolic activity are generally acidic and could alter pH considerably without a buffer system to counteract pH changes. The carbonic acid-bicarbonate buffer works throughout the body to maintain the pH of blood plasma close to 7.40. The body maintains the buffer by eliminating either the acid (carbonic acid) or the base (bicarbonate ions). The blood buffers, the lungs, and the kidneys represent the three defense systems against disturbances of pH homeostasis. Changes in carbonic acid concentration can be brought about within seconds via increased or decreased rate of breathing. The renal system, acts more slowly, but can permanently eliminate metabolic acids and regulate the levels of alkaline substances, controlling pH by either excreting or retaining ions.

### The Blood Buffer System

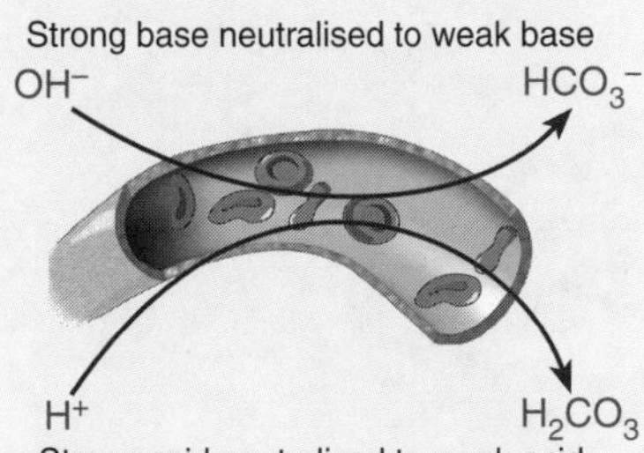


A buffer is able to resist changes to the pH of a fluid when either an acid or base is added to it. The bicarbonate ion ($HCO_3^-$) and its acid, carbonic acid ($H_2CO_3$), work in the following way:

$$H^+ + HCO_3^- \rightleftharpoons H_2CO_3$$

$$H_2CO_3 \rightleftharpoons H^+ + HCO_3^-$$

If a strong acid (such as HCl) is added to the system a weak acid is formed and thus the pH falls only slightly. Note that the blood also contains proteins, which contain basic and acidic groups that may act either as $H^+$ acceptors or donors to help maintain blood pH.

### The Respiratory System

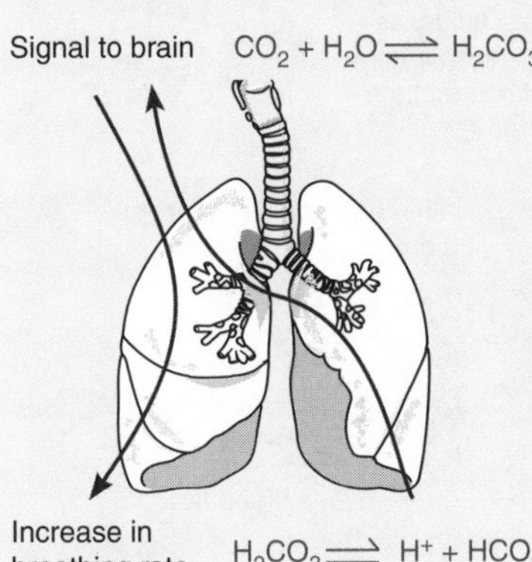


Carbon dioxide ($CO_2$) in the blood, an end-product of cellular respiration, forms carbonic acid ($H_2CO_3$) which dissociates to form $H^+$ and bicarbonate ($HCO_3^-$). This means that as $CO_2$ rises in the blood so too does the $H^+$ concentration. **Chemoreceptors** in the brain detect the rise in $H^+$ ions and increase the rate of breathing to expel the $CO_2$. Low levels of $CO_2$ have the effect of depressing the respiratory system so that $H^+$ builds up and the pH is once again restored.

### The Renal System

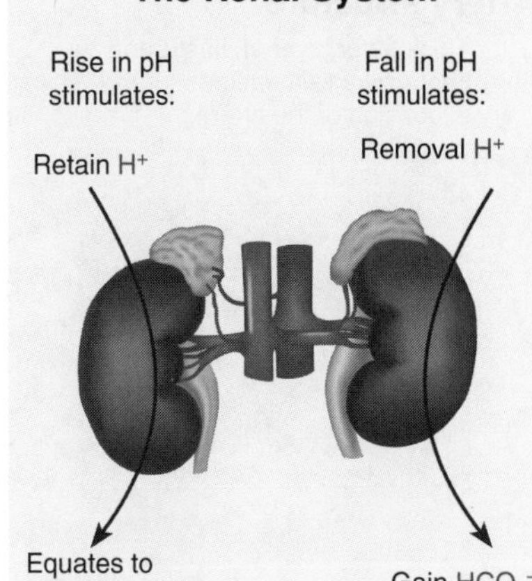


Recall that a net loss of $HCO_3^-$ effectively results in the gain of $H^+$. Bicarbonate is reabsorbed by the kidney tubules all the time so pH is regulated through retention or secretion of $H^+$. When blood pH rises, $H^+$ is retained by the tubule cells. When blood pH falls, $H^+$ is actively secreted into the kidney tubules. Urine pH normally varies from 4.5 to 8.0, reflecting the ability of the renal tubules to lose or retain ions to maintain the homeostasis of blood pH.

1. Explain why the blood must be kept at a pH between 7.35 and 7.45:

2. A drop in the blood pH to below 7.35 is called metabolic acidosis, even though the blood might still be at pH >7 and not strictly acidic. Describe how metabolic acidosis might arise:

3 (a) Describe how the blood buffer system maintains blood pH:

(b) Explain the effects of adding a base (e.g. ingestion of alkaline substances) to the system:

4. (a) Describe the respiratory response to excess $H^+$ in the blood:

(b) Explain where these $H^+$ ions come from:

(c) Describe how **respiratory acidosis** might arise:

5. Explain the role of the renal system in maintaining the pH of the blood:


**ISBN: 978-1-927173-12-1**


***Related activities:*** *Blood, The Physiology of the Kidney, Gas Exchange in Mammals*

# Blood Vessels

The blood vessels of the circulatory system connect the fluid environment of the body's cells to the organs that exchange gases, absorb nutrients, and dispose of wastes. In vertebrates, arteries carry blood away from the heart to the capillaries within the tissues. The large arteries leaving the heart branch repeatedly to form distributing arteries, which themselves divide to form small **arterioles** within the tissues and organs. Arterioles deliver blood to the capillaries connecting the arterial and venous systems. Capillaries enable the exchange of nutrients and wastes between the blood and tissues, and they form large networks, especially in tissues and organs with high metabolic rates. The structural differences between blood vessels are related to their functional roles. While vessels close to the heart exhibit all the layers typical of the vessel's type, one or more layers may be absent in vessels more distant from the heart. Capillaries, whose functional role is exchange, consist only of a thin endothelium.

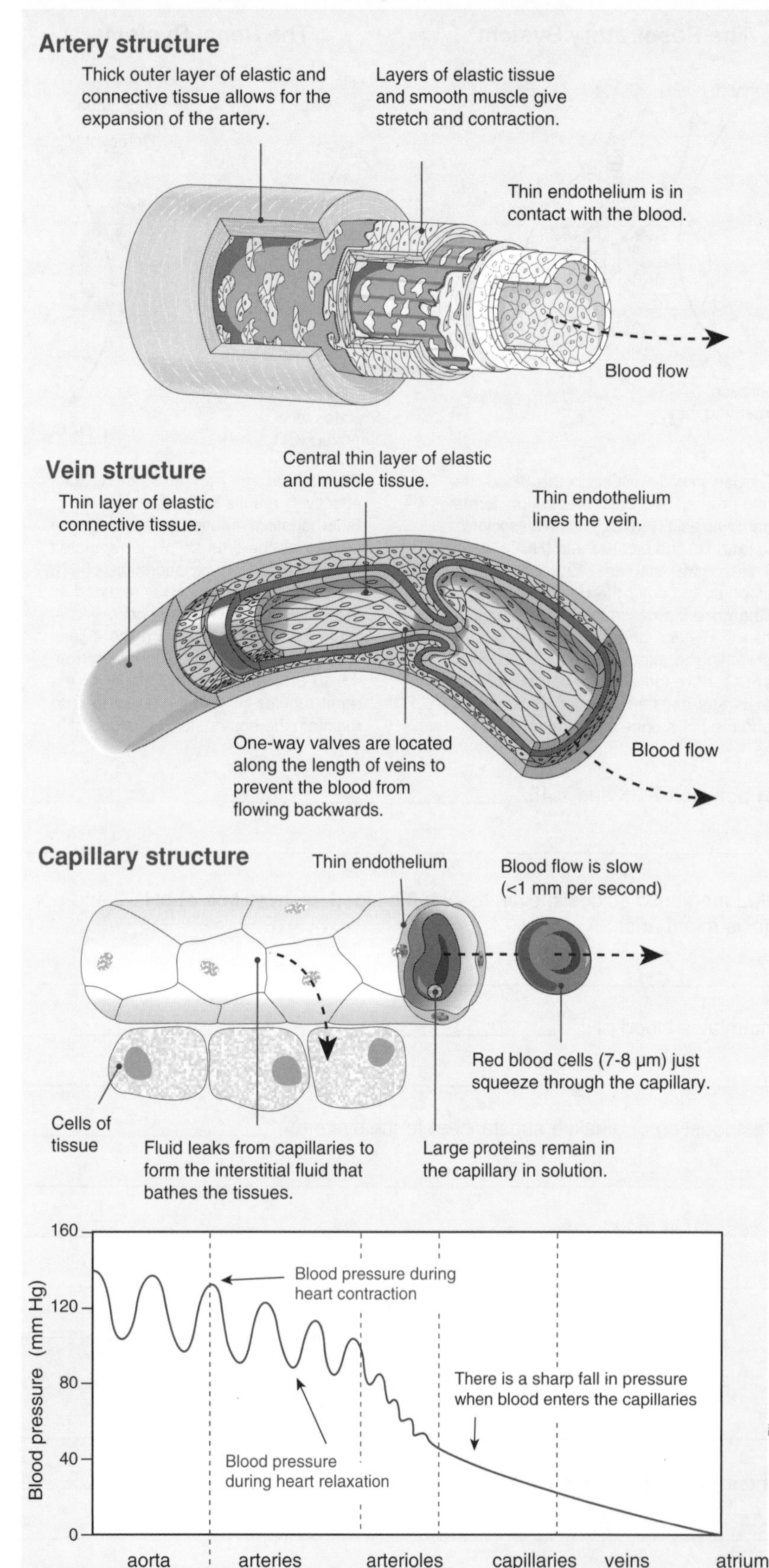

## Artery structure

Thick outer layer of elastic and connective tissue allows for the expansion of the artery.

Layers of elastic tissue and smooth muscle give stretch and contraction.

Thin endothelium is in contact with the blood.

Blood flow

## Vein structure

Thin layer of elastic connective tissue.

Central thin layer of elastic and muscle tissue.

Thin endothelium lines the vein.

One-way valves are located along the length of veins to prevent the blood from flowing backwards.

Blood flow

## Capillary structure

Thin endothelium

Blood flow is slow (<1 mm per second)

Red blood cells (7-8 μm) just squeeze through the capillary.

Cells of tissue

Fluid leaks from capillaries to form the interstitial fluid that bathes the tissues.

Large proteins remain in the capillary in solution.

Blood pressure (mm Hg)

160

120

80

40

0

Blood pressure during heart contraction

Blood pressure during heart relaxation

There is a sharp fall in pressure when blood enters the capillaries

aorta | arteries | arterioles | capillaries | veins | atrium

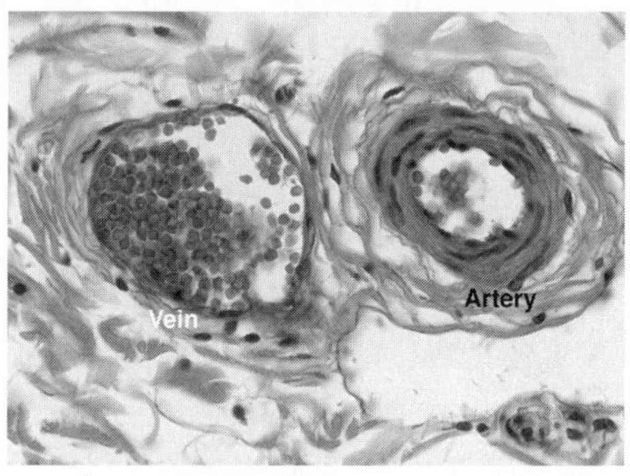


Arteries are made up of three layers; an inner layer of thin epthelium called the endothelium, a stretchy middle layer, and a thick outer layer. This structure enables arteries to withstand and maintain high blood pressure. **Veins** are similar in structure to arteries, but have less elastic and muscle tissue. Although veins are less elastic than arteries, they can still expand enough to adapt to changes in the pressure and volume of the blood passing through them.

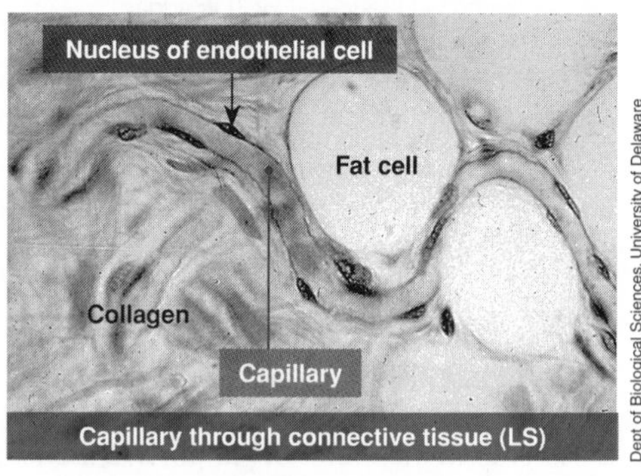


Dept of Biological Sciences, University of Delaware

Capillaries are very small blood vessels (4-10 μm diameter) made up of only a single layer of flattened (squamous) epithelial cells. Capillaries form a vast network of vessels that penetrate all parts of the body and are so numerous that no cell is more than 25 μm from any capillary. It is in the capillaries that the exchange of materials between the body cells and the blood takes place.

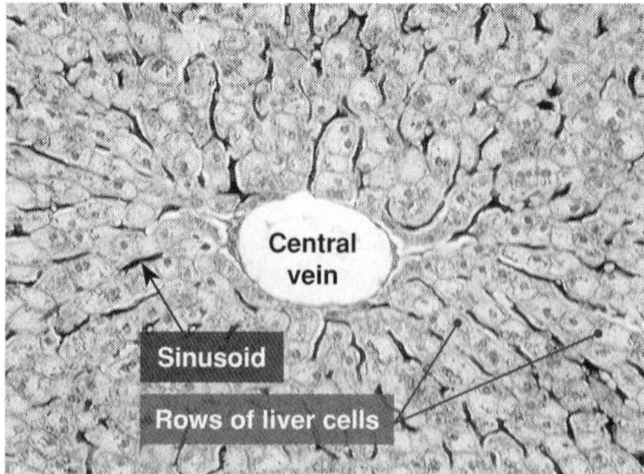


EII

Tiny blood vessels in dense organs, such as the liver (above), are called **sinusoids**. They are wider than capillaries and follow a more convoluted path through the tissue. Instead of the usual endothelial lining, they are lined with phagocytic cells. Sinusoids, like capillaries, transport blood from arterioles to venules.



***Related activities:*** *Capillary Networks*
***Weblinks:*** *Arteries, Veins*

***Periodicals:***
*Cunning plumbing*

**ISBN: 978-1-927173-12-1**

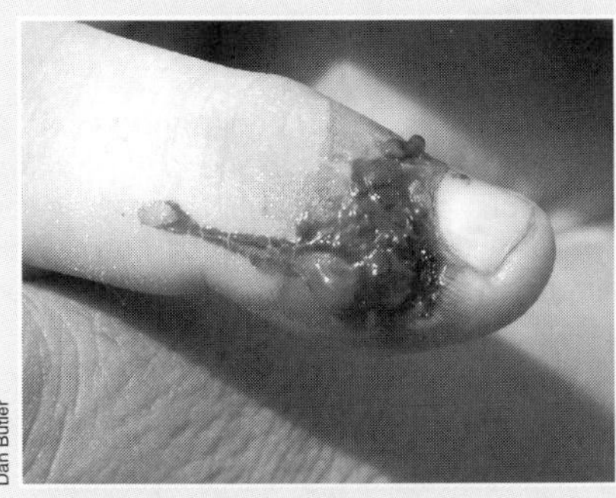


If a vein is cut, as in the severe finger wound shown above, the blood oozes out slowly in an even flow, and usually clots quickly as it leaves. In contrast, if a cut is made into an artery, the arterial blood spurts rapidly and requires pressure to staunch the flow.

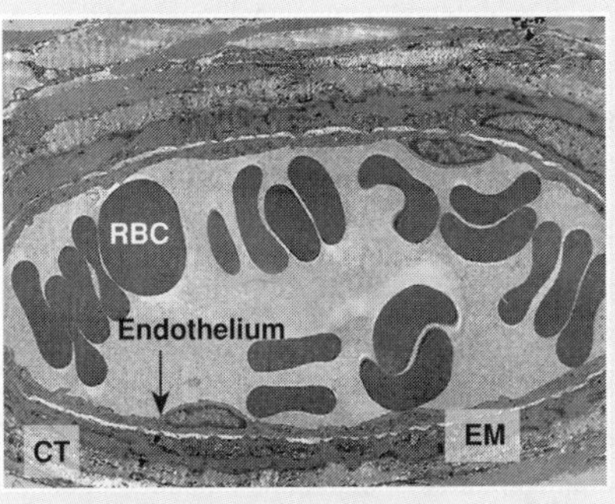


This TEM shows the structure of a typical vein. Note the red blood cells (RBC) in the lumen of the vessel, the inner layer of of epithelial cells (the endothelium), the central layer of elastic and muscle tissue (EM), and the outer connective tissue (CT) layer.

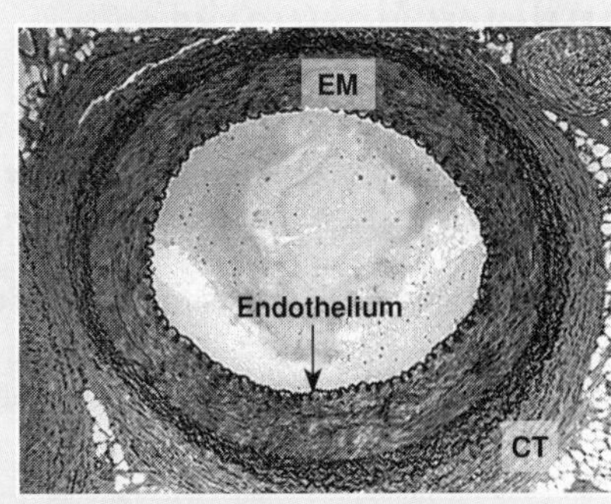


Arteries have a thick central layer of elastic and smooth muscle tissue (EM). Near the heart, arteries have more elastic tissue. This enables them to withstand high blood pressure. Arteries further from the heart have more smooth muscle; this helps them to maintain blood pressure.

1. Describe the contrasting structure of veins and arteries for each of the following properties:

   (a) Thickness of muscle and elastic tissue: ______

   (b) Size of the lumen (inside of the vessel): ______

2. Explain the reasons for the differences you have described above: ______

3. (a) Describe the structure of capillaries, explaining how it differs from that of veins and arteries: ______

   (b) Explain the reasons for these differences: ______

4. Compare the rate and force of blood flow in arteries, veins, and capillaries, explaining reasons for the differences:

5. Describe the role of the valves in assisting the veins to return blood back to the heart: ______

6. Explain why blood oozes from a venous wound, rather than spurting as it does from an arterial wound:

7. Explain why capillaries form dense networks in tissues with a high metabolic rate: ______


**ISBN: 978-1-927173-12-1**

# Capillary Networks

Capillaries form branching networks where exchanges between the blood and tissues take place. The flow of blood through a capillary bed is called **microcirculation**. In most parts of the body, there are two types of vessels in a capillary bed: the **true capillaries**, where exchanges take place, and a vessel called a **vascular shunt**, which connects the arteriole and venule at either end of the bed. The shunt diverts blood past the true capillaries when the metabolic demands of the tissue are low (e.g. vasoconstriction in the skin when conserving body heat). When tissue activity increases, the entire network fills with blood.

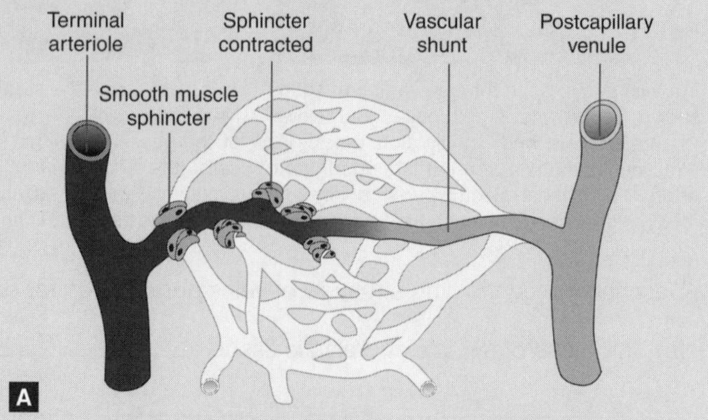

1. Describe the structure of a capillary network:

___________________________________________

___________________________________________

___________________________________________

___________________________________________

___________________________________________

___________________________________________

2. Explain the role of the smooth muscle sphincters and the vascular shunt in a capillary network:

___________________________________________

___________________________________________

___________________________________________

___________________________________________

___________________________________________

___________________________________________

3. (a) Describe a situation where the capillary bed would be in the condition labeled **A**:

___________________________________________

___________________________________________

___________________________________________

___________________________________________

(b) Describe a situation where the capillary bed would be in the condition labeled **B**:

___________________________________________

___________________________________________

___________________________________________

___________________________________________

4. Explain how a portal venous system differs from other capillary systems:

___________________________________________

___________________________________________

___________________________________________

___________________________________________

___________________________________________

___________________________________________



When the sphincters contract (close), blood is diverted via the vascular shunt to the postcapillary venule, bypassing the exchange capillaries.

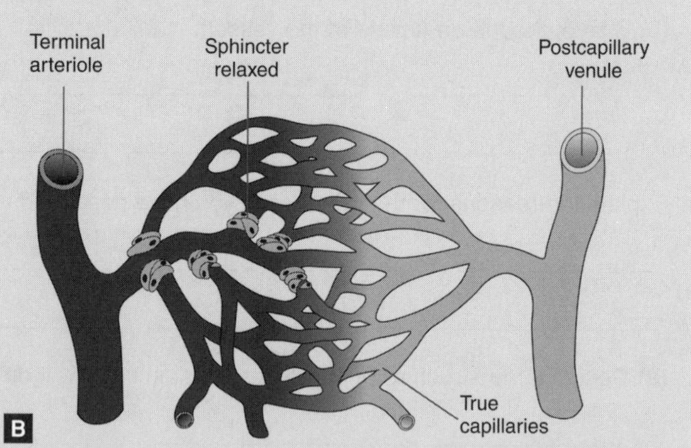

When the sphincters are relaxed (open), blood flows through the entire capillary bed allowing exchanges with the cells of the surrounding tissue.

### Connecting Capillary Beds

the role of portal venous systems

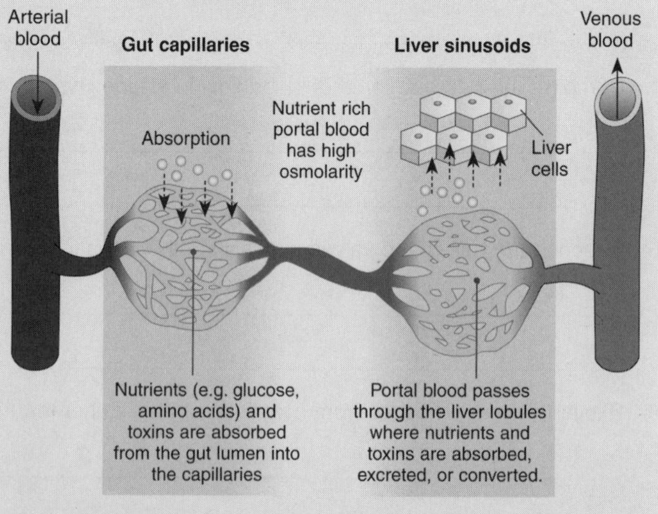

A portal venous system occurs when a capillary bed drains into another capillary bed through veins, without first going through the heart. Portal systems are relatively uncommon. Most capillary beds drain into veins which then drain into the heart, not into another capillary bed. The diagram above depicts the hepatic portal system, which includes both capillary beds and the blood vessels connecting them.



***Related activities:*** *Blood Vessels*
***Weblinks:*** *Microcirculation*


**ISBN: 978-1-927173-12-1**

# The Formation of Tissue Fluid

The network of capillaries supplying the body's tissues ensures that no substance has to diffuse far to enter or leave a cell. Substances exchanged first diffuse through the interstitial fluid (or tissue fluid), which surrounds and bathes the cells. As with all cells, substances can move into and out of the endothelial cells of the capillary walls in several ways; by diffusion, by cytosis, and through gaps where the membranes are not joined by tight junctions. Some fenestrated capillaries are also more permeable than others. These specialized capillaries are important where absorption or filtration occurs (e.g. in the intestine or the kidney). Because capillaries are leaky, fluid flows across their plasma membranes. Whether fluid moves into or out of a capillary depends on the balance between the blood (hydrostatic) pressure and the concentration of solutes at each end of a capillary bed.

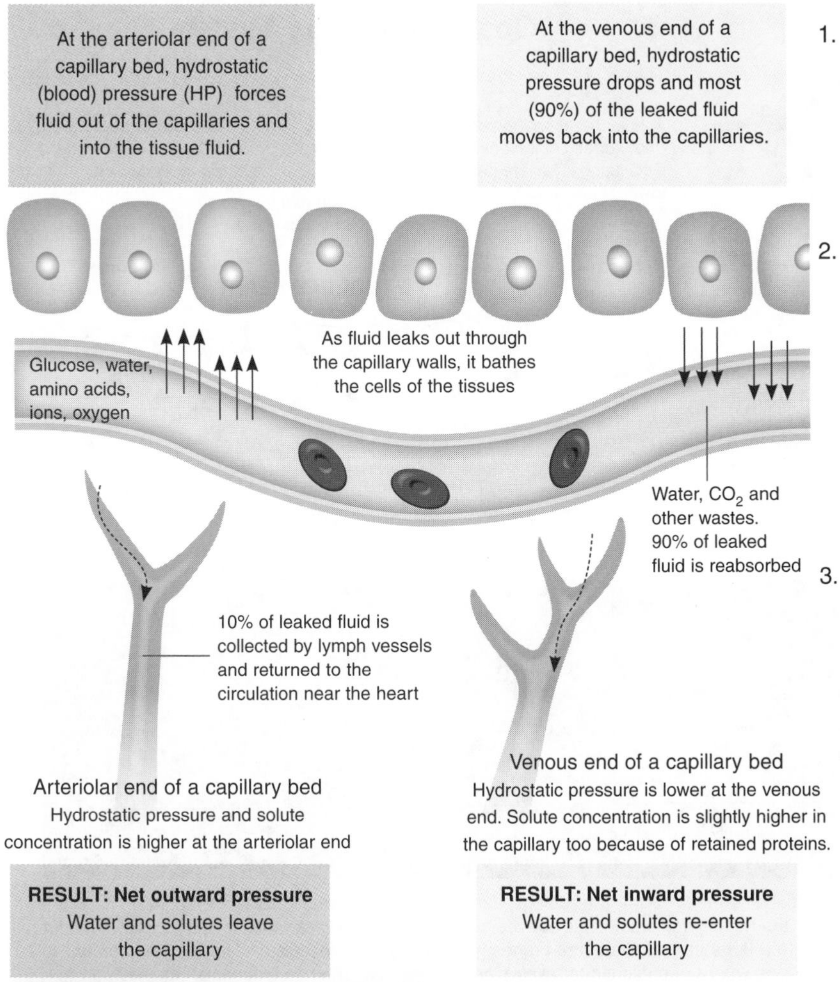

At the arteriolar end of a capillary bed, hydrostatic (blood) pressure (HP) forces fluid out of the capillaries and into the tissue fluid.

At the venous end of a capillary bed, hydrostatic pressure drops and most (90%) of the leaked fluid moves back into the capillaries.

Glucose, water, amino acids, ions, oxygen

As fluid leaks out through the capillary walls, it bathes the cells of the tissues

Water, $CO_2$ and other wastes. 90% of leaked fluid is reabsorbed

10% of leaked fluid is collected by lymph vessels and returned to the circulation near the heart

**Arteriolar end of a capillary bed**
Hydrostatic pressure and solute concentration is higher at the arteriolar end

**Venous end of a capillary bed**
Hydrostatic pressure is lower at the venous end. Solute concentration is slightly higher in the capillary too because of retained proteins.

**RESULT: Net outward pressure**
Water and solutes leave the capillary

**RESULT: Net inward pressure**
Water and solutes re-enter the capillary

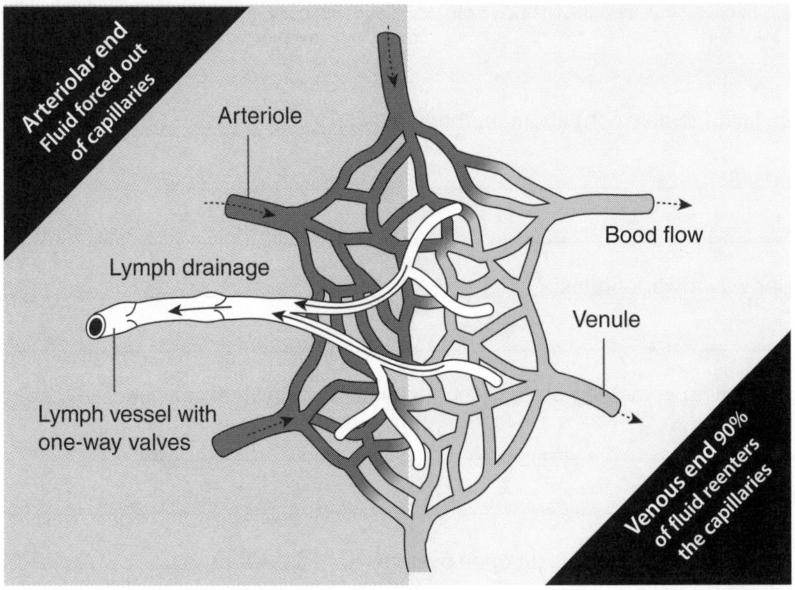

Arteriolar end
Fluid forced out of capillaries

Arteriole

Bood flow

Lymph drainage

Venule

Lymph vessel with one-way valves

Venous end 90% of fluid reenters the capillaries

As described above, not all the fluid reenters the capillaries at the venous end of the capillary bed. This extra fluid is collected by the lymphatic vessels, a network of vessels alongside the blood vessels. Once the fluid enters the lymphatic vessels it is called lymph. The lymphatic vessels drain into the subclavian vein near the heart. Lymph is similar to tissue fluid but has more lymphocytes.

1. What is the purpose of the tissue fluid?

_______________

_______________

_______________

2. Describe the features of capillaries that allow exchanges between the blood and other tissues:

_______________

_______________

_______________

_______________

3. Explain how hydrostatic (blood) pressure and solute concentration operate to cause fluid movement at:

(a) The arteriolar end of a capillary bed:

_______________

_______________

_______________

_______________

(b) The venous end of a capillary bed:

_______________

_______________

_______________

_______________

4. Describe the two ways in which tissue fluid is returned to the general circulation:

(a) _______________

_______________

_______________

_______________

(b) _______________

_______________

_______________

_______________


**ISBN: 978-1-927173-12-1**


***Related activities:*** *Blood Vessels, The Lymphatic System*
***Weblinks:*** *Microcirculation*

# Open Circulatory Systems

Two basic types of circulatory systems have evolved in animals. Many invertebrates have an **open circulatory system**, while vertebrates have a **closed circulatory system**, consisting of a heart and a network of tube-like vessels. The circulatory systems of arthropods are open but varied in complexity. Insects, unlike most other arthropods, do not use a circulatory system to transport oxygen, which is delivered directly to the tissues via the system of tracheal tubes. In addition to its usual transport functions, the circulatory system may also be important in hydraulic movements of the whole body (as in many mollusks) or its component parts (e.g. newly emerged butterflies expand their wings through hydraulic pressure).

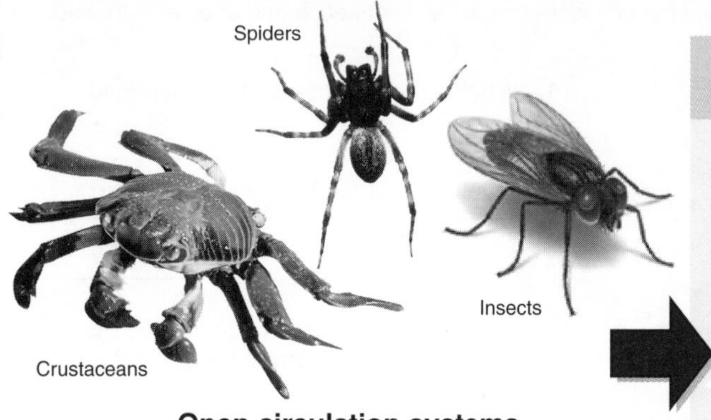


## Open circulation systems

Arthropods and mollusks (except cephalopods) have open circulatory systems in which the blood is pumped by a tubular, or sac-like, heart through short vessels into large spaces in the body cavity. The blood bathes the cells before reentering the heart through holes (**ostia**). Muscle action may assist the circulation of the blood.

## Open Circulatory System

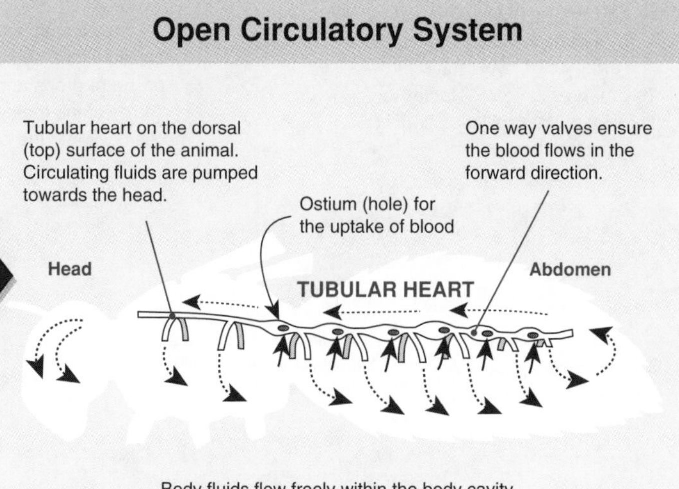


Body fluids flow freely within the body cavity

The circulatory system of crabs is best described as incompletely closed. The thoracic heart has three pairs of ostia a number of arteries, which leave the heart and branch extensively to supply various organs before draining into discrete channel-like sinuses.

In spiders, arteries from the dorsal heart empty the hemolymph into tissue spaces and then into a large ventral sinus that bathes the book lungs where gas exchange takes place. Venous channels conduct the hemolymph back to the heart.

The hemolymph occupies up to 40% of the body mass of an insect and is usually under low pressure due its lack of confinement in vessels. The circulation of the hemolymph is aided by body movements such as the ventilating movements of the abdomen.

1. Explain how an open circulatory system moves fluid (hemolymph) about the body: ____________________

2. Explain why arthropods do not bleed in a similar way to vertebrates: ____________________

3. Compare insects and decapod crustaceans (e.g. crabs) in the degree to which the circulatory system is closed: ____________________

4. (a) Explain why the crab's circulatory system is usually described as an open system: ____________________

   (b) Explain in what way this description is not entirely accurate: ____________________



***Related activities:*** *Closed Circulatory Systems*
***Weblinks:*** *Animal Circulatory Systems*


**ISBN: 978-1-927173-12-1**

# Closed Circulatory Systems

Closed circulatory systems are used by vertebrates, annelids (earthworms) and cephalopods (octopus and squid). The blood is pumped by a heart through a series of arteries and veins. Oxygen is transported around the body by the blood and diffuses through capillary walls into the body cells. Closed circulatory systems are useful for large, active animals where oxygen can not easily be transported to the interior of the body. They also allow the animal more control over the distribution of blood flow by contracting or dilating blood vessels. Closed systems are the most developed in vertebrates where a chambered heart pumps the blood into blood vessels at high pressure. The system can also be divided into two separate circuits, the pulmonary circuit taking up oxygen and the systemic circuit pumping oxygenated blood to the rest of the body.



## INVERTEBRATE CLOSED SYSTEMS

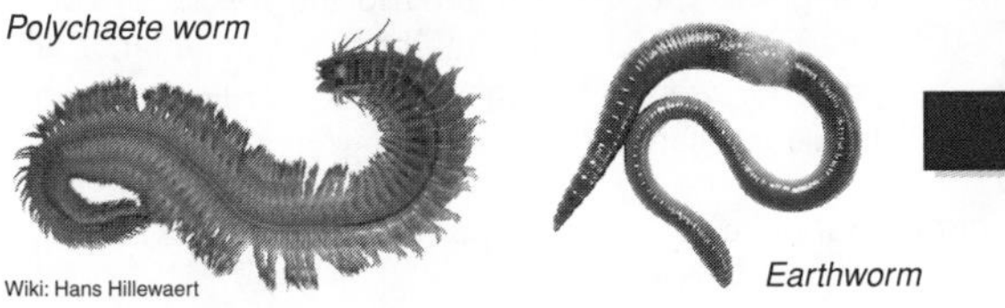


Contractle dorsal blood vessel
Aortic arches
Capillary networks
Head
Tail
Ventral blood vessel

The closed systems of many annelids (e.g. earthworms) circulate blood through a series of vessels before returning it to the heart. In annelids, the dorsal and ventral blood vessels are connected by lateral vessels in every segment (right). The dorsal vessel receives blood from the lateral vessels and carries it towards the head. The ventral vessel carries blood posteriorly and distributes it to the segmental vessels. The dorsal vessel is contractile and is the main method of propelling the blood, but there are also several contractile aortic arches ('hearts') which act as accessory organs for blood propulsion.

## VERTEBRATE CLOSED SYSTEMS

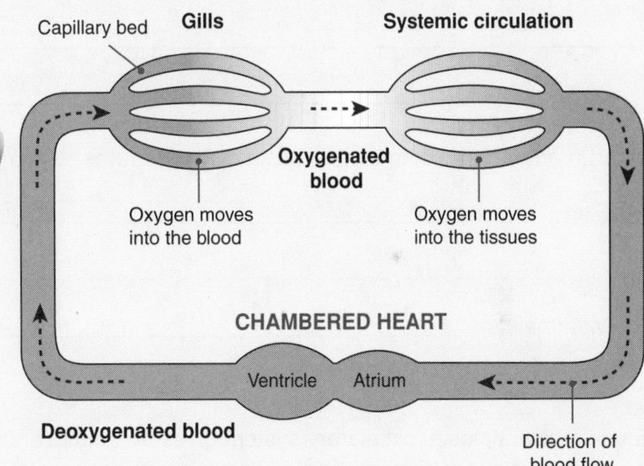


### Closed, single circuit systems

In closed circulation systems, the blood is contained within vessels and is returned to the heart after every circulation of the body. Exchanges between the blood and the fluids bathing the cells occurs by diffusion across capillaries. In single circuit systems, typical of fish, the blood goes directly from the gills to the body. The blood loses pressure at the gills and flows at low pressure around the body.

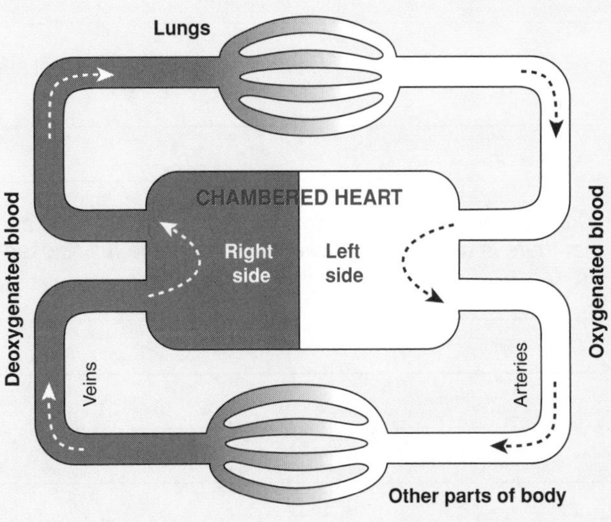


### Closed, double circuit systems

Double circulation systems occur in all vertebrates other than fish. The blood is pumped through a pulmonary circuit to the lungs, where it is oxygenated. The blood returns to the heart, which pumps the oxygenated blood, through a systemic circuit, to the body. In amphibians and most reptiles, the heart is not completely divided and there is some mixing of oxygenated and deoxygenated blood. In birds and mammals, the heart is fully divided and there is no mixing.


ISBN: 978-1-927173-12-1


***Related activities:*** *The Comparative Anatomy of the Heart*
***Weblinks:*** *Animal Circulatory Systems*

Photo: Hans Hillewaert, Wiki CC 2.5

## Cephalopod Mollusks: High Performing Invertebrates

The circulatory system is largely closed in all cephalopod mollusks (nautilus, cuttlefish, squid, and octopus). It has an extensive system of vessels making it the most complex and efficient system of all the mollusks, and enables cephalopods to be active, intelligent predators.

The circulatory system consists of one systemic heart, two branchial hearts, and blood vessels. The branchial hearts, which sit at the gill base, collect deoxygenated blood from all the body parts and direct it through the gills. The blood returns to the medial systemic heart and is pumped to the body via an anterior and posterior aorta, through smaller vessels and into tissue capillaries.

1. Describe the main difference between closed and open systems of circulation: ______________________

2. Describe where the blood flows to immediately after it has passed through the gills in a fish: ______________________

3. Describe where the blood flows immediately after it has passed through the lungs in a mammal: ______________________

4. Explain the higher functional efficiency of a double circuit system, relative to a single circuit system: ______________________

5. Hearts range from being simple contractile structures to complex chambered organs. Describe basic heart structure in:

   (a) Fish: ______________________

   (b) Mammals: ______________________

6. Explain how a closed circulatory system gives an animal finer control over the distribution of blood to tissues and organs:

7. Compare and contrast a vertebrate closed circulatory system with the circulatory system of an annelid: ______________________

8. "*Comparisons of the circulatory systems of insects, decapods (e.g. crabs), annelids, and cephalopod mollusks indicates that there is a gradient between fully open and fully closed circulatory systems*". Discuss this statement:


**ISBN: 978-1-927173-12-1**

# The Human Circulatory System

The blood vessels of the circulatory system form a vast network of tubes that carry blood away from the heart, transport it to the tissues of the body, and then return it to the heart. The arteries, arterioles, capillaries, venules, and veins are organized into specific routes to circulate the blood throughout the body. The figure below shows a number of the basic **circulatory routes** through which the blood travels. Humans, like all mammals have a double circulatory system: a **pulmonary circulation**, which carries blood between the heart and lungs, and a **systemic circulation**, which carries blood between the heart and the rest of the body. The systemic circulation has many subdivisions. Two important subdivisions are the coronary (cardiac) circulation, which supplies the heart muscle, and the **hepatic portal circulation**, which runs from the gut to the liver.

## Schematic Overview of the Human Circulatory System

Deoxygenated blood (colored gray below) travels to the right side of the heart via the vena cavae. The heart pumps the deoxygenated blood to the lungs where it releases carbon dioxide and receives oxygen. The oxygenated blood (colored white below) travels via the pulmonary vein back to the heart from where it is pumped to all parts of the body. The **venous system** (figure, left) returns blood from the capillaries to the heart. The **arterial system** (figure right) carries blood from the heart to the capillaries. **Portal systems** carry blood between two capillary beds.



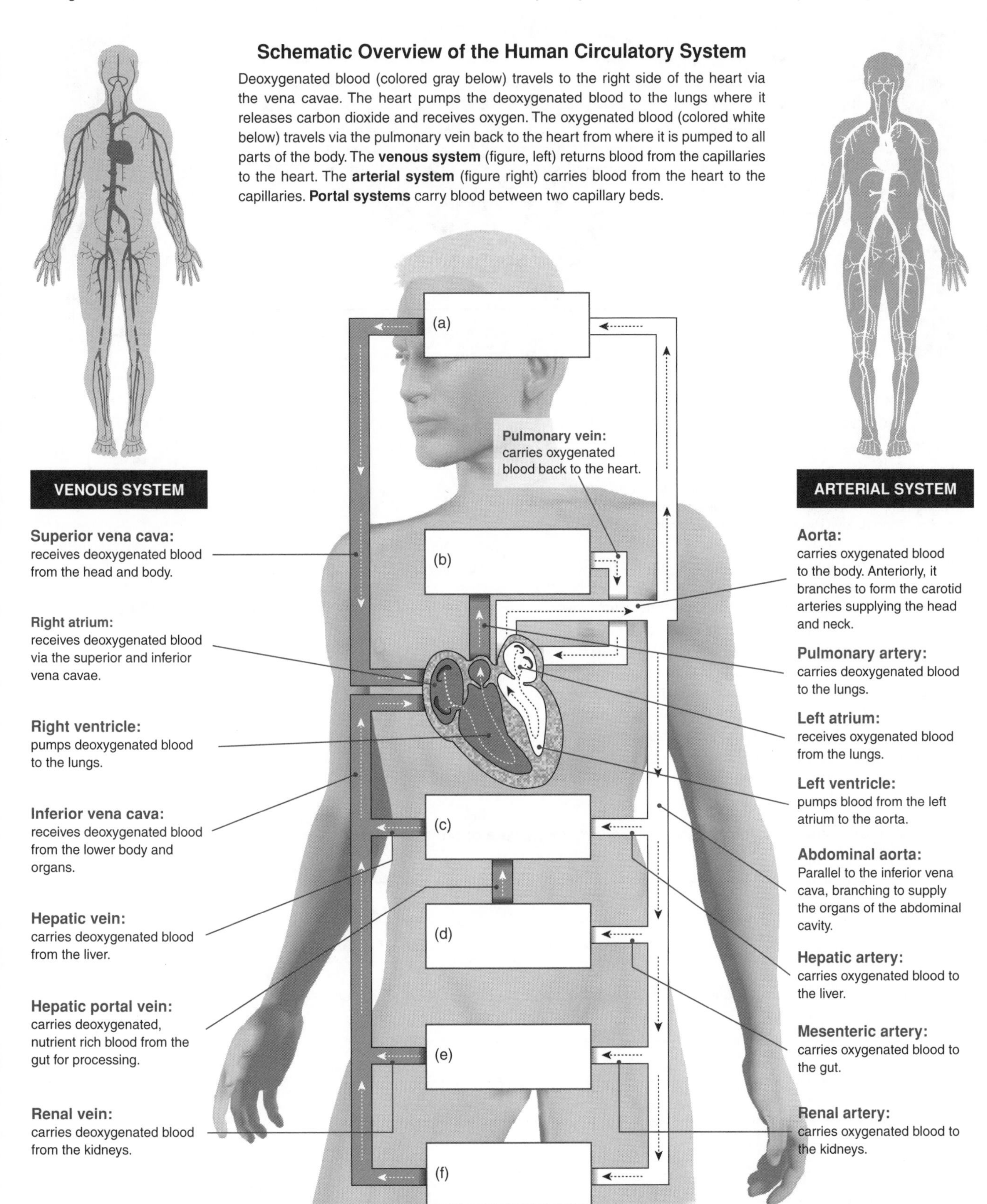

1. Complete the diagram above by labeling the boxes with the organs or structures they represent.

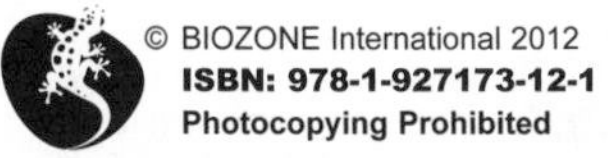


ISBN: 978-1-927173-12-1

# The Liver's Homeostatic Role

The liver is located just below the diaphragm and makes up 3-5% of body weight. It is the body's largest homeostatic organ and interacts with all the other systems of the body. It performs a vast number of functions including production of bile, storage and processing of nutrients, and detoxification of poisons and metabolic wastes. The liver has a **unique double blood supply** and up to 20% of the total blood volume flows through it at any one time. 30% of the blood flowing through the liver comes from the hepatic artery (oxygenated), and 70% comes from the hepatic portal vein (deoxygenated from the small intestine). This rich vascularization makes it the central organ for regulating activities associated with the blood and circulatory system. In spite of the complexity of its function, the liver tissue and the liver cells themselves are structurally relatively simple.

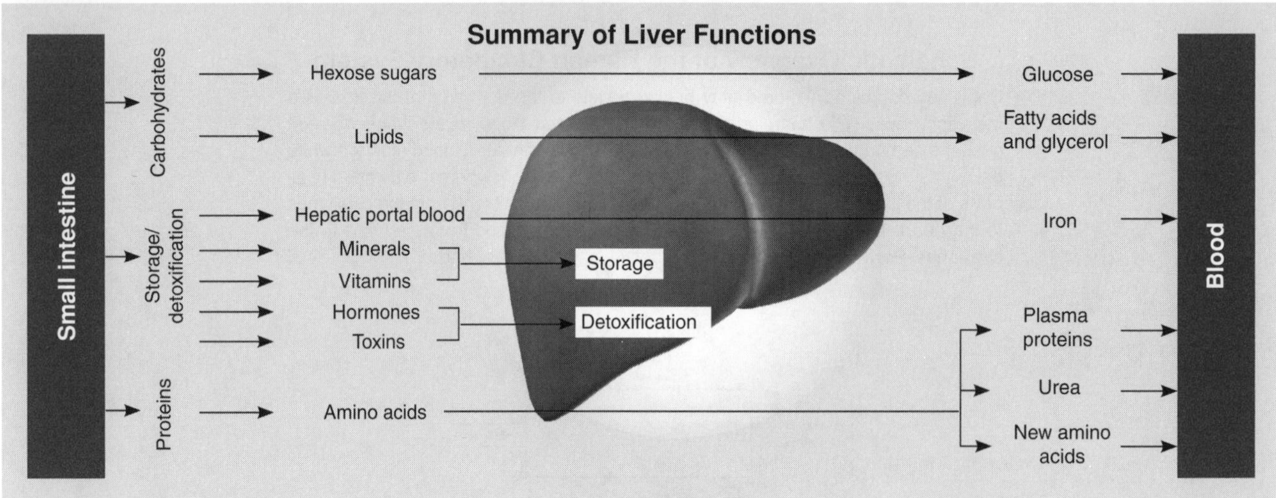


## Homeostatic Functions of the Liver

The liver is one of the largest and most complex organs in the body. It has a central role as an organ of homeostasis and performs many functions, particularly in relation to the regulation of blood composition. General functions of the liver are outlined below. Briefly summarized, the liver:

1. Secretes bile, important in emulsifying fats in digestion.
2. Metabolizes amino acids, fats, and carbohydrates (above).
3. Synthesizes glucose from non-carbohydrate sources when glycogen stores are exhausted (gluconeogenesis).
4. Stores iron, copper, and some vitamins (A, D, E, K, $B_{12}$).
5. Converts unwanted amino acids to urea (urea cycle, right).
6. Manufactures heparin and plasma proteins (e.g. albumin).
7. Detoxifies poisons or turns them into less harmful forms.
8. Some liver cells phagocytose worn-out blood cells.
9. Synthesizes cholesterol from acetyl coenzyme A.

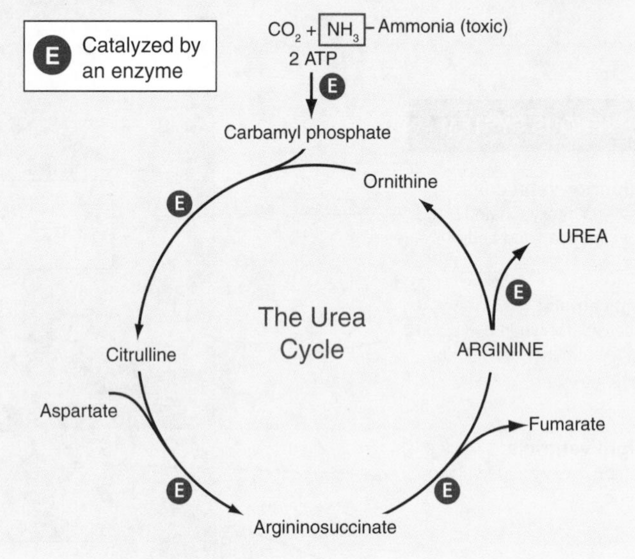

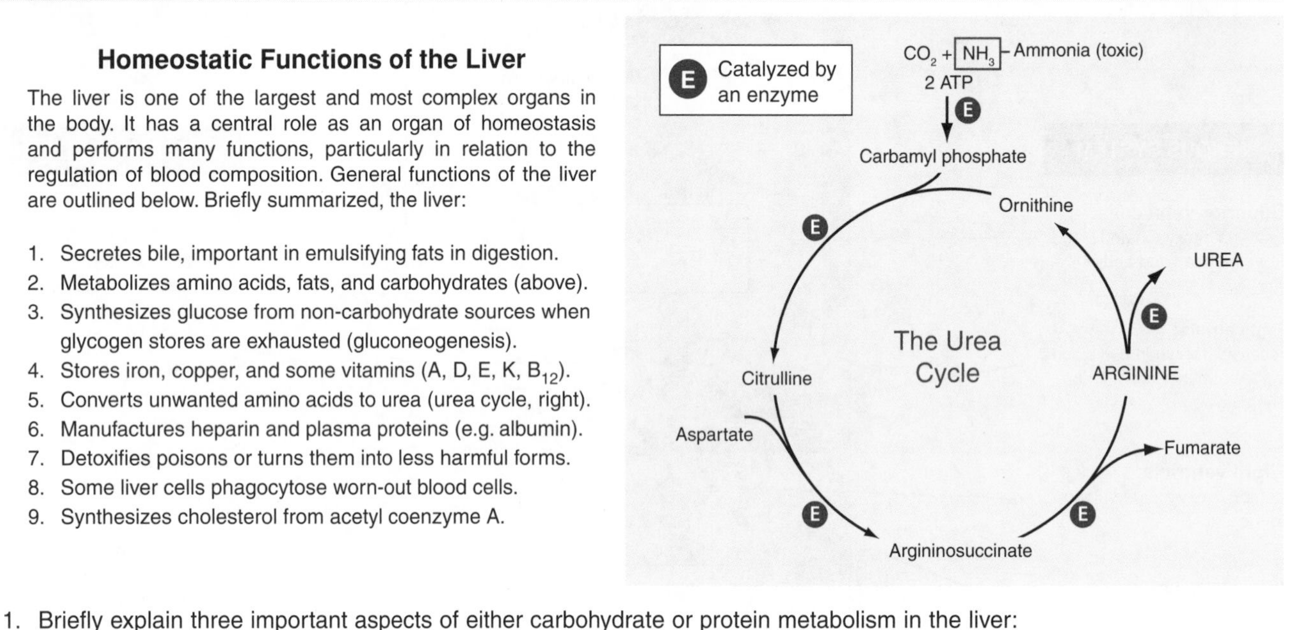

1. Briefly explain three important aspects of either carbohydrate or protein metabolism in the liver:

(a) ______________________________

(b) ______________________________

(c) ______________________________

2. Explain how the liver's interactions with other body systems make it central to homeostatic regulation:

______________________________



*Related activities:* Nitrogenous Wastes in Animals
*Weblinks:* Functions of the Liver

*Periodicals:* Metabolic powerhouse


**ISBN: 978-1-927173-12-1**

# The Comparative Anatomy of the Heart

In vertebrates, the heart shows a sequential increase in complexity from fish through to mammals. In fish, the heart is linear and contains two major chambers in series, and on the venous side there is an enlarged chamber or **sinus** on the vein (the sinus venosus). In mammals, the heart comprises four chambers - two pumps side by side - and there are large pressure differences between the pulmonary (lung) and systemic (body) circulations. The three chambered heart of amphibians reflects, in part, their incomplete shift to terrestrial life. Although the ventricle is undivided, a baffle-like spiral valve at the exit point of the ventricle helps to separate the arterial and venous flows and there is limited mixing of oxygenated and deoxygenated blood. The pulmonary circuit in amphibians also sends branches to the skin, reflecting its importance in oxygen uptake.

## Fish Heart

From rest of body
Sinus venosus
Atrium
Ventricle
Conus arteriosus
To gills
Ventral aorta
Atrioventricular valves
Sinoatrial valves
From rest of body

The fish heart is linear, with a sequence of three chambers in series (the conus may be included as a fourth chamber). Blood from the body first enters the heart through the sinus venosus, then passes into the atrium and the ventricle. A series of one-way valves between the chambers prevents reverse blood flow. Blood leaving the heart travels to the gills.

## Amphibian Heart

To lungs
To rest of body
From rest of body
From lungs
Right atrium
Left atrium
Ventricle (single chamber)

Amphibian hearts are three chambered. The atrium is divided into left and right chambers, but the ventricle lacks an internal dividing wall. Although this allows mixing of oxygenated and deoxygenated blood, the spongy nature of the ventricle reduces mixing. Amphibians are able to tolerate this because much of their oxygen uptake occurs across their moist skin, and not their lungs.

## Mammalian Heart

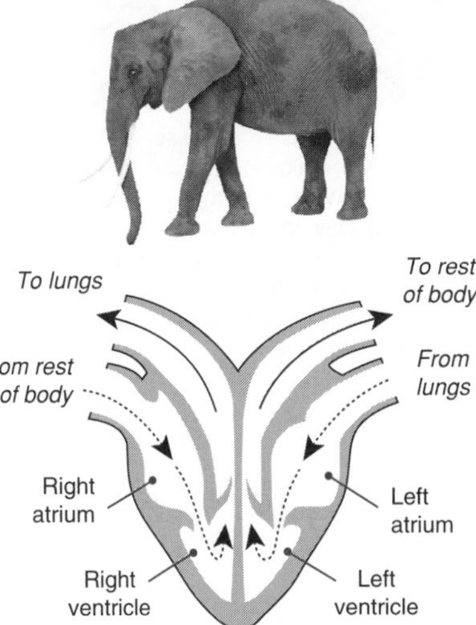


In birds and mammals, the heart is fully partitioned into two halves, resulting in four chambers. Blood circulates through two circuits, with no mixing of the two. Oxygenated blood from the lungs is kept separated from the deoxygenated blood returning from the rest of the body.

## Heart Size and Rate in Mammals

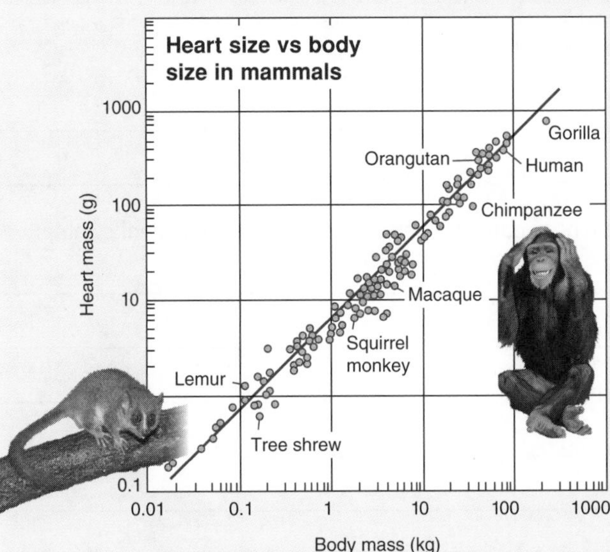


Heart size in mammals (above) increases with body size, but relative to body size, small and large mammals have the same heart size. Irrespective of body size, the size of the heart in mammals is 0.59% of body size. Humans and other primates fall within the range of other mammals.

*Adapted from Schmidt-Nielsen, 1979: Animal Physiology, Adaptation and Environment*

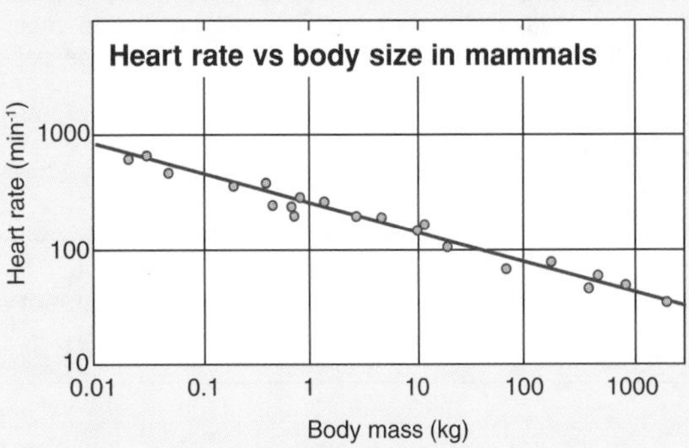


In contrast to the heart size-body size relationship, heart rate (above) is inversely related to body size. An elephant (3000 kg) has a resting pulse of around 25 beats per minute, compared with a 3 g shrew (the smallest mammal), which has a resting heart rate of over 600 beats per minute. The relationship is identical to that between body mass and oxygen consumption per unit body weight. The information from these two figures tell us that, in mammals:

1. The size of the heart (the pump) remains a constant percentage of body size, and...
2. The increase in heart rate in smaller mammals is in exact proportion to the need for oxygen.

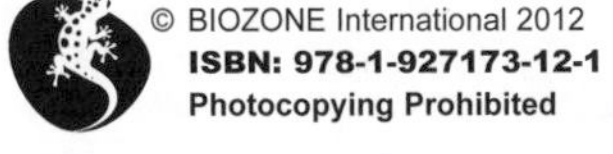


**ISBN: 978-1-927173-12-1**


***Related activities:*** *The Mammalian Heart*

# Evolution of the Heart and Circulatory Systems in Vertebrates

The structure of the vertebrate heart is a response to an enlarged body and an active lifestyle. Modifications to it reflect the major changes in lifestyle of the vertebrates, most importantly from living in water to living on land.

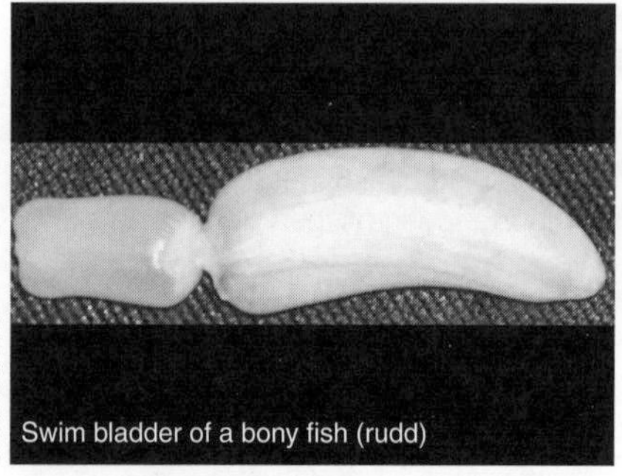

Uwe Gille CC 3.0

Swim bladder of a bony fish (rudd)

Amphibian (frog)

The development of accessory air breathing organs was a major step in the evolution of the heart. Outgrowths from the stomachs of fishes provided a way to increase oxygen supply in poorly oxygenated waters and adjust buoyancy. In fish, these outgrowths developed into the swim bladder. In ancestral amphibians, they developed into lungs.

Blood circulation in lungfish shows how a double circulatory system might have developed. Blood from the lungs follows a partially separated circuit. Modifications to the heart in **amphibians** produced three chambers in which blood from both the lungs and the body enter the single ventricle. Further modifications in early reptiles produced incomplete division within the ventricle.

Birds and mammals have four chambered hearts with a completely divided double circulatory system. The arrangement of their circulatory systems reflects their separate ancestries. Birds have lost what was the left hand part of the circulatory system in early reptiles. Mammals, on the other hand, have lost the right hand part of the system.

## Basic heart and circulatory patterns in vertebrates

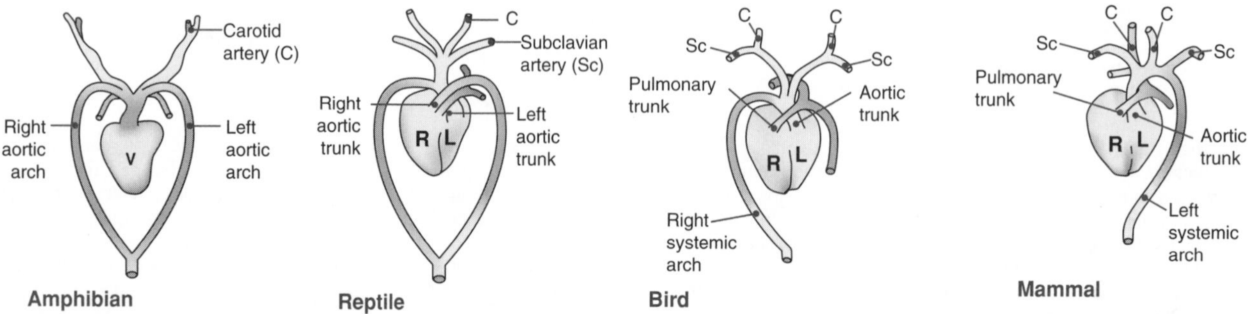


For an ancestral vertebrate, having a supplementary organ to provide air assisted survival in poorly oxygenated environments (such as swamps). During the evolution of lungs, the circulatory system began to take more blood through the lungs and less through the gills, which were eventually lost. The evolution of a double circulatory system, formed by separating the ventricle, allowed more efficient extraction and delivery of oxygen. This development was needed to meet the highly active lifestyles of birds and mammals. The Tbx5 gene encodes the Tbx5 transcription factor, and has a role in the development of the **septum**, the wall which divides the ventricle. A gradient of Tbx5 in the heart during embryonic development is needed for the septum to form. In embryonic lizards, Tbx5 is expressed evenly across the heart, there is no Tbx5 gradient, so no septum forms. Turtles have a weak gradient so a partial septum forms. In bird and mammals, there is a strong gradient between the left and right sides of the ventricle during embryonic development, so a full septum forms.

1. Describe the function of the aorta: ______________________________

2. Describe the function of the vena cava: ______________________________

3. Describe two advantages of possessing an accessory air breathing organ for a vertebrate living in an ancient swamp:

______________________________

______________________________

4. What evidence is there that birds and mammals evolved four chambered hearts independently: ______________________________

______________________________

______________________________

5. Discuss the structure of the heart in at least two vertebrate classes, relating features of the heart's structure and function to the animal's size, metabolic rate, or environment in each case

______________________________

______________________________

______________________________

______________________________


ISBN: 978-1-927173-12-1

# The Mammalian Heart

The heart is the centre of the human cardiovascular system. It is a hollow, muscular organ, weighing on average 342 grams. Each day it beats over 100,000 times to pump 3780 liters of blood through 100,000 kilometers of blood vessels. It comprises a system of four muscular chambers (two **atria** and two **ventricles**) that alternately fill and empty of blood, acting as a double pump. The left side pumps blood to the body tissues and the right side pumps blood to the lungs. The heart lies between the lungs, to the left of the body's midline, and it is surrounded by a double layered **pericardium** of tough fibrous connective tissue. The pericardium prevents overdistension of the heart and anchors the heart within the **mediastinum**.

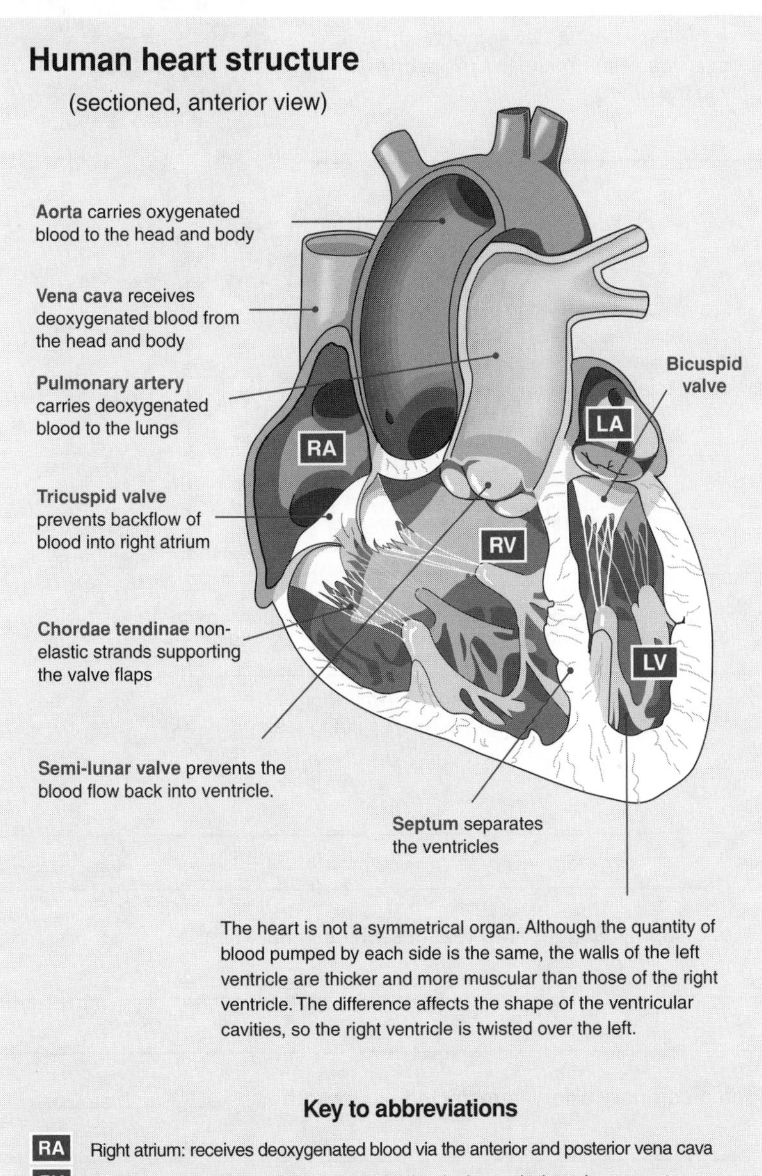


The heart is not a symmetrical organ. Although the quantity of blood pumped by each side is the same, the walls of the left ventricle are thicker and more muscular than those of the right ventricle. The difference affects the shape of the ventricular cavities, so the right ventricle is twisted over the left.

**Key to abbreviations**

- **RA** Right atrium: receives deoxygenated blood via the anterior and posterior vena cava
- **RV** Right ventricle: pumps deoxygenated blood to the lungs via the pulmonary artery
- **LA** Left atrium: receives blood returning to the heart from the lungs via the pulmonary veins
- **LV** Left ventricle: pumps oxygenated blood to the head and body via the aorta

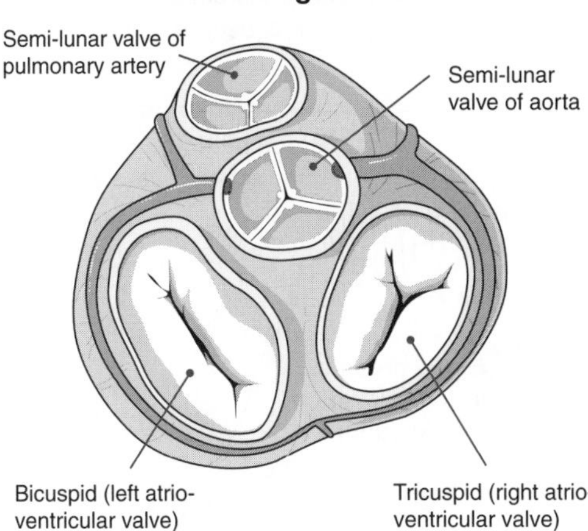


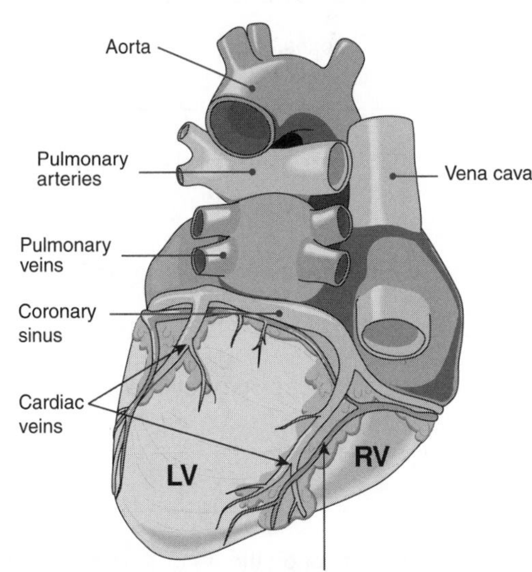


**Coronary arteries**: The high oxygen demands of the heart muscle are met by a dense capillary network. Coronary arteries arise from the aorta and spread over the surface of the heart supplying the cardiac muscle with oxygenated blood. Deoxygenated blood is collected by cardiac veins and returned to the right atrium via a large coronary sinus.

1. In the schematic diagram of the heart, below, label the four chambers and the main vessels entering and leaving them. The arrows indicate the direction of blood flow. Use large coloured circles to mark the position of each of the four valves.

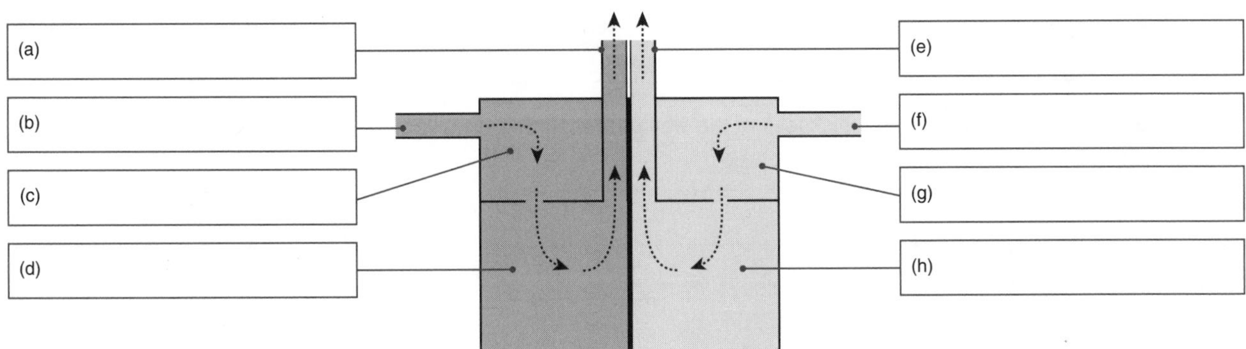



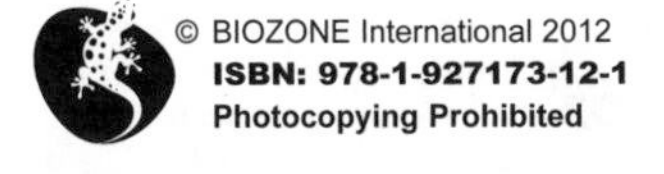

**ISBN: 978-1-927173-12-1**

## Pressure Changes and the Asymmetry of the Heart

The heart is not a symmetrical organ. The left ventricle and its associated arteries are thicker and more muscular than the corresponding structures on the right side. This asymmetry is related to the necessary pressure differences between the pulmonary (lung) and systemic (body) circulations (not to the distance over which the blood is pumped *per se*). The graph below shows changes in blood pressure in each of the major blood vessel types in the systemic and pulmonary circuits (the horizontal distance not to scale). The pulmonary circuit must operate at a much lower pressure than the systemic circuit to prevent fluid from accumulating in the alveoli of the lungs. The left side of the heart must develop enough "spare" pressure to enable increased blood flow to the muscles of the body and maintain kidney filtration rates without decreasing the blood supply to the brain.

aorta, 100 mg Hg

radial artery, 98 mg Hg

arterial end of capillary, 30 mg Hg

Pressure (mm Hg)

120, 100, 80, 60, 40, 20, 0

**Blood pressure during contraction (systole)**

**Blood pressure during relaxation (diastole)**

The greatest fall in pressure occurs when the blood moves into the capillaries, even though the distance through the capillaries represents only a tiny proportion of the total distance travelled.

aorta | arteries | **A** | capillaries | **B** | veins | vena cava | pulmonary arteries | **C** | **D** | venules | pulmonary veins

Systemic circulation horizontal distance not to scale

Pulmonary circulation horizontal distance not to scale

2. What is the purpose of the valves in the heart? ______________________

3. The heart is full of blood, yet it requires its own blood supply. Suggest two reasons why this is the case:

   (a) ______________________

   (b) ______________________

4. Predict the effect on the heart if blood flow through a coronary artery is restricted or blocked: ______________________

5. Identify the vessels corresponding to the letters **A-D** on the graph above:

   A: __________ B: __________ C: __________ D: __________

6. (a) Why must the pulmonary circuit operate at a lower pressure than the systemic system? ______________________

   (b) Relate this to differences in the thickness of the wall of the left and right ventricles of the heart: ______________________

7. What are you recording when you take a pulse? ______________________

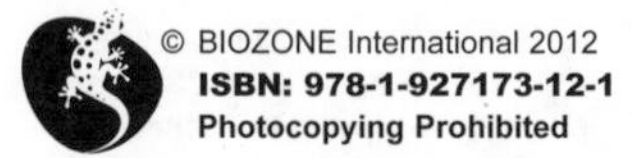


**ISBN: 978-1-927173-12-1**

# The Control of Heart Activity

Given adequate supplies of oxygen, the heart will continue to beat when removed from the body. Thus, the origin of the heart-beat is **myogenic**, i.e. contraction is an intrinsic property of the cardiac muscle itself. The heart-beat is regulated by a conduction system consisting of the pacemaker (**sinoatrial node** or SAN) and a specialized conduction system of **Purkyne tissue**. The pacemaker sets a basic rhythm for the heart, but this rate is influenced by the cardiovascular control center in the brainstem, which alters heart rate via parasympathetic and sympathetic nerve impulses. Changing the rate and force of heart contraction is the main mechanism for controlling cardiac output in order to meet changing demands.

## Generation of the Heartbeat

The basic rhythmic heartbeat is **myogenic**. The nodal cells (SAN and atrioventricular node) spontaneously generate rhythmic action potentials without neural stimulation. The normal resting rate of self-excitation of the SAN is about 50 beats per minute.

The amount of blood ejected from the left ventricle per minute is called the **cardiac output**. It is determined by the **stroke volume** (the volume of blood ejected with each contraction) and the **heart rate** (number of heart beats per minute).

**Cardiac muscle** responds to stretching by contracting more strongly. The greater the blood volume entering the ventricle, the greater the force of contraction. This relationship is known as **Starling's Law.**

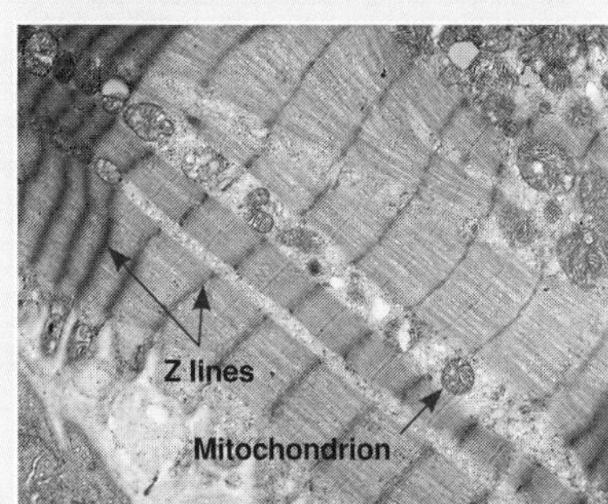


TEM of cardiac muscle showing striations in a fiber (muscle cell). Each fiber has one or two nuclei and many large mitochondria. Note the Z lines that delineate the contractile units (or sarcomeres) of the rod-like units (myofibrils) of the fiber. The fibers are joined by specialized electrical junctions called Intercalated discs, which allow impulses to spread rapidly through the heart muscle.

**Sinoatrial node** (SAN) is also called the **pacemaker**. It is a small mass of specialized muscle cells on the wall of the right atrium, near the entry point of the superior vena cava. The pacemaker initiates the cardiac cycle, spontaneously generating **action potentials** that cause the atria to contract. The SAN sets the basic heart rate, but this rate is influenced by hormones (e.g. epinephrine) and impulses from the autonomic nervous system.

**Atrioventricular node** (AVN) at the base of the atrium briefly delays the impulse to allow time for the atrial contraction to finish before the ventricles contract.

**Bundle of His** (atrioventricular bundle) A tract of conducting (Purkyne) fibers that distribute the action potentials over the ventricles causing ventricular contraction.

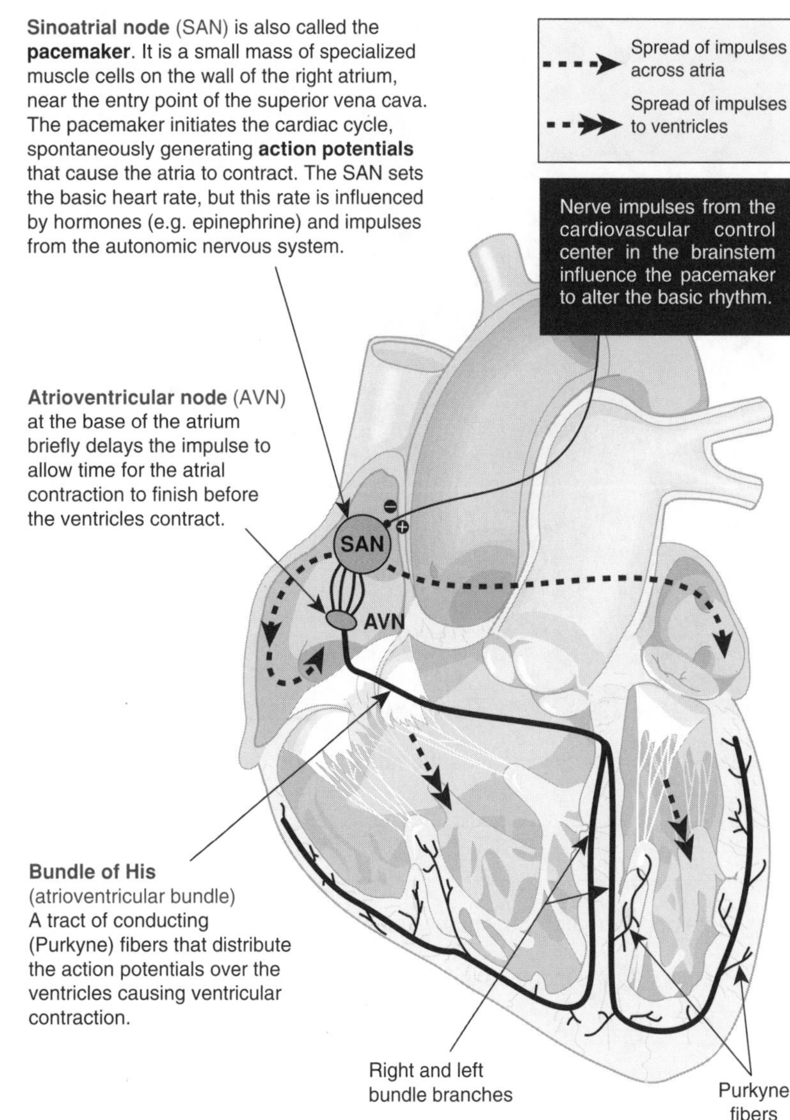


1. Describe the role of each of the following in heart activity:

   (a) The sinoatrial node: ______

   (b) The atrioventricular node: ______

   (c) The bundle of His: ______

   (d) Intercalated discs: ______

2. What is the significance of delaying the impulse at the AVN? ______

3. What is the advantage of the physiological response of cardiac muscle to stretching? ______

4. The heart-beat is intrinsic. Why is it important to be able to influence the basic rhythm via the central nervous system?

ISBN: 978-1-927173-12-1


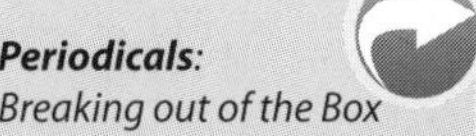

# Exercise and Blood Flow

Exercise promotes health by improving the rate of blood flow back to the heart (venous return). This is achieved by strengthening all types of muscle and by increasing the efficiency of the heart. During exercise blood flow to different parts of the body changes in order to cope with the extra demands of the muscles, the heart and the lungs.

1. The following table gives data for the **rate** of blood flow to various parts of the body at rest and during strenuous exercise. **Calculate** the **percentage** of the total blood flow that each organ or tissue receives under each regime of activity.

| Organ or tissue | At rest | | Strenuous exercise | |
|---|---|---|---|---|
| | $cm^3 min^{-1}$ | % of total | $cm^3 min^{-1}$ | % of total |
| Brain | 700 | 14 | 750 | 4.2 |
| Heart | 200 | | 750 | |
| Lung tissue | 100 | | 200 | |
| Kidneys | 1100 | | 600 | |
| Liver | 1350 | | 600 | |
| Skeletal muscles | 750 | | 12 500 | |
| Bone | 250 | | 250 | |
| Skin | 300 | | 1900 | |
| Thyroid gland | 50 | | 50 | |
| Adrenal glands | 25 | | 25 | |
| Other tissue | 175 | | 175 | |
| TOTAL | 5000 | 100 | 17 800 | 100 |

2. Explain how the body increases the rate of blood flow during exercise: ______________________

3. (a) State approximately how many times the total rate of blood flow increases between rest and exercise: __________

   (b) Explain why the increase is necessary: ______________________

4. (a) Identify which organs or tissues show no change in the rate of blood flow with exercise: __________

   (b) Explain why this is the case: ______________________

5. (a) Identify the organs or tissues that show the most change in the rate of blood flow with exercise: __________

   (b) Explain why this is the case: ______________________


ISBN: 978-1-927173-12-1

# Adaptations to High Altitude

Air at high altitude contains less oxygen than air at sea level. Air pressure decreases with altitude so the pressure (therefore amount) of oxygen in the air also decreases. Sudden exposure to an altitude of 2000 m causes breathlessness and above 7000 m most people would become unconscious. The effects of altitude on physiology are related to this lower oxygen availability. Humans and other animals can make some physiological adjustments to life at altitude. These changes are referred to **acclimatization** (below left), and are different to the evolutionary adaptations of high altitude populations, which are inherited.

## Mountain Sickness

Altitude sickness or mountain sickness is usually a mild illness associated with trekking to altitudes of 5000 meters or so. Common symptoms include headache, insomnia, poor appetite and nausea, vomiting, dizziness, tiredness, coughing and breathlessness. The best way to avoid mountain sickness is to ascend to altitude slowly (no more than 300 m per day above 3000 m). Continuing to ascend with mountain sickness can result in more serious illnesses: accumulation of fluid on the brain (cerebral edema) and accumulation of fluid in the lungs (pulmonary edema). These complications can be fatal if not treated with oxygen and a rapid descent to lower altitude.

### Physiological Adjustment to Altitude

| Effect | Minutes | Days | Weeks |
|---|---|---|---|
| Increased heart rate | ←→ | ←→ | |
| Increased breathing | ←→ | ←→ | |
| Concentration of blood | | ←→ | |
| Increased red blood cell production | | ←→ | ←→ |
| Increased capillary density | | | ←→ |

The human body can make adjustments to life at altitude. Some of these changes, e.g. increased breathing and heart rates, take place almost immediately. Other adjustments may take weeks (see above). These responses are all aimed at improving the rate of supply of oxygen to the body's tissues. When more permanent adjustments to physiology are made (increased blood cells and capillary networks), heart and breathing rates can return to normal.

## Adaptations to High Altitude

Studies on three high altitude human populations (Andean, Tibetan, and Ethiopian highlands) show three different adaptations for life at high altitude. Populations living in the Andean Altiplano in South America have higher concentrations of hemoglobin than populations living at sea level.

Tibetans have similar hemoglobin concentrations to human populations at sea level, but have an innately higher ventilation rate to compensate for the low oxygen pressure (15 L min$^{-1}$ compared with the 10.5 L min$^{-1}$ of an Andean). Their lungs also synthesize larger amounts of nitric oxide from the air, which increases the diameter of blood vessels, so that a lower oxygen content in the blood is offset by increased blood flow.

Tibetan porter

### Concentration of Hemoglobin in the Blood

| Group | Hemoglobin concentration (g 100 mL$^{-1}$) |
|---|---|
| U.S. sea level mean | 15.3 |
| Andean male | 19.2 |
| Andean female | 17.8 |
| Tibetan male | 15.6 |
| Tibetan female | 14.2 |
| Ethiopian highlands male | 15.9 |
| Ethiopian highlands female | 15.0 |

Recent studies on Ethiopian highlanders (altitude of >1500 m) have found that they do not experience **hypoxia** (lack of oxygen), yet have none of the physiological adaptations seen in the Andeans or Tibetans. Their hemoglobin concentration is no higher than sea level populations and they do not have a higher ventilation rate.

Tolerance of low oxygen in indigenous Ethiopian highlanders may be the result of changes to the hypoxia-inducible factor-1 (HIF-1) pathway. This pathway controls systemic and cellular responses to hypoxic conditions. Several genes are activated at low oxygen concentrations to control the HIF-1 pathway. Positive selection may be acting on these genes in Ethiopian highlanders.

1. (a) Describe the general effects of high altitude on the body: ______

   (b) Name the general term given to describe these effects: ______

2. (a) Name one short term physiological adaptation that humans make to high altitude: ______

   (b) Explain how this adaptation helps to increase the amount of oxygen the body receives: ______

3. Suggest why the altitude adaptations of indigenous highland populations is of great interest to evolutionary biologists: ______

ISBN: 978-1-927173-12-1


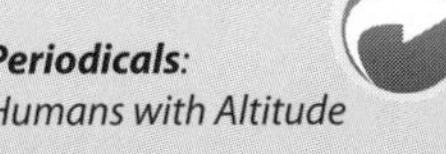

# Physiological Responses to Toxic Substances

Humans living in the twenty first century are exposed to a wide range of toxic substances every day. Some cause no noticeable effect unless encountered at high concentrations (e.g. carbon monoxide), others may build up in the body over many years before being noticeable e.g. lead) and others may cause death within moments of contact. Most of these substances are toxic because of the way they interact with the body's many physiological systems.

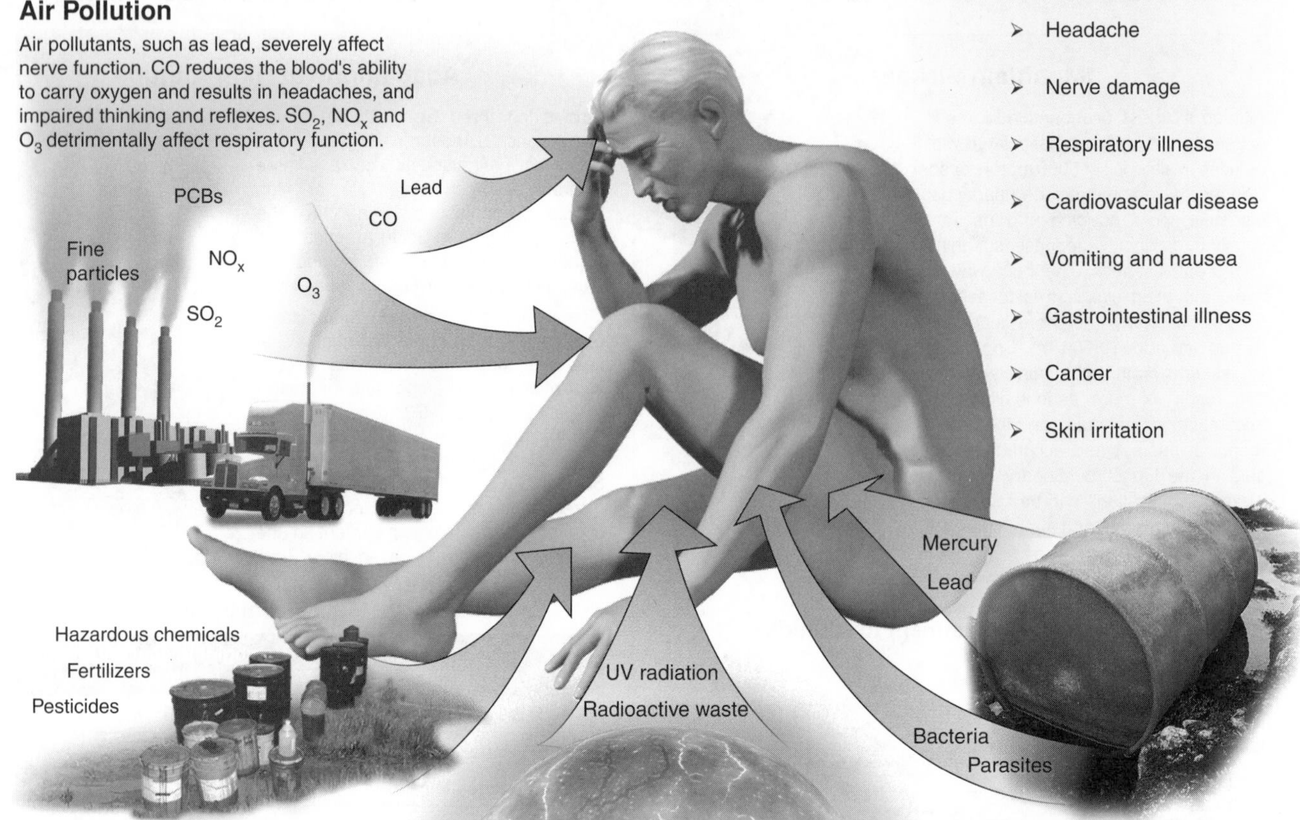


### Soil Contamination

Pesticides based on organophosphates are extremely toxic to humans and other mammals. Fertilizers can cause methemoglobinemia, cancer, and respiratory illness.

### Radiation

UV radiation from the sun causes thousands of cases of melanoma skin cancer every year, while radioactive waste can cause cancer and genetic defects in fetuses.

### Water Pollution

Heavy metals such as mercury and lead can cause nerve damage, while bacteria such as cholera or parasites such as giardia cause intestinal illness.

Removing lead based paint

Lead poisoning is one of the oldest known work place hazards, and there is no safe level of exposure. Lead mimics many metals used as cofactors by enzymes but inhibits their actions.

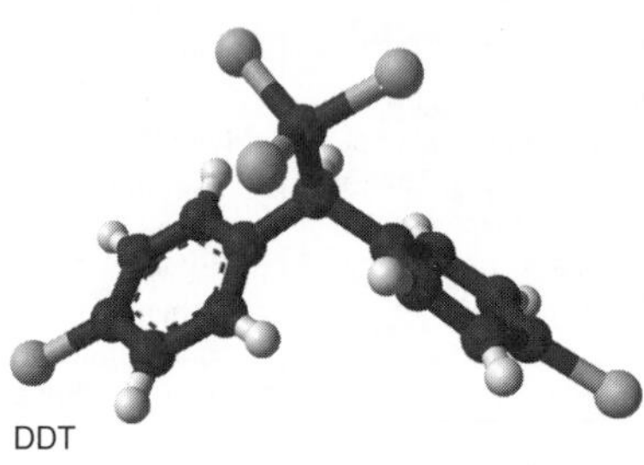

DDT

Some substances mimic hormones. DDT (now banned in many countries) is chemically similar enough to estrogen to trigger reproductive problems in animals. It can also cause cancer.

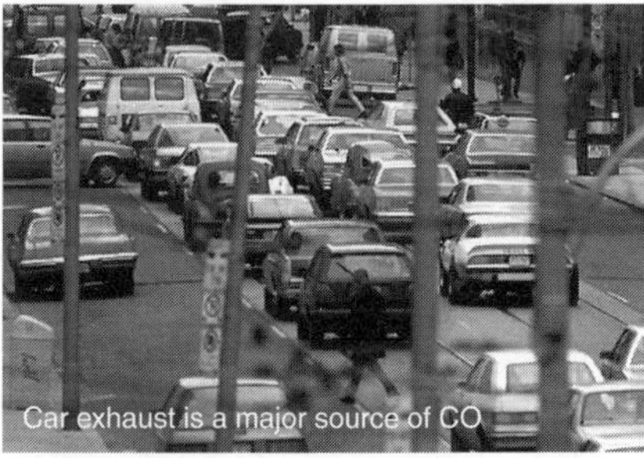

Car exhaust is a major source of CO

Hemoglobin has a greater affinity to CO than for $O_2$. CO is therefore preferentially transported in red blood cells. High concentrations inhibit $O_2$ transport, causing death.

1. Identify three types of pollutants to which people are often exposed and their effects on human health:

2. Discuss the social and economic costs of these pollutants:




ISBN: 978-1-927173-12-1

# KEY TERMS: Word Find

Use the clues below to find the relevant key terms in the WORD FIND grid

A O O S I N O A T R I A L N O D E T Z K I J B E V

T O I L L S R C L O S E D K H V B W D J V I V K E

R H T I S S U E F L U I D S W X F V Q G Y K O E O

I I C V E I N L M M I R Y O P E N P R B P I R H J

U Q J T Q P G G Q Z S A H A W E B X A Y T U T L B

M X T S Y S T E M I C C I R C U L A T I O N E A L

S U R F A C E A R E A V O L U M E I A D A G D R O

X X P B R E S P I R A T O R Y P I G M E N T U T O

Z H E M O L Y M P H B U F F E R F B Y Y H M B E D

S I N G L E P H O G Y G N S A B U L K F L O W R M

K R S O Q M V F K Z O A Y N V E N T R I C L E Y J

W D H T Z U Z B J N V T A O C A P I L L A R Y D I

A T R I O V E N T R I C U L A R N O D E O Y G O O

U M C H E M O R E C E P T O R S G N A F W B R U R

F K D P U L M O N A R Y C I R C U L A T I O N B F

S T O X I C F L R R L M H E A R T Z L Y M P H L Q

B Y A V J G E Z U V B Y B L O O D V E S S E L E O

A large blood vessel with a thick, muscled wall which carries blood away from the heart.

A chamber of the heart that receives blood from the body or lungs.

Circulatory fluid comprising numerous cell types, which moves respiratory gases and nutrients around the body.

Sensory structures that detect changes in hydrogen ion concentration in the blood and so respond to changes in pH.

A vein is an example of this.

The smallest blood vessel. May have a wall only one or two cells thick.

The combination of the interstitial fluid and blood in insects and other invertebrates.

Muscular organ used to pump blood about the body. In vertebrates this may have up to four chambers while in invertebrates it may consist only of specialized contractile blood vessels.

Specialized tissue between the right atrium and the right ventricle of the heart. It delays the delivery of impulse between the atria and ventricles.

A clear fluid contained within the lymphatic system. Similar in composition to the interstitial fluid.

Circulatory system in which blood travels from the heart to lungs and back before being pumped to the rest of the body.

Circulatory system in which the fluid (hemolymph) in the body cavity directly bathes the organs. Is not completely enclosed within vessels.

That part of a double circulatory system in air breathing vertebrates that transports blood from the heart to the lungs and back.

A substance that is damaging or poisonous to living things is called this.

A substance that is able to resist changes to pH.

A ratio that expresses the surface area of a structure relative to its volume.

That part of a double circulatory system that transports blood from the heart to the body and back.

The specialized cardiac cells that initiate the cardiac cycle, and set the basic heart rate. Also call the pacemaker.

A fluid derived from the blood plasma by leakage through capillaries. It bathes the tissues and is also called interstitial fluid.

Large blood vessel that returns blood to the heart.

A chamber of the heart that pumps blood into arteries.

Circulatory system in which blood travels from the heart to the gills and then to the body without first returning to the heart.

A substance present in blood that is able to bind oxygen for transport to cells, e.g. hemoglobin.

A circulatory system in which the blood is fully contained within vessels.

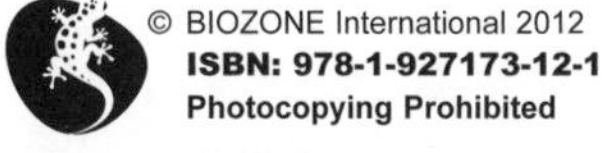


ISBN: 978-1-927173-12-1

Enduring Understanding

**2.D**

# Defense Mechanisms

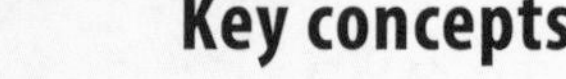

## Key concepts

- The body can distinguish self from non-self.
- Organisms defend themselves against pathogens using a variety of chemical defenses.
- Non-specific defenses target any foreign material.
- The immune response in mammals targets specific antigens and remembers previously encountered antigens.
- Some pathogens, through their specific mode of action, cause immune system failure.

## Key terms

active immunity
AIDS
antibody (=immunoglobulin)
allergies
antigen
autoimmune disease
B cell (=B lymphocyte)
cell-mediated immunity
clonal selection
disease
HIV
humoral immunity
immune response
immunity
immunological memory
infection
inflammation
interferon
leukocyte
lymphocyte
MHC antigens
monoclonal antibody
non-specific defense
passive immunity
pathogen
phagocyte
phagocytosis
primary response
secondary response
specific defense
T cell (=T lymphocyte)
vaccination (=immunization)

***Periodicals:***
*Listings for this chapter are on page 374*

***Weblinks:***
*www.thebiozone.com/weblink/AP2-3121.html*

*BIOZONE APP: Student Review Series Lymphatic System & Immunity*

## Essential Knowledge

☐ 1. Use the **KEY TERMS** to compile a glossary for this topic.

### Non-Specific Defenses *(2.D.4: a)* pages 169-170

☐ 2. Describe the range of **pathogens** that affect organisms and their distinguishing features with respect to **infection**, **disease**, and treatment. Include reference to viruses, bacteria, fungi, and protozoans. In a general way, understand how pathogens are transmitted between individuals.

☐ 3. Describe passive and **active defenses** in plants, recognizing chemical defenses operating at the cellular level after passive defenses have been breached.

☐ 4. Describe **non-specific defenses** in invertebrates, recognizing similarities with equivalent systems operating in vertebrates. Appreciate that invertebrates generally lack sophisticated pathogen-specific responses.

☐ 5. Describe non-specific defenses in vertebrates, e.g. **antimicrobial substances**, **inflammation**, and **phagocytosis**. Appreciate that non-specific defenses in vertebrates represent the first and second lines of defense against pathogens.

### The Mammalian Immune System *(2.D.4: b)* pages 171-188

☐ 6. Explain how the body distinguishes self from non-self and why this is important. Describe the consequences of the self-recognition system failing.

☐ 7. Describe the **immune response**, including the importance of both **specificity** and **memory**. Distinguish between naturally acquired and artificially acquired immunity and between **active** and **passive immunity**.

☐ 8. Describe **cell-mediated immunity** and **humoral** (antibody-mediated) immunity, identifying the specific white blood cells involved in each case.

☐ 9. Describe **clonal selection** and the basis of **immunological memory**. Explain how the immune system is able to respond to the large and unpredictable range of potential antigens.

☐ 10. Explain **antibody** production, including how B cells bring about humoral (antibody-mediated) immunity to specific **antigens**.

☐ 11. Explain the principles of **vaccination**, including reference to the **primary** and **secondary response** to infection and the role of these.

☐ 12. EXTENSION: Describe the applications of **monoclonal antibodies**.

### Immune Dysfunction and Disease pages 172, 189

☐ 13. Describe the consequences of inappropriate immune responses, e.g. in allergies and other **hypersensitivity** reactions (also see #6).

☐ 14. Describe the effects of **HIV** on the immune system, including the reduction in the number of active lymphocytes and the loss of immune function.

# Chemical Defenses In Animals

Living organisms are under constant attack from many different pathogens (see below). As a result, organisms have developed many ways to reduce infection by **pathogens** (organisms that cause disease). Plants and animals have a number of defenses, including **chemical defenses**, to destroy the pathogen. It was initially thought that invertebrates lacked specific immunity, but recent findings suggest that invertebrates do have specific immunity, but it is different to that seen in vertebrates. A comparison of the chemical defenses in vertebrates and invertebrates is given below.

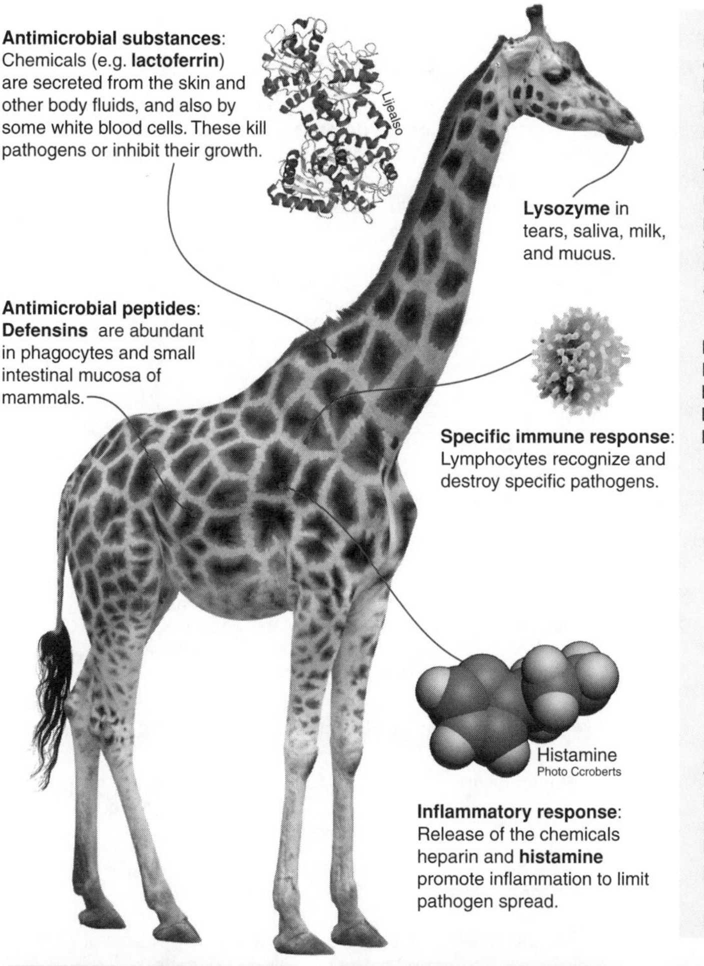


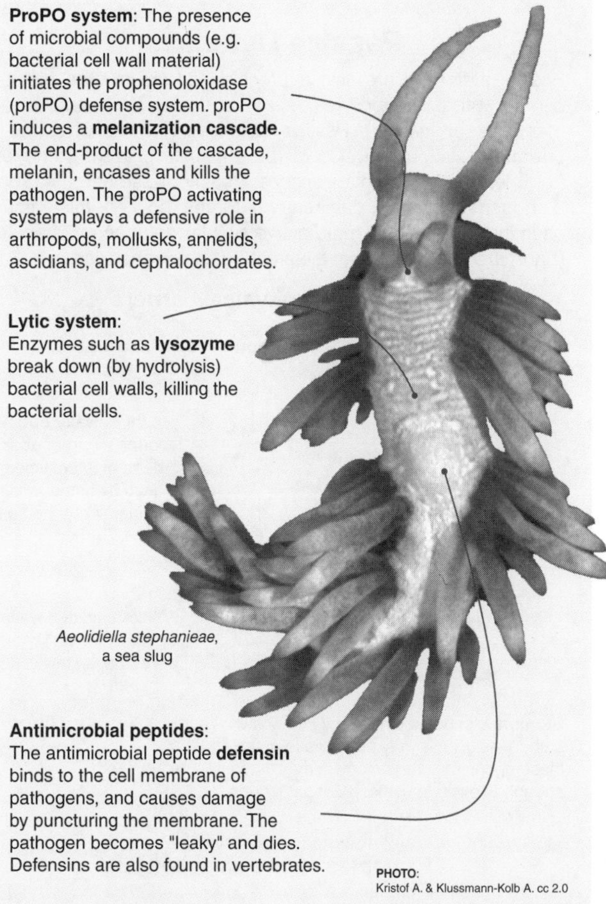


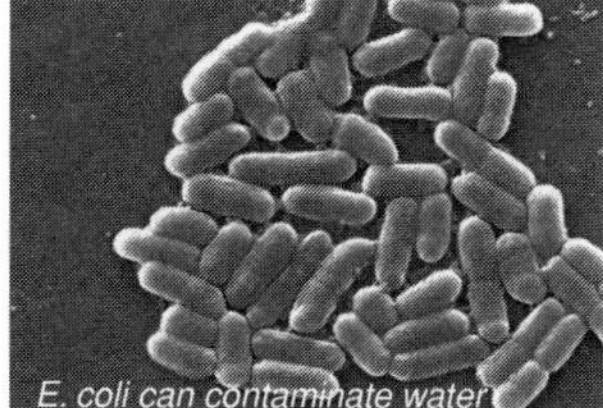


Pathogenic **bacteria** cause a wide range of diseases. Some types of *E. coli* are harmless gut residents, but others can cause disease.

Vertebrates and invertebrates are host to a range of arthropod vectors, which may carry pathogenic microorganisms on their mouthparts.

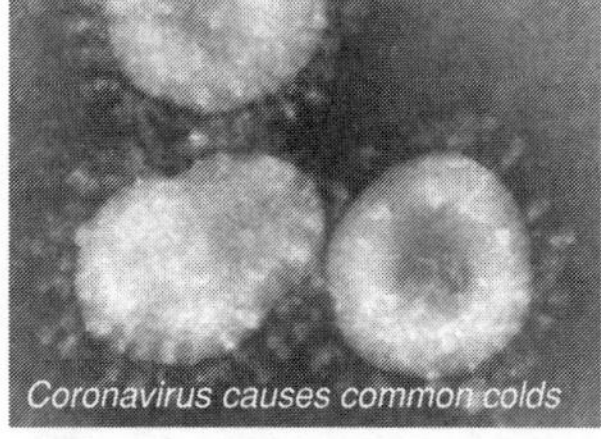


**Viruses** are species-specific intracellular parasites and cause many common diseases. They infect all other organisms, including bacteria.

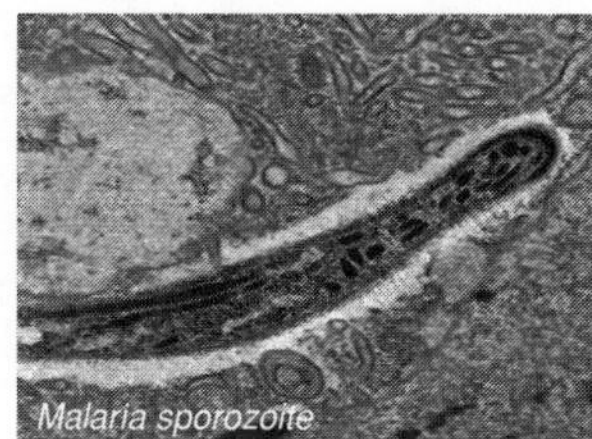


**Eukaryotic pathogens**: protozoa (e.g. the malaria parasite above), algae, fungi, and parasitic worms, may be pathogenic.

1. (a) Describe the advantage of having multiple (non-specific) defense responses: ______

(b) Describe a disadvantage of having only general (non-specific) defense responses: ______

2. Compare and contrast the non-specific defenses of vertebrate and invertebrate animals: ______

ISBN: 978-1-927173-12-1


**Related activities:** *Plant Defenses, Our Body's Defenses*

# Plant Defenses

Plants possess various biochemical and structural defense mechanisms, which protect them from infection and the activities of herbivorous animals that graze on plants or suck their sap. Some defense mechanisms are always present as part of the plant's basic make-up, while others are activated in response to an attack. **Passive defenses** take the form of physical (structural) and chemical barriers. **Active defenses** are produced in direct response to an infection or physical attack and act more specifically against the pathogen. Some plants produce chemicals that inhibit the growth of others nearby (**allelopathy**).

## Passive Defenses

Passive defenses are always present and are not the result of contact with pathogens (e.g. fungi) or grazers (e.g. herbivorous mammals or insects). Passive defenses may be **physical** or **chemical**. Various physical barriers help to prevent pathogens from penetrating the plant tissues. Despite these barriers, some pathogens may still gain entry into the host. If this occurs, chemical defenses or more active cellular defense mechanisms (right) are used to protect the plant against further damage.

### Examples of physical barriers

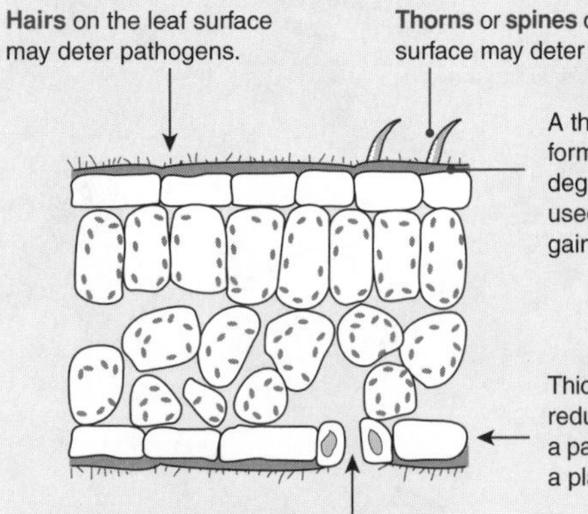


### Examples of chemical barriers

Plants can cover their surfaces with compounds that **inhibit** the development of **pathogens**.

Many plants have developed distasteful chemicals as a defense to deter insects and other grazers.

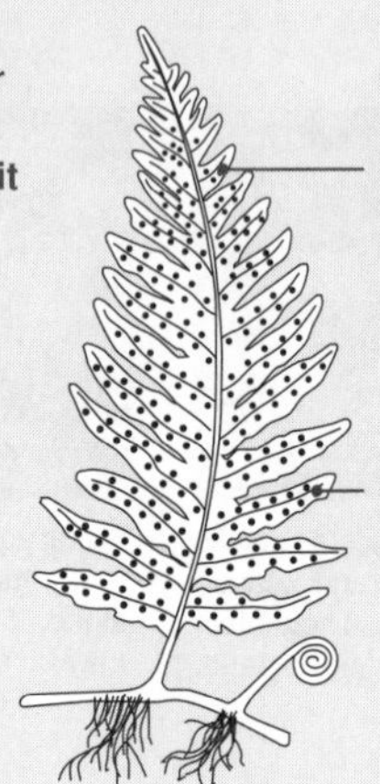


## Active Defenses

Once infected, a plant needs to respond actively to prevent any further damage. **Active defenses** are invoked only after the **pathogen** has been recognized, or after wounding or attack by a herbivore. Many plant defenses contribute to slowing pathogen growth without necessarily stopping it.

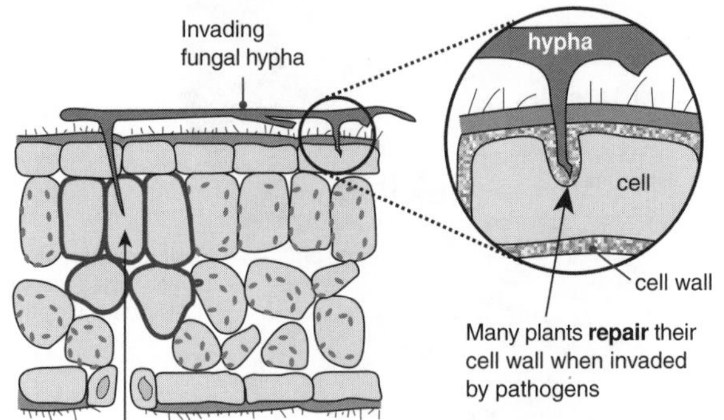


### Other cellular active defenses

| | |
|---|---|
| **Phytoalexins** | Antimicrobial substances that destroy a range of pathogens, e.g. by puncturing the cell wall or disrupting metabolism. |
| **Reactive oxygen levels** | An increase in reactive oxygen species (e.g. $H_2O_2$) in cells kills pathogens. |
| **Wound repair** | Infected areas are sealed off by layers of thickened cells called **cork cells.** |

Sealing off infected areas gives rise to abnormal swellings called galls (oak gall, above left; bulls-eye galls on a maple leaf, above right). These galls limit the spread of the parasite or the infection in the plant.

1. Distinguish between **passive** and **active** defense mechanisms: ______

2. How are **galls** effective in reducing the spread of infection in some plants? ______

3. What similarities are there between the active defense mechanisms of plants and the immune responses of animals? ______




**ISBN: 978-1-927173-12-1**

# Targets for Defense

In order for the body to present an effective defense against pathogens, it must first be able to recognize its own tissues (self). It must also ignore the normal **microflora** inhabiting our bodies and be able to deal with abnormal cells that periodically appear in the body and might develop into cancer. Failure of self/non-self recognition can lead to **autoimmune disorders**, in which the immune system mistakenly destroys its own tissues. The ability of the body to recognize its own molecules has implications for medical techniques such as tissue grafts, organ transplants, and blood transfusions. Incompatible tissues (correctly identified as foreign) are attacked by the body's immune system (rejected). Even a healthy pregnancy involves suppression of specific features of the self recognition system, allowing the mother to tolerate a nine month relationship with a foreign body (a fetus).

## The Body's Natural Microbiota

After birth, normal and characteristic microbial populations begin to establish themselves on and in the body. A typical human body contains 1 X $10^{13}$ body cells, yet harbors 1 X $10^{14}$ bacterial cells. These microorganisms establish more or less permanent residence but, under normal conditions, do not cause disease. In fact, this normal microflora can benefit the host by preventing the overgrowth of harmful pathogens. They are not found throughout the entire body, but are located in certain regions.

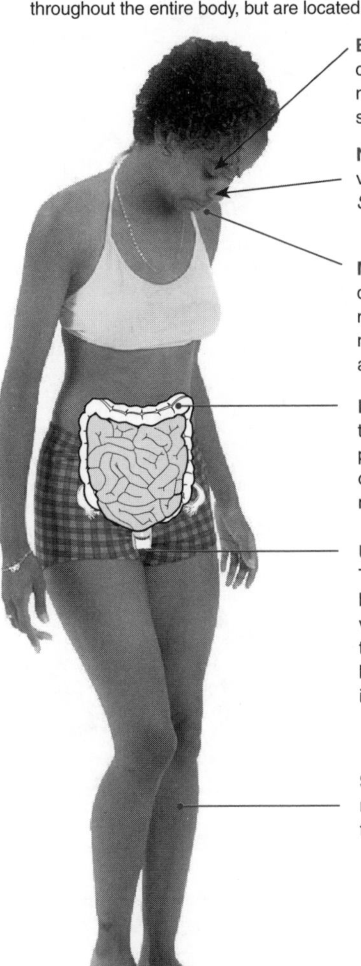

**Eyes**: The conjunctiva, a continuation of the skin or mucous membrane, contains a similar microbiota to the skin.

**Nose and throat**: Harbors a variety of microorganisms, e.g. *Staphylococcus* spp.

**Mouth**: Supports a large and diverse microbiota. It is an ideal microbial environment: high in moisture, warmth, and nutrient availability.

**Large intestine**: Contains the body's largest resident population of microbes because of its available moisture and nutrients.

**Urinary and genital systems**: The lower urethra in both sexes has a resident population; the vagina has a particular acid-tolerant population of microbes because of the low pH nature of its secretions.

**Skin**: Skin secretions prevent most of the microbes on the skin from becoming residents.

## Distinguishing Self from Non-Self

The human immune system achieves self-recognition through the **major histocompatibility complex** (MHC). This is a cluster of tightly linked genes on chromosome 6 in humans. These genes code for protein molecules (MHC antigens) that are attached to the surface of body cells. They are used by the immune system to recognize its own or foreign material. **Class I MHC** antigens are located on the surface of virtually all human cells, but **Class II MHC** antigens are restricted to macrophages and the antibody-producing B-lymphocytes.

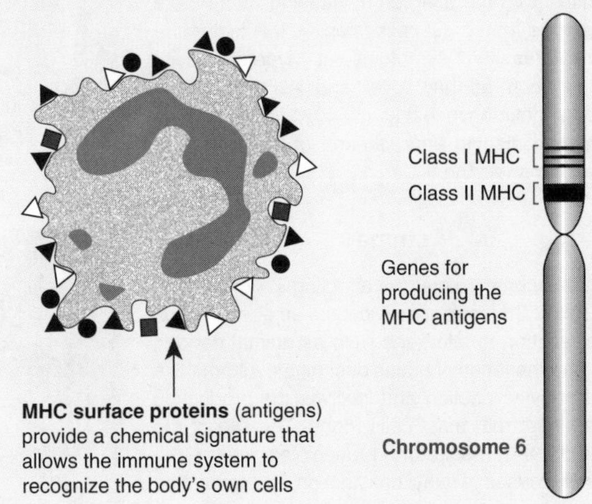


## Tissue Transplants

The MHC is responsible for the rejection of tissue grafts and organ transplants. Foreign MHC molecules are antigenic, causing the immune system to respond in the following way:

- T cells directly lyze the foreign cells.
- Macrophages are activated by T cells and engulf foreign cells.
- Antibodies are released that attack the foreign cell.
- The complement system injures blood vessels supplying the graft or transplanted organ.

To minimize this rejection, attempts are made to match the MHC of the organ donor to that of the recipient as closely as possible.

1. Explain why it is healthy to have a natural population of microbes on and inside the body: ________________

2. (a) Explain the nature and purpose of the **major histocompatibility complex** (MHC): ________________

   (b) Explain the importance of such a self-recognition system: ________________

3. Identify two situations when the body's recognition of 'self' is undesirable: ________________

Defense Mechanisms


**ISBN: 978-1-927173-12-1**

# Allergies and Hypersensitivity

Sometimes the immune system may overreact, or react to the wrong substances instead of responding appropriately. This is termed **hypersensitivity** and the immunological response leads to tissue damage rather than immunity. Hypersensitivity reactions occur after a person has been **sensitized** to an antigen. In some cases, this causes only localized discomfort, as in the case of hayfever. More generalized reactions (such as anaphylaxis from insect venom or drug injections), or localized reactions that affect essential body systems (such as asthma), can cause death through asphyxiation and/or circulatory shock.

### Hypersensitivity

**Hay fever** (allergic rhinitis) is an allergic reaction to airborne substances such as dust, molds, pollens, and animal fur or feathers. Allergy to wind-borne pollen is the most common, and certain plants (e.g. ragweed and privet) are highly allergenic. A person becomes **sensitized** when they form antibodies to harmless substances in the environment such as pollen or spores (steps 1-2 right). These substances, termed **allergens**, act as antigens to induce antibody production and an allergic response. Once a person is sensitized, the antibodies respond to further encounters with the allergen by causing the release of **histamine** from mast cells (step 4). It is histamine that mediates the symptoms of hypersensitivity reactions such as hay fever and asthma. These symptoms include wheezing and airway constriction, inflammation, itching and watering of the eyes and nose, and/or sneezing.

Eyewire

### Asthma

Asthma is a common disease affecting 9% of children in the United States. It usually occurs as a result of an allergic reaction to allergens such as animal dander, pollen, and the feces of house dust mites. Asthma is a hypersensitivity reaction and involves the production of histamines from mast cells (right). The site of the reaction is the respiratory bronchioles where the histamine causes airway constriction, accumulation of fluid and mucus, and inability to breathe. During an attack, breathing is labored and the chest cavity overexpands (photo, right). Asthma treatments work by suppressing the inflammatory response to prevent attacks and dilating the airways if an attack occurs. Asthma attacks are often triggered by environmental factors such as cold air, exercise, air pollutants, and viral infections. Recent evidence has also indicated the involvement of a bacterium, *Chlamydia pneumoniae*, in about half of all cases of asthma in susceptible adults.

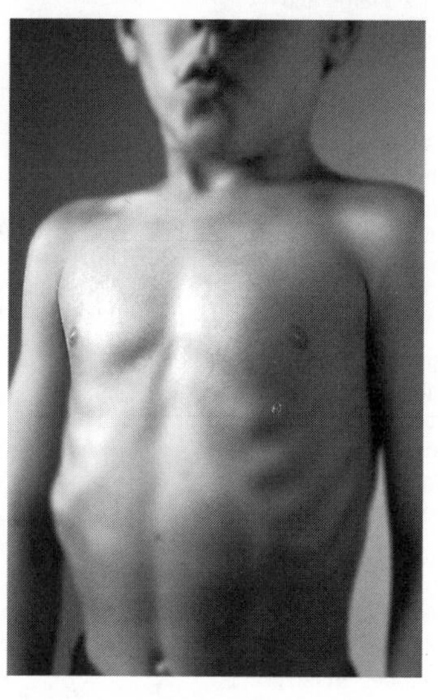

### The Basis of Hypersensitivity

1. B-cell encounters the allergen and differentiates into plasma cells

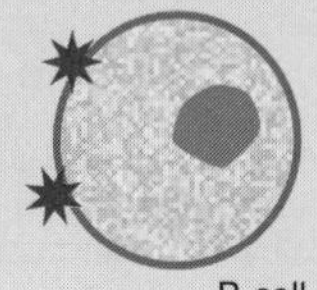


2. The plasma cell produces antibodies

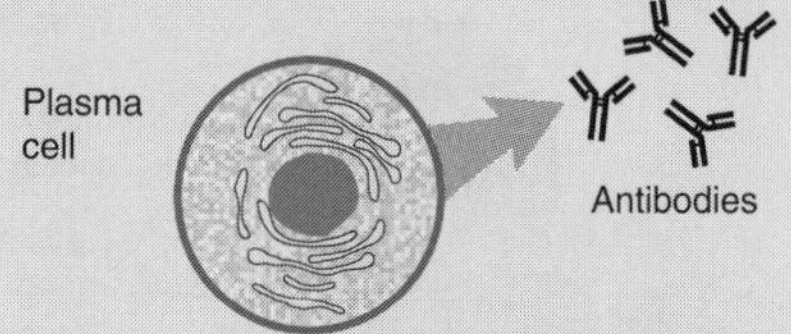


3. Antibodies bind to specific receptors on the surface of the mast cells

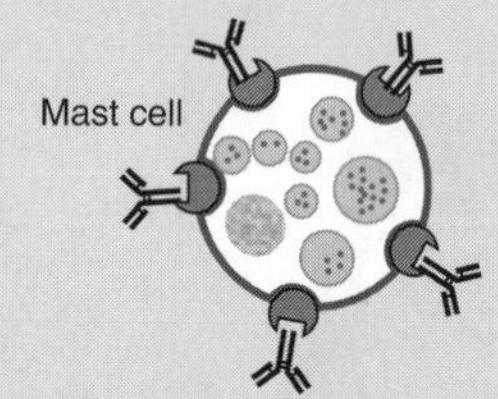


4. When the mast cell encounters the antigen again, it binds the antigen and releases histamine.

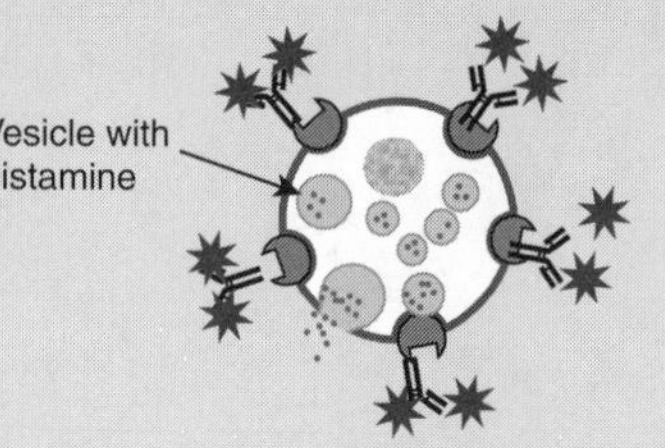


1. What is the role of histamine in hypersensitivity responses? ______________________________

2. What happens when a person becomes **sensitized** to an allergen? ______________________________

3. In what way is the hypersensitivity reaction a malfunction of the immune system? ______________________________

4. (a) What do **bronchodilators** do? ______________________________

   (b) Why are they used to treat asthma? ______________________________



***Related activities:*** *The Immune System*
***Weblinks:*** *What is Asthma?*

***Periodicals:***
*Misery for all seasons*


**ISBN: 978-1-927173-12-1**

# Our Body's Defenses

The human body has a tiered system of defenses to prevent or limit infection by pathogens. The first line of defense has a role in keeping microorganisms from entering the body. If this fails, a second line of defense targets any foreign bodies (including microbes) that manage to get inside. If microorganisms manage to evade this level of defense, the body's immune system provides a third line of (specific) defense. The ability to ward off disease through the various defense mechanisms is called **resistance**. The lack of resistance, or vulnerability to disease, is known as **susceptibility**. **Non-specific resistance** includes the first and second lines of defense. **Specific resistance** (the immune response) is specific to particular pathogens.

### 1st Line of Defense

The skin provides a physical barrier to the entry of pathogens. Healthy skin is rarely penetrated by microorganisms. Its low pH is unfavorable to the growth of many bacteria and its chemical secretions (e.g. sebum, antimicrobial peptides) inhibit growth of bacteria and fungi. Tears, mucus, and saliva also help to wash bacteria away.

### 2nd Line of Defense

A range of defense mechanisms operate inside the body to inhibit or destroy pathogens. These responses react to the presence of any pathogen, regardless of which species it is. White blood cells are involved in most of these responses.

It includes the **complement system** whereby plasma proteins work together to bind pathogens and induce an inflammatory responses to help fight infection.

### 3rd Line of Defense

Once the pathogen has been identified by the immune system, a specific response from white blood cells called lymphocytes occurs. These coordinate a range of specific responses to the pathogen.

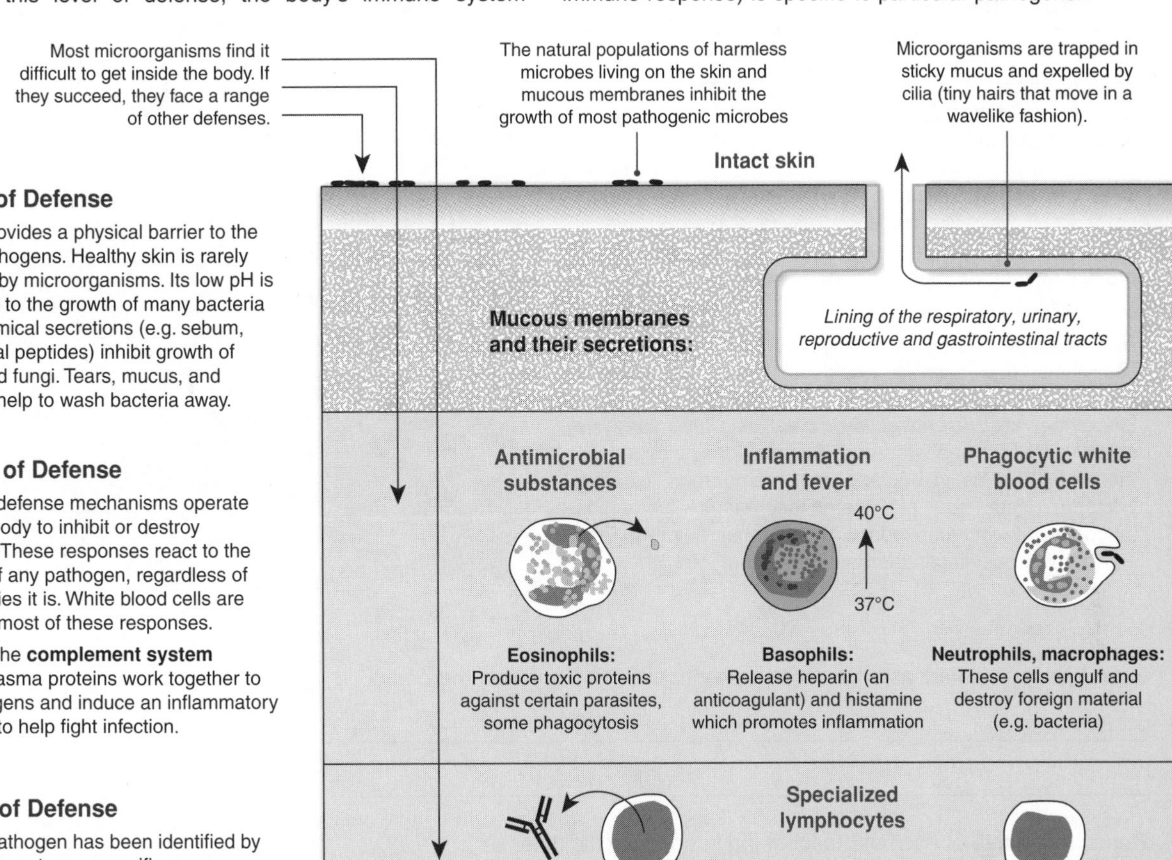




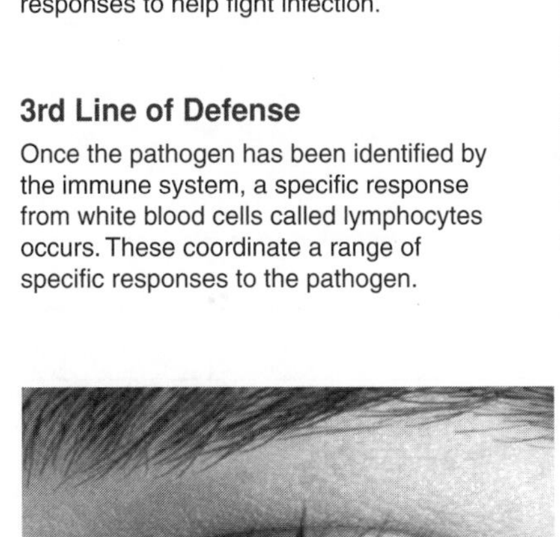

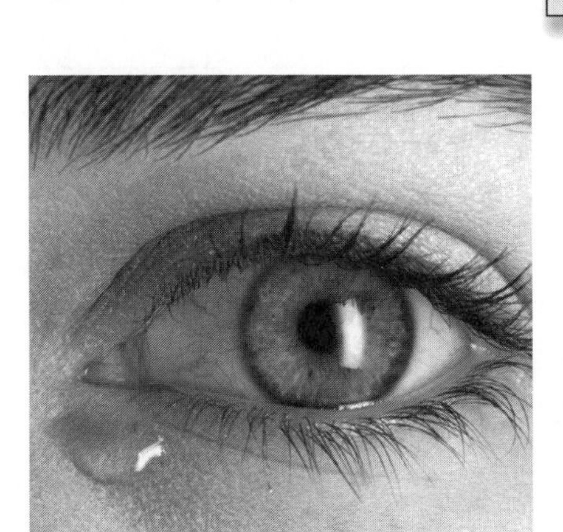

Tears contain antimicrobial substances as well as washing contaminants from the eyes.

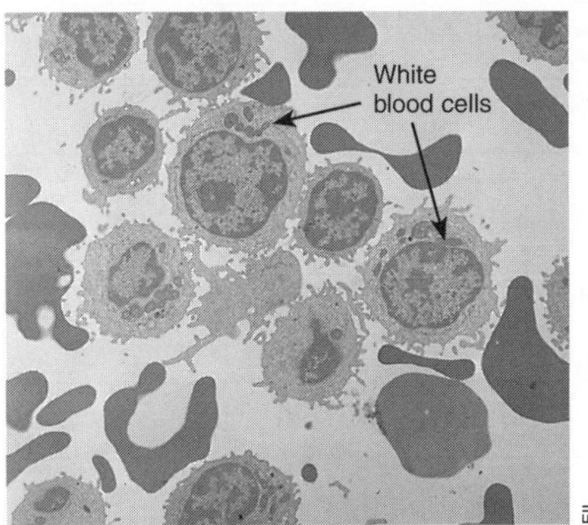


A range of white blood cells (the larger cells in the photograph) form the second line of defense.

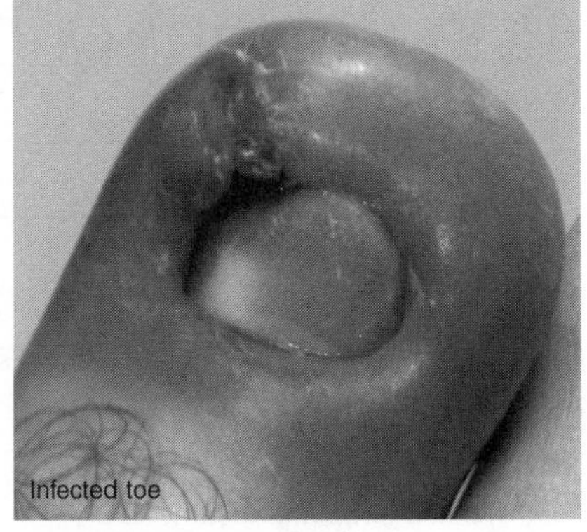


Inflammation is a localized response to infection characterized by swelling, pain, and redness.

1. Distinguish between specific and non-specific resistance:

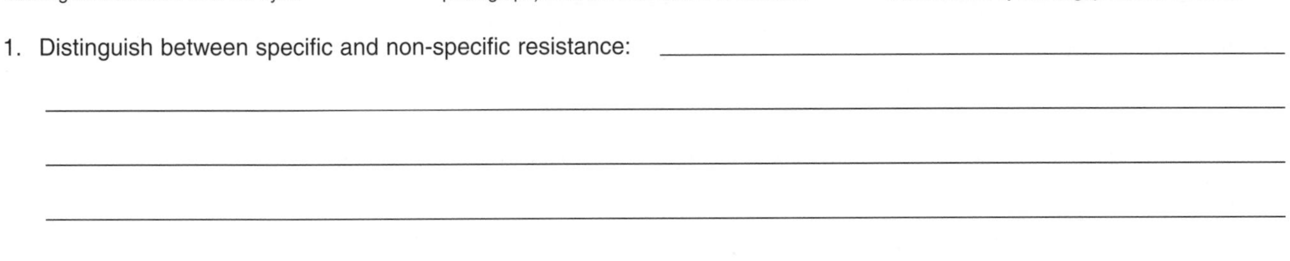

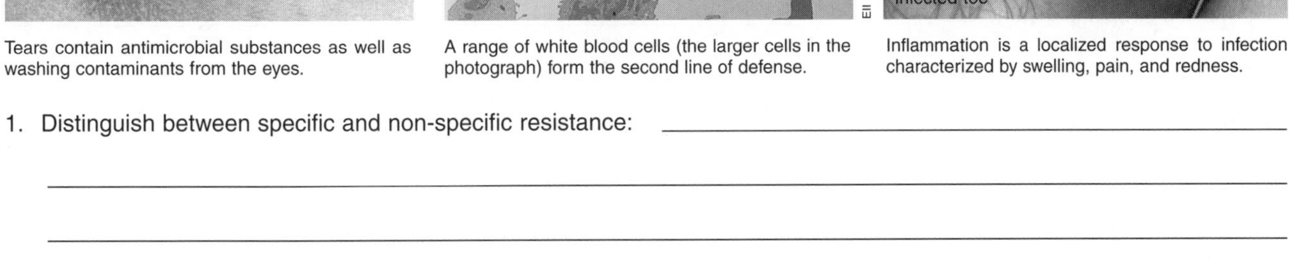

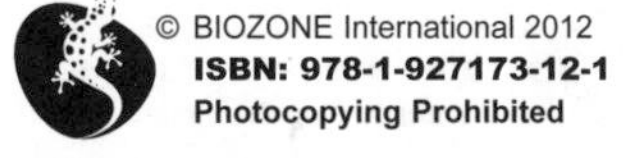


ISBN: 978-1-927173-12-1


***Related activities:*** *The Action of Phagocytes, Inflammation, The Immune System*

***Weblinks:*** *Immunoanimations*

## The Importance of the First Line of Defense

The skin is the largest organ of the body. It forms an important physical barrier against the entry of pathogens into the body. A natural population of harmless microbes live on the skin, but most other microbes find the skin inhospitable. The continual shedding of old skin cells (arrow, right) physically removes bacteria from the surface of the skin. Sebaceous glands in the skin (labelled right) produce sebum, which has antimicrobial properties, and the slightly acidic secretions of sweat inhibit microbial growth.

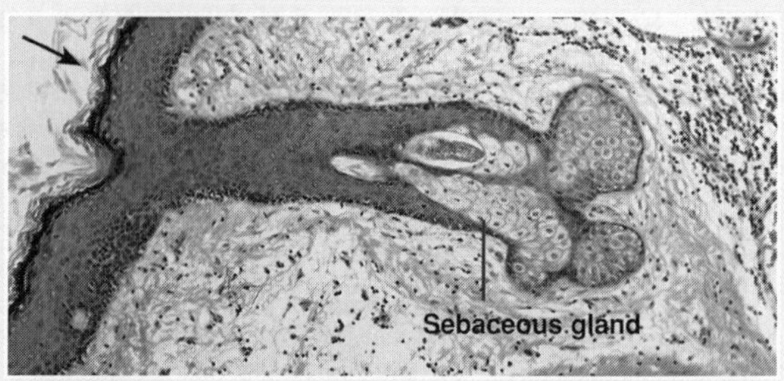


Cilia line the epithelium of the **nasal passage** (below right). Their wave-like movement sweeps foreign material out and keeps the passage free of microorganisms, preventing them from colonizing the body.

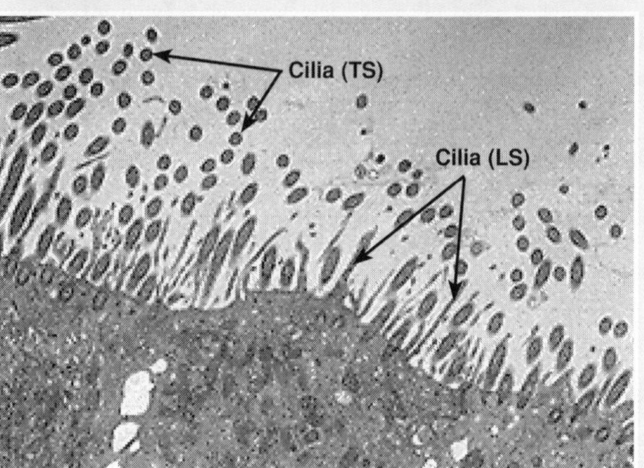


Antimicrobial chemicals are present in many bodily secretions. Tears, saliva, nasal secretions, and human breast milk all contain **lysozymes** and **phospholipases**. Lysozymes kill bacterial cells by catalyzing the hydrolysis of cell wall linkages, whereas phospholipases hydrolyze the phospholipids in bacterial cell membranes, causing bacterial death. Low pH gastric secretions also inhibit microbial growth, and reduce the number of pathogens establishing colonies in the gastrointestinal tract.

2. How does the skin act as a barrier to prevent pathogens entering the body? ______________________

3. Describe the role of each of the following in non-specific defense:

(a) Phospholipases: ______________________

(b) Cilia: ______________________

(c) Sebum: ______________________

4. Describe the functional role of each of the following defense mechanisms:

(a) Phagocytosis by white blood cells: ______________________

(b) Antimicrobial substances: ______________________

(c) Antibody production: ______________________

5. Explain the value of a three tiered system of defense against microbial invasion: ______________________


**ISBN: 978-1-927173-12-1**

# Blood Group Antigens

**Blood groups** classify blood according to the different marker molecules (**antigens**) on the surface of red blood cells (RBCs). These antigens determine the ability of RBCs to provoke an immune response. **Blood group typing** is essential for safe **blood transfusion**. Transfusion with incompatible blood types will cause clumping of the red blood cells and cell lysis (hemolysis). Although human RBCs have more than 500 known antigens, fewer than 30 (in 9 blood groups) are regularly tested for when blood is donated for transfusion. The **ABO** and **rhesus** (Rh) blood group antigens are the best known. Where the father of a baby is Rh-positive and the mother is Rh-negative, a second baby, if Rh-positive, will suffer from **hemolytic disease of the newborn**. This severe immune reaction is caused by the mother's acquired antibodies attacking the fetal blood cells.

| | Blood type A | Blood type B | Blood type AB | Blood type O |
|---|---|---|---|---|
| ***Antigens* present on the *red blood cells*** | antigen ***A*** | antigen ***B*** | antigens ***A*** and ***B*** | Neither antigen ***A*** nor ***B*** |
| ***Anti-bodies* present in the *plasma*** | Contains **anti-B** antibodies; but no antibodies that would attack its own antigen ***A*** | Contains **anti-A** antibodies; but no antibodies that would attack its own antigen ***B*** | Contains neither **anti-A** nor **anti-B** antibodies | Contains both **anti-A** and **anti-B** antibodies |

## Rh Blood Groups and HDN

Like the ABO grouping, the Rh system is based on antigens on the surfaces of red blood cells (RBCs). People whose RBCs have the Rh antigens on their surface are Rh⁺. Those who lack the Rh antigen are Rh⁻. About 84% of people in the US are Rh⁺. Normally, human plasma does not contain antibodies against the Rh antigen. However, if an Rh⁻ mother encounters Rh⁺ blood from the baby during delivery, her body will make antibodies against the antigen. If, in a second pregnancy, the fetus is Rh⁺, these antibodies will pass across the placenta and react with and destroy the baby's blood cells. This condition is called **hemolytic disease of the newborn** (HDN).

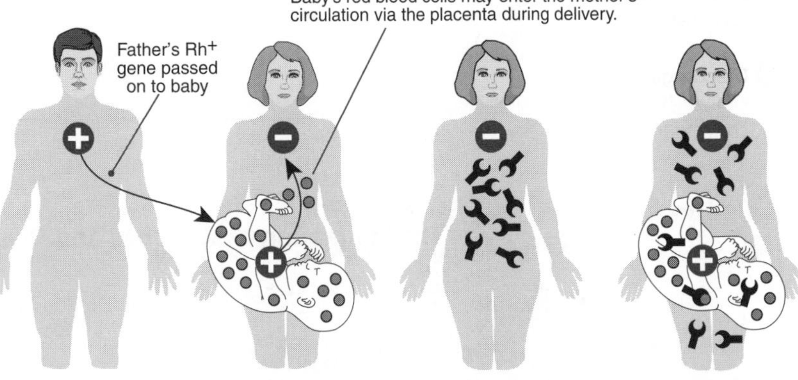


Father is Rh⁺ (positive)

**First pregnancy**
Rh⁻ mother is pregnant with Rh⁺ fetus. Rh⁺ antigens pass to the mother during labor.

After exposure to the Rh⁺ antigens, the mother makes anti-Rh antibodies.

**Second pregnancy**
Mother's anti-Rh antibodies cross the placenta to the fetal blood. An Rh⁺ baby will develop HDN.

1. Complete the table below to show the antibodies and antigens present in each blood group, and donor blood types:

| Blood Type | Freq. in US | | Antigen | Antibody | Can donate blood to: | Can receive blood from: |
|---|---|---|---|---|---|---|
| | Rh⁺ | Rh⁻ | | | | |
| **A** | 34% | 6% | A | anti-B | A, AB | A, O |
| **B** | 9% | 2% | | | | |
| **AB** | 3% | 1% | | | | |
| **O** | 38% | 7% | | | | |

2. Why are people with **blood type O⁻** sometimes called universal donors? ____________________

3. What causes **hemolytic disease of the newborn**? ____________________

**ISBN: 978-1-927173-12-1**


***Related activities:*** *Antibodies, Blood*
***Weblinks:*** *Blood Typing Game*

# The Action of Phagocytes

Human cells that ingest microbes and digest them by the process of **phagocytosis** are called **phagocytes**. All are types of white blood cells. During many kinds of infections, especially bacterial infections, the total number of white blood cells increases by two to four times the normal number. The ratio of various white blood cell types changes during the course of an infection.

## How a Phagocyte Destroys Microbes

**1 Detection**
Phagocyte detects microbes by the chemicals they give off and sticks the microbes to its surface.

**2 Ingestion**
The microbe is engulfed by the phagocyte wrapping pseudopodia around it to form a vesicle.

**3 Phagosome forms**
A phagosome (phagocytic vesicle) is formed, which encloses the microbes in a membrane.

**4 Fusion with lysosome**
Phagosome fuses with a lysosome (which contains powerful enzymes that can digest the microbe).

**5 Digestion**
The microbes are broken down by enzymes into their chemical constituents.

**6 Discharge**
Indigestible material is discharged from the phagocyte cell.

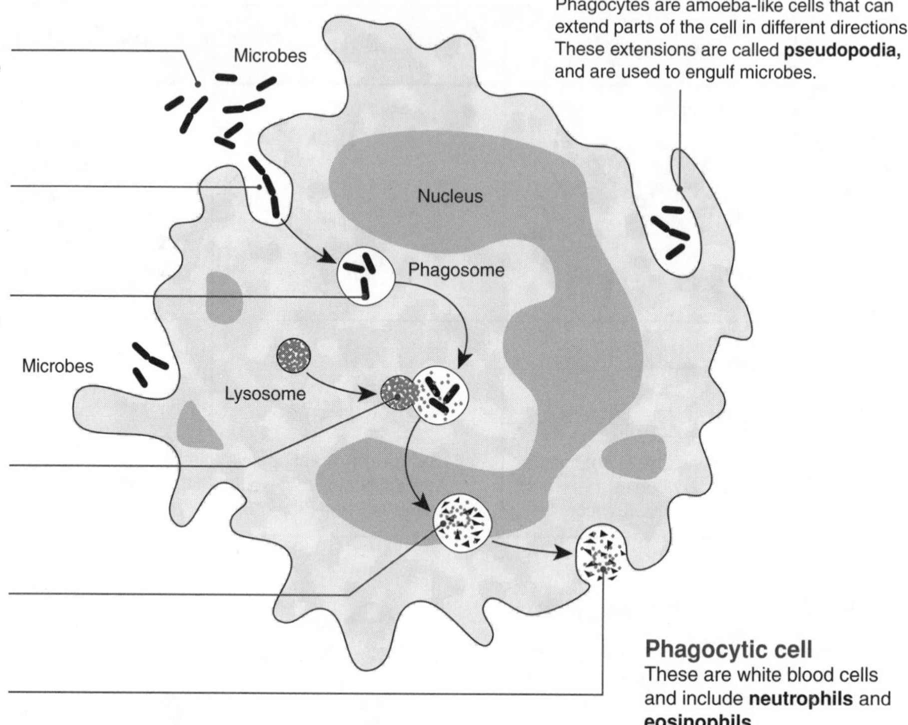


Phagocytes are amoeba-like cells that can extend parts of the cell in different directions. These extensions are called **pseudopodia,** and are used to engulf microbes.

**Phagocytic cell**
These are white blood cells and include **neutrophils** and **eosinophils**.

## The Interaction of Microbes and Phagocytes

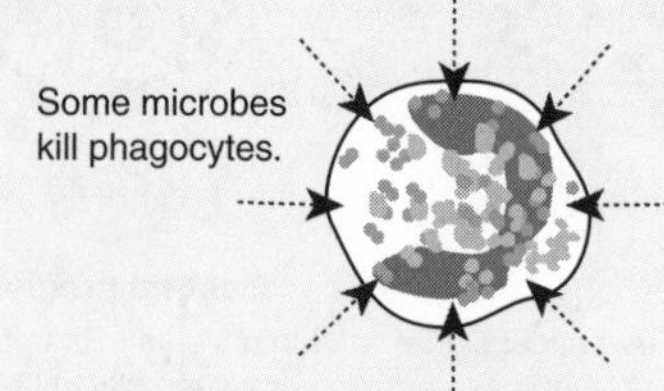


### Some microbes kill phagocytes

Some microbes produce toxins that can actually kill phagocytes, e.g. toxin-producing staphylococci and the dental plaque-forming bacteria *Actinobacillus*.

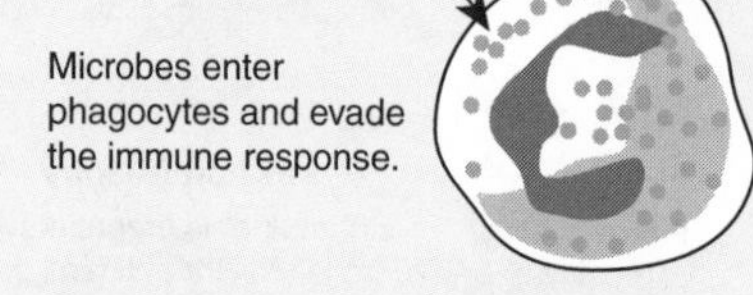


### Microbes evade immune system

Some microbes can evade the immune system by entering phagocytes. The microbes prevent fusion of the lysosome with the phagosome and multiply inside the phagocyte, almost filling it. Examples include *Chlamydia*, *Mycobacterium tuberculosis*, *Shigella*, and malarial parasites.

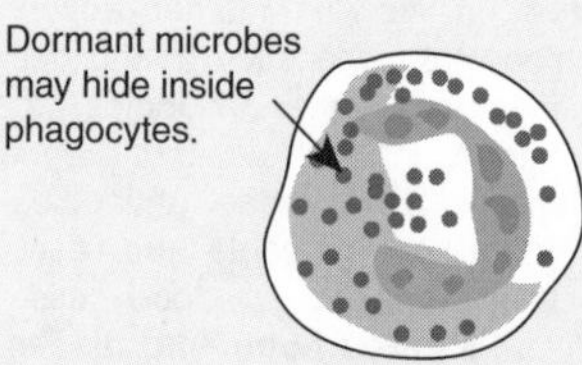


### Dormant microbes hide inside

Some microbes can remain dormant inside the phagocyte for months or years at a time. Examples include the microbes that cause brucellosis and tularemia.

1. Identify the white blood cells capable of phagocytosis: ____________________

____________________

2. Describe how a blood sample from a patient may be used to determine whether they have a microbial infection (without looking for the microbes themselves):

____________________

____________________

3. Explain how some microbes are able to overcome phagocytic cells and use them to their advantage:

____________________

____________________




**ISBN: 978-1-927173-12-1**

# Inflammation

Damage to the body's tissues can be caused by physical agents (e.g. sharp objects, heat, radiant energy, or electricity), microbial infection, or chemical agents (e.g. gases, acids and bases). The damage triggers a defensive response called **inflammation**. It is usually characterized by four symptoms: pain, redness, heat, and swelling. The inflammatory response is beneficial and has the following functions: (1) to destroy the cause of the infection and remove it and its products from the body; (2) if this fails, to limit the effects on the body by confining the infection to a small area; (3) replacing or repairing tissue damaged by the infection. The process of inflammation can be divided into three distinct stages. These are described below.

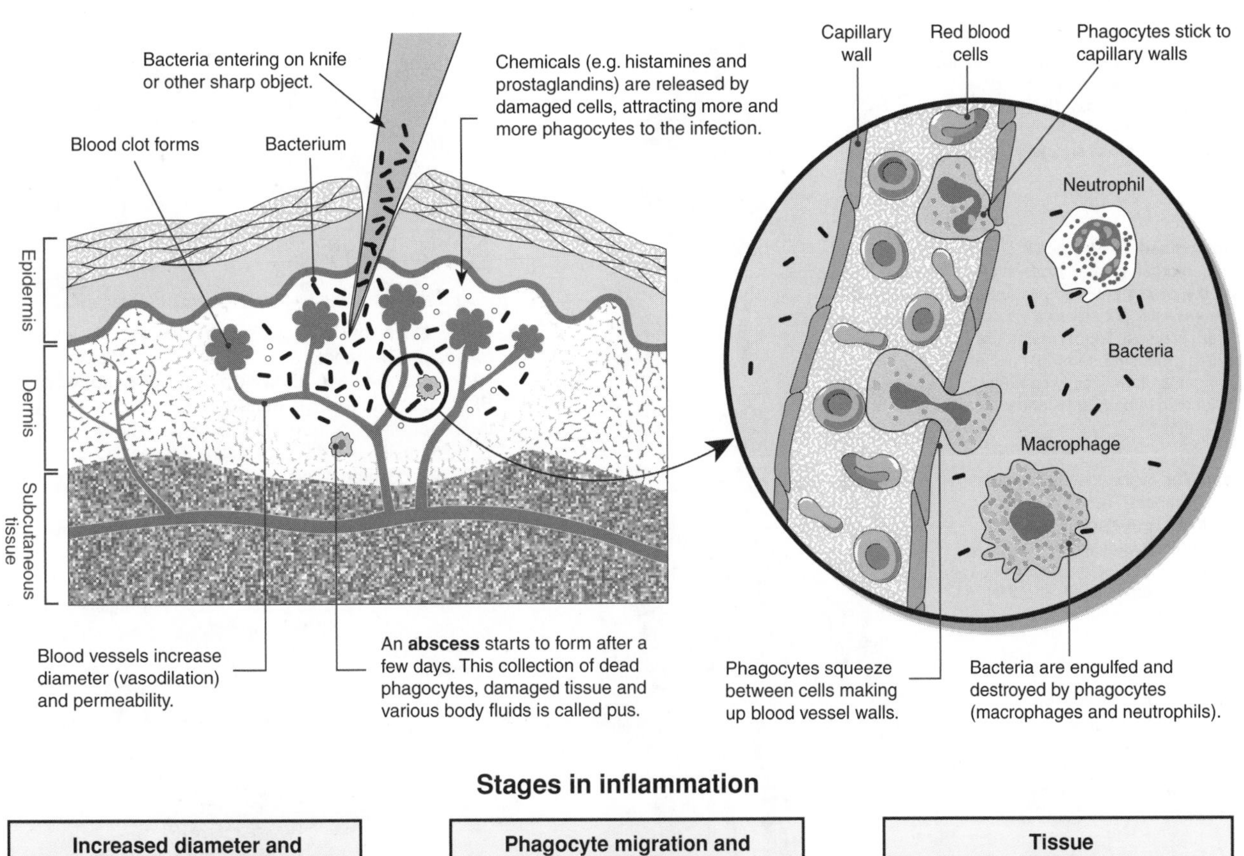


### Stages in inflammation

**Increased diameter and permeability of blood vessels**
Blood vessels increase their diameter and permeability in the area of damage. This increases blood flow to the area and allows defensive substances to leak into tissue spaces.

→

**Phagocyte migration and phagocytosis**
Within one hour of injury, phagocytes appear on the scene. They squeeze between cells of blood vessel walls to reach the damaged area where they destroy invading microbes.

→

**Tissue repair**
Functioning cells or supporting connective cells create new tissue to replace dead or damaged cells. Some tissue regenerates easily (skin) while others do not at all (cardiac muscle).

1. Outline the three stages of inflammation and identify the beneficial role of each stage:

(a) ______________________________

(b) ______________________________

(c) ______________________________

2. Identify two features of phagocytes important in the response to microbial invasion: ______________________________

3. State the role of histamines and prostaglandins in inflammation: ______________________________

4. Explain why pus forms at the site of infection: ______________________________

Defense Mechanisms

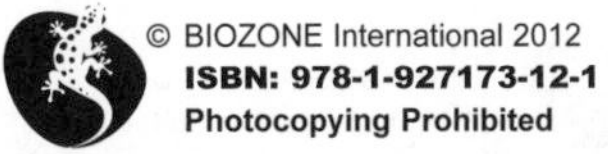

ISBN: 978-1-927173-12-1


***Related activities:*** *The Body's Defenses, The Action of Phagocytes*
***Weblinks:*** *Inflammation and Healing*

# The Lymphatic System

Fluid leaks out from capillaries and forms the tissue fluid, which is similar in composition to plasma but lacks large proteins. This fluid bathes the tissues, supplying them with nutrients and oxygen, and removing wastes. Some of the tissue fluid returns directly into the capillaries, but some drains back into the blood circulation through a network of lymph vessels. This fluid, called **lymph**, is similar to tissue fluid, but contains more leukocytes. Apart from its circulatory role, the lymphatic system also has an important function in the immune response. Lymph nodes are the primary sites where the destruction of pathogens and other foreign substances occurs. A lymph node that is fighting an infection becomes swollen and hard as the lymph cells reproduce rapidly to increase their numbers. The thymus, spleen, and bone marrow also contribute leukocytes to the lymphatic and circulatory systems.

**Tonsils**: Tonsils (and adenoids) comprise a collection of large lymphatic nodules at the back of the throat. They produce lymphocytes and antibodies and are well-placed to protect against invasion of pathogens.

**Thymus gland**: The thymus is a two-lobed organ located close to the heart. It is prominent in infants and diminishes after puberty to a fraction of its original size. Its role in immunity is to help produce **T cells** that destroy invading microbes directly or indirectly by producing various substances.

**Spleen**: The oval spleen is the largest mass of lymphatic tissue in the body, measuring about 12 cm in length. It stores and releases blood in case of demand (e.g. in cases of bleeding), produces mature **B cells**, and destroys bacteria by phagocytosis.

**Bone marrow**: Bone marrow produces red blood cells and many kinds of leukocytes: monocytes (and macrophages), neutrophils, eosinophils, basophils, and lymphocytes (B cells and T cells).

**Lymphatic vessels**: When tissue fluid is picked up by lymph capillaries, it is called **lymph**. The lymph is passed along lymphatic vessels to a series of lymph nodes. These vessels contain one-way valves that move the lymph in the direction of the heart until it is reintroduced to the blood at the subclavian veins.

**Disorders of the lymphatic system**: If the lymphatic system is not working properly, fluids can build up causing swelling (lymphedema). Other problems can result from infections, blockages, and cancer (e.g. Hodgkin's disease).

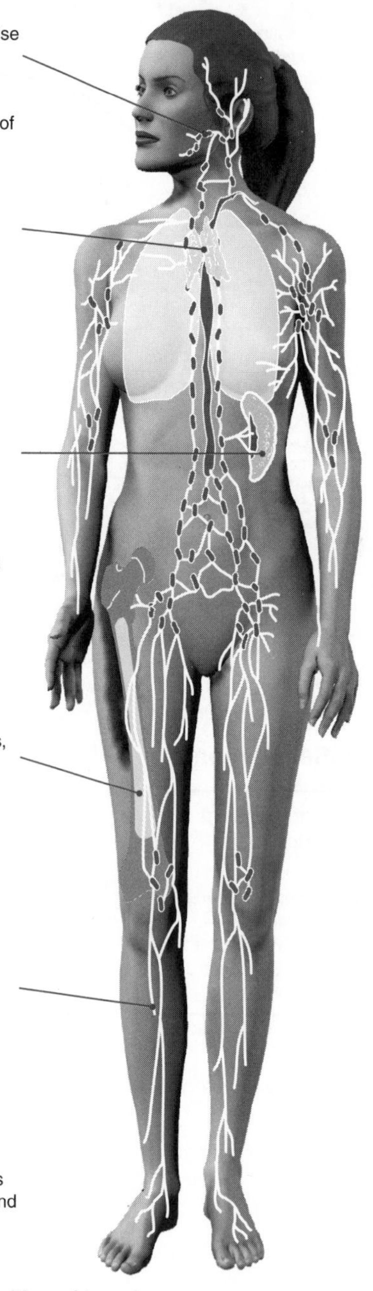

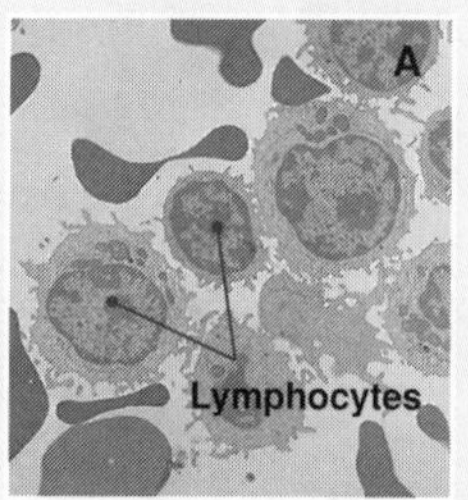


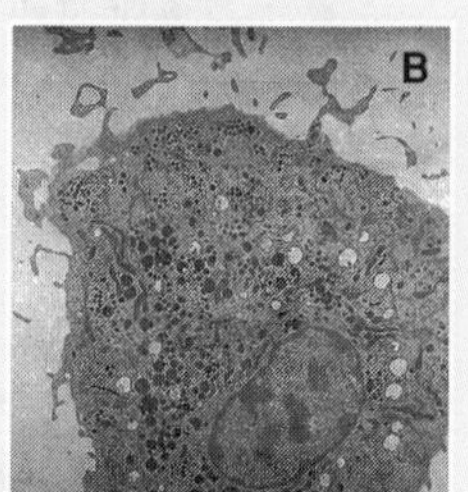


Photos: Eli

Many types of leukocytes are involved in internal defense. The photos above illustrate examples of leukocytes. Photo **A** shows a cluster of **lymphocytes**. Photo **B** shows a single **macrophage**: large, phagocytic cells that develop from monocytes and move from the blood to reside in many organs and tissues, including the spleen and lymph nodes.

### Lymph node

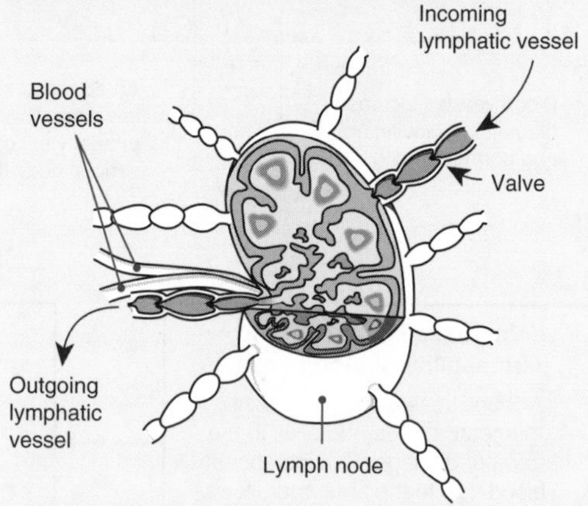


Lymph nodes are oval or bean-shaped structures, scattered throughout the body, usually in groups, along the length of lymphatic vessels. As lymph passes through the nodes, it filters foreign particles (including pathogens) by trapping them in fibers. Lymph nodes are also a "store" of **lymphocytes**, which may circulate to other parts of the body. Once trapped, macrophages destroy the foreign substances by phagocytosis. T cells may destroy them by releasing various products, and/or B cells may release antibodies that destroy them.

1. Briefly describe the composition of lymph: ______________________________

______________________________

2. Discuss the various roles of lymph: ______________________________

______________________________

______________________________

3. Describe one role of each of the following in the lymphatic system:

   (a) Lymph nodes: ______________________________

   (b) Bone marrow: ______________________________



***Related activities**: The Formation of Tissue Fluid, The Immune System*

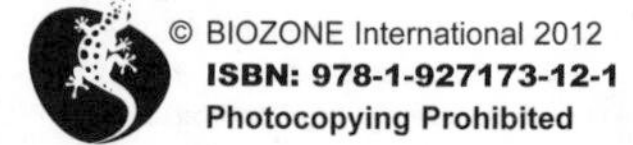


**ISBN: 978-1-927173-12-1**

# Acquired Immunity

We have natural or **innate resistance** to certain illnesses; examples include most diseases of other animal species. **Acquired immunity** refers to the protection an animal develops against certain types of microbes or foreign substances. Immunity can be acquired either passively or actively and is developed during an individual's lifetime. **Active immunity** develops when a person is exposed to microorganisms or foreign substances and the immune system responds. **Passive immunity** is acquired when antibodies are transferred from one person to another. Recipients do not make the antibodies themselves and the effect lasts only as long as the antibodies are present, usually several weeks or months. Immunity may also be **naturally acquired**, through natural exposure to microbes, or **artificially acquired** as a result of medical treatment.

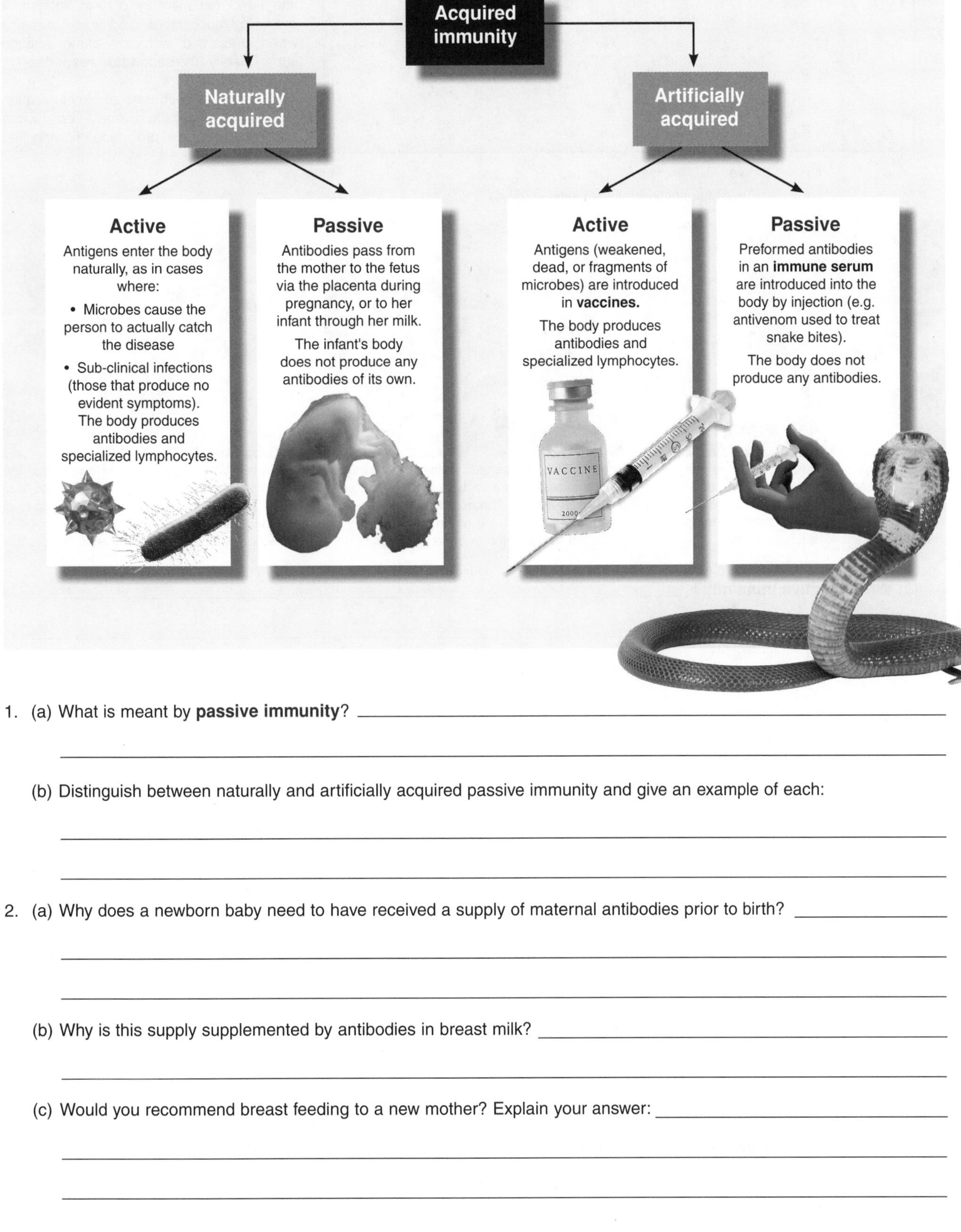




1. (a) What is meant by **passive immunity**? ______

   (b) Distinguish between naturally and artificially acquired passive immunity and give an example of each:

2. (a) Why does a newborn baby need to have received a supply of maternal antibodies prior to birth? ______

   (b) Why is this supply supplemented by antibodies in breast milk? ______

   (c) Would you recommend breast feeding to a new mother? Explain your answer: ______


ISBN: 978-1-927173-12-1


*Periodicals:* Hard to swallow, Immunology

**Related activities:** Antibodies, Vaccines and Vaccination

## Primary and Secondary Responses to Antigens

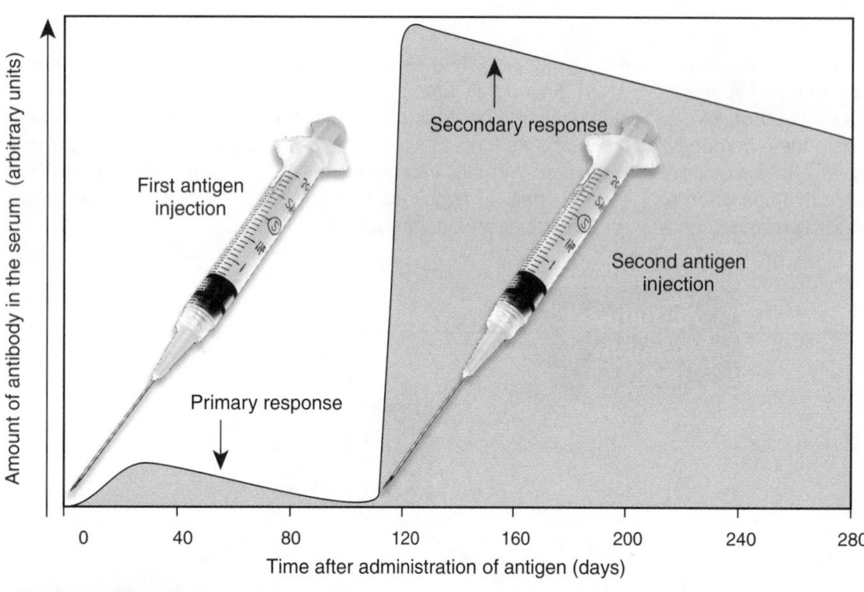


When the B cells encounter antigens and produce antibodies, the body develops **active immunity** against that antigen.

The initial response to antigenic stimulation, caused by the sudden increase in B cell clones, is called the **primary response**. Antibody levels as a result of the primary response peak a few weeks after the response begins and then decline. However, because the immune system develops an immunological memory of that antigen, it responds much more quickly and strongly when presented with the same antigen subsequently (the **secondary response**).

This forms the basis of immunization programmes where one or more booster shots are provided following the initial vaccination.

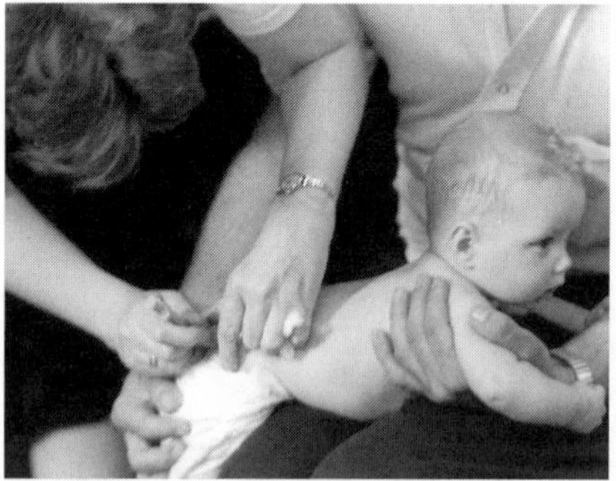

Vaccines against common diseases are given at various stages during childhood according to an immunization schedule. Vaccination has has resulted in the decline of some once-common childhood diseases, such as mumps.

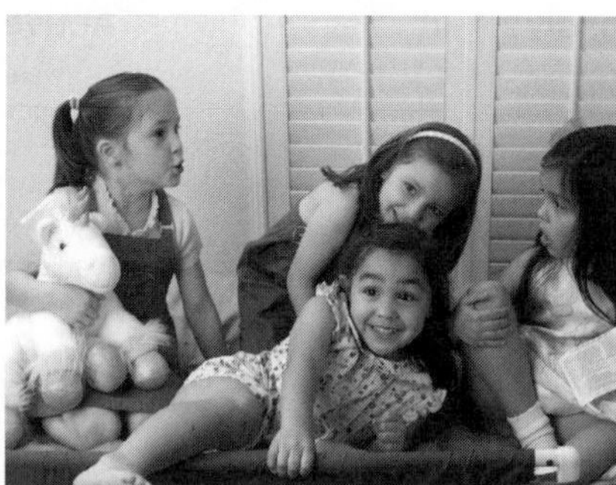

Many childhood diseases for which vaccination programmes exist are kept at a low level because of **herd immunity**. If most of the population is immune, those that are not immunized may be protected because the disease is uncommon.

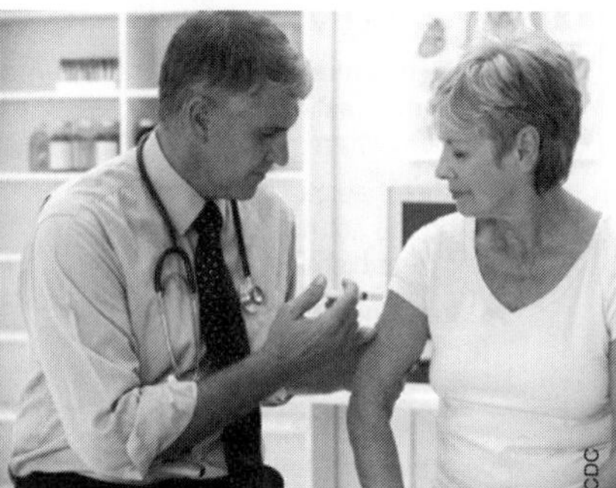


Most vaccinations are given in childhood, but adults may be vaccinated against a disease (e.g. TB, influenza) if they are in a high risk group (e.g. the elderly) or if they are travelling to a region in the world where a certain disease is prevalent.

3. (a) What is **active immunity**? ______

(b) Distinguish between naturally and artificially acquired active immunity and give an example of each: ______

4. (a) Describe two differences between the primary and secondary responses to presentation of an antigen: ______

(b) Why is the secondary response so different from the primary response? ______

5. (a) Explain the principle of herd immunity: ______

(b) Why are health authorities concerned when the vaccination rates for an infectious disease fall?


**ISBN: 978-1-927173-12-1**

# Vaccines and Vaccination

Vaccines operate on the principle that they alert the immune system to the presence of a pathogen by introducing harmless but recognizably foreign antigens against which the body can form antibodies. There are two basic types of vaccine: subunit vaccines and whole-agent vaccines. **Whole-agent vaccines** contain complete nonvirulent microbes, either **inactivated** (killed), or alive but **attenuated** (weakened). Attenuated viruses make very effective vaccines and often provide life-long immunity without the need for booster immunizations. Killed viruses are less effective and many vaccines of this sort have now been replaced by newer subunit vaccines. **Subunit vaccines** contain only the parts of the pathogen that induce the immune response. They are safer than attenuated vaccines because they cannot reproduce in the recipient, and they produce fewer adverse effects because they contain little or no extra material. There are several ways to make subunit vaccines but, in all cases, the subunit vaccine loses its ability to cause disease while retaining its antigenic properties. Some of the most promising vaccines under development consist of naked DNA which is injected into the body and produces an antigenic protein. The safety of DNA vaccines is uncertain but they show promise against rapidly mutating viruses such as influenza and HIV.

**Types of Vaccine**

VACCINE
2000ml

**Whole-Agent Vaccine**
Contains whole, **nonvirulent** microorganisms

**Subunit Vaccine**
Contains some part or product of microorganisms that can produce an immune response

**Inactivated (killed)**

FORMALIN

Viruses for vaccines may be inactivated with formalin or other chemicals. They present no risk of infection, e.g. most influenza vaccines, Salk polio vaccine.

**Attenuated (weakened)**

Mutated DNA

Attenuated viruses are usually strains in which mutations have accumulated during culture. These live viruses can back-mutate to a virulent form, e.g. MMR vaccine.

Many subunit vaccines are made with the addition of adjuvants: additional agents that increase the effectiveness of the vaccine. Adjuvants are not part of the pathogen but include irritants that increase the level of immune response.

**Recombinant vaccines**

Yeast makes viral proteins

Non-pathogenic microbes are genetically engineered to make a desired antigenic fraction, e.g. the hepatitis B vaccine.

**Toxoids**

Heat, iodine or formaldehyde

Toxins from bacteria are inactivated

Bacterial toxins inactivated by heat or chemicals. The toxoid stimulates the production of antibodies, e.g. diphtheria vaccine, tetanus vaccine.

**Conjugated vaccines**

Toxoid attached

Polysaccharide from pathogen

Polysaccharides from the pathogen are combined with proteins from other pathogens, e.g. a vaccine against *Haemophilus influenzae b*.

**Acellular vaccines**

Pieces of bacterial cells

Fragments of a whole-agent vaccine containing the desired antigens. Unable to cause infection, e.g. typhoid vaccine.

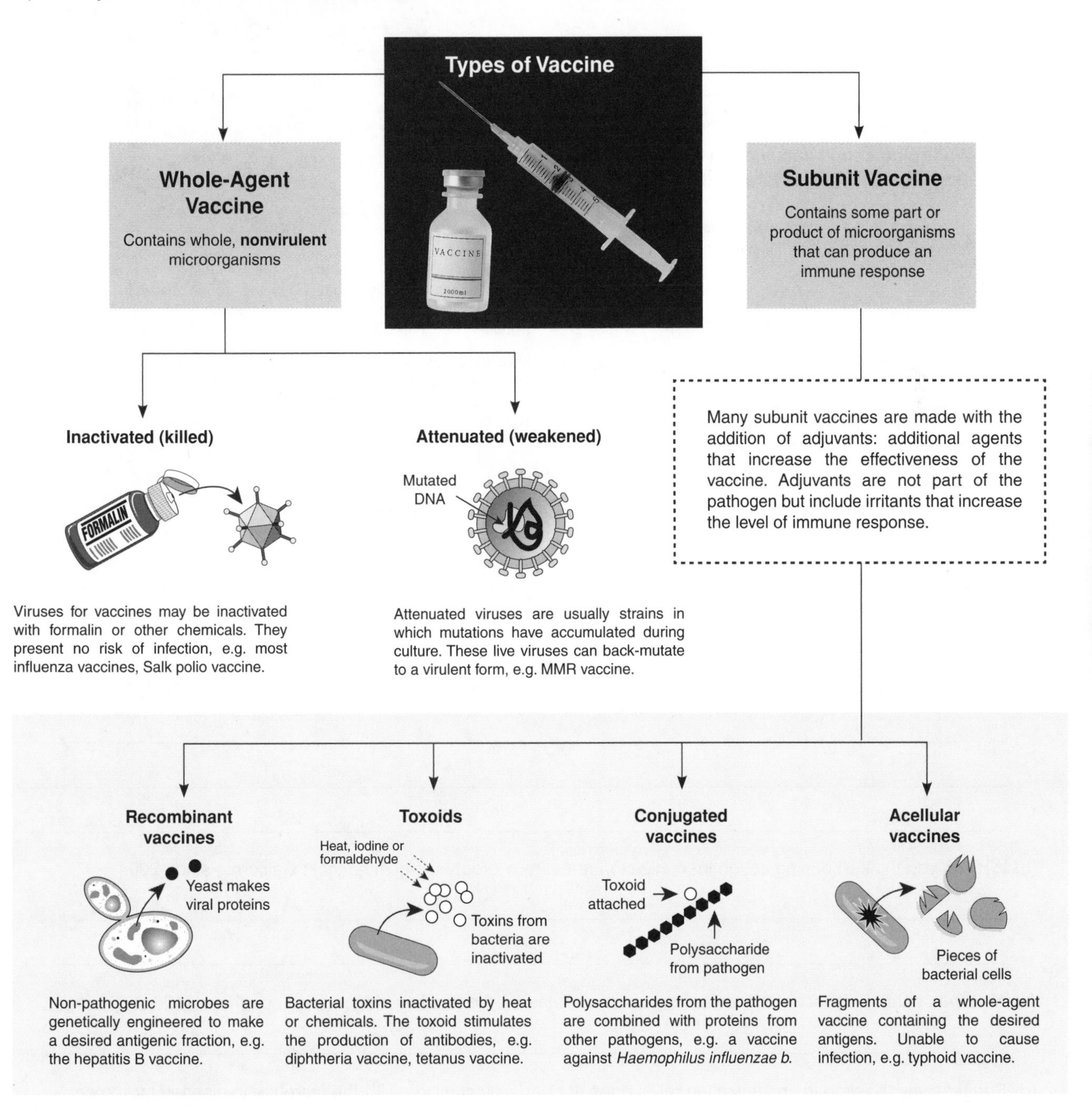

1. **Attenuated viruses** provide long term immunity to their recipients and generally do not require booster shots. Why do you think attenuated viruses provide such effective long-term immunity when inactivated viruses do not?

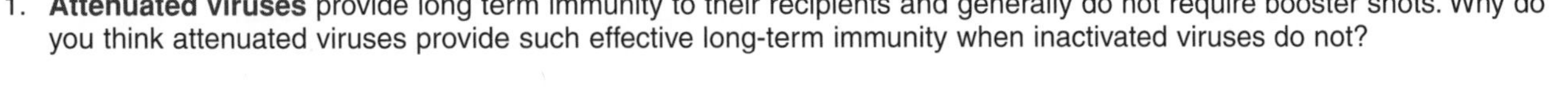


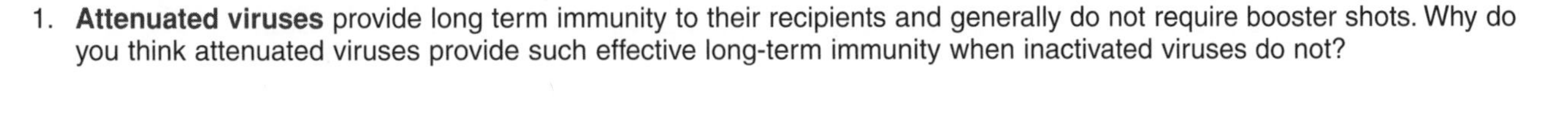

**ISBN: 978-1-927173-12-1**


***Periodicals:*** *Boosting vaccine power*

***Related activities:*** *Acquired Immunity*
***Weblinks:*** *Steps in Vaccine Development*

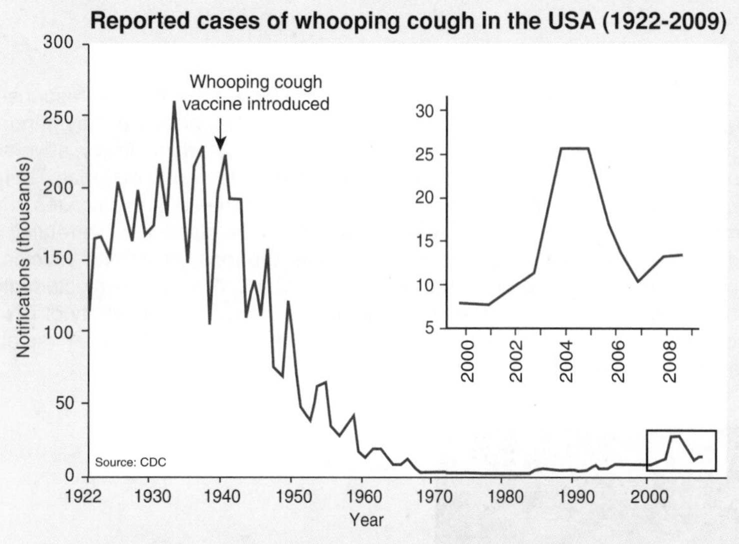


**Whooping cough vaccination rates vs reported cases for California (2000-2010)**

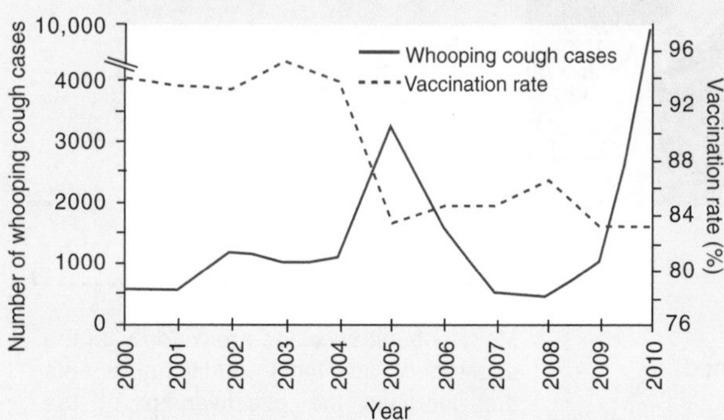


## Case Study: Whooping Cough

Whooping cough is caused by the bacterium *Bordetella pertussis*, and infection may last for two to three months. It is characterized by painful coughing spasms, and a cough that sounds like a "whoop". Severe coughing fits may be followed by periods of vomiting. Inclusion of the whooping cough vaccine into the US immunization schedule in the 1940s has greatly reduced the incidence rates of the disease (left).

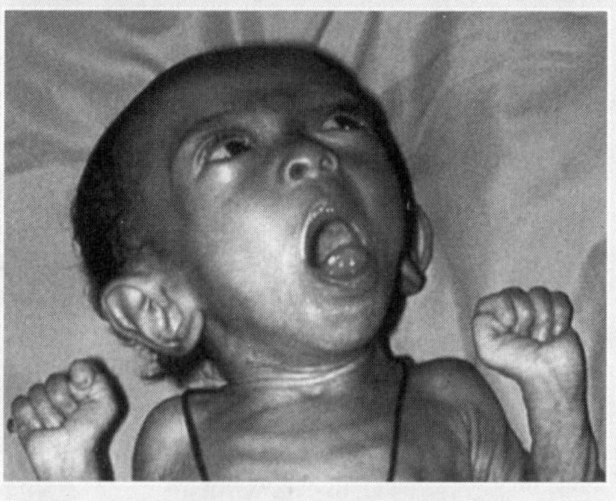


**Above**: Infants under six months of age are most at risk of developing complications or dying from whooping cough because they are too young to be fully protected by the vaccine. Ten infants died of whooping cough in California in 2010.

**Left**: In California, whooping cough vaccination rates have fallen amidst fears that it is responsible for certain health problems such as autism. As a result, rates of whooping cough have increased significantly since 2004. In 2010, over 9000 cases were reported, the highest level in 63 years.

2. How do high vaccination rates help to reduce the incidence of infectious disease? ____________

3. (a) Describe the effect of introducing the whooping cough vaccine into the immunization schedule in the US:

   (b) Why do you think whooping cough immunization rates have dropped significantly in California since 2004?

   (c) What has been the effect of the lower immunization rates on the number of whooping cough cases? ____________

   (d) Suggest why the drop in immunization rates does not perfectly coincide with the increase in disease incidence:

4. Originally the whooping cough vaccine was a whole agent vaccine. In the 1990s it started being manufactured as an acellular vaccine. What advantages does an acellular vaccine have over a whole agent vaccine?


ISBN: 978-1-927173-12-1

# The Immune System

The efficient internal defense provided by the immune system is based on its ability to respond specifically against a foreign substance and its ability to hold a memory of this response. There are two main components of the immune system: the humoral and the cell-mediated responses. They work separately and together to protect us from disease. The **humoral immune response** is associated with the serum (non-cellular part of the blood) and involves the action of **antibodies** secreted by B cell lymphocytes. Antibodies are found in extracellular fluids including lymph, plasma, and mucus secretions. The humoral response protects the body against circulating viruses, and bacteria and their toxins. The **cell-mediated immune response** is associated with the production of specialized lymphocytes called **T cells**. It is most effective against bacteria and viruses located within host cells, as well as against parasitic protozoa, fungi, and worms. This system is also an important defense against cancer, and is responsible for the rejection of transplanted tissue. Both B and T cells develop from stem cells located in the liver of fetuses and the bone marrow of adults. T cells complete their development in the thymus, whilst the B cells mature in the bone marrow.

## Lymphocytes and their Functions

B cells mature in the bone marrow in the shaft of the long bones (e.g. the femur). They migrate from here to the lymphatic organs.

**Stem cell**
Stem cells in the bone marrow and fetal liver give rise to T cells and B cells.

**Thymus gland**
The thymus gland is located above the heart on the wind pipe. It is large in infants but regresses with age. Immature T cells move to the thymus to mature before migrating to other lymphatic organs.

### B cell



B cells recognize and bind antigens. Each B cell recognizes one specific antigen. Helper T cells recognize specific antigens on B cell surfaces and induce their maturation and proliferation. A mature B cell may carry as many as 100,000 antigenic receptors embedded in its surface membrane. B cells defend against bacteria and viruses outside the cell and toxins produced by bacteria (free antigens).

*Differentiate into 2 kinds of cells:*



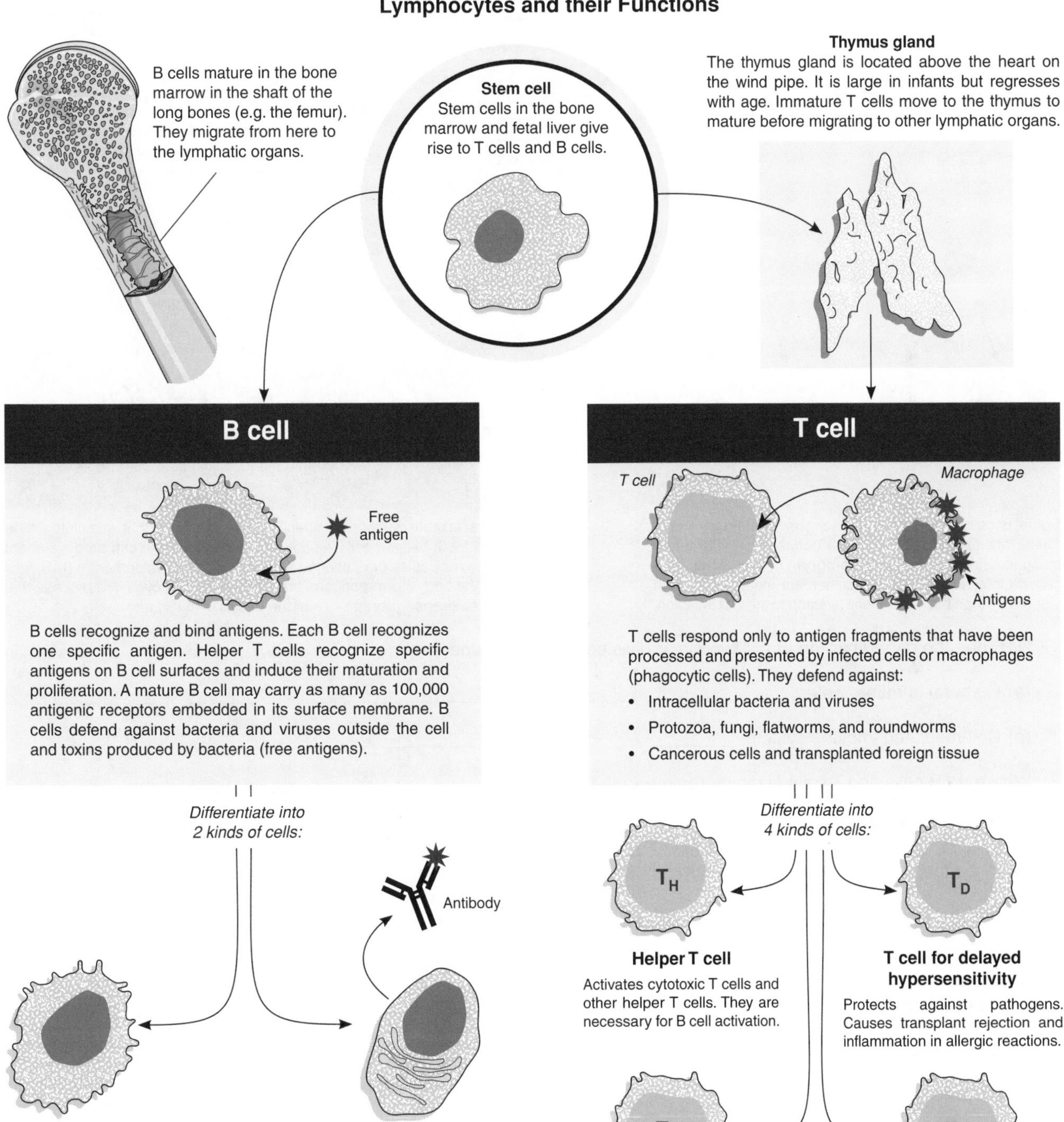

**Memory cells**
Some B cells differentiate into long-lived memory cells (see see Clonal Selection Theory, over page). When these cells encounter the same antigen again (even years or decades later), they rapidly differentiate into antibody-producing plasma cells.

**Plasma cells**
When stimulated by an antigen (see Clonal Selection Theory, over page), some B cells differentiate into plasma cells, which secrete antibodies into the blood system. The antibodies then inactivate the circulating antigens.

### T cell



T cells respond only to antigen fragments that have been processed and presented by infected cells or macrophages (phagocytic cells). They defend against:

- Intracellular bacteria and viruses
- Protozoa, fungi, flatworms, and roundworms
- Cancerous cells and transplanted foreign tissue

*Differentiate into 4 kinds of cells:*



**Helper T cell**
Activates cytotoxic T cells and other helper T cells. They are necessary for B cell activation.

**T cell for delayed hypersensitivity**
Protects against pathogens. Causes transplant rejection and inflammation in allergic reactions.

**Suppressor T cell**
Regulates immune response by turning it off when no more antigen is present.

**Cytotoxic T cell**
Destroys target cells on contact. Recognizes tumor (cancer) or virus infected cells by their surface (antigens and MHC markers).

ISBN: 978-1-927173-12-1


*Related activities:* Antibodies *Weblinks:* The Humoral Response, The Immune System Overview, Introducing...Specific Immunity

In 1955, the Australian **Sir Frank Macfarlane Burnet** proposed the **clonal selection theory** to explain how the immune system is able to respond to the large and unpredictable range of potential antigens in the environment. The diagram below describes **clonal selection** after antigen exposure for B cells. In the same way, a T cell stimulated by a specific antigen will multiply and develop into different types of T cells. Clonal selection and differentiation of lymphocytes provide the basis for **immunological memory**.

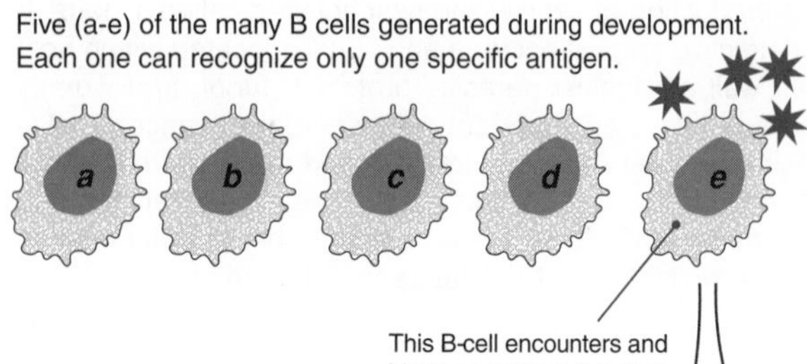

Five (a-e) of the many B cells generated during development. Each one can recognize only one specific antigen.
a
b
c
d
e
This B-cell encounters and binds an antigen. It is then stimulated to proliferate.

## Clonal Selection Theory

Millions of B cells form during development. Antigen recognition is randomly generated, so collectively they can recognize many antigens, including those that have never been encountered. Each B cell makes antibodies corresponding to the specific antigenic receptor on its surface. The receptor reacts only to that specific antigen. When a B cell encounters its antigen, it responds by proliferating and producing many clones all with the same kind of antibody. This is called clonal selection because the antigen selects the B cells that will proliferate.

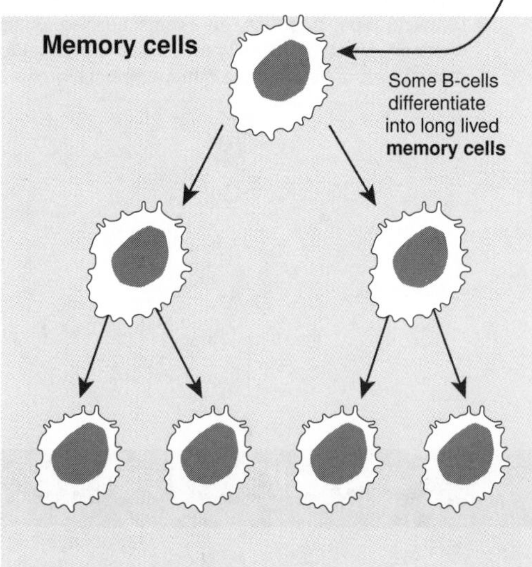

Memory cells
Some B-cells differentiate into long lived **memory cells**

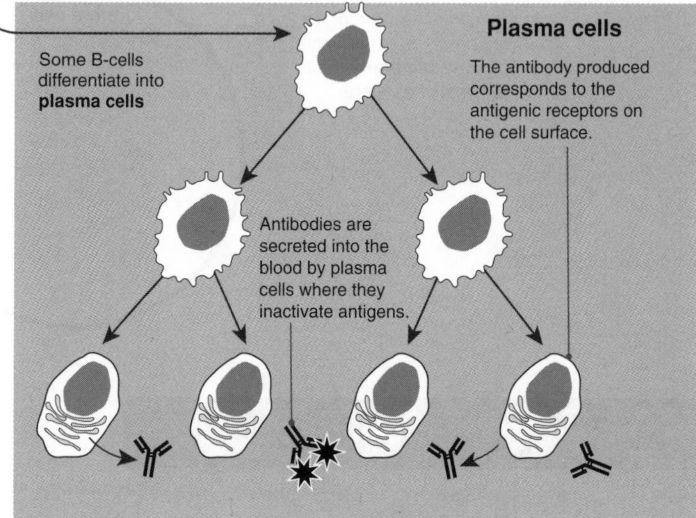

Plasma cells
Some B-cells differentiate into **plasma cells**
The antibody produced corresponds to the antigenic receptors on the cell surface.
Antibodies are secreted into the blood by plasma cells where they inactivate antigens.

Some B cells differentiate into long lived **memory cells**. These are retained in the lymph nodes to provide future immunity (**immunological memory**). In the event of a second infection, B-memory cells react more quickly and vigorously than the initial B-cell reaction to the first infection.

**Plasma cells** secrete antibodies specific to the antigen that stimulated their development. Each plasma cell lives for only a few days, but can produce about 2000 antibody molecules per second. Note that during development, any B cells that react to the body's own antigens are selectively destroyed in a process that leads to **self tolerance** (acceptance of the body's own tissues).

1. Describe the general action of the two major divisions in the immune system:

   (a) Humoral immune system: ______________________

   (b) Cell-mediated immune system: ______________________

2. Where do B cells and T cells originate (before maturing)? ______________________

3. (a) Where do B cells mature? ______________________ (b) Where do T cells mature? ______________________

4. Describe the function of each of the following cells in the immune system response:

   (a) Memory cells: ______________________

   (b) Plasma cells: ______________________

   (c) Helper T cells: ______________________

   (d) Suppressor T cells: ______________________

   (e) Delayed hypersensitivity T cells: ______________________

   (f) Cytotoxic T cells: ______________________

5. Explain the basis of **immunological memory**: ______________________

______________________

______________________

______________________

**ISBN: 978-1-927173-12-1**

# Antibodies

Antibodies and antigens play key roles in the response of the immune system. Antigens are foreign molecules that are able to bind to receptors and provoke a specific immune response. Antigens include potentially damaging microbes and their toxins (see below) as well as substances such as pollen grains, blood cell surface molecules, and the surface proteins on transplanted tissues. **Antibodies** (also called immunoglobulins) are proteins that are made in response to antigens. They are secreted into the plasma where they circulate and can recognize, bind to, and help to destroy antigens. There are five classes of **immunoglobulins**. Each plays a different role in the immune response (including destroying protozoan parasites, enhancing phagocytosis, protecting mucous surfaces, and neutralizing toxins and viruses). The human body can produce an estimated 100 million kinds of antibodies, recognizing many different antigens, including those it has never encountered. Each type of antibody is highly specific to only one particular antigen. The ability of the immune system to recognize and ignore the antigenic properties of its own tissues occurs early in development and is called **self-tolerance**. Exceptions occur when the immune system malfunctions and the body attacks its own tissues, causing an **autoimmune disorder**.

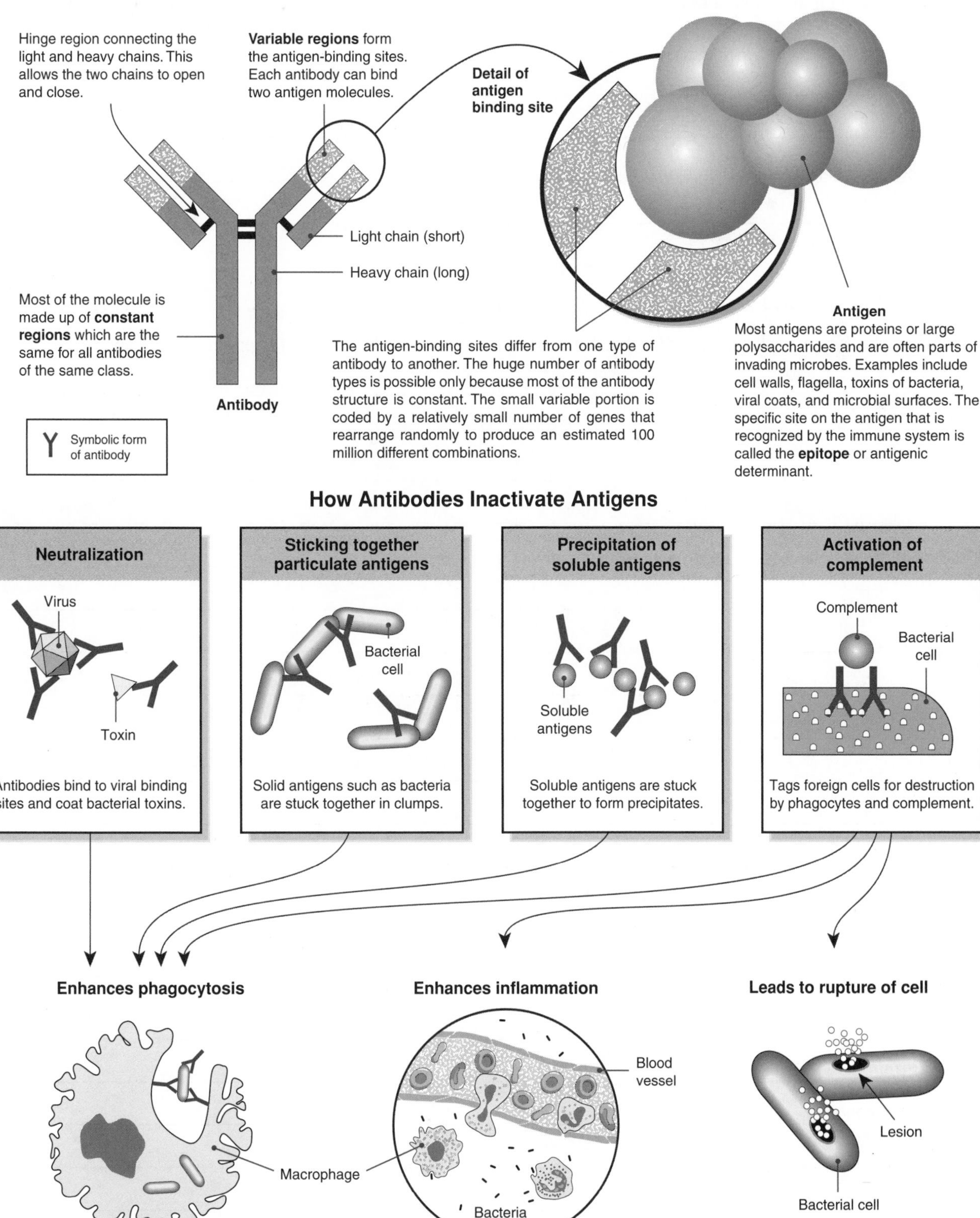



ISBN: 978-1-927173-12-1

1. Distinguish between an **antibody** and an **antigen**:

2. Describe the structure of an antibody, identifying the specific features of its structure that contribute to its function:

3. It is necessary for the immune system to clearly distinguish the body's own cells and proteins from foreign ones.

   (a) Why is this the case?

   (b) How does **self tolerance** develop?

   (c) What type of disorder results when this recognition system fails?

   (d) Describe two examples of disorders that are caused in this way, identifying what happens in each case:

4. Discuss the various ways in which antibodies inactivate antigens:

5. Explain how antibody activity enhances or leads to:

   (a) Phagocytosis:

   (b) Inflammation:

   (c) Bacterial cell lysis:

**ISBN: 978-1-927173-12-1**

# Monoclonal Antibodies

A **monoclonal antibody** is an artificially produced antibody that binds to and neutralizes only one specific protein (**antigen**). A monoclonal antibody binds an antigen in the same way that a normally produced antibody does. Monoclonal antibodies are used as diagnostic tools (e.g. detecting pregnancy) or to treat some types of cancer or autoimmune diseases. Therapeutic uses are still limited because the antibodies are produced are from non-human cells and can cause side effects. In the future, production of monoclonal antibodies from human cells will probably result in fewer side effects. Monoclonal antibodies are produced in the laboratory by stimulating the production of B-lymphocytes in mice injected with the antigen. These B-lymphocytes produce an antibody against a specific antigen. Once isolated, they are made to fuse with immortal tumor cells, and they can be cultured indefinitely in a suitable growing medium (below). Monoclonal antibodies are useful for three reasons: they are all the same (i.e. clones), they can be produced in large quantities, and they are highly specific.

## Making Monoclonal Antibodies



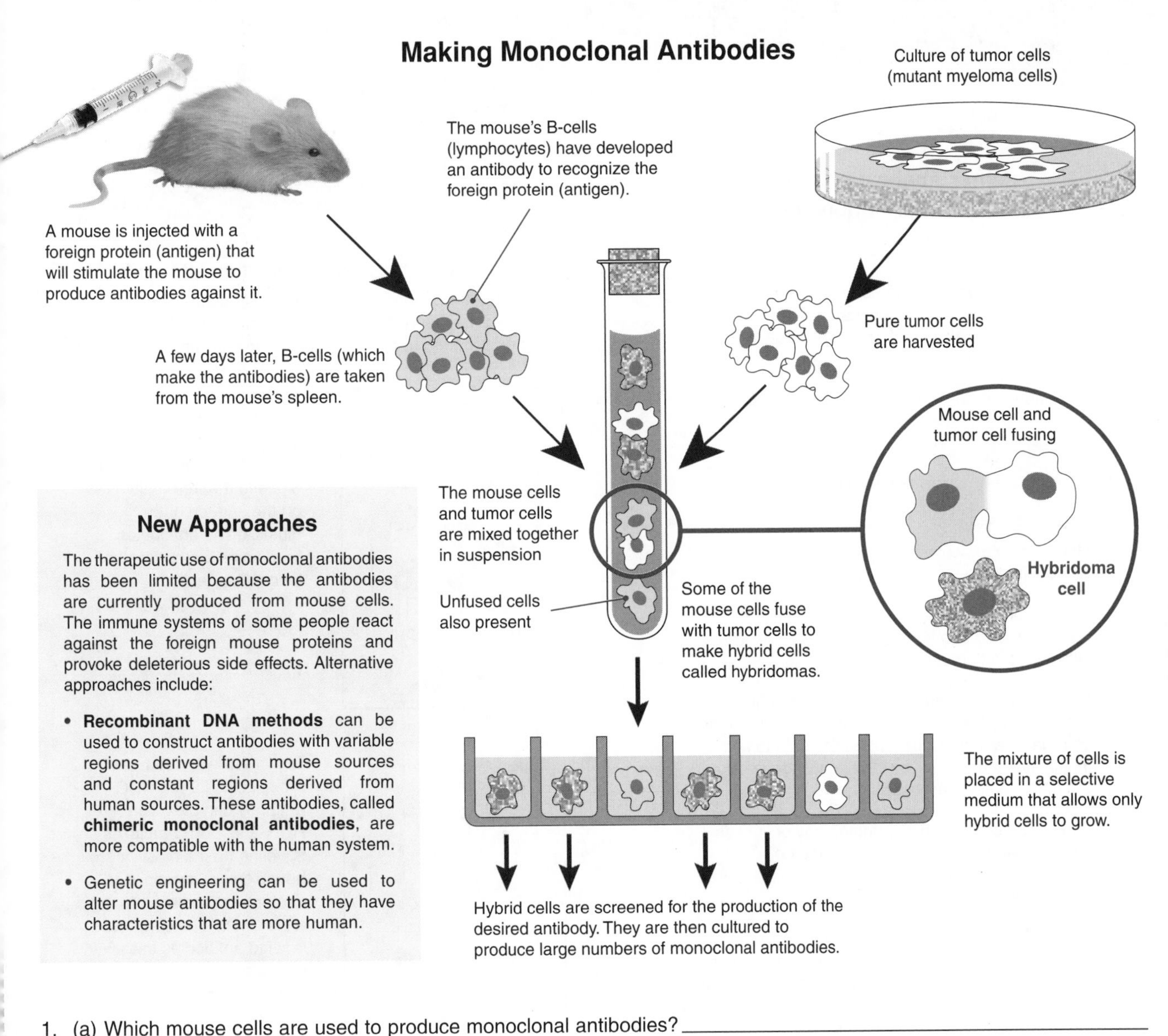

### New Approaches

The therapeutic use of monoclonal antibodies has been limited because the antibodies are currently produced from mouse cells. The immune systems of some people react against the foreign mouse proteins and provoke deleterious side effects. Alternative approaches include:

- **Recombinant DNA methods** can be used to construct antibodies with variable regions derived from mouse sources and constant regions derived from human sources. These antibodies, called **chimeric monoclonal antibodies**, are more compatible with the human system.
- Genetic engineering can be used to alter mouse antibodies so that they have characteristics that are more human.

1. (a) Which mouse cells are used to produce monoclonal antibodies? ______

   (b) What problem is associated with the use of mice to produce monoclonal antibodies? ______

2. Which characteristic of tumor cells allows an ongoing culture of antibody-producing lymphocytes to be made?

3. Describe four applications of monoclonal antibodies:

   (a) ______

   (b) ______

   (c) ______

   (d) ______

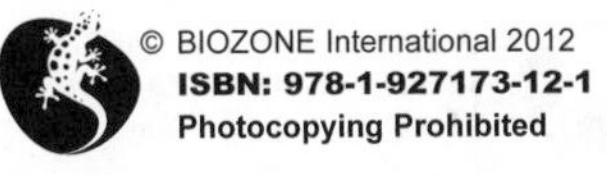

ISBN: 978-1-927173-12-1

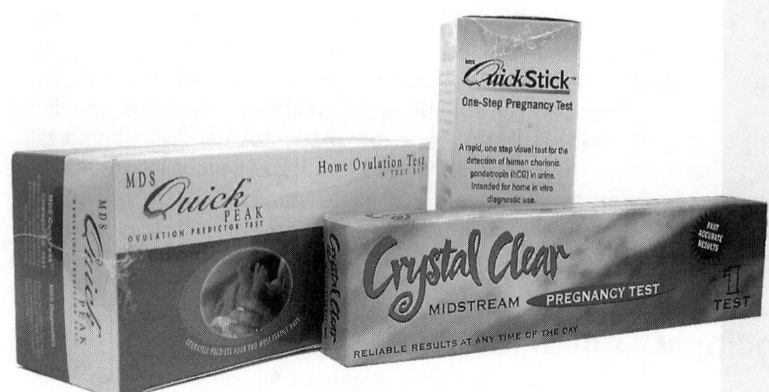


## Detecting Pregnancy using Monoclonal Antibodies

When a woman becomes pregnant, a hormone called **human chorionic gonadotropin** (HCG) is released. HCG accumulates in the bloodstream and is excreted in the urine. Antibodies can be produced against HCG and used in simple test kits (below) to determine if a woman is pregnant. Monoclonal antibodies are also used in other home testing kits, such as those for detecting ovulation time (far left).

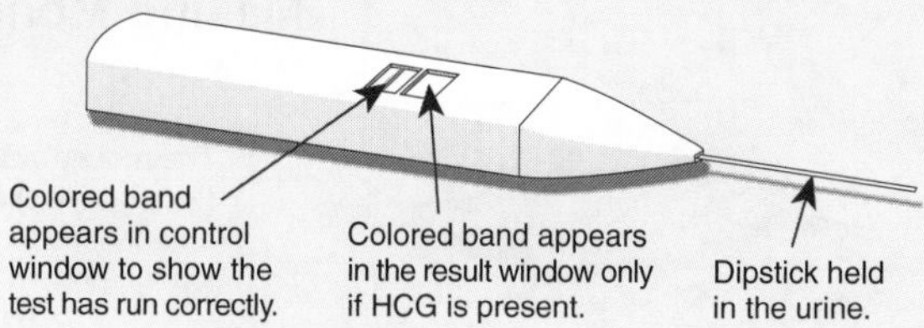


### How home pregnancy detection kits work

The test area of the dipstick (below) contains two types of antibodies: free monoclonal antibodies and capture monoclonal antibodies, bound to the substrate in the test window.

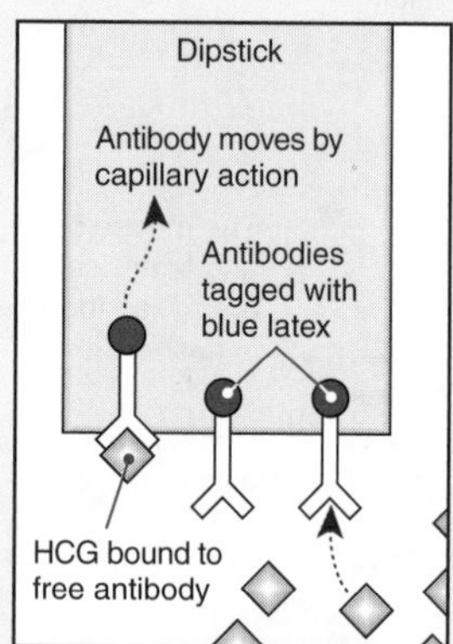


The free antibodies are specific for HCG and are color-labeled. HCG in the urine of a pregnant woman binds to the free antibodies on the surface of the dipstick. The antibodies then travel up the dipstick by capillary action.

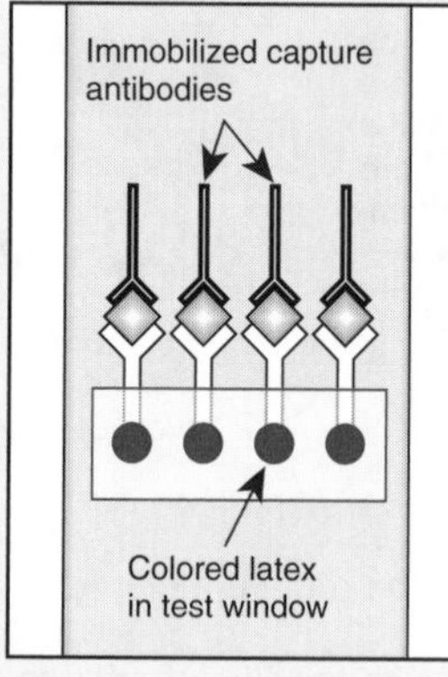


The capture antibodies are specific for the HCG-antibody complex. The HCG-antibody complexes traveling up the dipstick are bound by the immobilized capture antibodies, forming a sandwich. The color labeled antibodies then create a visible color change in the test window.

## Other Applications of Monoclonal Antibodies

**Diagnostic uses**

- Detecting the presence of pathogens such as *Chlamydia* and streptococcal bacteria, distinguishing between *Herpesvirus* I and II, and diagnosing AIDS.
- Measuring protein, toxin, or drug levels in serum.
- Blood and tissue typing.
- Detection of antibiotic residues in milk.

**Therapeutic uses**

- Neutralizing endotoxins produced by bacteria in blood infections.
- Used to prevent organ rejection, e.g. in kidney transplants, by interfering with the T cells involved with the rejection of transplanted tissue.
- Used in the treatment of some auto-immune disorders such as rheumatoid arthritis and allergic asthma. The monoclonal antibodies bind to and inactivate factors involved in the cascade leading to the inflammatory response.
- Immunodetection and immunotherapy of cancer. Newer methods specifically target the cell membranes of tumor cells, shrinking solid tumors without harmful side effects.
- Inhibition of platelet clumping, which is used to prevent reclogging of coronary arteries in patients who have undergone angioplasty. The monoclonal antibodies bind to the receptors on the platelet surface that are normally linked by fibrinogen during the clotting process.

4. For each of the following applications, suggest why an antibody-based test or therapy is so valuable:

(a) Detection of toxins or bacteria in perishable foods: ______________________________

______________________________

______________________________

(b) Detection of pregnancy without a doctor's prescription: ______________________________

______________________________

(c) Targeted treatment of tumors in cancer patients: ______________________________

______________________________


ISBN: 978-1-927173-12-1

# AIDS: Failures of Defense

AIDS (acquired immune deficiency syndrome) was first reported in the US in 1981. By 1983, the pathogen had been identified as **HIV** (human immunodeficiency virus) a virus that selectively infects **helper T cells**. It has since been established that HIV arose by the recombination of two simian viruses. It has probably been endemic in some central African regions for decades, as HIV has been found in blood samples from several African nations from as early as 1959. The disease causes a massive deficiency in the immune system. HIV is a **retrovirus** (RNA, not DNA) and is able to splice its genes into the host cell's chromosome. As yet, there is no cure or vaccine, and the disease has taken the form of a **pandemic**, spreading to all parts of the globe and killing more than a million people each year. HIV is constantly mutating and changing in ways which allow it to avoid detection, or adapt to the immune system. Its rapid rate of mutation is also why a successful vaccine has yet to be developed.

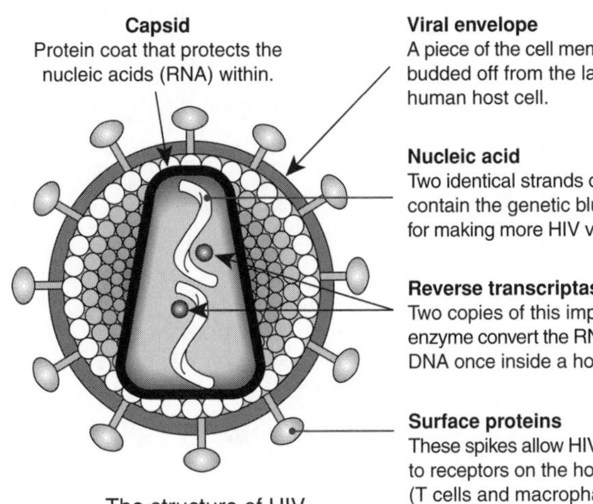


The structure of HIV

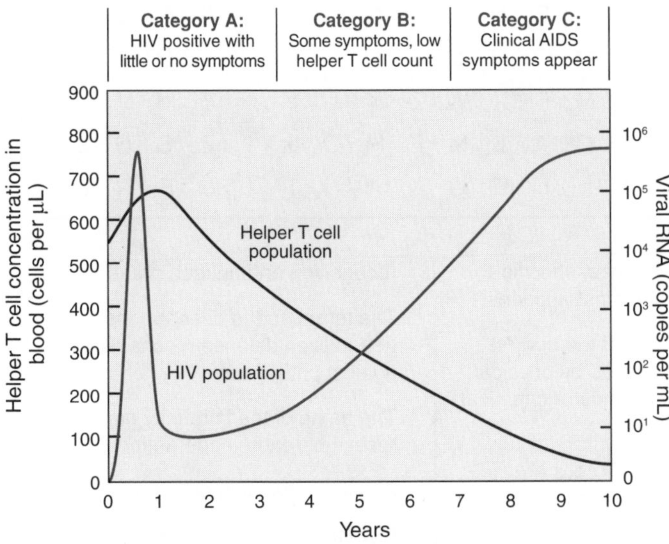


### The stages of an HIV infection

AIDS is only the end stage of a HIV infection. Shortly after the initial infection, HIV antibodies appear in the blood. The progress of infection has three clinical categories (shown in the graph above).

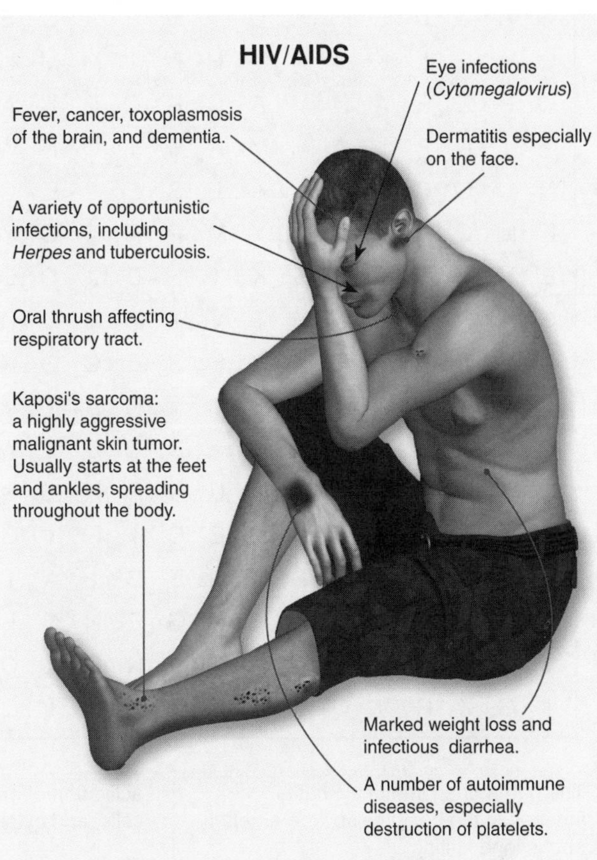


Individuals infected by HIV may have no symptoms, while medical examination may detect swollen lymph glands. Others may experience a short-lived illness when they first become infected. HIV infection is associated with a wide range of symptoms, which stem from secondary infections arising as a result of a **suppressed immune system** (low helper T cell numbers). These infections are from normally rare fungal, viral, and bacterial sources. Full blown AIDS can also feature rare forms of cancer.

Defense Mechanisms

1. Explain why HIV virus has such a debilitating effect on ability to fight disease: ______

______

______

2. Consult the graph above showing the stages of HIV infection (remember, HIV infects and destroys helper T cells).

   (a) Describe how the virus population changes with the progression of the disease: ______

   ______

   (b) How do the helper T cells respond to the infection? ______

   ______

   ______

ISBN: 978-1-927173-12-1




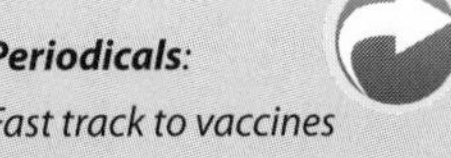

# KEY TERMS: Word Find

Use the clues below to find the relevant key terms in the WORD FIND grid.

| B | P | R | I | M | A | R | Y | R | E | S | P | O | N | S | E | M | H | C | A | P | E | I | Z | Y | M | L | Z | O | V |
|---|---|---|---|---|---|---|---|---|---|---|---|---|---|---|---|---|---|---|---|---|---|---|---|---|---|---|---|---|---|
| P | H | A | G | O | C | Y | T | E | A | T | K | U | K | I | S | C | X | D | I | A | M | S | F | C | J | F | L | I | Z |
| G | A | K | H | L | J | G | N | O | N | S | P | E | C | I | F | I | C | D | E | F | E | N | S | E | F | A | E | M | A |
| W | A | J | D | S | G | A | D | I | S | E | A | S | E | K | F | J | G | N | I | U | T | E | I | E | I | H | U | X | U |
| O | W | B | L | K | J | K | V | H | B | I | M | M | U | N | I | T | Y | T | I | K | C | W | J | X | F | Y | K | N | D |
| O | P | Z | E | A | Z | M | A | C | R | O | P | H | A | G | E | F | N | X | S | A | E | D | I | V | M | M | O | J | T |
| I | I | N | F | L | A | M | M | A | T | I | O | N | W | Y | E | G | F | K | S | I | L | V | W | N | F | J | C | Z | Y |
| N | U | O | C | L | O | N | A | L | S | E | L | E | C | T | I | O | N | P | N | D | L | J | L | N | E | U | Y | S | K |
| F | M | Q | M | U | H | L | Y | M | P | H | O | C | Y | T | E | G | R | N | L | S | X | C | B | V | K | A | T | C | G |
| E | V | T | K | K | S | S | K | I | N | B | J | G | O | E | W | O | F | S | W | G | B | B | C | E | L | L | E | K | E |
| C | M | P | M | D | U | Y | T | V | T | M | M | D | D | Z | P | A | C | T | I | V | E | I | M | M | U | N | I | T | Y |
| T | E | C | G | M | F | X | B | E | S | E | C | O | N | D | A | R | Y | R | E | S | P | O | N | S | E | A | S | H | G |
| I | C | X | Y | W | V | A | C | C | I | N | A | T | I | O | N | H | I | V | G | E | H | Z | T | W | W | B | P | X | G |
| O | S | H | T | F | A | N | T | I | B | O | D | Y | W | P | H | S | M | G | G | H | U | M | O | R | A | L | S | O | E |
| N | T | H | P | S | I | M | M | U | N | E | R | E | S | P | O | N | S | E | I | H | V | L | S | E | G | K | J | W | L |
| P | X | F | L | I | E | S | P | E | C | I | F | I | C | D | E | F | E | N | S | E | N | T | P | A | S | S | I | V | E |
| B | F | D | X | C | E | L | L | M | E | D | I | A | T | E | D | M | I | H | L | M | O | N | O | C | L | O | N | A | L |
| S | I | H | Q | P | A | T | H | O | G | E | N | N | L | B | F | M | F | X | W | Z | X | L | Y | U | Y | M | V | M | U |
| Y | U | Y | C | Y | J | I | M | M | U | N | O | L | O | G | I | C | A | L | M | E | M | O | R | Y | Z | G | G | Z | C |
| I | G | X | R | V | G | I | G | P | Y | G | K | I | Q | J | I | N | F | E | O | N | A | N | T | I | G | E | N | B | L |

The name of an immunity that is induced in the host itself by the antigen, and is long lasting.

A large white blood cell, found within tissues, produced by the differentiation of monocytes.

An immunity that is associated with the non-cellular part of the blood and involves the action of antibodies secreted by B cell lymphocytes.

A protein made in response to an antigen.

The ability of the immune system to respond rapidly in the future to antigens encountered in the past.

The acronym of a syndrome caused by a retrovirus that selectively infects helper T cells.

The term describing how the human body recognizes and defends itself against specific antigens.

A foreign molecule that is able to bind to an antibody and provoke a specific immune response.

A lymphocyte that makes antibodies against specific antigens.

The name of an immunity involving the activation of macrophages, specific T cells, and cytokines against antigens.

The defense response of the body's tissues to damage caused by physical agents, microbial infections or chemical agents.

A theory for how B and T cells are selected to target specific antigens invading the body.

A specific white blood cell involved in the adaptive immune response.

An abnormal condition of the body when bodily functions are impaired.

The abbreviation for the retrovirus which causes AIDS.

An antibody made by one type of immune cell that are all clones of a unique parent cell.

The term to describe the specific resistance of an organism to infection or disease.

A white blood cell.

The invasion of a host organism by a pathogen to the detriment of the host.

The abbreviation for the set of molecules displayed on cell surfaces that are responsible for lymphocyte recognition and antigen presentation.

The term used to describe the generalized defense mechanisms against pathogens.

The name of the immunity gained by the receipt of ready-made antibodies.

A disease-causing organism.

A white blood cell that destroys microbes by engulfing and digesting them.

The initial response of the immune system to exposure to an antigen.

This occurs when a range of defense mechanisms that operate inside the body are activated.

The term for a third line of defense against disease.

Lymphocyte responsible for the cellular immune response.

The gland with a role in the maturation of the T cells responsible for cell-mediated immunity.

The delivery of antigenic material to produce immunity to a disease.

ISBN: 978-1-927173-12-1

Enduring Understanding

2.E
3.E

# Timing, Coordination and Social Behavior

## Key concepts

- Organisms use environmental cues to coordinate life cycles and behavior.
- Biological rhythms have both an endogenous and an exogenous component.
- Many rhythmic behaviors involve the production of, and response to, chemical signals.
- Animals communicate using a variety of cues.
- Sociality offers many advantages to individual survival, reproduction, and population growth.

## Key terms

biological clock
biological rhythm
circadian rhythm
courtship
courtship
crepuscular
diurnal rhythm
dormancy
entrainment
environmental cue
environmental cycle
estivation
fitness
free running period
habituation
hibernation
homing
imprinting
insight learning
jet lag
long-day plant
migration
phase shift
pheromone
photoperiod
photoperiodism
phototropism
phytochrome
short-day plant
torpor
zeitgeber

## Essential Knowledge

☐ 1. Use the **KEY TERMS** to compile a glossary for this topic.

### Coordination of Physiology and Behavior *(2.E.2-2.E.3)* pages 192-218, 225

☐ 2. Describe and explain how plants coordinate physiological events, such as flowering and germination, with environmental cues. Examples include:
(a) **Phototropism** and **gravitropism** as directional growth responses. Explain how tropisms are adaptive in positioning the plant in a favorable environment.
(b) **Photoperiodism** and the role of **phytochrome**. Explain the advantages of photoperiodism to survival and fitness, e.g. in **long-day** and **short-day** plants.

☐ 3. Using examples, describe and explain how animals synchronize physiological processes with environmental cycles an cues. Examples (as below) could include **circadian rhythms**, **diurnal** and **nocturnal** activity patterns, and seasonal responses, such as breeding, **hibernation**, **torpor**, **estivation**, and **migration**. Explain why responses to environment cues, e.g **biological rhythms** and seasonal behaviors, are subject to strong selection pressure.

☐ 4. Describe and explain examples of an endogenous **circadian rhythm**. Using examples, e.g. **jet lag** in humans, explain the causes and consequences of disturbances to normal rhythmic behaviors.

☐ 5. Describe and explain **diurnal**, **nocturnal**, and **crepuscular** activity patterns.

☐ 6. Use examples to explain how simple organisms regulate physiological responses to the environment, e.g. **quorum sensing** and fruiting body formation in bacteria.

☐ 7. Distinguish between **innate** and **learned behavior**. Describe examples of innate and learned behaviors and the adaptive role of the behavior in each case. Examples include taxes (*sing.* **taxis**), kineses (*sing.* **kinesis**), **imprinting**, **habituation**, and **insight learning**.

☐ 8. Distinguish between **homing** and **migration**. Explain the adaptive value of homing and its dependence on environmental cues. Discuss the role of **navigation** in **migratory** and **homing** behavior in animals, including the mechanisms involved.

☐ 9. Describe and explain the role of **pheromones** in animal orientation, including their role as sex attractants and as signaling molecules to coordinate activities.

### Animal Communication *(3.E.1)* pages 219-224, 226-229

☐ 10. Explain how animals communicate information about changes in the environment and explain the survival value of such behaviors. Examples include bee dances, bird songs, territorial marking, pack behavior, and predator warnings.

☐ 11. Explain the survival value of responding appropriately to communicated information, e.g. in parent-offspring behaviors, in **courtship**, and in migration and foraging behaviors (also *2.E.3*).

☐ 12. Explain how cooperative behaviors, e..g herd behaviors, increase individual **fitness** and population survival. (also *2.E.3*)

***Periodicals:***
*Listings for this chapter are on page 375*

***Weblinks:***
*www.thebiozone.com/weblink/AP2-3121.html*

*BIOZONE APP: Student Review Series Populations & Interactions*

# Timing and Coordination in Simple Organisms

Survival in a changing environment requires the ability to synchronize various physiological functions with the environment. The timing and coordination of physiological functions requires the ability both to receive and act upon information from the environment, and to communicate with other individuals of the same species. Communication may be by chemicals secreted into the surrounding environment or by cell to cell contact. Often a physiological event is triggered by the density of a population reaching a certain critical level. For example, the pathogen *Pseudomonas aeruginosa* only becomes virulent when the population reaches a threshold that ensures a successful infection.

### Timing of Virulence in *Pseudomonas*

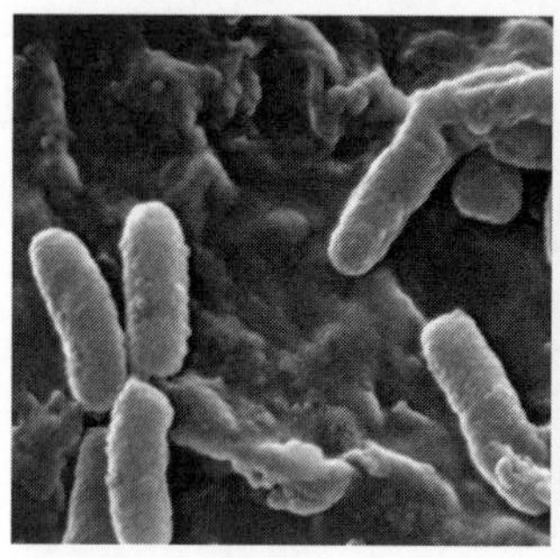

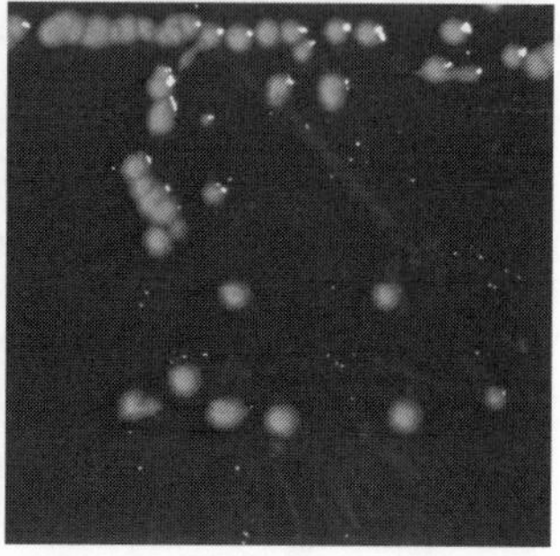

*Pseudomonas aeruginosa* is a common and usually relatively harmless soil bacterium. However, it is also an extremely versatile opportunistic pathogen and can cause fatal infections in people who have had at least part of their immune system compromised.

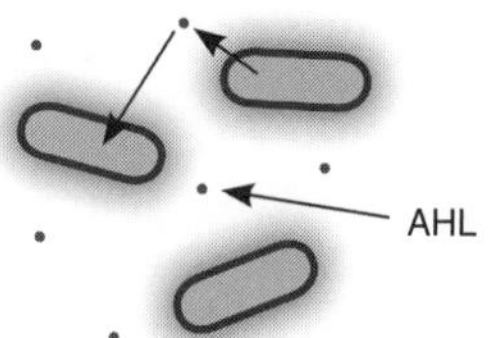


During its initial colonization phase, *P. aeruginosa* does not secrete toxins but devotes resources to increasing population size.

Individual bacteria produce signal molecules called acyl homoserine lactones (**AHL**s) that are received by others.

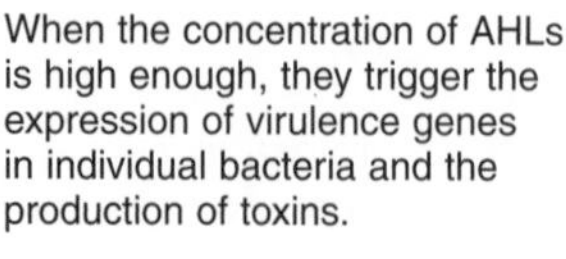

When the concentration of AHLs is high enough, they trigger the expression of virulence genes in individual bacteria and the production of toxins.

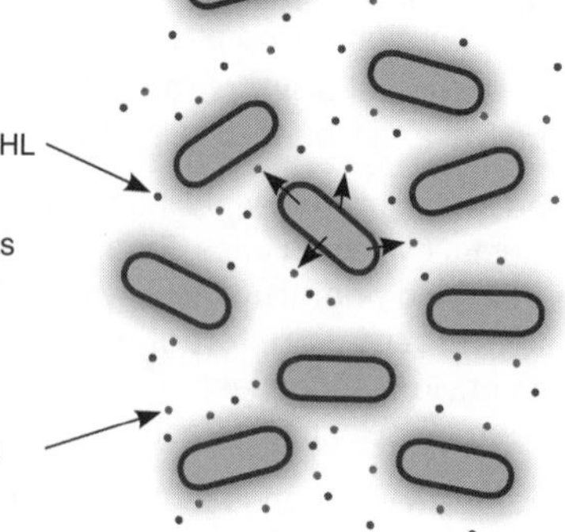


The switch to the virulent phase is regulated by quorum sensing. AHLs provide a way to coordinate the regulation of gene expression. When the population reaches a critical density *P. aeruginosa* suddenly switches to secreting an array of toxins and becomes highly virulent.

### Chemotaxis and Coordination in *Myxococcus*

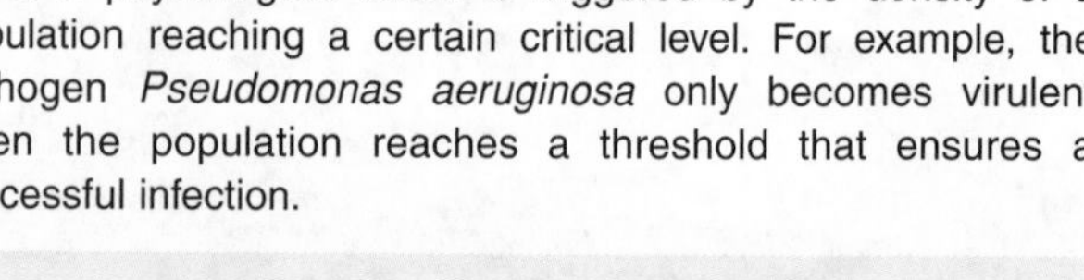

*Myxococcus xanthus* colonies contain rod shaped bacteria that show self organizing behavior in response to environmental cues.

The soil bacteria *Myxococcus xanthus* normally live in small colonies as a biofilm. Coordination between single cells in a colony can help in survival by the production of spores.

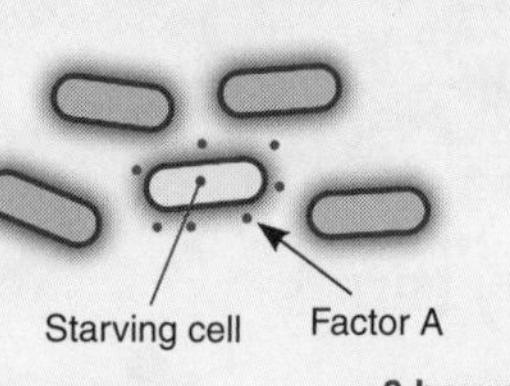


When resources dwindle, the cells begin to starve and signal to other bacteria using a chemical called **factor A.**

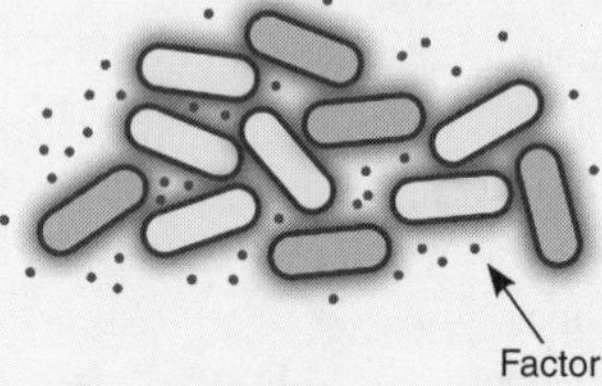


When factor A reaches high density (many cells are starving) the cells congregate and form a fruiting body, which can eventually contain about 100,000 cells.

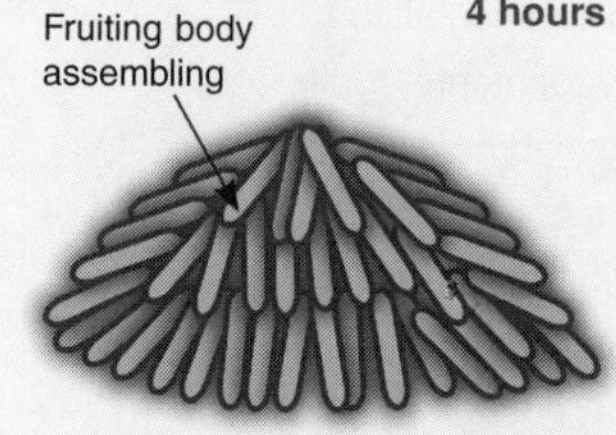


The fruiting body develops thousands of spores that are able to withstand the lack of resources. The spores may then be transported by wind or animals to new territory where they can germinate and produce a new colony.

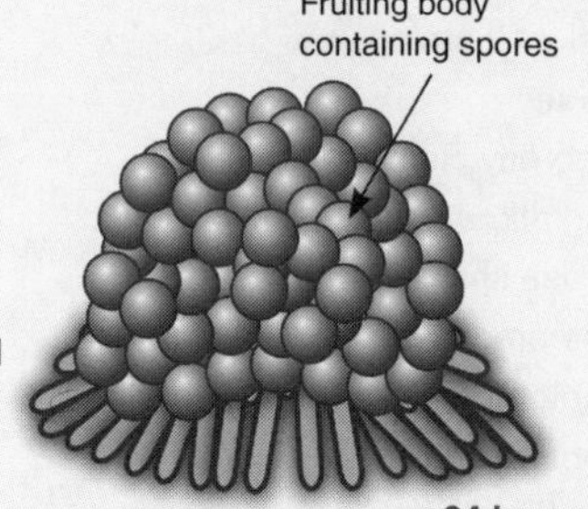


1. Explain how the following events are coordinated:

   (a) The timing of virulence in *Pseudomonas aeruginosa*: __________

   (b) The production of the fruiting body in *Myxococcus xanthus*: __________

2. Explain how coordinating these events gives a survival advantage to *P. aeruginosa* and *M. xanthus*:



**Weblinks:** About Myxococcus xanthus

**Periodicals:** Don't talk, reproduce


ISBN: 978-1-927173-12-1

# Plant Responses

Even though most plants are firmly rooted in the ground, they are still capable of responding and making adjustments to changes in their external environment. This ability is manifested chiefly in changing patterns of growth. These responses may involve relatively sudden physiological changes, as occurs in flowering, or a steady growth response, such as a **tropism**. Other responses of plants to the environment include nastic movements, circadian rhythms, photoperiodism, dormancy, and vernalization.

**TROPISMS**
Tropisms are growth responses made by plants to directional external stimuli, where the direction of the stimulus determines the direction of the growth response. A tropism may be positive (towards the stimulus), or negative (away from the stimulus). Common stimuli for plants include light, gravity, touch, and chemicals.

**LIFE CYCLE RESPONSES**
Plants use seasonal changes in the environment as cues for the commencement or ending of particular life cycle stages. Such changes are mediated by plant growth factors, such as phytochrome and gibberellin and enable the plant to avoid conditions unfavorable to growth or survival. Examples include flowering and other photoperiodic responses, dormancy and germination, and leaf fall.

**RAPID RESPONSES TO ENVIRONMENTAL STIMULI**
Plants are capable of quite rapid responses. Examples include the closing of stomata in response to water loss, opening and closing of flowers in response to temperature, and nastic responses. These responses may follow a circadian rhythm and are protective in that they reduce the plant's exposure to abiotic stress or grazing pressure.

**PLANT COMPETITION AND ALLELOPATHY**
Although plants are rooted in the ground, they can still compete with other plants to gain access to resources. Some plants produce chemicals that inhibit the growth of neighboring plants. Such chemical inhibition is called allelopathy. Plants also compete for light and may grow aggressively to shade out slower growing competitors.

**PLANT RESPONSES TO HERBIVORY**
Many plant species have responded to grazing or browsing pressure with evolutionary adaptations enabling them to survive constant cropping. Examples include rapid growth to counteract the constant loss of biomass (grasses), sharp spines or thorns to deter browsers (acacias, cacti), or toxins in the leaf tissues (eucalyptus).

1. Identify the stimuli to which plants typically respond and explain their significance: ______

2. Explain how plants benefit by responding appropriately to the environment: ______

3. Describe one adaptive response of plants to each of the following stressors in the environment:

   (a) Low soil water: ______

   (b) Falling autumn air temperatures: ______

   (c) High wind: ______

   (d) Browsing animals: ______

   (e) Low nighttime temperatures: ______


ISBN: 978-1-927173-12-1

# Tropisms and Growth Responses

**Tropisms** are plant growth responses to external stimuli, in which the stimulus direction determines the direction of the growth response. Tropisms are identified according to the stimulus involved, e.g. photo- (light), gravi- (gravity), hydro- (water), and may be positive or negative depending on whether the plant moves towards or away from the stimulus respectively. A tropism is distinguished from a **nastic response** by the directionality of the stimulus; nastic responses are independent of stimulus direction.

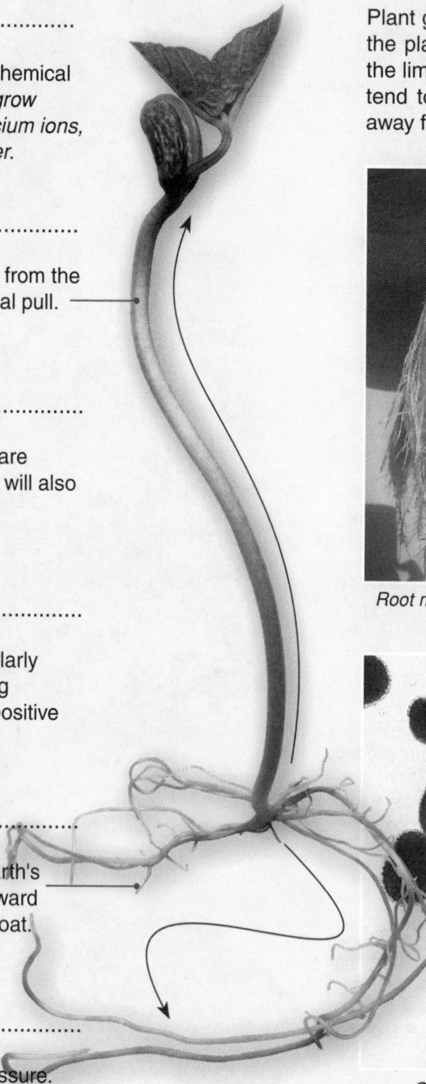

(a) ................................................

A positive growth response to a chemical stimulus. *Example: Pollen tubes grow towards a chemical, possibly calcium ions, released by the ovule of the flower.*

(b) ................................................

Stems and coleoptiles grow away from the direction of the Earth's gravitational pull.

(c) ................................................

Growth response to water. Roots are influenced primarily by gravity but will also grow towards water.

(d) ................................................

Growth responses to light, particularly directional light. Coleoptiles, young stems, and some leaves show a positive response.

(e) ................................................

Roots respond positively to the Earth's gravitational pull, and curve downward after emerging through the seed coat.

(f) ................................................

Growth responses to touch or pressure. Tendrils (modified leaves) have a positive coiling response stimulated by touch.

Plant growth responses are adaptive in that they position the plant body in a suitable growing environment, within the limits of the position in which it germinated. They also tend to reinforce each other. For example, shoots grow away from gravity and towards the light.

*Root mass in a hydroponically grown plant*

*Sweet pea tendrils*

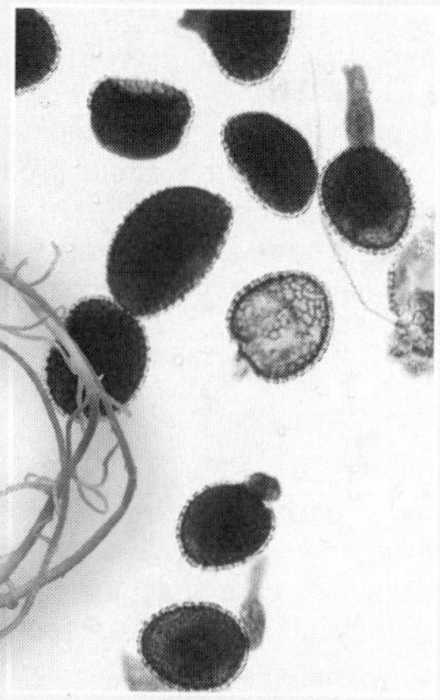

*Germinating pollen*

*Thale cress bending to the light*

Kristian Peters

1. Identify each of the plant tropisms described in (a)-(f) above. State whether the response is positive or negative.

2. Explain the difference between a **tropism** and a **nastic response** (nasty), and identify the adaptive value of each:

3. Describe the adaptive value of the following tropisms:

   (a) Positive gravitropism in roots: ____

   (b) Positive phototropism in coleoptiles: ____

   (c) Positive thigmomorphogenesis in weak stemmed plants: ____

   (d) Positive chemotropism in pollen grains: ____



***Related activities:*** *Investigating Phototropism, Investigating Gravitropism*
***Weblinks:*** *Plants in Motion*


**ISBN: 978-1-927173-12-1**

# Investigating Phototropism

Phototropism in plants was linked to a growth promoting substance in the 1920s. Early experiments investigating phototropism in severed coleoptiles gave evidence for the hypothesis that auxin was responsible for tropic responses in stems. These experiments (below) have been criticized as being too simplistic, although their conclusions have been shown to be valid. Auxins promote cell elongation and are inactivated by light. Thus, when a stem is exposed to directional light, auxin becomes unequally distributed either side of the stem. The stem responds to the unequal auxin concentration by differential growth, i.e. it bends. The mechanisms behind this response are now well understood.

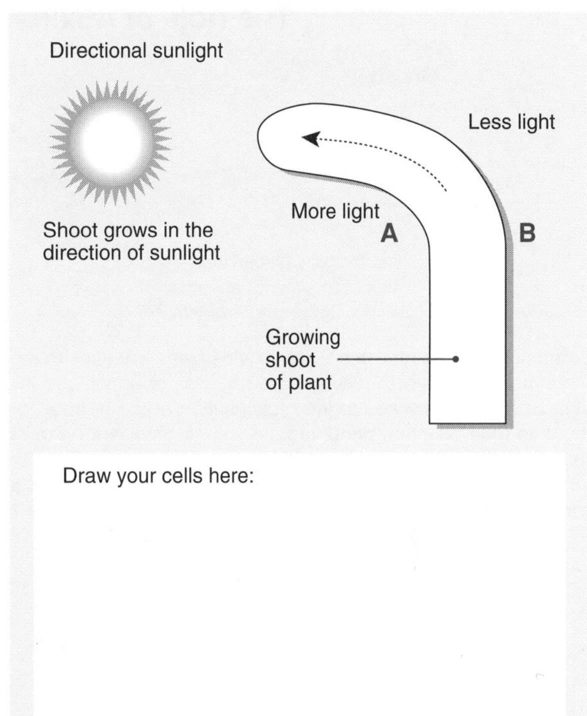


1. **Directional light:** A pot plant is exposed to direct sunlight near a window and as it grows, the shoot tip turns in the direction of the sun. When the plant was rotated, it adjusted by growing towards the sun in the new direction.

   (a) What hormone regulates this growth response?

   (b) What is the name of this growth response?

   (c) How do the cells behave to bring about this change in shoot direction at:

   Point **A**?

   Point **B**?

   (d) Which side (A or B) would have the highest hormone concentration and why?

   (e) Draw a diagram of the cells as they appear across the stem from point A to B (in the rectangle on the right).

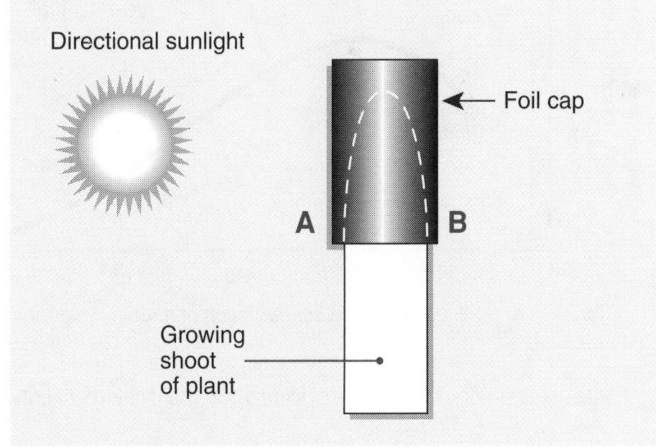


2. **Light excluded from shoot tip:** When a tin-foil cap is placed over the top of the shoot tip, light is prevented from reaching the shoot tip. When growing under these conditions, the direction of growth does not change towards the light source, but grows straight up. State what conclusion can you come to about the source and activity of the hormone that controls the growth response:

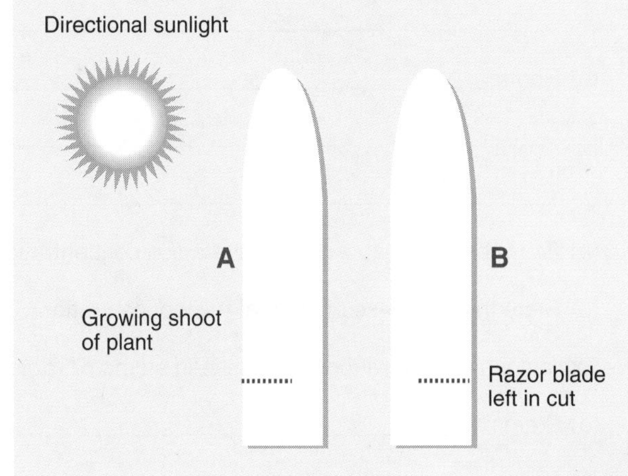

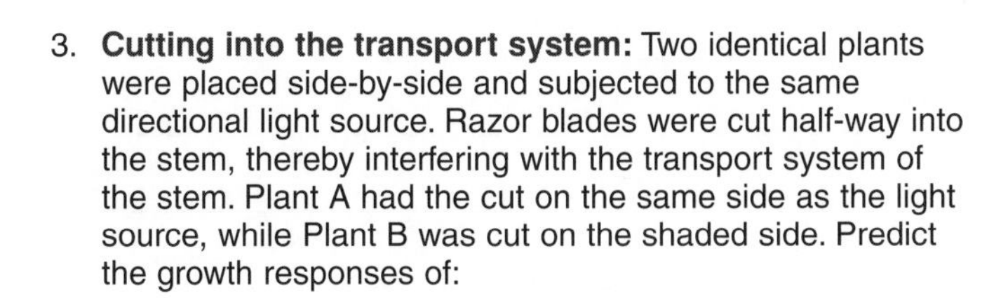

3. **Cutting into the transport system:** Two identical plants were placed side-by-side and subjected to the same directional light source. Razor blades were cut half-way into the stem, thereby interfering with the transport system of the stem. Plant A had the cut on the same side as the light source, while Plant B was cut on the shaded side. Predict the growth responses of:

   Plant **A**:

   Plant **B**:

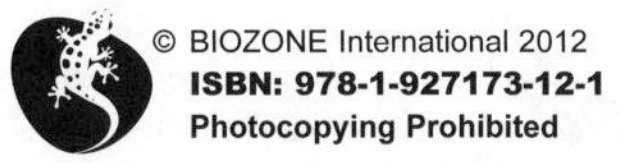


**ISBN: 978-1-927173-12-1**


***Related activities:*** *Tropisms and Growth Responses*
***Weblinks:*** *Plants in Motion*

# Investigating Gravitropism

Although the response of shoots and roots to gravity is well known, the mechanism behind it is not at all well understood. The importance of auxin as a plant growth regulator, as well as its widespread occurrence in plants, led to it being proposed as the primary regulator in the gravitropic response. The basis of auxin's proposed role in gravitropism (geotropism) is outlined below. The mechanism is appealing in its simplicity but, as noted below, has been widely criticized, and there is not a great deal of evidence to support it. Many of the early plant growth experiments (including those on phototropism) involved the use of coleoptiles. Their use has been criticized because the coleoptile (the sheath surrounding the young shoot of grasses) is a specialized and short-lived structure and is probably not representative of plant tissues generally.

## The Role of Auxins in Gravitropic Responses

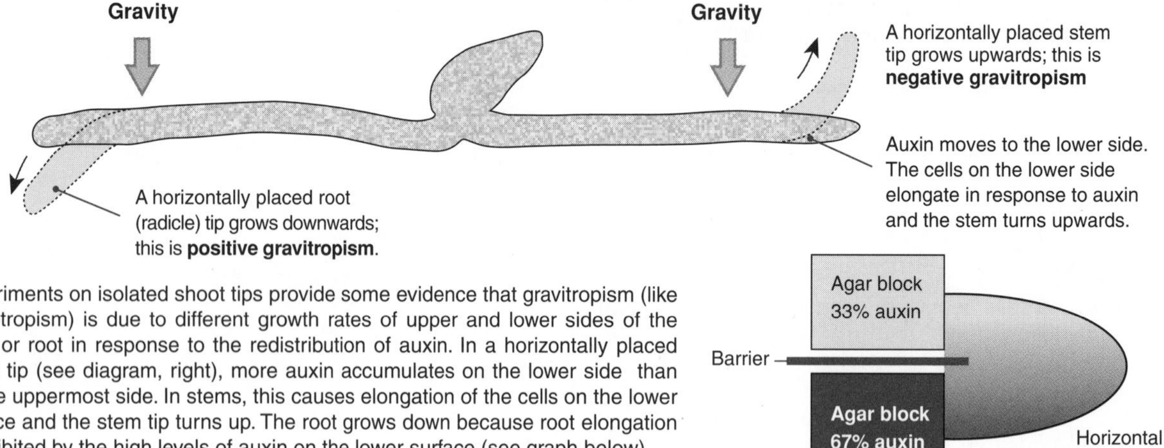


Experiments on isolated shoot tips provide some evidence that gravitropism (like phototropism) is due to different growth rates of upper and lower sides of the stem or root in response to the redistribution of auxin. In a horizontally placed shoot tip (see diagram, right), more auxin accumulates on the lower side than on the uppermost side. In stems, this causes elongation of the cells on the lower surface and the stem tip turns up. The root grows down because root elongation is inhibited by the high levels of auxin on the lower surface (see graph below).

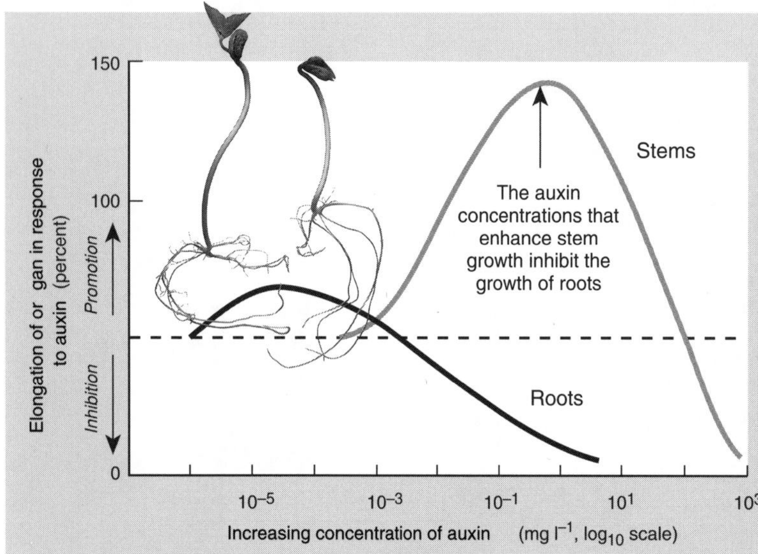


### Auxin Concentration and Root Growth

In a horizontally placed seedling, auxin moves to the lower side of the organ in both the stem and root. Whereas the stem tip grows upwards, the root tip responds by growing down. Root elongation is inhibited by the same level of auxin that stimulates stem growth (see graph left). The higher auxin levels on the lower surface cause growth inhibition there. The most elongated cells are then on the upper surface and the root turns down. This simple auxin explanation for the gravitropic response has been much criticized: the concentrations of auxins measured in the upper and lower surfaces of horizontal stems and roots are too small to account for the growth movements observed. Alternative explanations suggest that growth inhibitors are also somehow involved in the gravitropic response.

1. Explain the mechanism proposed for the role of auxin in the gravitropic response in:

   (a) Shoots (stems): ______________________

   (b) Roots: ______________________

2. (a) From the graph above, state the auxin concentration at which root growth becomes inhibited: ______________

   (b) State the response of stem at this concentration: ______________

3. Explain why the gravitropic response in stems or roots is important to the survival of a seedling:

   (a) Stems: ______________________

   (b) Roots: ______________________




ISBN: 978-1-927173-12-1

# Nastic Responses

As well as relatively slow growth responses, plants are also capable of quite rapid, reversible movements. These may be caused by localized turgor changes or the result of rapid cell growth in different parts of the plant. When the direction of the plant response is independent of the direction of the stimulus the response is said to be **nastic**. Types of nastic response include **thermonasty**, e.g. the opening and closing of tulip flowers to changes in air temperature, and **photonasty** e.g. the opening of evening-primrose flowers at dusk. Plant "sleep movements", where plants close up their flowers or lower their leaves at night, (**nyctinasties**) are specialized photonasties. The rapid collapse of the leaflets in the sensitive plant *Mimosa* (below) is termed **haptonasty**. The mechanisms operating in this response are also responsible for the leaf movements of the Venus flytrap.

The sensitive plant (*Mimosa pudica*) has long, compound leaves (leaves composed of small leaflets). When a leaf is touched, it collapses and its leaflets fold together. Strong disturbances cause the entire leaf to droop from its base. This response takes only a few seconds and is caused by a rapid loss of turgor pressure from the cells at the bases of the leaves and leaflets. The message that the plant has been disturbed is passed quickly around the plant by electrical signals (changes in membrane potential). Although the response could be likened to the action potential in animal nervous systems, it is not nearly as rapid. After the disturbance is removed, turgor is restored to the cells, and the leaflets will slowly return to their normal state.

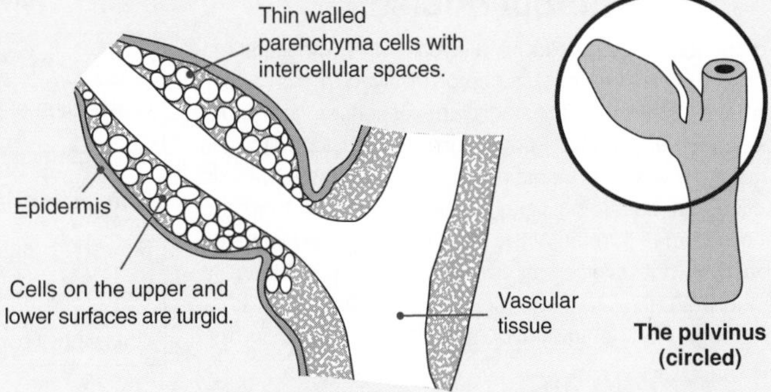


The compound leaves of *Mimosa* has joint-like thickenings, the **pulvini** (*sing.* pulvinus) at the bases of the petioles and also at the bases of each leaflet. These pulvini contain specialized parenchyma cells called **motor cells** which are involved in the rapid leaf movements.

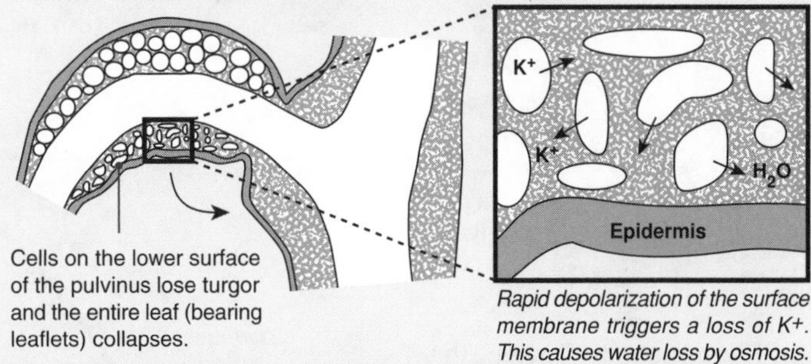


*Rapid depolarization of the surface membrane triggers a loss of K+. This causes water loss by osmosis.*

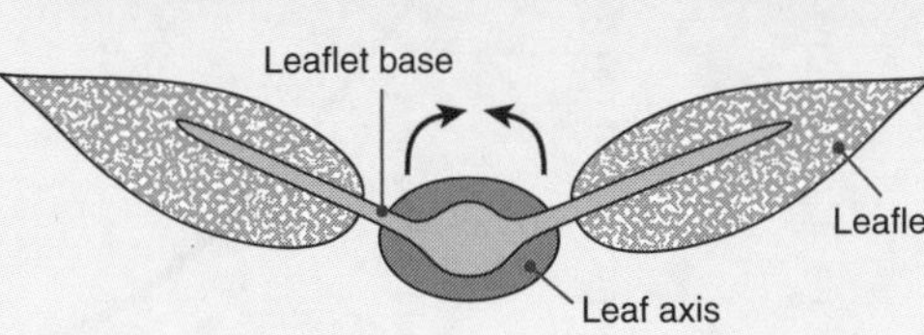


When disturbed, a change in membrane potential of the leaf cells is transmitted to the cells of the pulvinus. These cells respond by actively pumping potassium ions (and tannins) out of the cytoplasm. When water follows osmotically, there is a sudden loss of turgor. **NOTE**: The mechanism outlined above for the leaf bases, also operates at the leaflet bases, except that it is the cells on the **upper surface** of the pulvinus that lose turgor. Consequently, individual leaflets fold upwards, rather than drooping down (left).

1. Describe a possible adaptive advantage of the sensitivity response in plants such as *Mimosa*. Briefly explain your answer (HINT: leaf closure exposes thorns on the stem):

2. Describe how the sudden leaf movements are achieved in *Mimosa*:

3. Some plants use changes in turgor to lower their leaves in the evening and raise them again in the morning (so-called sleep movements). Explain how this behavior would benefit a plant:


ISBN: 978-1-927173-12-1

# Photoperiodism

**Photoperiodism** is the response of a plant to the relative lengths of daylight and darkness. Flowering is a photoperiodic activity; individuals of a single species will all flower at much the same time, even though their germination and maturation dates may vary. The exact onset of flowering varies depending on whether the plant is a short-day or long-day type (see the next page). Photoperiodic activities are controlled through the action of a pigment called **phytochrome**. Phytochrome acts as a signal for some biological clocks in plants and is also involved in other light initiated responses, such as germination, shoot growth, and chlorophyll synthesis. Plants do not grow at the same rate all of the time. In temperate regions, many perennial and biennial plants begin to shut down growth as autumn approaches. During unfavorable seasons, they limit their growth or cease to grow altogether. This condition of arrested growth is called **dormancy**, and it enables plants to survive periods of water scarcity or low temperature. The plant's buds will not resume growth until there is a convergence of precise environmental cues in early spring. Short days and long, cold nights (as well as dry, nitrogen deficient soils) are strong cues for dormancy. Temperature and daylength change seasonally in most parts of the world, so changes in these variables also influence many plant responses, including germination and flowering. In many plants, flowering is triggered only after a specific period of exposure to low winter temperatures. As described in the previous activity, this low-temperature stimulation of flowering is called **vernalization**.

## Photoperiodism

Photoperiodism is based on a system that monitors the day/night cycle. The photoreceptor involved in this, and a number of other light-initiated plant responses, is a blue-green pigment called **phytochrome**. Phytochrome is universal in vascular plants and has two forms: active and inactive. On absorbing light, it readily converts from the inactive form (**$P_r$**) to the active form (**$P_{fr}$**). **$P_{fr}$** predominates in daylight, but reverts spontaneously back to the inactive form in the dark. The plant measures daylength (or rather night length) by the amount of phytochrome in each form.

**Summary of phytochrome related activities in plants**

| Process | Effect of daylight | Effect of darkness |
|---|---|---|
| Conversion of phytochrome | Promotes | Promotes |
| Seed germination | Promotes | Inhibits |
| Leaf growth | Promotes | Inhibits |
| Flowering: long day plants | Promotes | Inhibits |
| Flowering: short day plants | Inhibits | Promotes |
| Chlorophyll synthesis | Promotes | Inhibits |

**Inactive phytochrome** — $P_r$

In natural light, $P_r$ converts rapidly to $P_{fr}$

In the dark, $P_{fr}$ reverts slowly back to $P_r$

**Active phytochrome** — $P_{fr}$

$P_{fr}$ may trigger the synthesis of specific enzymes in specific cells (see table above)

**Response**

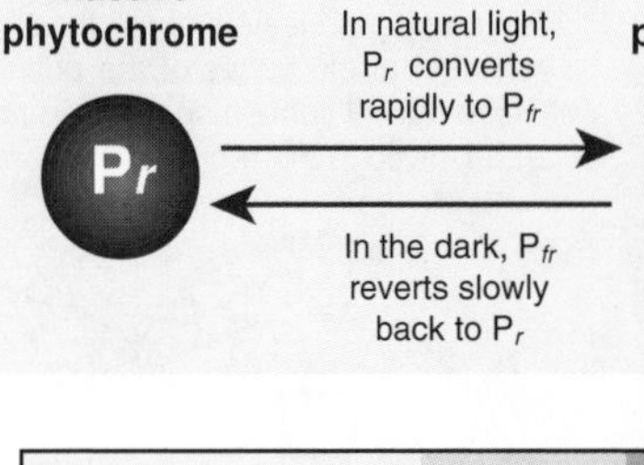

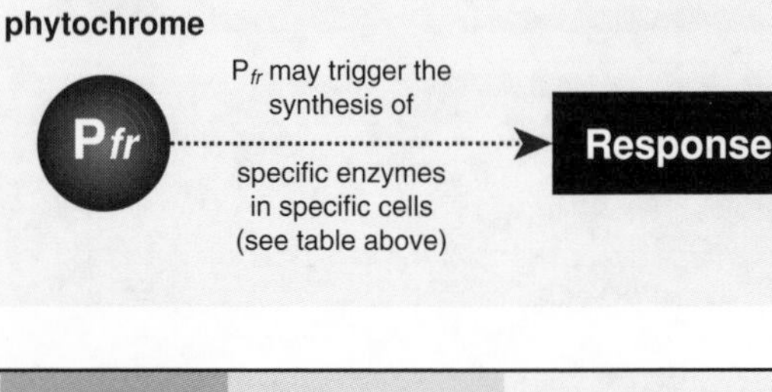

### Plant Responses to Daylength

The cycle of active growth and dormancy shown by temperate plants in the northern hemisphere is correlated with the number of daylight hours each day (right). In the southern hemisphere, the pattern is similar, but is six months out of phase. The duration of the periods may also vary on islands and in coastal regions because of the moderating effect of nearby oceans.

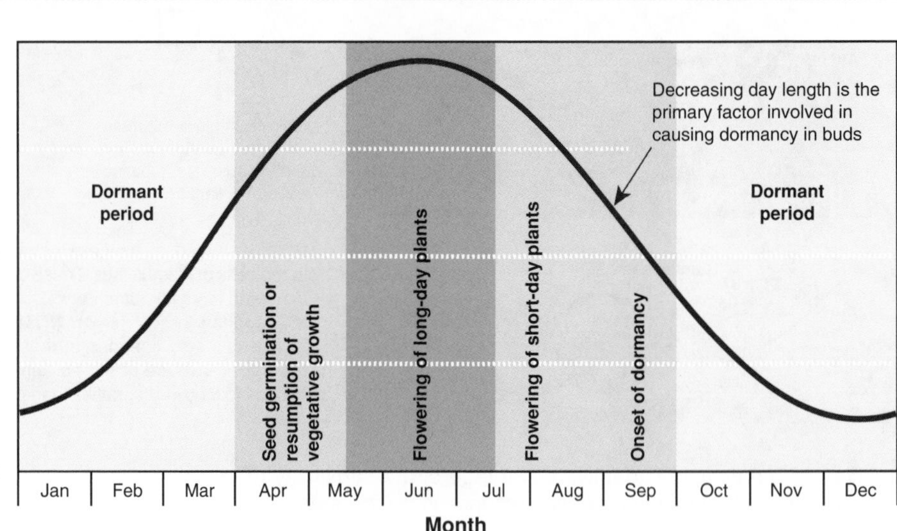


1. Describe two plant responses, initiated by exposure to light, which are thought to involve the action of phytochrome:

   (a) ______________________________

   (b) ______________________________

2. Discuss the role of phytochrome in a plant's ability to measure daylength: ______________________________

______________________________

______________________________

______________________________

______________________________




**ISBN: 978-1-927173-12-1**

## Photoperiodism in Plants

An experiment was carried out to determine the environmental cue that triggers flowering in 'long-day' and 'short-day' plants. The diagram below shows three different light regimes to which a variety of long-day and short-day plants were exposed.

0 ← hours → 24

| | |
|---|---|
| Long-day | Short night |
| Short-day | Long night |
| Short-day | Long \| night |

Long night interrupted by a short period exposed to light

### Long-day plants

When subjected to the light regimes on the right, the 'long-day' plants below flowered as indicated:

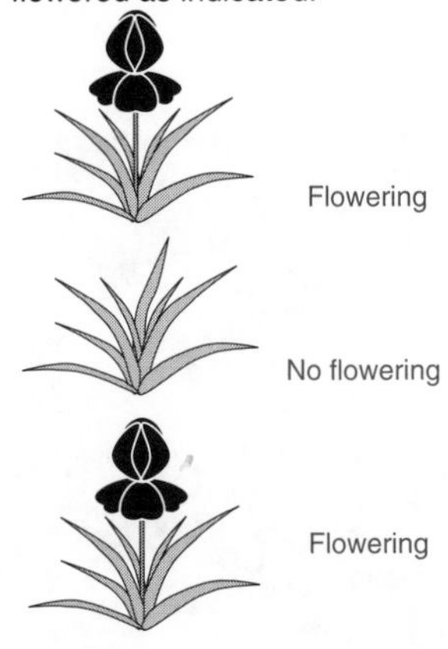


**Examples**: *lettuce, clover, delphinium, gladiolus, beets, corn, coreopsis*

### Short-day plants

When subjected to the light regimes on the left, the 'short-day' plants below flowered as indicated:

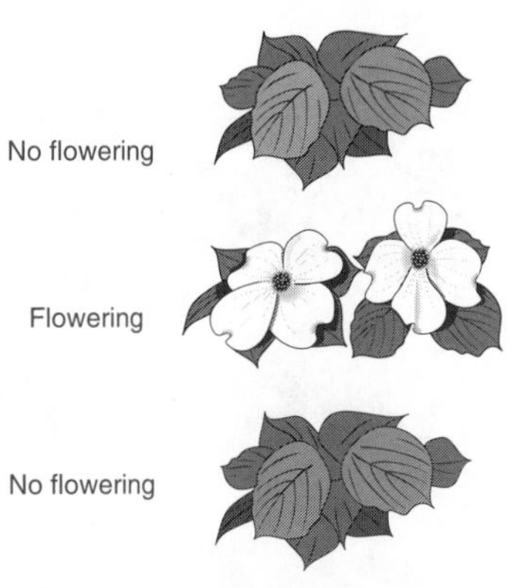


**Examples**: *potatoes, asters, dahlias, cosmos, chrysanthemums, pointsettias*

3. (a) What is the environmental cue that synchronizes flowering in plants? ______

   (b) Describe one biological advantage of this synchronization to the plants: ______

4. Discuss the role of environmental cues in triggering and breaking **dormancy** in plants: ______

5. Discuss the adaptive value of **dormancy** and **vernalization** in temperate climates: ______

6. Study the three light regimes above and the responses of short-day and long-day flowering plants to that light. From this observation, describe the most important factor controlling the onset of flowering in:

   (a) Short-day plants: ______

   (b) Long-day plants: ______

7. Using information from the experiment described above, discuss the evidence for the statement "*Short-day plants are really better described as long-night plants.*"




**ISBN: 978-1-927173-12-1**

# Plant Hormones and Timing Responses

**Plant hormones** are chemicals that act as signal molecules to regulate plant growth and responses. Alone or together, plant hormones target specific cells to cause a specific effect. Charles Darwin and his son Francis were the first to recognize the role of hormones when they identified the role of auxin in cell elongation. Since then, many other plant hormones have been identified. Many have roles in coordinating timing responses in plants including promoting and breaking seed dormancy, bolting (elongation), seed germination, and fruit ripening. These examples are presented below.

**Abscisic acid** (ABA) promotes **seed dormancy**, preventing seeds from germinating under unfavorable environmental conditions (e.g low soil temperature). **Gibberellins** have the opposite effect, **breaking dormancy** in seeds and promoting the growth of the embryo and emergence of the seedling.

**Ethylene** is a gaseous plant hormone with an important role in the ripening process of many fruits (such as these pears, above). Bananas are picked green and are artificially ripened after transport by being gassed with ethylene. Auxin and ethylene are believed to work together to promote **fruit fall**.

Deciduous plants shed their leaves every autumn in a process called **abscission**. The regulation of leaf abscission by plant hormones is not yet fully understood, but scientists believe that **auxin** (**IAA**) and **ethylene** work together to bring about leaf drop.

**Gibberellins** cause stem and leaf elongation by stimulating cell division and cell elongation. This results in **bolting**, in which an elongated flower stalk grows from the main stem of a plant. Bolting is common in *Brassica* plants, such as broccoli (above) and cabbage.

**Gibberellins** are required for **seed germination**. They stimulate cell division and cell elongation, allowing the root to penetrate the seed coat. In the brewing industry, they are used to speed up seed germination and ensure germination uniformity in the production of barley malt.

1. Why is it likely that the same hormones are responsible for both leaf fall (abscission) and fruit fall in plants?

______________________________________________

______________________________________________

2. Describe the role of **gibberellins** in stem elongation and in the germination of grasses such as barley:

______________________________________________

______________________________________________

______________________________________________

3. Why is it an advantage to pick fruit green and artificially ripen it with **ethylene** after shipping? ______________________

______________________________________________



***Related activities:*** *Plant Responses*
***Weblinks:*** *Seed Germination*


**ISBN: 978-1-927173-12-1**

# Biological Rhythms

**Environmental cues**, such as daylength, timing and the height of tides, temperature, and phase of the moon are often used by plants and animals to establish and maintain a pattern of activity. Regular environmental cues assist in survival by synchronizing important events in the life cycle of an organism; events such as pollination, mating, birth, germination, rearing of offspring, collection and storage of food reserves and body fat, and periods of torpor. **Biological rhythms** (biorhythms) in direct response to environmental stimuli are said to be **exogenous**, because the rhythm is controlled by an environmental stimulus that is external to the organism. Those rhythms that continue in the absence of external cues are said to be **endogenous**.

**Rhythmv** tidal.
**Period**: ~ 12.4 h (coincident with tidal flows).
**Examples**: In mud crabs, locomotion and feeding occurs when covered by tidal waters.

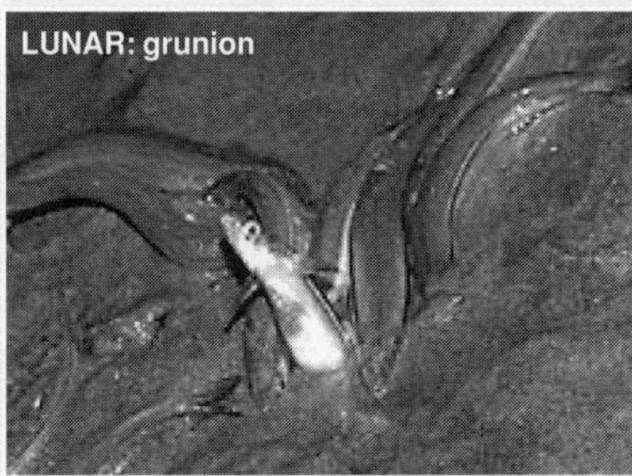


**Rhythm**: lunar.
**Period**: ~ 29.5 days (a month).
**Example**: In grunions, egg laying above the high water mark coincides with a new moon.

**Rhythm:** daily
**Period:** ~ 24 h.
**Examples**: In waxwings (*Bombycilla*), general activity and feeding occurs during daylight hours.

**Rhythm**: annual.
**Period**: ~ a year.
**Example**: In many domestic livestock and antelope species, young are born in spring.

**Rhythm**: annual.
**Period**: ~ a year.
**Example**: Northern bats (*Eptesicus nilssonii*) hibernate for 4-5 months during fall and winter.

**Rhythm**: intermittent.
**Period**: does not apply.
**Example**: In some shiners (a minnow-like fish), reproduction is triggered by flooding.

1. Use the examples provided above to determine a definition of each of the following terms describing biological rhythms. For each type of rhythm describe one other example to illustrate how the behavior follows the astronomical cycle:

   (a) **Daily** rhythm: ______

   Example: ______

   (b) **Lunar** rhythm: ______

   Example: ______

   (c) **Annual** rhythm: ______

   Example: ______

   (d) **Tidal** rhythm: ______

   Example: ______

2. For each of the examples below, describe an **environmental cue** that might be used to induce or maintain the activity:

   (a) Hedgehog's hibernation in winter: ______

   (b) Blackbird's foraging and social behavior during daylight: ______

   (c) Screech owl's activity of hunting for small mammals at night: ______

   (d) Coordinated flowering of plants in spring: ______

3. Explain what is meant by an exogenous rhythm: ______

______

Timing, Coordination & Social Behavior

ISBN: 978-1-927173-12-1


*Periodicals:*
*Times of our lives*

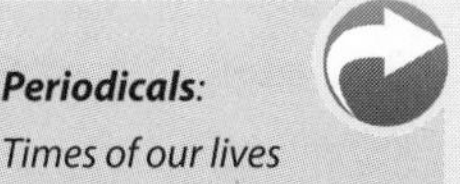

***Related activities:** Breeding Behavior*
***Weblinks:** Biological Clocks Animations*

# Biological Clocks

Animals and plants exhibit all kinds of regularly repeated behaviors. In animals, these rhythms include such behaviors as activity period, and times of reproduction and migration. In many cases, the rhythms are **exogenous** and are a direct response to environmental cues. However, many are thought to have an **endogenous** (internal) component. Such internal timing systems are often called **biological clocks**. Endogenous rhythms will continue even in the absence of environmental cues (although with a slightly different period than that observed in the natural environment). To remain in synchrony with the environment, biological clocks must be reset at regular intervals. This is the purpose of the external timekeeper or **zeitgeber**; the environmental cue that resets the biological clock. The process of regularly resetting the internal clock is known as **entrainment**.

## Functions of Biological Clocks

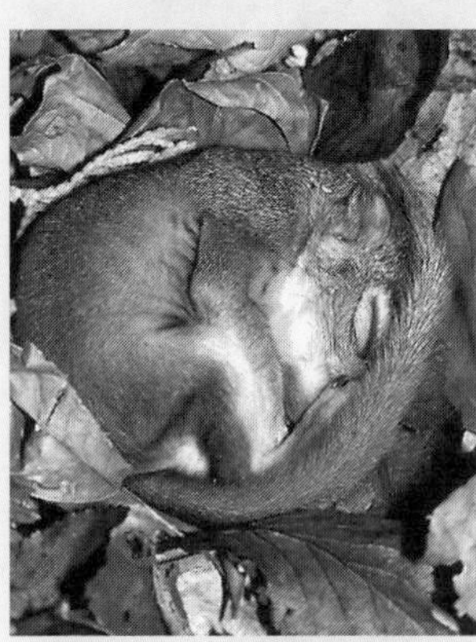

**Prediction** of and preparation for events in the environment (e.g. storing food reserves for hibernation).

**Synchronization** of social activities, migration, and reproduction. Animals congregate at breeding grounds at the same time of year.

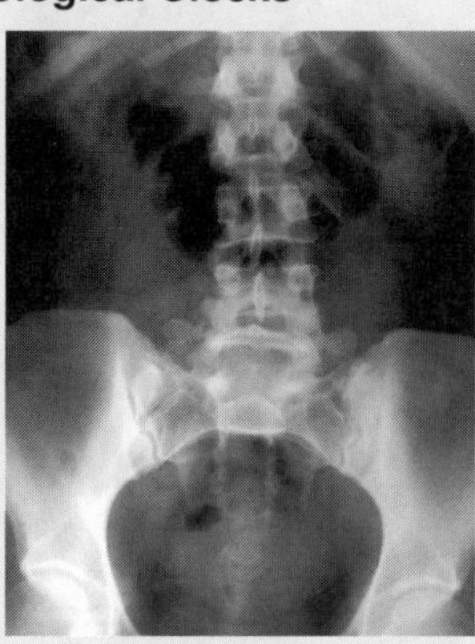

**Synchronization** of some internal processes such as those concerned with regulating the menstrual cycle and pregnancy.

**Time compensation** in navigation and suncompass orientation using a continuously consulted clock (e.g. honeybee food collection).

### The Human Biological Clock

For the majority of us, our biological clock runs at about a 25½ hour day. To keep it synchronized with our environment (the 24 hour-day cycle) it needs to be reset each day, reacting to outside stimuli such as light and dark, meal times, and exercise. The internal clock is made up of a collection of cells, called the **suprachiasmatic nucleus** (SCN), located in the midline of the brain, just behind the eyes. Light from the eyes stimulates the nerve pathways that connect with this biological clock and helps to reset it.

The **pineal gland** is an endocrine gland with extensive influences on the body. It secretes the hormone melatonin in the dark. Melatonin production is suppressed by bright light.

Melatonin

Eye

The suprachiasmatic nucleus is located in the hypothalamus and is directly connected to the eyes. It is about the size of a grain of rice and probably the dominant pacemaker for human biorhythms.

The SCN communicates with the nearby **pineal gland**, inducing it to begin or cease production of the sleep-inducing hormone, **melatonin**. Melatonin is produced in the dark, but not during the light. In this way, the light-dark cycle acts as the **zeitgeber** that resets the clock on a regular basis (entrains the endogenous rhythm) and forces it to take up the period of the environment.

### Circadian Disorders

Rapid, long distance air travel can lead to disruption of the normal sleep-wake cycle. When traveling across time zones, the body clock will not be synchronized with the destination time and must adjust to the new schedule. The misalignment of the body's rhythms is called **jet lag**. Typical symptoms include fatigue, insomnia, irritability, and grogginess.

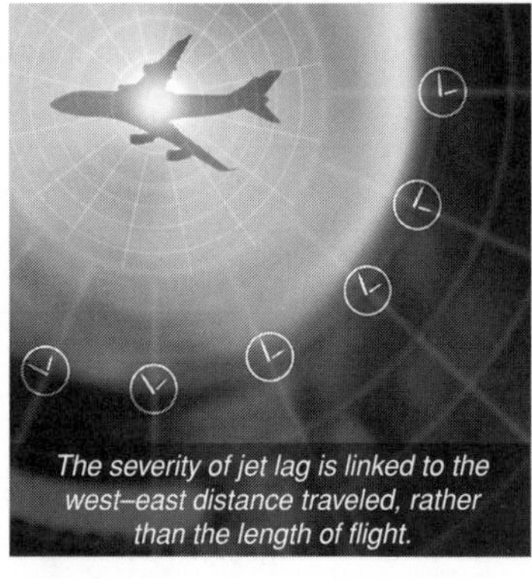

*The severity of jet lag is linked to the west–east distance traveled, rather than the length of flight.*

1. Naming the organisms, describe two examples of activities where preparation for environmental events is required:

   (a) Organism: ______________ Activity: ______________

   (b) Organism: ______________ Activity: ______________

2. (a) What is the purpose of a **zeitgeber**? ______________

   (b) Identify a common zeitgeber in animals: ______________

3. Explain why international travelers often take a **melatonin** supplement when they reach their destination:




**ISBN: 978-1-927173-12-1**

# Human Biorhythms (Biological Rhythms)

Humans exhibit a number of periodic changes in behaviour or physiology that are generated and maintained by a biological clock. In humans, the internal timing of production, release, and levels of various factors are closely aligned with our daily activities (below). These cycles are often called biorhythms, but they should not be confused with the pseudoscience by the same name, which proposes that rhythmic biological cycles, which begin at birth, can predict mental, physical and emotional activity.

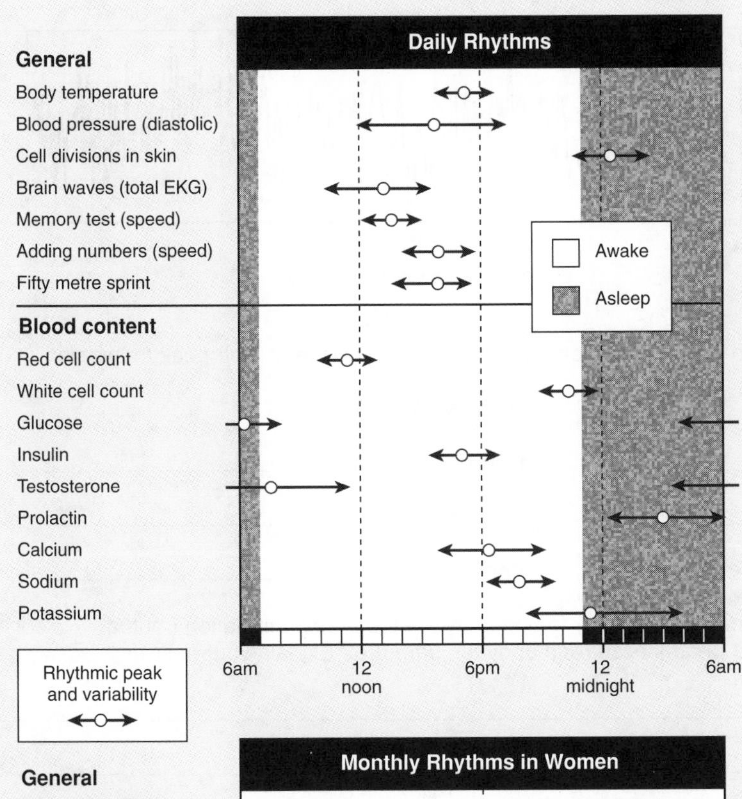

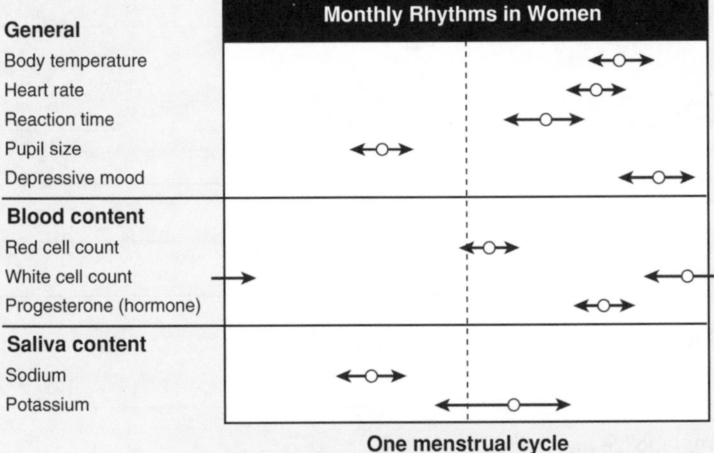


One menstrual cycle
(number of days varies with individual)

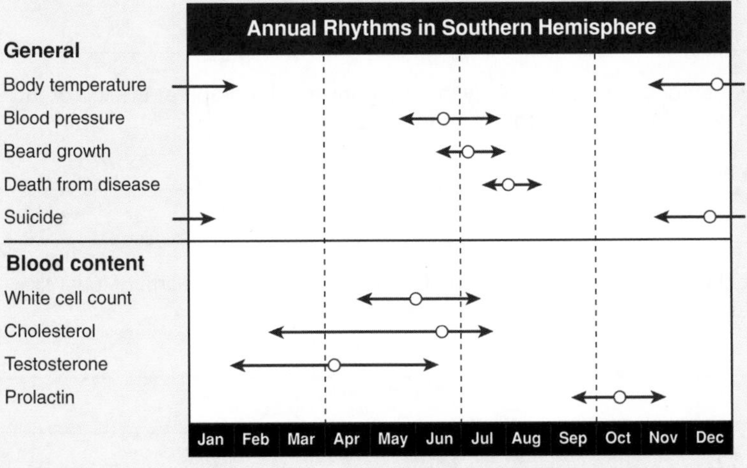


Months of the year

Source: Cycles of Nature – An Introduction to Biological Rhythms; Ahlgren, A. and Halberg, F.

## Selected Human Daily Rhythms

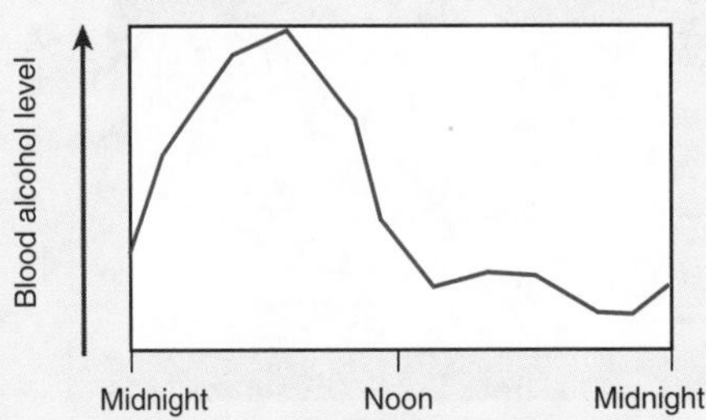


**Changes in susceptibility to alcohol**

The ability of the liver to metabolize (break down) alcohol changes with the time of day, being most efficient between 4 pm and 11 pm (the 'cocktail hour'). After midnight, the liver's activity slows down and alcohol starts to accumulate in the blood.

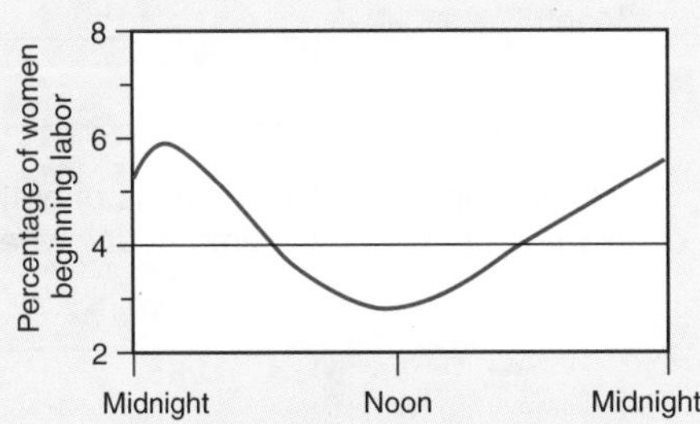


**Time of labor onset**

The graph above shows the percentage of a total of 200,000 women who began labor during each hour of the day. If labor were equally likely to begin at any hour, then 1/24 of the women (about 4%) would have begun each hour.

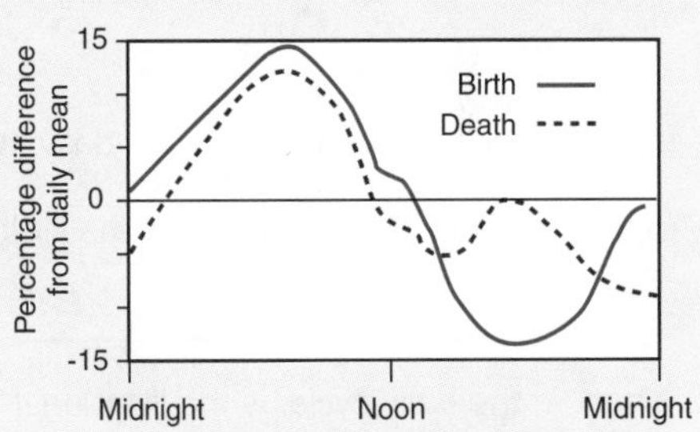


**Daily rhythms of birth and death**

The timing of births and deaths also seems to be tied to a daily cycle. The graph above shows how they vary in frequency compared to the daily mean.

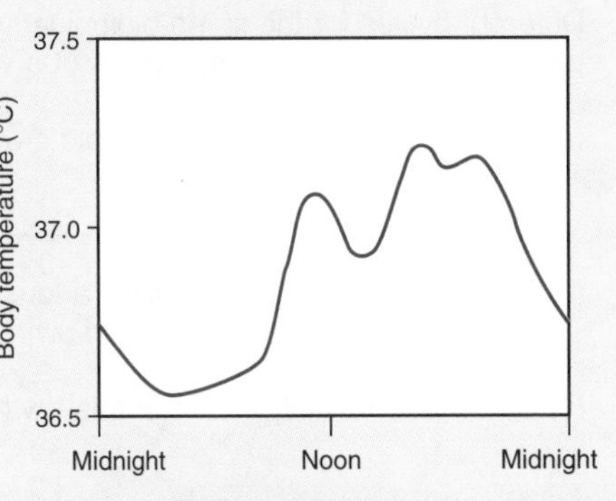


**Body temperature**

The amount of physical activity can affect body temperature; exercise will increase it. Data for the graph above was taken from volunteers lying in bed.

ISBN: 978-1-927173-12-1


*Periodicals:*
*Times of our lives*

***Related activities:*** *Biological Rhythms, Biological Clocks*
***Weblinks:*** *Biological Clocks Animation*

### Rhythms in a Long Duration Cave Occupation

A woman spent four months isolated underground in a cave so that her biological rhythms could be studied in the absence of the normal day/night environmental cues. The air temperature inside the cave remained constant over this time. Her body temperature was measured 3 times a day for 4 months.

Body temperature (°C)
38.0
37.0
36.0
0
5
10
15
Weeks in isolation underground

1. Study the three charts on the previous page and determine when each of the following are at their peak for women:

   (a) White blood cell count

   Daily: ______________________

   Monthly: ______________________

   Annually: ______________________

   (b) Body temperature

   Daily: ______________________

   Monthly: ______________________

   Annually: ______________________

2. Study the graphs for 'daily rhythms of birth and death' and 'body temperature'. Is there any correlation (mutual relationship) between the time of day when most deaths occur and body temperature? Explain your answer:

3. Imagine you are about to sit a theory exam.

   (a) At what time of the day should you sit the exam to provide your best performance? ______________________

   (b) Explain your reasoning: ______________________

4. (a) When does insulin peak in the daily cycle? ______________________

   (b) Explain why the production of insulin peaks at this time of day: ______________________

5. When, in the daily cycle, is the liver least able to metabolize alcohol? ______________________

6. Describe how beard growth is correlated with the seasons of the year: ______________________

7. Different cycles for the same biological variable can occur simultaneously in an organism. The graph at the top of this page shows the body temperature of a woman who stayed underground for four months.

   (a) What is represented by the rapid swings in body temperature? ______________________

   (b) The same graph also displays a longer undulating rhythm with a period of several weeks. Draw a line of 'best fit' through the middle of the 'rapid swings' to show this longer rhythm.

   (c) Describe the likely cause of this rhythm: ______________________

   (d) Explain why this experiment was carried out in a cave environment: ______________________


ISBN: 978-1-927173-12-1

# Recording Animal Activity

The activity patterns of two animals were recorded by students in an animal behavior course. Devices set up to monitor activity were linked to a computer using electronic sensors. The recorded activity patterns are displayed in the **actograms** below. The activity recordings for these exercises were kindly provided by Dr. Bob Lewis, Department of Zoology, University of Auckland.

**Mouse A**

**Mouse B**

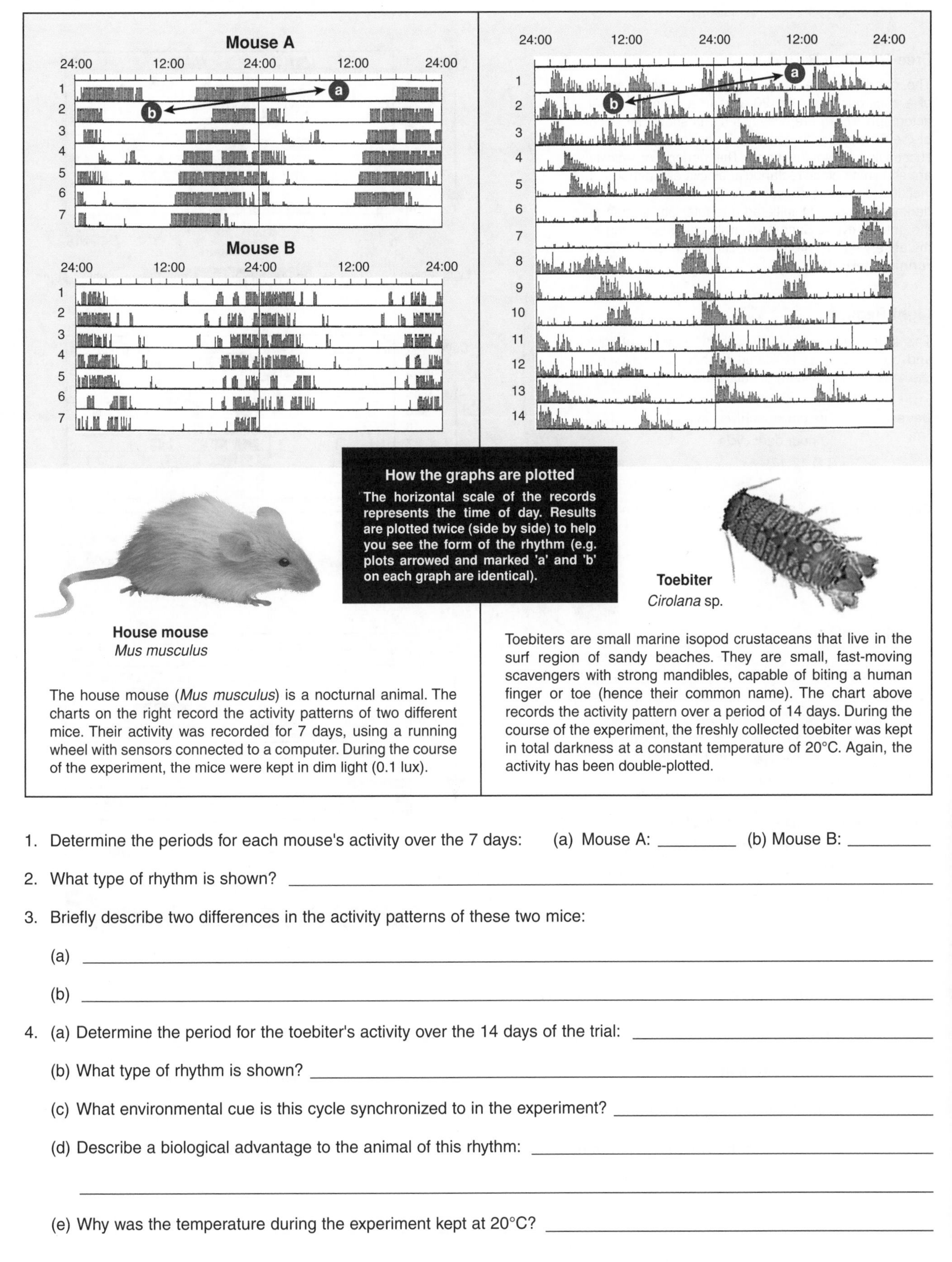

**How the graphs are plotted**

The horizontal scale of the records represents the time of day. Results are plotted twice (side by side) to help you see the form of the rhythm (e.g. plots arrowed and marked 'a' and 'b' on each graph are identical).

**House mouse**
*Mus musculus*

The house mouse (*Mus musculus*) is a nocturnal animal. The charts on the right record the activity patterns of two different mice. Their activity was recorded for 7 days, using a running wheel with sensors connected to a computer. During the course of the experiment, the mice were kept in dim light (0.1 lux).

**Toebiter**
*Cirolana* sp.

Toebiters are small marine isopod crustaceans that live in the surf region of sandy beaches. They are small, fast-moving scavengers with strong mandibles, capable of biting a human finger or toe (hence their common name). The chart above records the activity pattern over a period of 14 days. During the course of the experiment, the freshly collected toebiter was kept in total darkness at a constant temperature of 20°C. Again, the activity has been double-plotted.

1. Determine the periods for each mouse's activity over the 7 days: (a) Mouse A: ________ (b) Mouse B: ________
2. What type of rhythm is shown? ________
3. Briefly describe two differences in the activity patterns of these two mice:
   (a) ________
   (b) ________
4. (a) Determine the period for the toebiter's activity over the 14 days of the trial: ________
   (b) What type of rhythm is shown? ________
   (c) What environmental cue is this cycle synchronized to in the experiment? ________
   (d) Describe a biological advantage to the animal of this rhythm: ________
   (e) Why was the temperature during the experiment kept at 20°C? ________
   (f) Identify a cue on the shore to which the rhythm could be synchronized: ________



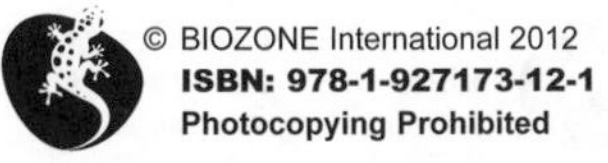


ISBN: 978-1-927173-12-1

# Rhythms in Cockroaches

Cockroaches are nocturnal insects. The experiments described below investigated the periodicity of their behavior under controlled conditions. The first experiment determined the **free-running period** (the periodicity of the cycle in the absence of environmental cues). The results of the second experiment illustrate the process of **entrainment** of the rhythm.

### Free-Running Period

The charts on the right record the activity rhythm of a cockroach kept for 20 days in a running wheel actograph. There are activity records shown for each of the 20 days, with each day's record presented in succession down the page. The periods of activity are shown as black rectangular blocks and periods of inactivity shown as no rectangle. The onset of constant darkness on day 11 initiated a regular phase-shift in the activity cycle. Such an activity cycle occurring in the absence of environmental cues is called the **free-running period**.

### Light Regime

This is the term used to describe the cycles of light and darkness. It is indicated by bars of 'light' and 'darkness' at the bottom of each table:

**Days 1-10:** The cockroach was in a 12 hour light / 12 hour dark cycle (**LD 12:12**)

**Days 11-20:** The cockroach was in constant darkness for these 10 days (**DD**)

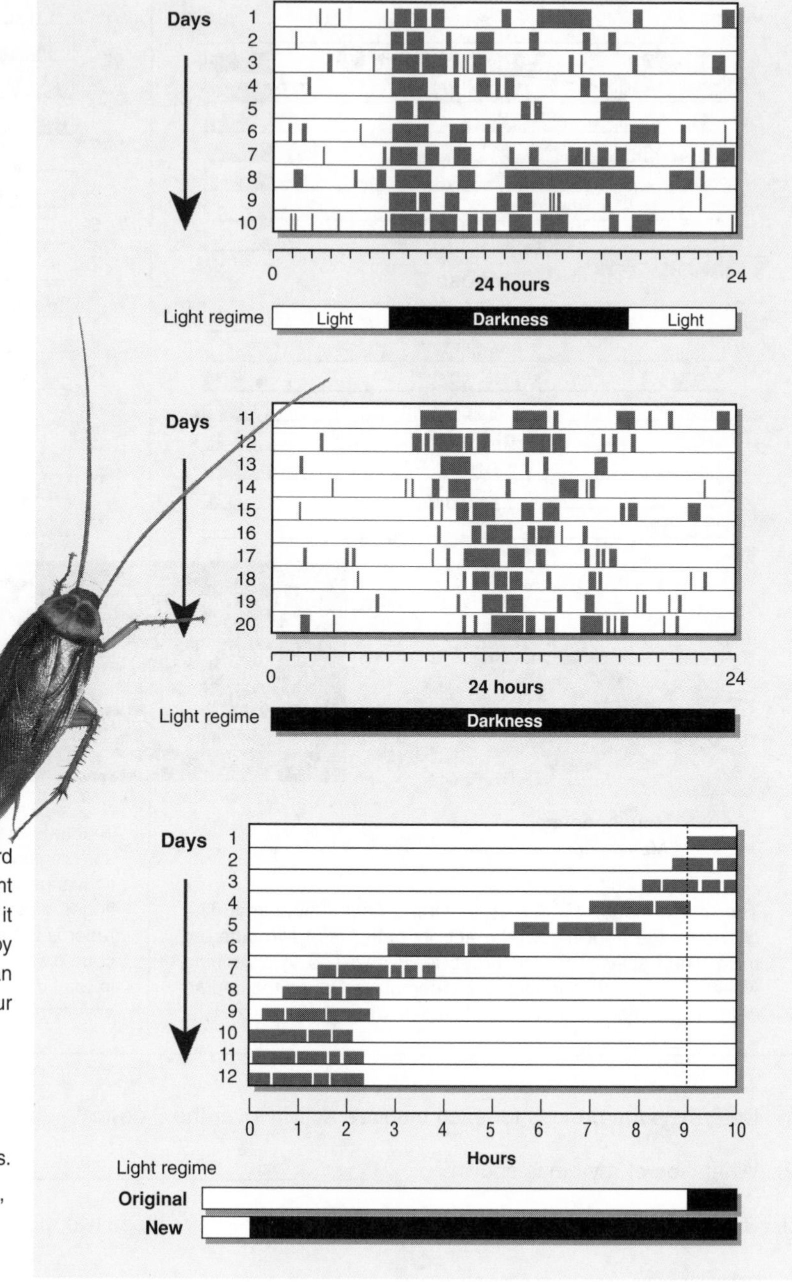


### Entrainment

The chart on the right shows the activity record of a cockroach. It is being retrained to a new light cycle, which occurs nine hours earlier than the one it had been experiencing previously. The process by which the endogenous rhythm is synchronized to an environmental cue, such as a 12 hour light / 12 hour dark cycle, is termed **entrainment**.

Entrainment usually has the following features:

- A **phase-shift** (a new starting position) for the commencement of activity is gradual, without jumps.
- As the activity gets nearer the new lights-out signal, the daily phase-shift is reduced.

1. What type of rhythm is displayed in the first 10 days? __________

2. In what part of the light/darkness cycle was the cockroach most active? __________

3. Describe the **activity pattern** displayed by the cockroach: __________

4. Determine the **free-running period** (in hours) displayed between days 11-20: __________

5. Discuss the effect of **entrainment** on the **free-running period**, including its adaptive value in the natural environment:


ISBN: 978-1-927173-12-1




**Related activities:** *Recording Animal Activity*

# Hibernation

Many animals **hibernate** during winter to conserve energy while food supplies are limited. During hibernation, the animal enters a state of inactivity and reduced metabolic rate. The basal metabolic rate can drop to 2-4% of normal, conserving the body's reserves during the hibernation period. Body temperature, breathing rate, and heart rate are all significantly decreased during hibernation. The animal is also less responsive to environmental stimuli. Hibernation periods vary between species (days, weeks, or months). Some animals exhibit reduced metabolic activity during the summer months, this is called **estivation**.

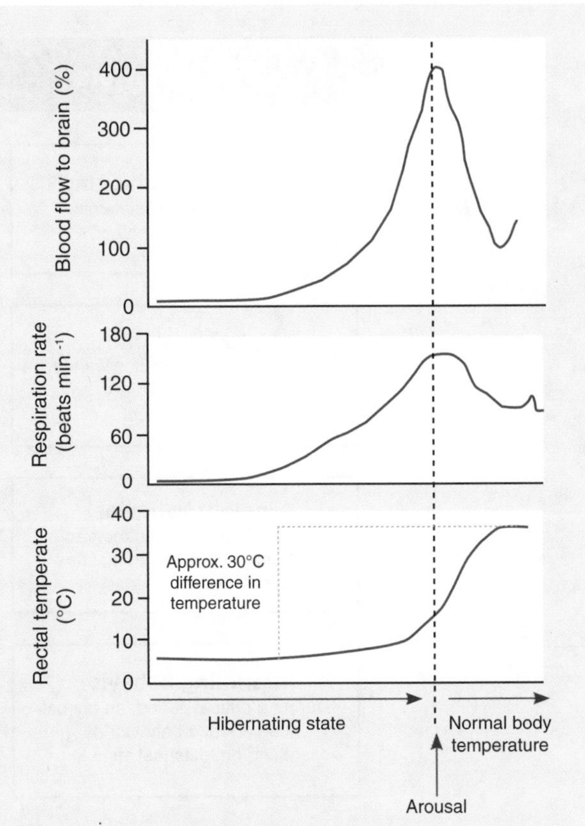


Source: Osborne http://www.asahikawa-med.ac.jp/dept/mc/phys1/profiles/osborne.html

Animals prepare for hibernation at the onset of specific **environmental cues**. Common hibernation cues include a shortage of food, shorter daylight periods, and cooler temperatures.

Many animals, such as this chipmunk (left), build up their energy stores by eating large quantities of food prior to hibernating. The excess energy is stored as fat deposits, which provide the energy to carry out metabolic activities during hibernation.

**Obligate hibernators** (e.g. rodents such as ground squirrels) drop their body temperature significantly, and cannot be awoken by external stimuli. However, they are periodically awoken by internal stimuli, so that maintenance functions can occur. **Facultative hibernators** (e.g. bats and bears) can be aroused by external stimuli and their period of inactivity is often described as **torpor**, rather than true hibernation. Arousal carries an energetic cost. Large animals, such as bears, do not drop their body temperature as much as smaller animals. Their body temperature drops only a few degrees because raising it from a very low temperature would take too long and would be too energy-expensive.

*The data (left) shows metabolic activity and temperature in golden hamsters during hibernation. Note the large difference in the animal's temperature (~30°C) between hibernation and normal body temperature. Metabolic activity (blood flow to the brain and respiration rate) significantly decreases during hibernation, increases to a maximum during arousal, and tapers off once normal body temperature is achieved. The elevated metabolic rate observed during the arousal period speeds up arousal and rapidly clears waste products from the body.*

1. (a) What is **hibernation**? ____________________

   (b) What are the survival advantages of hibernation? ____________________

   (c) What are the common environmental cues triggering hibernation? ____________________

   (d) What energetic risks might be associated with adopting a strategy of facultative hibernation? ____________________

2. (a) What happens to the body temperature of the golden hamster during hibernation? ____________________

   (b) Why does this change in temperature occur? ____________________

   (c) Explain why blood flow to the brain and respiration rate may peak during arousal from hibernation:

Timing, Coordination & Social Behavior


ISBN: 978-1-927173-12-1

# The Components of Behavior

Behavior in animals can be attributed to two components: **innate behavior** that has a genetic basis, and **learned behavior**, which results from the experiences of the animal. Together they combine to produce the total behavior exhibited by the animal. It should also be noted that experience can modify innate behavior. Animals behave in fixed, predictable ways in many situations. The innate behavior follows a classical pathway called a **fixed-action pattern** (FAP) where an innate behavioral programme is activated by a stimulus or **releaser** to direct some kind of behavioral response. Innate behaviors are generally adaptive and are performed for a variety of reasons. Learning, which involves the modification of behavior by experience, occurs in various ways.

## Innate Behaviors

**Reflex behavior**
A stimulus induces an automatic involuntary and stereotyped response. Reflex behaviors are protective, e.g the reflex escape behavior (caridoid response) of decapod crustaceans.

**Kinesis**
Random movement of an animal in which the rate of movement is related to the intensity of the stimulus, but not to its direction.

**Taxis**
A movement in response to the direction of a stimulus. Movements towards a stimulus are positive while those away from a stimulus are negative.

**Stereotyped behavior**
Occurs when the same response is given to the same stimulus on different occasions. This behavior shows fixed patterns of coordinated movements called fixed action patterns.

*The complex behavior patterns exhibited by an animal*

## Learned Behaviors

**Classical conditioning**
Animals may associate one stimulus with another.

**Habituation**
Response to a stimulus wanes when it is repeated with no apparent effect.

**Insight behavior**
Correct behavior on the first attempt where the animal has no prior experience.

**Imprinting behavior**
During a critical period, an animal can adopt a behavior by latching on to its first stimulus.

**Operant conditioning**
Also called trial and error learning; an animal is rewarded or punished after chance behavior.

### Fixed Action Patterns

A releaser (sometimes called a sign stimulus) is a signal that elicits a specific response from members of the same species. The releaser triggers the operation of an **innate releasing mechanism** in the brain that mediates a predictable behavioral response called the **fixed-action pattern** (FAP). FAPs are spontaneous, stereotyped, and indivisible. Once begun, a FAP runs to completion and is independent of learning.

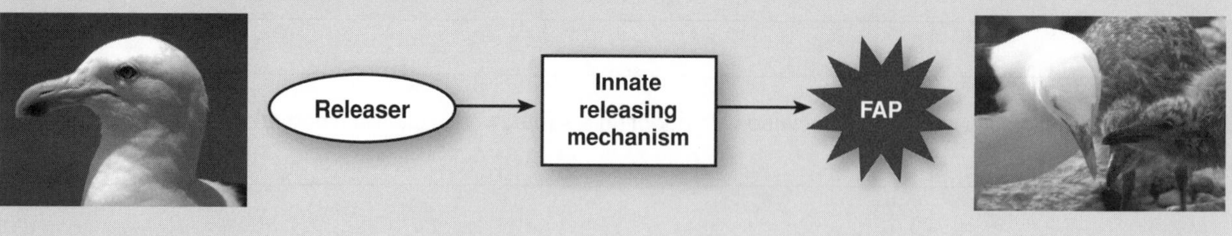


1. Distinguish between innate and learned behaviors: ______________________________

______________________________

2. (a) Explain the role of releasers in innate behaviors: ______________________________

______________________________

(b) Name a releaser for a fixed action pattern and the animal involved, and describe the behavior elicited:

______________________________

______________________________



*Related activities:* *Simple Behaviors, Learned Behavior*


ISBN: 978-1-927173-12-1

# Simple Behaviors

Taxes and kineses are adaptive locomotory behaviors and involve movements in response to external stimuli such as gravity, light, chemicals, touch, and temperature. A **kinesis** (*pl.* kineses) is a non-directional response to a stimulus in which the speed of movement or the rate of turning is proportional to the stimulus intensity. Kineses do not involve orientation directly to the stimulus and are typical of many invertebrates and protozoa. In contrast, **taxes** involve orientation and movement in response to a directional stimulus or a gradient in stimulus intensity. Taxes often involve moving the head (bearing sensory receptors) from side to side until the sensory input from both sides is equal (a klinotaxic response). Many taxic responses are complicated by a simultaneous response to more than one stimulus e.g. fish orientate dorsal side up in response to both light and gravity. Orientation responses are always classed according to whether they are towards the stimulus (positive) or away from it (negative).

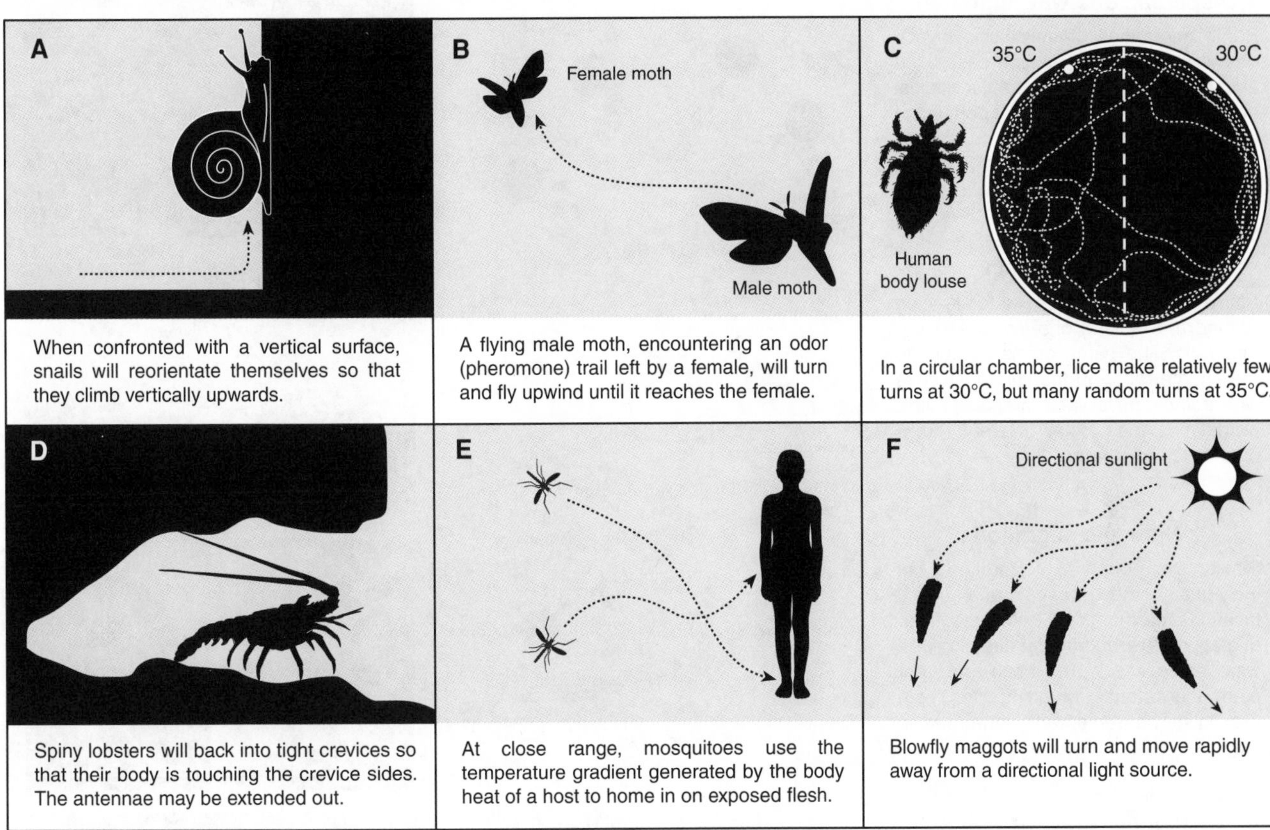


A: When confronted with a vertical surface, snails will reorientate themselves so that they climb vertically upwards.

B: A flying male moth, encountering an odor (pheromone) trail left by a female, will turn and fly upwind until it reaches the female.

C: In a circular chamber, lice make relatively few turns at 30°C, but many random turns at 35°C.

D: Spiny lobsters will back into tight crevices so that their body is touching the crevice sides. The antennae may be extended out.

E: At close range, mosquitoes use the temperature gradient generated by the body heat of a host to home in on exposed flesh.

F: Blowfly maggots will turn and move rapidly away from a directional light source.

1. Distinguish between a **kinesis** and a **taxis**, describing examples to illustrate your answer: ______________________

2. Giving an example, describe the adaptive value of simple orientation behaviors such as kineses: ______________________

3. Name the physical stimulus for each of the following **prefixes** used in naming orientation responses:

   (a) Gravi- ____________ (b) Hydro- ____________ (c) Thigmo- ____________

   (d) Photo- ____________ (e) Chemo- ____________ (f) Thermo- ____________

4. For each of the above examples (**A**-**F**), describe the orientation response. Indicate whether the response is positive or negative (e.g. positive phototaxis):

   (a) **A:** ____________ (d) **D:** ____________

   (b) **B:** ____________ (e) **E:** ____________

   (c) **C:** ____________ (f) **F:** ____________

5. Suggest what temperature body lice "prefer", given their response in the chamber (in C): ______________________




ISBN: 978-1-927173-12-1


***Related activities:*** *The Components of Behavior, Pheromones*

# Migration Patterns

In many animals, migration is an important response to environmental change. True migrations are those where animals travel from one well-defined region to another, for a specific purpose such as overwintering, breeding, or seeking food. Migrations often involve very large distances and usually involve a return journey. They are initiated by the activity of internal clocks or timekeepers in response to environmental cues such as change in temperature or daylength. Some mass movements of animals are not truly migrations in that they do not involve a return journey and they are governed by something other than an internal biological clock (e.g. depletion of a food resource). Such movements are best described as dispersals and are typical of species such as 'migratory' locusts and some large African mammals which relocate in response to changing food supplies.

### Dispersal: one-way migration

Some migrations of animals involve a one-way movement. In such cases, the animal does not return to its original home range. This is typical of population dispersal. This often occurs to escape deteriorating habitats and to colonise new ones.

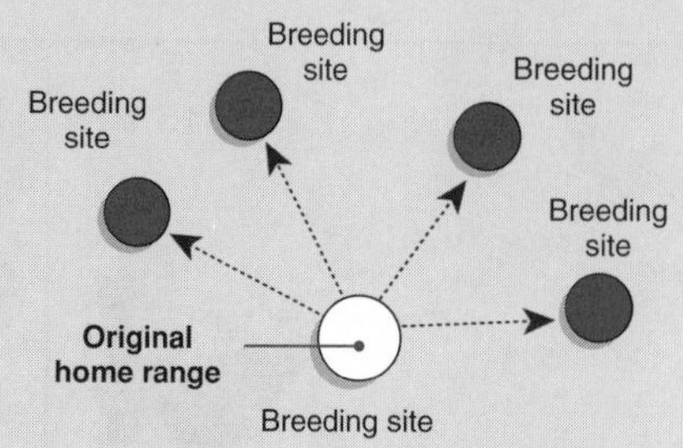

### Return migration

Animals that move to a winter feeding ground are making one leg of a return migration. The same animals return to their home range in the spring which is where they have their breeding sites. Sometimes they follow different routes on the return journey.

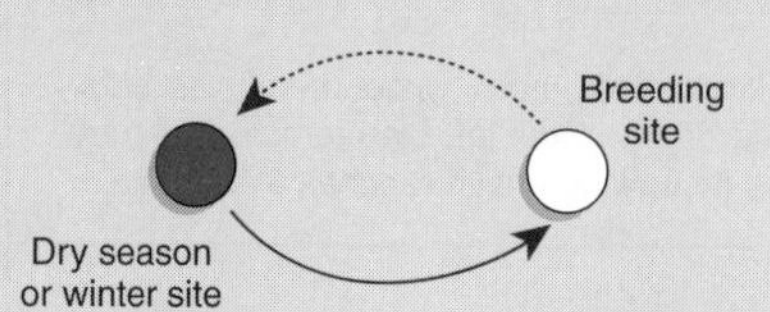

### Nomadic migration

Similar to one-way migration but individuals may breed at several locations during their lifetimes. These migrations are apparently directionless, with no set pattern. Each stopover point is a potential breeding site. There may also be temporary non-breeding stopovers for the winter or dry season.

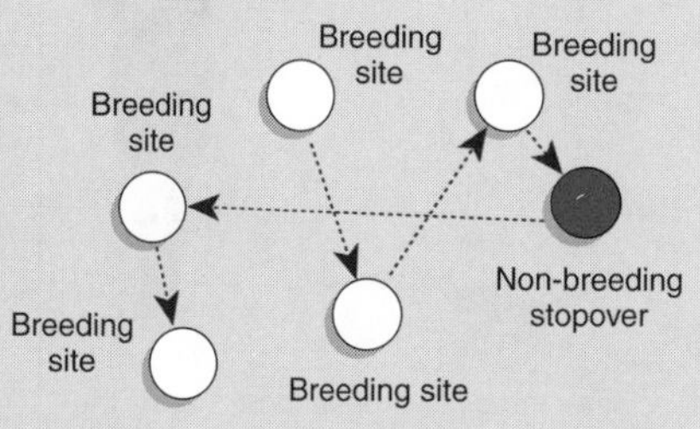

### Remigration circuits

In some populations, the return leg of a migration may have stopovers and may be completed by one or more subsequent generations. In addition to winter or dry season areas, there may be stops at feeding areas by juveniles or adults. Also included are closed circuits where animals die after breeding.

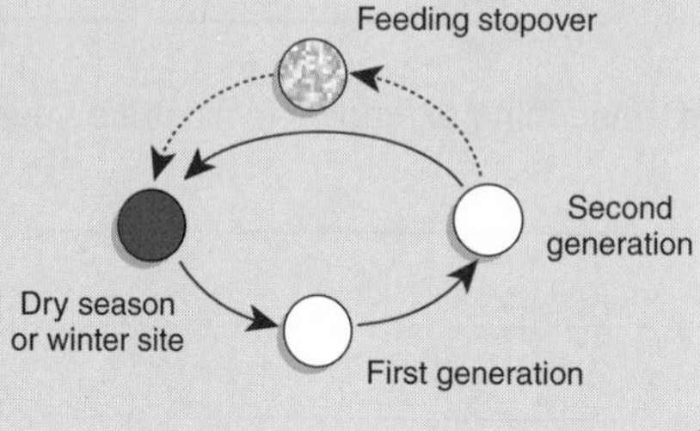

1. Giving an example, describe the conditions under which nomadic migration might be necessary:

2. Identify which of the above forms of migration would lead to further dispersal of a population. Explain your answer:

3. Describe an environmental cue important in the regular migratory behavior of a named species:

4. Discuss the adaptive value of migratory behavior:




**ISBN: 978-1-927173-12-1**

# Animal Migrations

There are many animals that move great distances at different times of the year or at certain stages in their life cycle. Genetically programmed behavior causes them to seek out more favorable conditions when triggered by an environmental cue (e.g. change of season). The cycles are often linked to breeding activity and the location of plentiful food resources at different times of the year. Migration is not without costs: migrating animals must expend huge amounts of energy and the journey itself is often hazardous.

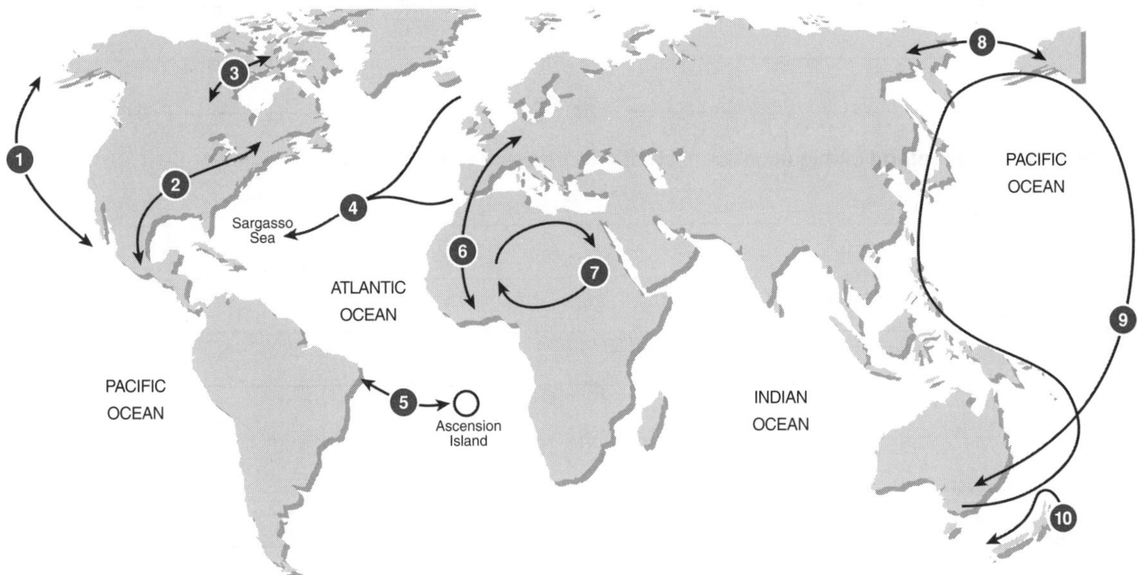


**Migratory locusts** are found in desert regions such as northern and eastern Africa, the Middle East and Australia. When the locust populations expand but food is limited, the developing insects change into the voracious migrating form of the species. Their migration is, more correctly, a dispersal.

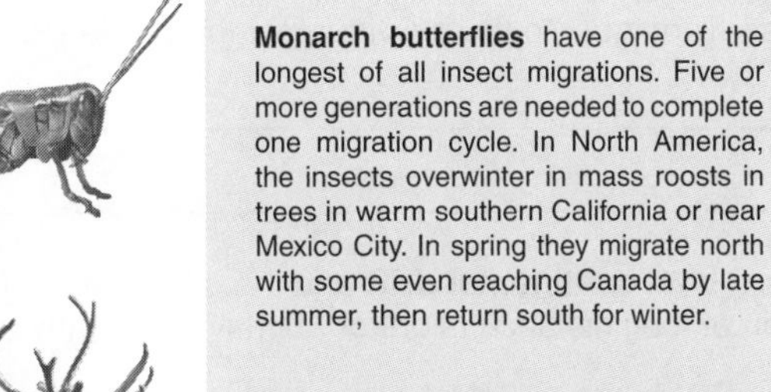

3000 km

**Caribou** spend the winter feeding in the coniferous forests in central Canada. In the spring they begin the northward trek to the tundra of the Barren Lands, within the Arctic circle; a distance of some 1000 km. There they give birth to their calves in the relative safety of the open tundra.

750 to 1000 km

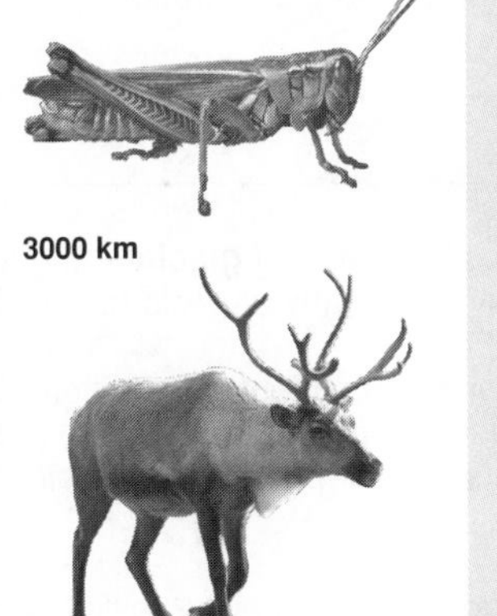

A number of **shearwater** species (including mutton birds) breed in Australia and islands around New Zealand, then migrate northwards with the onset of the southern winter to the north and northeast Pacific. The return journey across the eastern Pacific is assisted directly by the NE trade winds.

11,000 to 13,500 km

**Polar bears** can cover distances of up to 1600 km walking across ice from Alaska, USA to set up winter dens across the Bering Strait in Siberia.

1600 km

**Green turtles** swim vast distances showing amazing navigational skills. They return from the coasts of South America to the beach of their own spawning on Ascension Island to lay eggs.

3000 km

**Monarch butterflies** have one of the longest of all insect migrations. Five or more generations are needed to complete one migration cycle. In North America, the insects overwinter in mass roosts in trees in warm southern California or near Mexico City. In spring they migrate north with some even reaching Canada by late summer, then return south for winter.

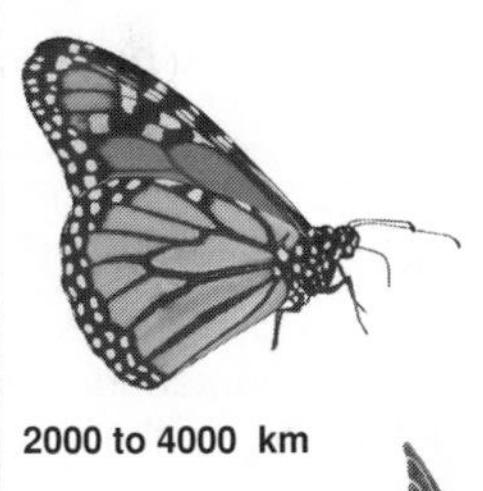

2000 to 4000 km

The **European swift** is one of 140 species of bird that follow one of Europe's most important migratory routes from northern Spain to North Africa and beyond. Swifts breed throughout Europe, and migrate to regions south of the Sahara after breeding. Swifts feed on the wing and the onset of the migration is thought to be triggered by the lack of nutritious airborne insects.

3000 to 12,000 km

A number of **whale** species that include the **humpback** and **gray whale**s follow an annual migration pattern. In summer the whales feed in the krill-rich waters of the polar regions. In winter they move closer to the equator to give birth to young conceived the previous year and to mate again. They seldom feed in transit.

7000 km

**European eels** undertake an outstanding migration across the northern Atlantic ocean to spawn in the Sargasso Sea off the coast of Florida. The larvae that hatch from the eggs gradually drift back across to Europe, a migration which takes several years. Eventually they enter estuaries and move upriver where they feed, grow and mature.

3000 km

In New Zealand and elsewhere, **spiny lobsters** periodically make migrations of many hundreds of kilometers. The movement is predominantly against the prevailing current. It is thought to be an attempt to compensate for the long-term downstream movement of the population as planktonic larvae are swept in one direction by the ocean currents.

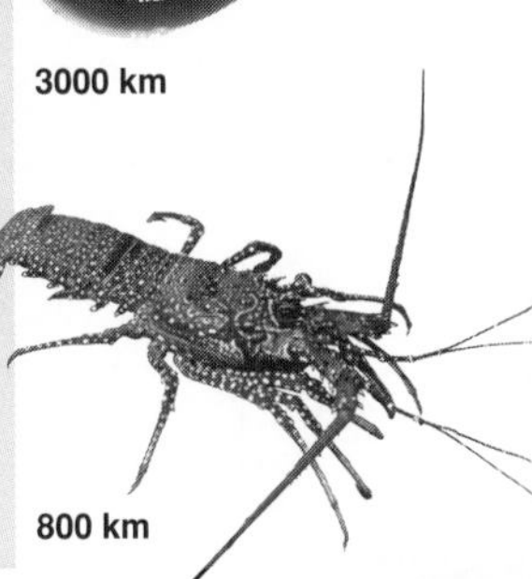

800 km

1. Match up the ten **migration routes** on the previous page with each of the animals below:

(a) Migratory locust: ________________ (f) Monarch butterfly: ________________

(b) Caribou: ________________ (g) European swift: ________________

(c) Shearwater: ________________ (h) Humpback whale: ________________

(d) Polar bear: ________________ (i) European eels: ________________

(e) Green turtle: ________________ (j) Spiny lobster: ________________

2. Describe a **biological advantage** of migration for each of the organisms listed below.

(a) Migratory locusts: ________________

(b) Monarch butterflies: ________________

(c) Humpback whale: ________________

(d) Spiny lobster: ________________

3. Describe two general disadvantages (to the individual) of long distance migration:

(a) ________________

(b) ________________

4. It has been suggested that both **continental drift** and the succession of **glacials** and **interglacials** have been major influences in the evolution of long distance migratory behavior in many animals (particularly birds).

**Continental drift**: About 50 million years ago the pattern of the continents was different, with some closer (e.g. Africa and Eurasia) with others further apart. Many of the bird species present at this time, at least at the group (family) level, would be identifiable today. The movement of the continents to their present day positions takes place at speeds of several centimeters a year.

**Glacial periods**: There have been a number of glacial periods over the last 2 million years, two of which occurred in the last 150,000 years. The cycle of glacial periods affected the geographical distribution of habitats, causing them to expand and contract, as well as moving them further away from the polar regions and then back again with each cycle. The speed of some of the temperature changes associated with the cycles may have been quite rapid: over tens of years, rather than hundreds. This must have benefited those species most that had a flexibility in their migratory strategy.

Discuss how each of these events may have influenced the development of migratory behavior:

(a) Continental drift: ________________

(b) Ice ages: ________________


ISBN: 978-1-927173-12-1

# Migratory Navigation in Birds

**Navigation** is the process by which an animal uses various cues to determine its position in reference to a particular goal. Migrating birds must be able to know their flight direction and when they have reached their destination (goal). They use a wide range of environmental stimuli to provide navigational cues. These include stellar and solar cues, visible landscape features, prevailing wind direction, low frequency sounds generated by winds, polarized light, the Earth's magnetic field, gravitational 'contours', and the smell of pungent sea bird colonies or the sweet smell of meadows. Some of these are examined below:

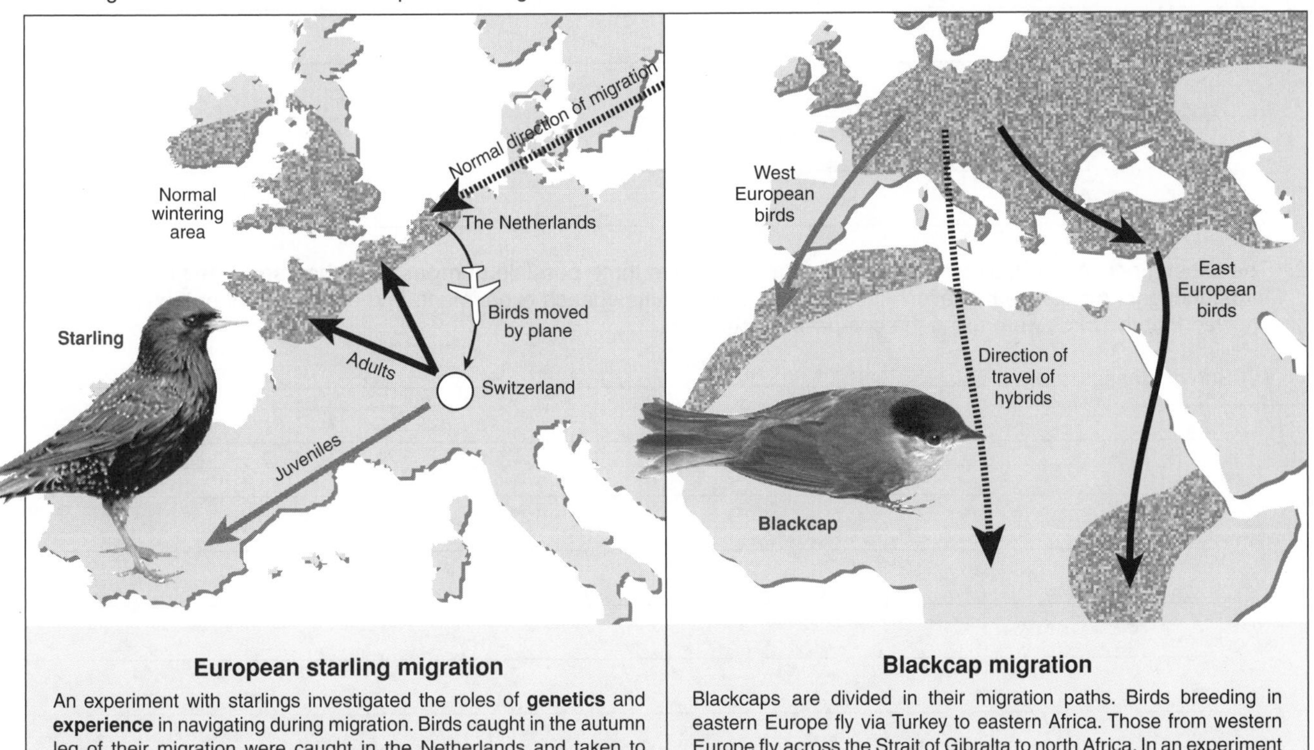


### European starling migration

An experiment with starlings investigated the roles of **genetics** and **experience** in navigating during migration. Birds caught in the autumn leg of their migration were caught in the Netherlands and taken to Switzerland and released. The juveniles, which had not migrated before, flew to Spain. The more experienced birds reached their winter homes in France, Britain and Ireland.

### Blackcap migration

Blackcaps are divided in their migration paths. Birds breeding in eastern Europe fly via Turkey to eastern Africa. Those from western Europe fly across the Strait of Gibraltar to north Africa. In an experiment to test their genetic memory for navigation, birds from both populations were crossed. The hybrids tried to fly south but on a course that takes them over the Alps and the widest part of the Mediterranean.

### Sun compass

Experiments have been carried out to investigate the existence of a sun compass and its importance for daytime migrations. Caged birds were placed in circular enclosures with four windows. Mirrors were used to alter the angle at which light entered the enclosures. At migration time, in natural conditions, these birds clearly showed a preferred flight direction (left). When mirrors bent the suns rays through 90°, the birds turned their preferred direction (middle and right).

### Magnetic compass

An experiment that investigated the possibility of a magnetic compass being used by migratory birds used magnetic coils to mimic the Earth's magnetic field. The birds detect magnetic north, the direction of their spring migration. When the magnetic field was twisted so that north was in the east-southeast position, the birds kept their original path for the first two nights. By the third night, they had detected the change and altered their path accordingly.

### Star compass

An experiment that investigated the use of star positions in the night sky used an ink pad at the base of a cone of blotting paper. Nocturnal migrants flutter in their preferred direction of travel as the amount of ink shows. In a planetarium that projected the real sky, Indigo Buntings located the Pole Star and used it to find north, the direction of their spring migration. When the sky in the planetarium was rotated 90° counter-clockwise the birds altered their direction accordingly. Simulating a cloudy night, the obscured sky confused the birds.

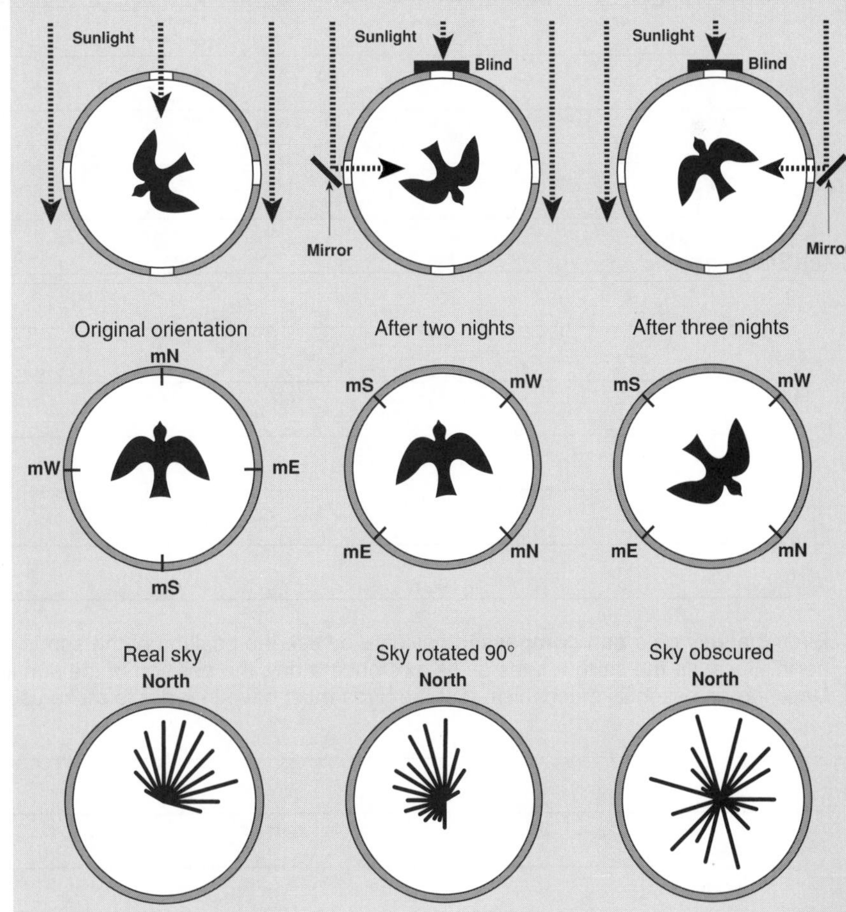

ISBN: 978-1-927173-12-1


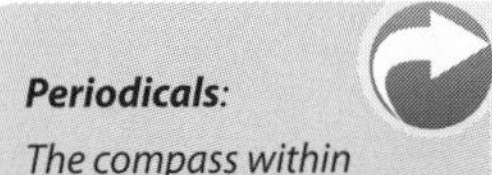

*Periodicals:*
*The compass within*

***Related activities:** Animal Migrations*
***Weblinks:** Bird Migration*

1. With reference to the information on the **European starling** and **blackcap** migrations (previous page), discuss the contributions of innate behavior (genetic programming) and learned behavior on navigation in these migratory birds:

2. The experiments described on the previous page investigate three possible **compass mechanisms** used by migratory birds during long distance migration flights. Discuss the results of each experiment, and explain whether the experiment showed that the birds were using the compass mechanism:

   (a) Sun compass:

   (b) Magnetic compass:

   (c) Star compass:

3. Birds that rely on a **sun compass** to navigate by use the position of the sun in the sky as a reference point to determine north. Because the earth rotates on its axis once a day, the position of the sun in the sky is constantly changing. Describe an essential mechanism that the birds must have in order to make use of this type of compass:

ISBN: 978-1-927173-12-1

# Homing in Insects

Unlike migration, which is the periodic movement from one location and climate to another, **homing** refers to the ability of an animal to return to its home site or locality after being displaced. **Navigation** is involved in both migration and homing, but homing behavior often relies on the recognition of familiar landmarks, especially where the distances involved are relatively short.

The **digger wasp**, *Philanthus triangulum*, builds a nest burrow in sand. It captures and paralyzes an insect grub as a food supply for the wasp's larvae to feed on during development. The paralyzed grub is then taken back to the wasp's underground nest, where it lays its eggs in the still living body. In a well-known experiment to test the homing behavior of this wasp, a research scientist by the name of Tinbergen, carried out the following 2-step procedure. (After Tinbergen, 1951. The Study of Instinct. Oxford University Press, London)

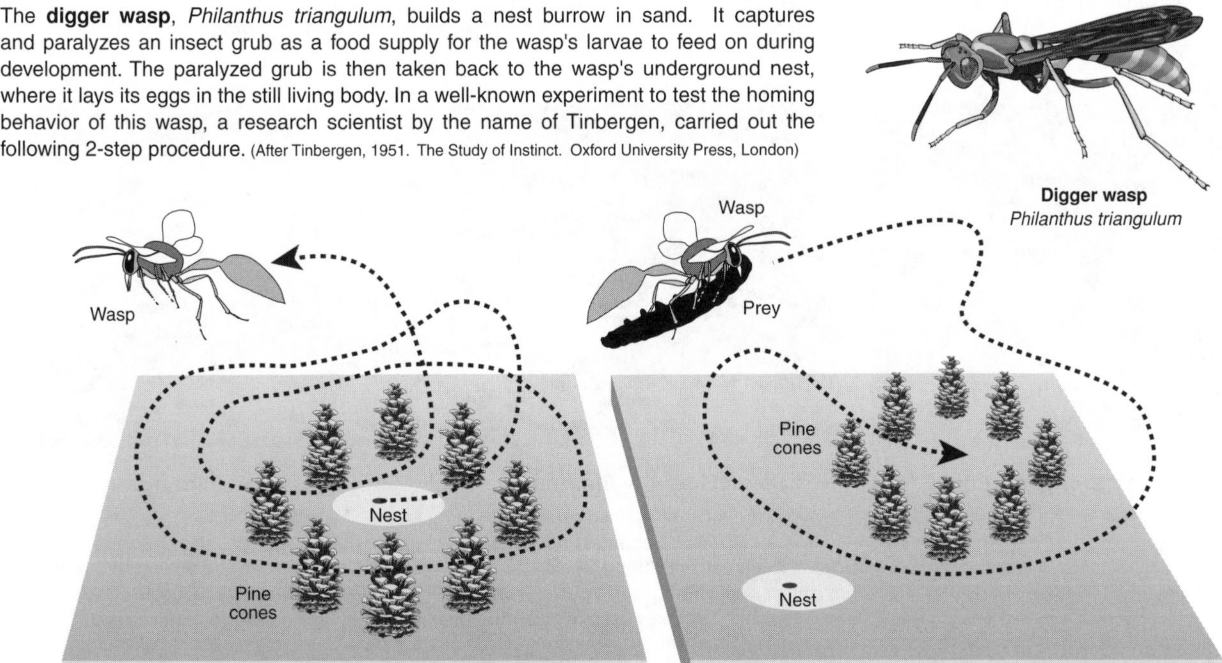


**Step 1: Orientation Flight**

While the female wasp was in the burrow, Tinbergen placed a circle of pine cones around the entrance. When she emerged, the wasp reacted to the situation by carrying out a wavering orientation flight before flying off.

**Step 2: Return Flight**

During her absence, the pine cones were moved away from the burrow leading to the nest. Returning to the nest with prey, the wasp orientated to the circle of pine cones, and not the entrance to the nest.

### Foraging Ant Trails

A foraging ant is able to leave a chemical trail so that it may return to its nest after having searched widely for food sources. The ant secretes a pheromone from its abdomen, that other ants will follow. Soon many ants will follow a well- established trail to and from the food. Later, when the food is gone, the ants stop secreting the pheromone, and the chemical soon diffuses from the area. The trail finally vanishes and the movement of the ants becomes random again as they search for new food sources.

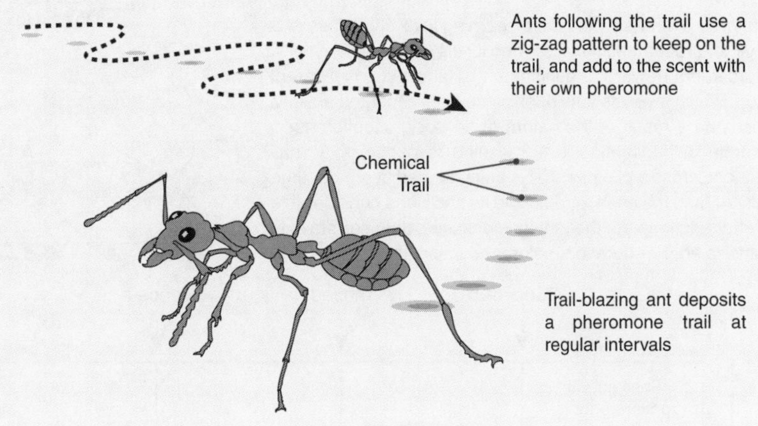


1. Describe a method of navigation used by each insect listed below in the following contexts:

   (a) Honeybees relocating a rich food source (flowers): ______

   (b) Moths finding a mate at night: ______

   (c) Digger wasps locating their nest: ______

   (d) Ants following other foraging ants: ______

   (e) Mosquito locating a meal (blood from a mammal): ______

2. Explain how **homing** is different from migration: ______

______

______



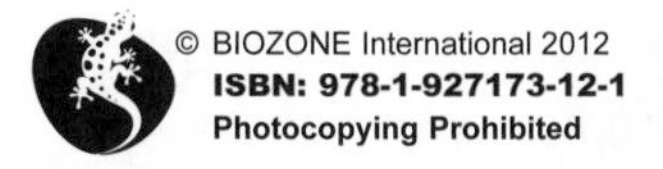


ISBN: 978-1-927173-12-1


Related activities: Pheromones

# Learned Behavior

Learning describes a relatively permanent modification of behavior that occurs as a result of practice or experience. Learning is a critical process that affects the behavior of animals of all ages, across many taxa. Learning behaviors vary widely. The simplest are **habituation** and **imprinting**, when an animal learns to make a particular response only to one type of animal or object. Like most behaviors, they are adaptive in that they enhance survival by ensuring appropriate responses in the given environment. More complex behaviors arise through **conditioning** and observational learning, such as imitation. Latent learning and insight are not readily demonstrated but have been shown experimentally in a range of species.

Digger wasp, *Philanthus triangulum*, with prey



### Filial (Parent) Imprinting

**Filial imprinting** is the process by which animals develop a social attachment. It differs from most other kinds of learning (including other types of imprinting), in that it normally can occur only at a specific time during an animal's life. This **critical period** is usually shortly after hatching (about 12 hours) and may last for several days. Ducks and geese have no innate ability to recognize *mother* or even their own species. They simply respond to, and identify as mother, the first object they encounter that has certain characteristics.

### Sexual Identity Imprinting

Individuals learn to direct their sexual behavior at some stimulus objects, but not at others. Termed **sexual imprinting**, it may serve as a species identifying and species isolating mechanism. The mate preferences of birds have been shown to be imprinted according to the stimulus they were exposed to (other birds) during early rearing. Sexual imprinting generally involves longer periods of exposure to the stimulus than filial imprinting (*see left*).

### Latent Learning

Latent learning describes an association made without reinforcement and expressed later. Tinbergen and Kruyt tested latent learning in predatory digger wasps by providing landmarks around the wasps' nest burrows. On emerging, the wasps survey the area and fly off to forage. If the landmarks were removed or rearranged, the returning wasps became disorientated, supporting the conclusion that the returning wasps use the entire configuration of landmarks as a guide to relocating the burrow.

### Habituation

Habituation is a very simple type of learning involving a loss of a response to a repeated stimulus when it fails to provide any form of reinforcement (reward or punishment). Habituation is different to fatigue, which involves loss of efficiency in a repeated activity, and arises as a result of the nature of sensory reception itself. An example of habituation is the waning response of a snail attempting to cross a platform that is being tapped at regular time intervals. At first, the snail retreats into its shell for a considerable period after each tap. As the tapping continues, the snail stays in its shell for a shorter duration, before resuming its travel.

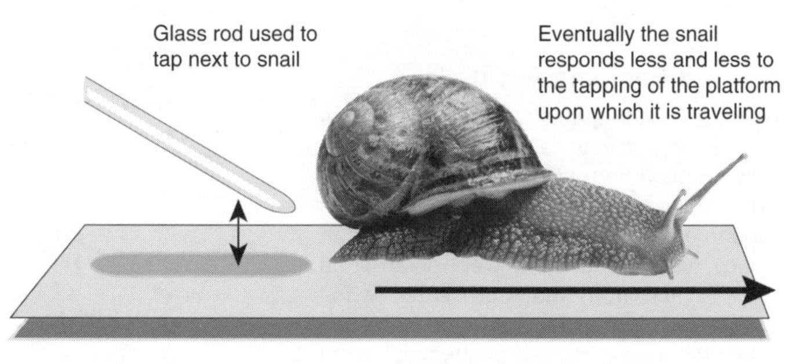


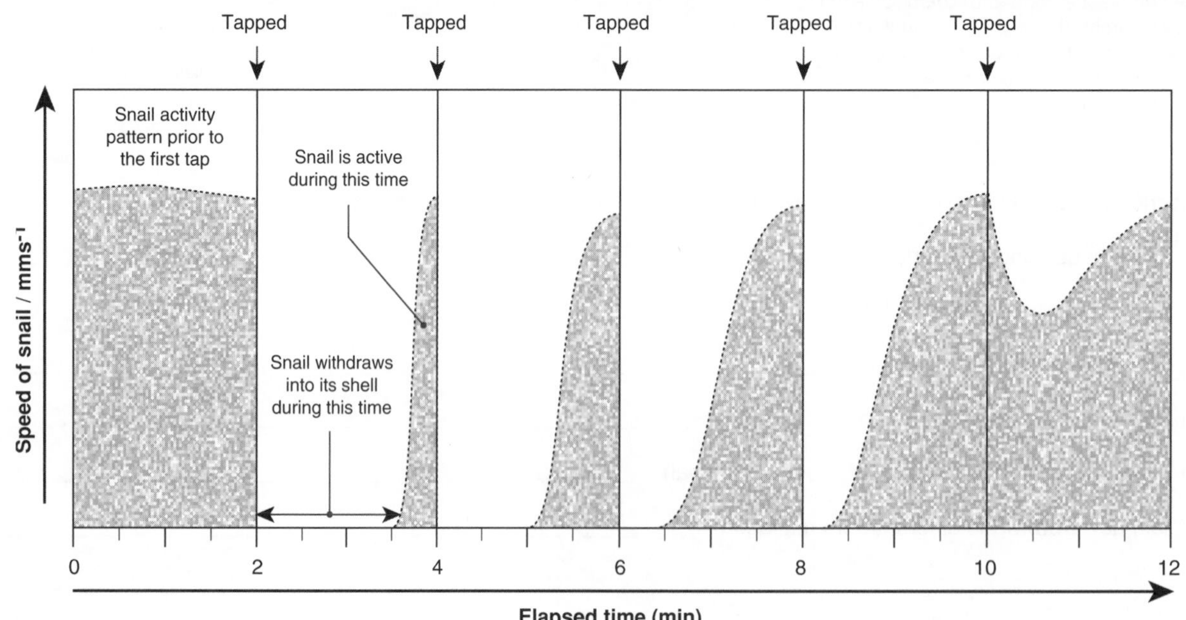




**Related activities:** *The Components of Behavior, Homing in Insects, Learning to Sing*
**Weblinks:** *Insight Learning in Ravens*


ISBN: 978-1-927173-12-1

## Operant Conditioning

Operant conditioning describes situations where an animal learns to associate a particular behavior with a reward. The appearance of the reward is dependent on the appearance of the behavior. Burrhus Skinner studied operant conditioning using an apparatus he invented called a Skinner box (below). When an animal (usually a pigeon or rat) pushed a particular button it was rewarded with food. The animals learned to associate the pushing of the button with obtaining food (the reward). The behavioral act that leads the animal to push the button in the first place is thought to be generated spontaneously (by accident or curiosity). This type of learning is also called instrumental learning and it is the predominant learning process found in animals.

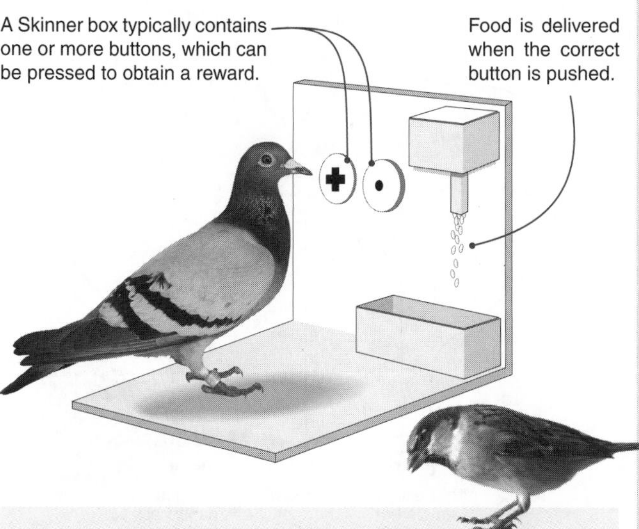


### Operant Learning in Sparrows

Common house sparrows provide a good example of operant conditioning. They have learned to gain access to restaurants and cafés through automatic doors by triggering the motion sensors that control their opening. The birds will flutter in front of the sensor until the door opens, or perch on top of the sensor and lean over until it is triggered. Presumably, after accidentally triggering the sensor and gaining access, the birds learned which behaviors will bring them a reward.

## Insight Behavior

Insight behavior involves using reason to form conclusions or solve a new problem. It is not based on past experiences of a similar problem, but does involve linking together isolated experiences from different problems to reach an appropriate response. Insight learning is common in higher mammals (e.g. apes) and there is good evidence for its occurrence in many other animals including dogs, pigeons and ravens.

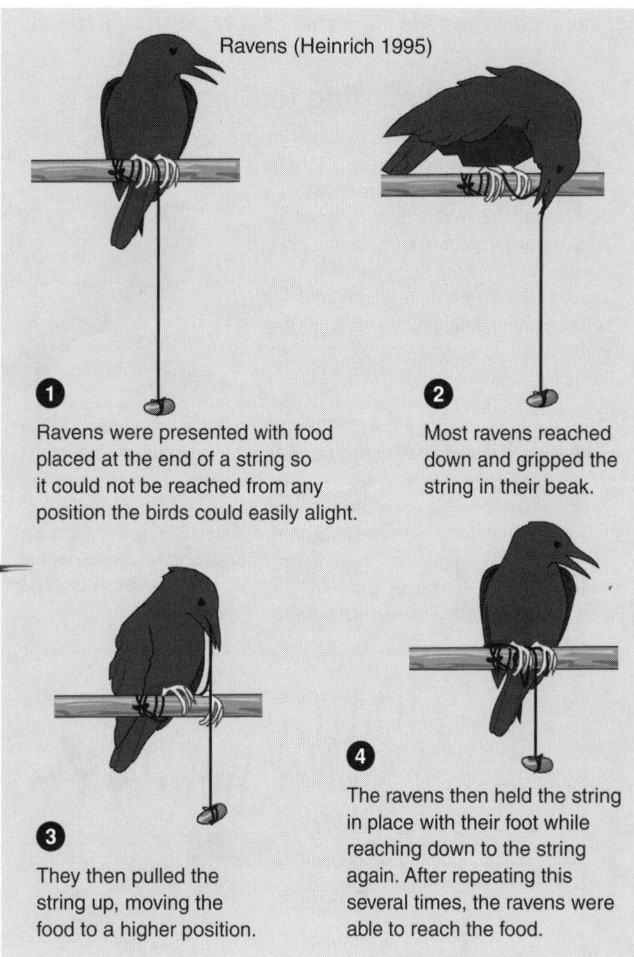


1. Explain the adaptive value of **filial imprinting**: ______________________

2. In what way is **habituation** adaptive? ______________________

3. Suggest when **latent learning** might be important to an animal's survival: ______________________

4. (a) Describe the basic features of **operant conditioning**: ______________________

   (b) Explain why operant conditioning is likely to be the predominant learning process in animals: ______________________

5. Explain why it is difficult to prove conclusively that an animal is using **insight learning** when solving a given problem: ______________________




ISBN: 978-1-927173-12-1

# Learning to Sing

Song birds use vocalizations (songs and calls) as a way to communicate, establish territories, and attract mates. The characteristics of a song may also be an indicator of fitness, as it has been shown that parasites and disease affect the song produced. While singing is instinctive, learning to sing the correct song is a learned behavior, and without it, a song bird is unlikely to gain a territory or a mate. Analyses of many bird species show that there are at least two major strategies for song development: (1) imitation of other birds, particularly adults of the same species, and (2) invention or improvisation. These strategies overlie the genetic template for the song learning process. The window during which a song can be learned varies between species. In some species, the inherited song pattern can be modified by learning only during early life. In others, the song is modified according to experience for at least another year, and some (e.g. blackbirds) modify their songs throughout life.

## Learning to Sing

The songs of different bird species vary but are generally characteristic of the species. The structure of bird song is studied using a technique called **acoustic spectroscopy**, a technique which produces a graphical representation of the sounds being made. This enables song patterns to be compared between individuals and has led to experimental work to establish how birds learn song and how much of the song is genetically determined.

*Blackbirds modify their songs throughout their life*

The sound spectrographs of **chaffinch** song (below) illustrate how the final song that is produced can be altered by exposure to the songs of same-species individuals during the first three months when the song is learned. The upper trace shows the song of a normal male, while the lower trace is the song of a male reared in isolation from the nest. The isolated male's song is the right pitch and relatively normal in length, but it is simpler and lacks the acoustic 'flourish' at the end (arrow), which is typical of the chaffinch's song.

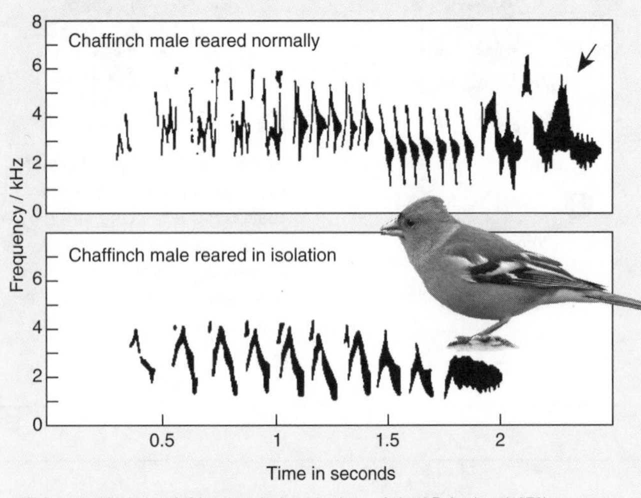


*All data modified after A. Manning: An Introduction to Animal Behaviour (1979)*

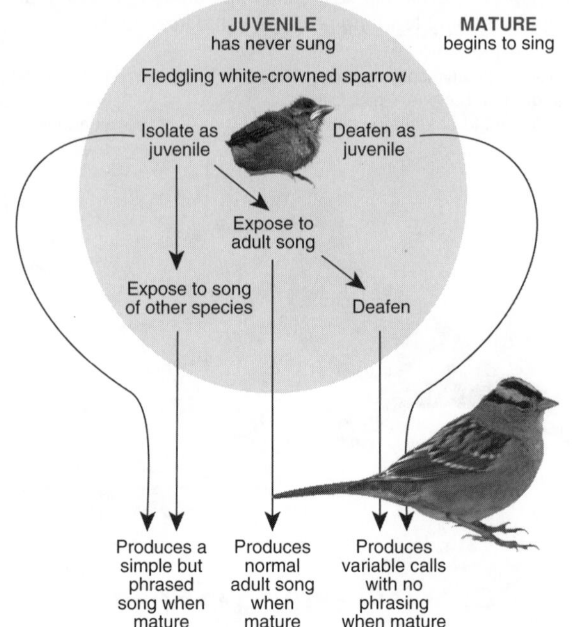


The diagram above summarizes experiments investigating song development in **white-crowned sparrows**. This small finch has a wide range on the Pacific Coast of America and birds from different regions have different **song dialects**. The experiments found:

- Isolated males will eventually sing similar and simplified versions of the normal song, regardless of which region they come from.
- Isolated males can be trained to sing their own dialect by playing them tape recorded songs of birds from their region.
- After 4 months of age, birds are unreceptive to further learning.
- A bird needs to be able to hear itself in order to produce the normal song. It requires auditory feedback to adjust the notes.
- Once birds have learned their normal adult song, they can continue to sing normally even if deafened.

1. From the studies of song development in white-crowned sparrows, describe the evidence for the following statements:

   (a) The inherited song pattern can be modified by learning only in the early stages of life: __________

   (b) Young birds need to hear themselves sing to produce a normal song as adults: __________

   (c) The song produced by an adult bird is determined by a genetic predisposition and what it hears when it is a juvenile:

2. Discuss the possible evolutionary significance of modifying a basic (genetically determined) song pattern by learning:



***Related activities:*** *Animal Communication, Behavior and Species Recognition*
***Weblinks:*** *Starling Talk*


**ISBN: 978-1-927173-12-1**

# Animal Communication

**Communication** (the transmission of (understood) information) between animals of the same species is essential to their survival and reproductive success. Effective communication relies on being correctly understood so the messages conveyed between animals are often highly **ritualized** and therefore not easily misinterpreted. Communication involves a range of signals, which commonly include visual, chemical, auditory, and tactile perception. Which of these signals is adopted will depend on the activity pattern and habitat of the animal. Visual displays, for example, are ineffective at night or in heavy undergrowth.

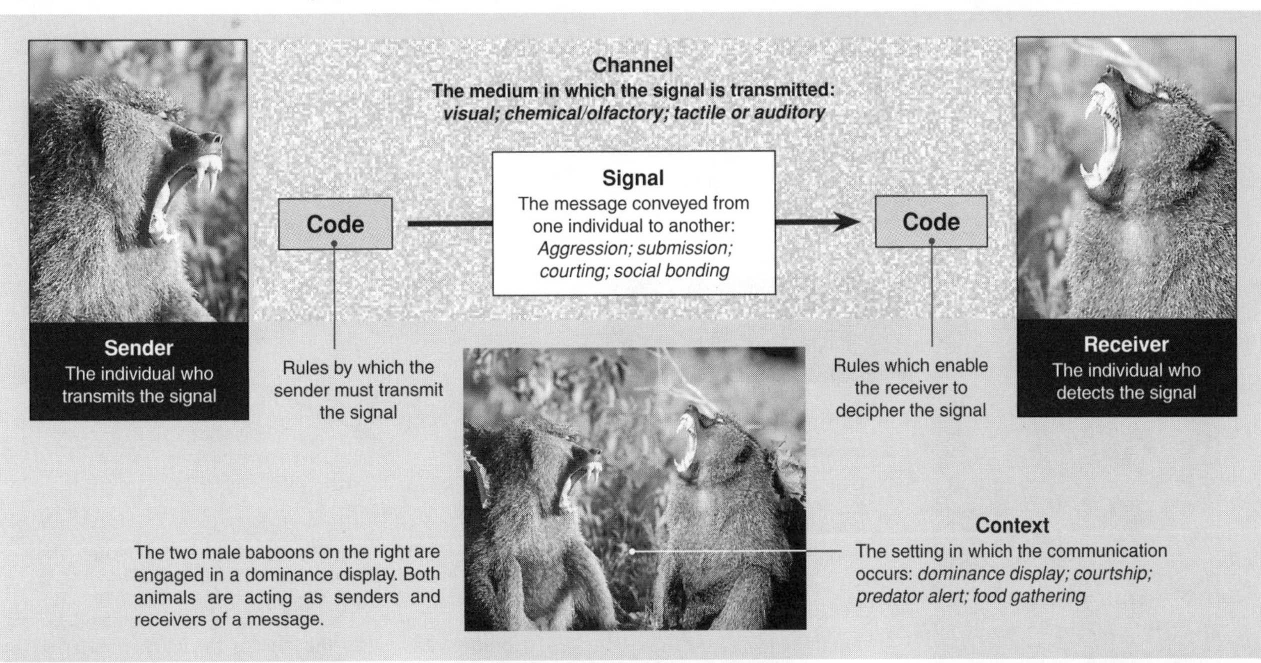


The two male baboons on the right are engaged in a dominance display. Both animals are acting as senders and receivers of a message.

RA

### Olfactory Messages

Some animals produce special scents that are carried considerable distance by the wind. This may serve to advertise for a potential mate, or warn neighboring competitors to keep out of a territory. In some cases, mammals use their urine and feces to mark territorial boundaries. Sniffing genitals is common among mammals.

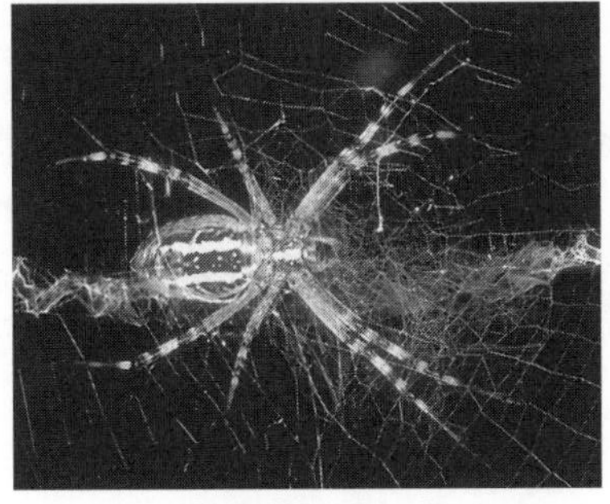

### Tactile Messages

The touching of one animal by another may be a cooperative interaction or an aggressive one. Grooming behavior between members of a primate group communicates social bonding. Vibrations sent along a web by a male spider communicates to a potential female mate not to eat him.

### Auditory Messages

Sound may be used to communicate over great distances. Birds keep rivals away and advertise for mates with birdsong. Fin whales are able to send messages over thousands of kilometers of ocean. Calls made by mammals may serve to attract mates, keep in touch with other members of a group or warn away competitors.

## Visual Messages

Many animals convey information to other members of the species through their body coverings and adornment, as well as through gestures and body language. Through visual displays, it is possible to deliver threat messages, show submission, attract a mate and even exert control over a social group.

### Warning

Animals may communicate a warning to other animals through visual displays. Many wasp species (like those above) have brightly colored black and yellow markings to tell potential predators that they risk being stung if attacked.

### Deception

Animals may seek to deceive other animals about their identity. As an alternative to camouflage, animals may use visual markings that startle or deter potential predators. The eye spots on this moth may confuse a predator.

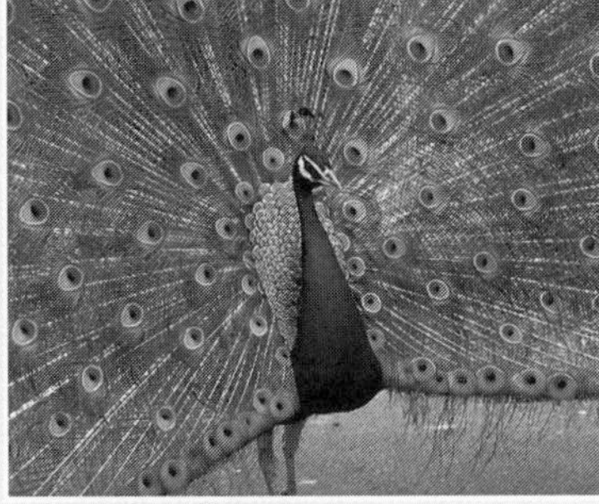

### Attraction

Some animals produce a stunning visual display in order to attract a mate. The plumage of some bird species can be extremely colorful and elaborate, such as the peacock (above), the birds of paradise, and the lyrebird.

**ISBN: 978-1-927173-12-1**


***Related activities:*** *Pheromones, Cooperation and Survival*
***Weblinks:*** *Norway Rat Behavior Repertoire*

# The Fight or Flight Response

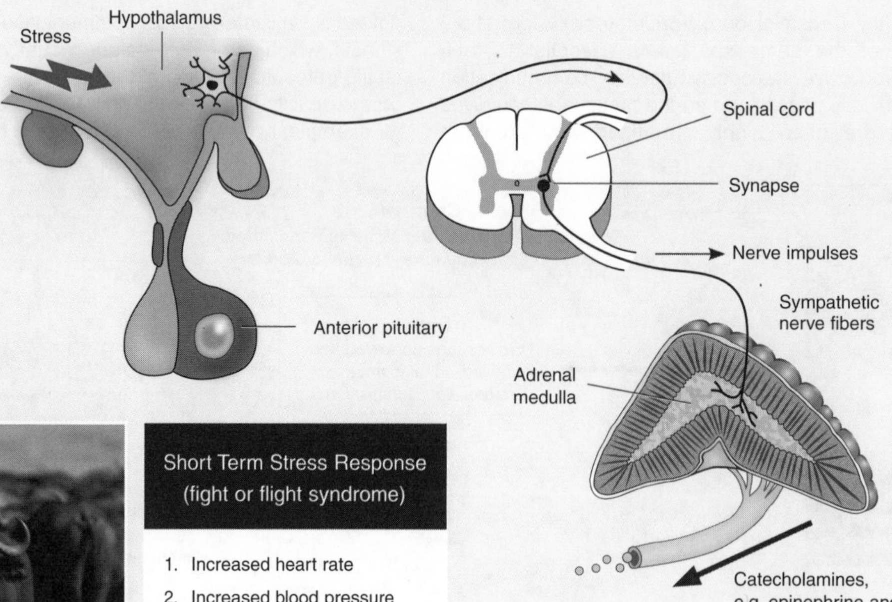


**Short Term Stress Response (fight or flight syndrome)**

1. Increased heart rate
2. Increased blood pressure
3. Liver converts glycogen to glucose; blood glucose levels increase
4. Dilation of bronchioles
5. Blood flow to gut and kidney reduced

   Blood flow to muscles and brain increased
6. Increased metabolic rate

A threat from another animal does not always cause an immediate fight or flight reaction. There may be a period of heightened awareness, during which each animal interprets behavioral signals and cues from the other before they decide to take action (fight or flight). The heightened awareness of one animal will be received by the rest of the group, potentially changing the behavior of the group as a whole (e.g. one individual's nervousness may unsettle the group to the point where a stampede occurs).

Internal changes or external cues can change the way an animal behaves in certain situations. A good example is the **fight or flight response**. When an animal is subjected to stress (e.g. being stalked by a predator) the way the animal reacts is controlled by complex hormonal and nervous interactions of the **hypothalamus, pituitary** and **adrenal glands**. The stress response is triggered through sympathetic stimulation of the central medulla region of the adrenal glands. This stimulation causes the release of catecholamines (**epinephrine** and **norepinephrine**). These physiological changes occur as part of the short term stress response so animals operate at peak efficiency when endangered, competing, or whenever a high level of performance is required.

1. (a) Explain why much of animal communication is ritualized: ____________________

   ____________________

   (b) Discuss the benefits to animals of effective communication over both long and short distances: ____________________

   ____________________

   ____________________

   ____________________

2. Describe and explain the communication methods best suited to nocturnal animals in a forest habitat: ____________________

   ____________________

   ____________________

   ____________________

3. Explain how the fight or flight response prepares an animal to react to a threatening or potentially dangerous situation:

   ____________________

   ____________________

   ____________________

   ____________________


ISBN: 978-1-927173-12-1

# Social Organization

All behavior appears to have its roots in the underlying genetic program of the individual. These innate behaviors may be modified by interactions of the individual with its environment, such as the experiences it is exposed to and its opportunities for learning. The behavioral adaptations of organisms affect their fitness (their ability to survive and successfully reproduce) and so are the products of natural selection. A behavior that leads to greater reproductive success should become more common in a species over time. Few animal species lead totally solitary lives. Many live in cooperative groups for all or part of their lives. Social animals comprise groups of individuals of the same species, living together in an organized fashion. They divide resources and activities between them and are mutually dependent (i.e. they do not survive or successfully reproduce outside the group).

Tigers are solitary and territorial animals, living and hunting alone. A male will remain with a female for 3-5 days during the mating season. A female may have 3 or 4 cubs which will stay with their mother for more than 2 years.

Many invertebrates (e.g. hermit crabs) are solitary animals, with occasional, random encounters. Some animals may be drawn together at feeding sites. Wind or currents may also cause aggregations.

Schooling fish and herds of mammals are examples of animals that form groups of a loose association. There is no set structure or hierarchy to the group. The grouping is often to provide mutual protection.

Family groups may consist of one or more parents with offspring of various ages. The relationship between parents may be a temporary, seasonal one or may be life-long.

Some insects (e.g. ants, termites, some wasp and many bee species) form colonies. The social structure of these colonies ranges from simple to complex, and may involve castes that provide for division of labor.

Primates such as chimpanzees and baboons have evolved complex social structures. Organized in terms of dominance hierarchies, higher ranked animals within the group have priority access to food and other resources.

### Advantages of large social groupings

1. Protection from physical factors
2. Protection from predators
3. Assembly for mate selection
4. Locating and obtaining food
5. Defense of resources against other groups
6. Division of labor amongst specialists
7. Richer learning environment
8. Population regulation

### Possible disadvantages of large social groupings

1. Increased competition between group members for resources as group size increases.
2. Increased chance of the spread of diseases and parasites.
3. Interference with reproduction (e.g. cheating in parental care; infanticide by non-parents).

1. Briefly describe two ways in which behavior may be passed on between generations:

   (a) ______________________________

   (b) ______________________________

2. Explain how large social groupings confer an advantage by providing:

   (a) Richer learning environment: ______________________________

   ______________________________

   ______________________________

   ______________________________


ISBN: 978-1-927173-12-1


*Periodicals:*
*Kiss and make up,*
*All for one*

***Related activities:*** *Cooperation and Survival*
***Weblinks:*** *The African Lion Agonistic Behavior*

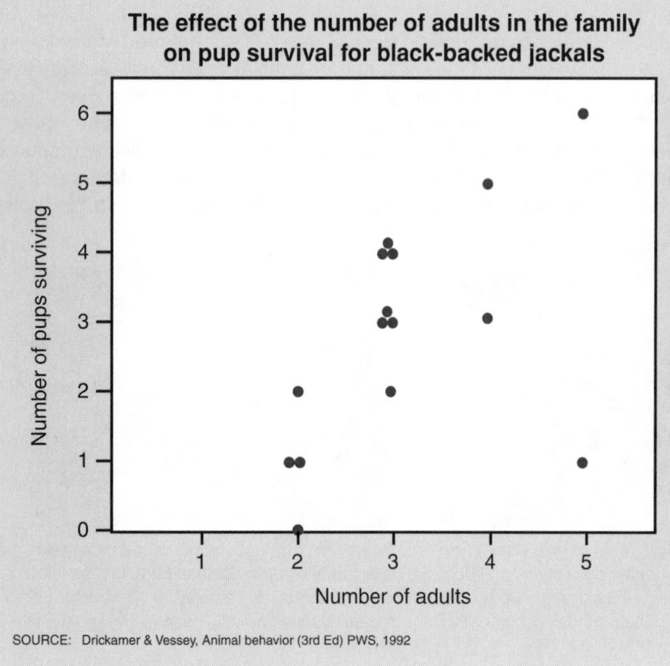


SOURCE: Drickamer & Vessey, Animal behavior (3rd Ed) PWS, 1992

**Black-backed jackal** *(Canis mesomelas)*

Black-backed jackals live in the brushland of Africa. Monogamous pairs (single male and female parents) hunt cooperatively, share food and defend territories. Offspring from the previous year's litter frequently help rear their siblings by regurgitating food for the lactating mother and for the pups themselves. The pup survival results of 15 separate jackal groups are shown in the graph on the left.

(b) Division of labor among specialists: ______

(c) Assembly for mate selection: ______

3. The graph at the top of this page shows how the survival of black-backed jackal pups is influenced by the number of adult helpers in the group.

   (a) Draw an approximate 'line of best fit' on the graph (by eye) and describe the general trend: ______

   (b) Describe two ways in which additional adult helpers may increase the survival prospects of pups:

4. Explain how a social behavior that is beneficial to individuals in a species may become more common over time:


ISBN: 978-1-927173-12-1

# Behavior and Species Recognition

Many of the behaviors observed in animals are associated with reproduction, reflecting the importance of this event in an individual's life cycle. Many types of behavior are aimed at facilitating successful reproduction. These include **courtship behaviors**, which may involve attracting a mate to a particular breeding site. Courtship behaviors are aimed at reducing conflict between the sexes and are often stereotyped or **ritualized**. They rely on sign stimuli to elicit specific responses in potential mates. In addition, there are other reproductive behaviors which are associated with assessing the receptivity of a mate, defending mates against competitors, and rearing the young. Behavioral (ethological) differences between species are a type of **prezygotic** isolating mechanism to help preserve the uniqueness of a species gene pool.

## Courtship and Species Recognition

Accurate species recognition when choosing a mate is vital for successful reproduction and species survival. Failure to choose a mate of the same species would result in reproductive failure or hybrid offspring which are infertile or unable to survive. Birds exhibit a wide range of species-specific courtship displays to identify potential mates of the same species who are physiologically ready to reproduce. They may use simple visual or auditory stimuli, or complex stimuli involving several modes of communication specific to the species.

Peacock courtship (left) involves a visually elaborate tail display to attract female attention. The male raises and fans his tail to display the bright colors and eye-spot patterns. Peahens tend to mate with peacocks displaying the best quality tail display which includes the quantity, size and distribution of eye-spots.

Bird song is an important behavioral isolation method for many species including eastern and western meadowlarks. Despite the fact that they look very similar and share the same habitat, they have remained as two separate species. Differences between the songs of the two species enables them to recognize individuals of their own species and mate only with them. This maintains the species isolation.

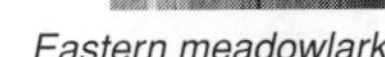

*Eastern meadowlark*

Some species use chemical cues as mating signals and to determine mate choice. The crested auklet (left) secretes aldehydes which smell like tangerines. Birds rub their bills in the scented nape of a partner during courtship. This "ruff-sniff" behavior allows mate evaluation based on chemical potency. A potential partner might be seen as fitter and more attractive if it produces more aldehydes, because the chemical repels ectoparasites.

## Courtship Behavior is a Necessary Precursor to Successful Mating

**Courtship** behavior occurs as a prelude to mating. One of its functions is to synchronize the behaviors of the male and female so that mating can occur, and to override attack or escape behavior. Here, a male greater frigatebird calls, spreads its wings, and inflates its throat pouch to court a female.

In many bird and arthropod species, the male will provide an offering, such as food, to the female. These **rituals** reduce aggression in the male and promote appeasement behavior by the female. For some **monogamous** species, e.g. the blue-footed boobies (left), the pairing begins a long term breeding partnership.

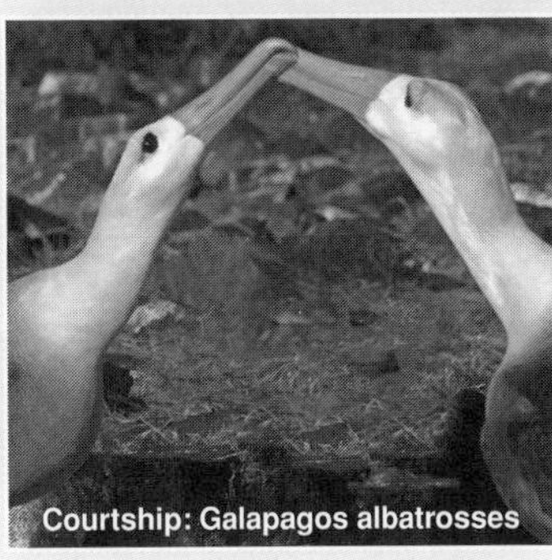

Courtship: Galapagos albatrosses

Although courtship rituals may be complex, they are very stereotyped and not easily misinterpreted. Males display, usually through exaggerated physical posturing, and the females then select their mates. Courtship displays are species specific and may include ritualized behavior such as dancing, feeding, and nest-building.

1. (a) Suggest why courtship behavior may be necessary prior to mating: ______

   (b) Explain why courtship behavior is often ritualized and involves stereotyped displays: ______

2. In terms of species continuity, explain the significance of courtship behavior in species recognition: ______

ISBN: 978-1-927173-12-1

# Breeding Behavior

In the animal world, humans are unusual in being sexually receptive most of the time. Most animals breed on an annual basis and show no reproductive behavior outside the breeding season. Photoperiod is an important cue for triggering the onset of breeding behavior in many species. Species with a relatively short gestation (e.g. most birds) respond to increasing daylength, whereas those with a long gestation (e.g. large mammals) respond to decreasing daylength. The short time period that most sexually reproducing animals have in which to breed creates strong selective pressure for clearly understood patterns of behavior that improve the chances of reproductive success. A breeding scenario typical for many insects is described below.

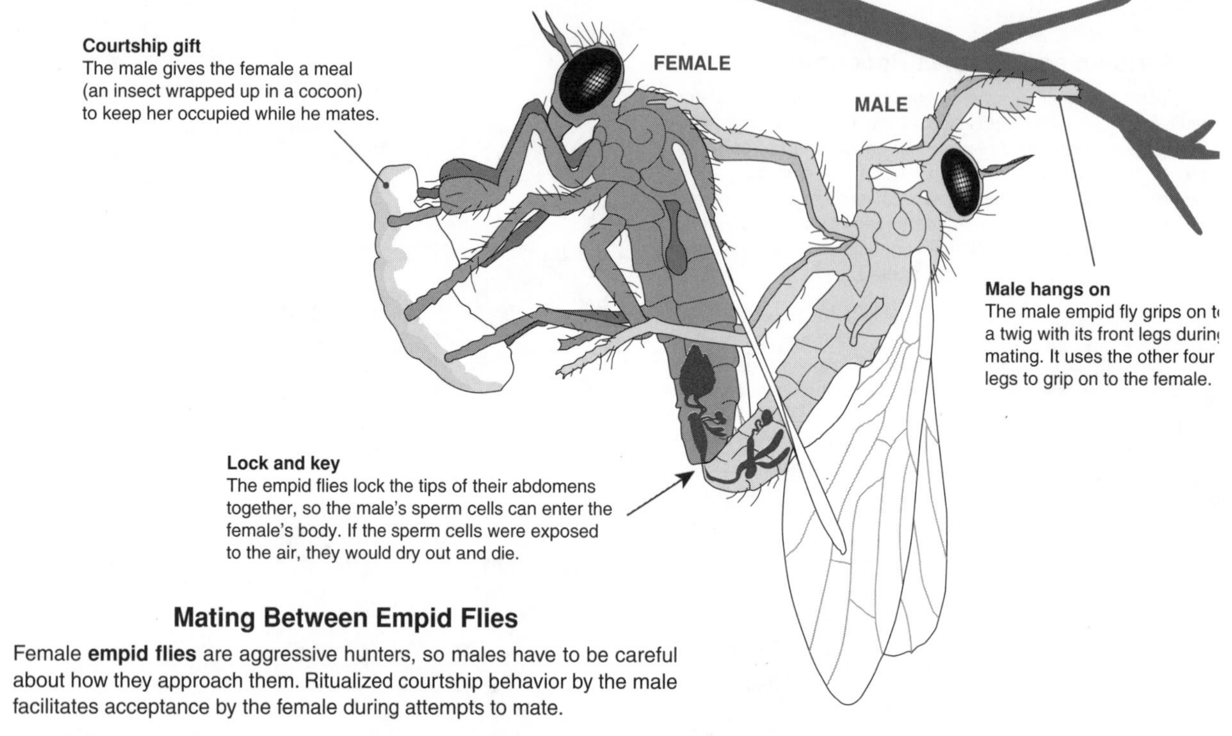


### Mating Between Empid Flies

Female **empid flies** are aggressive hunters, so males have to be careful about how they approach them. Ritualized courtship behavior by the male facilitates acceptance by the female during attempts to mate.

## Territories and Reproductive Behavior

A territory is any area that is defended against members of the same species. The territory, which usually contains access to valuable resources, is usually defended by acts of aggression or ritualized signals (e.g. vocal, visual or chemical signals).

Resource availability often determines territory size and the population becomes spread out accordingly. Gannet territories are relatively small, with hens defending only the area they can reach while sitting on their nest.

Topi, a lek species

Lek systems involve gatherings of males for competitive mating displays. In a lek system, the only area defended by a male is the space where mating occurs. Females compete for the highest ranked males, as judged by their position on the lek.

1. Describe two aspects of mating behavior in empid flies that help to ensure successful mating:

   (a) ______________________________

   (b) ______________________________

2. Explain the role of photoperiod in determining the onset of reproductive behavior in many species: ______________________________

3. Describe the role a territory could play in an individual's reproductive success: ______________________________






ISBN: 978-1-927173-12-1

# Pheromones

A **pheromone** is a chemical produced by an animal and released into the external environment where it has an effect on the physiology or behavior of members of the same species. Hundreds of pheromones, some of which are sex attractants, are known. They are especially common amongst insects and mammals, and commonly relate to reproductive behavior. Many mammals, including canids and all members of the cat family, commonly use scent marking to mark territorial boundaries and to advertise their sexual receptivity to potential mates. Other mammals, including rabbits, release a mammary pheromone that triggers nursing behavior in the young. Pheromones are also used as signaling molecules in social insects such as bees, wasps, and ants. They may be used to mark a scent trail to a food source or to signal alarm. Species specific odor cues are now widely used as bait in traps when controlling insect or mammalian pests or to capture animals for study.

## Pheromones and Animal Communication

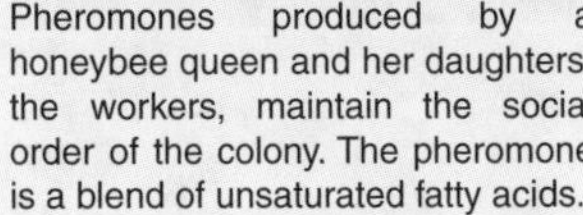

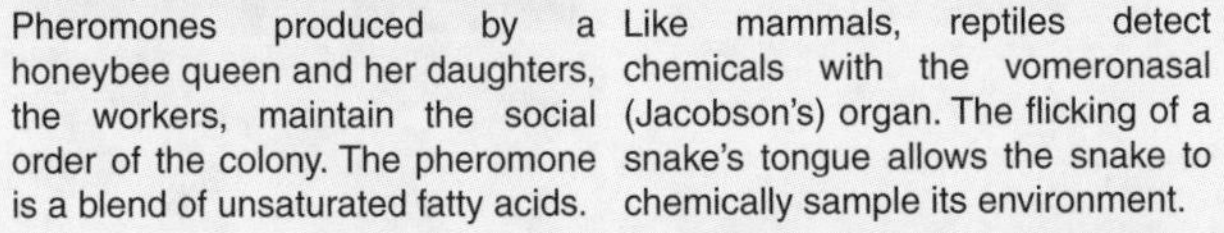

Pheromones produced by a honeybee queen and her daughters, the workers, maintain the social order of the colony. The pheromone is a blend of unsaturated fatty acids.

Like mammals, reptiles detect chemicals with the vomeronasal (Jacobson's) organ. The flicking of a snake's tongue allows the snake to chemically sample its environment.

In mammals, pheromones are used to signal sexual receptivity and territorial presence, or to synchronize group behavior. Pheromone detection relies on the vomeronasal organ (VNO), an area of receptor tissue in the nasal cavity. Mammals use a flehmen response, in which the upper lip is curled up, to better expose the VNO to the chemicals of interest.

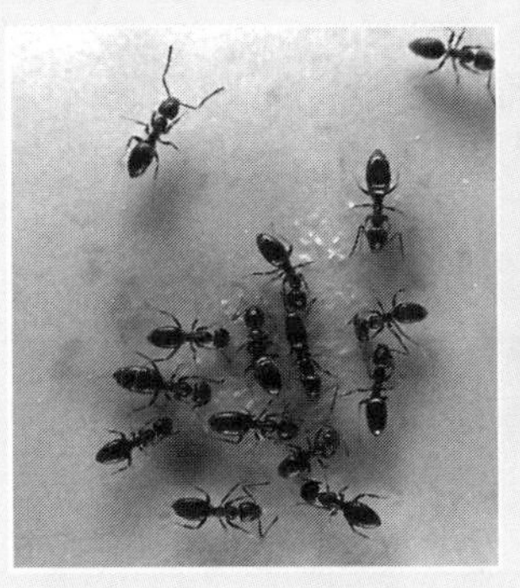

Communication in ants and other social insects occurs through detection of pheromones. Foraging ants will leave a trail along the ground which other ants will follow and reinforce until the food source is depleted. Ants also release alarm substances, which will send other ants in the vicinity into an attack frenzy. These signals dissipate rapidly if not reinforced.

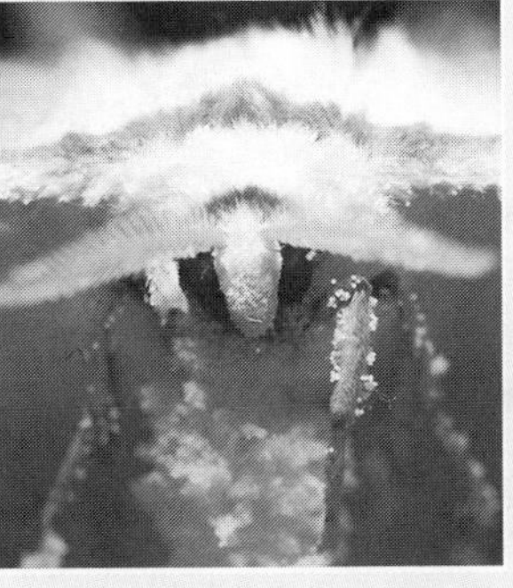

Photo courtesy of Cereal Research Centre, AAFC

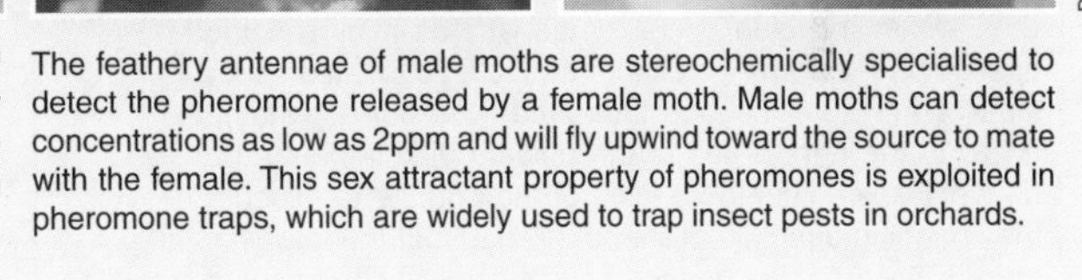

The feathery antennae of male moths are stereochemically specialised to detect the pheromone released by a female moth. Male moths can detect concentrations as low as 2ppm and will fly upwind toward the source to mate with the female. This sex attractant property of pheromones is exploited in pheromone traps, which are widely used to trap insect pests in orchards.

1. (a) Distinguish between hormones and **pheromones**: ______

   (b) Explain the significance of pheromones being species specific: ______

2. Giving examples, briefly describe the role of pheromones in three aspects of animals behavior:

   (a) ______

   (b) ______

   (c) ______

3. From what you know of pheromone activity, suggest how a pheromone trap would operate to control an insect pest:

______




ISBN: 978-1-927173-12-1


***Related activities:*** *Simple Behaviors*
***Weblinks:*** *Cichlid Response to Pheromones*

# Honeybee Communication

Animals communicate with each other about where to find sources of food. Honeybees use the **waggle dance**, a series of figure of eight movements, to tell other bees the direction and distance to food or water sources. Honeybees navigate using a sun compass, so honeybees communicate the direction and distance to food relative to the current position of the sun.

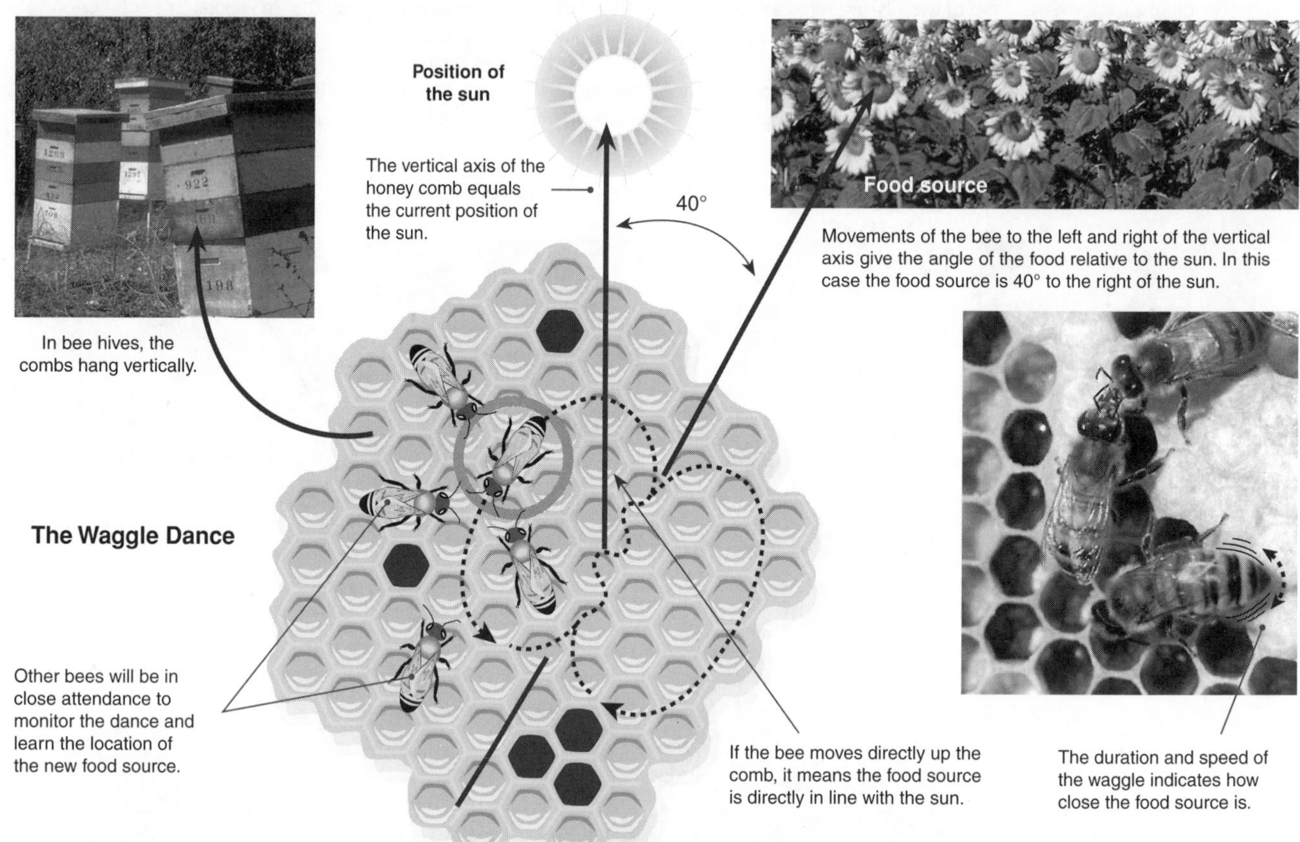


### The Waggle Dance

Bees communicate the direction and distance of the food source through the waggle dance (above). If food is located directly in line with the sun, the communicator (bee within the blue circle) demonstrates it by running directly up the comb. To direct bees to food located either side of the sun, the bee introduces the corresponding angle to the right or left of the upward direction into the dance. Bees adjust the angles of their dance to account the changing direction of the sun throughout the day. This means directions to the food source are still correct even though the sun has changed positions.

### The Round Dance

If the food source is very close (less than 50 m) the honeybee will perform a round dance. The honeybee's round dance stimulates other workers to leave the hive and search within 50 m for a food source (see right).

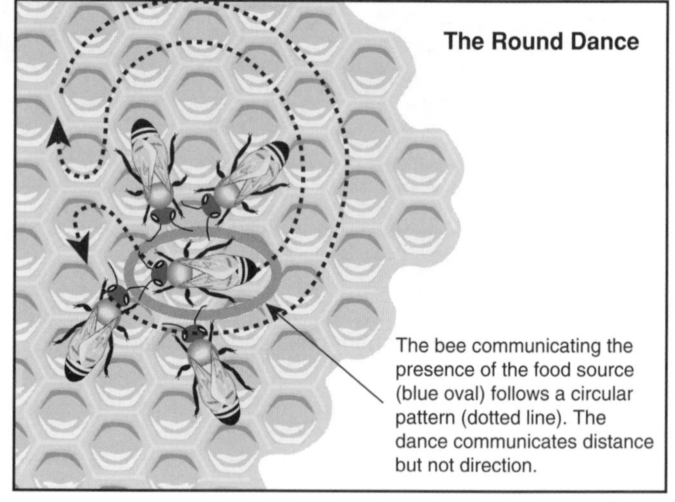


1. Name the environmental reference used by honeybees to orientate for navigation: ______________________

2. Describe how a bee communicates the proximity of a food source in the waggle dance: ______________________

3. Explain how the bee compensates for the time it takes between finding the food and delivering its message to the hive: ______________________

4. Describe the circumstances under which the round dance is used: ______________________



***Related activities:** Animal Communication* ***Weblinks:** Honeybee Waggle Dance, Using Vector Calculus to Communicate*

***Periodicals:** Show me the honey*


**ISBN: 978-1-927173-12-1**

# Cooperation and Survival

**Cooperative behavior** both within and between species can aid survival. Individuals may cooperate with each other for many reasons: for mutual defense and protection, to enhance food acquisition, or to rear young. To explain the evolution of cooperative behavior, it has been suggested that individuals benefit their own survival or the survival of their genes (offspring) by cooperating. **Kin selection** is a form of selection that favors altruistic (self-sacrificing) behavior towards relatives. In this type of behavior an individual will sacrifice its own opportunity to reproduce for the benefit of its close relatives. Individuals may also cooperate and behave altruistically if there is a chance that the "favor" may be returned at a later time. **Altruistic behavior** towards non-relatives is usually explained in terms of trade-offs, where individuals weigh up the costs and benefits of helpful behavior. Cooperation will evolve in systems where, in the long term, individuals all derive some benefit.

**Gray wolves**

Many mammalian predators live in well organized social groups. These are formed for the purposes of cooperative hunting and defense, and they facilitate offspring survival within the entire group. In the gray wolves above, territories are marked by scent. Howling promotes group bonding and helps to keep neighboring packs away.

Mammals lack the enzymes required to break down cellulose. However, a mutualistic relationship exists between many herbivores and the microbes in their gut, which enables cellulose to be digested. In ruminants, the rumen microflora break down the cellulose and the ruminant obtains energy from the fatty acids released by the microbes.

**Waxeye**

©Dr. M. Soper

The males of many species help their mates collect enough food to meet reproductive needs. In some species, especially amongst birds, non-breeding individuals, e.g. older siblings, may assist in rearing the offspring by protecting or feeding them. This type of altruism may arise through **kin selection**.

**Bee pollinating a flower**

A mutualistic relationship exists between plants and their pollinators. Most pollinators receive food in the form of pollen or nectar from the plant, and in return, the pollen of the plant is transferred by the pollinating organism to other flowers to ensure fertilization occurs.

**Meerkats**

South African meerkats live in communities in earth burrows. They are vulnerable to attack from land and aerial predators (especially vultures). The group maintains a constant surveillance by posting sentinels to warn the rest of the group of danger.

JW

Cooperative (mutualistic) associations can occur between different species. Cape buffalos are warned of approaching predators by cattle egrets and maribou storks which in turn feed on insects disturbed by the buffalo as it grazes.

1. Using examples, discuss the difference between **altruistic behavior** and **kin selection**:

2. Explain (in evolutionary terms) why an animal would raise the offspring of a close relative rather than their own:

3. Discuss how cooperative interactions enhance both individual and population survival:

Timing, Coordination & Social Behavior

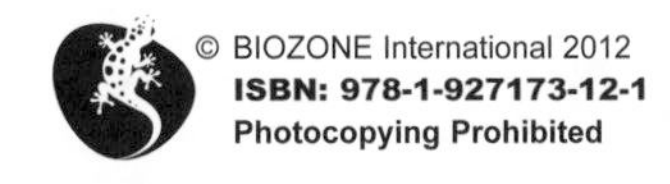


**ISBN: 978-1-927173-12-1**


*Periodicals:* *Relative distance*

***Related activities:*** *Animal Communication, Social Organization*

# Cooperative Defense and Attack

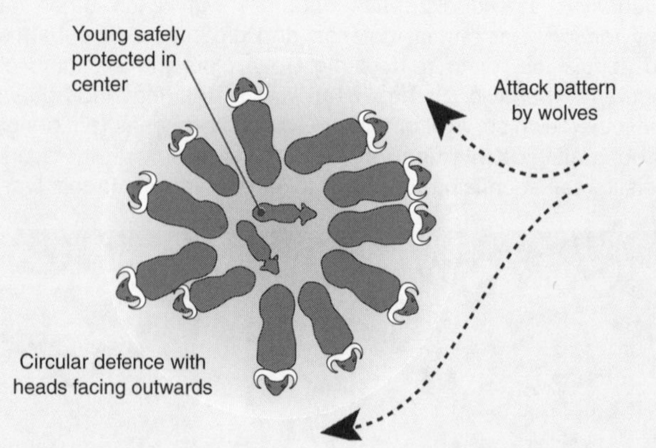


**Group defense in musk oxen:** In the Siberian steppes, which are extensive grasslands, a large grazing animal like the musk ox must find novel ways of protecting itself from predators. There is often no natural cover to help with defense, so they must make their own barrier in the form of a defensive circle. When wolves (their most common predator) attack, they shield the defenseless young inside the circle. Lone animals have little chance of surviving an attack as wolves hunt in packs.

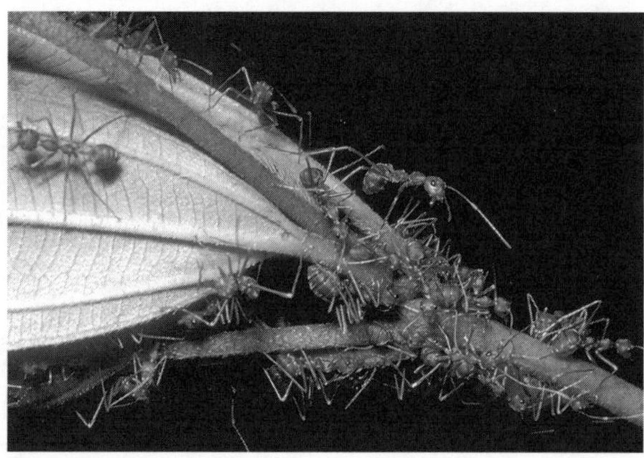

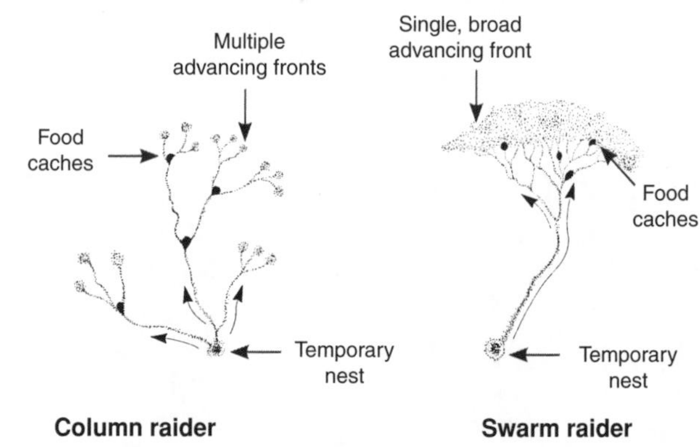


**Army ants foraging:** There are two species of army ant that have quite different raiding patterns: *Eciton hamatum* whose columns go in many directions and *Eciton burchelli*, which is a swarm-raider, forming a broad front. Both species cache food at various points along the way (dark patches in the diagram above). Through group cooperation, the tiny ants are able to subdue prey much larger than themselves, even managing to kill and devour animals such as lizards and small mammals. This would not be possible if they hunted as individuals.

1. Describe a benefit of the cooperative interaction for each of the following species:

   (a) Musk oxen: ______

   (b) Army ants: ______

2. Sheep need to spend most of their day feeding on grass. They form mobs both naturally in the wild as well as on farms.

   (a) Explain why sheep form mobs: ______

   (b) Explain how this might enhance an individual sheep's ability to feed: ______



*Related activities:* *Cooperation and Survival, Cooperative Food Gathering*


**ISBN: 978-1-927173-12-1**

# Cooperative Food Gathering

**Humpback whales**: The two whales pictured above are feeding near the surface. They swim below a school of fishes and confuse them by emitting a stream of small bubbles. They then swim upward in a spiral pattern with the mouth open, closing it as they break the surface. Water is squeezed out of the mouth, through a sieve of baleen plates, trapping the fish. By fishing cooperatively in this way, several whales herd the fish more effectively.

**Pelicans fishing**: Group hunting behavior in pelicans enables the birds to herd together large quantities of fish and facilitates the scoop-beak fishing method. Groups of five to ten birds gather in shallow water. They swim in horseshoe formation closing almost to a complete circle to trap the fish. They plunge their beaks into the water exactly at the same time to catch the fish. A lone pelican's fishing success may not be as good.

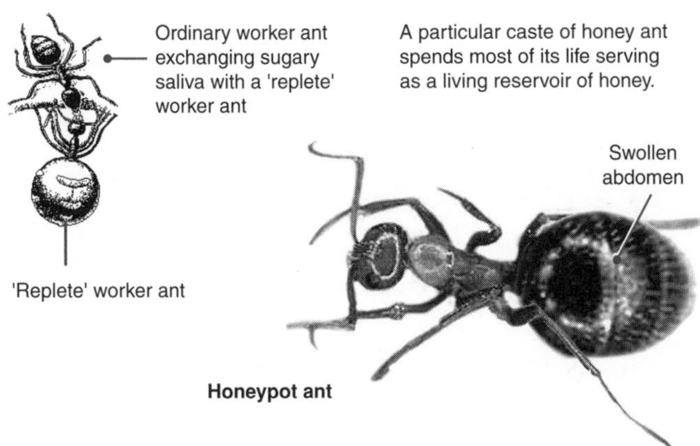


**Honeypot ants**: Honeypot ants of central Australia have a special group of workers called 'repletes'. These never leave the nest, but stay in underground galleries where they serve as vessels for storing a rich food supply. Regular workers that have been foraging for honeydew and nectar return to the nest where they regurgitate food from their crops to feed the replete. The replete will continue to accept these offerings until its abdomen has swollen to the size of a pea (normally it is the size of a grain of rice). The repletes become so swollen that their movements are restricted to clinging to the gallery ceiling where many hundreds of them hang in a row. When the dry season arrives and food supplies become scarce, workers return to the repletes, coaxing them to regurgitate droplets of honey.

**Lions hunting**: Lions hunt on the savannah grasslands of East Africa. In this terrain, the sparse distribution of trees creates a great advantage for the fast moving prey of the lions (e.g. antelope). They can detect approaching lions easily, raise the alarm, and escape. Unlike solitary big cats, such as the leopard, that hunt in forest environments, lions must work as a team and use a strategy to trap the prey. Solitary lions have poor hunting success. When lions sight a herd of prey, several lionesses hide downwind. Others circle upwind and stampede the herd towards the lionesses waiting to attack. Lions must be careful not get injured in the hunt. A solitary, injured lion that cannot hunt will almost certainly starve. Cooperation when hunting reduces the risk of injury, and provides group support when it does.

1. Discuss the energetic and survival benefits of cooperative food gathering: ______________________

2. Using an example, describe how diversity of roles within a population for food gathering can aid survival: ______________________

Timing, Coordination & Social Behavior

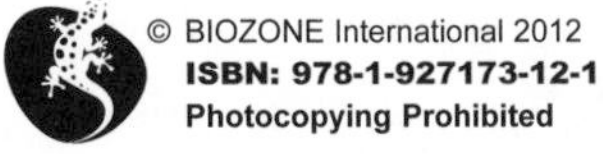

ISBN: 978-1-927173-12-1

# KEY TERMS: Mix and Match

*INSTRUCTIONS: Test your vocabulary by matching each term to its definition, as identified by its preceding letter code.*

abscisic acid (ABA) ..........
auxin ..........
cooperative behavior ..........
courtship ..........
environmental cue ..........
ethylene ..........
gibberellin ..........
gravitropism ..........
hibernation ..........
homing ..........
innate behavior ..........
insight ..........
kinesis (pl. kineses) ..........
learned behavior ..........
magnetic compass ..........
migration ..........
nastic response (nasty) ..........
navigation ..........
operant learning ..........
pheromone ..........
photoperiodism ..........
phototropism ..........
social behavior ..........
sun compass ..........
taxis (pl. taxes) ..........
tropism ..........

**A** Behavior which uses reasoning to correctly perform a task on the first attempt. There is no prior experience or knowledge.

**B** Stimulus from the environment used by animals and plants to establish and maintain a pattern of activity, e.g. day length, temperature.

**C** A directional growth response to gravity in plants.

**D** The recognition of land marks and use of environmental stimuli to find and arrive at a location.

**E** Genetically scripted behavior, which is displayed even when the animal is kept in isolation. It is usually inflexible, and a given stimulus triggers a specific stereotyped response.

**F** Non-directional response to a stimulus in which the speed or rate of turning is proportional to the stimulus intensity.

**G** A plant hormone (phytohormone) responsible for apical dominance, phototropism and cell elongation.

**H** The movement of a population of animals. Classic examples involve the movement to and return from the location.

**I** The working together (either knowingly or unknowingly) of individuals in order to reach a common goal, e.g. the gain of resources to enhance survival.

**J** An orientation movement by an animal in response to a directional stimulus.

**K** A situation where an animal learns to associate a particular behavioral act with a reward.

**L** Behavior that is directed towards or taking place between members of the same species.

**M** The physiological reaction of organisms to the presence and absence of light.

**N** The ability of an animal to return to its home site after being displaced.

**O** A growth response in plants to directional light.

**P** A state of inactivity and reduced metabolic rate, often in response to food shortages.

**Q** Plant hormone that acts as a growth inhibitor, promoting closing of stomata and seed dormancy.

**R** A plant response that is independent of the direction of the stimulus.

**S** Plant hormone that promotes fruit ripening and leaf fall.

**T** Plant hormone with many effects including delay of senescence and leaf fall, breaking dormancy, promotion of secondary growth.

**U** The use of the movement of the sun to help with navigation. Normally used in conjunction with an internal clock.

**V** A chemical produced by an individual and released into the environment that has an effect on the behavior of another individual of the same species.

**W** Any process in an animal in which its behavior becomes consistently modified as a result of experience.

**X** Behavior that occurs as a prelude to mating. One function is to override attack or escape behaviors.

**Y** Magnetic perception of direction. In some animals (e.g. pigeons), this consists of grains of magnetite within cells that align to the magnetic field of the Earth, helping them to navigate.

**Z** A directional growth response in plants either towards (positive) or away (negative) from a stimulus.

ISBN: 978-1-927173-12-1

Enduring Understanding

**3.E**

# Nervous Systems and **Responses**

## Key concepts

- Neurons are electrically excitable cells capable of transmitting impulses over a considerable distance.
- Action potentials are discrete all-or-nothing impulses.
- Action potentials are transmitted across chemical synapses by diffusion of a neurotransmitter.
- Sensory receptors act as biological transducers.
- Reflexes are central to simple behaviors.
- In muscle, the movement of actin filaments against myosin filaments creates contraction. ATP is required.

## Key terms

acetylcholine
actin
action potential
biological transducer
brain
cholinergic synapse
contraction
dopamine
electrochemical gradient
inhibition
integration
motor neuron
muscle tone
myelin
myelinated neuron
myofilament
myosin
neuromuscular junction
neuron
neurotransmitter
nodes of Ranvier
non-myelinated neuron
postsynaptic
presynaptic
reflex
refractory period
resting potential
saltatory conduction
Schwann cells
sense organ
sensory neuron
sensory receptor
serotonin
skeletal muscle
stimulus (pl. stimuli)
summation
synapse
synaptic integration

## Essential Knowledge

□ 1. Use the **KEY TERMS** to compile a glossary for this topic.

### Transmission of Nerve Impulses *(3.E.2: a-c)* pages 233-238

□ 2. Recognize **neurons** as the basic structural unit of nervous systems. Describe neurons as electrically excitable cells capable of responding to signal input.

□ 3. Using diagrams, describe the structure and functions of **sensory neurons** and **motor neurons**. Using a simple **reflex** as an example, explain how neurons enable detection, generation, transmission, and **integration** of signal information.

□ 4. Explain how the **resting potential** of a neuron is established and maintained. Include reference to the differential permeability of the membrane to $Na^+$ and $K^+$, and the generation of an **electrochemical gradient**.

□ 5. Explain how an **action potential** (nerve impulse) is generated. Explain the all-or-nothing nature of the impulse and the significance of the **refractory period** in producing discrete impulses. Interpret graphs of the voltage changes occurring during the generation and transmission of an action potential.

□ 6. Explain how an action potential is propagated along a myelinated nerve by **saltatory conduction**. Include reference to the voltage-gated ion channels and the roles of **myelin** and the **nodes of Ranvier**.

### System Integration and Response *(3.E.2: d)* pages 232, 239-262

□ 7. Describe the structure and function of **synapses** in nervous systems, including their role in unidirectionality, and **integration**. Describe transmission of an action potential across a **cholinergic synapse** by diffusion of **neurotransmitter**.

□ 8. Explain the role of neurotransmitters, including **dopamine** and **serotonin**, in behavior. Explain how **drugs** can mimic or block neurotransmitter activity.

□ 9. Describe regional specialization in the vertebrate **brain**. Compare the brain structure of different vertebrates and explain reasons for the differences.

□ 10. Describe the **stimuli** to which animals respond and describe diversity in the sense organs responsible for receiving stimuli in different taxa.

□ 11. Using examples, explain how sensory receptors receive and respond to stimuli. Explain the role of sense organs as **biological transducers**.

□ 12. Using an example, describe how nerves, muscles, and the skeleton interact to bring about movement.

□ 13. Describe the ultrastructure of skeletal muscle **fibers**, identifying the **sarcomere** and **myofibrils**, and the composition and arrangement of the **(myo)filaments**.

□ 14. Explain the sequence of events in **contraction** of skeletal muscle, from the arrival of an action potential at the motor end plate. Include reference to the role of **actin** and **myosin filaments**, ATP, the **sarcoplasmic reticulum**, and **calcium ions**.

□ 15. EXTENSION: Explain how muscles are equipped with sensory structures (e.g. **muscle spindle organ**) that provide feedback to regulate **muscle tone**.

**Periodicals:**
*Listings for this chapter are on page 375*

**Weblinks:**
*www.thebiozone.com/weblink/AP2-3121.html*

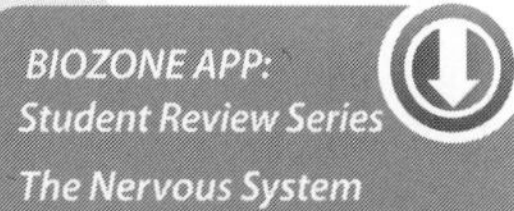

# Detecting Changing States

A **stimulus** is any physical or chemical change in the environment capable of provoking a response in an organism. Animals respond to stimuli in order to survive. This response is adaptive; it acts to maintain the organism's state of homeostasis. Stimuli may be either external (outside the organism) or internal (within its body). Some of the stimuli to which humans and other mammals respond are described below, together with the sense organs that detect and respond to these stimuli. Note that sensory receptors respond only to specific stimuli. The sense organs an animal possesses therefore determine how it perceives the world.

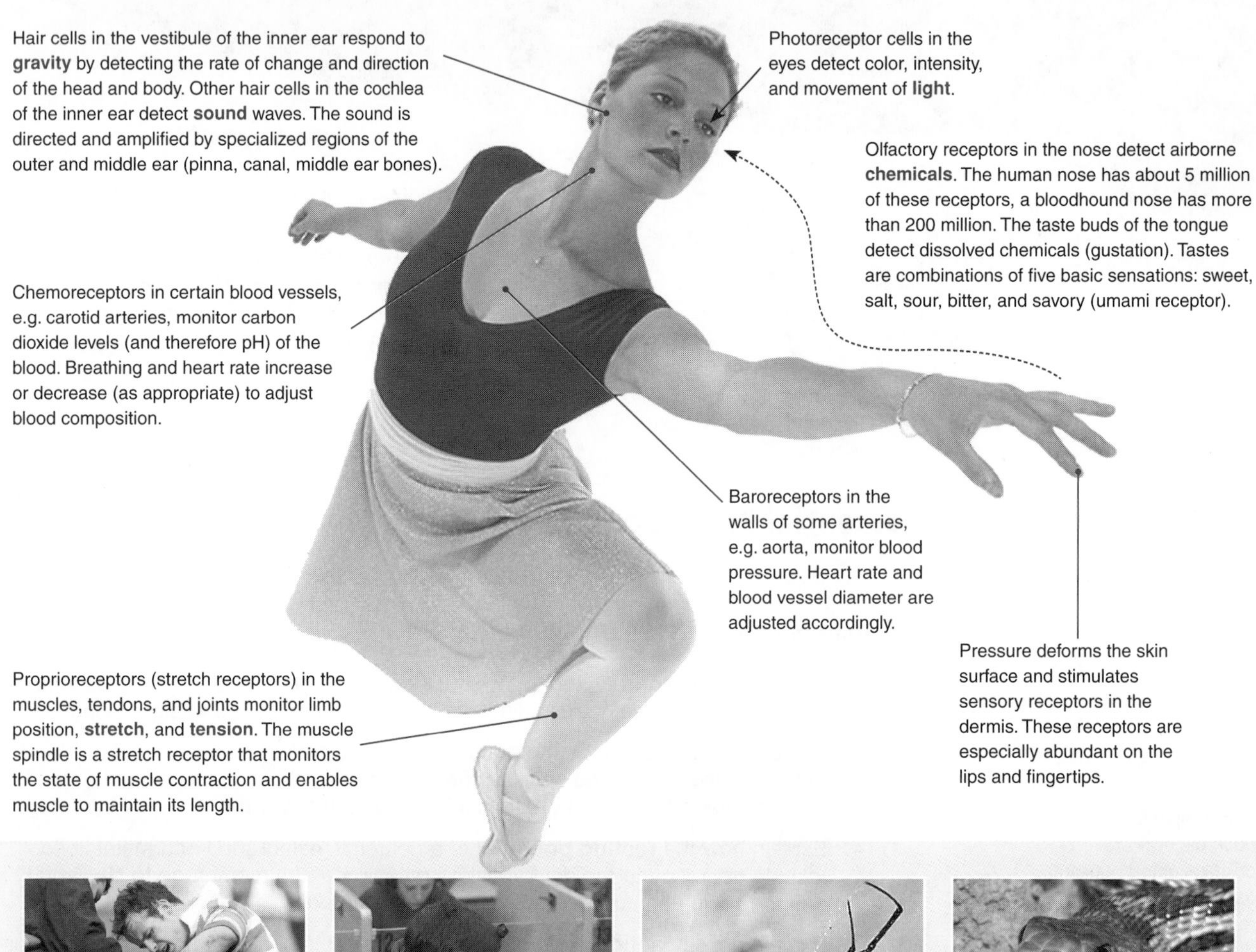

Hair cells in the vestibule of the inner ear respond to **gravity** by detecting the rate of change and direction of the head and body. Other hair cells in the cochlea of the inner ear detect **sound** waves. The sound is directed and amplified by specialized regions of the outer and middle ear (pinna, canal, middle ear bones).

Photoreceptor cells in the eyes detect color, intensity, and movement of **light**.

Olfactory receptors in the nose detect airborne **chemicals**. The human nose has about 5 million of these receptors, a bloodhound nose has more than 200 million. The taste buds of the tongue detect dissolved chemicals (gustation). Tastes are combinations of five basic sensations: sweet, salt, sour, bitter, and savory (umami receptor).

Chemoreceptors in certain blood vessels, e.g. carotid arteries, monitor carbon dioxide levels (and therefore pH) of the blood. Breathing and heart rate increase or decrease (as appropriate) to adjust blood composition.

Baroreceptors in the walls of some arteries, e.g. aorta, monitor blood pressure. Heart rate and blood vessel diameter are adjusted accordingly.

Pressure deforms the skin surface and stimulates sensory receptors in the dermis. These receptors are especially abundant on the lips and fingertips.

Proprioreceptors (stretch receptors) in the muscles, tendons, and joints monitor limb position, **stretch**, and **tension**. The muscle spindle is a stretch receptor that monitors the state of muscle contraction and enables muscle to maintain its length.

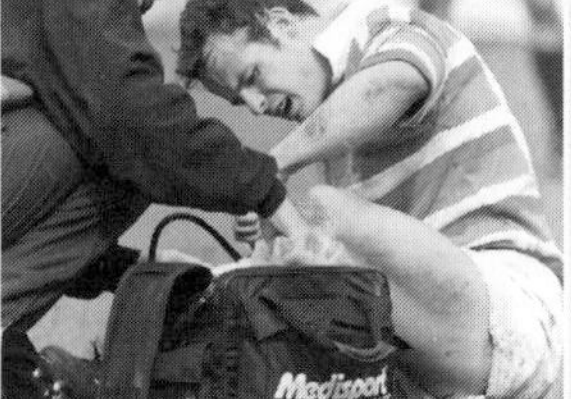


Pain and temperature are detected by simple nerve endings in the skin. Deep tissue injury is sometimes felt on the skin as referred pain.

Humans rely heavily on their hearing when learning to communicate; without it, speech and language development are more difficult.

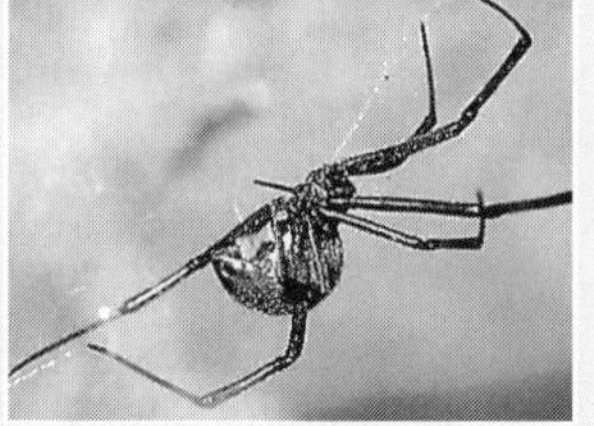

The vibration receptors in the limbs of arthropods are sensitive to movement: either sound or vibration (as caused by struggling prey).

The chemosensory Jacobson's organ in the roof of the mouth of reptiles (e.g. snakes) enables them to detect chemical stimuli.

Breathing and heart rates are regulated in response to sensory input from chemoreceptors.

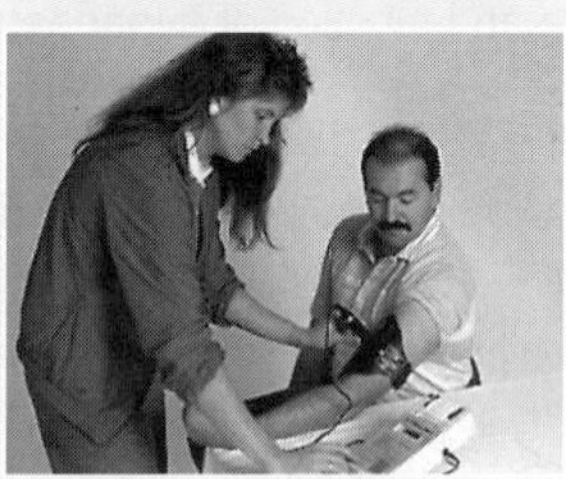

Baroreceptors and osmoreceptors act together to keep blood pressure and volume within narrow limits.

Many insects, such as these ants, rely on chemical sense for location of food and communication.

Jacobson's organ is also present in mammals and is used to detect sexual receptivity in potential mates.

1. Provide a concise definition of a stimulus: ______________________________

______________________________________________

2. (a) Name one external stimulus and its sensory receptor: ______________________

(b) Name one internal stimulus and its sensory receptor: ______________________




ISBN: 978-1-927173-12-1

# Nervous Regulation in Vertebrates

An essential feature of living organisms is their ability to coordinate their activities in response to environmental stimuli. The vertebrate plan is a good model for studying the basics of nervous regulation. Vertebrates detect and respond to environmental change through the nervous and endocrine systems. These two systems are quite different structurally, but interact to coordinate behavior and physiology. The **nervous system** is the body's control and communication center. It has three functions: to detect stimuli, interpret them, and initiate appropriate responses. It comprises millions of **neurons** (nerve cells), which are specialized to transmit information in the form of electrochemical impulses (action potentials). The nervous system forms a signaling network with branches carrying information to and from specific target tissues. Impulses can be transmitted rapidly over considerable distances and although it comprises millions of neural connections, its plan (below) is quite simple.

## Coordination by the Nervous System

The vertebrate nervous system consists of the **central nervous system** (brain and spinal cord), and the nerves and receptors outside it (**peripheral nervous system**). Sensory input to receptors comes via stimuli. Information about the effect of a response is provided by feedback mechanisms so that the system can be readjusted. The basic organization of the nervous system can be simplified into a few key components: the sensory receptors, a central nervous system processing point, and the effectors which bring about the response (below):

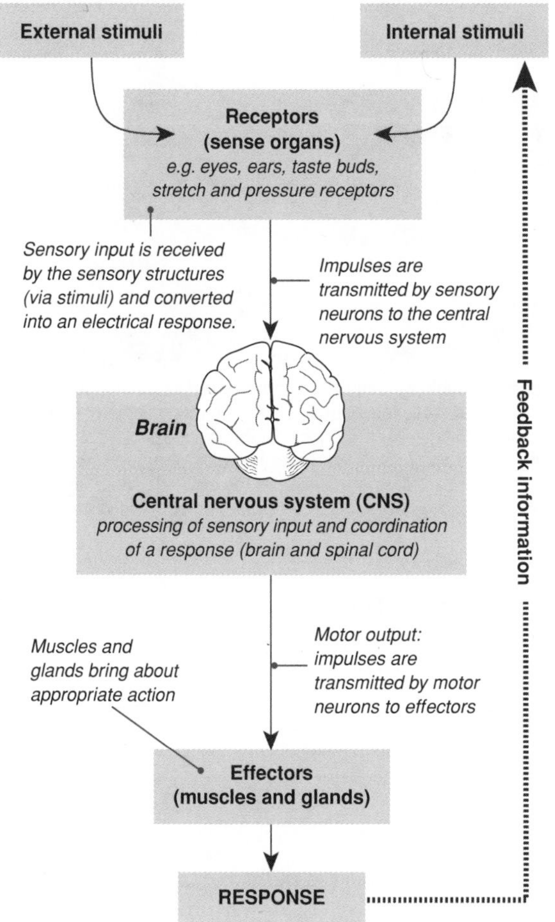

In the example above, the frisbee's approach is perceived by the eye. The motor cortex of the brain integrates the sensory message. Coordination of hand and body orientation is brought about through motor neurons to the muscles.

**Comparison of nervous and hormonal control**

| | Nervous control | Hormonal control |
|---|---|---|
| **Communication** | *Impulses across synapses* | *Hormones in the blood* |
| **Speed** | *Very rapid (within a few milliseconds)* | *Relatively slow (over minutes, hours, or longer)* |
| **Duration** | *Short term and reversible* | *Longer lasting effects* |
| **Target pathway** | *Specific (through nerves) to specific cells* | *Hormones broadcast to target cells everywhere* |
| **Action** | *Causes glands to secrete or muscles to contract* | *Causes changes in metabolic activity* |

1. Identify the three basic components of a nervous system and describe their role:

(a) ______________________________

(b) ______________________________

(c) ______________________________

2. Comment on the significance of the differences between the speed and duration of nervous and hormonal controls:

______________________________

______________________________

______________________________


ISBN: 978-1-927173-12-1


***Related activities**: The Mammalian Nervous System*
***Weblinks**: Nervous System Animation*

# Reflexes

A reflex is an automatic response to a stimulus involving a small number of neurons and a central nervous system (CNS) processing point (usually the spinal cord, but sometimes the brain stem). This type of circuit is called a **reflex arc**. Reflexes permit rapid responses to stimuli. They are classified according to the number of CNS synapses involved; **monosynaptic reflexes** involve only one CNS synapse (e.g. knee jerk reflex), **polysynaptic reflexes** involve two or more (e.g. pain withdrawal reflex). Both are spinal reflexes. The pupil reflex (opening and closure of the pupil) is an example of a cranial reflex.

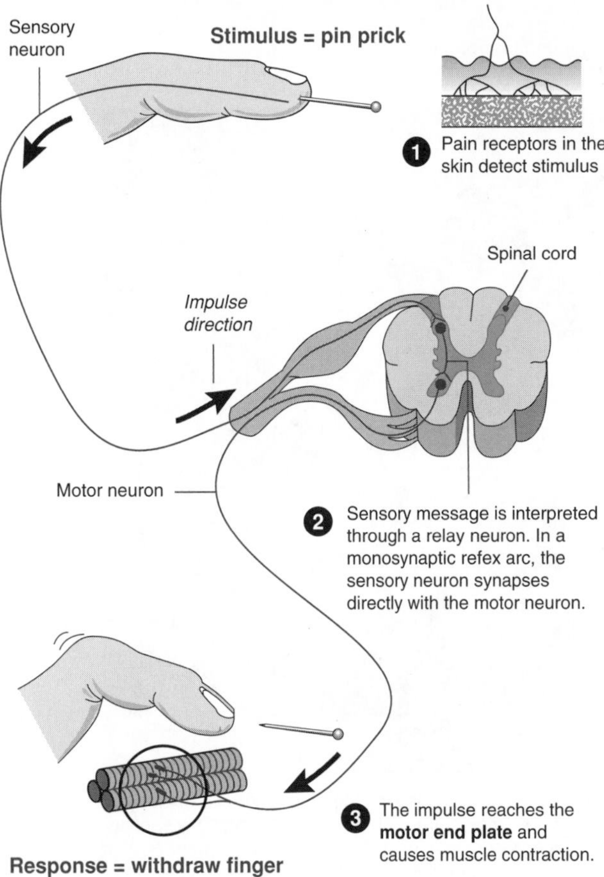


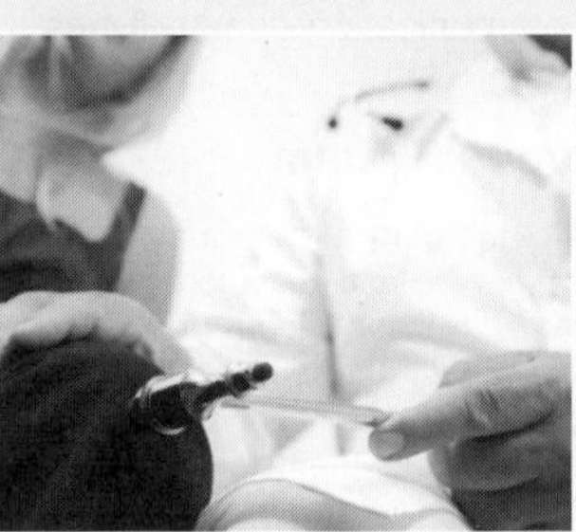

The patella (knee jerk) reflex is a simple deep tendon reflex used to test the function of the femoral nerve and spinal cord segments L2-L4. It helps to maintain posture and balance when walking.

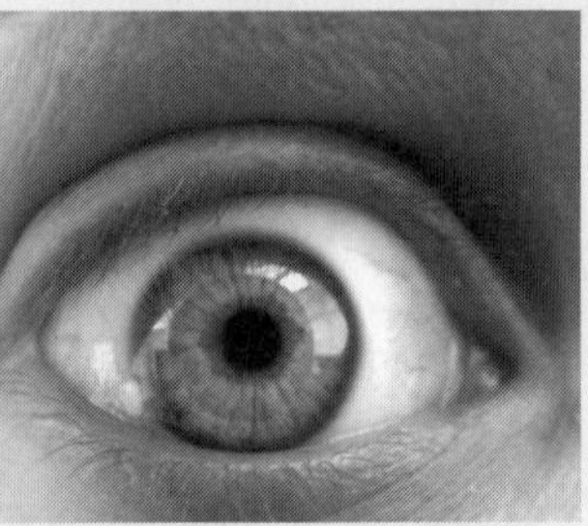

The pupillary light reflex refers to the rapid expansion or contraction of the pupils in response to the intensity of light falling on the retina. It is a polysynaptic cranial reflex and can be used to test for brain death.

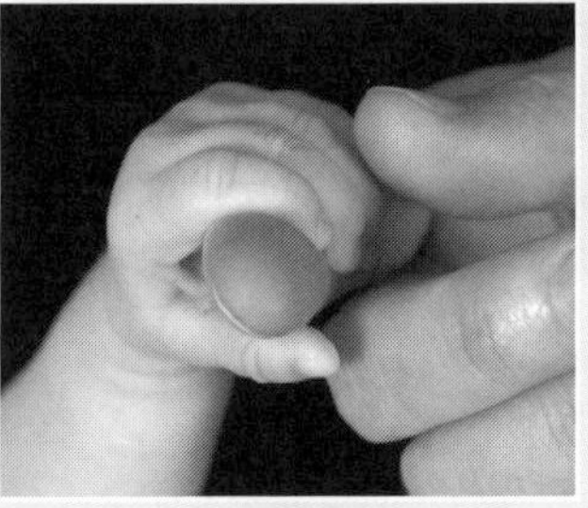

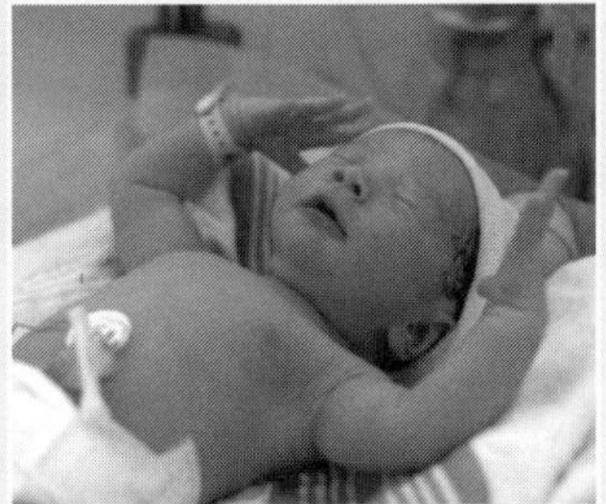

Normal newborns exhibit a number of primitive reflexes in response to particular stimuli. These reflexes disappear within a few months of birth as the child develops. Primitive reflexes include the grasp reflex (above left) and the startle or Moro reflex (above right) in which a sudden noise will cause the infant to throw out its arms, extend the legs and head, and cry. The rooting and sucking reflexes are other examples of primitive reflexes.

1. Why are reasoning and conscious thought not necessary or desirable features of reflex behaviours?

2. Distinguish between a spinal reflex and a cranial reflex and give an example of each:

3. (a) Distinguish between a monosynaptic and a polysynaptic reflex arc and give an example of each:

   (b) Which would produce the most rapid response, given similar length sensory and motor pathways? Explain:

4. (a) With reference to examples, describe the adaptive value of primitive reflexes in newborns:

   (b) Why are newborns tested for the presence of these reflexes?



**Related activities:** *Nervous Regulation in Vertebrates*


**ISBN: 978-1-927173-12-1**

# Neuron Structure and Function

As described earlier, homeostasis depends on the nervous system detecting, interpreting, and responding appropriately to both internal and external stimuli. Many of these responses are involuntary and are achieved through **reflexes**. Information, in the form of electrochemical impulses, is transmitted along nerve cells (**neurons**) from receptors to effectors. The speed of impulse conduction depends primarily on the axon diameter and whether or not the axon is **myelinated**. Within the tolerable physiological range, an increase in temperature also increases the speed of impulse conduction. In cool environments, impulses travel faster in endothermic than in ectothermic vertebrates. Neurons typically consist of a cell body, dendrites, and an axon (below). The principle behind increasing impulse speed through **saltatory conduction** is described overleaf.

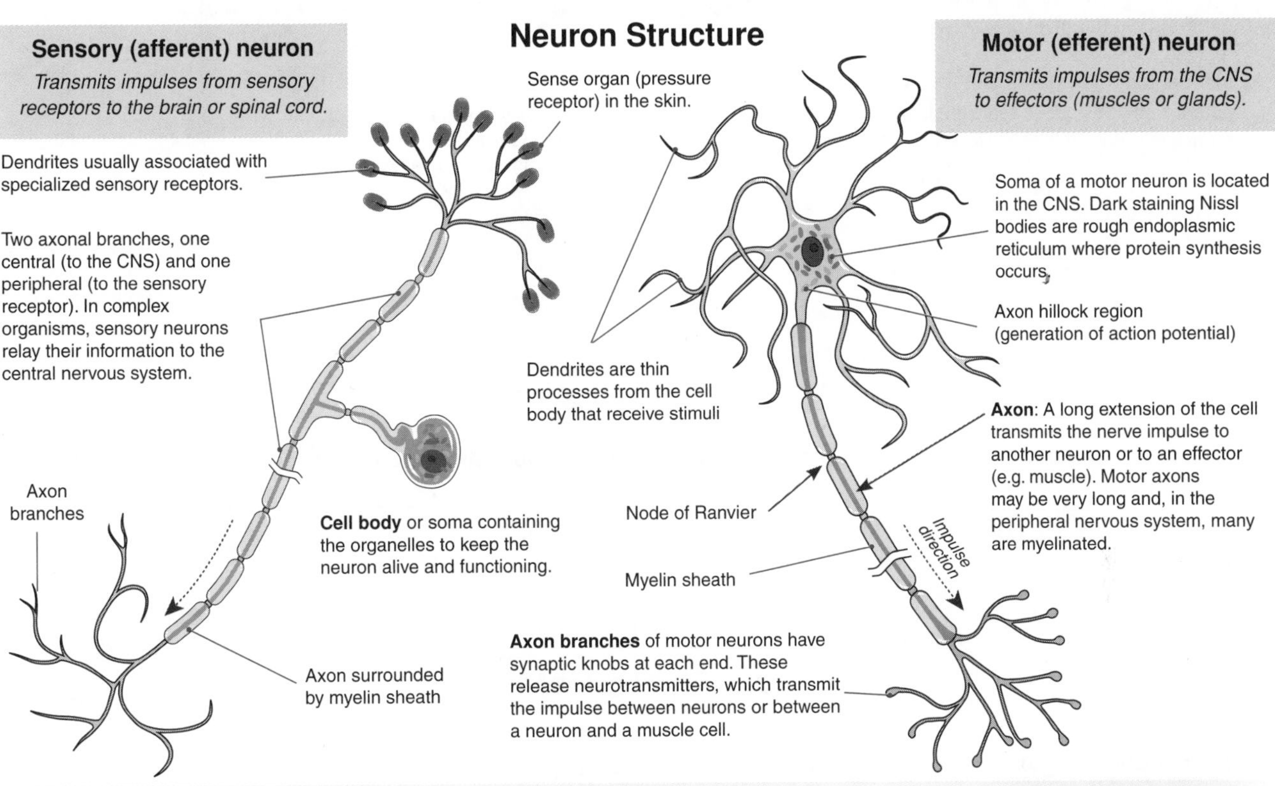


Where conduction speed is important, the axons of neurons are sheathed within a lipid and protein rich substance called **myelin**. Myelin is produced by **oligodendrocytes** in the central nervous system (CNS) and by **Schwann cells** in the peripheral nervous system (PNS). At intervals along the axons of myelinated neurons, there are gaps between neighboring Schwann cells and their sheaths. These are called **nodes of Ranvier**. Myelin acts as an insulator, increasing the speed at which nerve impulses travel because it prevents ion flow across the neuron membrane and forces the current to "jump" along the axon from node to node.

## Myelinated Neurons

Diameter: 1-25 μm

Conduction speed: 6-120 $ms^{-1}$

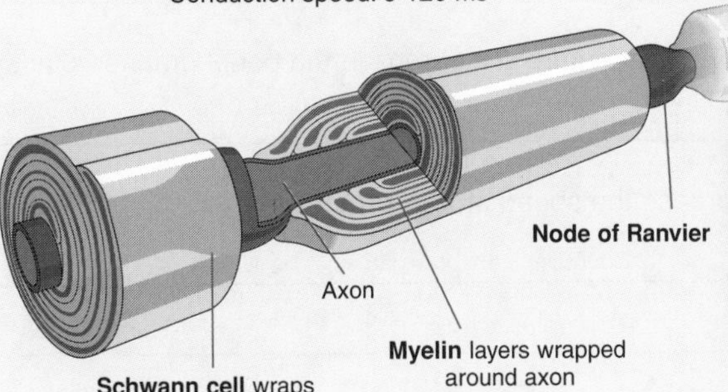


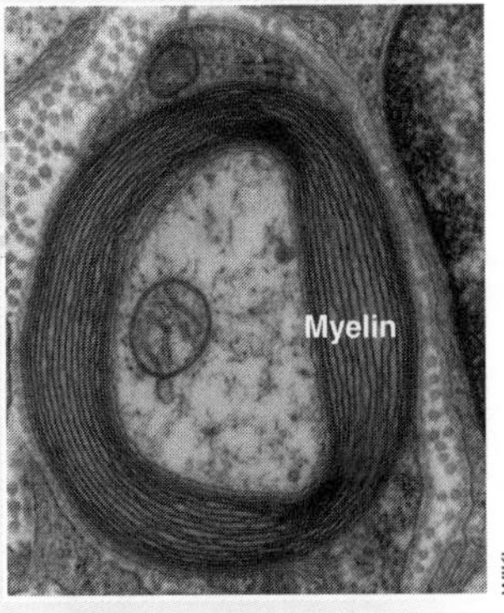


TEM cross section through a myelinated axon

**Non-myelinated axons** are relatively more common in the CNS where the distances travelled are less than in the PNS. Here, the axons are encased within the cytoplasmic extensions of oligodendrocytes or Schwann cells, rather than within a myelin sheath. The speed of impulse conduction is slower than in myelinated neurons because the nerve impulse is propagated along the entire axon membrane, rather than jumping from node to node as occurs in myelinated neurons. Conduction speeds are slower than in myelinated neurons, although they are faster in larger neurons (there is less ion leakage from a larger diameter axon).

## Non-myelinated Neurons

Diameter: <1 μm in vertebrates

Conduction speed: 0.2-0.5 $ms^{-1}$

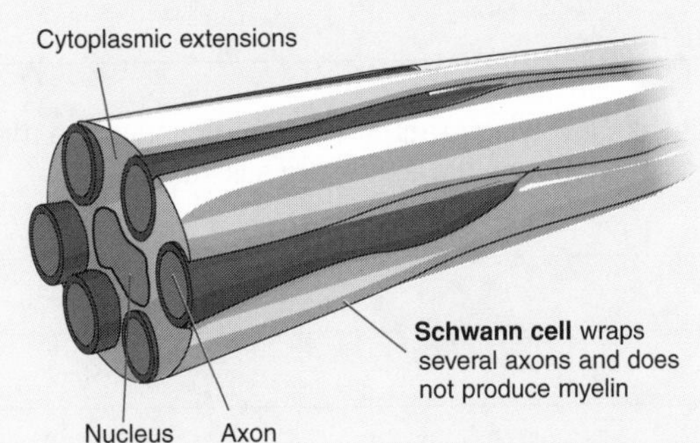


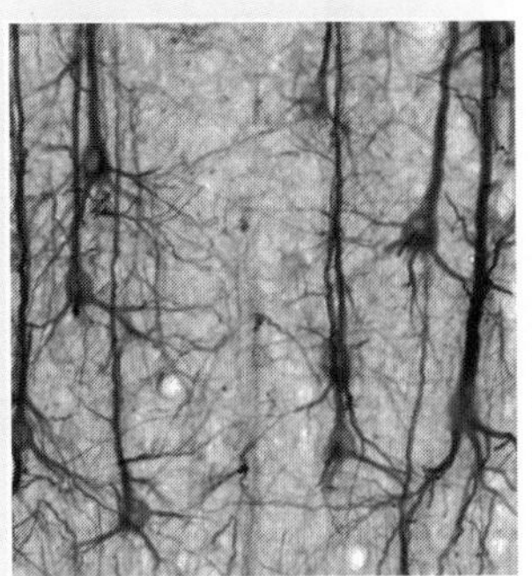


Unmyelinated pyramidal neurons of the cerebral cortex

ISBN: 978-1-927173-12-1


*Related activities:* Nervous Regulation in Vertebrates

*Weblinks:* Unipolar and Multipolar Neurons

Axon myelination is a characteristic feature of vertebrate nervous systems and it enables them to achieve very rapid speeds of nerve conduction. Myelinated neurons conduct impulses by **saltatory conduction**, a term that describes how the impulse jumps along the fiber. In saltatory conduction, only the nodes of Ranvier are involved in action potential generation. In myelinated (insulated) regions, there is no leakage of ions across the neuron membrane and the action potential at one node is sufficient to trigger an action potential in the next node.

Apart from increasing the speed of the nerve impulse, the myelin sheath helps in reducing energy expenditure because the area of depolarization is decreased (and therefore also the number of sodium and potassium ions that need to be pumped to restore the resting potential).

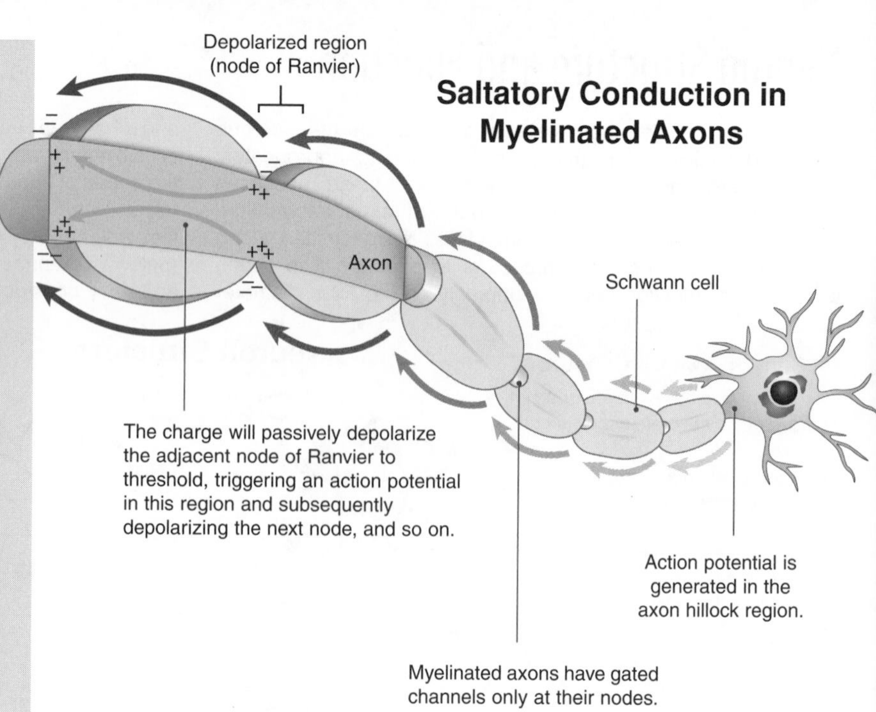


1. Complete the missing panels of the following table summarizing structural and functional differences between neurons:

| | Sensory neuron | Interneuron | Motor neuron |
|---|---|---|---|
| Structure | | Short dendrites, long or short axon | |
| Location | Dendrites outside the spinal cord, cell body in spinal ganglion | Entirely within the CNS | Dendrites and cell body in the spinal cord; axon outside the spinal cord. |
| Function | | Connect sensory & motor neurons | |

2. (a) What is the function of myelination in neurons?

   (b) What cell type is responsible for myelination in the CNS?

   (c) What cell type is responsible for myelination in the PNS?

   (d) Why is myelination typically a feature of neurons in the peripheral nervous system?

3. How does myelination increase the speed of nerve impulse conduction?

4. (a) Describe the adaptive advantage of faster conduction of nerve impulses:

   (b) Why does increasing the axon diameter also increase the speed of impulse conduction?

5. Multiple sclerosis (MS) is a disease involving progressive destruction of the myelin sheaths around axons. Why does MS impair nervous system function even though the axons are still intact?


ISBN: 978-1-927173-12-1

# Action Potentials

The plasma membranes of cells, including neurons, contain **sodium-potassium ion pumps** which actively pump sodium ions ($Na^+$) out of the cell and potassium ions ($K^+$) into the cell. The action of these ion pumps in neurons creates a separation of charge (a potential difference or voltage) either side of the membrane and makes the cells **electrically excitable**. It is this property that enables neurons to transmit electrical impulses. The **resting state** of a neuron, with a net negative charge inside, is maintained by the sodium-potassium pumps, which actively move two $K^+$ into the neuron for every three $Na^+$ moved out (below left). When a nerve is stimulated, a brief increase in membrane permeability to $Na^+$ temporarily reverses the membrane polarity (a depolarization). After the nerve impulse passes, the sodium-potassium pump restores the resting potential. The depolarization is propagated along the axon by local current in non-myelinated fibers and by **saltatory conduction** in myelinated fibers. Impulses pass from neuron to neuron by crossing junctions called **synapses**.

### The Resting Neuron

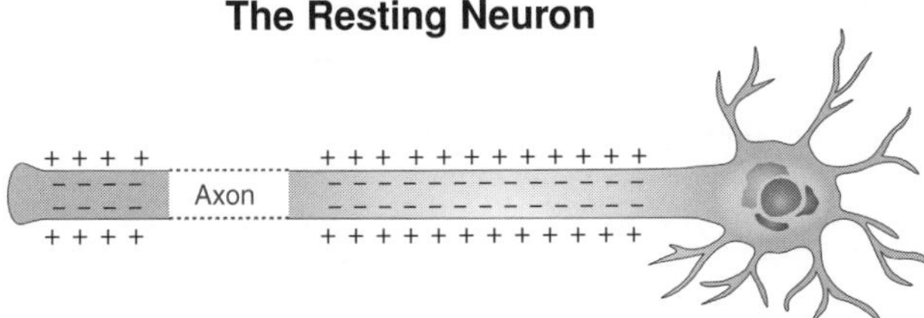


When a neuron is not transmitting an impulse, the inside of the cell is negatively charged relative to the outside and the cell is said to be electrically polarised. The potential difference (voltage) across the membrane is called the **resting potential**. For most nerve cells this is about -70 mV. Nerve transmission is possible because this membrane potential exists.

### The Nerve Impulse

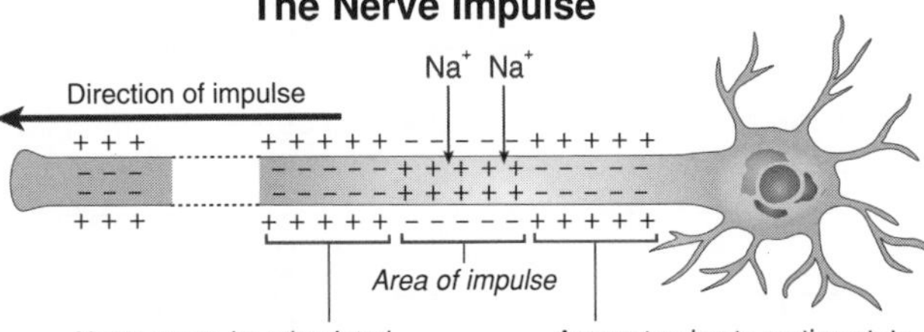


When a neuron is stimulated, the distribution of charges on each side of the membrane briefly reverses. This process of **depolarization** causes a burst of electrical activity to pass along the axon of the neuron as an **action potential**. As the charge reversal reaches one region, local currents depolarize the next region and the impulse spreads along the axon.

### The Action Potential

Membrane potential: +50 mV, 0, -50 mV (Threshold potential), -70 mV (Resting potential)

Elapsed time in milliseconds: 0, 1, 2, 3, 4, 5

The depolarization in an axon can be shown as a change in membrane potential (in millivolts). A stimulus must be strong enough to reach the **threshold potential** before an action potential is generated. This is the voltage at which the depolarization of the membrane becomes unstoppable.

The action potential is **all or nothing** in its generation and because of this, impulses (once generated) always reach threshold and move along the axon without attenuation. The resting potential is restored by the movement of potassium ions ($K^+$) out of the cell. During this **refractory period**, the nerve cannot respond, so nerve impulses are discrete.

Voltage-Gated Ion Channels and the Course of an Action Potential

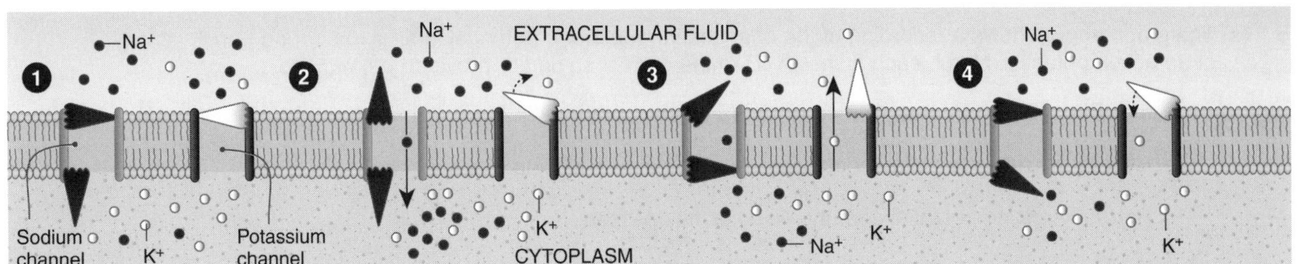


**Resting state**: Voltage activated $Na^+$ and $K^+$ channels are closed.

**Depolarization**: Voltage activated $Na^+$ channels open and there is a rapid influx of $Na^+$ ions. The interior of the neuron becomes positive relative to the outside.

**Repolarization**: Voltage activated $Na^+$ channels close and the $K^+$ channels open; $K^+$ moves out of the cell, restoring the negative charge to the cell interior.

**Returning to resting state**: Voltage activated $Na^+$ and $K^+$ channels close to return the neuron to the resting state.

ISBN: 978-1-927173-12-1


*Periodicals:* *Refractory period*

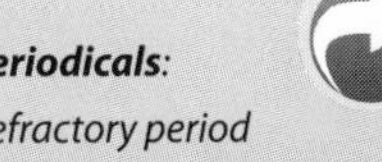

*Related activities: Chemical Synapses*
*Weblinks: Nerve Action Potential, Neurobiology*

1. (a) Describe a defining feature of neurons: ______________________

   ______________________

   (b) Explain how the supporting cells of nervous tissue (e.g. Schwann cells) differ from neurons: ______________________

   ______________________

2. Explain how an action potential is able to pass along a neuron: ______________________

   ______________________

   ______________________

   ______________________

3. Explain how the refractory period influences the direction in which an impulse will travel: ______________________

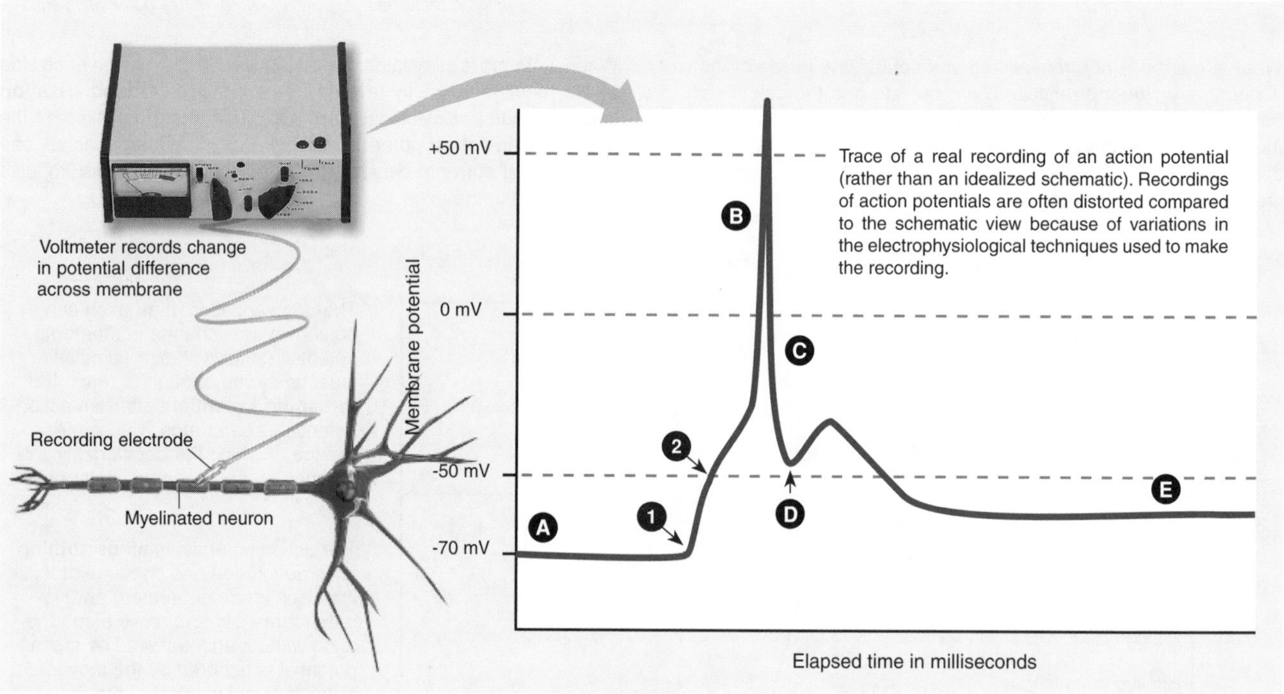

   ______________________

   ______________________

Voltmeter records change in potential difference across membrane

Recording electrode

Myelinated neuron

Membrane potential

+50 mV

0 mV

-50 mV

-70 mV

A B C D E 1 2

Trace of a real recording of an action potential (rather than an idealized schematic). Recordings of action potentials are often distorted compared to the schematic view because of variations in the electrophysiological techniques used to make the recording.

Elapsed time in milliseconds

4. Action potentials themselves are indistinguishable from each other. Explain how the nervous system is able to interpret the impulses correctly and bring about an appropriate response:

   ______________________

   ______________________

   ______________________

5. (a) The graph above shows a recording of the changes in membrane potential in an axon during transmission of an action potential. Match each stage (**A-E**) to the correct summary provided below.

   - [ ] Membrane depolarization (due to rapid $Na^+$ entry across the axon membrane.
   - [ ] Hyperpolarization (an overshoot caused by the delay in closing of the $K^+$ channels.
   - [ ] Return to resting potential after the stimulus has passed.
   - [ ] Repolarization as the $Na^+$ channels close and slower $K^+$ channels begin to open.
   - [ ] The membrane's resting potential.

   (b) Explain what is happening at point **1** on the graph: ______________________

   (c) Explain what is happening at point **2** on the graph: ______________________

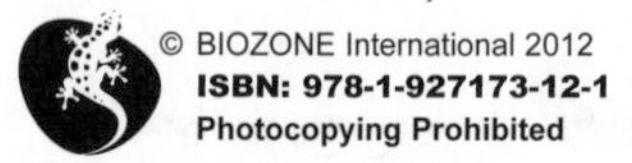


**ISBN: 978-1-927173-12-1**

# Chemical Synapses

Action potentials are transmitted between neurons across synapses: junctions between the end of one axon and the dendrite or cell body of a receiving neuron. **Chemical synapses** are the most widespread type of synapse in nervous systems. The axon terminal is a swollen knob, and a small gap separates it from the receiving neuron. The synaptic knobs are filled with tiny packets of chemicals called **neurotransmitters**. Transmission involves the diffusion of the neurotransmitter across the gap, where it interacts with the receiving membrane and causes an electrical response. The response of a receiving cell to the arrival of a neurotransmitter depends on the nature of the cell itself, on its location in the nervous system, and on the neurotransmitter involved. Synapses that release acetylcholine (ACh) are termed **cholinergic**. In the example below, ACh causes membrane depolarization and the generation of an action potential (termed excitation or an excitatory response).

## A Cholinergic Synapse

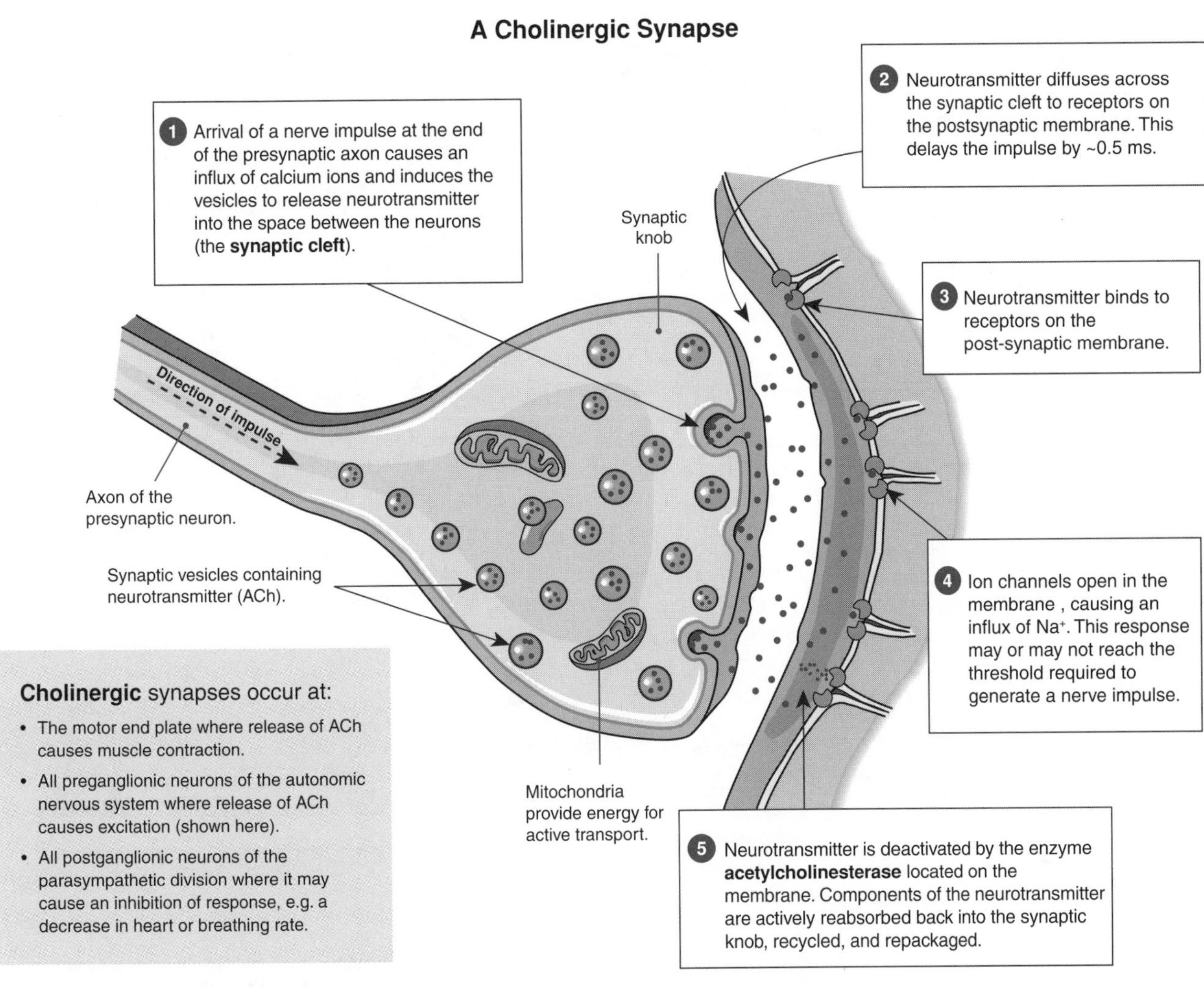


1. Explain what is meant by a **synapse**: ______

2. Explain what causes the release of neurotransmitter into the synaptic cleft: ______

3. State why there is a brief delay in transmission of an impulse across the synapse: ______

4. (a) State how the neurotransmitter is deactivated: ______

   (b) Explain why it is important for the neurotransmitter substance to be deactivated soon after its release: ______

5. Consult a reference source to identify one function of acetylcholine in the nervous system: ______

6. Suggest one factor that might influence the strength of the response in the receiving cell: ______


ISBN: 978-1-927173-12-1

# Encoding Information

Sensory receptors are specialized to detect stimuli and respond by producing an electrical discharge. In this way they act as **biological transducers**, converting the energy from a stimulus into an electrochemical signal. Stimulation of a sensory receptor cell results in an electrical impulse with specific properties. The frequency of impulses produced by the receptor cell encodes information about the strength of the stimulus; a stronger stimulus produces more frequent impulses. Sensory receptors also show **sensory adaptation** and will cease responding to a stimulus of the same intensity. The simplest sensory receptors consist of a single sensory neuron (e.g. free nerve endings). More complex sense cells form synapses with their sensory neurons (e.g. taste buds). Sensory receptors are classified according to the stimuli to which they respond (for example, photoreceptors respond to light). The response of a simple **mechanoreceptor**, the Pacinian corpuscle, to a stimulus (pressure) is described below.

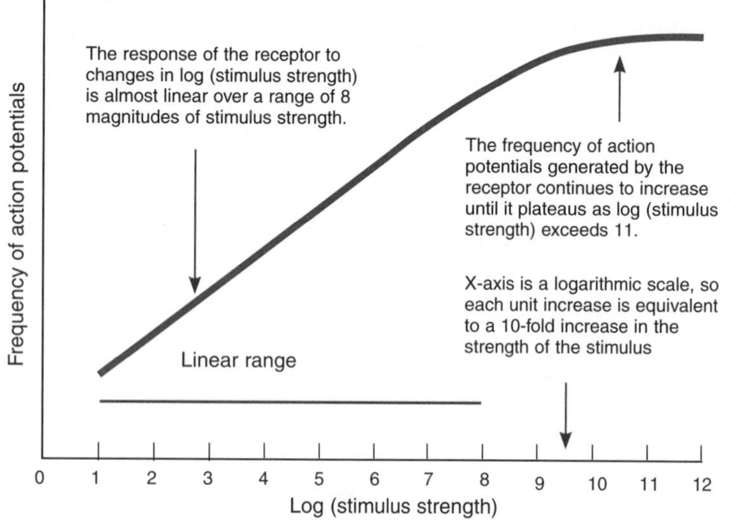


Receptors can use variation in action potential frequency to encode stimulus strengths that vary by nearly 11 orders of magnitude.

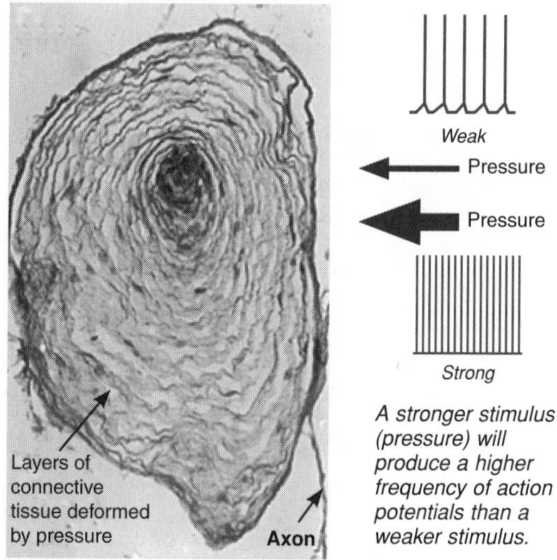


A Pacinian corpuscle (above), illustrating the many layers of connective tissue. Pacinian corpuscles are rapidly adapting receptors; they fire at the beginning and end of a stimulus, but do not respond to unchanging pressure.

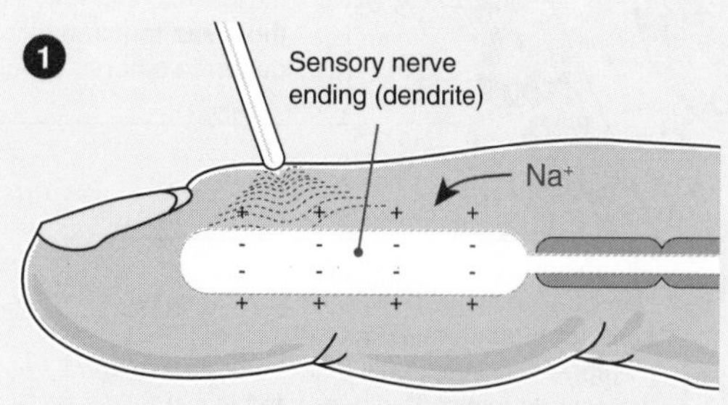


Deforming the corpuscle leads to an increase in the permeability of the nerve to sodium. $Na^+$ diffuses into the nerve ending creating a localized depolarization. This depolarization is called a **generator potential**.

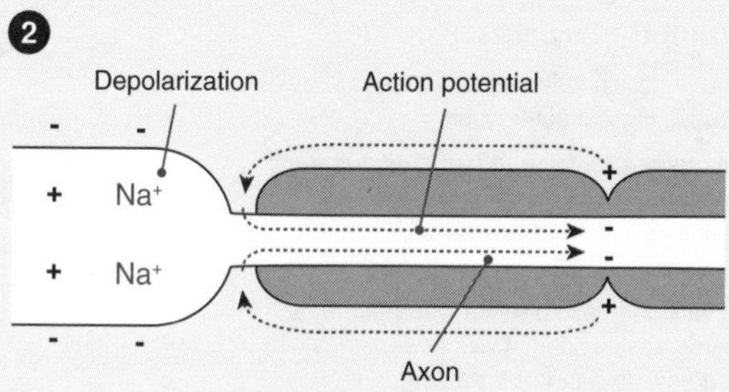


A volley of **action potentials** is triggered once the generator potential reaches or exceeds a **threshold value**. These action potentials are conducted along the sensory axon. A strong stimulus results in a high frequency of impulses.

1. Explain why sensory receptors are termed '**biological transducers**': ____________________

2. Explain the significance of linking the magnitude of a sensory response to stimulus intensity: ____________________

3. Explain the physiological importance of sensory adaptation: ____________________

4. Suggest why a simple mechanoreceptor, such as the Pacinian corpuscle, does not fire action potentials unless a stimulus of threshold value is reached:

____________________



***Related activities:*** *Action Potentials*
***Weblinks:*** *Neuron Information Coding and Transfer*

***Periodicals:***
*Infinite sensation*

**ISBN: 978-1-927173-12-1**

# Integration at Synapses

Synapses play a pivotal role in the ability of the nervous system to respond appropriately to stimulation and to adapt to change. The nature of synaptic transmission allows the **integration** (interpretation and coordination) of inputs from many sources. These inputs need not be just excitatory (causing depolarization). Inhibition results when the neurotransmitter released causes negative chloride ions (rather than sodium ions) to enter the postsynaptic neuron. The postsynaptic neuron then becomes more negative inside (hyperpolarized) and an action potential is less likely to be generated. At synapses, it is the sum of **all** inputs (excitatory and inhibitory) that leads to the final response in a postsynaptic cell. Integration at synapses makes possible the various responses we have to stimuli. It is also the most probable mechanism by which learning and memory are achieved.

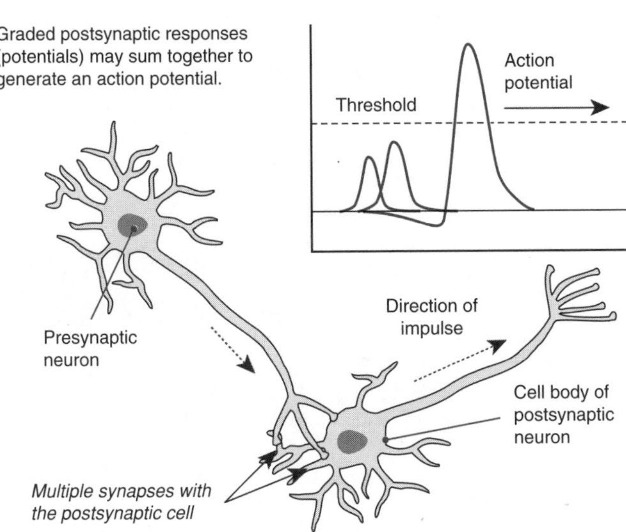


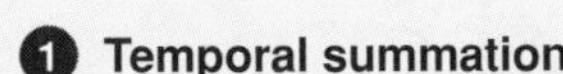


## Synapses and Summation

Nerve transmission across chemical synapses has several advantages, despite the delay caused by neurotransmitter diffusion. Chemical synapses transmit impulses in one direction to a precise location and, because they rely on a limited supply of neurotransmitter, they are subject to fatigue (inability to respond to repeated stimulation). This protects the system against overstimulation.

Synapses also act as centers for the **integration** of inputs from many sources. The response of a postsynaptic cell is sometimes graded and not strong enough on its own to generate an action potential. However, because the strength of the response is related to the amount of neurotransmitter released, subthreshold responses can sum to produce a response in the post-synaptic cell. This additive effect is termed **summation**. Summation (below) can be **temporal** (in time) or **spatial** (in space). A neuromuscular junction (photo below) is a specialized form of synapse between a motor neuron and a skeletal muscle fiber. Functionally, it is similar to any excitatory cholinergic synapse.

### 1 Temporal summation

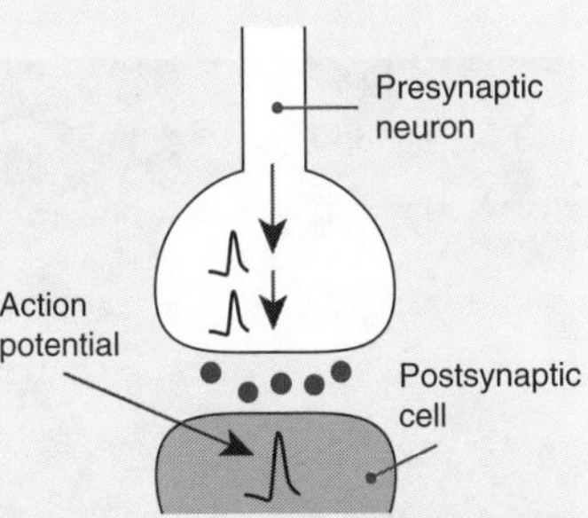


Several impulses may arrive at the synapse in quick succession from a single axon. The individual responses are so close together in time that they sum to reach threshold and produce an action potential in the postsynaptic neuron.

### 2 Spatial summation

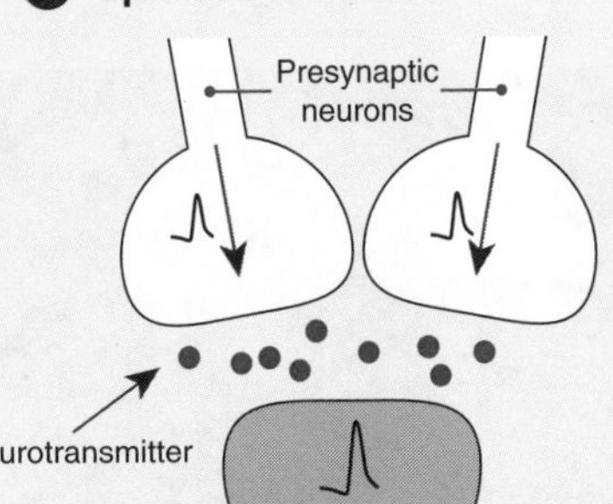


Individual impulses from spatially separated axon terminals may arrive **simultaneously** at different regions of the same postsynaptic neuron. The responses from the different places sum to reach threshold and produce an action potential.

### 3 Neuromuscular junction

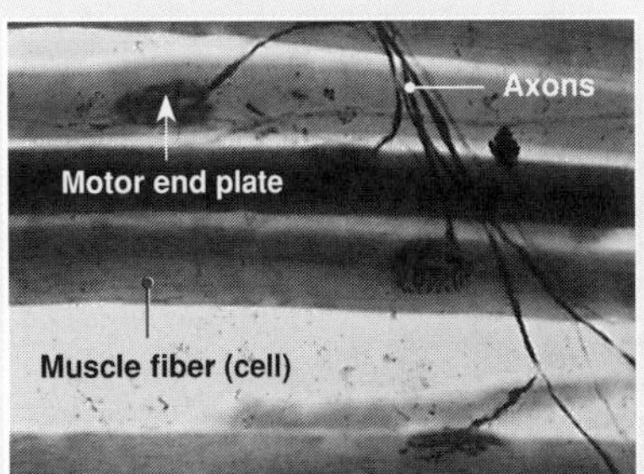


The arrival of an impulse at the neuromuscular junction causes the release of acetylcholine from the synaptic knobs. This causes the muscle cell membrane (sarcolemma) to depolarize, and an action potential is generated in the muscle cell.

1. Explain the purpose of nervous system integration: ______________________

2. (a) Explain what is meant by **summation**: ______________________

   (b) In simple terms, distinguish between temporal and spatial summation: ______________________

3. Describe two ways in which a neuromuscular junction is similar to any excitatory cholinergic synapse:

   (a) ______________________

   (b) ______________________


ISBN: 978-1-927173-12-1


*Related activities: Chemical Synapses*

# Drugs at Synapses

Synapses in the peripheral nervous system are classified by the type of neurotransmitter they release. **Cholinergic** synapses release **acetylcholine (Ach)**, while adrenergic synapses release **epinephrine** (adrenaline) or norepinephrine (noradrenaline). Postsynaptic receptors are also classified by the type of neurotransmitters they bind. Cholinergic receptors all bind acetylcholine, but they can also bind **drugs** that mimic Ach. Drugs act on the nervous system by mimicking (**agonists**) or blocking (**antagonists**) the activity of neurotransmitters. Because of the small amounts of chemicals involved in synaptic transmission, drugs that affect the activity of neurotransmitters, or their binding sites, can have powerful effects in small doses.

## Drugs at Cholinergic Synapses

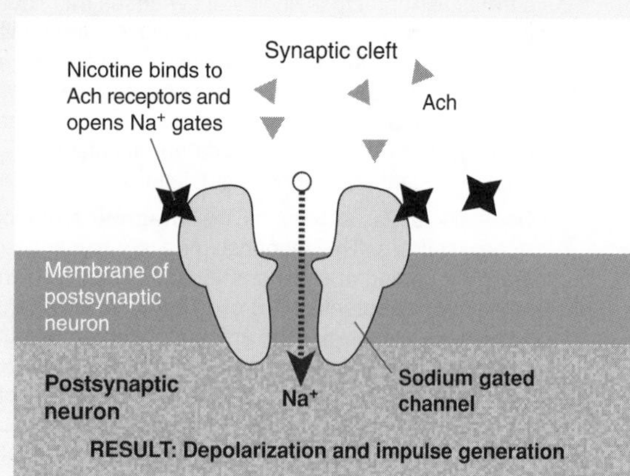


**Nicotine** acts as a **direct agonist** at nicotinic synapses. Nicotine binds to and activates acetylcholine (Ach) receptors on the postsynaptic membrane. This opens sodium gates, leading to a sodium influx and membrane depolarization. Some agonists work indirectly by preventing Ach breakdown. Such drugs are used to treat elderly patients with Alzheimer's.

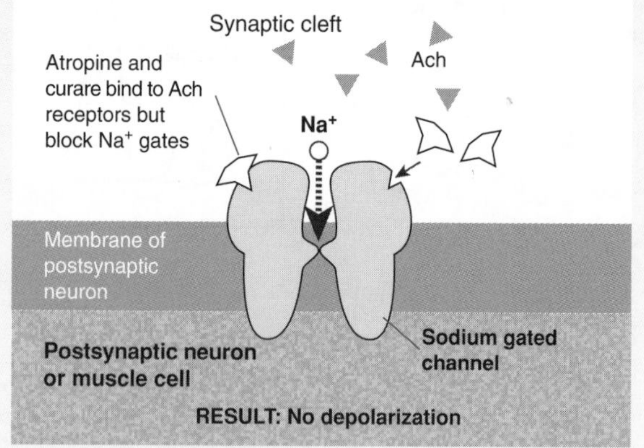


**Atropine** and **curare** act as **antagonists** at some cholinergic synapses. These molecules compete with Ach for binding sites on the postsynaptic membrane, and block sodium influx so that impulses are not generated. If the postsynaptic cell is a muscle cell, muscle contraction is prevented. In the case of curare, this causes death by flaccid paralysis.

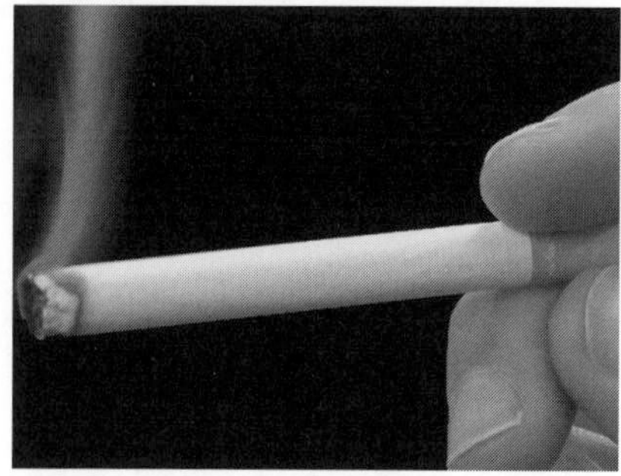

**Nicotine** is the highly addictive substance in cigarettes. It acts on the nicotinic acetylcholine receptors, increasing the levels of several neurotransmitters, including dopamine. Dopamine produces feelings of euphoria and relaxation. These feelings reinforce nicotine consumption, and create nicotine addiction.

Fly agaric (*Amanita muscaria*)

**Muscarine**, a compound found in several types of mushrooms, binds to muscarinic acetylcholine receptors. Muscarine is used to treat a number of medical conditions (e.g. glaucoma), but consumption of the mushrooms can deliver a fatal overdose of muscarine.

Eastern green mamba

Mamba snake venom contains a number of neurotoxins including **dendrotoxins**. These small peptide molecules act as acetylcholine receptor antagonists (blocking muscarinic receptors). They have many effects including disrupting muscle contraction.

1. Providing an example of each, outline two ways in which drugs can act at a cholinergic synapse:

   (a) ______

   (b) ______

2. Explain why atropine and curare are described as direct antagonists: ______

3. Suggest why curare (carefully administered) is used during abdominal surgery: ______



***Related activities:*** *Chemical Synapses*
***Weblinks:*** *The Science of Addiction*


**ISBN: 978-1-927173-12-1**

# The Vertebrate Brain

The vertebrate brain develops as an expansion of the anterior end of the neural tube in embryos. The forebrain, midbrain, and hindbrain can be seen very early in development, with further differentiation as development continues. The brains of fish and amphibians are relatively unspecialized, with a rudimentary cerebrum and cerebellum. The reptiles show the first real expansion of the cerebrum, with the gray matter external in the cortex. In the birds and mammals, the brain is relatively large, with well developed cerebral and cerebellar regions. In primitive vertebrates, the cerebral regions act primarily as olfactory centres. In higher vertebrates, the cerebrum takes over the many of the functions of other regions of the brain (e.g. the optic lobes), becoming the primary integration center of the brain. The cerebellum also becomes more important as locomotor and other muscular activities increase in complexity. The relative sizes of different regions of vertebrate brains are shown below.

## Vertebrate Brains

All vertebrates, from fish and amphibians, to humans and other mammals, have brains with the same basic structure. The brain develops from a hollow tube and comprises the forebrain, midbrain, and hindbrain (which runs into the spinal cord). During the course of vertebrate evolution some parts of the brain (e.g. the medulla) have remained largely unchanged, retaining their primitive functions. Other parts (e.g. the cerebrum of the forebrain) have expanded and taken on new functions.

**Key to Brain Regions and their Functions**

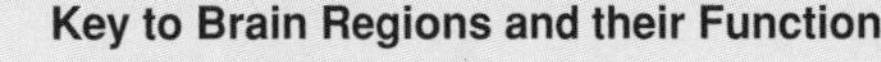

- **Olfactory bulb**: Receives and processes olfactory signals.
- **Cerebrum**: Behavior, complex thought and reasoning.
- **Cerebellum**: Center for controlling movement and balance.
- **Medulla**: Reflex functions and relay for sensory information.
- **Optic lobe**: Receives and processes visual information.

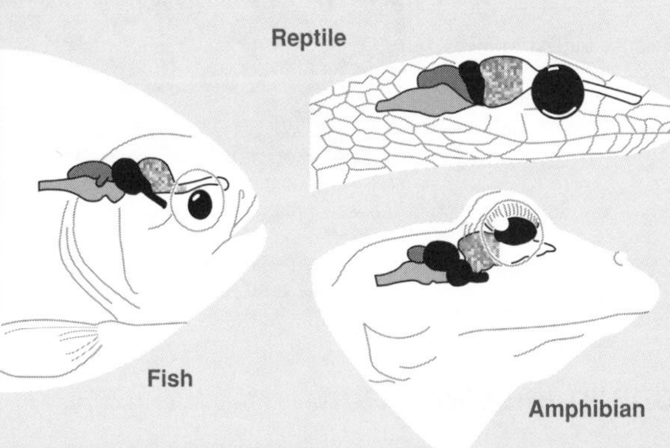


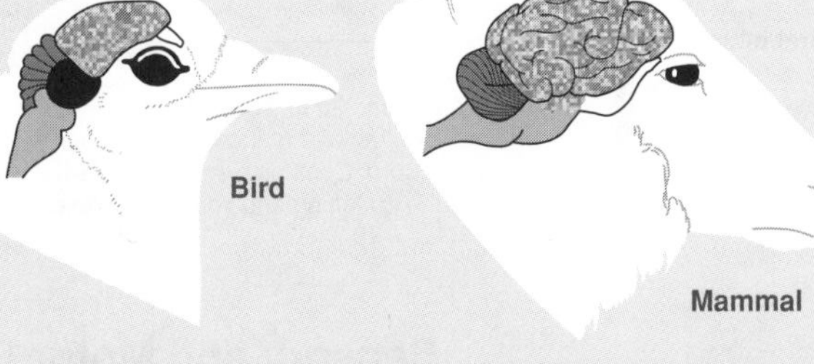


The medulla (part of the hindbrain) in fish, amphibians, and reptiles is prominent. It relays sensory information to or from parts of the brain associated with sensory processing. In these groups both the optic lobe (visual processing) and the centers associated with processing olfactory information (in the thalamus) are also large.

In vertebrates other than mammals, a major part of the midbrain is associated with the analysis of vision, and the optic lobes are very large. In mammals, the analysis of vision is a function of the forebrain. The forebrain itself has changed dramatically in size during vertebrate evolution. In birds most of the cerebrum is associated with complex behavior. In mammals, there has been an progressive increase in the size and importance of the cerebrum; particularly the parts associated with complex thought, reasoning, and communication.

1. (a) Describe one major difference between the brain structure of mammals and other vertebrates:

   (b) Suggest what brain structure can tell us about the sensory perception of an animal:

2. Discuss the trends in brain development during the course of vertebrate evolution and relate these to changes to changes in lifestyle and behavior:


ISBN: 978-1-927173-12-1


***Related activities:*** *The Human Brain*

# The Human Brain

The brain is one the largest organs in the body. It is protected by the skull, the **meninges**, and the **cerebrospinal fluid** (CSF). The brain is the body's control center. It receives a constant flow of sensory information, but responds only to what is important at the time. Some responses are very simple (e.g. cranial reflexes), whilst others require many levels of processing. The human brain is noted for its large, well developed cerebral region, with its prominent folds (**gyri**) and grooves (**sulci**). Each cerebral hemisphere has an outer region of gray matter and an inner region of white matter, and is divided into four lobes by deep sulci or fissures. These lobes: temporal, frontal, occipital, and parietal, correspond to the bones of the skull under which they lie.

## Primary Structural Regions of the Brain

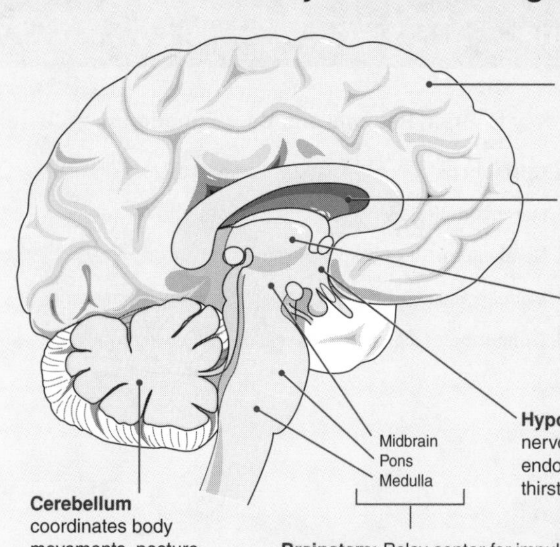


**Cerebrum:** Divided into two cerebral hemispheres. Many, complex roles. It contains sensory, motor, and association areas, and is involved in memory, emotion, language, reasoning, and sensory processing.

**Ventricles:** Cavities containing the CSF, which absorbs shocks and delivers nutritive substances.

**Thalamus** is the main relay center for all sensory messages that enter the brain, before they are transmitted to the cerebrum.

**Hypothalamus** controls the autonomic nervous system and links nervous and endocrine systems. Regulates appetite, thirst, body temperature, and sleep.

**Cerebellum** coordinates body movements, posture, and balance.

**Brainstem:** Relay center for impulses between the rest of the brain and the spinal cord. Controls breathing, heartbeat, and the coughing and vomiting reflexes.

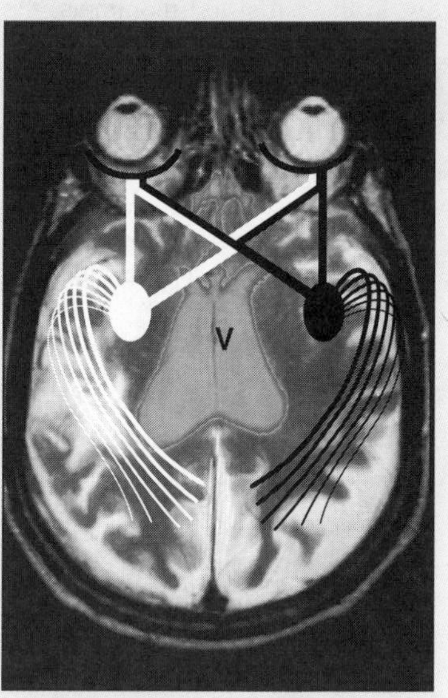


MRI scan of the brain viewed from above. The visual pathway has been superimposed on the image. Note the crossing of some sensory neurons to the opposite hemisphere and the fluid filled ventricles (V) in the center.

## Sensory and Motor Regions in the Cerebrum

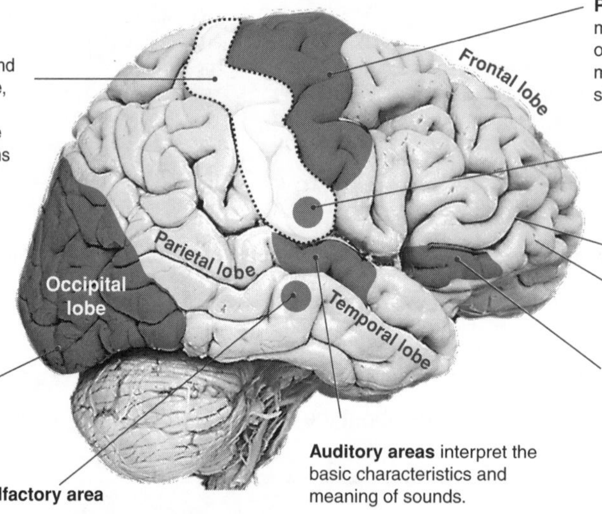


**Primary somatic sensory area** receives sensations from receptors in the skin, muscles and viscera, allowing recognition of pain, temperature, or touch. Sensory information from receptors on one side of the body crosses to the opposite side of the cerebral cortex where conscious sensations are produced. The size of the sensory region for different body parts depends on the number of receptors in that particular body part

**Visual areas** within the occipital lobe receive, interpret, and evaluate visual stimuli. In vision, each eye views both sides of the visual field but the brain receives impulses from left and right visual fields separately (see photo caption above). The visual cortex combines the images into a single impression or perception of the image.

**Olfactory area**

**Primary motor area** controls muscle movement. Stimulation of a point one side of the motor area results in muscular contraction on the opposite side of the body.

**Primary gustatory area** interprets sensations related to taste.

Sulci (grooves)

Gyri (elevated folds)

**Language areas**: The motor speech area (Broca's area) is concerned with speech production. The sensory speech area (Wernicke's area) is concerned with speech recognition and coherence.

**Auditory areas** interpret the basic characteristics and meaning of sounds.

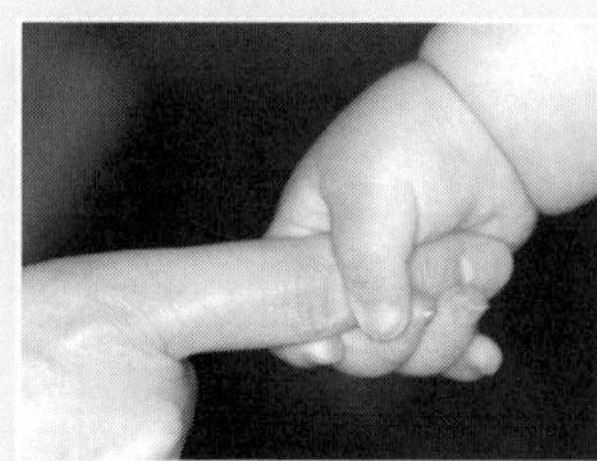

Touch is interpreted in the primary somatic sensory area. The fingertips and the lips have a relatively large amount of area devoted to them.

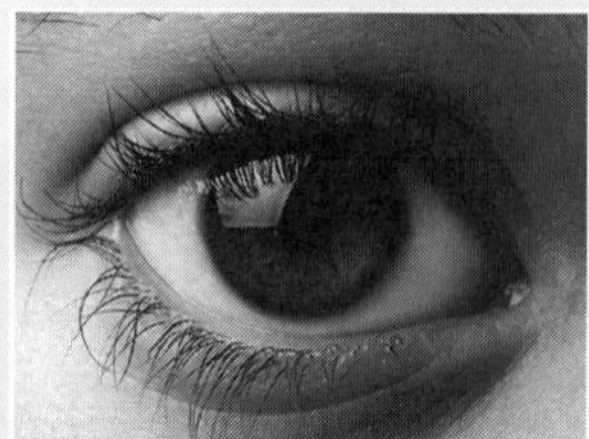

Humans rely heavily on vision. The importance of this **special sense** in humans is indicated by the large occipital region of the brain.

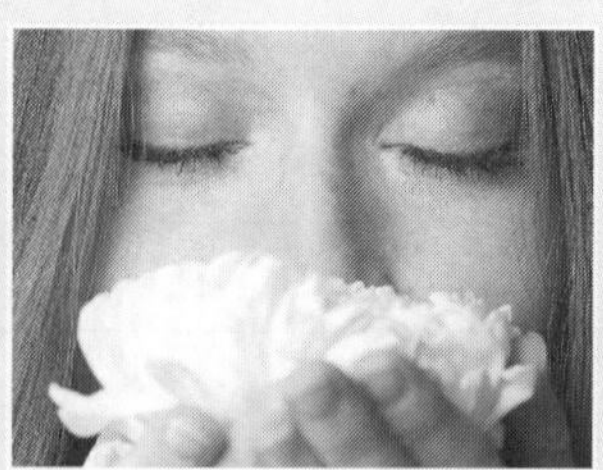

The olfactory tract connects the olfactory bulb with the cerebral hemispheres where olfactory information is interpreted.

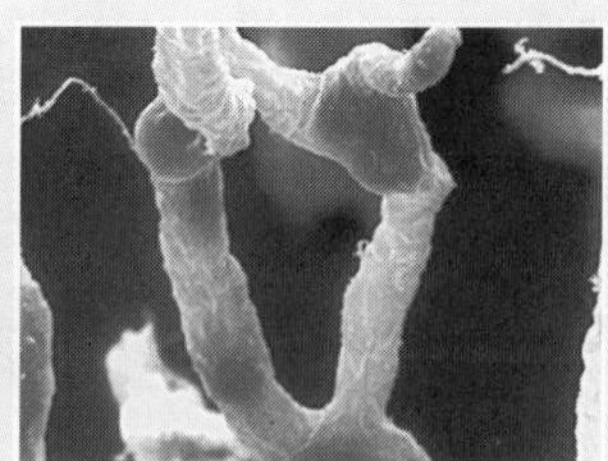


The endothelial tight junctions of the capillaries supplying the brain form a protective **blood-brain barrier** against toxins and infection.



***Related activities:*** *The Mammalian Nervous System, Imaging the Brain*
***Weblinks:*** *Inside the Brain: An Interactive Tour*

**ISBN: 978-1-927173-12-1**

# The Ventricles and CSF

The delicate nervous tissue of the brain and spinal cord is protected against damage by the **bone** of the skull and vertebral column, the membranes overlying the brain (the **meninges**), and the watery but nutritive **cerebrospinal fluid** (CSF), which lies between the inner two of the meningeal layers.

The meninges are collectively three membranes: a tough double-layered outer **dura mater**, a web-like middle **arachnoid mater**, and an inner delicate **pia mater** that adheres to the surface of the brain. The CSF is formed from the blood by clusters of capillaries on the roof of each of the brain's ventricles (choroid plexuses). The CSF is constantly circulated through the ventricles of the brain (and into the spinal cord), returning to the blood via specialized projections of the middle meningeal layer (the arachnoid).

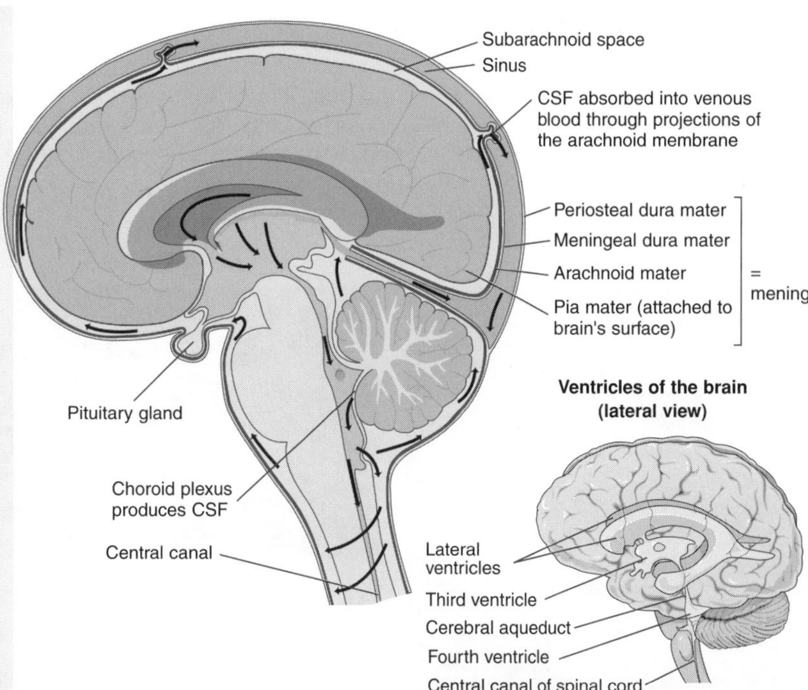


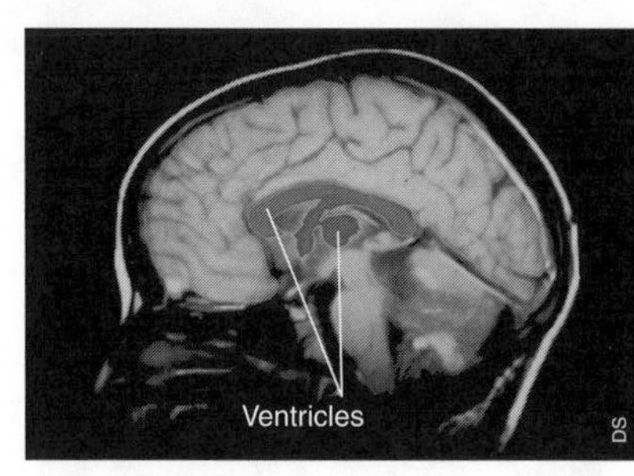


If the passages that normally allow the CSF to exit the brain become blocked, the CSF accumulates within the brain's ventricles causing a condition called hydrocephalus

The accumulated fluid can be seen in this MRI scan.

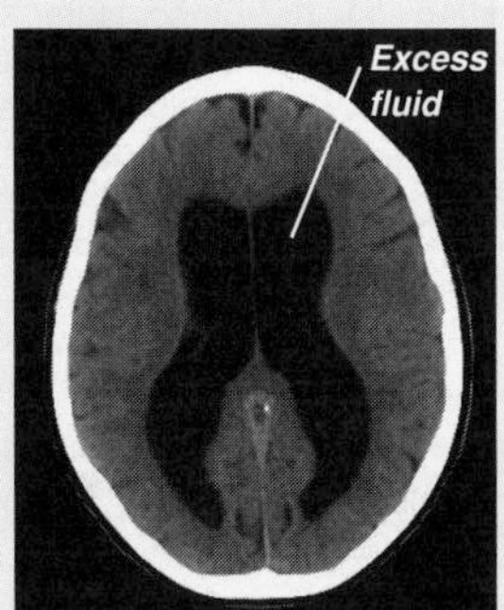


MRI scanning is a powerful technique to visualize the structure and function of the body. It provides much greater contrast between the different soft tissues than computerized tomography (CT) does, making it especially useful in neurological (brain) imaging, especially for indicating the presence of tumors or fluid, and showing up abnormalities in blood supply. In the scan pictured right, the fluid within the lateral and third ventricles is clearly visible.

Ventricles

DS

1. For each of the following bodily functions, identify the region(s) of the brain involved in its control:

(a) Breathing and heartbeat: ____________________

(b) Memory and emotion: ____________________

(c) Posture and balance: ____________________

(d) Autonomic functions: ____________________

(e) Visual processing: ____________________

(f) Body temperature: ____________________

(g) Language: ____________________

(h) Muscular movement: ____________________

2. Explain how the brain is protected against physical damage and infection: ____________________

3. (a) Describe where CSF is produced and how the CSF returns to the blood: ____________________

(b) Explain the consequences of blocking this return flow of CSF: ____________________

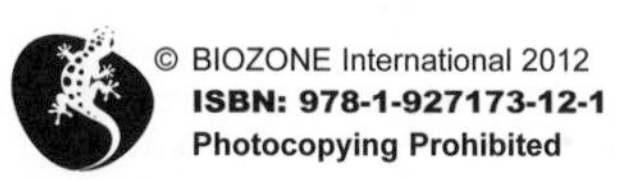


**ISBN: 978-1-927173-12-1**

# Imaging the Brain

**Neuroimaging** uses imaging techniques to study the brain. **Structural imaging** (e.g. **CT scans, MRI**) is used to determine the structure of the brain, and can be used to diagnose the presence of diseases, such as tumors, or damage resulting from injury. **Functional imaging** (e.g. **fMRI**) measures the change in blood flow related to neural activity in the brain. The brain activity is visualized as light colors on the scan image. Functional imaging can be used to diagnose metabolic disorders and small scale lesions (as seen in Alzheimer's patients).

## CT Scans

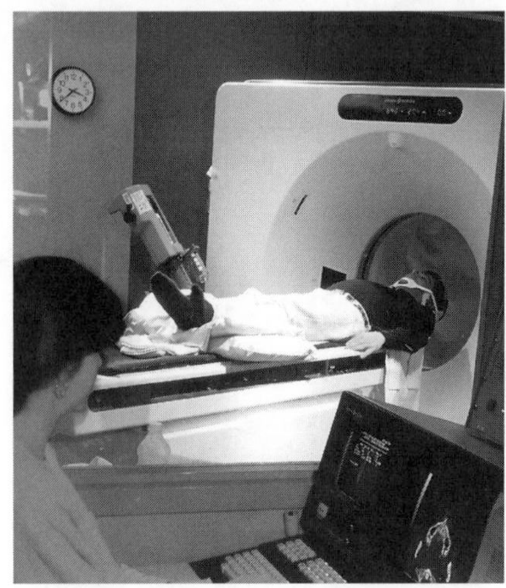

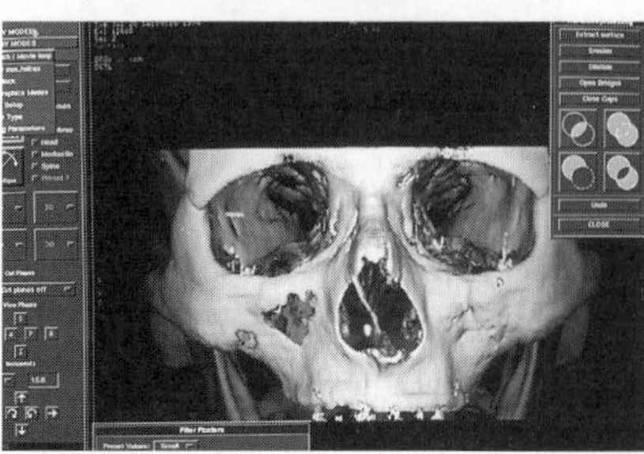

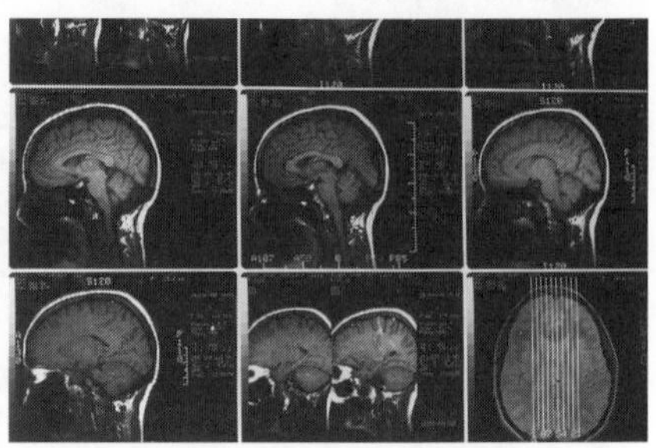

**Computed axial tomography** (CT or CAT scan) involves taking multiple X-ray images of an object from many angles. The images are fitted together by computer software to produce a high resolution three dimensional model showing both the inside and outside of the object. CT can detect a 1% difference in tissue density, so tumors can be detected early. Resolution can be improved by injecting iodine based contrasting chemicals during the scan. CT images are used to diagnose cancer, vascular disease, spinal problems, determine bone density, and organ injury. A disadvantage of CT scanning is that it exposes the patient to moderate doses of radiation.

## MRI Scans

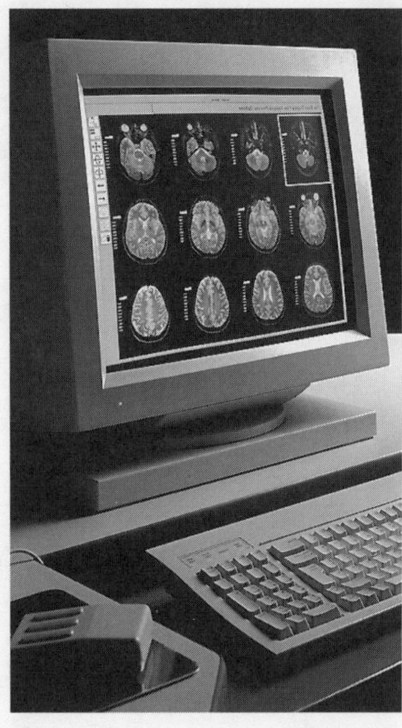

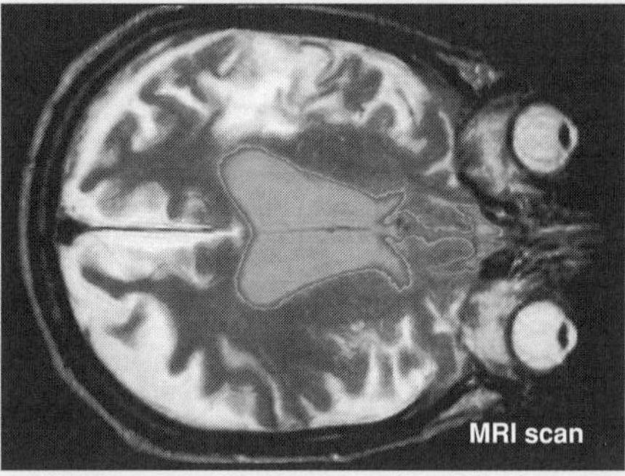


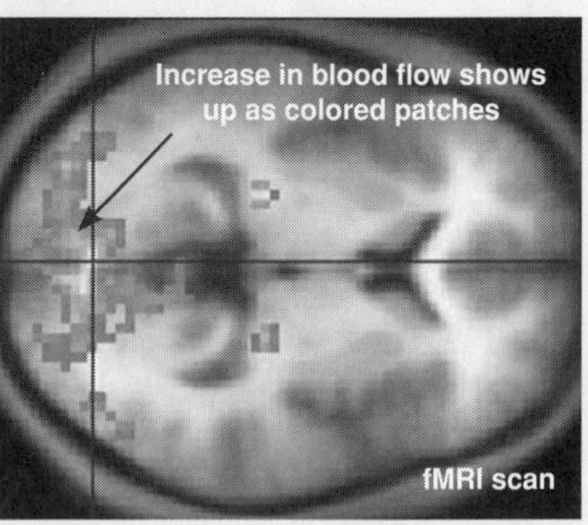


**Magnetic resonance imaging** (MRI) uses a powerful magnetic field and radio waves to produce images of the body's organs and structures. MRI images are very detailed and allow doctors to examine the body's organs (e.g. brain), tissues (e.g. breast) and blood vessels (including arterial blood flow). Diagnostically MRI is used to detect cancer, diagnose heart problems, find cardiovascular obstructions, diagnose diseases of the digestive system and causes of infertility. There is no radiation involved (unlike CT scans), but the high magnetic field means people with pacemakers or metal body parts cannot enter the MRI machine.

Functional MRI (**fMRI**) maps brain activity by mapping blood flow in the brain as it is working. The blood flow can be related to the neural activity of different regions of the brain. fMRI allows scientists to see which parts of the brain are active during different activities (e.g. hearing or recalling events). Normal brain activity can be compared with a damaged brain (e.g. Alzheimer's patients) to determine which areas of the brain are affected. These studies may help find a treatment in the future.

1. (a) Describe how CT scans are used to image the brain: ______________________________

______________________________

______________________________

(b) Describe the main disadvantage of CT scans: ______________________________

2. (a) Describe the advantages of MRI over CT imaging: ______________________________

(b) Explain how fMRI may benefit people with Alzheimer's disease in the future: ______________________________

______________________________

______________________________

(c) Describe a disadvantage of MRI: ______________________________




ISBN: 978-1-927173-12-1

# The Malfunctioning Brain: Alzheimer's

**Alzheimer's disease** (AD) is a disabling neurological disorder affecting 5.4 million Americans. Although its causes are largely unknown, people with a family history of Alzheimer's have a greater risk, implying that a genetic factor is involved. Some of the cases of Alzheimer's with a familial (inherited) pattern involve a mutation of the gene for amyloid precursor protein (APP), found on chromosome 21 and nearly all people with Down syndrome (trisomy 21) who live into their 40s develop the disease. The gene for the protein apoE, which has an important role in lipid transport, degeneration and regulation in nervous tissue, is also a risk factor that may be involved in modifying the age of onset. Sufferers of Alzheimer's have trouble remembering recent events and they become confused and forgetful. In the later stages of the disease, people with Alzheimer's become very disorientated, lose past memories, and may become paranoid and moody. Dementia and loss of reason occur at the end stages of the disease. The effects of the disease are irreversible and it has no cure.

## The Effects of Alzheimer's Disease

Alzheimer's is associated with accelerated loss of neurons, particularly in regions of the brain that are important for memory and intellectual processing, such as the cerebral cortex and hippocampus. The disease has been linked to abnormal accumulations of protein-rich **amyloid** plaques and tangles, which invade the brain tissue and interfere with synaptic transmission.

**Cerebral cortex:** Conscious thought, reasoning, and language. Alzheimer's sufferers show considerable loss of function from this region.

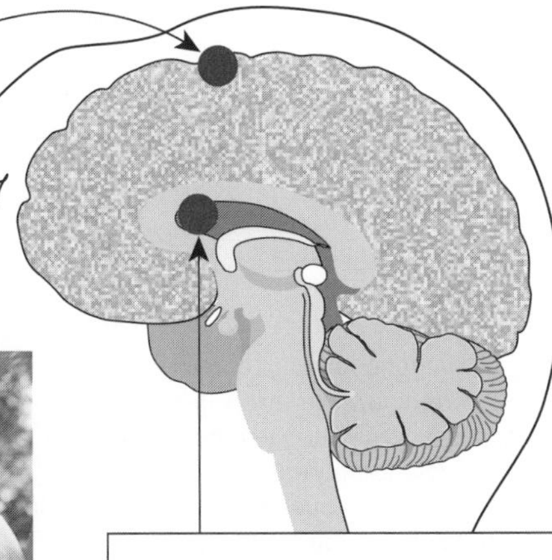

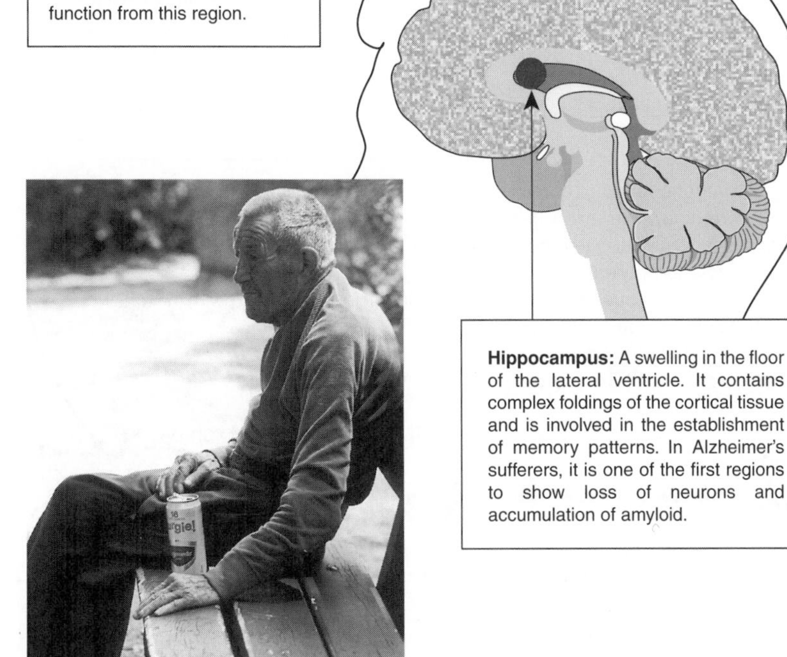

**Hippocampus:** A swelling in the floor of the lateral ventricle. It contains complex foldings of the cortical tissue and is involved in the establishment of memory patterns. In Alzheimer's sufferers, it is one of the first regions to show loss of neurons and accumulation of amyloid.

It is not uncommon for Alzheimer's sufferers to wander and become lost and disorientated.

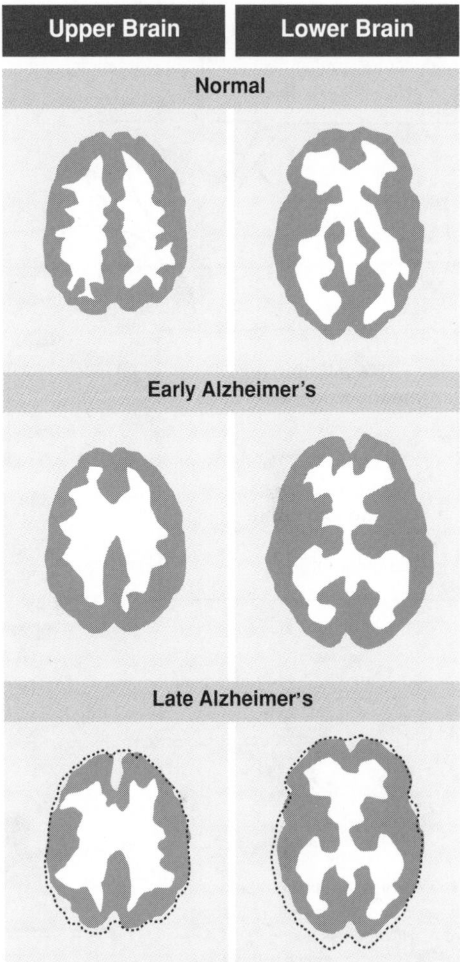


The brain scans above show diminishing brain function in certain areas of the brain in Alzheimer's sufferers. Note, particularly in the two lower scans, how much the brain has shrunk (original size indicated by the dotted line). White areas indicate brain activity.

1. Describe the biological basis behind the degenerative changes associated with Alzheimer's disease:

___

___

2. Describe the evidence for the Alzheimer's disease having a genetic component in some cases:

___

___

3. Some loss of neuronal function occurs normally as a result of ageing. Identify the features distinguishing Alzheimer's disease from normal age related loss of neuronal function:

___

___

___

**ISBN: 978-1-927173-12-1**


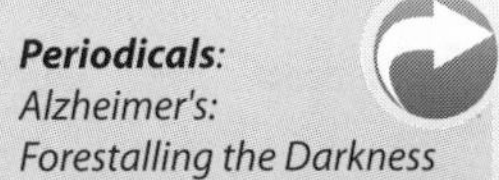

***Periodicals:***
*Alzheimer's: Forestalling the Darkness*

***Related activities:*** *The Human Brain, Imaging the Brain*

# Dopamine and Behavior

The role of genes in some aspects of human behavior, although controversial, is becoming more apparent as research is carried out on more genes and their functions. The dopamine receptor D4 (DRD4) has received a lot of research attention, because of its apparent link to substance dependence and sensation seeking. Dopamine pathways are neural pathways in the brain which transmit the neurotransmitter dopamine from one region of the brain to another. Dopamine has many functions in the brain, including important roles in behavior and cognition, motivation and reward, mood, attention, and learning. In cases of deficiency, dopamine cannot be administered as a drug itself, because it cannot cross the blood-brain barrier.

### Dopamine Pathways in the Brain

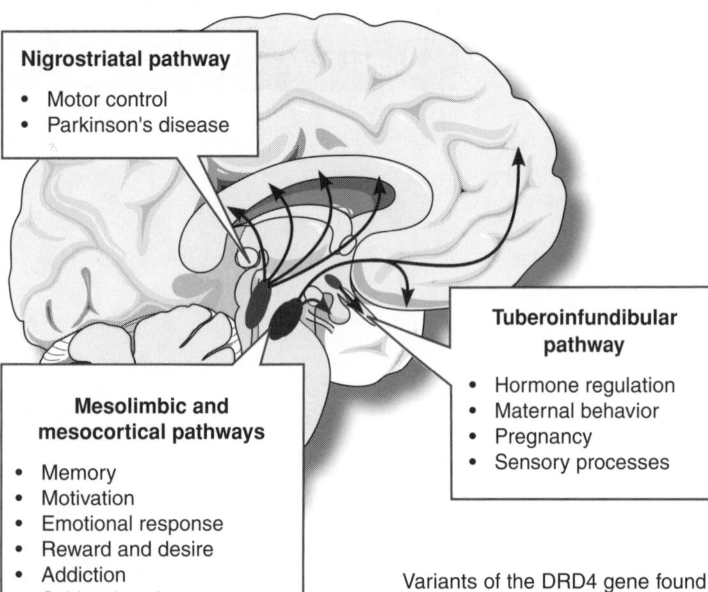


**Effects of dopamine in the mesolimbic and mesocortical pathways**

| Level | Effect |
|---|---|
| Very low | Inability to focus. Associated with Attention Deficient Hyperactivity Disorder (ADHD). |
| Low | Addiction: the euphoric feeling caused by certain drugs is similar to the effects of dopamine and can leave the body wanting more. |
| Normal | Focus constant. |
| High | Hyperstimulation. Focus becomes narrowed to highly specific objects. Increase in perception of senses. |
| Very high | Paranoia and hallucination. Extreme situations can lead to schizophrenia. |

Variants of the DRD4 gene found of chromosome 11 have been much studied. A section of the gene known as exon III has a repeating section of 48 base pairs. This section is typically repeated between 2 and 10 times. The number of repeats is divided into short repeats (from 2-5 repeats) and long repeats (from 6-10 repeats). Long repeats cause the receptor to be inefficient in binding with dopamine, leaving the body dopamine deficient.

The occurrence of long repeats has been linked with higher frequencies of ADHD and risk taking. "DRD4 knock-out" mice (mice without the DRD4 gene) show a deficiency in exploration of novel situations.

### Structure of DRD4 gene

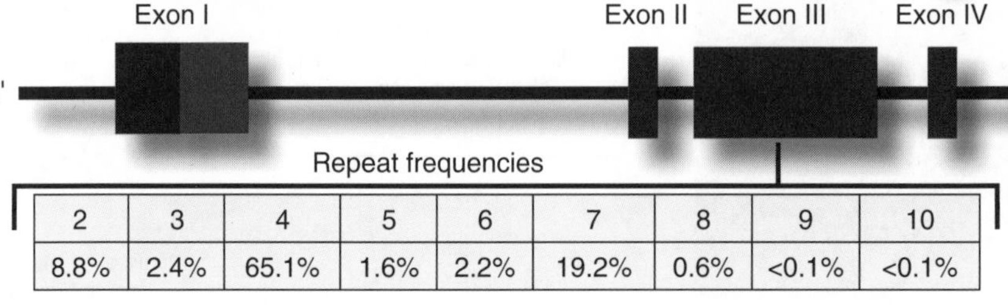


| 2 | 3 | 4 | 5 | 6 | 7 | 8 | 9 | 10 |
|---|---|---|---|---|---|---|---|---|
| 8.8% | 2.4% | 65.1% | 1.6% | 2.2% | 19.2% | 0.6% | <0.1% | <0.1% |

1. Explain why it is difficult to attribute behavior to a single gene: ______

2. Contrast the effects of various levels of dopamine in the brain on behavior. ______

3. Explain the link between the DRD4 gene and behavior: ______

4. Discuss how this link helps us understand other aspects of human behavior: ______




**ISBN: 978-1-927173-12-1**

# Chemical Imbalances in the Brain

The brain uses chemicals (**neurotransmitters**) to transmit messages between nerve cells. Neurotransmitters are released from presynaptic neurons and diffuse across the synaptic cleft to postsynaptic neurons to cause a specific effect. Many brain disorders result from disturbances to natural levels of specific neurotransmitters, and can lead to the failure of specific neural pathways. Sometimes the pathways can be restored using drugs that either replace or boost levels of specific neurotransmitters.

## Parkinson's Disease

Patients with **Parkinson's disease** show decreased stimulation in the motor cortex of the brain. This results from reduced dopamine production in the substantia nigra region (right) where dopamine is produced. This is usually the result of the death of nerve cells. Symptoms, slow physical movement and spasmodic tremors, often don't begin to appear until a person has lost 70% of their dopamine-producing cells.

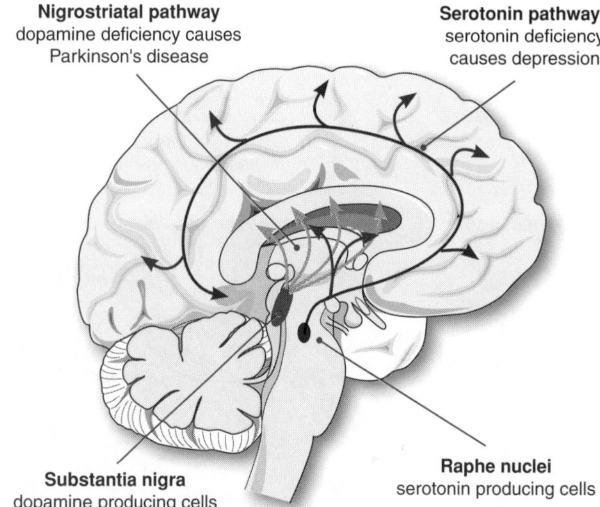


## Treating Parkinson's Disease

**Parkinson's disease** is caused by reduced dopamine production and low dopamine levels in the brain pathways involved with movement. Treatments for Parkinson's have focused on increasing the body's dopamine levels. Dopamine is unable to cross the blood-brain barrier, so cannot be administered as a treatment. However, **L-dopa** is a dopamine precursor that can cross the blood-brain barrier and enter the brain. Once in the brain, it is converted to dopamine. L-dopa has been shown to reduce some of the symptoms of Parkinson's disease.

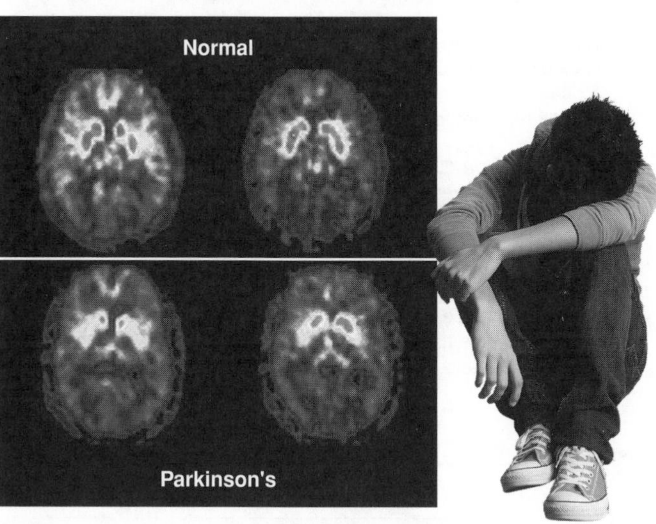


Positron emission tomography (PET) measures the activity of dopamine neurons in the substantia nigra area of the brain. Parkinson's patients (lower panel) show reduced activity in the dopamine neurons compared with normal patients.

## Depression

A person with **depression** (left) experiences prolonged periods of extremely low mood, including low self esteem, regret, guilt, and feelings of hopelessness. Depression may be caused by a mixture of environmental factors (e.g. stress) and biological factors (e.g. low **serotonin** production by the raphe nuclei in the brain, above).

## Treating Depression

Recognition of the link between **serotonin** and **depression** has resulted in the development of **antidepressant drugs** that alter serotonin levels. Monoamine oxidase inhibitors (MAOI) are commonly used antidepressants that increase serotonin levels by preventing its breakdown in the brain. Newer drugs, called Selective Serotonin Re-uptake Inhibitors (SSRIs), stop serotonin re-uptake by presynaptic cells. This increases the levels of extracellular serotonin, making more available to bind to the postsynaptic cells, and stabilizing serotonin levels in the brain. SSRIs have fewer side effects than other antidepressants because they specifically target serotonin and no other neurotransmitters.

1. Describe the function of a neurotransmitter: ____________________

2. Describe the pharmacological cause of the following diseases and identify the major symptom of each:

   (a) Parkinson's disease: ____________________

   (b) Depression: ____________________


ISBN: 978-1-927173-12-1

*Periodicals:* Circuit training, Disco inferno

*Related activities:* Dopamine and Behavior
*Weblinks:* The Effect of Ecstasy

## Recreational Drugs

Most recreational drugs act by altering brain function (i.e. they are psychotropic). Recreational drugs have different mechanisms of action and produce a variety of effects including hallucinations and feelings of euphoria and intimacy. Most recreational drugs are illegal because of their extreme and unpredictable effects, and all can be deadly if an overdose is taken. Recreational drugs are especially dangerous when taken together or taken in conjunction with alcohol.

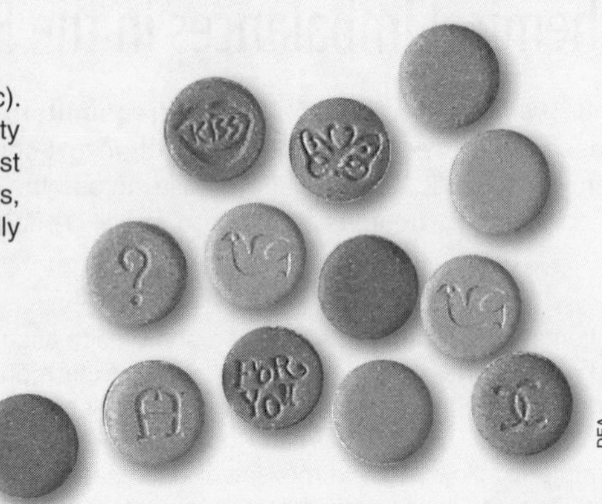


**Ecstasy** or **MDMA** (right) is an illegal drug that produces feelings of euphoria. Some psychologists have suggested that MDMA may have therapeutic benefits, such as relieving post-traumatic stress disorder or the anxiety associated with terminal diseases such as cancer.

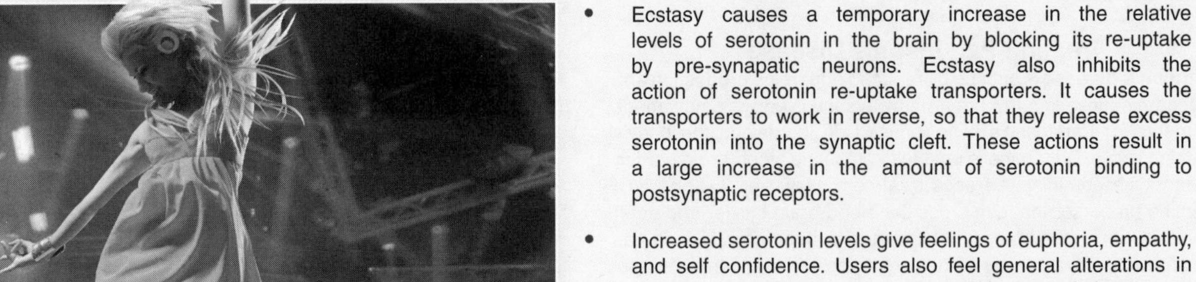

Ecstasy is a stimulant and is popular in dance clubs. It can be deadly in the hot and crowded environment. An overdose can cause heart failure and difficulty breathing, induce coma, and cause death.

- Ecstasy causes a temporary increase in the relative levels of serotonin in the brain by blocking its re-uptake by pre-synapatic neurons. Ecstasy also inhibits the action of serotonin re-uptake transporters. It causes the transporters to work in reverse, so that they release excess serotonin into the synaptic cleft. These actions result in a large increase in the amount of serotonin binding to postsynaptic receptors.
- Increased serotonin levels give feelings of euphoria, empathy, and self confidence. Users also feel general alterations in consciousness, increased energy, and increased alertness.
- Serotonin is involved in temperature regulation and vasoconstriction. Elevated levels can cause body temperature to rise (hyperthermia). This can lead to rapid organ failure and death if not treated quickly.
- Serotonin is produced at a fixed rate so spikes in serotonin secretion can eventually lead to depletion of serotonin in the brain. This imbalance may last for weeks, causing feelings similar to depression.
- Prolonged ecstasy use may permanently damage the brain's ability to produce serotonin. Long term effects may include memory problems, development of psychiatric disorders, appetite loss, and depression.

3. Describe how L-dopa is used to help treat Parkinson's disease: ______________________

4. Explain how the actions of antidepressive reduce the feelings of depression: ______________________

5. Explain how ecstasy and SSRIs increase extracellular serotonin levels: ______________________

6. Explain the long term effect of ecstasy use on the brain: ______________________


ISBN: 978-1-927173-12-1

# The Structure of the Eye

The eye is a complex and highly sophisticated sense organ specialized to detect light. The adult eyeball is about 25 mm in diameter. Only the anterior one-sixth of its total surface area is exposed; the rest lies recessed and protected by the **orbit** into which it fits. The eyeball is protected and given shape by a fibrous tunic. The posterior part of this structure is the **sclera** (the white of the eye), while the anterior transparent portion is the **cornea**, which covers the colored iris.

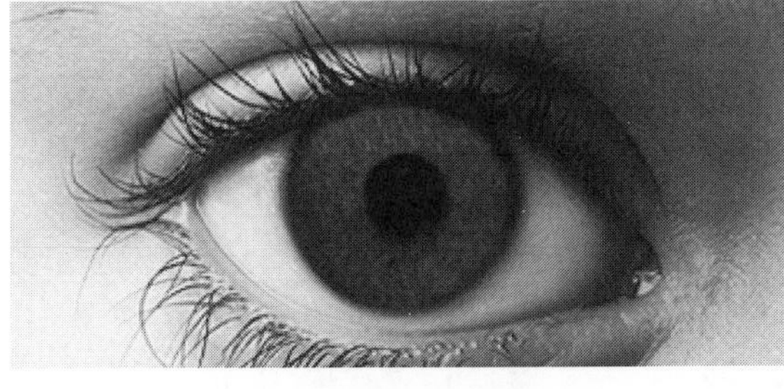

### Forming a Visual Image

Before light can reach the photoreceptor cells of the retina, it must pass through the cornea, aqueous humor, pupil, lens, and vitreous humor. For vision to occur, light reaching the photoreceptor cells must form an image on the retina. This requires **refraction** of the incoming light, **accommodation** of the lens, and **constriction** of the pupil.

The anterior of the eye is concerned mainly with **refracting** (bending) the incoming light rays so that they focus on the retina. Most refraction occurs at the cornea. The lens adjusts the degree of refraction to produce a sharp image. **Accommodation** adjusts the eye for near or far objects. Constriction of the pupil narrows the diameter of the hole through which light enters the eye, preventing light rays entering from the periphery.

The point at which the nerve fibers leave the eye as the optic nerve, is the **blind spot** (the point at which there are no photoreceptor cells). Nerve impulses travel along the optic nerves to the visual processing areas in the cerebral cortex. Images on the retina are inverted and reversed by the lens but the brain interprets the information it receives to correct for this image reversal.

### The Structure and Function of the Mammalian Eye

The human eye is essentially a three layered structure comprising an outer fibrous layer (the sclera and cornea), a middle vascular layer (the choroid, ciliary body, and iris), and inner **retina** (neurons and **photoreceptor cells**). The shape of the eye is maintained by the fluid filled cavities (aqueous and vitreous humors), which also assist in light refraction. Eye color is provided by the pigmented iris. The iris also regulates the entry of light into the eye through the contraction of circular and radial muscles.

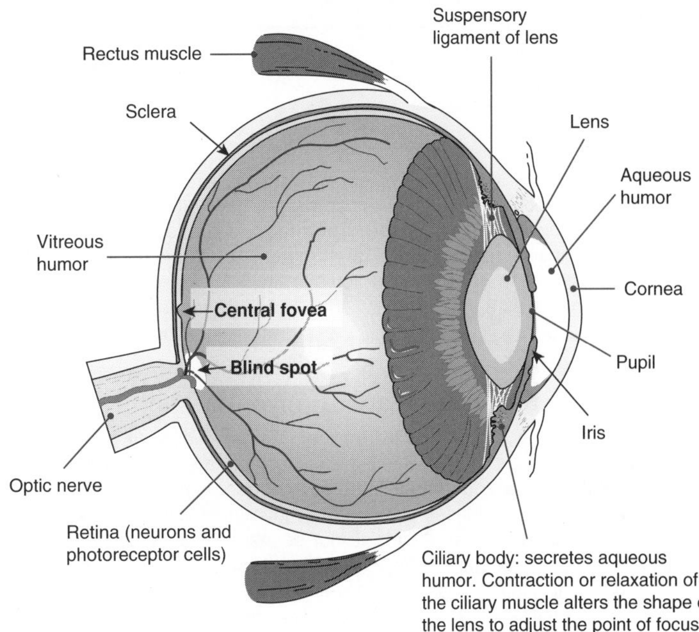


1. Identify the function of each of the structures of the eye listed below:

   (a) Cornea: ______________________________

   (b) Ciliary body: ______________________________

   (c) Iris: ______________________________

2. (a) The first stage of vision involves forming an image on the retina. Explain what this involves: ______________________________

   (b) Explain how accommodation is achieved: ______________________________


ISBN: 978-1-927173-12-1

The lens of the eye has two convex surfaces (biconvex). When light enters the eye, the lens bends the incoming rays towards each other so that they intersect at the focal point on the central fovea of the retina. By altering the curvature of the lens, the focusing power of the eye can be adjusted. This adjustment of the eye for near or far vision is called **accommodation** and it is possible because of the elasticity of the lens. For some people, the shape of the eyeball or the lens prevents convergence of the light rays on the central fovea, and images are focused in front of, or behind, the retina. Such visual defects (below) can be corrected with specific lenses. As we age, the lens loses some of its elasticity and, therefore, its ability to accommodate. This inability to focus on nearby objects due to loss of lens elasticity is a natural part of ageing and is called far sight.

## Normal vision

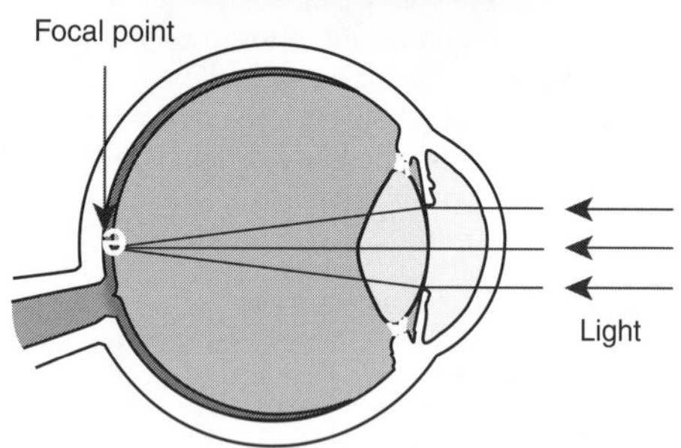


In normal vision, light rays from an object are bent sufficiently by the cornea and lens, and converge on the central fovea. A clear image is formed. Images are focused upside down and mirror reversed on the retina. The brain automatically interprets the image as right way up.

## Accommodation for near and distant vision

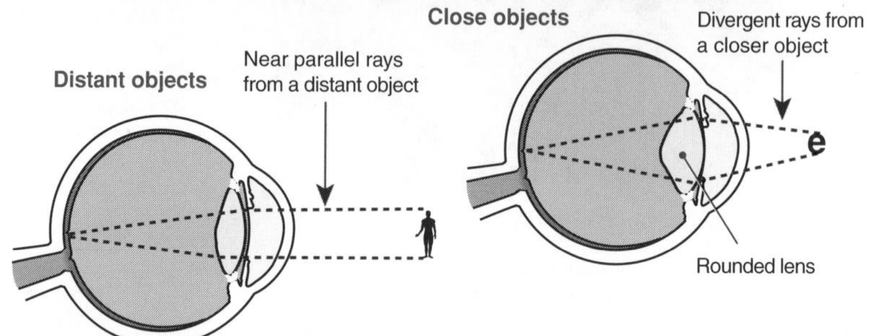


The degree of refraction occurring at each surface of the eye is precise. The light rays reflected from an object 6 m or more away are nearly parallel to one another. Those reflected from near objects are divergent. The light rays must be refracted differently in each case so that they fall exactly on the central fovea. This is achieved through adjustment of the shape of the lens (accommodation). Accommodation from distant to close objects occurs by rounding the lens to shorten its focal length, since the image distance to the object is essentially fixed.

### Short sightedness (myopia)

Myopia (top row, right) results from an elongated eyeball or a thickened lens. Left uncorrected, distant objects are focused in front of the retina and appear blurred. To correct myopic vision, concave (negative) lenses are used to move the point of focus backward to the retina. Myopia is not necessarily genetic, nor is it necessarily caused by excessive close work, as was once thought, although myopia does seem to be more prevalent amongst those living in very confined spaces (e.g. people working and living in submarines).

### Long sightedness (hypermetropia)

Long sightedness (bottom row, right) results from a shortened eyeball or from a lens that is too thin. Left uncorrected, light is focused behind the retina and near objects appear blurred. Mild or moderate hypermetropia, which occurs naturally in young children, may be overcome by **accommodation**. In more severe cases, corrective lenses are used to bring the point of focus forward to produce a clear image. This is achieved using a convex (positive) lens.

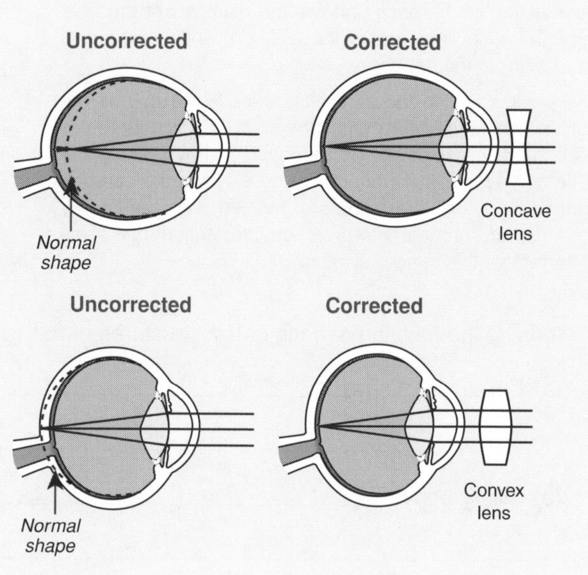


3. (a) Describe the function of the pupil: ______

   (b) Suggest why control of pupil diameter would be under reflex control: ______

   ______

   ______

4. With respect to formation of the image, describe what is happening in:

   (a) Short sighted people: ______

   (b) Long sighted people: ______

5. In general terms, describe how the use of lenses corrects the following problems associated with vision:

   (a) Myopia: ______

   (b) Hypermetropia: ______


ISBN: 978-1-927173-12-1

# The Physiology of Vision

Vision involves essentially two stages: formation of the image on the retina), and generation and conduction of nerve impulses. When light reaches the retina, it is absorbed by the photosensitive pigments associated with the membranes of the photoreceptor cells (the rods and cones). The pigment molecules are altered by the absorption of light in such a way as to lead to the generation of nerve impulses. It is these impulses that are conducted via nerve fibers to the visual processing center of the cerebral cortex.

## Structure and Function of the Retina

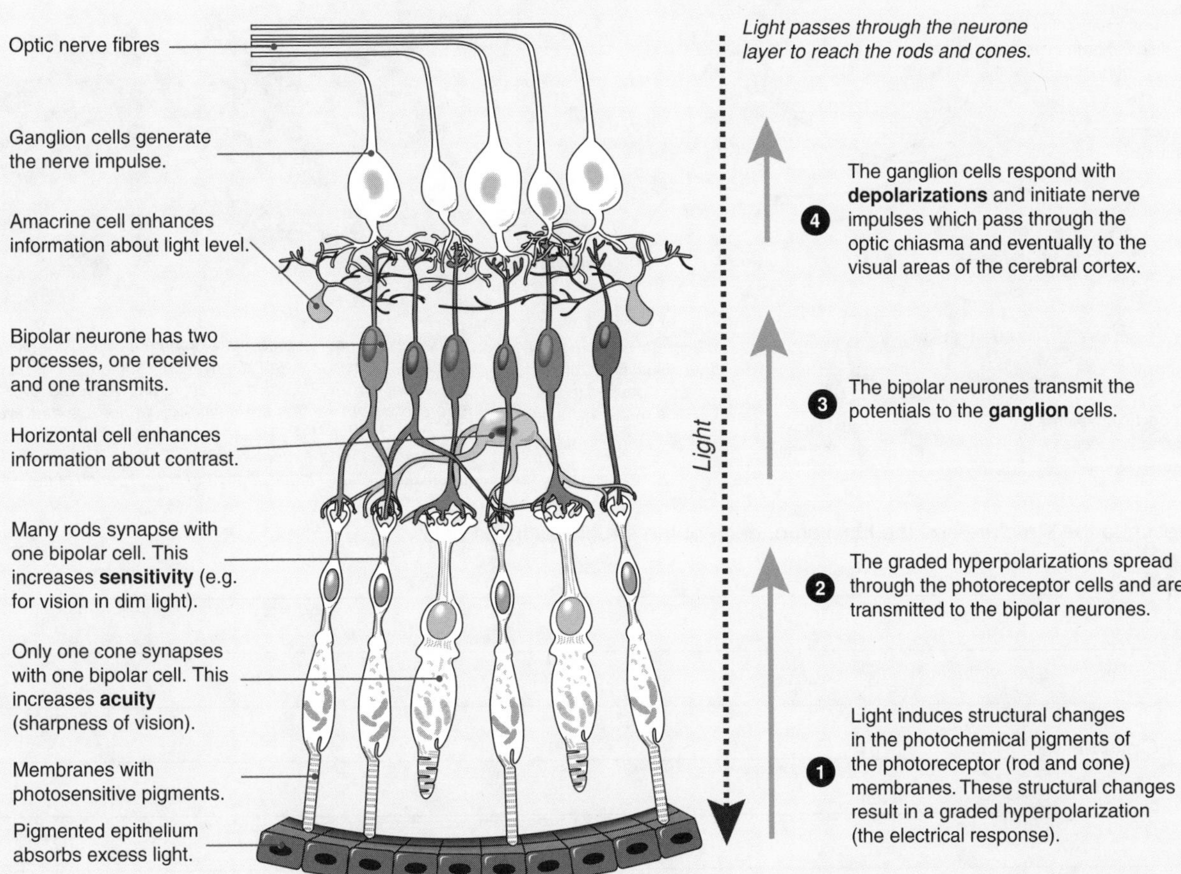


The photoreceptor cells of the mammalian retina are the **rods** and **cones**. Rods are specialized for vision in dim light, whereas cones are specialized for color vision and high visual acuity. Cone density and visual acuity are greatest in the **central fovea** (rods are absent here). After an image is formed on the retina, light impulses must be converted into nerve impulses. The first step is the development of graded potentials (in this case, hyperpolarizations). The graded changes in membrane conductance spread through the cells of the retina to the **ganglion cells**, which initiate nerve impulses. These pass through the optic chiasma and eventually to the visual areas of the cerebral cortex. The frequency and pattern of impulses in the optic nerve conveys information about the changing visual field.

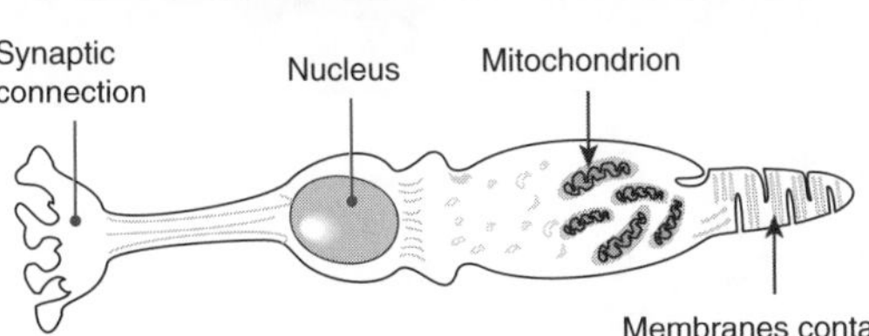


Each **cone** synapses with only one bipolar cell giving high acuity.

## The Basis of Trichromatic Vision

There are three classes of cones, each with a maximal response in either short (blue), intermediate (green) or long (yellow-green) wavelength light (below). The yellow-green cone is also sensitive to the red part of the spectrum and is often called the red cone. The differential responses of the cones to light of different wavelengths provides the basis of trichromatic color vision.

**Cone response to light wavelengths**

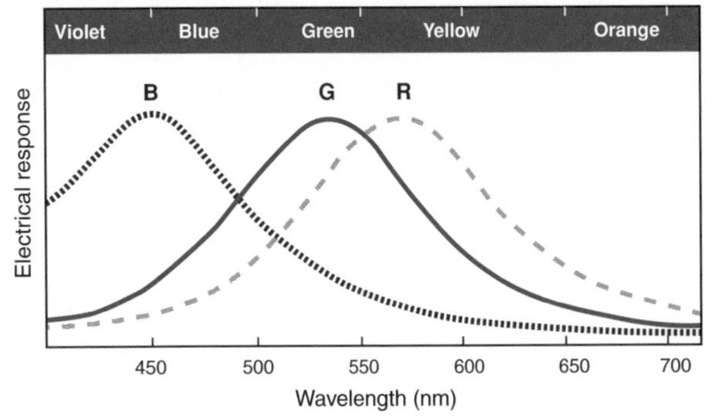

ISBN: 978-1-927173-12-1


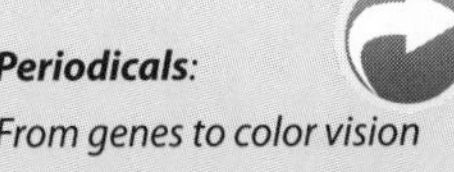

# Experimental Evidence for how Vision Develops

Cells in the visual cortex are arranged in columns (**ocular dominance columns**), which alternate between receiving information from the right and the left eyes. These columns, and the axons that stimulate them, are present at birth. If an eye is deprived of visual input, the axons supplying the corresponding columns have reduced capacity and eventually fail to function.

David Hubel and Torsten Wiesel established that a **critical window** (between 4-8 weeks of age) exists for developing a mature, fully functioning visual cortex (the area of the brain that processes visual stimuli). If the brain does not receive visual stimuli during the critical period, neural connections are not made, and normal visual pathways do not develop.

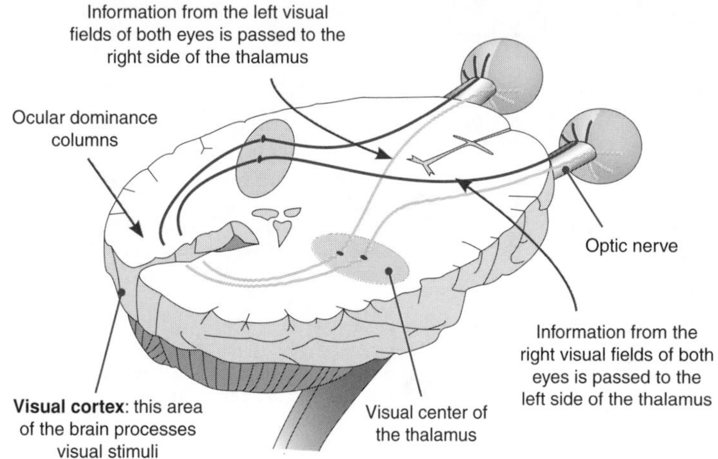

*Hubel and Wiesel carried out experiments on infant kittens and monkeys to see if visual deprivation affected neural activity in the visual cortex. Visual deprivation was achieved by stitching one eyelid shut so that the animal could not use that eye. They discovered that the visual cortex developed physiological abnormalities and atrophied (wasted away) if the animals were deprived of visual stimuli during the critical period of development.*

1. Describe the structure and the function of each of the structures listed below:

   (a) Retina: ______

   (b) Optic nerve: ______

2. Contrast the structure of the blind spot and the central fovea: ______

3. Complete the table below, comparing the features of rod and cone cells:

| Feature | Rod cells | Cone cells |
|---|---|---|
| Visual pigment(s): | | |
| Visual acuity: | | |
| Overall function: | | |

4. Account for the differences in acuity and sensitivity between rod and cone cells: ______

5. Explain the evidence supporting the importance of critical windows in brain development: ______


**ISBN: 978-1-927173-12-1**

# Hearing

Most animals respond to sound and so have receptors for the detection of sound waves. In mammals, these receptors are organized into hearing organs called ears. Sound is produced by the vibration of particles in a medium and it travels in waves that can pass through solids, liquids, or gases. The distance between wave 'crests' determines the frequency (pitch) of the sound. The absolute size (amplitude) of the waves determines the intensity or loudness of the sound. Sound reception in mammals is the role of **mechanoreceptors**: tiny hair cells in the cochlea of the inner ear. The hair cells are very sensitive and are easily damaged by prolonged exposure to high intensity sounds. Gradual hearing loss with age is often caused by the cumulative loss of sensory hair cell function, especially at the higher frequencies. Such hearing loss is termed perceptive deafness.

## The Human Ear

Ear canal
Ear drum
Vestibular apparatus (senses balance and body position)
Auditory nerve
Cochlea
Round window
Oval window
Malleus (hammer)
Incus (anvil)
Stapes (stirrup)
Ear ossicles (bones)
Eustachian tube
Pinna focuses sound waves

In mammals, sound waves are converted to pressure waves in the inner ear. The ears of mammals use mechanoreceptors (sensory hair cells) to change the pressure waves into nerve impulses. The mammalian ear contains not only the organ of hearing, the cochlea, but all the specialized structures associated with gathering, directing, and amplifying the sound. The cochlea is a tapered, coiled tube, divided lengthwise into **three fluid filled canals**. The cochlea is shown below, unrolled to indicate the way in which sound waves are transmitted through the canals to the sensory cells. The mechanisms involved in hearing are outlined in a simplified series of steps. In mammals, the inner ear is also associated with the organ for detecting balance and position (the vestibular apparatus), although this region is not involved in hearing.

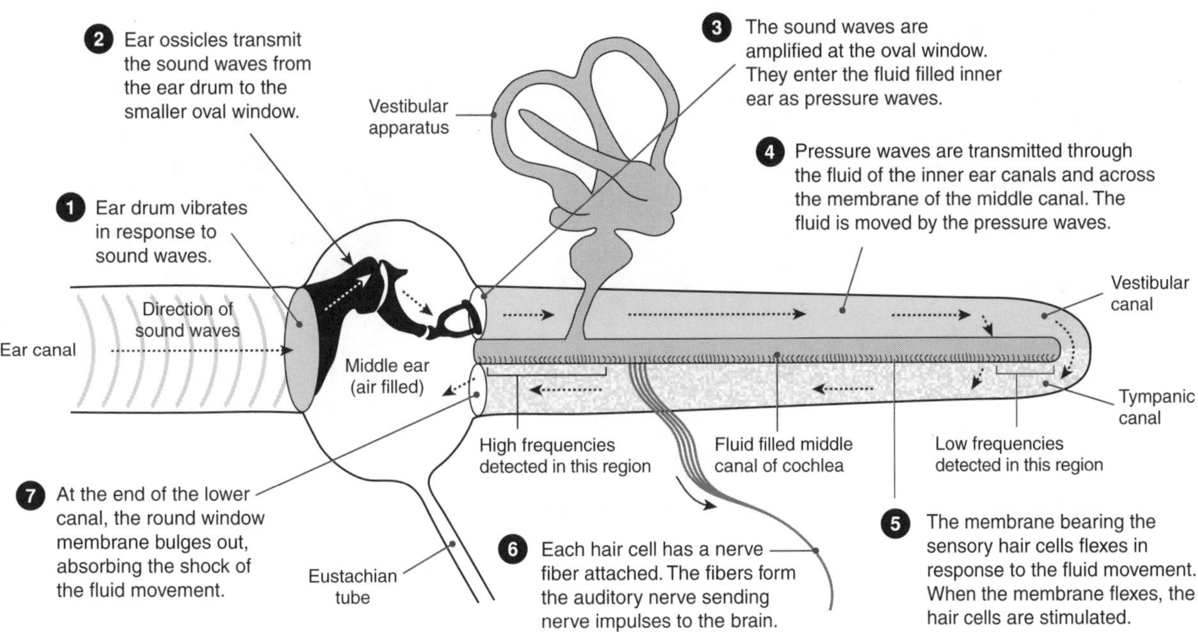


1. In a short sentence, outline the role of each of the following in the reception and response to sound:

   (a) The ear drum: 

   (b) The ear ossicles: 

   (c) The oval window: 

   (d) The sensory hair cells: 

   (e) The auditory nerve: 

2. What is the significance of the inner ear being fluid filled?


ISBN: 978-1-927173-12-1


*Related activities: Encoding Information*
*Weblinks: Hearing, Sound Waves and the Cochlea, Hair Cell Transduction*

# Taste and Smell

**Chemosensory receptors** are responsible for our sense of smell (**olfaction**) and taste (**gustation**). The receptors for smell and taste both respond to chemicals, either carried in the air (smell) or dissolved in a fluid (taste). In humans and other mammals, these are located in the nose and tongue respectively. Each receptor type is basically similar: they are collections of receptor cells equipped with chemosensory microvilli or cilia. When chemicals stimulate their membranes, the cells respond by producing nerve impulses that are transmitted to the appropriate region of the cerebral cortex for interpretation.

## Taste (Gustation)

The organs of taste are the **taste buds**, which are located on the tongue. Most of the taste buds on the tongue are located on raised protrusions of the tongue surface called **papillae**. Each bud is flask-like in shape, with a pore opening to the surface of the tongue enabling molecules and ions dissolved in saliva to reach the receptor cells inside. Each taste bud is an assembly of 50-150 taste cells. These connect with nerves that send messages to the gustatory region of the brain. There are five basic taste sensations. **Salty** and **sour** operate through ion channels, while **sweet**, **bitter**, and **umami** (savory) operate through membrane signaling proteins. These taste sensations are found on all areas of the tongue although some regions are more sensitive than others.

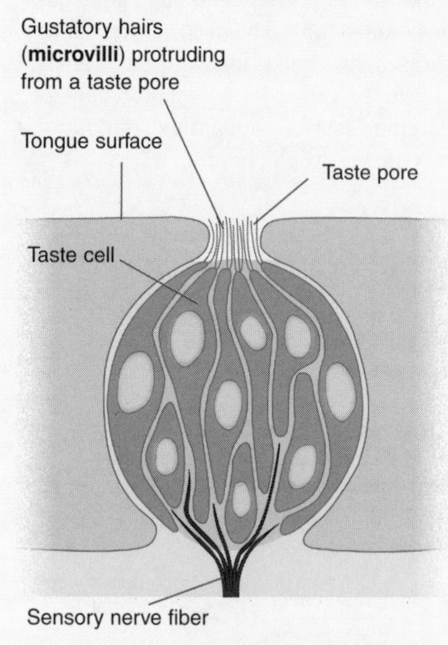


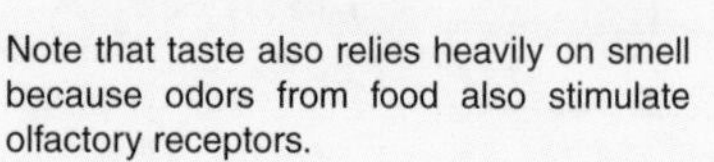


Note that taste also relies heavily on smell because odors from food also stimulate olfactory receptors.

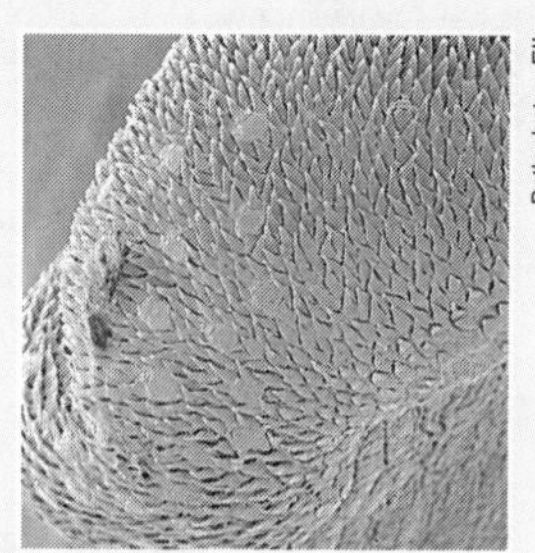


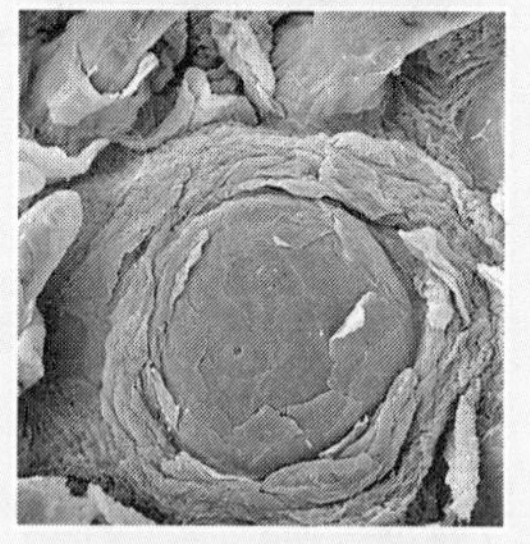

**Above**: SEMs of the surface of the tongue (top) and close up of one of the papillae (below).

## Smell (Olfaction)

In humans, the receptors for smell are located at the top of the nasal cavity. The receptors are specialized hair cells that detect airborne molecules and respond by sending nerve impulses to the olfactory centre of the brain. Unlike taste receptors, olfactory receptors can detect many different odors. However, they quickly adapt to the same smell and will cease to respond to it. This phenomenon is called **sensory adaptation**.

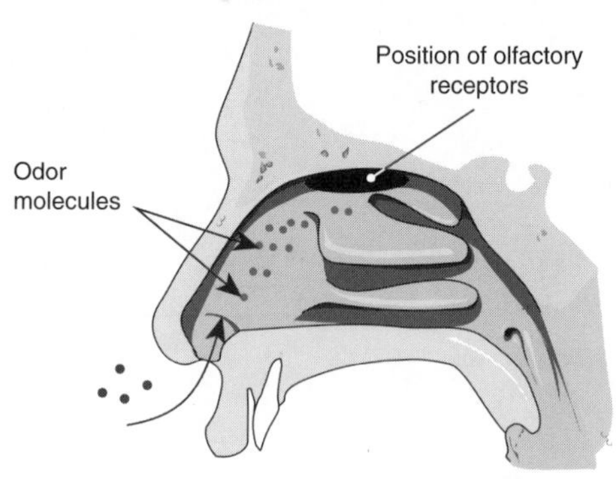


***Detail of olfactory membrane***

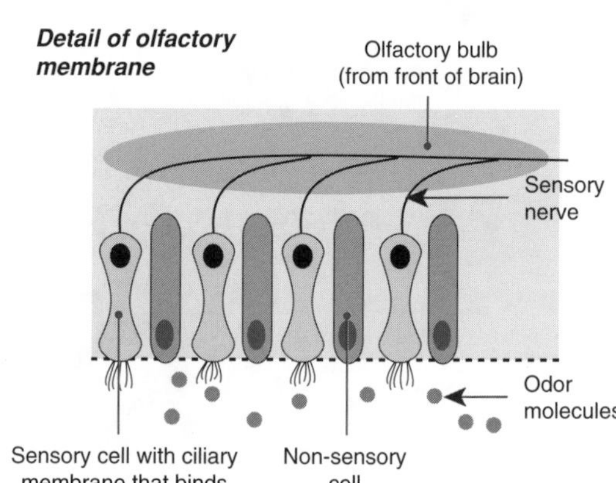


1. Describe the basic mechanism by which chemical sense operates: ______________________________

______________________________

______________________________

2. Take a deep breath of a non-toxic, pungent substance such as perfume. Take a sniff of the substance at 10 second intervals for about 1 minute. Make a record of how strongly you perceived the smell to be at each time interval. Use a scale of 1 to 6: **1.** *Very strong;* **2.** *Quite strong;* **3.** *Noticeable;* **4.** *Weak;* **5.** *Very faint;* **6.** *Could not detect.*

| Time | Strength | Time | Strength | Time | Strength |
|---|---|---|---|---|---|
| 10 s | ______ | 30 s | ______ | 50 s | ______ |
| 20 s | ______ | 40 s | ______ | 1 min | ______ |

3. (a) Explain what happened to your sense of smell over the time period: ______________________________

(b) State the term that describes this phenomenon: ______________________________

(c) Describe the adaptive advantage of this phenomenon: ______________________________

______________________________



***Related activities:*** *Encoding Information*
***Weblinks:*** *Sense of Taste, Sense of Smell, Olfactory Receptor Stimulation*

***Periodicals:***
*Exquisite sense*

**ISBN: 978-1-927173-12-1**

# The Mechanics of Movement

We are familiar with the many different bodily movements achievable through the action of muscles. Contractions in which the length of the muscle shortens in the usual way are called **isotonic contractions**: the muscle shortens and movement occurs. When a muscle contracts against something immovable and does not shorten the contraction is called **isometric**. Skeletal muscles are attached to bones by tough connective tissue structures called **tendons**. They always have at least two attachments: the **origin** and the **insertion**. They create movement of body parts when they contract across **joints**. The type and degree of movement achieved depends on how much movement the joint allows and where the muscle is located in relation to the joint. Some common types of body movements are described below (left panel). Because muscles can only pull and not push, most body movements are achieved through the action of opposing sets of muscles (below, right panel).

## The Action of Antagonistic Muscles

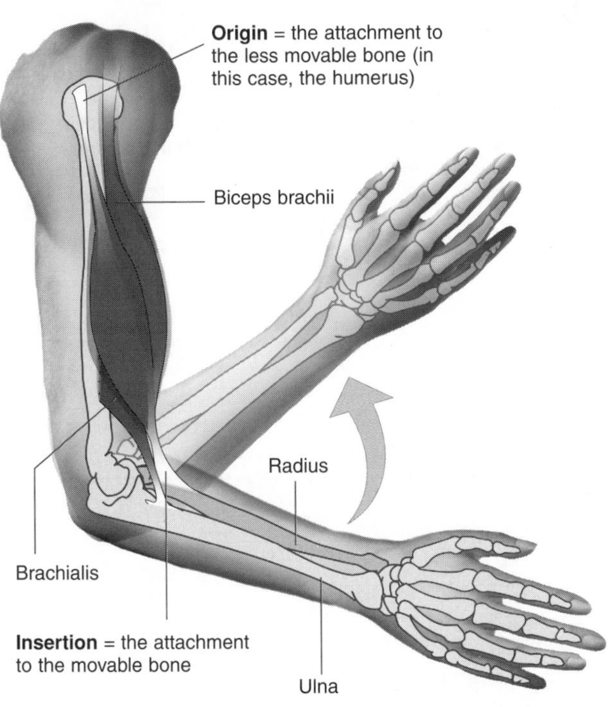


Two muscles are involved in flexing the forearm. The **brachialis**, which underlies the biceps brachii and has an origin half way up the humerus, is the **prime mover**. The more obvious **biceps brachii**, which is a two headed muscle with two origins and a common insertion near the elbow joint, acts as the synergist. During contraction, the insertion moves towards the origin.

The skeleton works as a system of levers. The joint acts as a **fulcrum** (or pivot), the muscles exert the **force**, and the weight of the bone being moved represents the **load**. The flexion (bending) and extension (unbending) of limbs is caused by the action of **antagonistic muscles**. Antagonistic muscles work in pairs and their actions oppose each other. During movement of a limb, muscles other than those primarily responsible for the movement may be involved to fine tune the movement.

Every coordinated movement in the body requires the application of muscle force. This is accomplished by the action of agonists, antagonists, and synergists. The opposing action of agonists and antagonists (working constantly at a low level) also produces muscle tone. Note that either muscle in an antagonistic pair can act as the agonist or **prime mover**, depending on the particular movement (for example, flexion or extension).

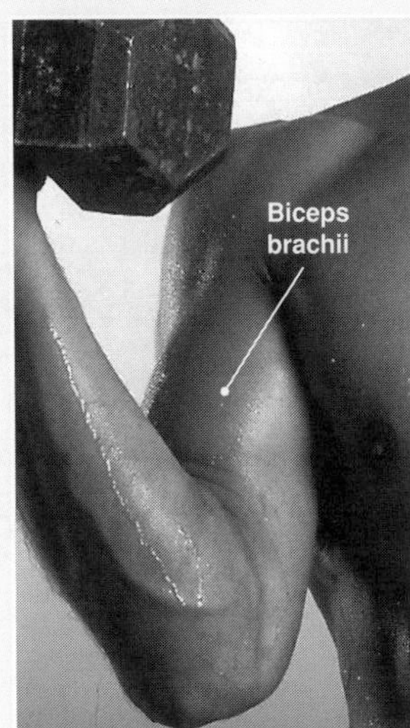


**Agonists** or prime movers: muscles that are primarily responsible for the movement and produce most of the force required.

**Antagonists:** muscles that oppose the prime mover. They may also play a protective role by preventing over-stretching of the prime mover.

**Synergists:** muscles that assist the prime movers and may be involved in fine-tuning the direction of the movement.

During flexion of the forearm (left) the **brachialis** muscle acts as the prime mover and the **biceps brachii** is the synergist. The antagonist, the **triceps brachii** at the back of the arm, is relaxed. During extension, their roles are reversed.

### Movement at Joints

The synovial joints of the skeleton allow free movement in one or more planes. The articulating bone ends are separated by a joint cavity containing lubricating synovial fluid. Two types of synovial joint, the shoulder ball and socket joint and the hinge joint of the elbow, are illustrated below.

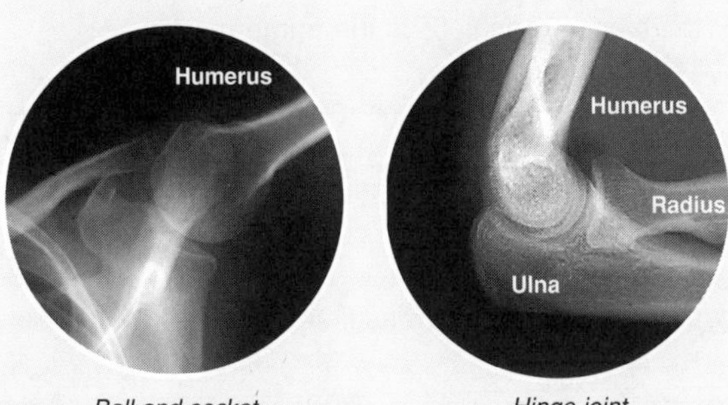


*Ball and socket* *Hinge joint*

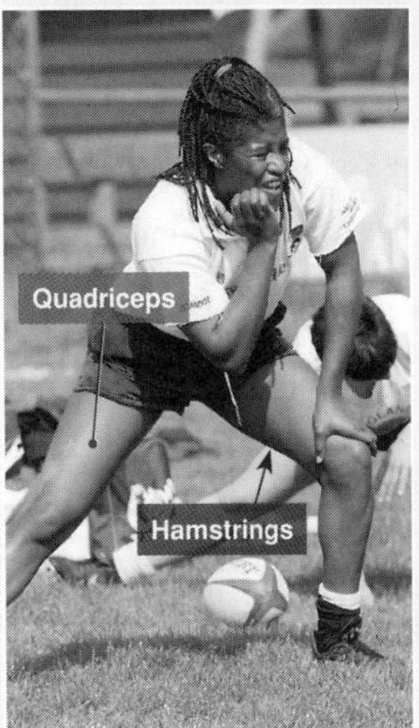


Movement of the upper leg is achieved through the action of several large groups of muscles, collectively called the **quadriceps** and the **hamstrings**.

The hamstrings are actually a collection of three muscles, which act together to flex the leg.

The quadriceps at the front of the thigh (a collection of four large muscles) opposes the motion of the hamstrings and extends the leg.

When the prime mover contracts forcefully, the antagonist also contracts very slightly. This stops overstretching and allows greater control over thigh movement.


**ISBN: 978-1-927173-12-1**

## Types of Body Movement

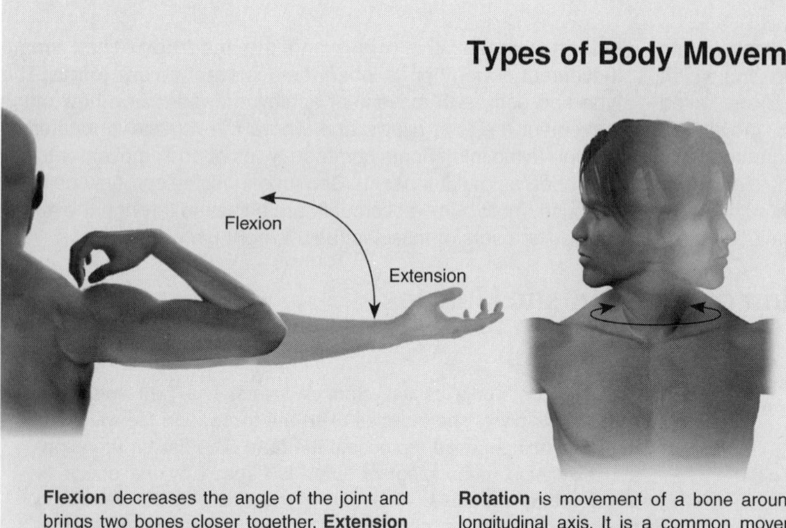


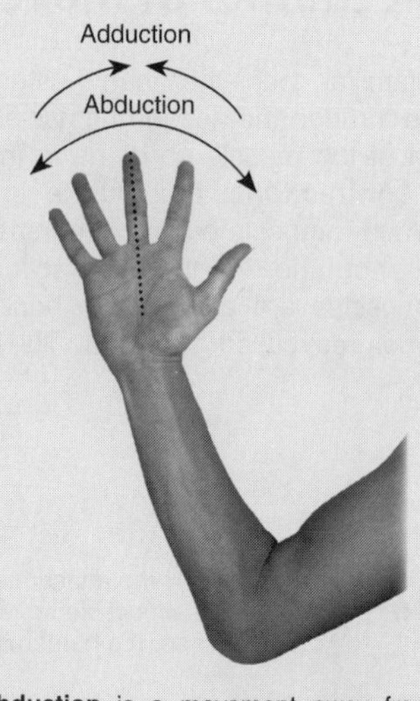


**Flexion** decreases the angle of the joint and brings two bones closer together. **Extension** is its opposite. Extension more than 180° is called **hyperextension**.

**Rotation** is movement of a bone around its longitudinal axis. It is a common movement of ball and socket joints and the movement of the atlas around the axis.

**Abduction** is a movement away from the midline, whereas **adduction** describes movement towards the midline. The terms also apply to opening and closing the fingers.

1. Describe the role of each of the following muscles in moving a limb:

   (a) Prime mover: ___

   (b) Antagonist: ___

   (c) Synergist: ___

2. Explain why the muscles that cause movement of body parts tend to operate as antagonist pairs: ___

3. Describe the relationship between muscles and joints Using appropriate terminology, explain how antagonistic muscles act together to raise and lower a limb:

4. Explain the role of joints in the movement of body parts: ___

5. (a) Identify the insertion for the biceps brachii during flexion of the forearm: ___

   (b) Identify the insertion of the brachialis muscle during flexion of the forearm: ___

   (c) Identify the antagonist during flexion of the forearm: ___

   (d) Given its insertion, describe the forearm movement during which the biceps brachialis is the prime mover:

6. (a) Describe a forearm movement in which the brachialis is the antagonist: ___

   (b) Identify the prime mover in this movement: ___

7. (a) Describe the actions that take place in the neck when you nod your head up and down as if saying "yes":

   (b) Describe the action being performed when a person sticks out their thumb to hitch a ride: ___


**ISBN: 978-1-927173-12-1**

# Muscle Structure and Function

There are three kinds of muscle tissue: **skeletal, cardiac**, and **smooth** muscle, each with a distinct structure. The muscles used for posture and locomotion are skeletal (voluntary) muscles and are largely under conscious control. Their distinct appearance is the result of the regular arrangement of contractile elements within the muscle cells. Muscle fibers are innervated by the branches of motor neurons, each of which terminates in a specialized cholinergic synapse called the **neuromuscular junction** (or motor end plate). A motor neuron and all the fibers it innervates (which may be a few or several hundred) are called a **motor unit**.

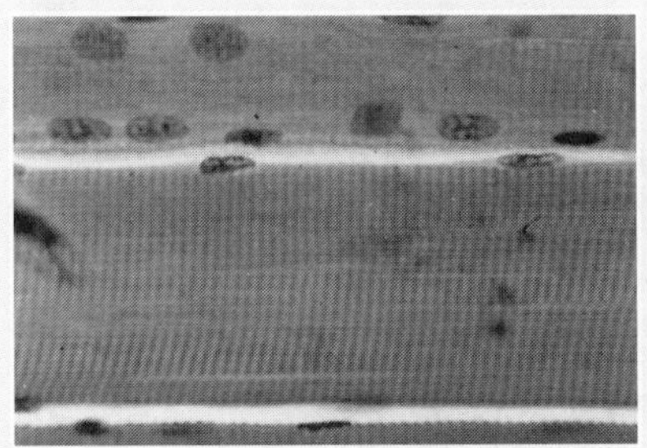


**Skeletal muscle**

Also called striated or striped muscle. It has a banded appearance under high power microscopy. Sometimes called voluntary muscle because it is under conscious control. The cells are large with many nuclei at the edge of each cell.

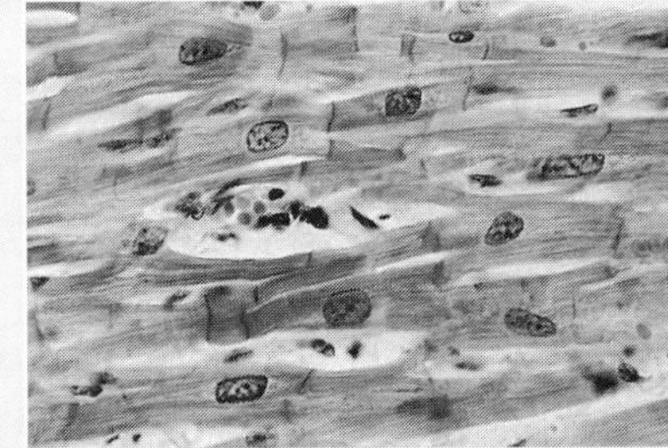

**Cardiac muscle**

Specialized striated muscle that does not fatigue. Cells branch and connect with each other to assist the passage of nerve impulses through the muscle. Cardiac muscle is not under conscious control (it is involuntary).

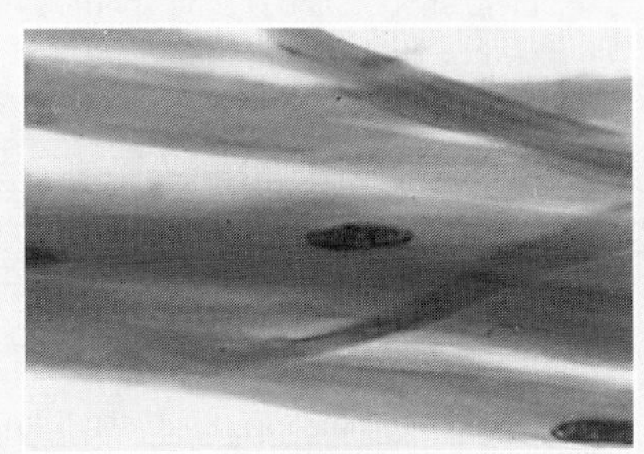


**Smooth muscle**

Also called involuntary muscle because it is not under conscious control. Contractile filaments are irregularly arranged so the contraction is not in one direction as in skeletal muscle. Cells are spindle shaped with one central nucleus.

## Structure of Skeletal Muscle

Skeletal muscle is organized into bundles of muscle cells or **fibers**. Each fiber is a single cell with many nuclei and each fiber is itself a bundle of smaller **myofibrils** arranged lengthwise. Each myofibril is in turn composed of two kinds of **myofilaments** (thick and thin), which overlap to form light and dark bands. It is the alternation of these light and dark bands which gives skeletal muscle its striated or striped appearance. The **sarcomere**, bounded by the dark Z lines, forms one complete contractile unit.

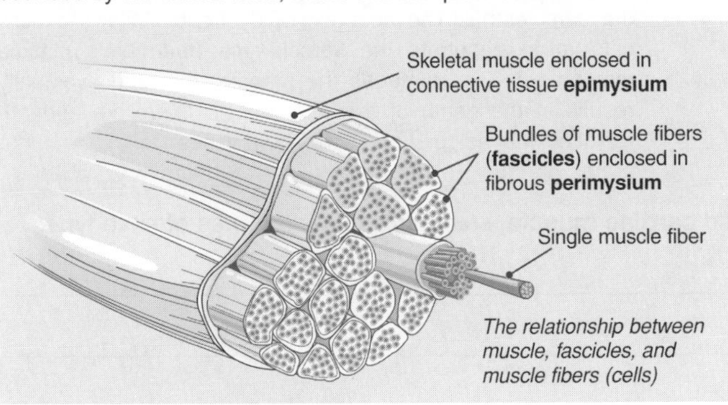


*The relationship between muscle, fascicles, and muscle fibers (cells)*

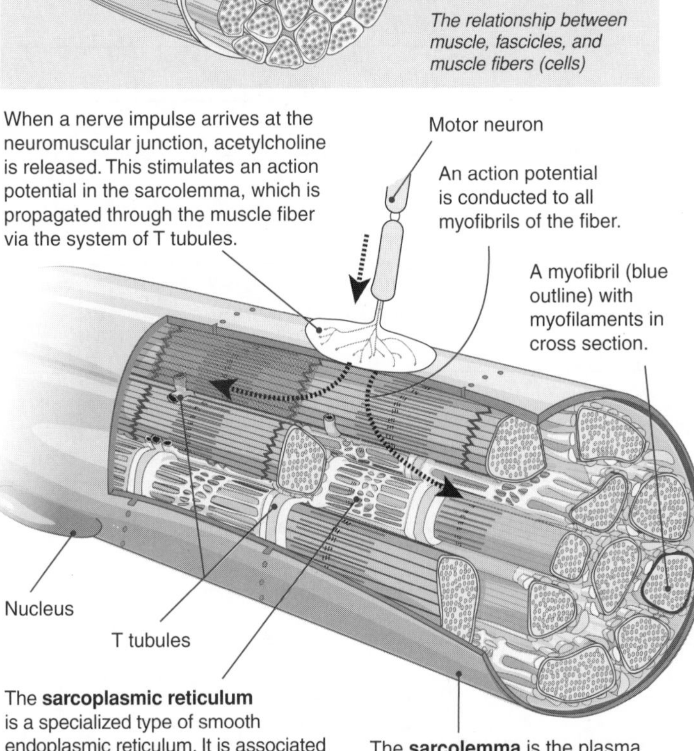


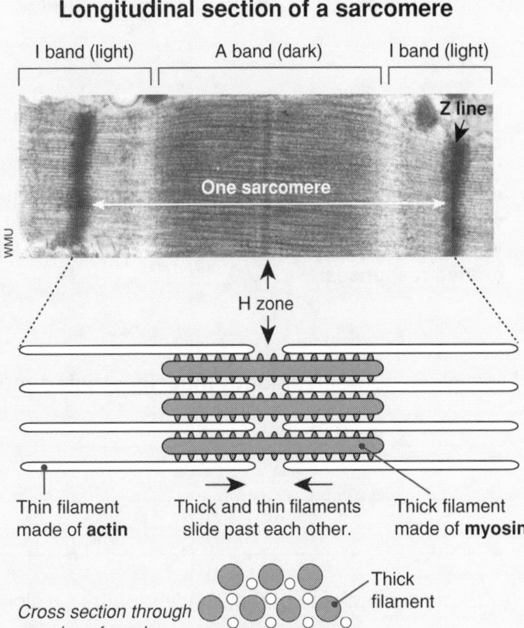


The photograph of a sarcomere (above) shows the banding pattern arising as a result of the arrangement of thin and thick filaments. It is represented schematically in longitudinal section and cross section.

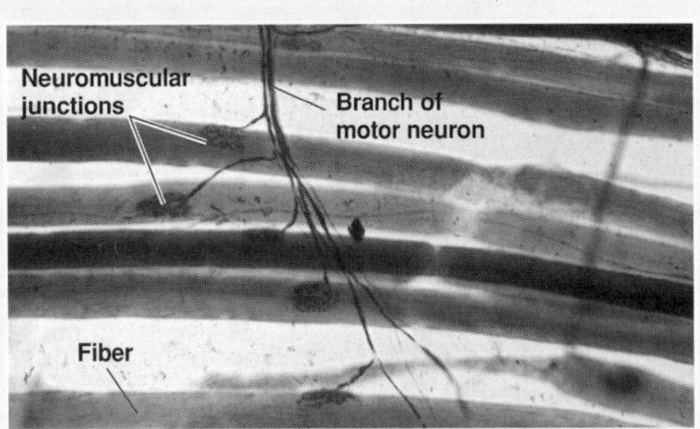


Above: Axon terminals of a motor neuron supplying a muscle. The branches of the axon terminate on the sarcolemma of a fiber at regions called the neuromuscular junction. Each fiber receives a branch of an axon, but one axon may supply many muscle fibers.

## The Banding Pattern of Myofibrils

Within a myofibril, the thin filaments, held together by the **Z lines**, project in both directions. The arrival of an action potential sets in motion a series of events that cause the thick and thin filaments to slide past each other. This is called **contraction** and it results in shortening of the muscle fiber and is accompanied by a visible change in the appearance of the myofibril: the I band and the sarcomere shorten and H zone shortens or disappears (below).

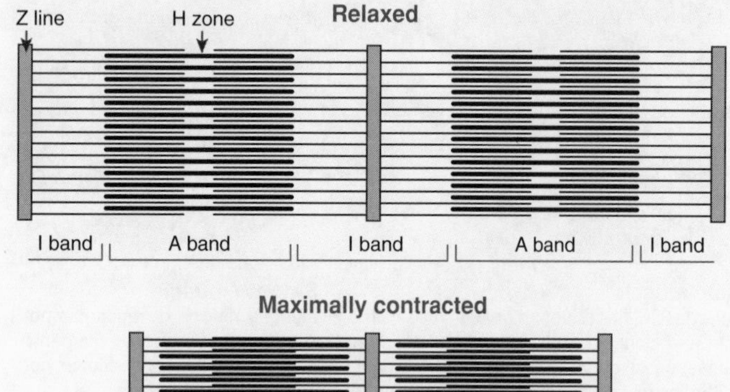


The response of a single muscle fiber to stimulation is to contract maximally or not at all; its response is referred to as the **all-or-none law** of muscle contraction. If the stimulus is not strong enough to produce an action potential, the muscle fiber will not respond. However skeletal muscles as a whole are able to produce varying levels of contractile force. These are called **graded responses**.

## When Things Go Wrong

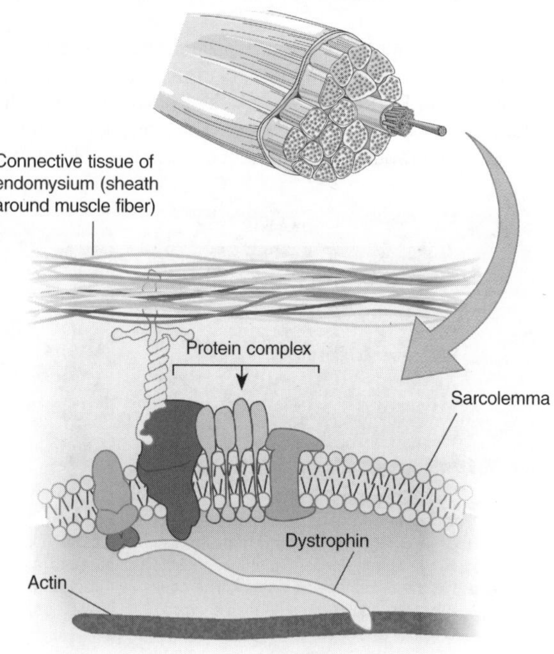


Duchenne's muscular dystrophy is an X-linked disorder caused by a mutation in the gene DMD, which codes for the protein **dystrophin**. The disease causes a rapid deterioration of muscle, eventually leading to loss of function and death. It is the most prevalent type of muscular dystrophy and affects only males. Dystrophin is an important structural component within muscle tissue and it connects muscles fibers to the extracellular matrix through a protein complex on the sarcolemma. The absence of dystrophin allows excess calcium to penetrate the sarcolemma (the fiber's plasma membrane). This damages the sarcolemma, and eventually results in the death of the cell. Muscle fibers die and are replaced with adipose and connective tissue.

1. Distinguish between **smooth muscle**, **striated muscle**, and **cardiac muscle**, summarizing the features of each type:

2. (a) Explain the cause of the banding pattern visible in striated muscle:

   (b) Explain the change in appearance of a myofibril during contraction with reference to the following:

   The I band:

   The H zone:

   The sarcomere:

3. Describe the purpose of the connective tissue sheaths surrounding the muscle and its fascicles:

4. Explain what is meant by the all-or-none response of a muscle fiber:

5. Explain why the inability to produce **dystrophin** leads to a loss of muscle function:

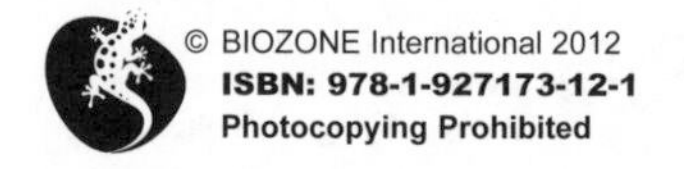


ISBN: 978-1-927173-12-1

# The Sliding Filament Theory

The previous activity described how muscle contraction is achieved by the thick and thin muscle filaments sliding past one another. This sliding is possible because of the structure and arrangement of the thick and thin filaments. The ends of the thick myosin filaments are studded with heads or **cross bridges** that can link to the thin filaments next to them. The thin filaments contain the protein actin, but also a regulatory protein complex. When the cross bridges of the thick filaments connect to the thin filaments, a shape change moves one filament past the other. Two things are necessary for cross bridge formation: calcium ions, which are released from the **sarcoplasmic reticulum** when the muscle receives an action potential, and ATP, which is hydrolyzed by ATPase enzymes on the myosin. When cross bridges attach and detach in sarcomeres throughout the muscle cell, the cell shortens. Although a muscle fiber responds to an action potential by contracting maximally, skeletal muscles as a whole can produce varying levels of contractile force. These **graded responses** are achieved by changing the frequency of stimulation (**frequency summation**) and by changing the number and size of motor units recruited (**multiple fiber summation**). Maximal contractions of a muscle are achieved when nerve impulses arrive at the muscle at a rapid rate and a large number of motor units are active at once.

### The Sliding Filament Theory

Muscle contraction requires calcium ions ($Ca^{2+}$) and energy (in the form of ATP) in order for the thick and thin filaments to slide past each other. The steps are:

1. The binding sites on the **actin** molecule (to which myosin 'heads' will locate) are blocked by a complex of two protein molecules: tropomyosin and troponin.
2. Prior to muscle contraction, ATP binds to the heads of the myosin molecules, priming them in an erect high energy state. Arrival of an action potential causes a release of $Ca^{2+}$ from the sarcoplasmic reticulum. The $Ca^{2+}$ binds to the troponin and causes the blocking complex to move so that the myosin binding sites on the actin filament become exposed.
3. The heads of the cross-bridging myosin molecules attach to the binding sites on the actin filament. Release of energy from the hydrolysis of ATP accompanies the cross bridge formation.
4. The energy released from ATP hydrolysis causes a change in shape of the myosin **cross bridge**, resulting in a bending action (*the power stroke*). This causes the actin filaments to slide past the myosin filaments towards the centre of the sarcomere.
5. (Not illustrated). Fresh ATP attaches to the myosin molecules, releasing them from the binding sites and repriming them for a repeat movement. They become attached further along the actin chain as long as ATP and $Ca^{2+}$ are available.

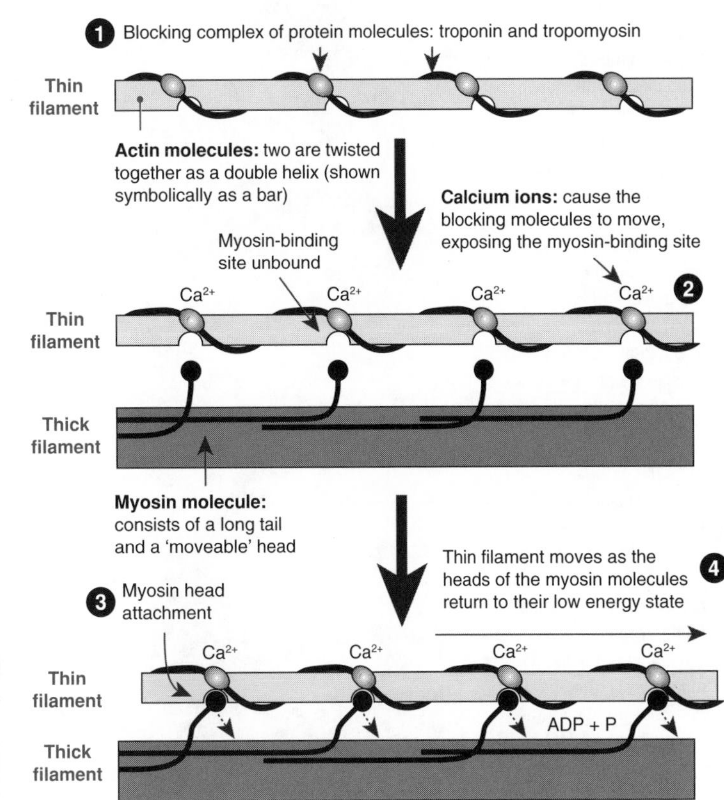


1. Match the following chemicals with their functional role in muscle movement (draw a line between matching pairs):

   (a) Myosin • Bind to the actin molecule in a way that prevents myosin head from forming a cross bridge

   (b) Actin • Supplies energy for the flexing of the myosin 'head' (power stroke)

   (c) Calcium ions • Has a moveable head that provides a power stroke when activated

   (d) Troponin-tropomyosin • Two protein molecules twisted in a helix shape that form the thin filament of a myofibril

   (e) ATP • Bind to the blocking molecules, causing them to move and expose the myosin binding site

2. Describe the two ways in which a muscle as a whole can produce contractions of varying force:

   (a) ______________________________

   ______________________________

   (b) ______________________________

   ______________________________

3. (a) Identify the two things necessary for cross bridge formation: ______________________________

   (b) Explain where each of these comes from: ______________________________

   ______________________________


ISBN: 978-1-927173-12-1


***Related activities:*** *Muscle Structure and Function*
***Weblinks:*** *Muscle Cell Contraction, Sliding Filament*

# Muscle Tone and Posture

Even when we consciously relax a muscle, a few of its fibers at any one time will be involuntarily active. This continuous and passive partial contraction of the muscles is responsible for **muscle tone** and is important in maintaining **posture**. The contractions are not visible but they are responsible for the healthy, firm appearance of muscle. The amount of muscle contraction is monitored by sensory receptors in the muscle called **muscle spindle organs**. These provide the sensory information necessary to adjust movement as required. Abnormally low muscle tone (**hypotonia**) can arise as a result of traumatic or degenerative nerve damage, so that the muscle no longer receives the innervation it needs to contract. The principal treatment for these disorders is physical therapy to help the person compensate for the neuromuscular disability.

We are usually not aware of the skeletal muscles that maintain posture, although they work almost continuously making fine adjustments to maintain body position. Both posture and functional movements of the body are highly dependent on the strength of the body's core (the muscles in the pelvic floor, belly, and mid and lower back). The core muscles stabilize the thorax and the pelvis and lack of core strength is a major contributor to postural problems and muscle imbalances.

Physical therapy is a branch of health care concerned with maintaining or restoring functional movement throughout life. Loss of muscle tone and strength can develop as a result of aging, disease, or trauma. As a result of not being used, muscles will **atrophy**, losing both mass and strength. Although the type of physical therapy depends on the problem, it usually includes therapeutic exercise to help restore mobility and strength, and prevent or slow down the loss of muscle tissue.

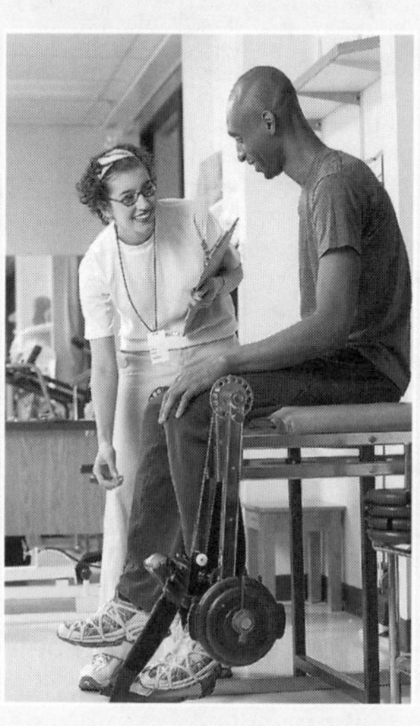

## The Role of the Muscle Spindle

Changes in length of a muscle are monitored by the muscle spindle organ, a stretch receptor located within skeletal muscle, parallel to the muscle fibers themselves. The muscle spindle is stimulated in response to sustained or sudden stretch on the central region of its specialized intrafusal fibers. Sensory information from the muscle spindle is relayed to the spinal cord. The motor response brings about adjustments to the degree of stretch in the muscle. These adjustments help in the coordination and efficiency of muscle contraction. Muscle spindles are important in the maintenance of muscle tone, postural reflexes, and movement control, and are concentrated in muscles that exert fine control over movement.

**Motor nerves** send impulses to adjust the degree of contraction in the intrafusal and extrafusal fibers.

**Sensory nerves** monitor stretch in the non-contractile region of the spindle and send impulses to the spinal cord.

Striated appearance of contractile elements

Nucleus of muscle fiber

The **muscle spindle organ** comprises special **intrafusal fibers** which lie parallel to the muscle fibers within a lymph-filled capsule. Only the regions near the end can contract.

The spindle is surrounded by the muscle fibers of the skeletal muscle.

1. (a) Explain what is meant by muscle tone: ______

   (b) Explain how this is achieved: ______

2. (a) Explain the role of the muscle spindle organ: ______

   (b) With reference to the following, describe how the structure of the muscle spindle organ is related to its function:

   Intrafusal fibers lie parallel to the extrafusal fibers: ______

   Sensory neurons are located in the non-contractile region of the organ: ______

   Motor neurons synapse in the extrafusal fibers and the contractile region of the intrafusal fibers: ______




**ISBN: 978-1-927173-12-1**

# KEY TERMS: Crossword



Complete the crossword below, which will test your understanding of key terms in this chapter and their meanings

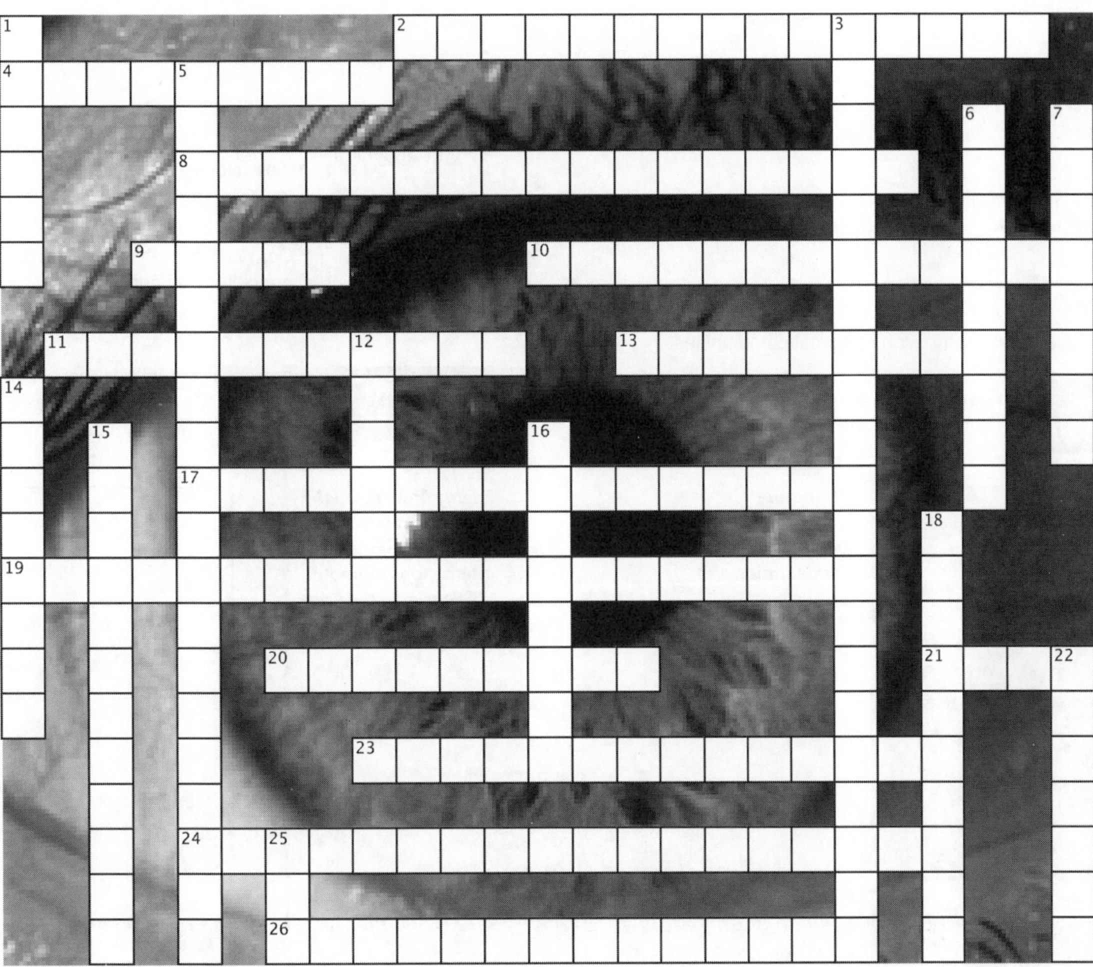

## Clues Across

2. A self propagating nerve impulse. (2 words: 6, 9)
4. Motor nerves carry impulses from the central nervous system to these.
8. These chemicals relay signals between a neuron and another cell.
9. Along with the spinal cord, this organ comprises the central nervous system.
10. Specialized striated muscle found only in the heart. It does not fatigue. (2 words: 7, 6)
11. These synapses release acetylcholine.
13. This region connects the brain to the spinal cord. (2 words: 5, 4)
17. The potential difference across the cell membrane of a neuron when there is no impulse passing. (2 words: 7, 9)
19. A sensory receptor that is specialized to detect stimuli and respond by producing an electrical discharge. (2 words: 10, 10)
20. The sense of smell.
21. The sense organs that respond to light.
23. Another name for striated muscle. (2 words: 8, 6)
24. The membrane potential to which a membrane must be depolarized to initiate an action potential (2 words: 9, 9)
26. A temporary change in membrane potential caused by influx of sodium ions.

## Clues Down

1. A cell specialized to transmit electrical impulses.
3. The site at which an axon terminal contacts a muscle cell. (2 words: 13, 8)
5. The portion of the nervous system that comprises the brain and spinal cord. (3 words: 7, 7, 6)
6. A complete contractile unit in skeletal muscle.
7. A specialized cell that detects stimuli and responds by producing a nerve impulse.
12. The region of the eye containing the photoreceptors.
14. The largest and most anterior part of the human brain.
15. Type of muscle lining the alimentary canal, blood vessels, and many hollow organs. (2 words: 6, 6)
16. This neurotransmitter can affect behavior and mood.
18. The region of the human brain which coordinates body movement, posture and balance.
22. The gap between neighboring neurons or between a neuron and an effector.
25. A photoreceptor found in the retina, specialized for vision in dim light.

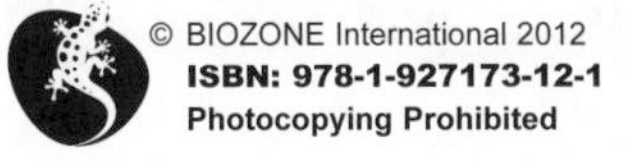


ISBN: 978-1-927173-12-1

# Ecology

## Important in this section ...

- *Understand ecosystem structure and function.*
- *Understand population dynamics and how population changes can be modeled.*
- *Understand how human activity can influence ecosystem stability and accelerate change.*

## The Nature of Ecosystems

| | |
|---|---|
| Biomes and environmental gradients | • The world's biomes<br>• Distribution and physical gradients |
| Habitat and niche | • Habitat as a component of niche<br>• Niche and adaptation |
| Sampling communities | • Sampling communities<br>• Quadrats, transects, mark-recapture |
| Community strutcure | • Stratification<br>• Zonation<br>• Distribution in animal communities |
| **Connect 4.A.5 with 2.D.1c** | *Community composition is affected by complex biotic and abiotic interactions* |

*The nature and stability of populations, communities, and ecosystems are determined by biotic and abiotic interactions.*

## Energy Flow and Nutrient Cycles

| | |
|---|---|
| Energy flow | • Food chains and webs<br>• Production and trophic efficiency<br>• Quantifying energy flow<br>• Energy flow in unusual systems |
| Nutrient cycles | • Principles of nutrient cycling<br>• The role of the microbial community<br>• The hydrologic cycle<br>• The nitrogen cycle |
| Human influences | • Comparing energy flows in natural and agricultural systems<br>• Nitrogen pollution |
| **Connect 2.A.2 with 4.A.6** | *Autotrophs capture free energy to power the production of organic molecules. Food webs depend on their activities.* |

*Energy flows through ecosystems but matter is recycled. Biotic interactions promote efficiencies in the use of energy and matter.*

*Populations exhibit properties not shown by individuals. Interactions between individuals contribute to these properties.*

## Populations

| | |
|---|---|
| Features of populations | • Density and distribution<br>• Age structure and survivorship<br>• Population regulation<br>• Models of population growth |
| Interactions within and between populations | • Competition and niche differentiation<br>• Predator-prey interactions<br>• Symbiotic relationships<br>• Human demographics |
| Field work | • Methods for studying communities |
| **Connect 4.B.3 with 4.A.5** | *Population interactions influence distribution and abundance and their effects can be modeled.* |
| **Connect 4.A.5 with 3.E.1** | *Natural selection favors those attributes that increase individual fitness and enhance population survival.* |

*Ecosystems may change as a result of natural events. Human activity can accelerate change at global and local levels.*

## The Diversity and Stability of Ecosystems

| | |
|---|---|
| Stability and diversity | • What is biodiversity?<br>• Threats to biodiversity<br>• Disturbance and community structure<br>• Ecological succession |
| Human influences on ecosystem structure and function | • Global warming and its effects<br>• Ice sheet melting<br>• Ocean acidification<br>• Climate change models<br>• Tropical deforestation<br>• The impact of alien species<br>• Species responses to change |
| **Connect 4.A.6 with 2.D.3** | *Human activities have an impact on natural ecosystems that can disrupt the system's dynamic homeostasis.* |

Enduring Understanding

2.D
4.A

# The Nature of Ecosystems

## Key concepts

- The world's biomes are distributed according to broad zones of climate, altitude, and latitude.
- Environmental gradients and biotic interactions contribute to community diversity.
- Distributional variation in communities can be described and quantified.
- The realized niche of a species is a function of its tolerance range and biotic interactions.

## Key terms

abiotic (=physical) factor
adaptation
altitude
belt transect
biome
biotic factor
community
distribution
ecological niche
ecosystem
environmental gradient
habitat
latitude
line transect
limiting factor
microclimate
microhabitat
niche breadth
physical (=abiotic) factor
population
quadrat
stratification
tolerance range
transect
zonation

## Essential Knowledge

☐ 1. Use the **KEY TERMS** to compile a glossary for this topic.

### Community Structure *(2.D.1: c)* pages 266-68, 273-78 281-85, 292-93

☐ 2. Explain the effect of **latitude**, **altitude**, and **rainfall** in determining the distribution of world **biomes**. Describe the components of an **ecosystem**.

☐ 3. Understand what is meant by **habitat** and **tolerance range**. Recognize habitat as one component of the **ecological niche**. Distinguish between the **fundamental** and the **realized niche** and describe the factors affecting **niche breadth**. Describe and explain the relationship of niche breadth to community diversity.

☐ 4. Describe examples to show how **biotic** and **abiotic** (physical) **factors** influence the structure and stability of **populations**, **communities**, and **ecosystems**. Examples could include **environmental gradients** and community diversity (as below), trophic relationships in food chains and webs, habitat diversity and reproductive success, and suppression of population density by competition.

☐ 5. Explain how gradients in abiotic factors can occur over relatively short distances, e.g. on a rocky shore, or in a forest, desert, or lake. Explain the role of these environmental gradients in influencing species **distribution** and in contributing to community diversity.

☐ 6. Describe examples of distributional variation in communities, e.g. **zonation** or **stratification**. Explain how these patterns arise and how they increase community diversity.

### Sampling Communities *(4.A.5: a)* pages 269-72, 279-80, 286-87, 289-91

☐ 7. Describe techniques used to quantify distributional variation in communities, e.g. **belt transects** and **line transects** using point sampling or **quadrats**. Appreciate advantages and limitations of different methods with respect to sampling time, cost, and suitability to habitat.

☐ 8. Explain how community structure is described in terms of **species diversity** and the composition of species populations.

### Adaptation to Environment *(4.A.6: g)* pages 273-75

☐ 9. Describe **adaptations** for survival in a given niche. Use examples to illustrate how the adaptations of organisms are related to obtaining and using energy and matter in a particular environment.

☐ 10. Understand that adaptations are the result of evolutionary changes to the species, but not to individuals within their own lifetimes.

*Periodicals:*
*Listings for this chapter are on page 375*

*Weblinks:*
*www.thebiozone.com/weblink/AP2-3121.html*

*BIOZONE APP:*
*Student Review Series Introduction to Ecosystems, The Ecological Niche*

# Biomes

The Earth's **biomes** are the largest geographically based biotic communities that can be conveniently recognized. These are large areas where the vegetation type shares a particular suite of physical requirements. Biomes have characteristic features, but the boundaries between them are not distinct. The same biome may occur in widely separated regions of the world wherever the climatic and soil conditions are similar. Terrestrial biomes are recognized for all the major climatic regions of the world. They are classified by their predominant vegetation type.

## Earth's Climate and Biomes

Biomes are closely related to the major air cells that circle the Earth and are reflected in the Northern and Southern Hemispheres.

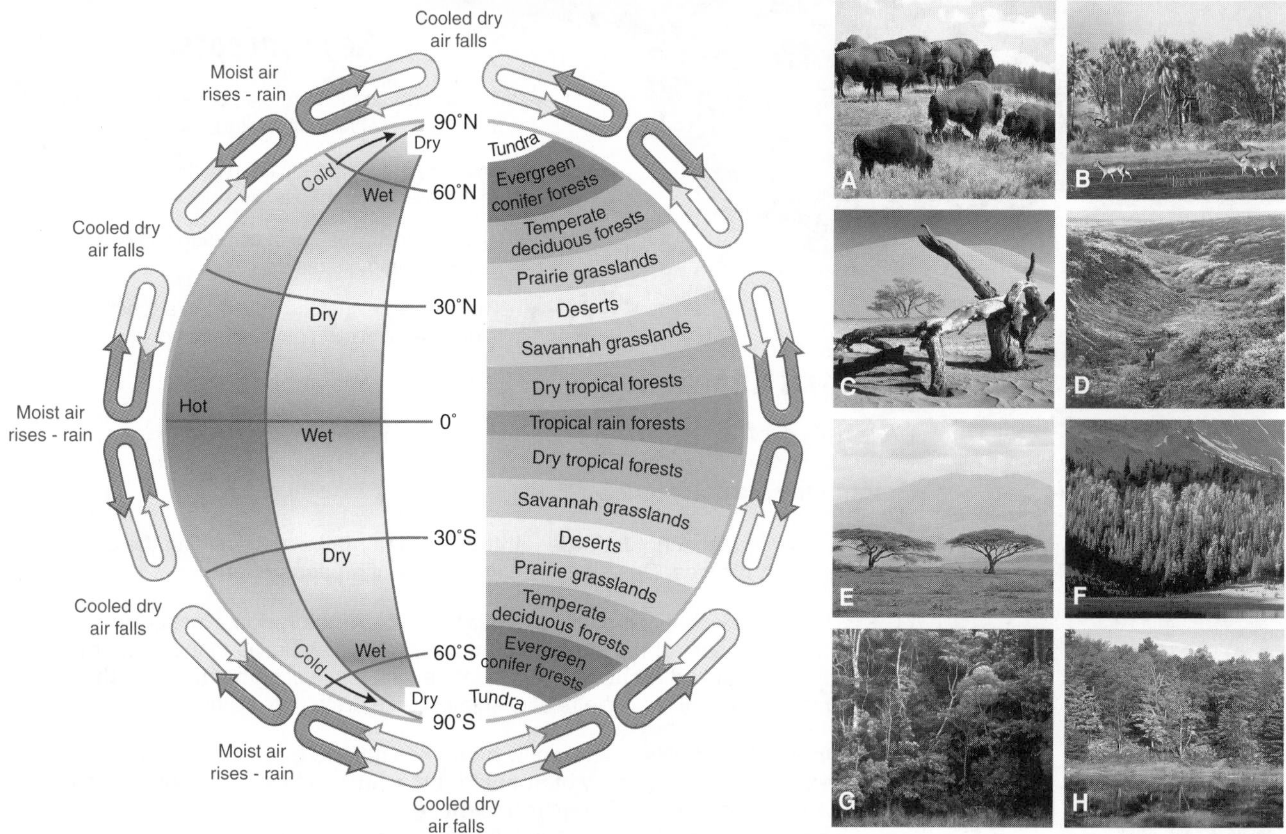


## Biomes and Landscapes

Climate is heavily modified by the landscape. Where there are large mountain ranges, wind is deflected upwards causing rain on the windward side and a **rain shadow** on the leeward side. The biome that results from this is considerably different from the one that may have appeared with no wind deflection. Large expanses of ocean and flat land also change the climate by modifying air temperatures and the amount of rainfall.

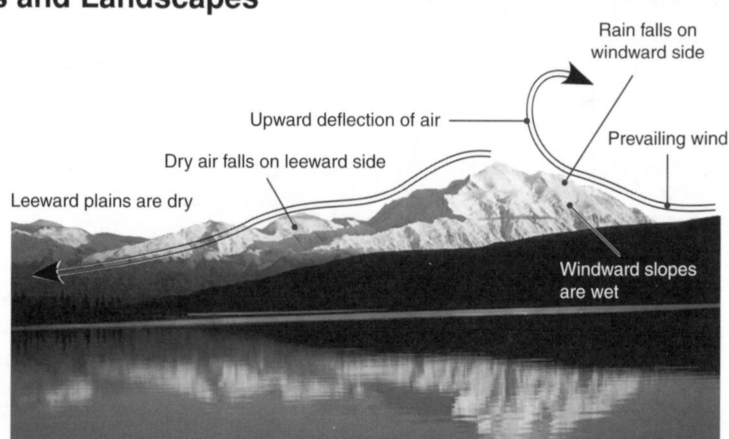


1. Match the lettered biome images (above, right) with the appropriate biome name:

(a) Tundra: ______________________

(b) Temperate deciduous forest: ______________________

(c) Deserts: ______________________

(d) Dry tropical forests: ______________________

(e) Evergreen conifer forest: ______________________

(f) Prairie grasslands: ______________________

(g) Savannah grasslands: ______________________

(h) Tropical rain forests: ______________________

2. Identify which abiotic factor(s) limit the extent of temperate deciduous forests: ______________________

______________________________________________



***Related activities**: Components of an Ecosystem*
***Weblinks**: The World's Biomes*

***Periodicals**:*
*Grasslands*

**ISBN: 978-1-927173-12-1**

# Components of an Ecosystem

The concept of the ecosystem was developed to describe the way groups of organisms are predictably found together in their physical environment. A community comprises all the organisms within an ecosystem. The structure and function of a community is determined by the physical (abiotic) and biotic factors, which determine species distribution and survival.

### What is the Biosphere?

The biosphere is a narrow belt around the Earth containing all the Earth's living organisms. It extends from the bottom of the oceans to the upper atmosphere. Broad scale life-zones or **biomes** within the biosphere are characterized according to the predominant vegetation. Within these biomes, ecosystems form natural units comprising the non-living, physical environment (the **atmosphere**, **water**, and **soil**) and the community.

**Physical Environment Abiotic Factors:**

**Atmosphere**
- Wind speed and direction
- Light intensity and quality
- Precipitation and humidity
- Air temperature

**Soil**
- Nutrient availability
- Soil moisture and pH
- Composition
- Temperature

**Water**
- Dissolved nutrients
- pH and salinity
- Dissolved oxygen
- Temperature

**Community: Biotic Factors**
- **Producers** • Consumers
- Detritivores • Decomposers

**Interact in the community as**: Competitors, parasites, pathogens, symbionts, predators, herbivores

123Rf Tomas Sobek



1. Distinguish clearly between a community and an ecosystem: ______________________

2. Distinguish between biotic and abiotic factors: ______________________

3. Use one or more of the following terms to describe each of the features of a beech community listed below:
**Terms**: *population, community, ecosystem, physical factor.*

(a) All the beech trees present: ____________ (c) All the organisms present: ____________

(b) The entire forest: ____________ (d) The humidity: ____________

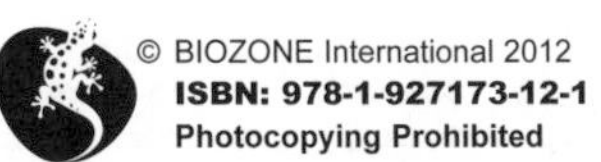


ISBN: 978-1-927173-12-1


*Periodicals:* *Getting to grips with ecology*

***Weblinks:*** *racerocks.com*

# Habitat

The environment in which a species population (or an individual organism) lives (including all the physical and biotic factors) is termed its **habitat**. Within a prescribed habitat, each species population has a range of tolerance to variations in its physical and chemical environment. Within the population, individuals will have slightly different tolerance ranges based on small differences in genetic make-up, age, and health. The wider an organism's tolerance range for a given abiotic factor (e.g. temperature or salinity), the more likely it is that the organism will be able to survive variations in that factor. Species **dispersal** is also strongly influenced by **tolerance range**. The wider the tolerance range of a species, the more widely dispersed the organism is likely to be. As well as a tolerance range, organisms have a narrower **optimum range** within which they function best. This may vary from one stage of an organism's development to another or from one season to another. Every species has its own optimum range. Organisms will usually be most abundant where the abiotic factors are closest to the optimum range.

## Habitat Occupation and Tolerance Range

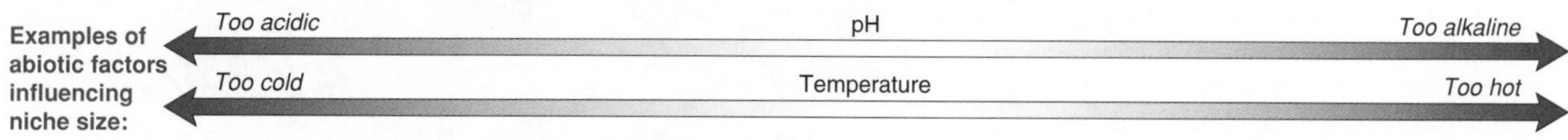


The law of tolerances states that "*for each abiotic factor, a species population (or organism) has a tolerance range within which it can survive. Toward the extremes of this range, that abiotic factor tends to limit the organism's ability to survive*".

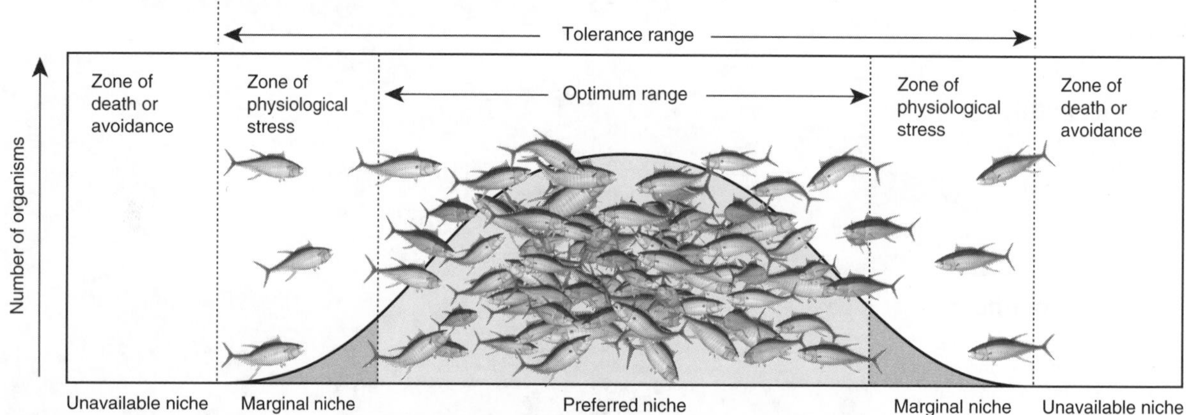


## The Scale of Available Habitats

A habitat may be vast and relatively homogeneous, as is the open ocean. Barracuda (above) occur around reefs and in the open ocean where they are aggressive predators.

For non-mobile organisms, such as the fungus above, a suitable habitat may be defined by the particular environment in a relatively tiny area, such as on this decaying log.

For microbial organisms, such as the bacteria and protozoans of the ruminant gut, the habitat is defined by the chemical environment within the rumen (R) of the host animal, in this case, a cow.

1. Explain how an organism's habitat occupation relates to its tolerance range: ______________________

______________________

______________________

2. (a) Identify the range in the diagram above in which most of the species population is found. Explain why this is the case:

______________________

______________________

(b) Describe the greatest constraints on an organism's growth and reproduction within this range: ______________________

______________________

3. Describe some probable stresses on an organism forced into a marginal niche: ______________________

______________________

______________________



**Related activities:** *The Ecological Niche*


**ISBN: 978-1-927173-12-1**

# Sampling Communities

In most field studies, it is not possible to measure or count every member of a population. Instead, the population is **sampled** in a way that provides a fair (unbiased) representation of the organisms present and their distribution. This is usually achieved through **random sampling**, a technique in which every possible sample of a given size has the same chance of selection. Most practical exercises in community ecology involve collecting or recording data from living organisms. The technique you use to sample must be appropriate to the community being studied and the information you want to obtain. You must also think about the time and equipment available, the organisms involved, and the impact your study might have on the environment.

## Quantifying the Diversity of Ecosystems

Reef community:
high density, clumped distribution

The methods we use to sample communities and their constituent populations must be appropriate to the ecosystem being investigated. Communities in which the populations are at low density and have a random or clumped distribution will require a different sampling strategy to those where the populations are uniformly distributed and at higher density. There are many sampling options, each with advantages and drawbacks for particular communities. How would you assess aspects (e.g. species abundance or distribution) of the reef community above?

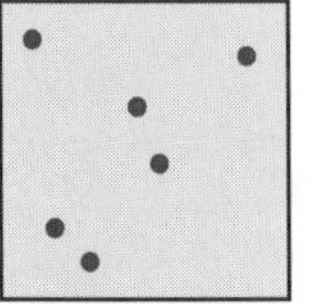

*Random point sampling*

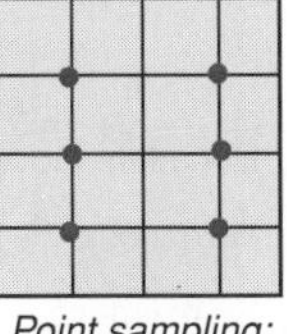

*Point sampling: systematic grid*

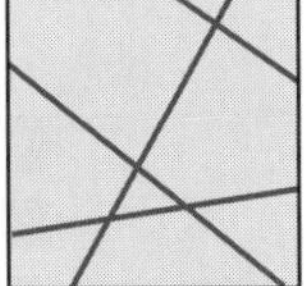

*Line and belt transects*

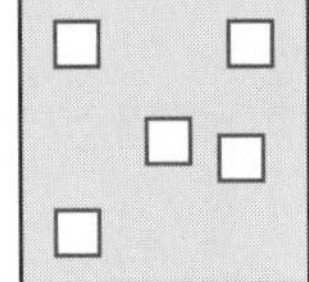

*Random quadrats*

Photo: www.coastal planning.net

Marine ecologists use quadrat sampling to estimate biodiversity prior to works such as dredging.

RA

Line transects are appropriate to estimate biodiversity along an environmental gradient.

Tagging has been used for more than 30 years to follow the migration of monarch butterflies. The photograph here depicts an older tagging method, which has largely been replaced by a tag on the underside of the hindwing (inset). The newer method results in better survival and recapture rates and interferes less with flight.

## Which Sampling Method?

Field biologists take a number of factors into consideration when deciding on a sampling method for a chosen population or community. The benefits and drawbacks of some common methods are outlined below:

**Point sampling** is time efficient and good for determining species abundance and community composition. However, organisms in low abundance may be missed.

**Transects** are well suited to determining changes in community composition along an environmental gradient but can be time consuming to do well.

**Quadrats** are also good for assessments of community diversity and composition but are largely restricted to plants and immobile animals. Quadrat size must also be appropriate for the organisms being sampled.

**Mark and recapture** is useful for highly mobile species which are otherwise difficult to record. However, it is time consuming to do well. **Radio-tracking** offers an alternative to mark and recapture and is now widely used in conservation to study the movements of both threatened species and pests.

## Sensors and Meters

Various meters can be used to quantify aspects of the physical environment, including the pH, temperature, light levels, and turbidity. Meters that measure single factors have now largely been replaced by multi-purpose meters.

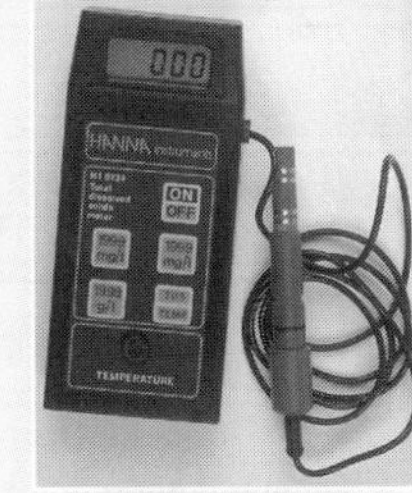

**Total dissolved solids (TDS) meter**: Measures the content of dissolved solids (as ions) in water in mg $L^{-1}$ giving an indication of water quality. The probe measures the conductivity of the water to approximate the level of TDS. TDS can also be measured gravimetrically by evaporating a sample leaving the residue behind.

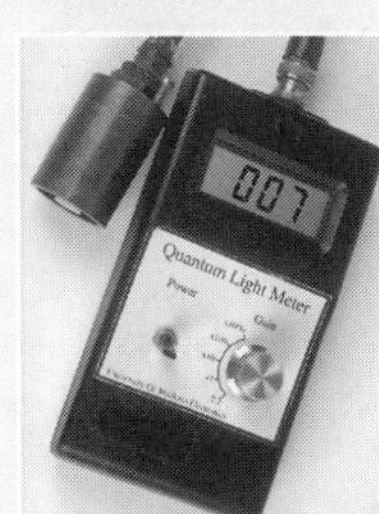

**Quantum light meter**: Measures light intensity levels but not light quality (wavelength). Light levels can change dramatically from a forest floor to its canopy. A light meter provides a quantitative measure of these changes, many of which are not detectable with our own visual systems.

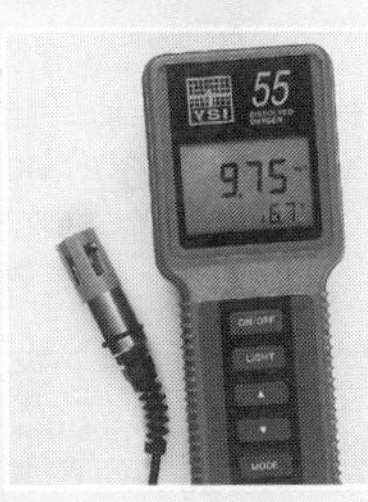

**Dissolved oxygen meter**: This measures the amount of oxygen dissolved in water (as mg $L^{-1}$), which gives and indication of water quality and suitability to support organisms such as fish. The **Winkler** method uses a titration of $MnSO_4$, KI, and $K_2S_2O_3$ to determine the concentration of $O_2$.

The Nature of Ecosystems

ISBN: 978-1-927173-12-1


***Related activities:*** *Quadrat Sampling, Transect Sampling, Mark and Recapture*
***Weblinks:*** *Ecological Sampling Methods*

# Using Dataloggers in Field Studies

Usually, when we collect information about populations in the field, we also collect information about the physical environment. This provides important information about the local habitat and can be useful in assessing habitat preference. With **dataloggers,** collecting this information is straightforward.

Dataloggers are electronic instruments that record measurements over time. They are equipped with a microprocessor, data storage facility, and sensor. Different sensors are used to measure a range of variables in water or air. The datalogger is connected to a computer, and software is used to set the limits of operation (e.g. the sampling interval) and initiate the logger. The logger is then disconnected and used remotely to record and store data. When reconnected to the computer, the data are downloaded, viewed, and plotted. Dataloggers make data collection quick and accurate, and they enable prompt data analysis.

Datalogger photos courtesy of PASCO

Dataloggers fitted with sensors are portable and easy to use in a wide range of aquatic (left) and terrestrial (right) environments. Different variables can be measured by changing the sensor attached to the logger.

1. Why do we **sample** populations? ______________________________

2. Describe a sampling technique that would be appropriate for determining each of the following:

   (a) The percentage cover of a plant species in pasture: ______________________________

   (b) The density and age structure of a plankton population: ______________________________

   (c) Change in community composition from low to high altitude on a mountain: ______________________________

3. Why is it common practice to also collect information about the physical environment when sampling populations?

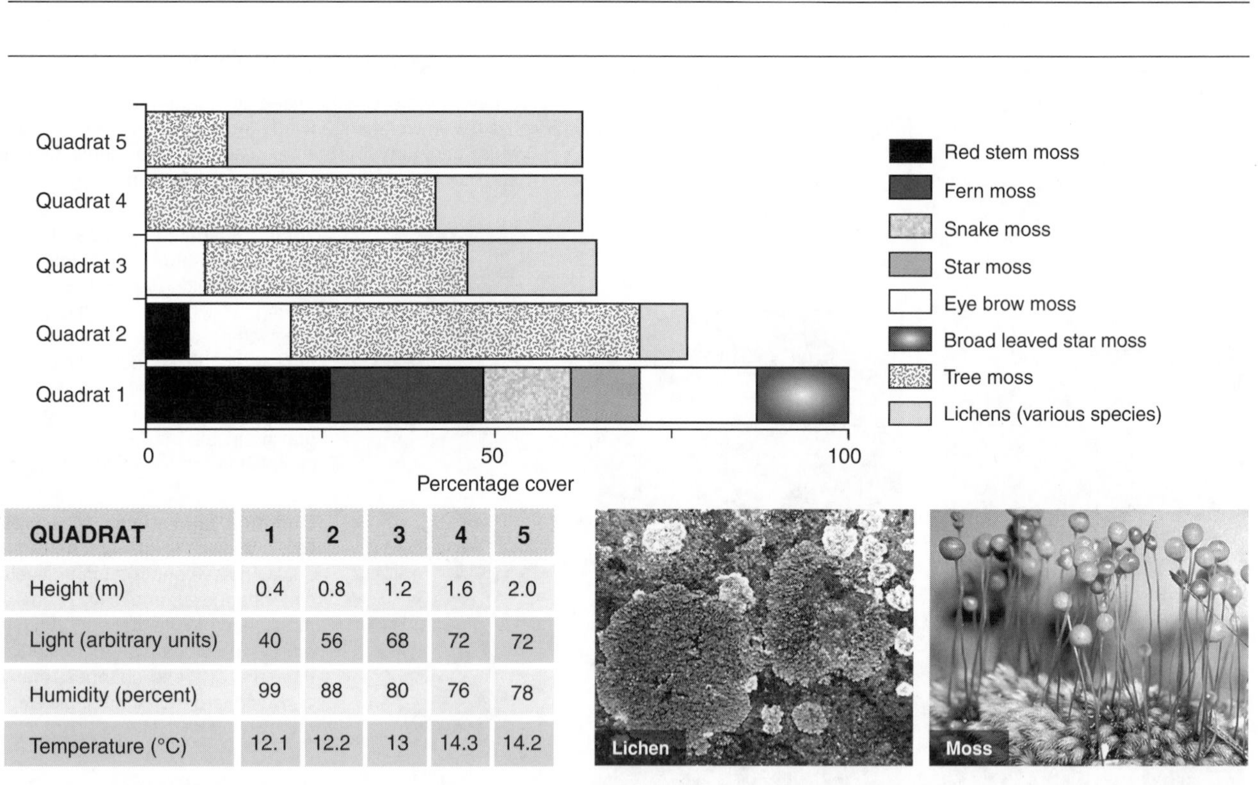


| QUADRAT | 1 | 2 | 3 | 4 | 5 |
|---|---|---|---|---|---|
| Height (m) | 0.4 | 0.8 | 1.2 | 1.6 | 2.0 |
| Light (arbitrary units) | 40 | 56 | 68 | 72 | 72 |
| Humidity (percent) | 99 | 88 | 80 | 76 | 78 |
| Temperature (°C) | 12.1 | 12.2 | 13 | 14.3 | 14.2 |

4. The figure (above) shows the changes in vegetation cover along a 2 m vertical transect up the trunk of an oak tree (*Quercus*). Changes in the physical factors light, humidity, and temperature along the same transect were also recorded. From what you know about the ecology of mosses and lichens, account for the observed vegetation distribution:


ISBN: 978-1-927173-12-1

# Quadrat Sampling

**Quadrat sampling** is a method by which organisms in a certain proportion (sample) of the habitat are counted directly. It is used when the organisms are too numerous to count in total. It can be used to estimate population **abundance** (number), **density, frequency of occurrence**, and **distribution**. Quadrats may be used without a transect when studying a relatively uniform habitat. In this case, the quadrat positions are chosen randomly using a random number table.

The general procedure is to count all the individuals (or estimate their percentage cover) in a number of quadrats of known size and to use this information to work out the abundance or percentage cover value for the whole area. The number of quadrats used and their size should be appropriate to the type of organism involved (e.g. grass vs tree).

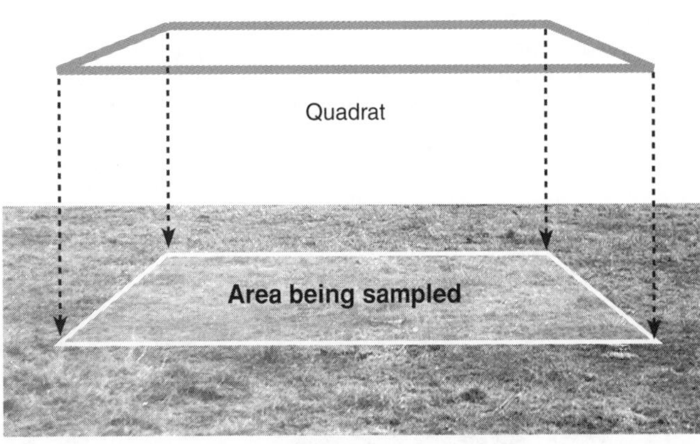


$$\text{Estimated average density} = \frac{\text{Total number of individuals counted}}{\text{Number of quadrats X area of each quadrat}}$$

### Guidelines for Quadrat Use:

1. The **area of each quadrat** must be known exactly and ideally quadrats should be the same shape. The quadrat does not have to be square (it may be rectangular, hexagonal etc.).
2. **Enough quadrat samples** must be taken to provide results that are representative of the total population.
3. The **population of each quadrat** must be known exactly. Species must be distinguishable from each other, even if they have to be identified at a later date. It has to be decided beforehand what the count procedure will be and how organisms over the quadrat boundary will be counted.
4. The size of the quadrat should be appropriate to the organisms and habitat, e.g. a large size quadrat for trees.
5. The quadrats must be **representative of the whole area.** This is usually achieved by **random sampling** (right).

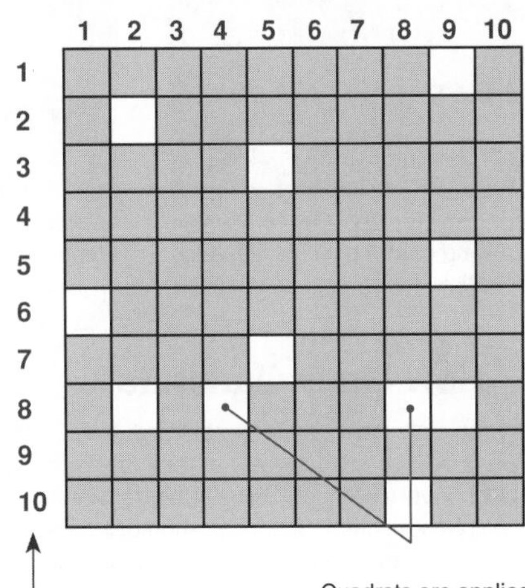


The area to be sampled is divided up into a grid pattern with indexed coordinates

Quadrats are applied to the predetermined grid on a random basis. This can be achieved by using a random number table.



### Sampling a centipede population

A researcher by the name of Lloyd (1967) sampled centipedes in Wytham Woods, near Oxford in England. A total of 37 hexagon–shaped quadrats were used, each with a diameter of 30 cm (see diagram on right). These were arranged in a pattern so that they were all touching each other. Use the data in the diagram to answer the following questions.

1. Determine the average number of centipedes captured per quadrat:

___

2. Calculate the estimated average density of centipedes per square meter (remember that each quadrat is 0.08 square meters in area):

___

3. Looking at the data for individual quadrats, describe in general terms the distribution of the centipedes in the sample area:

___

___

4. Describe one factor that might account for the distribution pattern:

___

___

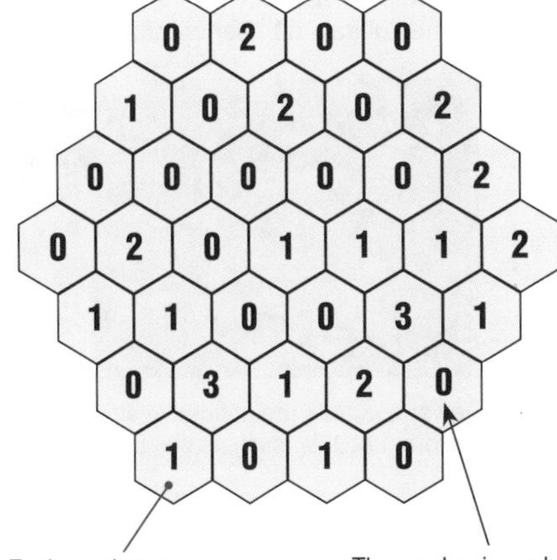


Each quadrat was a hexagon with a diameter of 30 cm and an area of 0.08 square meters.

The number in each hexagon indicates how many centipedes were caught in that quadrat.

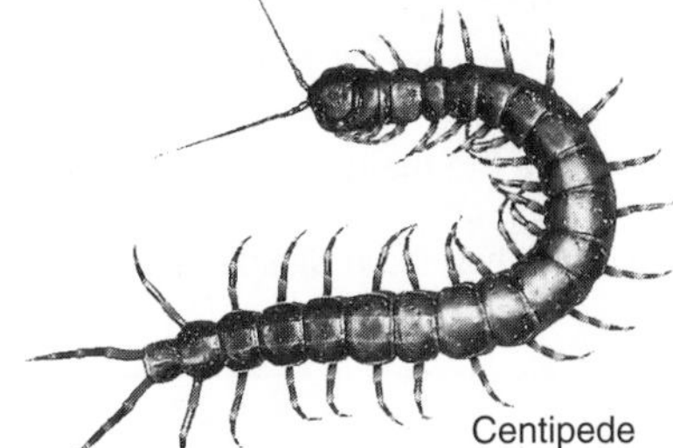

**ISBN: 978-1-927173-12-1**

# Quadrat-Based Estimates

The simplest description of a plant community in a habitat is a list of the species that are present. This qualitative assessment of the community has the limitation of not providing any information about the **relative abundance** of the species present. Quick estimates can be made using **abundance scales**, such as the ACFOR scale described below. Estimates of percentage cover provide similar information. These methods require the use of **quadrats**. Quadrats are used extensively in plant ecology. This activity outlines some of the common considerations when using quadrats to sample plant communities.

## What Size Quadrat?

Quadrats are usually square, and cover 0.25 $m^2$ (0.5 m x 0.5 m) or 1 $m^2$, but they can be of any size or shape, even a single point. The quadrats used to sample plant communities are often 0.25 $m^2$. This size is ideal for low-growing vegetation, but quadrat size needs to be adjusted to habitat type. The quadrat must be large enough to be representative of the community, but not so large as to take a very long time to use.

A quadrat covering an area of 0.25 $m^2$ is suitable for most low growing plant communities, such as this alpine meadow, fields, and grasslands.

Larger quadrats (e.g.1 $m^2$) are needed for communities with shrubs and trees. Quadrats as large as 4 m x 4 m may be needed in woodlands.

Small quadrats (0.01 $m^2$ or 100 mm x 100 mm) are appropriate for lichens and mosses on rock faces and tree trunks.

## How Many Quadrats?

As well as deciding on a suitable quadrat size, the other consideration is how many quadrats to take (the sample size). In species-poor or very homogeneous habitats, a small number of quadrats will be sufficient. In species-rich or heterogeneous habitats, more quadrats will be needed to ensure that all species are represented adequately.

### Determining the number of quadrats needed

- Plot the cumulative number of species recorded (on the y axis) against the number of quadrats already taken (on the x axis).
- The point at which the curve levels off indicates the suitable number of quadrats required.

Fewer quadrats are needed in species-poor or very uniform habitats, such as this bluebell woodland.

## Describing Vegetation

Density (number of individuals per unit area) is a useful measure of abundance for animal populations, but can be problematic in plant communities where it can be difficult to determine where one plant ends and another begins. For this reason, plant abundance is often assessed using **percentage cover**. Here, the percentage of each quadrat covered by each species is recorded, either as a numerical value or using an abundance scale such as the ACFOR scale.

### The ACFOR Abundance Scale

**A** = Abundant (30% +)

**C** = Common (20-29%)

**F** = Frequent (10-19%)

**O** = Occasional (5-9%)

**R** = Rare (1-4%)

The ACFOR scale could be used to assess the abundance of species in this pasture. Abundance scales are subjective, but it is not difficult to determine which abundance category each species falls into.

1. Describe one difference between the methods used to assess species abundance in plant and in animal communities:

2. What is the main consideration when determining appropriate quadrat size?

3. What is the main consideration when determining number of quadrats?

4. Explain two main disadvantages of using the ACFOR abundance scale to record information about a plant community:

 (a)

 (b)



**Related activities:** *Quadrat Sampling*

**Weblinks:** *Ecological Sampling Methods*


**ISBN: 978-1-927173-12-1**

# The Ecological Niche

The **ecological niche** describes the functional position of a species in its ecosystem. It is a description of how an organism uses available resources and alters those resources for other species. The full range of environmental conditions (biological and physical) under which a species can exist describes its **fundamental niche**. Interactions with other organisms usually force a species to occupy a **realized niche** that is narrower than this. When two species utilise some of the same resources (e.g. food items of a particular size) they will **compete**. Sometimes, one species will exclude another from an area. Such interactions can be very important in determining patterns of species distribution and abundance in a community.

## The Ecological Niche

**Physical conditions**
- Substrate
- Light and humidity
- Altitude and aspect
- Salinity
- pH
- Exposure
- Temperature
- Depth

**Adaptations for**
- Locomotion
- Activity (day/night)
- Tolerance range
- Self defense
- Defense of range
- Predator avoidance
- Reproduction
- Feeding

**Adaptations** are the **structural**, **physiological**, and **behavioral** features that enable the organism to exploit the resources of the habitat.

**Resources offered by the habitat**
- Food sources
- Shelter from climatic extremes
- Mating sites
- Nesting sites
- Predator avoidance

The physical conditions influence the habitat and the habitat provides the organism with resources and opportunities.

Resource availability is affected by the presence of other organisms and interactions with them: competition, predation, parasitism and disease.

**Other organisms present**

## Competition and Niche Size

Realised niche of species A

Possible tolerance range

### The realized niche

The tolerance range represents the potential (**fundamental**) niche a species could exploit. The actual or realized niche of a species is narrower than this because of competition with other species.

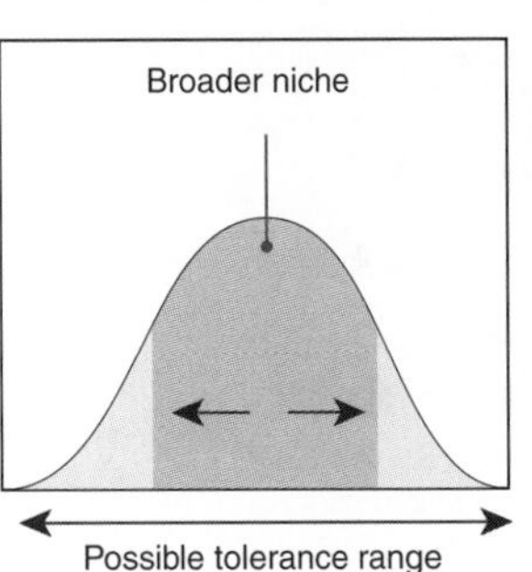


### Intraspecific competition

Competition is strongest between individuals of the same species, because their resource needs exactly overlap. When intraspecific competition is intense, individuals are forced to exploit resources in the extremes of their tolerance range. This leads to expansion of the realized niche.

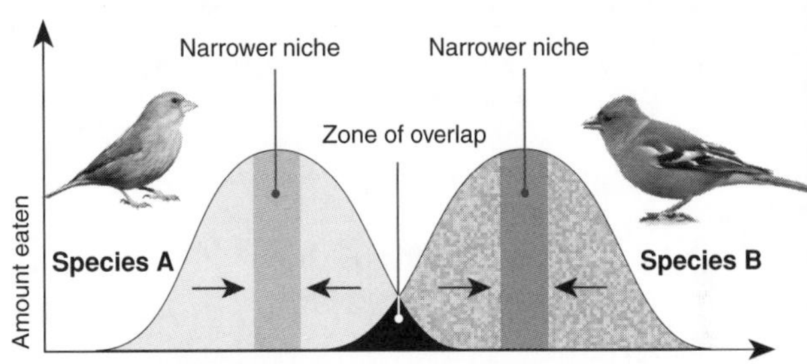


### Interspecific Competition

If two (or more) species compete for some of the same resources, their resource use curves will overlap. Within the zone of overlap, resource competition will be intense and selection will favor niche specialization so that one or both species occupy a narrower niche.

1. (a) In what way could the realized niche be regarded as flexible? ______

   (b) What factors might further reduce the extent of the realized niche? ______

2. Explain how differences in the niche requirements of organisms can contribute to community patterns such as **zonation**: ______



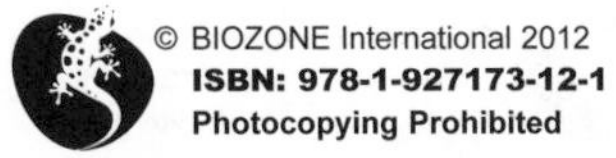


**ISBN: 978-1-927173-12-1**


***Related activities:*** *Niche Differentiation, Interspecific Competition, Intraspecific Competition*

# Adaptation and Niche

The adaptive features that evolve in species are the result of selection pressures on them through the course of their evolution. These features enable an organism to function most effectively in its niche, enhancing its exploitation of its environment and therefore its survival. The examples below illustrate some of the adaptations of two species: a British placental mammal and a migratory Arctic bird. Note that adaptations may be associated with an animal's structure (morphology), its internal physiology, or its behavior.

## Northern or Common Mole
### (*Talpa europaea*)

*Head-body length: 113-159 mm, tail length: 25-40 mm, weight range: 70-130 g.*

Moles (photos above) spend most of the time underground and are rarely seen at the surface. Mole hills are the piles of soil excavated from the tunnels and pushed to the surface. The cutaway view above shows a section of tunnels and a nest chamber. Nests are used for sleeping and raising young. They are dug out within the tunnel system and lined with dry plant material.

The northern (common) mole is a widespread insectivore found throughout most of Britain and Europe, apart from Ireland. They are found in most habitats but are less common in coniferous forest, moorland, and sand dunes, where their prey (earthworms and insect larvae) are rare. They are well adapted to life underground and burrow extensively, using enlarged forefeet for digging. Their small size, tubular body shape, and heavily buttressed head and neck are typical of burrowing species.

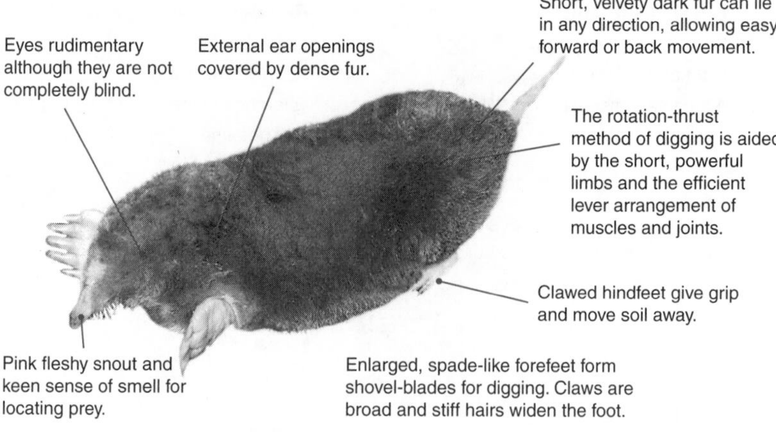


**Habitat and ecology**: Moles spend most of their lives in underground tunnels. Surface tunnels occur where their prey is concentrated at the surface (e.g. land under cultivation). Deeper, permanent tunnels form a complex network used repeatedly for feeding and nesting, sometimes for several generations.

**Senses and behavior**: Keen sense of smell but almost blind. Both sexes are solitary and territorial except during breeding. Life span about 3 years. Moles are prey for owls, buzzards, stoats, cats, and dogs. Their activities aerate the soil and they control many soil pests. Despite this, they are regularly trapped and poisoned as pests.

## Snow Bunting
### (*Plectrophenax nivalis*)

The snow bunting is a small ground feeding bird that lives and breeds in the Arctic and sub-Arctic islands. Although migratory, snow buntings do not move to traditional winter homes but prefer winter habitats that resemble their Arctic breeding grounds, such as bleak shores or open fields of northern Britain and the eastern United States. Snow buntings have the unique ability to molt very rapidly after breeding. During the warmer months, the buntings are a brown color, changing to white in winter (right). They must complete this color change quickly, so that they have a new set of feathers before the onset of winter and before migration. In order to achieve this, snow buntings lose as many as four or five of their main flight wing feathers at once, as opposed to most birds, which lose only one or two.

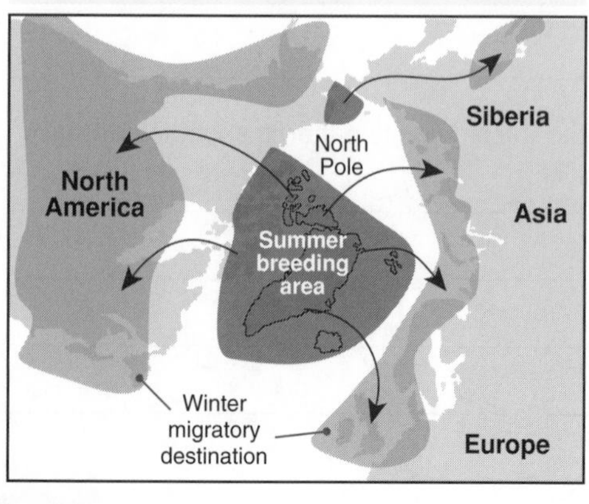


Very few small birds breed in the Arctic, because most small birds lose more heat than larger ones. In addition, birds that breed in the brief Arctic summer must migrate before the onset of winter, often traveling over large expanses of water. Large, long winged birds are better able to do this. However, the snow bunting is superbly adapted to survive in the extreme cold of the Arctic region.

**Habitat and ecology**: Widespread throughout Arctic and sub-Arctic Islands. Active throughout the day and night, resting for only 2-3 hours in any 24 hour period. Snow buntings may migrate up to 6000 km but are always found at high latitudes.

**Reproduction and behavior**: The nest, which is concealed amongst stones, is made from dead grass, moss, and lichen. The male bird feeds his mate during the incubation period and helps to feed the young.



**Related activities:** *The Ecological Niche, Niche Differentiation*

***Periodicals:*** *Fatal attraction*

**ISBN: 978-1-927173-12-1**

Carnivorous plants, such as the pitcher plant and Venus fly trap (below), obtain at least some of their nutrients by consuming protozoans or small animals (e.g. insects). The plants are photosynthetic, but live in nitrogen-poor environments and their nitrogen requirements are met by the digestion of animal tissue. In order to trap organisms, the leaves of carnivorous plants have been modified. Traps can be active or passive depending on whether leaf movement is involved in capturing the prey.

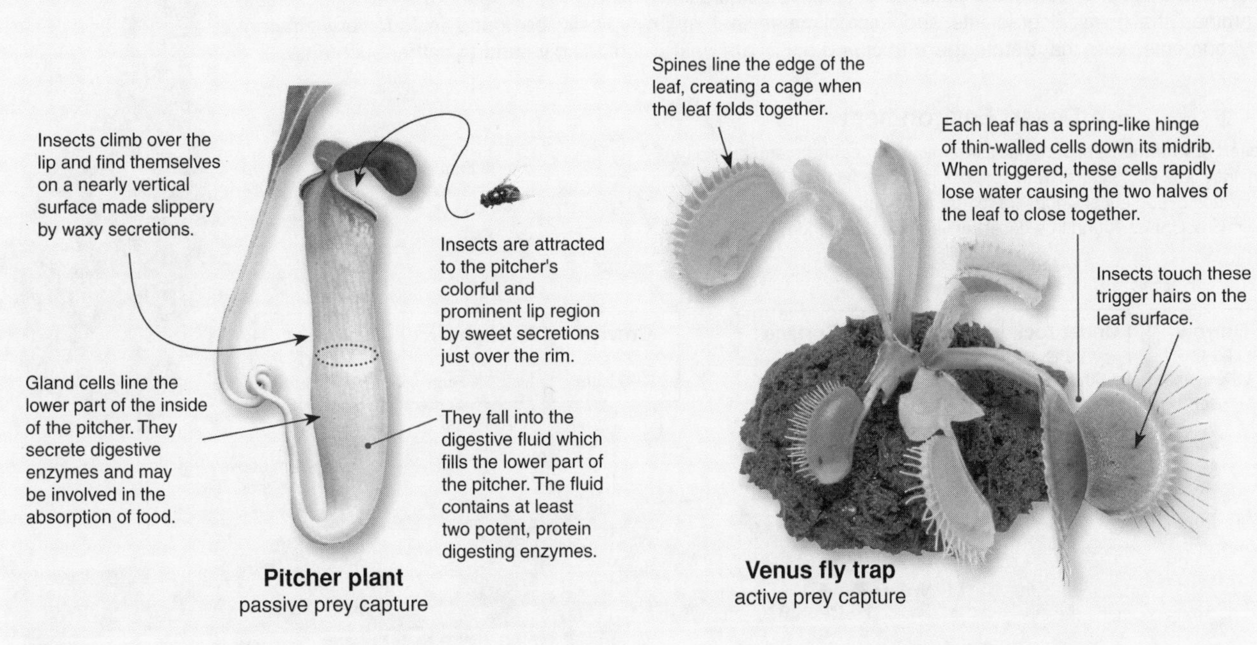


1. Describe a structural, physiological, and behavioral adaptation of the **common mole**, explaining how each aids survival:

   (a) Structural adaptation: ______

   (b) Physiological adaptation: ______

   (c) Behavioral adaptation: ______

2. Describe a structural, physiological, and behavioral adaptation of the **snow bunting**, explaining how each aids survival:

   (a) Structural adaptation: ______

   (b) Physiological adaptation: ______

   (c) Behavioral adaptation: ______

3. For the carnivorous plants above, describe the modifications to the leaf that enable prey capture:

   (a) **Pitcher plant**: ______

   (b) **Venus fly trap**: ______

4. Suggest why plants such as the Venus fly trap do not compete well with other plants in high nutrient environments:

______




ISBN: 978-1-927173-12-1

# Physical Factors and Gradients

Gradients in abiotic factors are found in almost every environment; they influence habitats and **microclimates**, and determine patterns of species distribution. This activity, covering the next four pages, examines the physical gradients and microclimates that might typically be found in four very different environments. Note that **dataloggers** (pictured right), are being increasingly used to gather such data.

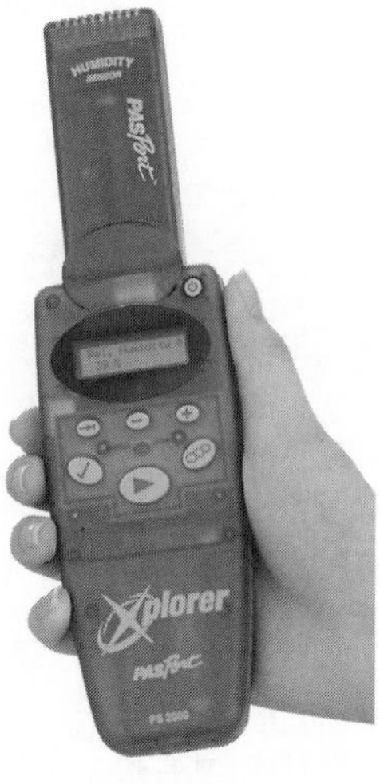

## A Desert Environment

Desert environments experience extremes in temperature and humidity, but they are not uniform with respect to these factors. This diagram illustrates hypothetical values for temperature and humidity for some of the microclimates found in a desert environment at midday.

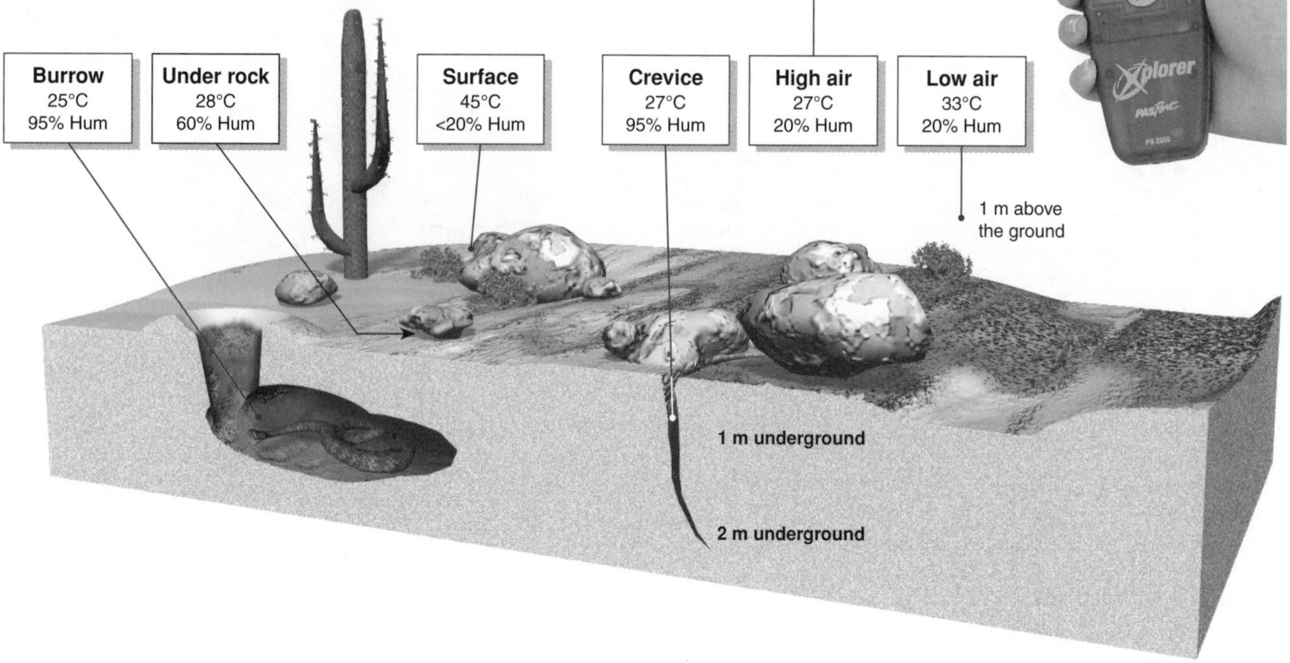

1. Distinguish between **climate** and **microclimate**:

2. Study the diagram above and describe the general conditions where high humidity is found:

3. Identify the three microclimates that a land animal might exploit to avoid the extreme high temperatures of midday:

4. Describe the likely consequences for an animal that was unable to find a suitable microclimate to escape midday sun:

5. Describe the advantage of high humidity to the survival of most land animals:

6. Describe the likely changes to the temperature and relative humidity that occur during the night:




ISBN: 978-1-927173-12-1

# Physical Factors in a Forest

We have seen how environmental gradients can occur with altitude and with horizontal distance along a shore, but they can also arise as a result of vertical distance from the ground. In a forest, the light quantity and quality, humidity, wind speed, and temperature change gradually from the canopy to the forest floor. These changes give rise to a layered or stratified community in which different species occupy different vertical positions in the forest according to their particular tolerances.

Canopy

**Light:** 70%
**Wind:** 15 kmh$^{-1}$
**Humid:** 67%

**Light:** 50%
**Wind:** 12 km h$^{-1}$
**Humid:** 75%

**Light:** 12%
**Wind:** 9 km h$^{-1}$
**Humid:** 80%

**Light:** 6%
**Wind:** 5 km h$^{-1}$
**Humid:** 85%

**Light:** 1%
**Wind:** 3 km h$^{-1}$
**Humid:** 90%

**Light:** 0%
**Wind:** 0 km h$^{-1}$
**Humid:** 98%

Leaf litter

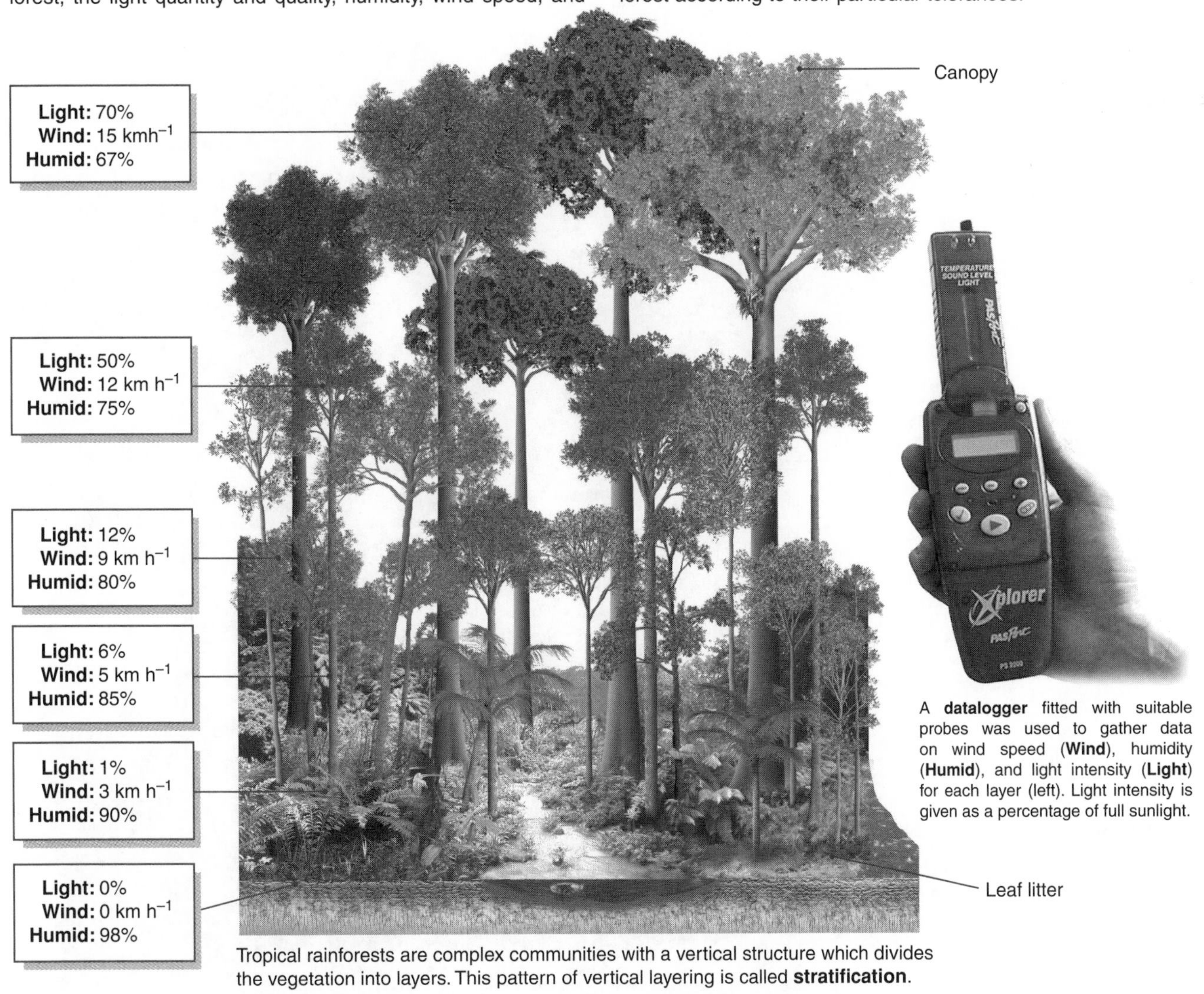

A **datalogger** fitted with suitable probes was used to gather data on wind speed (**Wind**), humidity (**Humid**), and light intensity (**Light**) for each layer (left). Light intensity is given as a percentage of full sunlight.

Tropical rainforests are complex communities with a vertical structure which divides the vegetation into layers. This pattern of vertical layering is called **stratification**.

1. Describe the general trend from the canopy to the leaf litter for each of the following:

   (a) Light intensity: ______________________

   (b) Wind speed: ______________________

   (c) Humidity: ______________________

2. Explain why each of these factors changes as the distance from the canopy increases:

   (a) Light intensity: ______________________

   (b) Wind speed: ______________________

   (c) Humidity: ______________________

3. What other feature of light, other than intensity, will also change with distance from the canopy and why?

______________________

______________________

**ISBN: 978-1-927173-12-1**


***Related activities:*** *Stratification in a Forest*

# Stratification in a Forest

Forest communities throughout the world show a pattern of vertical layering called **stratification**. Stratification produces a heterogeneous environment, providing greater habitat diversity and opportunity for a greater number of niches. Fallen logs and cavities in trees also add to vertical structure and enhance biodiversity. The forest composition itself (the actual species present) depends on many factors including altitude, light levels, soil type, drainage, and the past history of the area.

**Canopy**
The canopy intercepts most of the direct sunlight. Canopy trees that grow taller than the canopy layer are called emergents.

**Subcanopy**
Sometimes called the understory. This lower level of smaller trees is not always present.

**Epiphytes and lianes**
Epiphytes are plants that have no contact with the soil but grow in crevices in the branches and trunks of larger trees. Lianes are rooted in the ground, but clamber into the canopy. This layer includes ferns and orchids.

**Shrub layer**
A layer of plants 1-3 m tall. Includes seedlings less than 1 m tall and shade adapted, low growing plants such as ferns.

**Ground layer**
Includes mosses, fungi, lichens, dead leaves, and debris. This layer may also incorporate some of the plants from the shrub layer (ferns and shrubs).

1. Using examples, explain why a forest with a strong pattern of stratification might provide a greater diversity of habitats than a forest without such a vertical structure:

2. Predict the impact of deliberate removal (logging) of emergents and large canopy trees on the community composition. Consider how logging alters the physical environment and how existing and colonizing species might respond to this:




**ISBN: 978-1-927173-12-1**

# Transect Sampling

A **transect** is a line placed across a community of organisms. Transects are usually carried out to provide information on the **distribution** of species in the community. This is of particular value in situations where environmental factors that change over the sampled distance. This change is called an **environmental gradient** (e.g. up a mountain or across a seashore). The usual practice for small transects is to stretch a string between two markers. The string is marked off in measured distance intervals, and the species at each marked point are noted. The sampling points along the transect may also be used for the siting of quadrats, so that changes in density and community composition can be recorded. Belt transects are essentially a form of continuous quadrat sampling. They provide more information on community composition but can be difficult to carry out. Some transects provide information on the vertical, as well as horizontal, distribution of species (e.g. tree canopies in a forest).





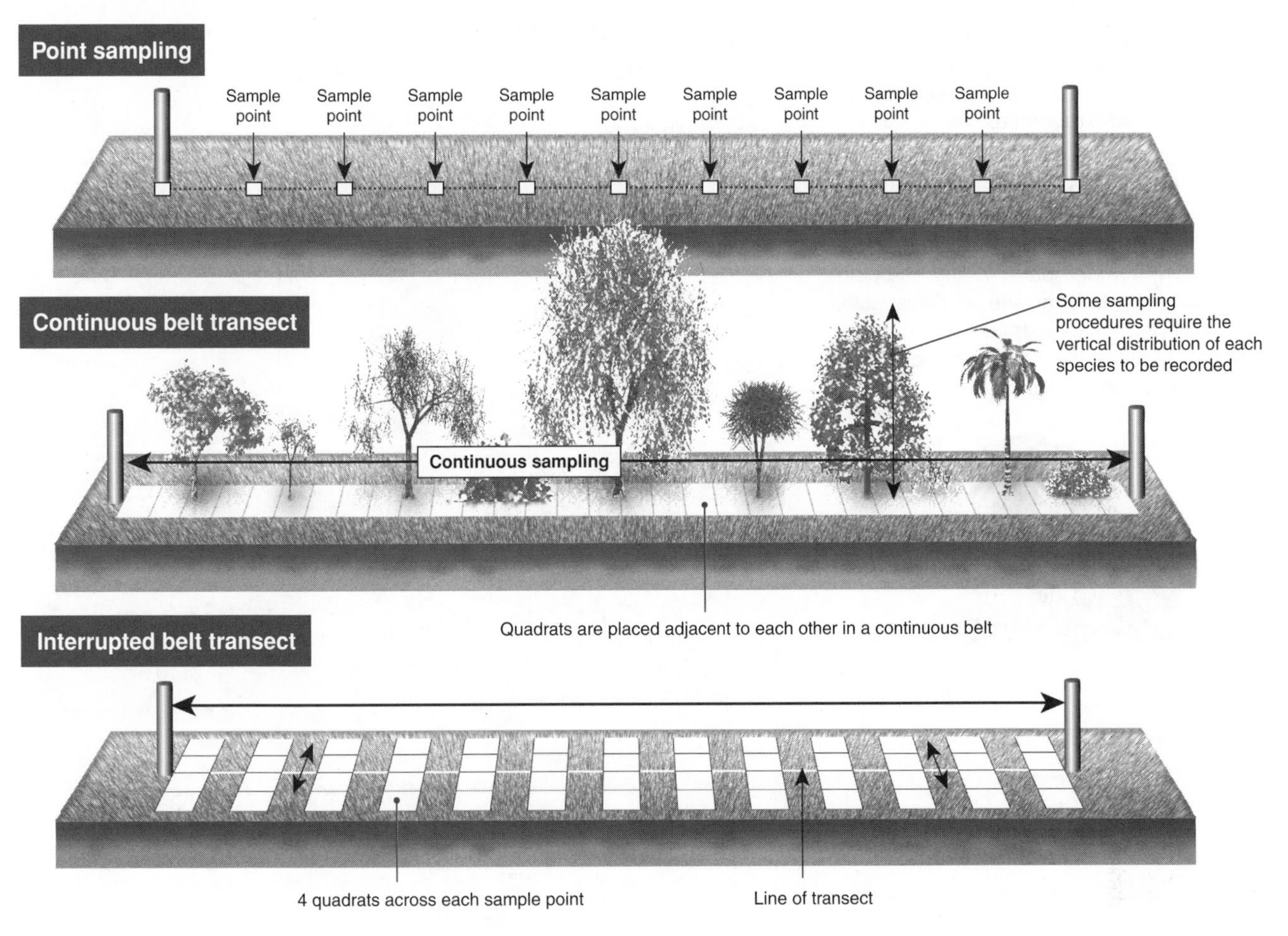

1. Belt transect sampling uses quadrats placed along a line at marked intervals. In contrast, point sampling transects record only the species that are touched or covered by the line at the marked points.

   (a) Describe one disadvantage of belt transects: ______

   (b) Explain why line transects may give an unrealistic sample of the community in question: ______

   (c) Explain how belt transects overcome this problem: ______

   (d) Describe a situation where the use of transects to sample the community would be inappropriate: ______

2. Explain how you could test whether or not a transect sampling interval was sufficient to accurately sample a community: ______

ISBN: 978-1-927173-12-1

Kite graphs are an ideal way in which to present distributional data from a belt transect (e.g. abundance or percentage cover along an environmental gradient). Usually, they involve plots for more than one species. This makes them good for highlighting probable differences in habitat preference between species. Kite graphs may also be used to show changes in distribution with time (e.g. with daily or seasonal cycles).

3. The data on the right were collected from a rocky shore field trip. Periwinkles from four common species of the genus *Littorina* were sampled in a continuous belt transect from the low water mark, to a height of 10 m above that level. The number of each of the four species in a 1 $m^2$ quadrat was recorded.

   Plot a **kite graph** of the data for all four species on the grid below. Be sure to choose a scale that takes account of the maximum number found at any one point and allows you to include all the species on the one plot. Include the scale on the diagram so that the number at each point on the kite can be calculated.

## Field data notebook

### Numbers of periwinkles (4 common species) showing vertical distribution on a rocky shore

| Height above low water (m) | Periwinkle species: *L. littorea* | *L. saxatalis* | *L. neritoides* | *L. littoralis* |
|---|---|---|---|---|
| 0-1 | 0 | 0 | 0 | 0 |
| 1-2 | 1 | 0 | 0 | 3 |
| 2-3 | 3 | 0 | 0 | 17 |
| 3-4 | 9 | 3 | 0 | 12 |
| 4-5 | 15 | 12 | 0 | 1 |
| 5-6 | 5 | 24 | 0 | 0 |
| 6-7 | 2 | 9 | 2 | 0 |
| 7-8 | 0 | 2 | 11 | 0 |
| 8-9 | 0 | 0 | 47 | 0 |
| 9-10 | 0 | 0 | 59 | 0 |

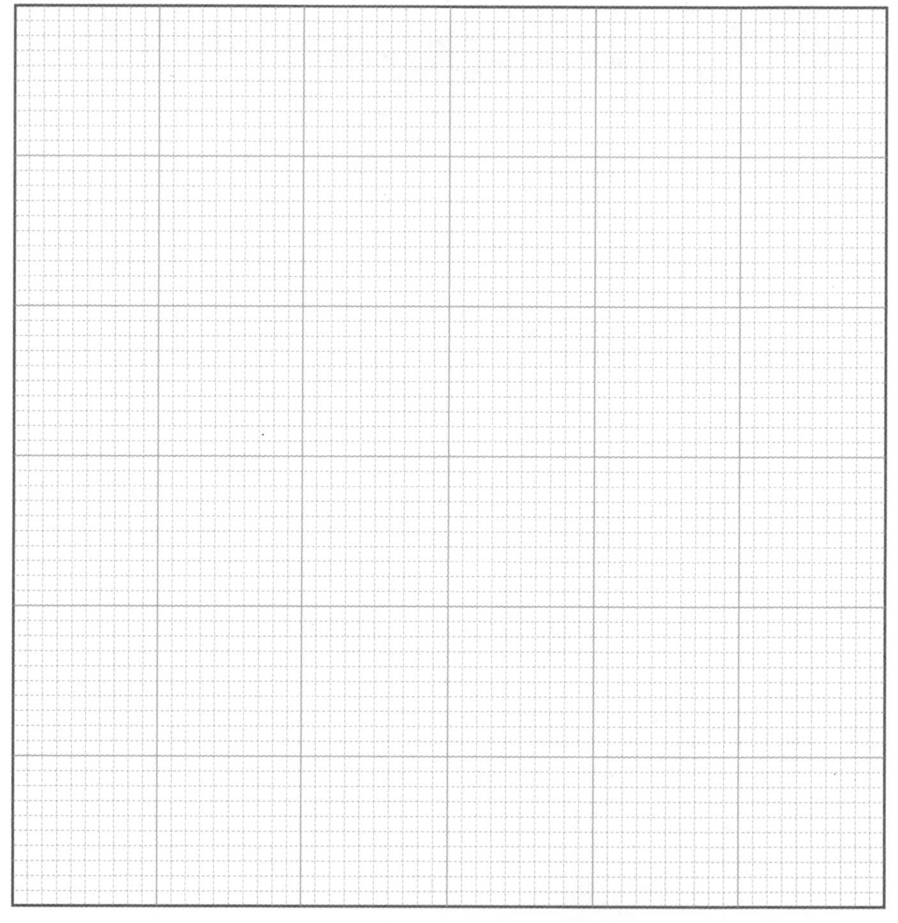

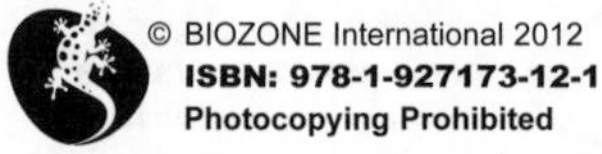


**ISBN: 978-1-927173-12-1**

# Community Change with Altitude

The Kosciusko National Park lies on the border between Victoria and New South Wales. In 1959, a researcher by the name of A.B. Costin carried out a detailed sampling of a transect between Berridale and the summit of Mt. Kosciusko. The map on the right shows the transect as a dotted line representing a distance of some 50-60 km.

The distribution of the various plant species up the slope of Mount Kosciusko is affected by changes in the physical conditions with increasing altitude. Below are two diagrams, one showing the profile of the transect showing changes in vegetation and soil types with increasing altitude. The diagram below the profile shows the changes in temperature and rainfall (precipitation) with altitude.

The low altitude soil around Berridale has low levels of organic matter supporting dry tussock grassland vegetation. The high altitude alpine soils are rich in organic matter, largely due to slow decay rates.

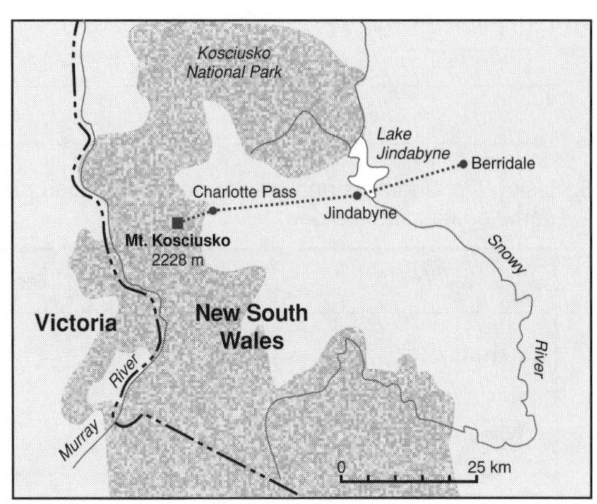


## Profile of Mount Kosciusko

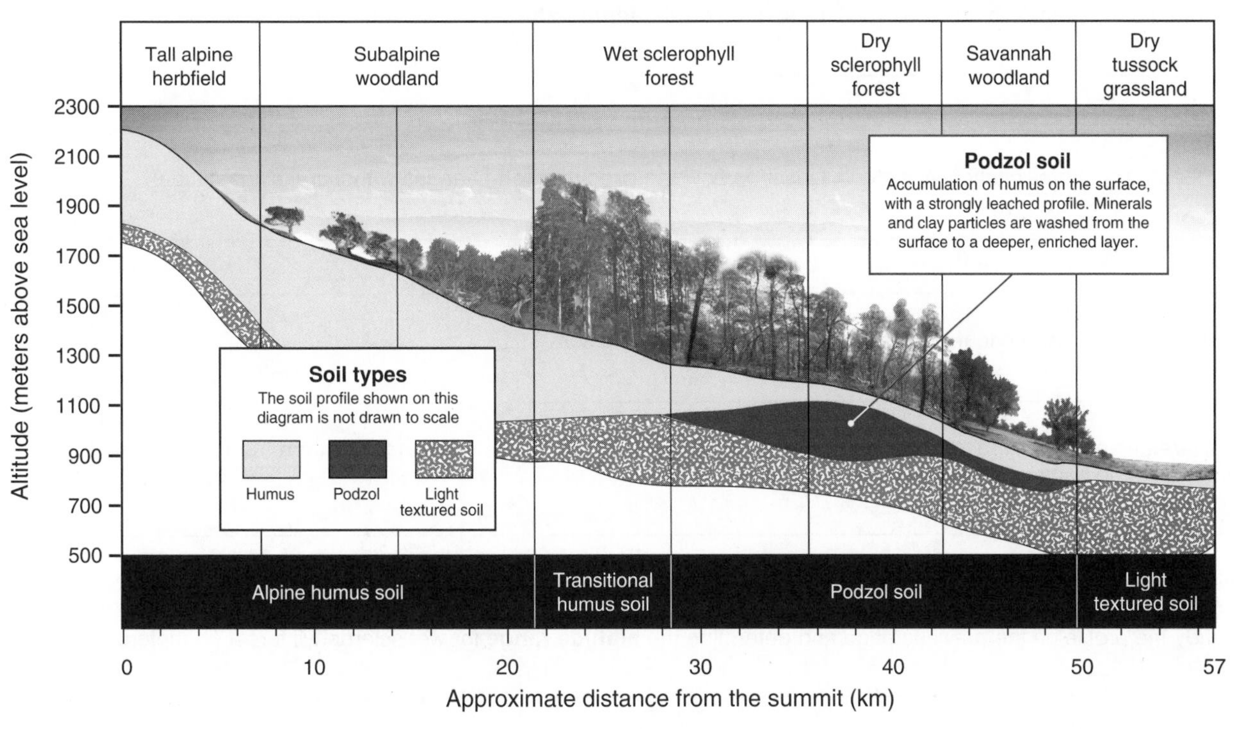


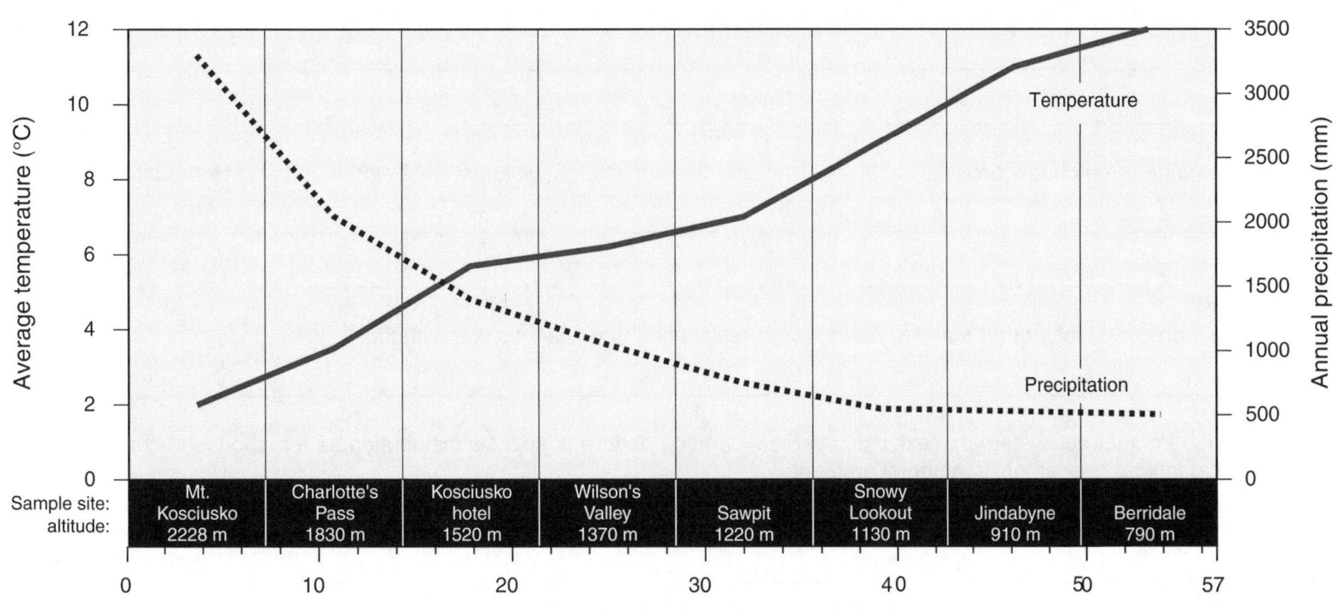




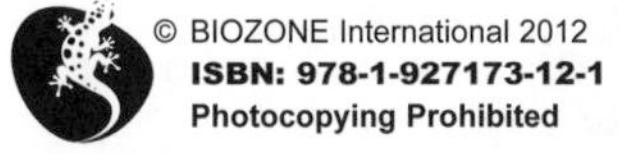

**ISBN: 978-1-927173-12-1**

1. Calculate the vertical distance (change in altitude) in meters, between Berridale and Mount Kosciusko: ______________

2. Name the three physical factors that are illustrated on the diagrams on the previous page:

3. Using the diagrams and graphs on the previous page, describe the following physical measurements for the three sample sites listed below:

| | Altitude (m) | Temperature (°C) | Precipitation (mm) | Soil type |
|---|---|---|---|---|
| **Berridale:** | | | | |
| **Wilson's Valley:** | | | | |
| **Mt Kosciusko:** | | | | |

4. Study the graph of temperature vs altitude. Describe how the **temperature** changes with increasing altitude:

5. Study the graph of precipitation vs altitude. Describe how the **precipitation** changes with increasing altitude:

6. Suggest a reason why the leaf litter is slow to decay in the alpine soil: ______________

7. The different vegetation types are distributed on the slopes of Mt. Kosciusko in a banded pattern from low altitude to the summit. Give the name of this kind of distribution pattern:

8. Wet sclerophyll forest is found part way up the slope of Mt. Kosciusko.

   (a) Study the profile on the previous page and determine the **altitude range** for wet sclerophyll forest (in meters):

   (b) Describe the probable physical factor that prevents the wet sclerophyll forest not being found at a lower altitude:

   (c) Describe the probable physical factor that prevents the wet sclerophyll forest not being found at a higher altitude:

9. Name a physical factor other than temperature or precipitation that changes with altitude:

10. Describe another ecosystem that exhibits a vertical banding pattern of species distribution as a response to changing physical factors over an environmental gradient:


**ISBN: 978-1-927173-12-1**

# Physical Factors on a Rocky Shore

Gradients in abiotic factors are found in almost every environment. They influence habitats and create **microclimates**, and are important in determining community patterns. This activity examines the physical gradients and microclimates that might be found on a rocky shore, where the community often shows clear zones of species distribution. Use this activity as a background to the activities covering shoreline zonation and the niches of intertidal organisms.

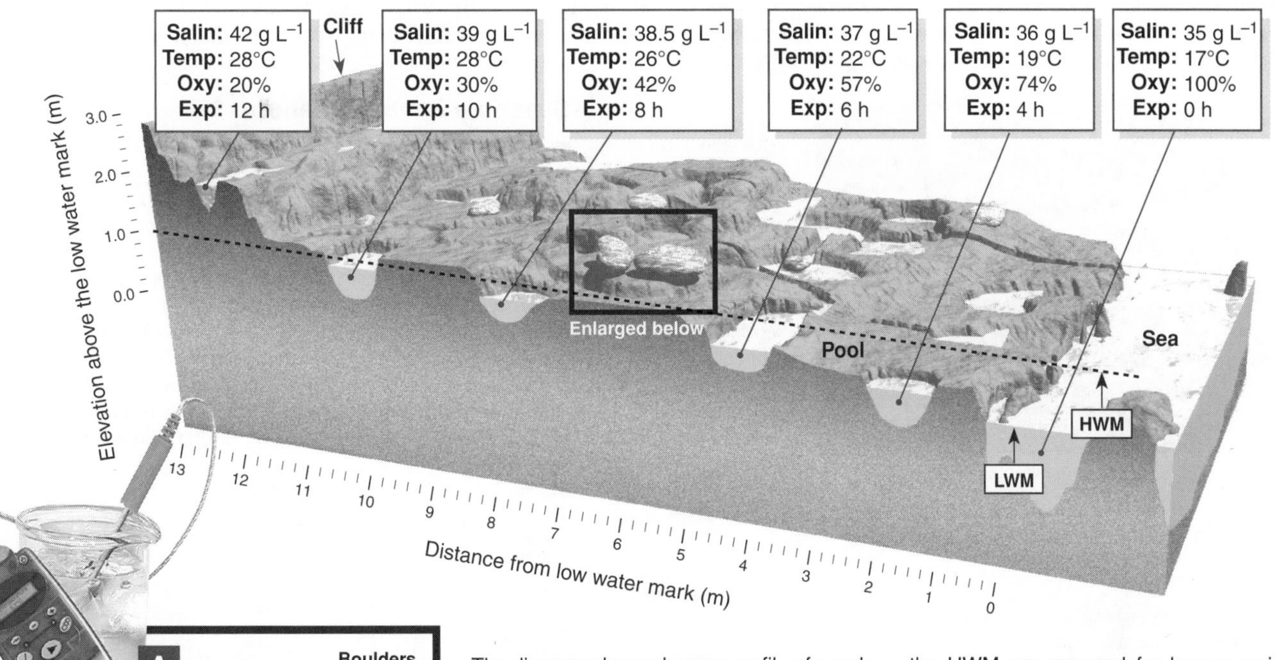


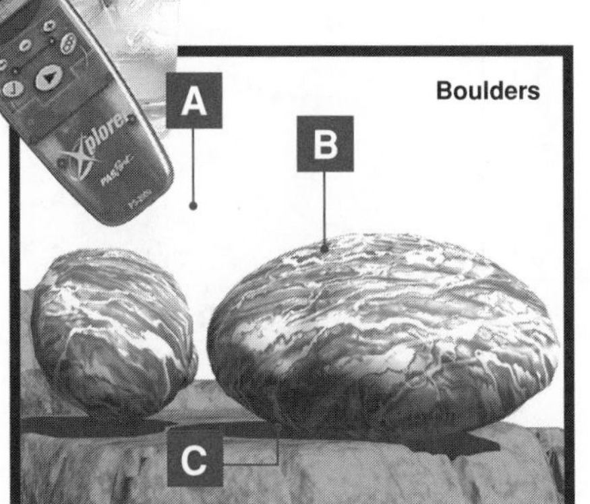


The diagram above shows a profile of a rock platform at low tide. The **high water mark** (HWM) shown is the average height of the spring tide. In reality, the high tide level varies with the phases of the moon. The **low water mark** (LWM) is an average level subject to the same variations. The rock pools vary in size, depth, and position on the platform. They are isolated at different elevations, trapping water from the ocean for time periods that may be relatively brief or up to 10-12 hours duration. Pools near the HWM are exposed for longer periods of time than those near the LWM. The difference in exposure times results in some of the physical factors exhibiting a **gradient**; the factor's value gradually changes over a horizontal and/or vertical distance. Physical factors sampled in the pools include salinity, or the amount of dissolved salts (g) per liter (**Salin**), temperature (**Temp**), dissolved oxygen compared to that of open ocean water (**Oxy**), and exposure, or the amount of time isolated from the ocean water (**Exp**).

1. Describe the environmental gradient (general trend) from the low water mark (LWM) to the high water mark (HWM) for:

   (a) Salinity: ______

   (b) Temperature: ______

   (c) Dissolved oxygen: ______

   (d) Exposure: ______

2. Rock pools above the normal high water mark (HWM), such as the uppermost pool in the diagram above, can have wide extremes of salinity. What abiotic conditions might cause these pools to have very high salinity?

   ______

3. (a) The inset diagram (above, left) is an enlarged view of two boulders on the rock platform. How might the abiotic factors listed below differ at each of the labelled points **A**, **B**, and **C**?

   Mechanical force of wave action: ______

   Surface temperature when exposed: ______

   (b) State the term given to these localized variations in physical conditions: ______


ISBN: 978-1-927173-12-1

# Shoreline Zonation

**Zonation** refers to the division of an ecosystem into distinct zones that experience similar abiotic conditions. In a more global sense, zonation may also refer to the broad distribution of vegetation according to latitude and altitude. Zonation is particularly clear on a rocky seashore, where assemblages of different species form a banding pattern approximately parallel to the waterline. This effect is marked in temperate regions where the prevailing weather comes from the same general direction.

Exposed shores show the clearest zonation. On sheltered rocky shores there is considerable species overlap and it is only on the upper shore that distinct zones are evident. Rocky shores exist where wave action prevents the deposition of much sediment. The rock forms a stable platform for the secure attachment of organisms such as large seaweeds and barnacles. Sandy shores are less stable than rocky shores and the organisms found there are adapted to the more mobile substrate.

Tide pools on a rocky shore, Santa Cruz, California

## Seashore Zonation Patterns

The zonation of species distribution according to an environmental gradient is well shown on rocky shorelines. In the United States, exposed rocky shores occur along much of the coastline. Variations in low and high tide affect zonation, and in areas with little tidal variation, zonation is restricted. High on the shore, some organisms may be submerged only at spring high tide. Low on the shore, others may be exposed only at spring low tide. There is a gradation in extent of exposure and the physical conditions associated with this. Zonation patterns generally reflect the vertical movement of seawater. Sheer rocks can show marked zonation as a result of tidal changes with little or no horizontal shift in species distribution. The profiles below compare generalized zonation patterns on an exposed rocky shore (left profile) and an exposed sandy shore (right profile). **SLT** = Spring low tide mark, **MLT** = Mean low tide mark, **MHT** = Mean high tide mark, **SHT** = Spring high tide mark.

### Key to species

1 Lichen: sea ivory
2 Small periwinkle *Littorina neritoides*
3 Lichen *Verrucaria maura*
4 Rough periwinkle *Littorina saxatilis*
5 Common limpet *Patella vulgaris*
6 Laver *Porphyra*
7 Spiral wrack *Fucus spiralis*
8 Australian barnacle
9 Common mussel *Mytilus edulis*
10 Common whelk *Buccinum undatum*
11 Grey topshell *Gibbula cineraria*
12 Carrageen (Irish moss) *Chondrus crispus*
13 Thongweed *Himanthalia elongata*
14 Toothed wrack *Fucus serratus*
15 Dabberlocks *Alaria esculenta*
16 Common sandhopper
17 Sandhopper *Bathyporeia pelagica*
18 Common cockle *Cerastoderma edule*
19 Lugworm *Arenicola marina*
20 Sting winkle *Ocinebra erinacea*
21 Common necklace shell *Natica alderi*
22 Rayed trough shell *Mactra corallina*
23 Sand mason worm *Lanice conchilega*
24 Sea anemone *Halcampa*
25 Pod razor shell *Ensis siliqua*
26 Sea potato *Echinocardium* (a heart urchin)

*Note: Where several species are indicated within a single zonal band, they occupy the entire zone, not just the position where their number appears.*

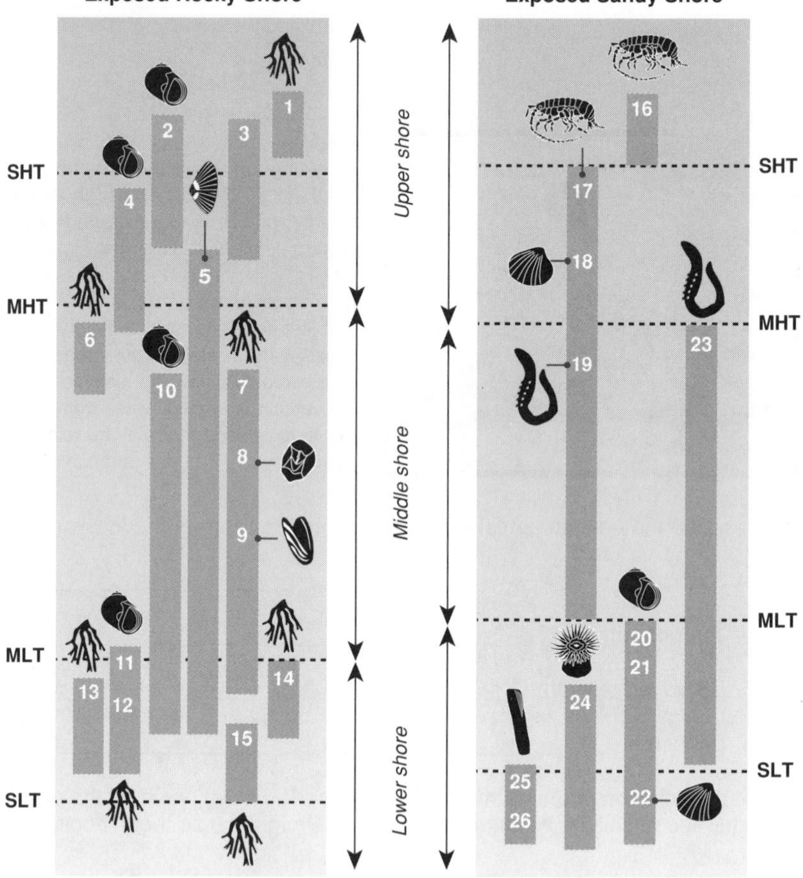


1. (a) Suggest why the time of exposure above water is a major factor controlling species distribution on a rocky shore:

(b) Identify two other abiotic factors that might influence species distribution on a rocky shore:

(c) Identify two biotic factors that might influence species distribution on a rocky shore:

2. Describe the zonation pattern on a rocky shore:



*Related activities: Physical Factors on a Rocky Shore*
*Weblinks: Tide Pool Ecology*


ISBN: 978-1-927173-12-1

# Seaweed Zonation

Zonation patterns in the brown seaweeds of the genus *Fucus* provide an ideal opportunity to combine both quantitative and qualitative work. In determining patterns of distribution and abundance for *Fucus* species, it is also useful to make qualitative observations about the size, vigor, and degree of desiccation of specimens at different points on the shore. These observations provide biological information which can help you to explain the patterns of distribution and abundance you find. *Fucus* is a genus of marine brown algae, commonly called wracks, which are found in the midlittoral zone of rocky seashores (i.e. the zone between the low and high levels). Three *Fucus* species are the dominant seaweeds on many rocky shores along the north east coast of North America and the Californian coast, where they form distinct zones along the shore. The species' distribution is governed by the extent to which each species is tolerant of exposure to air. This activity summarizes a study of their ecology.

Spiral wrack (*Fucus spiralis*)

Andreas Trepte

A group of students made a study of a rocky shore dominated by three species of wrack: spiral wrack, bladder wrack, and serrated wrack.

Their aim was to investigate the distribution of three *Fucus* species in the midlittoral zone and relate this to the size and **vigor (V)** of the seaweeds and the degree of **desiccation (D)** evident.

Bladder wrack (*F. vesiculosus*)

Thalli

Stemonitis

Serrated wrack (*F. serratus*)

Stemonitis

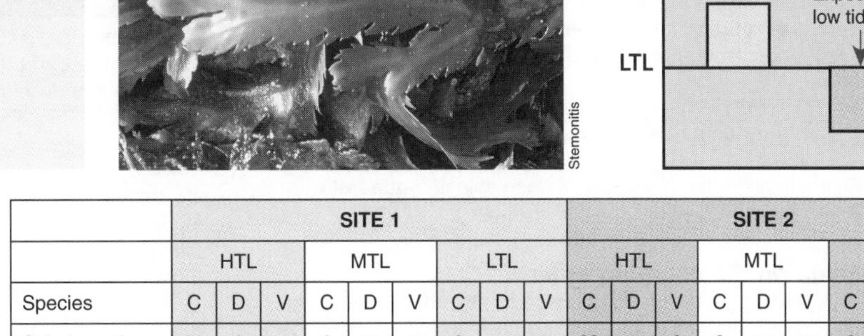

Site 1
Site 2
Covered at high tide only
HTL
$50\ cm^3$
Upper midlittoral
Equally covered and exposed
MTL
Lower midlittoral
Exposed at low tide only
LTL
Lower littoral

**Procedure**

Three $50\ cm^3$ quadrats were positioned from the LTL to the HTL at two locations on the shoreline as shown in the diagram (far right). An estimate of **percentage cover** (**C**) of each species of *Fucus* was made for each sample. Information on vigor and degree of desiccation was collected at the same time.

Qualitative data were collected as simple scores:

- \+ = vigorous with large thalli; no evidence of dessication
- 0 = less vigorous with smaller thalli; some evidence of dessication
- – = small, poorly grown thalli; obvious signs of desiccation

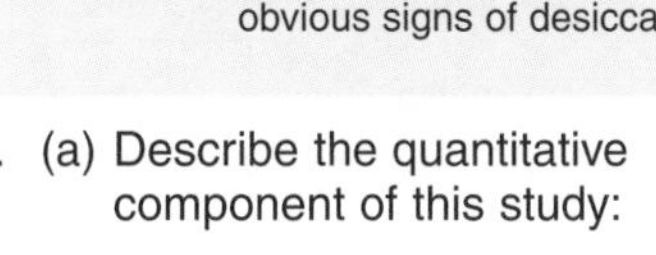

1. (a) Describe the quantitative component of this study:

   (b) Describe the qualitative component of this study:

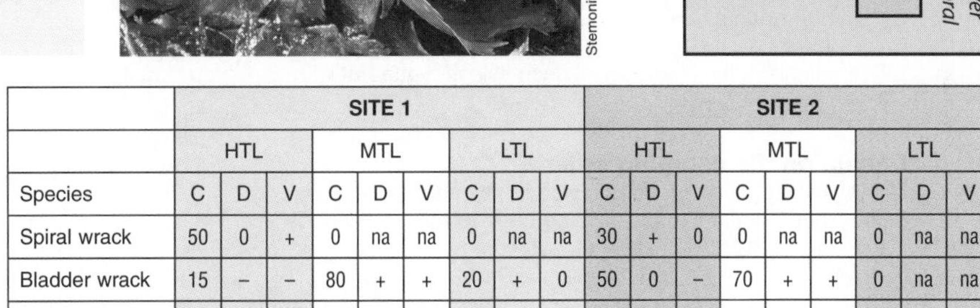

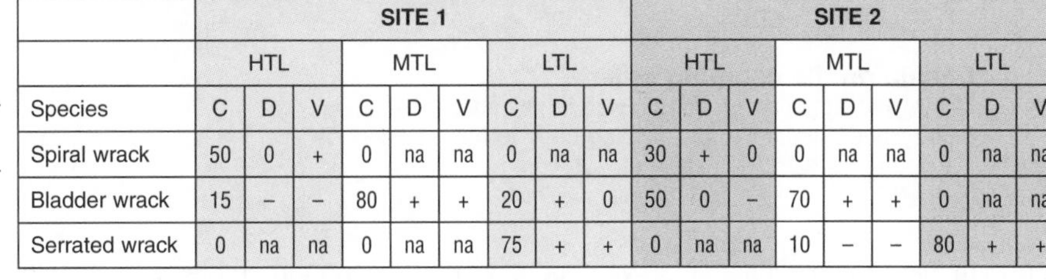

| | SITE 1 | | | | | | | | | SITE 2 | | | | | | | | |
|---|---|---|---|---|---|---|---|---|---|---|---|---|---|---|---|---|---|---|
| | HTL | | | MTL | | | LTL | | | HTL | | | MTL | | | LTL | | |
| Species | C | D | V | C | D | V | C | D | V | C | D | V | C | D | V | C | D | V |
| Spiral wrack | 50 | 0 | + | 0 | na | na | 0 | na | na | 30 | + | 0 | 0 | na | na | 0 | na | na |
| Bladder wrack | 15 | – | – | 80 | + | + | 20 | + | 0 | 50 | 0 | – | 70 | + | + | 0 | na | na |
| Serrated wrack | 0 | na | na | 0 | na | na | 75 | + | + | 0 | na | na | 10 | – | – | 80 | + | + |

2. The results of the quadrat survey are tabulated above. On a separate sheet, plot a column graph of the percentage coverage of each species at each position on the shore and at sites 1 and 2. Staple it to this page.

3. Relate the distribution pattern to the changes in degree of desiccation and in size and vigor of the seaweed thalli:

4. Suggest why the position of the quadrats was staggered for the two sites and describe a disadvantage of this design:


ISBN: 978-1-927173-12-1


*Related activities:* Shoreline Zonation

# Sampling a Rocky Shore Community

The diagram on the opposite page represents an area of seashore with its resident organisms. The distribution of coralline algae and four animal species are shown. This exercise is designed to prepare you for planning and carrying out a similar procedure to practically investigate a natural community. It is desirable, but not essential, that students work in groups of 2–4.

1. **Decide on the sampling method**
For the purpose of this exercise, it has been decided that the populations to be investigated are too large to be counted directly and a quadrat sampling method is to be used to estimate the average density of the four animal species as well as that of the algae.

2. **Mark out a grid pattern**
Use a ruler to mark out 3 cm intervals along each side of the sampling area (area of quadrat = 0.03 x 0.03 m). **Draw lines** between these marks to create a 6 x 6 grid pattern (total area = 0.18 x 0.18 m). This will provide a total of 36 quadrats that can be investigated.

3. **Number the axes of the grid**
Only a small proportion of the possible quadrat positions are going to be sampled. It is necessary to select the quadrats in a random manner. It is not sufficient to simply guess or choose your own on a 'gut feeling'. The best way to choose the quadrats randomly is to create a numbering system for the grid pattern and then select the quadrats from a random number table. Starting at the *top left hand corner*, **number the columns** and **rows** from 1 to 6 on each axis.

4. **Choose quadrats randomly**
To select the required number of quadrats randomly, use random numbers from a random number table. The random numbers are used as an index to the grid coordinates. Choose 6 quadrats from the total of 36 using table of random numbers provided for you at the bottom of the next page. Make a note of which column of random numbers you choose. Each member of your group should choose a different set of random numbers (i.e. different column: A–D) so that you can compare the effectiveness of the sampling method.

Column of random numbers chosen: ______

NOTE: Highlight the boundary of each selected quadrat with colored pen/highlighter.

5. **Decide on the counting criteria**
Before the counting of the individuals for each species is carried out, the criteria for counting need to be established.

There may be some problems here. You must decide before sampling begins as to what to do about individuals that are only partly inside the quadrat. Possible answers include:

(a) Only counting individuals that are completely inside the quadrat.
(b) Only counting individuals with a clearly defined part of their body inside the quadrat (such as the head).
(c) Allowing for 'half individuals' (e.g. 3.5 barnacles).
(d) Counting an individual that is inside the quadrat by half or more as one complete individual.

**Discuss the merits and problems** of the suggestions above with other members of the class (or group). You may even have counting criteria of your own. Think about other factors that could cause problems with your counting.

6. **Carry out the sampling**
Carefully examine each selected quadrat and **count the number of individuals** of each species present. Record your data in the spaces provided on the next page.

7. **Calculate the population density**
Use the combined data TOTALS for the sampled quadrats to estimate the average density for each species by using the formula:

$$\text{Density} = \frac{\text{Total number in all quadrats sampled}}{\text{Number of quadrats sampled} \times \text{area of a quadrat}}$$

Remember that a total of 6 quadrats are sampled and each has an area of 0.0009 $m^2$. The density should be expressed as the number of individuals *per square meter (no.* $m^{-2}$*)*.

| | | | |
|---|---|---|---|
| Plicate barnacle: | | Snakeskin chiton: | |
| Oyster: borer | | Coralline algae: | |
| Limpet: | | | |

8. (a) In this example the animals are not moving. Describe the problems associated with sampling moving organisms. Explain how you would cope with sampling these same animals if they were really alive and very active:

(b) Carry out a direct count of all 4 animal species and the algae for the whole sample area (all 36 quadrats). Apply the data from your direct count to the equation given in (7) above to calculate the actual population density (remember that the number of quadrats in this case = 36):

Barnacle: ______ Oyster borer: ______ Chiton: ______ Limpet: ______ Algae: ______

Compare your estimated population density to the actual population density for each species:




**ISBN: 978-1-927173-12-1**

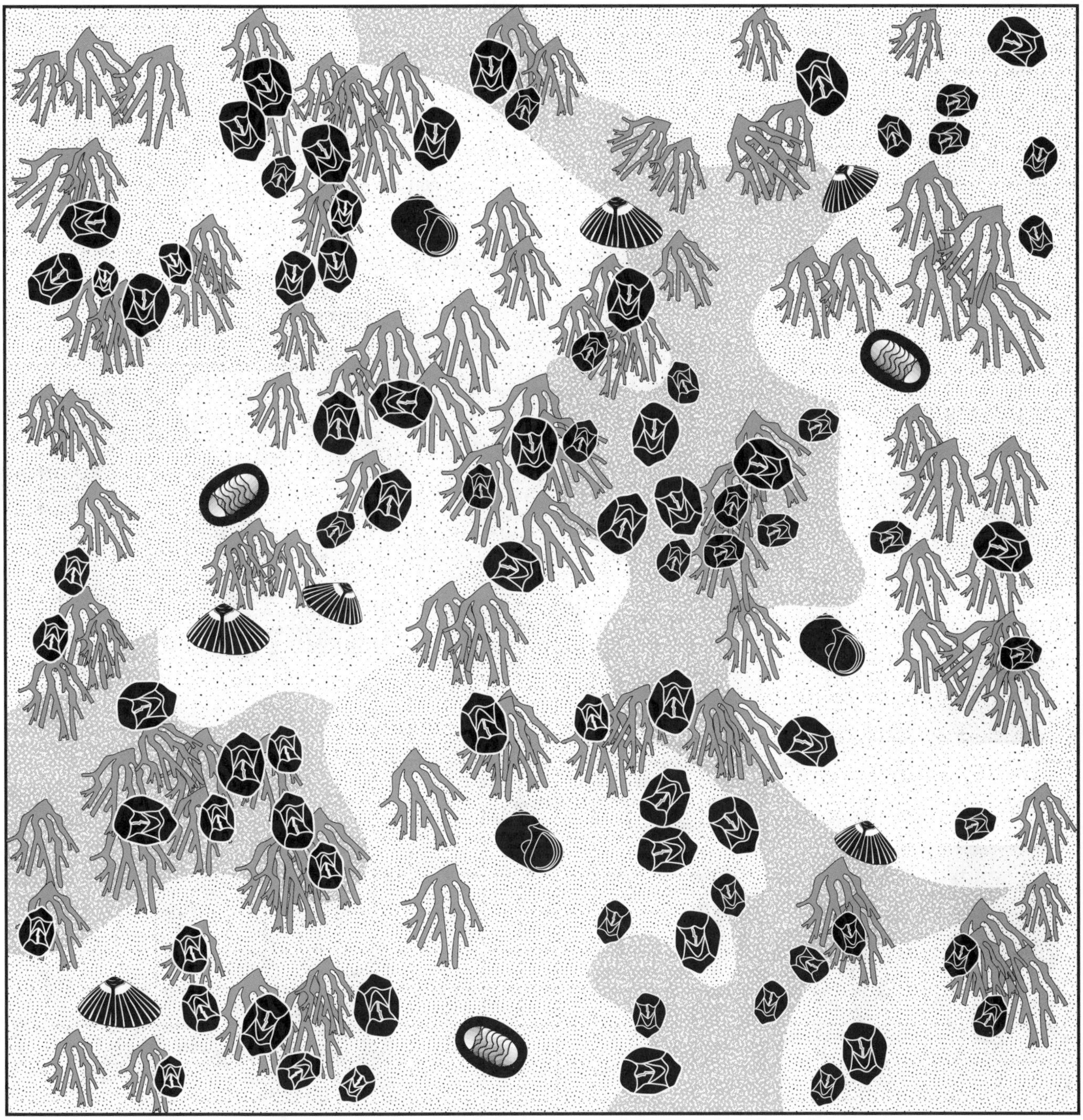

| Coordinates for each quadrat | Plicate barnacle | Oyster borer | Snakeskin chiton | Limpet | Coralline algae |
|---|---|---|---|---|---|
| 1: | | | | | |
| 2: | | | | | |
| 3: | | | | | |
| 4: | | | | | |
| 5: | | | | | |
| 6: | | | | | |
| TOTAL | | | | | |

**Table of random numbers**

| A | B | C | D |
|---|---|---|---|
| 22 | 31 | 62 | 22 |
| 32 | 15 | 63 | 43 |
| 31 | 56 | 36 | 64 |
| 46 | 36 | 13 | 45 |
| 43 | 42 | 45 | 35 |
| 56 | 14 | 31 | 14 |

The table above has been adapted from a table of random numbers from a statistics book. Use this table to select quadrats randomly from the grid above. Choose one of the columns (A to D) and use the numbers in that column as an index to the grid. The first digit refers to the row number and the second digit refers to the column number. To locate each of the 6 quadrats, find where the row and column intersect, as shown below:

Example: 52 refers to the 5th row and the 2nd column

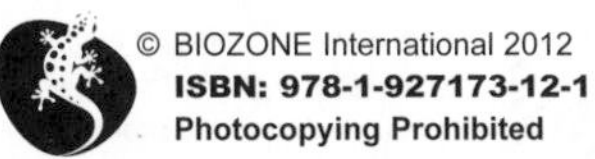


ISBN: 978-1-927173-12-1

# Interspecific Competition and Zonation

In naturally occurring populations, direct competition between different species (**interspecific competition**) is usually less intense than intraspecific competition because coexisting species have evolved slight differences in their realized niches, even though their fundamental niches may overlap. This has been well documented in barnacle species (below). However, when two species with very similar niche requirements are brought into direct competition through the introduction of a foreign species, one usually benefits at the expense of the other, which is excluded. The introduction of foreign species is a factor in the competitive displacement and decline of many native species. These introductions can irreversibly alter community structure and function. It can be difficult to show that competition from one species has forced the decline of another, but it is often inferred if the range of the native species contracts and that of the introduced competitor shows a corresponding increase.

## Competitive Exclusion in Barnacles

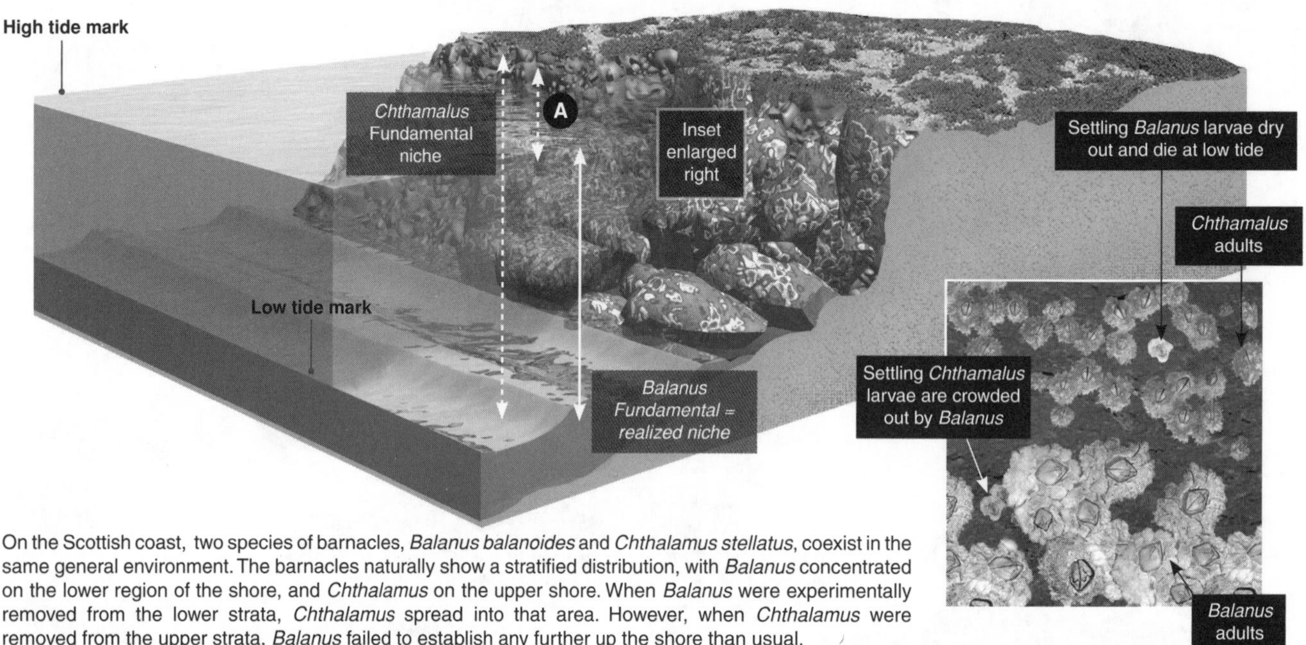


On the Scottish coast, two species of barnacles, *Balanus balanoides* and *Chthalamus stellatus*, coexist in the same general environment. The barnacles naturally show a stratified distribution, with *Balanus* concentrated on the lower region of the shore, and *Chthalamus* on the upper shore. When *Balanus* were experimentally removed from the lower strata, *Chthalamus* spread into that area. However, when *Chthalamus* were removed from the upper strata, *Balanus* failed to establish any further up the shore than usual.

**Left**: *Epopella plicata*, the plicate barnacle, is a large barnacle and abundant throughout New Zealand on shores with some wave exposure.

**Right**: *Chamaesipho columna*, the columnar barnacle, will extend higher on the shore than *Epopella plicata*, but the zones of the two species overlap and they compete for space and position.

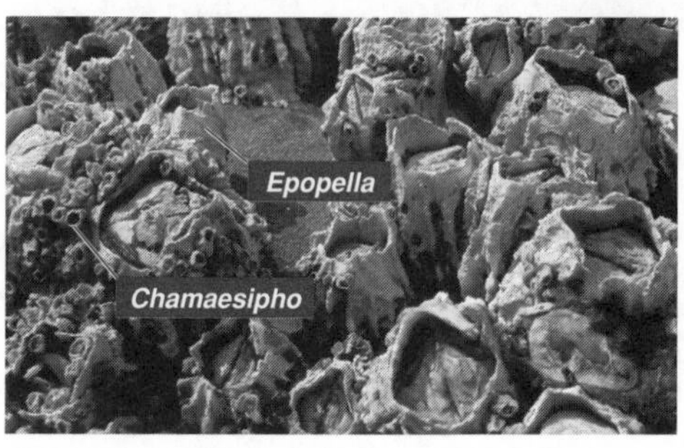


Photos: Conrad Pilditch

1. (a) In the example of the barnacles (above), describe what is represented by the zone labelled with the arrow A:

   (b) Explain the evidence for the barnacle distribution being the result of competitive exclusion:

2. Using examples, explain how biotic and abiotic factors can act together to contribute to a pattern of species distribution:



**Related activities:** Species Interactions, Intraspecific Competition


**ISBN: 978-1-927173-12-1**

# Field Study of a Rocky Shore

Many biological investigations require data to be gathered from natural communities in the field. Recording the physical aspects of the site from which the data are collected (e.g. sheltered sandy beach) allows it to be compared with other sites. Collected data may include the total number of plants or animals at a site, the number per square meter of sample area, or some other aspect of a plant or animal's niche. The investigation below compares the animals found on an exposed and a sheltered rocky shore.

**Sample site A**: Exposed rocky shore. Frequent heavy waves and high winds. Smooth rock face with few boulders and relatively steep slope towards the sea.

Coastline

Prevailing direction of wind and swell

1 km

**Sample site B**: Sheltered rocky shore. Small, gentle waves and little wind. Jagged rock face with large boulders and shallower slope leading to the sea.

### The Aim

To investigate the differences in the abundance of intertidal animals on an exposed rocky shore and a sheltered rocky shore.

### Background

The composition of rocky shore communities is strongly influenced by the shore's physical environment. Animals that cling to rocks must keep their hold on the substrate while being subjected to intense wave action and currents. However, the constant wave action brings high levels of nutrients and oxygen. Communities on sheltered rocky shores, although encountering less physical stress, may face lower nutrient and oxygen levels.

To investigate differences in the abundance of intertidal animals, students laid out 1 $m^2$ quadrats at regular intervals along one tidal zone at two separate but nearby sites: a rocky shore exposed to wind and heavy wave action and a rocky shore with very little heavy wave action. The animals were counted and their numbers in each quadrat recorded.

The Nature of Ecosystems


ISBN: 978-1-927173-12-1

1. Underline an appropriate hypothesis for this field study from the possible hypotheses below:

   (a) Wave action causes differences in communities on rocky shores.

   (b) The topography of the coastline affects rocky shore communities.

   (c) The communities of intertidal animals differ between exposed rocky shores and sheltered rocky shores.

   (d) Water temperature affects rocky shore communities.

2. During the field study, students counted the number of animals in each quadrat and recorded them in a notebook. In the space below, tabulate the data to show the total number of each species and the mean number of animals per quadrat:

Field data notebook

Count per quadrat. Quadrats 1 $m^2$

| Site A | 1 | 2 | 3 | 4 | 5 | 6 | 7 | 8 |
|---|---|---|---|---|---|---|---|---|
| Brown barnacle | 39 | 38 | 37 | 21 | 40 | 56 | 36 | 41 |
| Oyster borer | 6 | 7 | 4 | 3 | 7 | 8 | 9 | 2 |
| Columnar barnacle | 6 | 8 | 14 | 10 | 9 | 12 | 8 | 11 |
| Plicate barnacle | 50 | 52 | 46 | 45 | 56 | 15 | 68 | 54 |
| Ornate limpet | 9 | 7 | 8 | 10 | 6 | 7 | 6 | 10 |
| Radiate limpet | 5 | 6 | 4 | 8 | 6 | 7 | 5 | 6 |
| Black nerite | 7 | 7 | 6 | 8 | 4 | 6 | 8 | 9 |
| **Site B** | | | | | | | | |
| Brown barnacle | 7 | 6 | 7 | 5 | 8 | 5 | 7 | 7 |
| Oyster borer | 2 | 3 | 1 | 3 | 2 | 2 | 1 | 1 |
| Columnar barnacle | 56 | 57 | 58 | 55 | 60 | 47 | 58 | 36 |
| Plicate barnacle | 11 | 11 | 13 | 10 | 14 | 9 | 9 | 8 |
| Rock oyster | 7 | 8 | 8 | 6 | 2 | 4 | 8 | 6 |
| Ornate limpet | 7 | 8 | 5 | 6 | 5 | 7 | 9 | 3 |
| Radiate limpet | 13 | 14 | 11 | 10 | 14 | 12 | 9 | 13 |
| Black nerite | 6 | 5 | 3 | 1 | 4 | 5 | 2 | 3 |


ISBN: 978-1-927173-12-1

3. Use the grid below to draw a column graph of the mean number of species per 1 $m^2$ at each sample site. Remember to include a title, correctly labelled axes, and a key.

4. (a) Explain why more brown barnacles and plicate barnacles were found at site A:

(b) Explain why more oyster borers were found at site A:

5. Which species was entirely absent from site A?

6. A student wrote the following discussion of the field study. If you read it carefully, you can see that it restates the results of the study, but falls short of discussing them. Revise it to include explanatory detail that might account for the results:

We investigated the difference in communities between an exposed rocky shore and a sheltered rocky shore. The sample site A was more exposed than the second sample site with bigger waves and stronger winds. The animals we sampled were those that attach themselves to the rock surface. Quadrats were used to count the numbers of animals present. It was found that the brown barnacle and the plicate barnacles were the most common animal on the exposed rocky shore. Their numbers were reduced on the sheltered shore, but the columnar barnacle was more prevalent. Rock oysters were found only at site B. The abundance of the other animals varied only slightly except the oyster borer which was more abundant at site A.



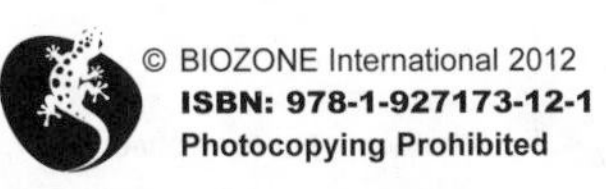

ISBN: 978-1-927173-12-1

# Physical Factors in a Small Lake

**Oxbow lakes** are formed from old river meanders which have been cut off and become isolated from the main channel following a change of the river's course. They are shallow (about 2-9 m deep) but often deep enough to develop temporary, but relatively stable, temperature gradients from top to bottom (below). Oxbows are commonly very productive and this can influence values for abiotic factors such as dissolved oxygen and light penetration, which can vary widely both with depth and proximity to the shore. Typical values for water temperature (**Temp**), dissolved oxygen (**Oxygen**), and light penetration as a percentage of the light striking the surface (**Light**) are indicated below.

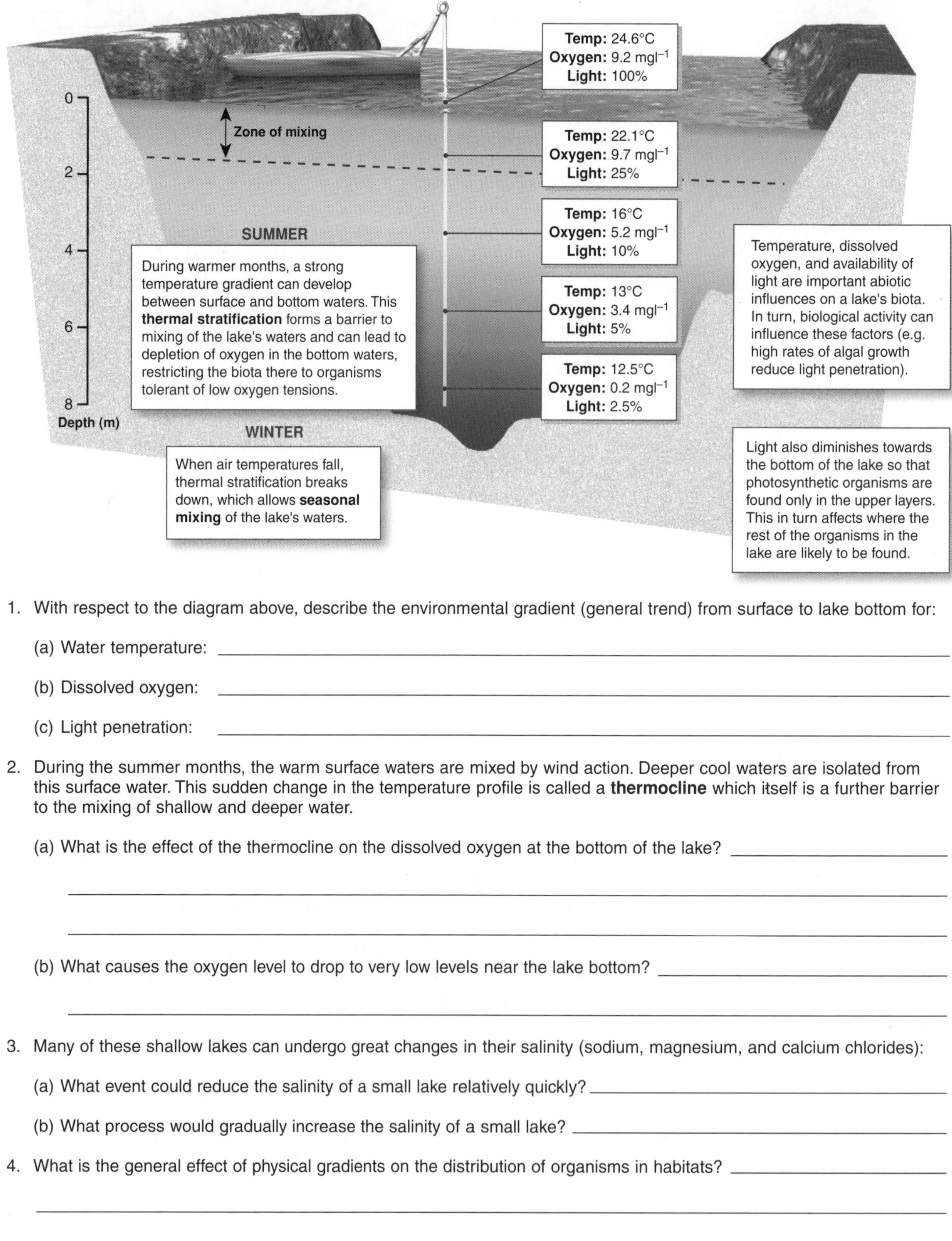


1. With respect to the diagram above, describe the environmental gradient (general trend) from surface to lake bottom for:

   (a) Water temperature: ____________________

   (b) Dissolved oxygen: ____________________

   (c) Light penetration: ____________________

2. During the summer months, the warm surface waters are mixed by wind action. Deeper cool waters are isolated from this surface water. This sudden change in the temperature profile is called a **thermocline** which itself is a further barrier to the mixing of shallow and deeper water.

   (a) What is the effect of the thermocline on the dissolved oxygen at the bottom of the lake? ____________________

   ____________________

   ____________________

   (b) What causes the oxygen level to drop to very low levels near the lake bottom? ____________________

   ____________________

3. Many of these shallow lakes can undergo great changes in their salinity (sodium, magnesium, and calcium chlorides):

   (a) What event could reduce the salinity of a small lake relatively quickly? ____________________

   (b) What process would gradually increase the salinity of a small lake? ____________________

4. What is the general effect of physical gradients on the distribution of organisms in habitats? ____________________

   ____________________

   ____________________

   ____________________



***Related activities:*** *Vertical Distribution in a Lake Community*


**ISBN: 978-1-927173-12-1**

# Vertical Distribution in a Lake Community

Lake communities often show seasonal and daily changes in their patterns of distribution in the water column. A lake's water column is not uniform. Light intensity, light quality, temperature, and oxygen level may all show considerable variation with depth and these changes affect the movements and distribution of the plankton community. The distribution of animal plankton (zooplankton) in winter is often very different to their distribution in summer. Even on a daily basis, zooplankton populations will migrate vertically in the water column. The graphs below show two examples of how distribution can change over time.

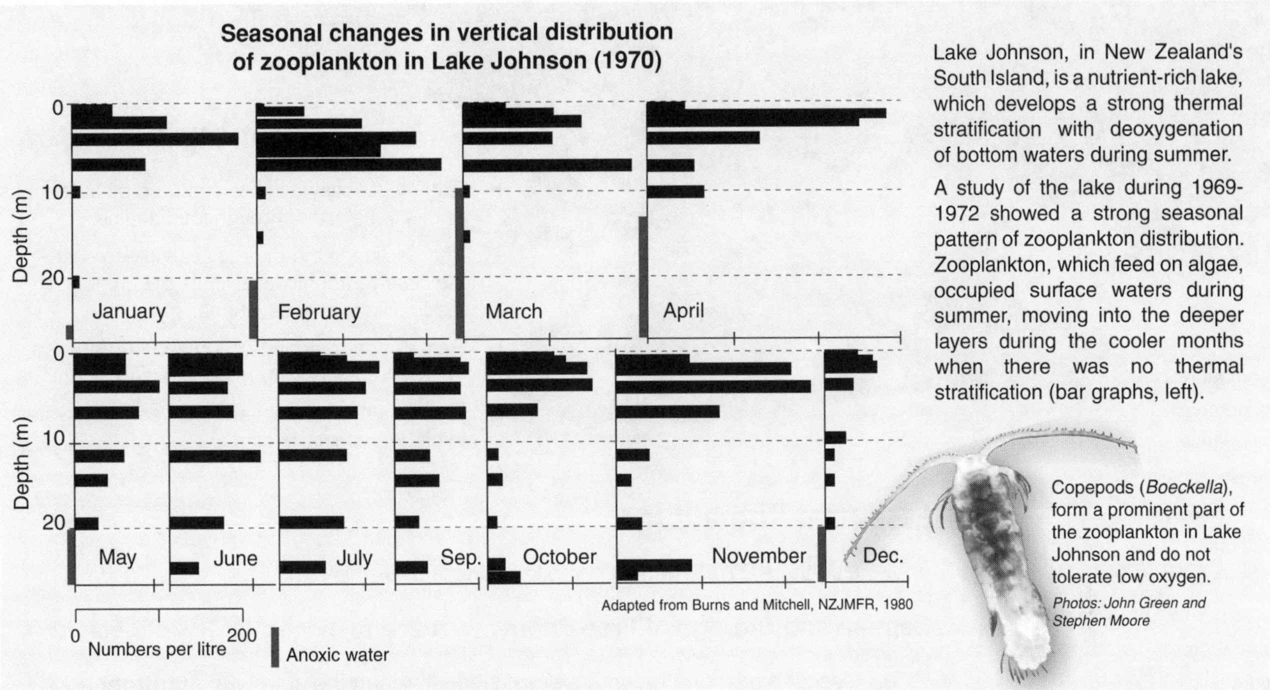


Lake Johnson, in New Zealand's South Island, is a nutrient-rich lake, which develops a strong thermal stratification with deoxygenation of bottom waters during summer.

A study of the lake during 1969-1972 showed a strong seasonal pattern of zooplankton distribution. Zooplankton, which feed on algae, occupied surface waters during summer, moving into the deeper layers during the cooler months when there was no thermal stratification (bar graphs, left).

Copepods (*Boeckella*), form a prominent part of the zooplankton in Lake Johnson and do not tolerate low oxygen.

*Photos: John Green and Stephen Moore*

### Daily pattern of vertical distribution of adult *Daphnia galeata* in Fuller Pond, Connecticut

*Daphnia* adults are prey for golden shiners, a small carnivorous fish. Golden shiners are visual predators and eat larger, adult *Daphnia*. They require enough light to see and capture their prey (graph, right). The kite graphs below show how *Daphnia* in Fuller Pond migrate in the water column over a 24 hour period.

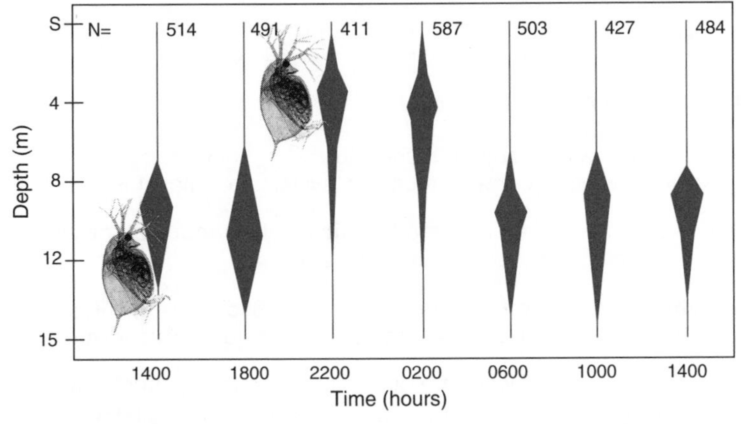


### Predation rate at different light intensities

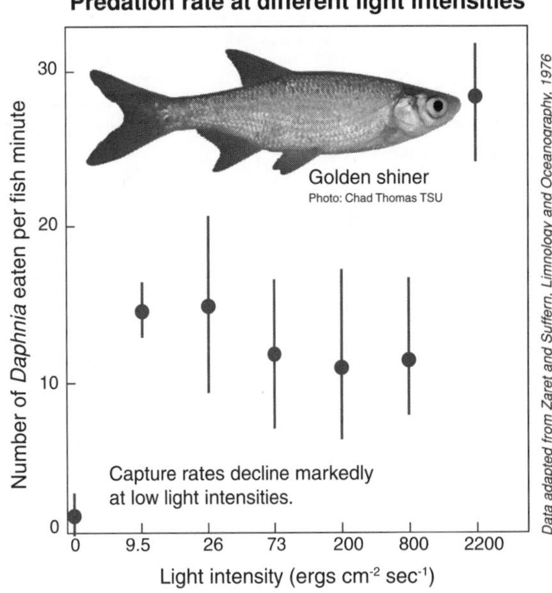


1. For the data relating to the Lake Johnson zooplankton community:

   (a) Describe the seasonal pattern of zooplankton distribution: ______

   (b) Identify the factor that appears to govern this pattern: ______

   (c) Is it biotic or abiotic? ______

2. What evidence is there, from the data presented, to indicate that *Daphnia* migrate in the water column to avoid predation?

   ______

   ______

3. Using a separate sheet of paper, use the information in this activity to write a 200 word essay discussing the role of **abiotic factors** and **predation** in the daily and seasonal distribution patterns of organisms in a lake community.



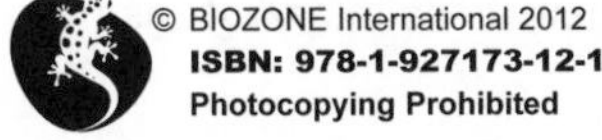


**ISBN: 978-1-927173-12-1**

Enduring Understanding

**2.A**
**4.A**

# Energy Flow and Nutrient Cycles

## Key concepts

- Energy in ecosystems is captured by autotrophs and is transferred through food chains.
- The decline in the energy available to each successive trophic level limits the number of feeding links in ecosystems.
- Rates of primary production and efficiency of energy transfers vary regionally and globally.
- Nutrients move within and between ecosystems in biogeohemical cycles.

## Key terms

autotroph
carbon cycle
chemoautotroph
consumers
decomposer
detritivore
ecological pyramid
food chain
food web
gross primary production
heterotroph
hydrologic cycle
net primary production
nitrogen cycle
nutrient cycle
photoautotroph
primary consumer
primary productivity
saprotroph
secondary consumer
ten percent rule
trophic efficiency
trophic group
trophic level

## Essential Knowledge

☐ 1. Use the **KEY TERMS** to compile a glossary for this topic.

### Capture and Storage of Free Energy *(2.A.2: a-b)* pages 11-12, 295-96, 300-01

☐ 2. Describe how energy enters ecosystems through the activity of **autotrophs**. Recognize autotrophs as **producers** in ecosystems. Distinguish between **photoautotrophs** (photosynthetic organisms) and **chemoautotrophs** (chemosynthetic organisms) in terms of their source of free energy.

☐ 3. Describe how **heterotrophs** obtain their free energy. Recognize heterotrophs as **consumers** in ecosystems. Distinguish between different types of consumers, including **detritivores**, and **saprotrophs** (saprophytes or decomposers). Outline the role of each of these **trophic groups** in energy transfer and nutrient cycling.

☐ 4. Recall how energy capturing processes use different types of electron acceptors (e.g. $NADP^+$ in photosynthesis, oxygen in cellular respiration).

### Flow of Energy in Ecosystems *(4.A.6: a-d)* pages 11-12, 297-308

☐ 5. Explain what the energy laws mean with respect to energy conversions in ecosystems. Explain how energy flows through ecosystems but matter is recycled.

☐ 6. Describe how energy is transferred between **trophic levels** in **food chains** and **food webs**. Comment on the efficiency of energy transfers.

☐ 7. Construct **food chains** and a **food web** for a community. Show how the organisms interact through their feeding relationships and assign them to **trophic levels**.

☐ 8. Understand how food chains and webs are dependent on **primary productivity**. Explain how primary productivity is quantified and explain contrasting patterns of productivity and **trophic efficiency**.

☐ 9. Describe energy flow quantitatively using an energy flow diagram. Include reference to trophic levels, direction of energy flow, processes involved in energy transfer, energy sources, and energy sinks.

☐ 10. Describe food chains quantitatively using **ecological pyramids**. Construct or interpret pyramids of energy, numbers, or biomass for different communities. Identify the relationship between each of these types of pyramids and their corresponding food chains and webs. Explain the shape of pyramid.

### Movement of Matter in Ecosystems *(4.A.6: a-d, f)* pages 309-16, 357-67

☐ 11. Describe the role of **nutrient cycles** in ecosystems. Use specific examples, e.g. the **carbon cycle**, **nitrogen cycle**, or **hydrologic cycle** to show how nutrients are exchanged within and between ecosystems, moving between the atmosphere, the Earth's crust, water, and organisms.

☐ 12. Discuss how humans intervene in nutrient cycles and how these interventions may occur on local, regional, and global scales.

**Periodicals:**
*Listings for this chapter are on page 375*

**Weblinks:**
*www.thebiozone.com/weblink/AP2-3121.html*

*BIOZONE APP: Student Review Series Communities*

# Plants as Producers

Life on earth is solar-powered; it runs on energy from the sun. Plants, algae, and some bacteria capture this solar energy and convert it into sugars. They achieve this through a process called **photosynthesis**. Each year these organisms produce more than 200 billion tonnes of food. The chemical energy stored in this food fuels the reactions that sustain life (metabolism). Producers are **autotrophs** (self-feeding). Organisms that cannot make their own food are **heterotrophs** and rely on producers either directly or indirectly for their energy. The photosynthesis that occurs in the oceans is vital to the Earth's functioning, providing oxygen and absorbing carbon dioxide. The oceans cover nearly three quarters of the globe. The evaporation from oceans provides most of the Earth's rainfall and ocean temperatures have a major effect on the world's climate. Despite its importance, humans have harvested the ocean heavily and used it to dump waste. Only in recent years have we realized the consequences of this.

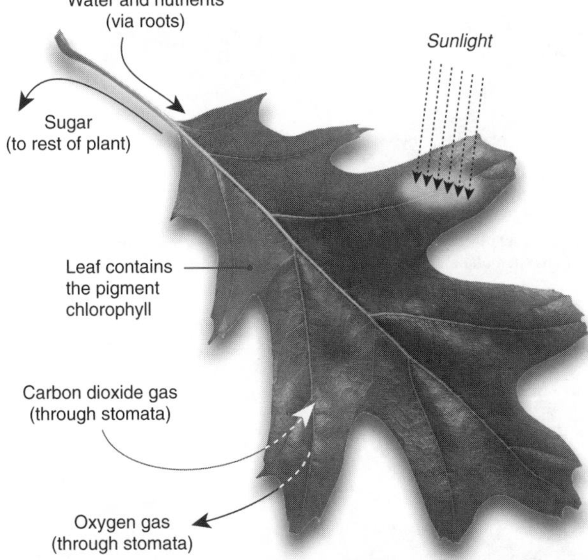


## Requirements for Photosynthesis

In order to produce their own food, plants need only a few raw materials, light energy from the sun, and the pigment chlorophyll, which is contained in chloroplasts (in the leaves and stems of higher plants). Photosynthesis is summarized in the chemical equation below. It is important to note that this equation is a deceptively simple summary of a more involved process. Photosynthesis is not a single process but two complex processes (the light dependent and light independent reactions) each with multiple steps.

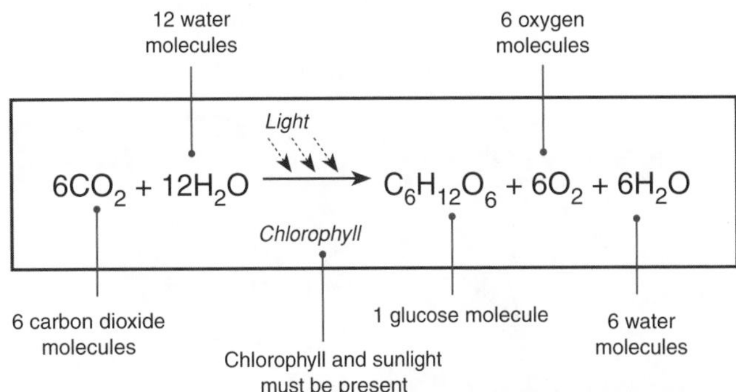

The photosynthesis of marine algae supplies a substantial portion of the world's oxygen. The oceans also act as sinks for absorbing large amounts of carbon dioxide.

Macroalgae, like this giant kelp, are important marine producers. Algae living near the ocean surface get access to light used in photosynthesis (the red wavelength).

On land, vascular plants (such as trees with transport vessels) are the main producers of food. Plants at different levels in a forest receive different intensity and quality of light.

1. Write the overall chemical equation for photosynthesis using:

   (a) Words: ______________________________

   (b) Chemical symbols: ______________________________

2. Describe three things of fundamental biological importance provided by photosynthesis:

   (a) ______________________________

   (b) ______________________________

   (c) ______________________________

3. Explain how light limits the distribution of algae in the ocean: ______________________________

   ______________________________

4. Predict one probable effect of deforestation on the Earth's climate and/or level of the gases oxygen and carbon dioxide:

   ______________________________

ISBN: 978-1-927173-12-1

# Modes of Nutrition

The way in which living organisms obtain their source of energy and carbon is termed their nutritional mode. There is a great diversity in nutritional modes amongst different phyla, which has resulted in organisms developing numerous adaptations to obtain and use energy. The prokaryotes (bacteria) show the greatest variety in terms of the range of organic and inorganic compounds used as energy sources. The diagram below illustrates the classification of nutritional modes in living organisms. The diagram simplifies the real situation and concentrates on the diversity within the eukaryotic groups.

## Nutritional Patterns in Organisms

Living organisms can be classified according to their source of energy and carbon. According to their **energy source**, organisms are classified as either **phototrophs** (using light as their main energy source) or **chemotrophs** (using inorganic or organic compounds for energy). As a carbon source, **autotrophs** (self-feeders) use carbon dioxide, and **heterotrophs** (feeders on others) need an organic carbon source. Most organisms are either photoautotrophs, chemoautotrophs, or chemoheterotrophs. Prokaryotes show a huge variety of nutritional modes. Many are photo- or chemoautotrophs (chemosynthetic) but a large number are chemoheterotrophs (as are animals, fungi, and many protists). For many, the energy and carbon source is glucose.

## Autotrophic Nutrition

**Photoautotrophs**: Photosynthetic bacteria, cyanobacteria, algae, plants.

**Chemoautotrophs**: Sulfur, hydrogen, iron, and nitrifying bacteria.

## Heterotrophic Nutrition (Chemoheterotrophs*)

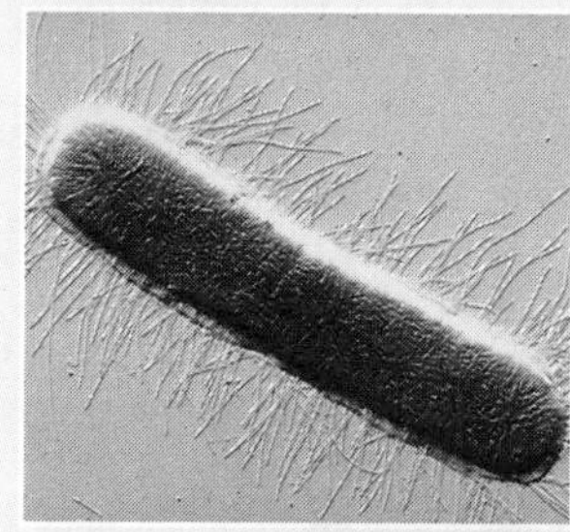


Most of the bacteria with which we are familiar are chemoheterotrophs.

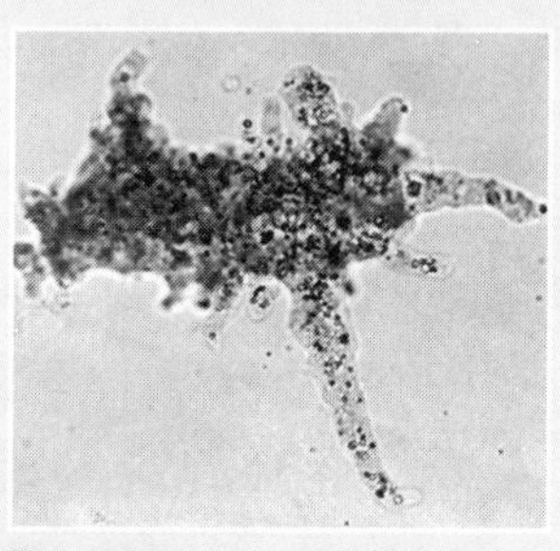


**Protozoans**, such as *Amoeba*, engulf food particles by phagocytosis.

** A few bacterial groups are photoheterotrophic.*

Fungi may be saprophytic (below), parasitic, or mutualistic.

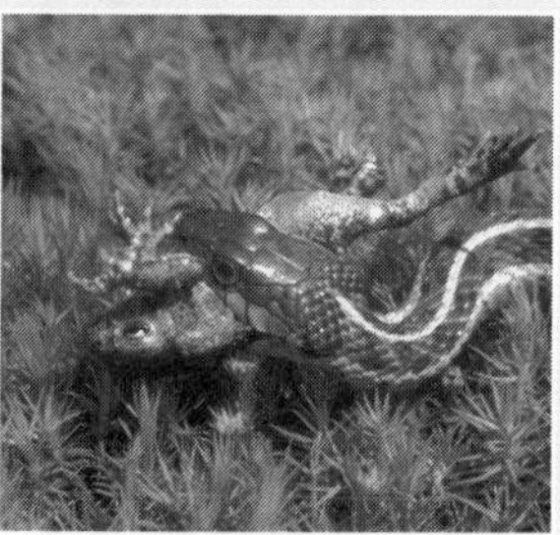

Feeding provides animals with a carbon and energy source: glucose.

## Nutritional Modes of Heterotrophs

Heterotrophic organisms feed on organic material in order to obtain the energy and nutrients they require. They depend either directly on other organisms (dead or alive), or their by-products (e.g. feces, cell walls, or food stores). There are three principal modes of heterotrophic nutrition: saprotrophic, parasitic, and holozoic. Within the animal phyla, holozoic nutrition is the most common nutritional mode.

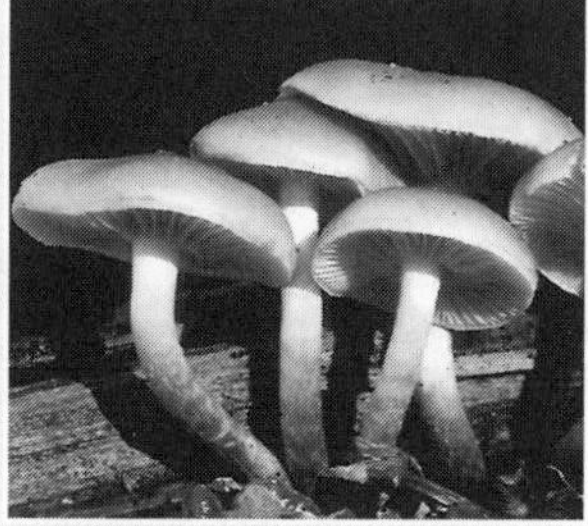

Most fungi and many bacteria are saprotrophs. They are decomposers, obtaining nutrition from the extra-cellular digestion of dead organic material.

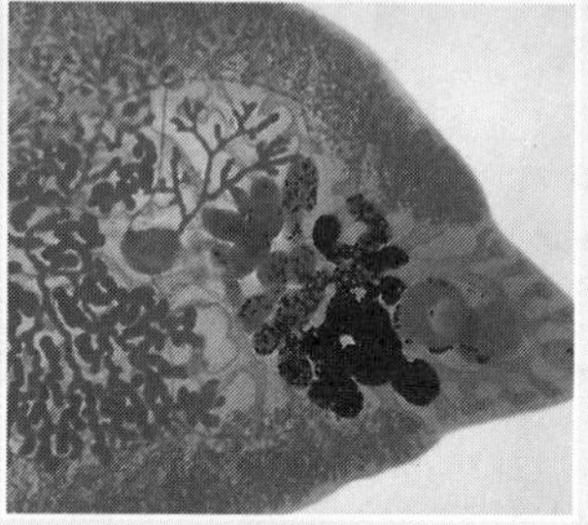


Parasites, e.g. flukes, live on or within their living host organism for part or all of their life. Bacteria, fungi, protists, and animals all have parasitic representatives.

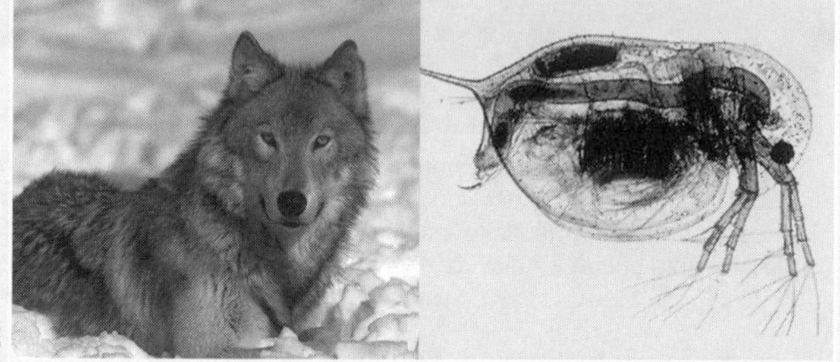


Holozoic means to ingest solid organic material from the bodies of other organisms. It is the main feeding mode of animals, although a few specialized plants may obtain some nutrients this way. Holozoic animals are classified according to the form of the food they take in: small or large particles, or fluid.

1. Discuss the differences in nutritional mode between photoautotrophs, chemoautotrophs, and chemoheterotrophs:

2. Explain how saprotrophs differ from parasites:



***Related activities:*** *Plants as Producers*


**ISBN: 978-1-927173-12-1**

# Food Chains

Every ecosystem has a trophic structure: a hierarchy of feeding relationships that determines the pathways for energy flow and nutrient cycling. Species are assigned to trophic levels on the basis of their sources of nutrition. The first trophic level (**producers**), ultimately supports all other levels. The consumers are those that rely on producers for their energy. Consumers are ranked according to the trophic level they occupy (first order, second order, etc.). The sequence of organisms, each of which is a source of food for the next, is called a **food chain**. Food chains commonly have four links but seldom more than six. Those organisms whose food is obtained through the same number of links belong to the same trophic level. Note that some consumers (particularly top carnivores and omnivores) may feed at several different trophic levels, and many primary consumers eat many plant species. The different food chains in an ecosystem therefore tend to form complex webs of interactions (food webs).

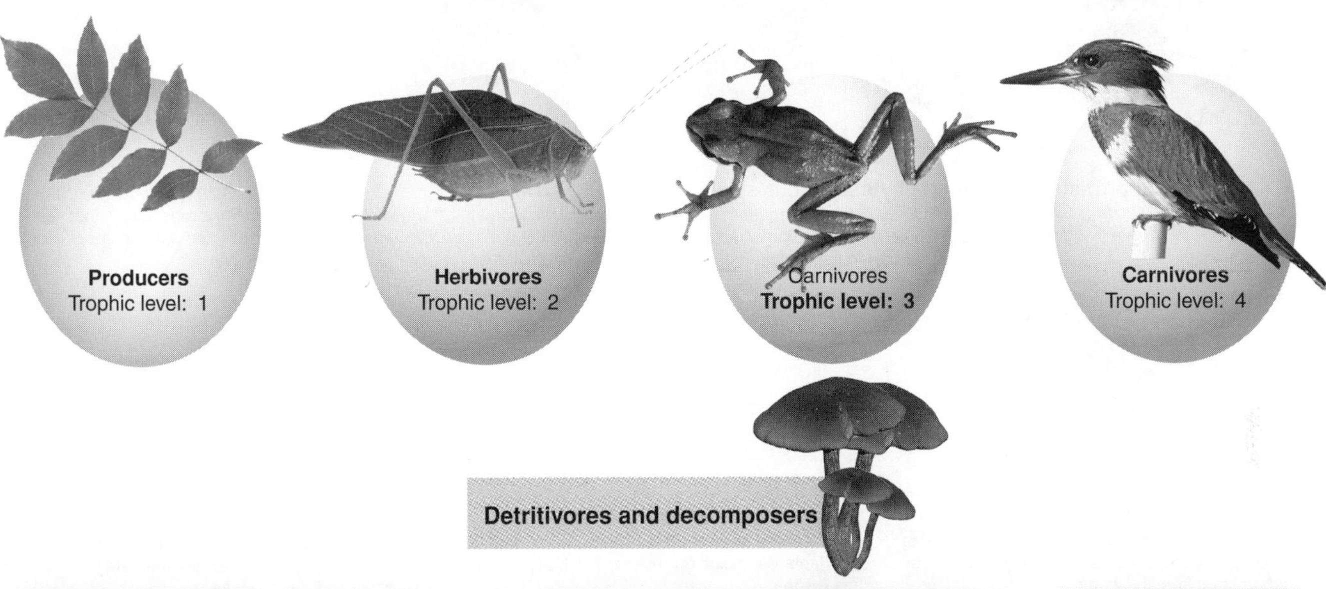

Ladybug eating aphid

**Producers** (algae, green plants, and some bacteria) make their own food using simple inorganic carbon sources (e.g. $CO_2$). Sunlight is the most common energy source for this process.

**Consumers** (animals, non-photosynthetic protists, and some bacteria) rely on other living organisms or organic particulate matter for their energy and their source of carbon. First order consumers, such as aphids (left), feed on producers. Second (and higher) order consumers, such as ladybugs (center) eat other consumers. **Detritivores** consume (ingest and digest) detritus (decomposing organic material) from every trophic level. In doing so, they contribute to decomposition and the recycling of nutrients. Common detritivores include wood-lice, millipedes (right), and many terrestrial worms.

**Decomposers** (fungi and some bacteria) obtain their energy and carbon from the extracellular breakdown of dead organic matter (DOM). Decomposers play a central role in nutrient cycling.

The diagram above represents the basic elements of a food chain. In the questions below, you are asked to add to the diagram the features that indicate the flow of energy through the community of organisms.

1. (a) State the original energy source for this food chain: ______________________

   (b) Draw arrows on the diagram above to show how the energy flows through the organisms in the food chain. Label each arrow with the process involved in the energy transfer. Draw arrows to show how energy is lost by respiration.

2. (a) Describe what happens to the **amount** of energy available to each successive trophic level in a food chain:

   ______________________

   (b) Explain why this is the case: ______________________

   ______________________

3. Explain what you could infer about the tropic level(s) of the kingfisher, if it was found to eat both katydids and frogs:

   ______________________



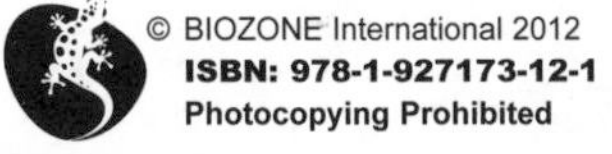

# Constructing a Food Web

Every ecosystem has a **trophic structure**: a hierarchy of feeding relationships which determines the pathways for energy flow and nutrient cycling. Species are assigned to trophic levels on the basis of their sources of nutrition, with the first trophic level (the **producers**), ultimately supporting all other (consumer) levels. Consumers are ranked according to the trophic level they occupy, although some consumers may feed at several different trophic levels. The sequence of organisms, each of which is a source of food for the next, is called a **food chain**. The different food chains in an ecosystem are interconnected to form a complex web of feeding interactions called a **food web**. In the example of a lake ecosystem below, your task is assemble the organisms into a food web in a way that illustrates their trophic status and their relative trophic position(s).

## Feeding Requirements of Lake Organisms

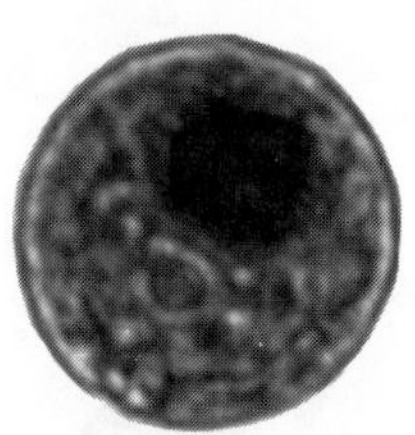

**Autotrophic protists**
*Chlamydomonas* (above), *Euglena*
Two of the many genera that form the phytoplankton.

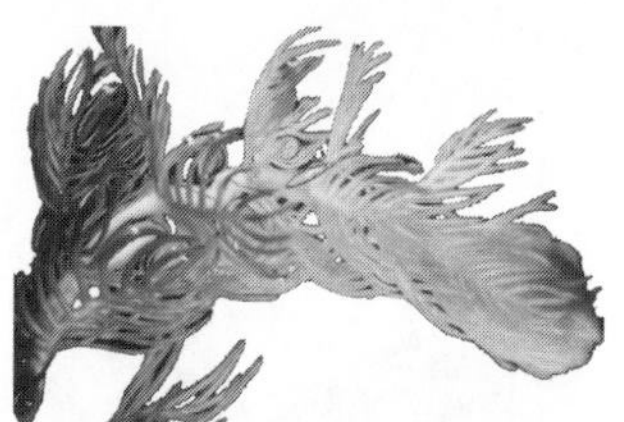

**Macrophytes** (various species)
A variety of flowering aquatic plants are adapted for being submerged, free-floating, or growing at the lake margin.

**Detritus**
Decaying organic matter from within the lake itself or it may be washed in from the ake margins.

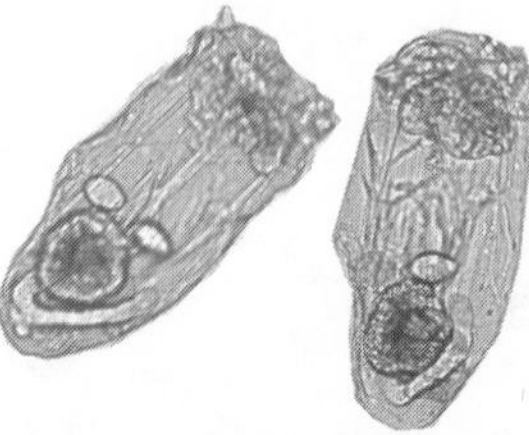

***Asplanchna*** (planktonic rotifer)
A large, carnivorous rotifer that feeds on protozoa and young zooplankton (e.g. small *Daphnia*).

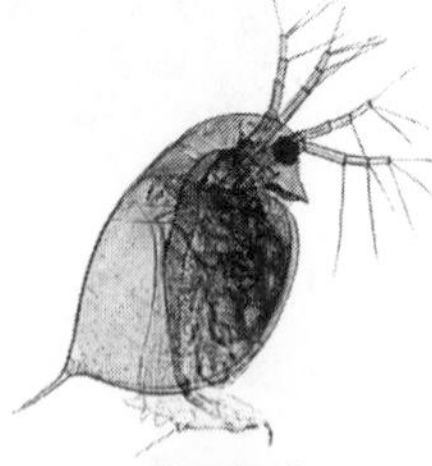

***Daphnia***
Small freshwater crustacean that forms part of the zooplankton. It feeds on planktonic algae by filtering them from the water with its limbs.

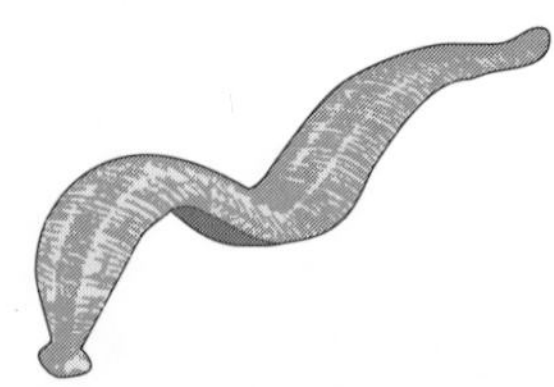

**Leech** (*Glossiphonia*)
Leeches are fluid feeding predators of smaller invertebrates, including rotifers, small pond snails and worms.

**Three-spined stickleback** (*Gasterosteus*)
A common fish of freshwater ponds and lakes. It feeds mainly on small invertebrates such as *Daphnia* and insect larvae.

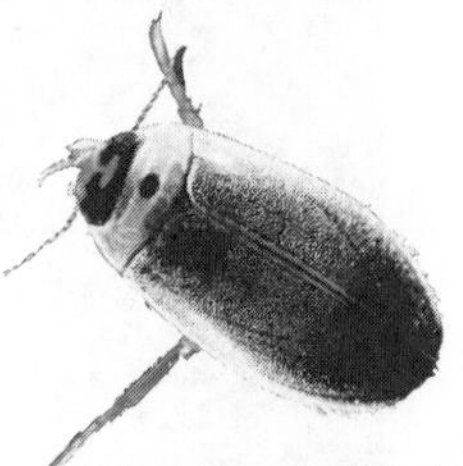

**Diving beetle** (*Dytiscus*)
Diving beetles feed on aquatic insect larvae and adult insects blown into the lake community. The will also eat organic detritus collected from the bottom mud.

**Carp** (*Cyprinus*)
A heavy bodied freshwater fish that feeds mainly on bottom living insect larvae and snails, but will also take some plant material (not algae).

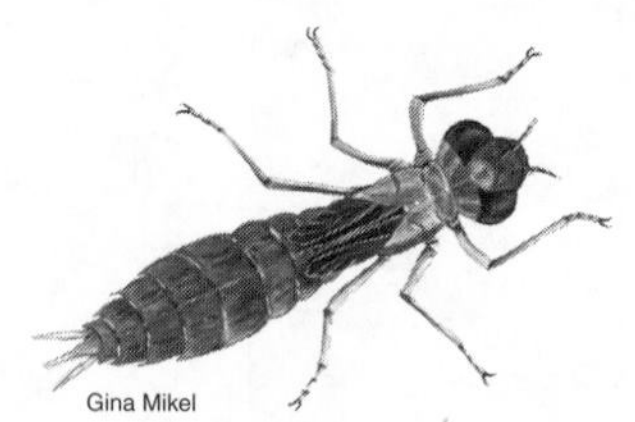


**Dragonfly** larva
Large aquatic insect larvae that are voracious predators of small invertebrates including *Hydra*, *Daphnia*, other insect larvae, and leeches.

**Great pond snail** (*Limnaea*)
Omnivorous pond snail, eating both plant and animal material, living or dead, although the main diet is aquatic macrophytes.

**Herbivorous water beetles** (e.g. *Hydrophilus*)
Feed on water plants, although the young beetle larvae are carnivorous, feeding primarily on small pond snails.

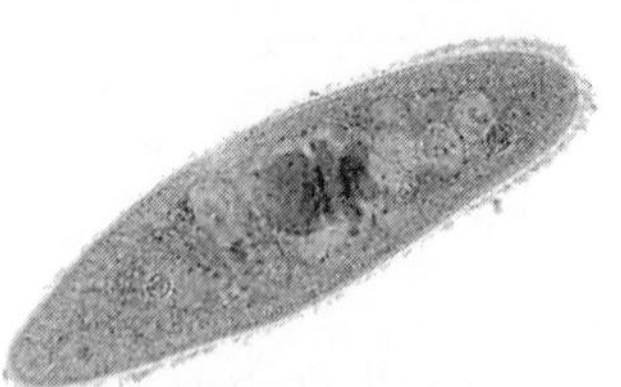

**Protozan** (e.g. *Paramecium*)
Ciliated protozoa such as *Paramecium* feed primarily on bacteria and microscopic green algae such as *Chlamydomonas*.

**Pike** (*Esox lucius*)
A top ambush predator of all smaller fish and amphibians. They are also opportunistic predators of rodents and small birds.

**Mosquito larva** (*Culex* spp.)
The larvae of most mosquito species, e.g. *Culex*, feed on planktonic algae and small protozoa before passing through a pupal stage and undergoing metamorphosis into adult mosquitoes.

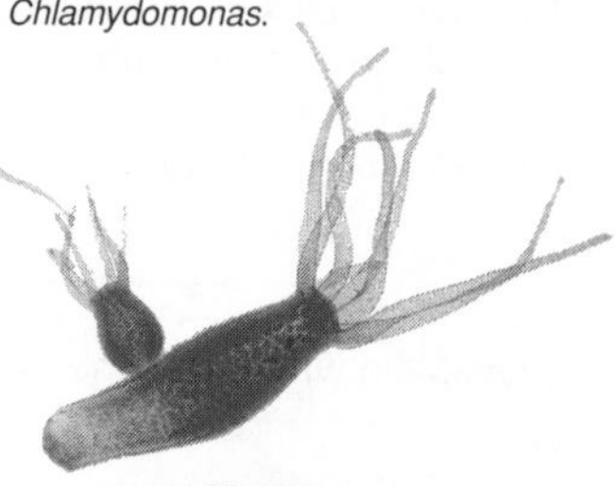

***Hydra***
A small carnivorous cnidarian that captures small prey items, e.g. rotifers, *Daphnia*, and insect larvae, using its stinging cells on the tentacles.

**ISBN: 978-1-927173-12-1**

1. From the information provided for the lake food web components on the previous page, construct **ten** different **food chains** to show the feeding relationships between the organisms. Some food chains may be shorter than others and most species will appear in more than one food chain. An example has been completed for you.

Example 1: Macrophyte ⟶ Herbivorous water beetle ⟶ Carp ⟶ Pike

(a) ______________________________

(b) ______________________________

(c) ______________________________

(d) ______________________________

(e) ______________________________

(f) ______________________________

(g) ______________________________

(h) ______________________________

(i) ______________________________

(j) ______________________________

2. (a) Use the food chains created above to help you to draw up a **food web** for this community. Use the information supplied to draw arrows showing the flow of **energy** between species (only energy **from** the detritus is required).

(b) Label each species to indicate its position in the food web, i.e. its trophic level (**T1, T2, T3, T4, T5**). Where a species occupies more than one trophic level, indicate this, e.g. **T2/3**:

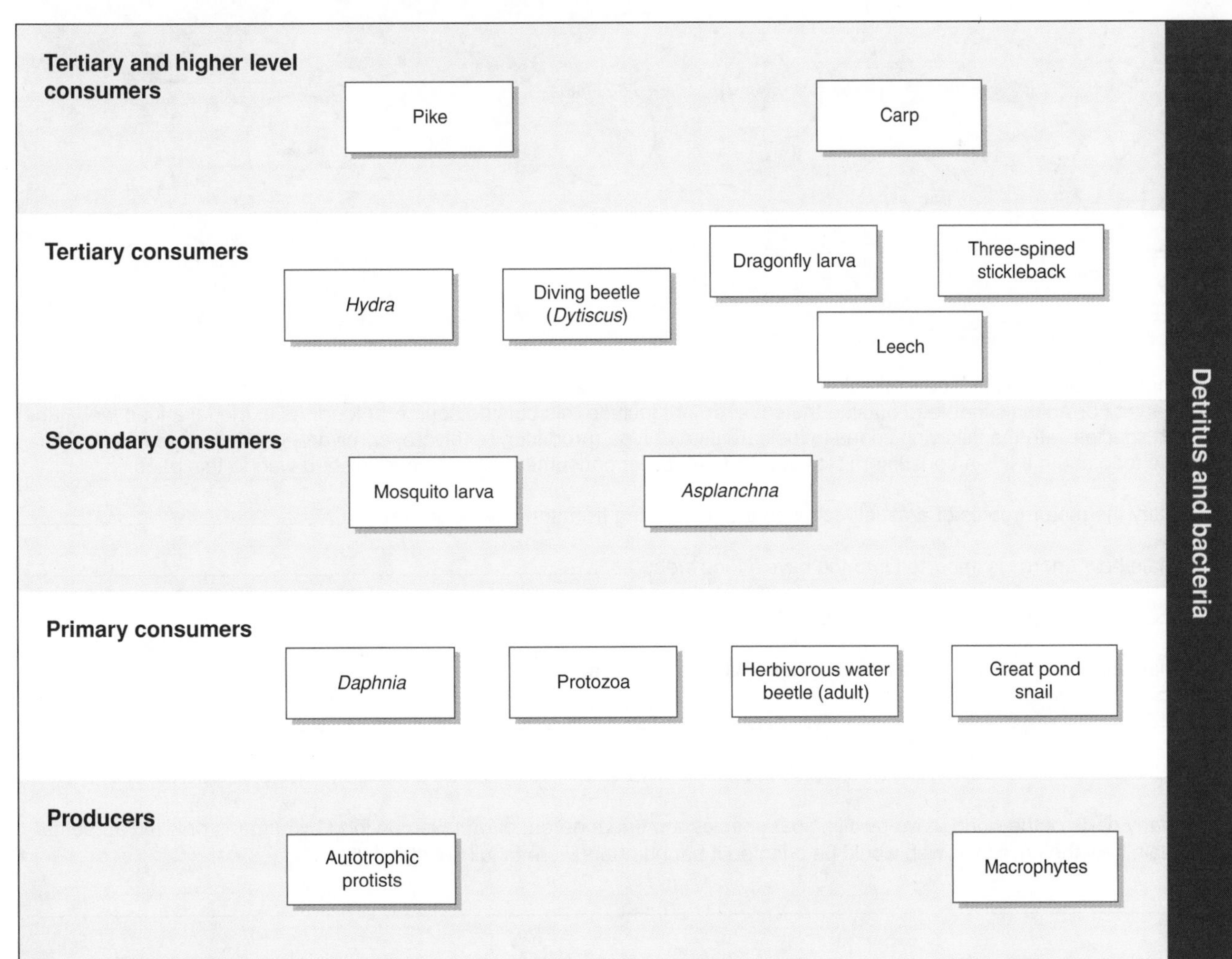



ISBN: 978-1-927173-12-1

# Cave Food Webs

A cave is a barren environment, without the light that normally sustains most ecosystems. Despite this, a wide range of animals inhabit caves and are specially adapted to that particular environment. Other animals, such as bats, do not live permanently in the cave but use it as a roosting or resting and breeding area, safe from many predators and other dangers. The food webs of caves are very fragile and based on few resources. **Around the entrance** of the cave, the owl (**1**) preys on the mouse (**2**) which itself feeds on the vegetation outside the cave. The owl and the mouse leave droppings that support the cave dung beetle (**3**) and the millipede (**4**). The cave cricket (**5**) scavenges dead birds and mammals near the cave entrance. The harvestman (**6**) is a predator of the dung beetle, the millipede, and the cricket. **Inside the cave**, the horseshoe bat (**7**) roosts and breeds in safety, leaving the cave to feed outside on slow flying insects. The bats produce vast quantities of guano (droppings). The guano is eaten by the blind cave beetle (**8**), the millipede (**4**) and the springtail (**9**). These invertebrates are hunted by the predatory cave spider (**10**). Occasionally, in tropical caves, snakes (not shown) may enter the cave and feed on bats. **In underground pools**, the bat guano supports the growth of bacteria (**11**). Flatworms (**12**) and isopods (**13**) feed on the bacteria and themselves are eaten by the blind cave shrimp (**14**). The blind cave fish (**15**) is the top predator, feeding on isopods and the blind cave shrimps.



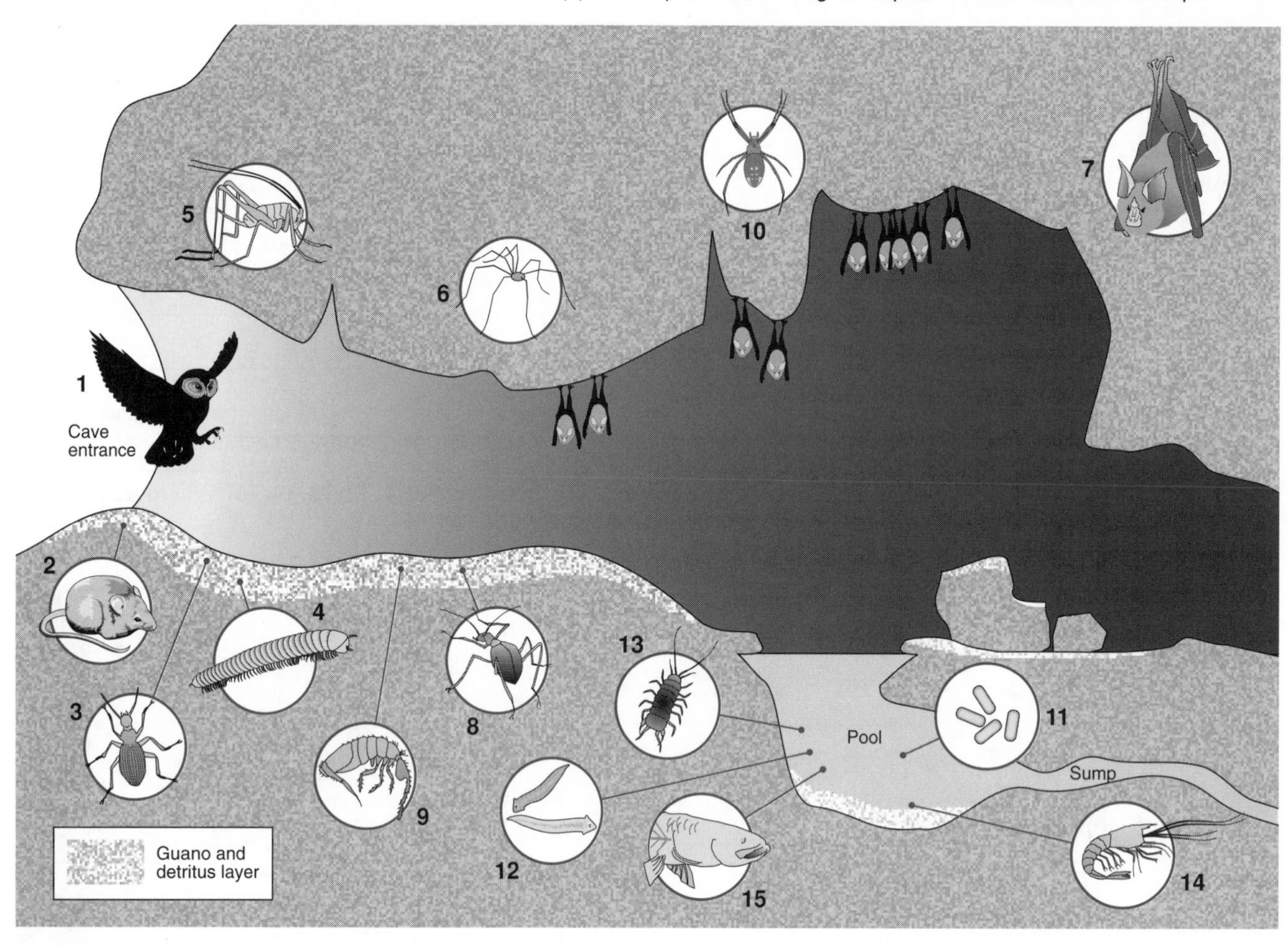

1. Using the previous lake food web activity as a guide, construct a food web for the cave ecosystem on a separate sheet of paper. For animals that feed outside the cave, do not include this outside source of food. As in the lake food web, label each species with the following codes to indicate its diet type (producer, herbivore, carnivore, omnivore) and its position in the food chain if it is a consumer (1st, 2nd, 3rd, 4th order consumer). Staple your finished web to this page.

2. Identify the major trophic of a usual food web that is missing from the cave food web: ________________

3. Explain how energy is imported into the cave's food web: ________________

4. Explain how energy from the cave ecosystem might be removed: ________________

5. In many parts of the world, cave-dwelling bat species are endangered, often taken as food by humans or killed as pests. Explain how the cave food web would be affected if bat numbers were to fall substantially:



**_Related activities:_** _Constructing a Food Web_

**ISBN: 978-1-927173-12-1**

# The Darkest Depths

Deep sea hydrothermal vents occur at around 2000 m depth and where tectonic plates meet. Buckling in the plates causes fault lines to form where water can move down into the crust before being heated and ejected at temperatures of up to 350°C. Temperatures this high are possible because the pressure of the ocean prevents the water from boiling into steam. The high water temperature dissolves minerals from crustal rocks and, when they reach the surface, they precipitate into formations called **black smokers** - chimneys of minerals that may reach 60m high.

Hydrothermal vents are the site of unique communities. At this depth, no light penetrates and the amount of organic debris falling from above is minimal because much of it has been used up by the time it reaches the bottom. The organisms living here are restricted in their movement; only a few tens of metres from the mouth of the vent, the water temperature plummets to barely above freezing. The degree of isolation has resulted in the evolution of a unique fauna.

Photo: NOAA

Photo: US federal govt

The water spewing from the hydrothermal vents is rich in minerals and the bacteria living there have evolved to use these to manufacture food. These chemosynthetic bacteria use oxygen and hydrogen sulfide (highly toxic to most organisms) to build organic molecules. They are the producers on which the vent community is based. They form thick mats around the vents or float in aggregations resembling snow storms.

Tube worms, one of the larger organisms in these communities, provide shelter for the bacteria and benefit from the products of bacterial **chemosynthesis**. Vent mussels also have bacteria living within them and have abandoned a filter-feeding lifestyle to form a mutualistic relationship with bacteria in their tissues. Blind shrimps and crabs scavenge on decaying material and the bacterial mats. Octopi and fish also make up part of the food web, preying on smaller animals. Of most interest to scientists is the Pompeii worm. It can withstand temperatures of 80°C; higher than any other complex organism. The hairy coat that covers it is, in fact, mats of bacteria on which the worm feeds.

1. Describe the environmental conditions found around deep sea hydrothermal vents: ______________________

______________________

______________________

2. Explain reasons for the uniqueness of the vent communities: ______________________

______________________

______________________

3. Discuss the relationships between the organisms of the vent community and use the information to construct a basic food web in the space provided:

______________________

______________________

______________________

______________________

______________________

______________________

______________________

______________________

______________________

______________________

______________________



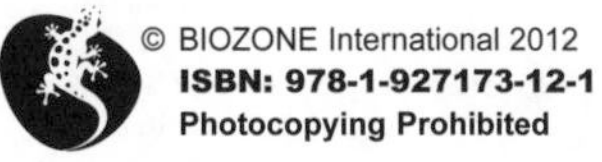


ISBN: 978-1-927173-12-1


***Related activities**: Constructing a Food Web*

# Measuring Primary Productivity

The energy entering an ecosystem is determined by the rate at which producers can convert sunlight energy or inorganic compounds into chemical energy. Photosynthesis by vascular plants accounts for most of the energy entering a terrestrial ecosystem. The total energy fixed by a plant in photosynthesis is the **gross primary production** (**GPP**) and it is usually expressed as J $m^{-2}$ (or kJ $m^{-2}$), or as g $m^{-2}$. However, a portion of this energy is required by the plant for respiration. Subtracting respiration from GPP gives the **net primary production** (**NPP**). Measurement of this energy is important because is represents the energy available to consumers in the ecosystem and thus how much biomass the ecosystem can support (in general, the higher the productivity, the higher the biomass of the ecosystem). Some important measurements of productivity are given below:

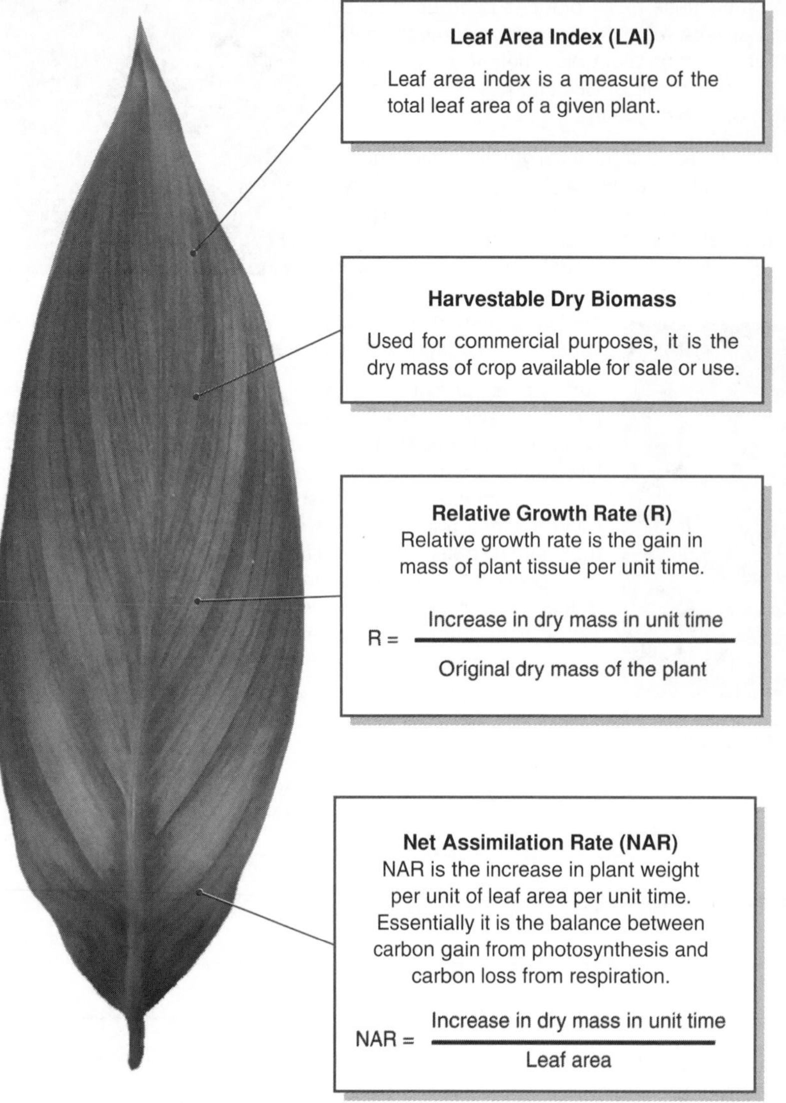


### Measuring Productivity

Measuring gross primary productivity (GPP) can be difficult due to the effect of on-going respiration, which uses up some the organic material produced (glucose). One method for measuring GPP is to measure the difference in production between plants kept in the dark and those in the light.

A simple method for measuring GPP in phytoplankton is illustrated below:

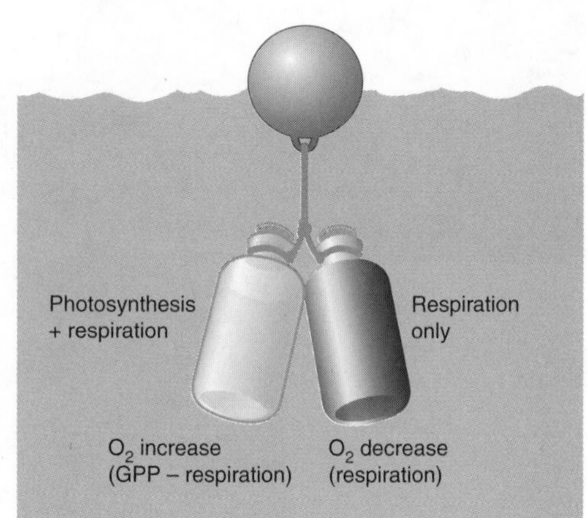


Two bottles are lowered into the ocean or lake to a specified depth, filled with water, and then stoppered. One bottle is transparent, the other is opaque. The $O_2$ concentration of the water surrounding the bottles is measured and the bottles are left for a specified amount of time. The phytoplankton in the transparent bottle will photosynthesize, increasing the $O_2$ concentration, and respire, using some of that $O_2$. The phytoplankton in the opaque bottle will only respire. The final measured difference in $O_2$ between the bottles gives the amount of $O_2$ produced by the phytoplankton in the specified time (including that used for respiration).

The amount of $O_2$ used allows us to determine the amount of glucose produced and therefore the GPP of the phytoplankton.

1. Suggest how the LAI might influence the rate of primary production: ______________________

______________________

______________________

2. Estimating the NPP is relatively simple: all the plant material (including root material) from a measured area (e.g. 1 $m^2$) is collected and dried (at 105°C) until it reaches a constant mass. This mass, called the **standing crop**, is recorded (in kg $m^{-2}$). The procedure is repeated after some set time period (e.g. 1 month). The difference between the two calculated masses represents the estimated NPP:

(a) Suggest why this procedure only provides an estimate of NPP: ______________________

______________________

(b) State what extra information would be required in order to express the standing crop value in kJ $m^{-2}$: ______________________

______________________

(c) Suggest what information would be required in order to calculate the GPP: ______________________

______________________



*Related activities: Production and Trophic Efficiency*

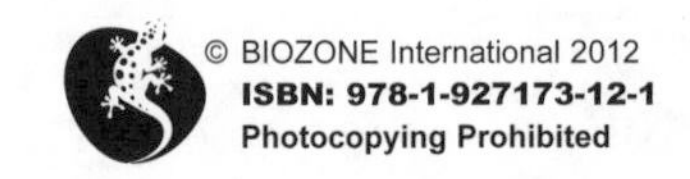


ISBN: 978-1-927173-12-1

# Production and Trophic Efficiency

The energy entering ecosystems is fixed by producers at a rate that is dependent on **limiting factors** such as temperature and the availability of light, water, and nutrients. This energy to converted to **biomass** (the mass of biological material) by **anabolic reactions**. The **rate** of biomass production, or **net primary productivity**, is the biomass produced per area per unit time. **Trophic efficiency** refers to the efficiency of energy transfer from one trophic level to the next. The trophic efficiencies of herbivores can vary widely, depending on how much of the producer biomass is consumed and **assimilated** (incorporated into new biomass). In some natural ecosystems this can be surprisingly high. Humans intervene in natural energy flows by simplifying the system and reducing the number of transfers occurring between trophic levels.

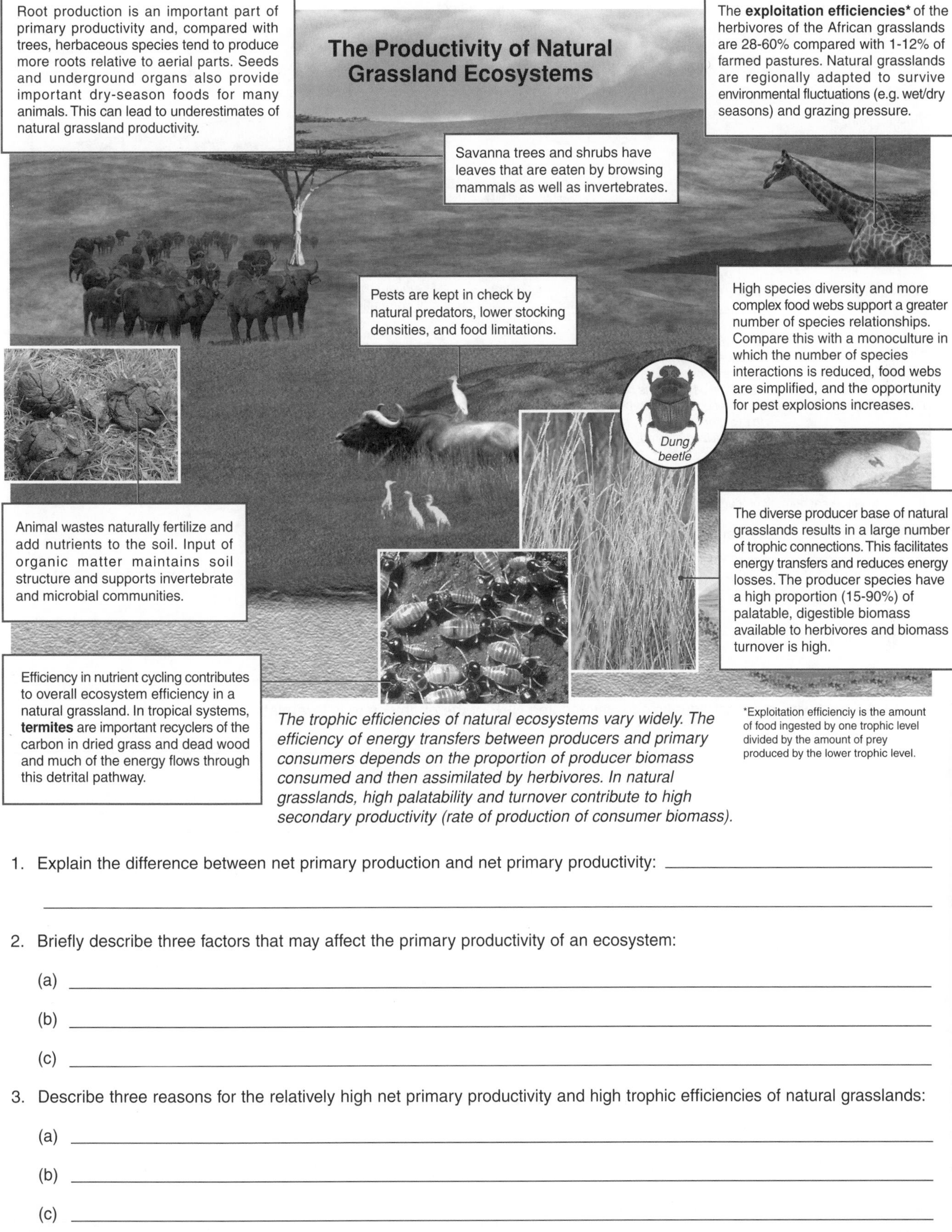


*The trophic efficiencies of natural ecosystems vary widely. The efficiency of energy transfers between producers and primary consumers depends on the proportion of producer biomass consumed and then assimilated by herbivores. In natural grasslands, high palatability and turnover contribute to high secondary productivity (rate of production of consumer biomass).*

*Exploitation efficienciy is the amount of food ingested by one trophic level divided by the amount of prey produced by the lower trophic level.

1. Explain the difference between net primary production and net primary productivity: ____________________

____________________

2. Briefly describe three factors that may affect the primary productivity of an ecosystem:

(a) ____________________

(b) ____________________

(c) ____________________

3. Describe three reasons for the relatively high net primary productivity and high trophic efficiencies of natural grasslands:

(a) ____________________

(b) ____________________

(c) ____________________




**ISBN: 978-1-927173-12-1**

## Measuring Productivity

The gross primary productivity of an ecosystem will depend on the capacity of the producers to capture and fix carbon in organic compounds. In most ecosystems, this is limited by constraints on photosynthesis (availability of light, nutrients, or water for example). The net primary productivity is then determined by how much of this goes into plant biomass per unit time, after respiratory needs are met. This will be the amount available to the next trophic level. It is difficult to measure productivity, but it is often estimated from the harvestable dry biomass or standing crop (the net primary production).

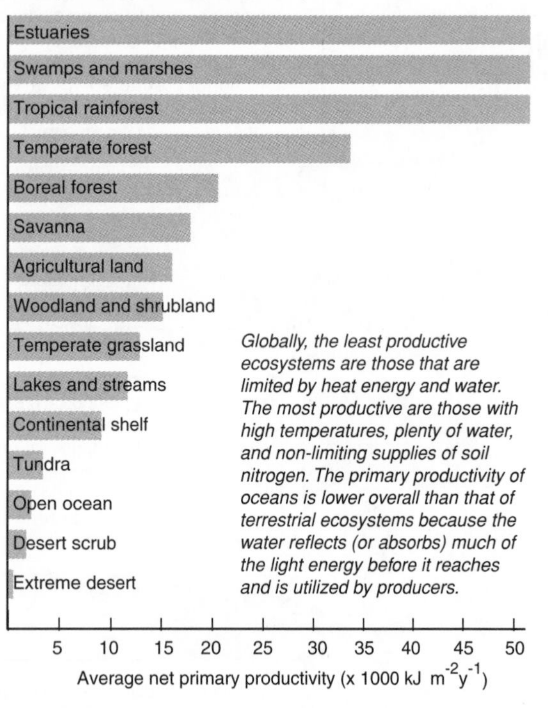


## Agriculture and Productivity

Increasing net productivity in agriculture (increasing yield) is a matter of manipulating and maximizing energy flow through a reduced number of trophic levels. On a farm, the simplest way to increase the net primary productivity is to produce a monoculture. Monocultures reduce competition between the desirable crop and weed species, allowing crops to put more energy into biomass. Other agricultural practices designed to increase productivity in crops include pest (herbivore) control and spraying to reduce disease. Higher productivity in feed-crops also allows greater secondary productivity (e.g. in livestock). Here, similar agricultural practices make sure the energy from feed-crops is efficiently assimilated by livestock.

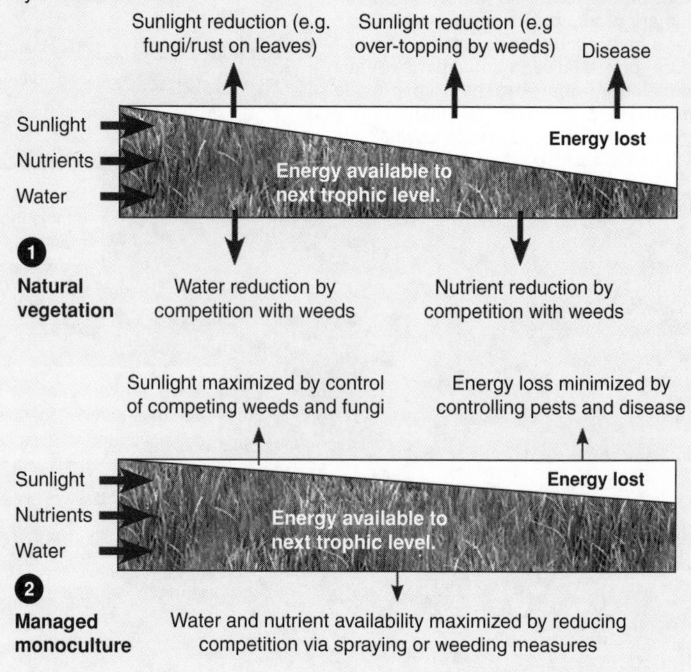


4. With reference to the bar graph above:

   (a) Suggest why tropical rainforests are among the most productive terrestrial ecosystems, while tundra and desert ecosystems are among the least productive:

   (b) Suggest why, amongst aquatic ecosystems, the NPP of the open ocean is low relative to that of coastal systems:

5. (a) How could a farmer maximize the net primary productivity of a particular crop?

   (b) How could a farmer maximize the productivity of their livestock?

6. Explain the contrasting net primary productivities of intensive agricultural land and extreme desert:


ISBN: 978-1-927173-12-1

# Quantifying Energy Flow in an Ecosystem

The flow of energy through an ecosystem can be measured and analyzed. It provides some idea as to the energy trapped and passed on at each trophic level. Each trophic level in a food chain or web contains a certain amount of biomass: the dry weight of all organic matter contained in its organisms. Energy stored in biomass is transferred from one trophic level to another (by eating, defecation etc.), with some being lost as low-grade heat energy to the environment in each transfer. Three definitions are useful:

- **Gross primary production**: The total of organic material produced by plants (including that lost to respiration).
- **Net primary production**: The amount of biomass that is available to consumers at subsequent trophic levels.
- **Secondary production**: The amount of biomass at higher trophic levels (consumer production). Production figures are sometimes expressed as rates (productivity).

The percentage of energy transferred from one trophic level to the next is called the **trophic efficiency** or ecological efficiency. It varies between 5% and 20% and is a measure of the efficiency of energy transfer. An average figure of 10% is often used (called the **ten percent rule**). The path of energy flow in an ecosystem depends on its characteristics. In a tropical forest, most of the primary production enters the detrital and decomposer food chains. However, in an intensively grazed pasture more than half the primary production may enter the grazing food chain.

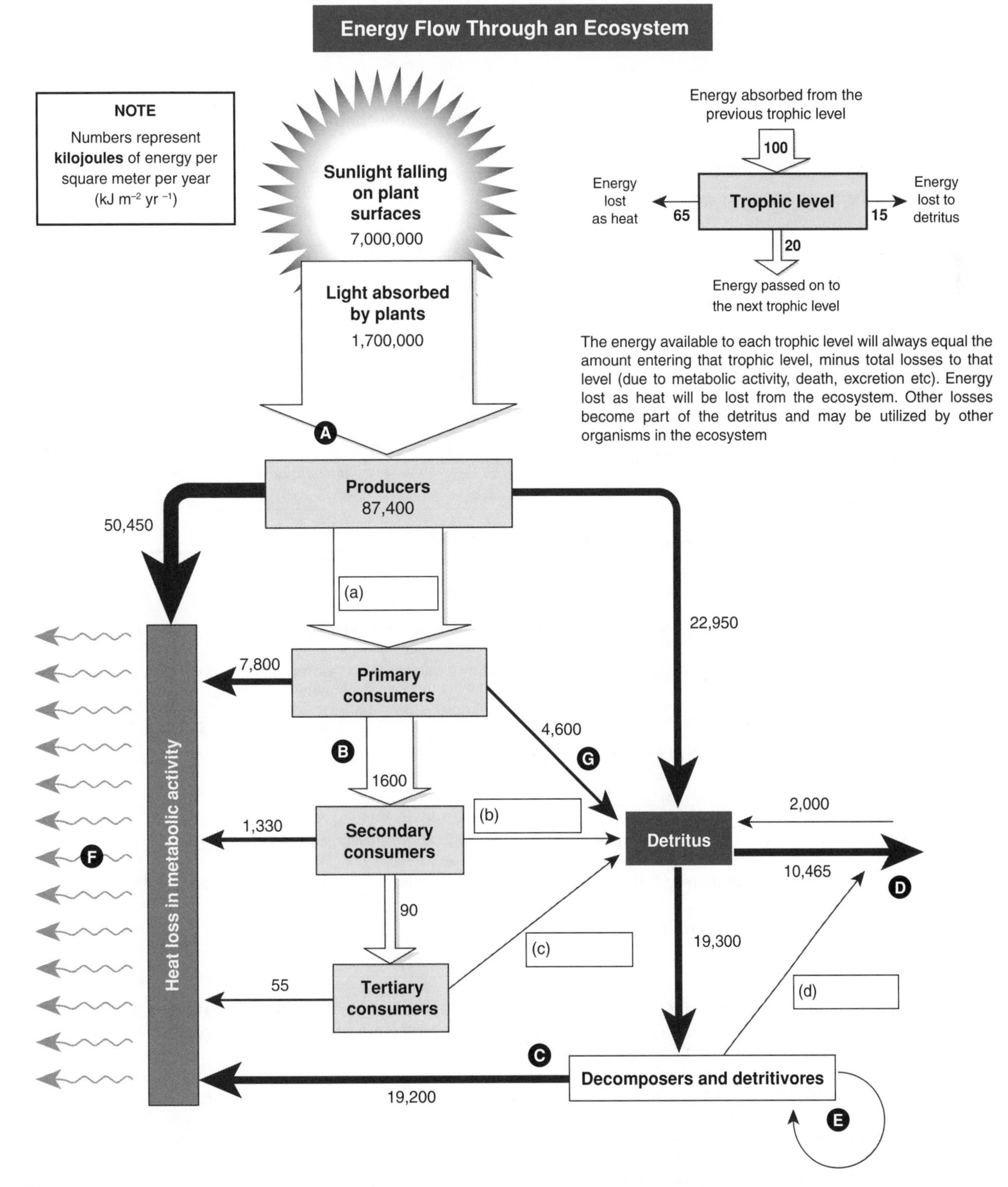

## Energy Flow Through an Ecosystem

**NOTE**
Numbers represent **kilojoules** of energy per square meter per year ($kJ\ m^{-2}\ yr^{-1}$)

Energy absorbed from the previous trophic level
100
Trophic level
Energy lost as heat 65
Energy lost to detritus 15
20 Energy passed on to the next trophic level

The energy available to each trophic level will always equal the amount entering that trophic level, minus total losses to that level (due to metabolic activity, death, excretion etc). Energy lost as heat will be lost from the ecosystem. Other losses become part of the detritus and may be utilized by other organisms in the ecosystem

Sunlight falling on plant surfaces 7,000,000
Light absorbed by plants 1,700,000
A
Producers 87,400
50,450
(a)
22,950
Primary consumers
7,800
4,600
G
B
1600
Secondary consumers
1,330
(b)
Detritus
2,000
10,465
D
90
(c)
19,300
Tertiary consumers
55
(d)
Heat loss in metabolic activity
F
C
Decomposers and detritivores
19,200
E

1. Study the diagram on the previous page illustrating energy flow through a hypothetical ecosystem. Use the example at the top of the page as a guide to calculate the missing values (a)–(d) in the diagram. Note that the sum of the energy inputs always equals the sum of the energy outputs. Place your answers in the spaces provided on the diagram.

2. What is the original source of energy for this ecosystem? ________________

3. Identify the processes occurring at the points labelled **A – G** on the diagram:

A. ________________ E. ________________

B. ________________ F. ________________

C. ________________ G. ________________

D. ________________

4. (a) Calculate the percentage of light energy falling on the plants that is absorbed at point **A**:

Light absorbed by plants ÷ sunlight falling on plant surfaces x 100 = ________________

(b) What happens to the light energy that is not absorbed? ________________

5. (a) Calculate the percentage of light energy absorbed that is actually converted (fixed) into producer energy:

Producers ÷ light absorbed by plants x 100 = ________________

(b) How much light energy is absorbed but not fixed: ________________

(c) Account for the difference between the amount of energy absorbed and the amount actually fixed by producers:

6. Of the total amount of energy **fixed** by producers in this ecosystem (at point **A**) calculate:

(a) The total amount that ended up as metabolic waste heat (in kJ): ________________

(b) The percentage of the energy fixed that ended up as waste heat: ________________

7. (a) State the groups for which detritus is an energy source: ________________

(b) How could detritus be removed or added to an ecosystem? ________________

8. Under certain conditions, decomposition rates can be very low or even zero, allowing detritus to accumulate:

(a) From your knowledge of biological processes, what conditions might slow decomposition rates?

(b) What are the consequences of this lack of decomposer activity to the energy flow? ________________

(c) Add an additional arrow to the diagram on the previous page to illustrate your answer. ________________

(d) Describe three examples of materials that have resulted from a lack of decomposer activity on detrital material:

9. The **ten percent rule** states that the total energy content of a trophic level in an ecosystem is only about one-tenth (or 10%) that of the preceding level. For each of the trophic levels in the diagram on the preceding page, determine the amount of energy passed on to the next trophic level as a percentage:

(a) Producer to primary consumer: ________________

(b) Primary consumer to secondary consumer: ________________

(c) Secondary consumer to tertiary consumer: ________________

**ISBN: 978-1-927173-12-1**

# Ecological Pyramids

The trophic levels of any ecosystem can be arranged in a pyramid shape. The first trophic level is placed at the bottom and subsequent trophic levels are stacked on top in their 'feeding sequence'. Ecological pyramids can illustrate changes in the numbers, biomass (weight), or energy content of organisms at each level. Each of these three kinds of pyramids tell us something different about the flow of energy and materials between one trophic level and the next. The type of pyramid you choose in order to express information about the ecosystem will depend on what particular features of the ecosystem you are interested in and, of course, the type of data you have collected.

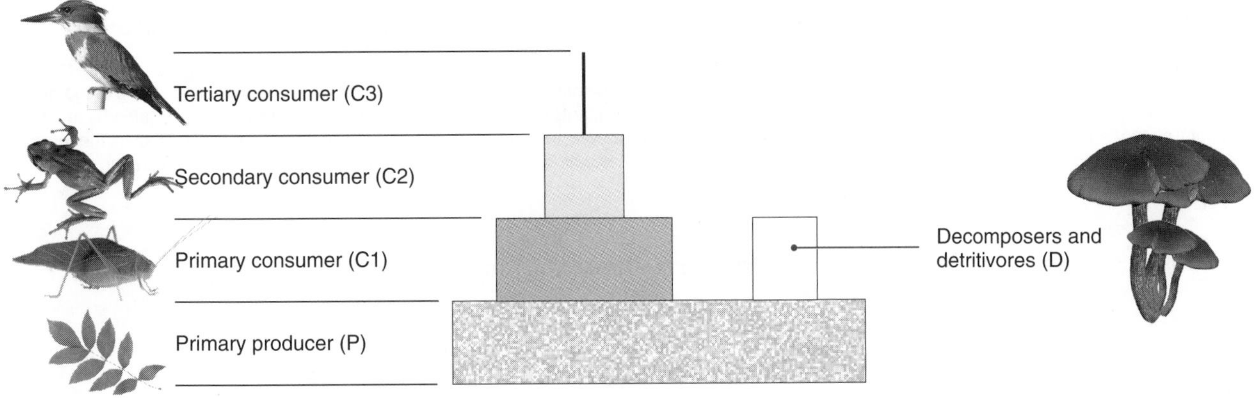


The generalized ecological pyramid pictured above shows a conventional pyramid shape, with a large number (or biomass) of producers forming the base for an increasingly small number (or biomass) of consumers. Decomposers are placed at the level of the primary consumers and off to the side. They may obtain energy from many different trophic levels and so do not fit into the conventional pyramid structure. For any particular ecosystem at any one time (e.g. the forest ecosystem below), the shape of this typical pyramid can vary greatly depending on whether the trophic relationships are expressed as numbers, biomass or energy

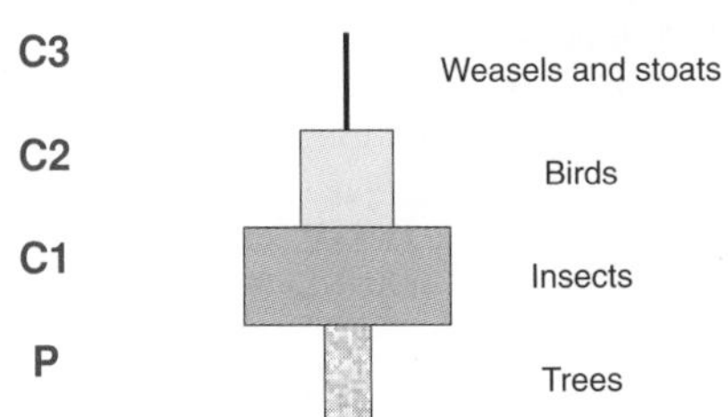


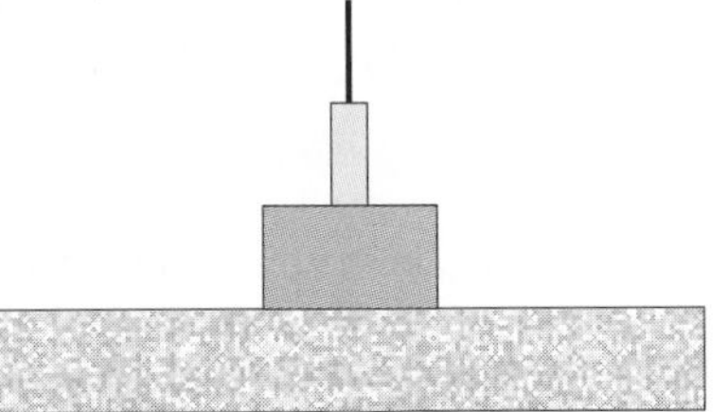

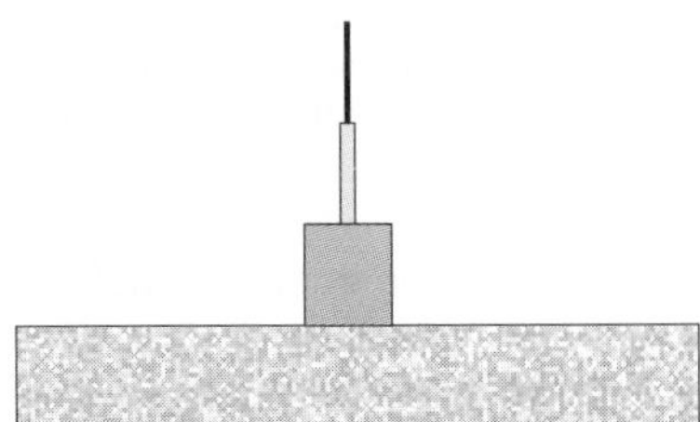

### Numbers in a forest community

Pyramids of numbers display the number of individual organisms at each trophic level. The pyramid above has few producers, but they may be of a very large size (e.g. trees). This gives an 'inverted pyramid', although not all pyramids of numbers are like this.

### Biomass in a forest community

Biomass pyramids measure the 'weight' of biological material at each trophic level. Water content of organisms varies, so 'dry weight' is often used. Organism size is taken into account, allowing meaningful comparisons of different trophic levels.

### Energy in a forest community

Pyramids of energy are often very similar to biomass pyramids. The energy content at each trophic level is generally comparable to the biomass (i.e. similar amounts of dry biomass tend to have about the same energy content).

1. What do each of the following types of ecological pyramids measure?

   (a) Number pyramid: ______

   (b) Biomass pyramid: ______

   (c) Energy pyramid: ______

2. What is the advantage of using a biomass or energy pyramid rather than a pyramid of numbers to express the relationship between different trophic levels?

3. How can a forest community with relatively few producers (see next page) support a large number of consumers?

ISBN: 978-1-927173-12-1


***Related activities:*** *Food Chains, Quantifying Energy Flow in an Ecosystem*

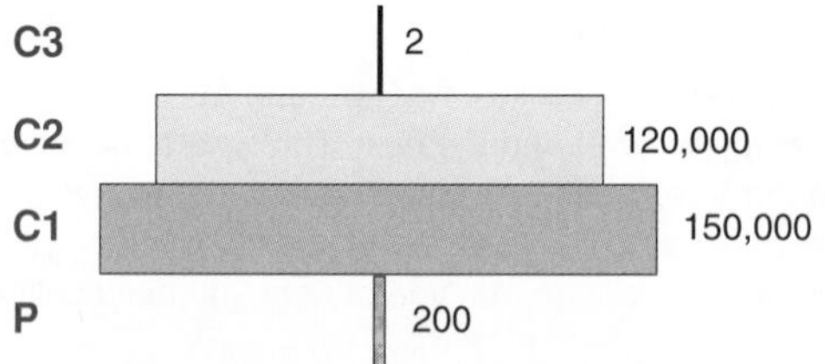


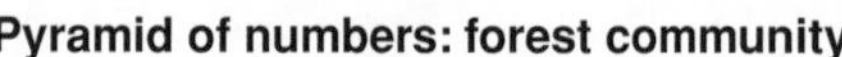

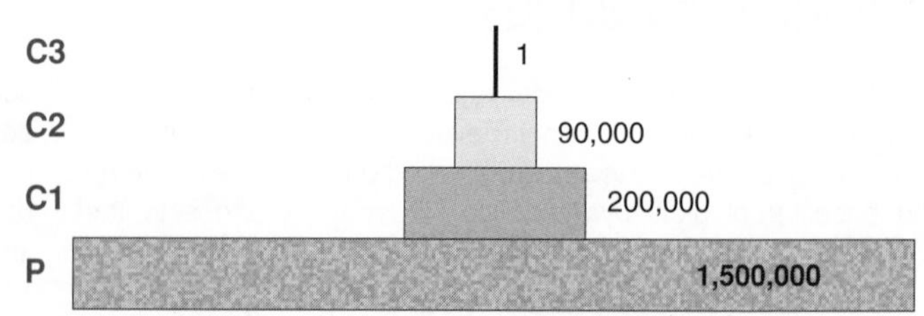


**Pyramid of numbers: forest community**

In a forest community a few producers may support a large number of consumers. This is due to the large size of the producers; large trees can support many individual consumer organisms. The example above shows the numbers at each trophic level for an oak forest in England, in an area of 10 $m^2$.

**Pyramid of numbers: grassland community**

In a grassland community a large number of producers are required to support a much smaller number of consumers. This is due to the small size of the producers. Grass plants can support only a few individual consumer organisms and take time to recover from grazing pressure. The example above shows the numbers at each trophic level for a derelict grassland area (10 $m^2$) in Michigan, United States.

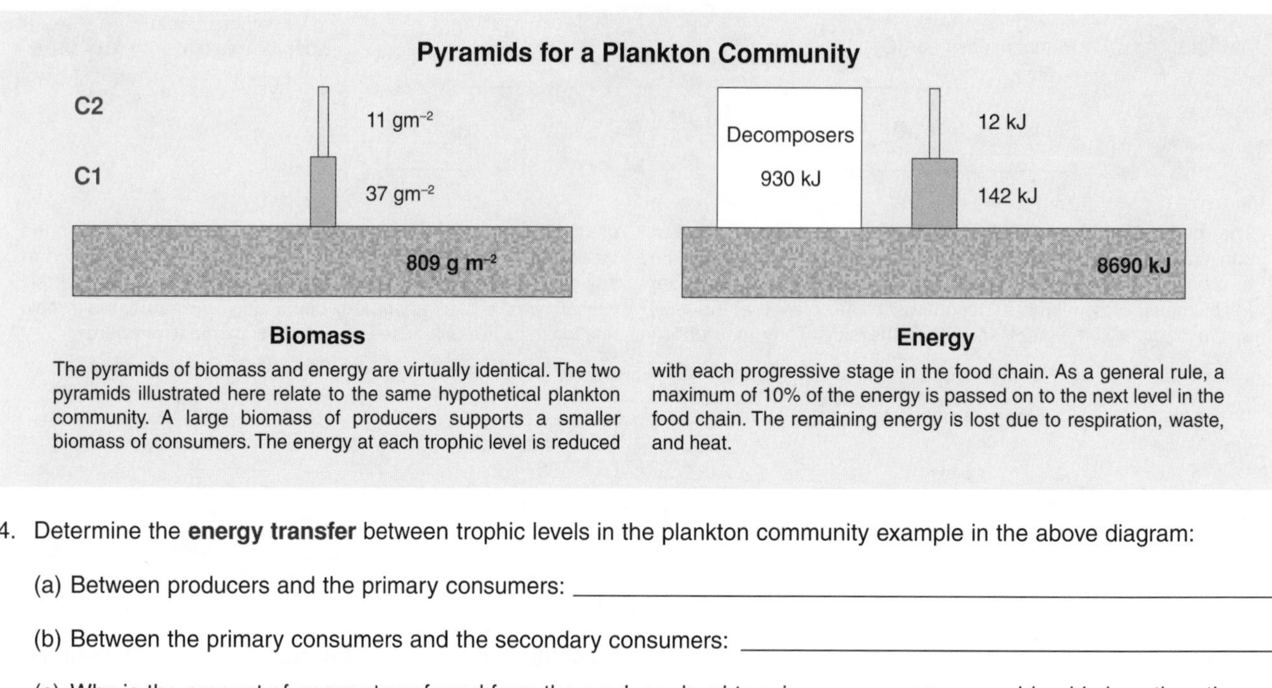


**Biomass** **Energy**

The pyramids of biomass and energy are virtually identical. The two pyramids illustrated here relate to the same hypothetical plankton community. A large biomass of producers supports a smaller biomass of consumers. The energy at each trophic level is reduced with each progressive stage in the food chain. As a general rule, a maximum of 10% of the energy is passed on to the next level in the food chain. The remaining energy is lost due to respiration, waste, and heat.

4. Determine the **energy transfer** between trophic levels in the plankton community example in the above diagram:

   (a) Between producers and the primary consumers: __________

   (b) Between the primary consumers and the secondary consumers: __________

   (c) Why is the amount of energy transferred from the producer level to primary consumers considerably less than the expected 10% that occurs in many other communities?

   (d) After the producers, which trophic group has the greatest energy content? __________

   (e) Give a likely explanation why this is the case: __________

## An unusual biomass pyramid

The biomass pyramids of some ecosystems appear rather unusual with an inverted shape. The first trophic level has a lower biomass than the second level. What this pyramid does not show is the rate at which the producers (algae) are reproducing in order to support the larger biomass of consumers.

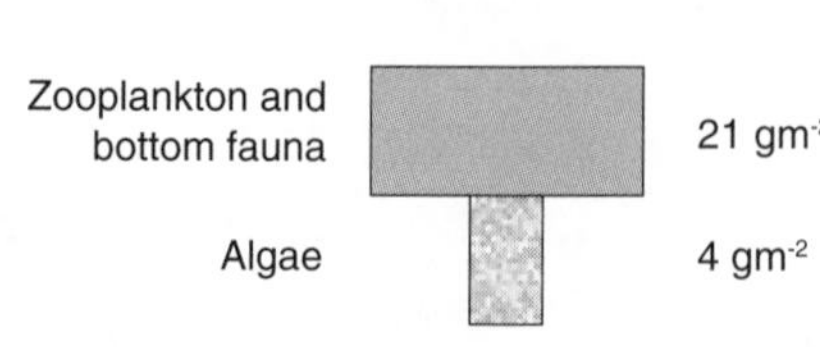


**Biomass**

5. Give a possible explanation of how a small biomass of producers (algae) can support a larger biomass of consumers (zooplankton):

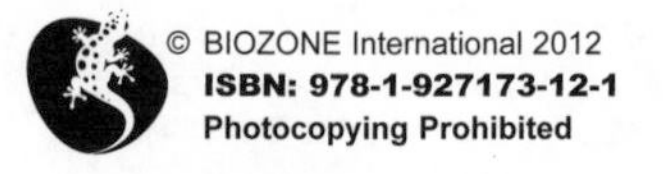


ISBN: 978-1-927173-12-1

# Nutrient Cycles

Nutrient cycling is an important part of every ecosystem. Elements essential for the efficient operation of living systems move through the environment through the processes of uptake and deposition. Commonly, nutrients must be in an ionic (rather than elemental) form in order for plants and animals to have access to them. Some bacteria have the ability to convert elemental forms of nutrients, such as sulfur, into ionic forms and so play an important role in making nutrients available to plants and animals.

### Essential Nutrients

| Macronutrient | Common form | Function |
|---|---|---|
| Carbon (C) | $CO_2$ | Organic molecules |
| Oxygen (O) | $O_2$ | Respiration |
| Hydrogen (H) | $H_2O$ | Cellular hydration |
| Nitrogen (N) | $N_2$, $NO_3^-$, $NH_4^+$ | Proteins, nucleic acids |
| Potassium (K) | $K^+$ | Principal ion in cells |
| Phosphorus (P) | $H_2PO_4^-$, $HPO_4^{2-}$ | Nucleic acids, lipids |
| Calcium (Ca) | $Ca^{2+}$ | Membrane permeability |
| Magnesium (Mg) | $Mg^{2+}$ | Chlorophyll |
| Sulfur (S) | $SO_4^{2-}$ | Proteins |
| **Micronutrient** | **Common form** | **Function** |
| Iron (Fe) | $Fe^{2+}$, $Fe^{3+}$ | Chlorophyll, blood |
| Manganese (Mn) | $Mn^{2+}$ | Enzyme activation |
| Molybdenum (Mo) | $MoO_4^-$ | Nitrogen metabolism |
| Copper (Cu) | $Cu^{2+}$ | Enzyme activation |
| Sodium (Na) | $Na^+$ | Ion in cells |
| Silicon (Si) | $Si(OH)_4$ | Support tissues |

### Tropical Rainforest

### Temperate Woodland

The speed of nutrient cycling can vary markedly. Some nutrients are cycled slowly, others quickly. The environment and diversity of an ecosystem can also have a large effect on the speed at which nutrients are recycled.

## The Role of Organisms in Nutrient Cycling

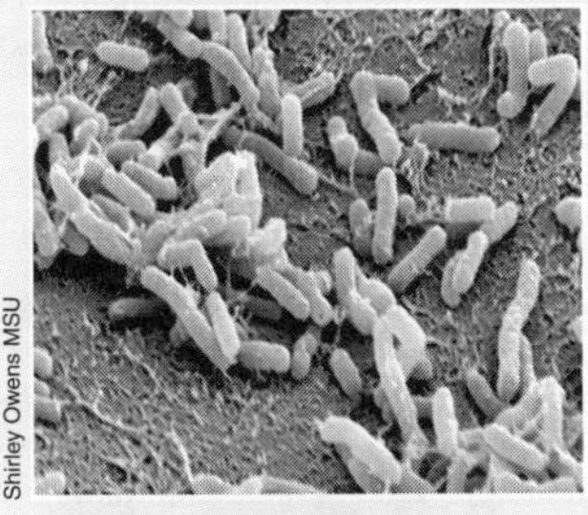

Shirley Owens MSU

**Bacteria**

Bacteria play an essential role in all nutrient cycles. They have the ability to act as saprophytes, decomposing material, but are also able to convert nutrients from inaccessible to biologically accessible forms.

**Fungi**

Fungi are saprophytes and play a critical role in decomposing organic material, returning nutrients to the soil or converting them into forms accessible to plants and animals.

**Plants**

Plants have an important role in absorbing many nutrients from the soil and making them directly available to browsing animals. They also add their own decaying matter to soils.

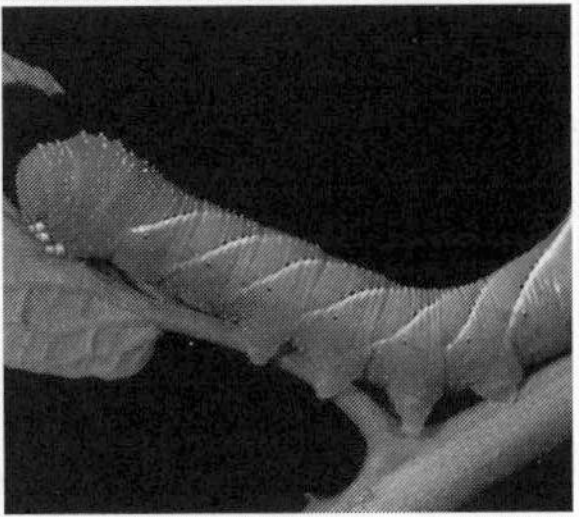

**Animals**

Animals utilize and break down materials from bacteria, plants, and fungi and return the nutrients to soils and water via their wastes and when they die.

1. Describe the role of each of the following in nutrient cycling:

   (a) Bacteria: ____________________

   (b) Fungi: ____________________

   (c) Plants: ____________________

   (d) Animals: ____________________

2. Why are soils in tropical rainforests nutrient deficient relative to soils in temperate woodlands? ____________________

3. Distinguish between macronutrients and micronutrients: ____________________


ISBN: 978-1-927173-12-1


***Related activities:*** *Energy Inputs and Outputs*

# The Hydrologic Cycle

The **hydrologic cycle** (water cycle), collects, purifies, and distributes the Earth's fixed supply of water. The main processes in this water recycling are described below. Besides replenishing inland water supplies, rainwater causes erosion and is a major medium for transporting dissolved nutrients within and among ecosystems. On a global scale, evaporation (conversion of water to gaseous water vapor) exceeds precipitation (rain, snow etc.) over the oceans. This results in a net movement of water vapor (carried by winds) over the land. On land, precipitation exceeds evaporation. Some of this precipitation becomes locked up in snow and ice, for varying lengths of time. Most form surface and groundwater systems that flow back to the sea, completing the major part of the cycle. Living organisms, particularly plants, participate to varying degrees in the water cycle. Over the sea, most of the water vapor is due to evaporation alone. However on land, about 90% of the vapor results from plant transpiration. Animals (particularly humans) intervene in the cycle by utilizing the resource for their own needs.

Transport overland: net movement of water vapor by wind

**Precipitation** **(rain, sleet, hail, snow, fog)**

**Condensation** conversion of gaseous water vapor into liquid water

Rain clouds

Precipitation to land

**Evaporation** from inland lakes and rivers

**Transpiration**

**Evaporation** from the land

**Precipitation**

Precipitation over the ocean

**Evaporation**

Evaporation from the ocean

Transpiration from plants

**Surface runoff** (rapid)

Water locked up in snow and ice

Lakes

**Rivers**

**Infiltration**: movement of water into soil

Aquifers: groundwater storage areas

**Ocean storage** 97% of total wate

**Percolation**: downward flow of water

**Groundwater movement (slow)**

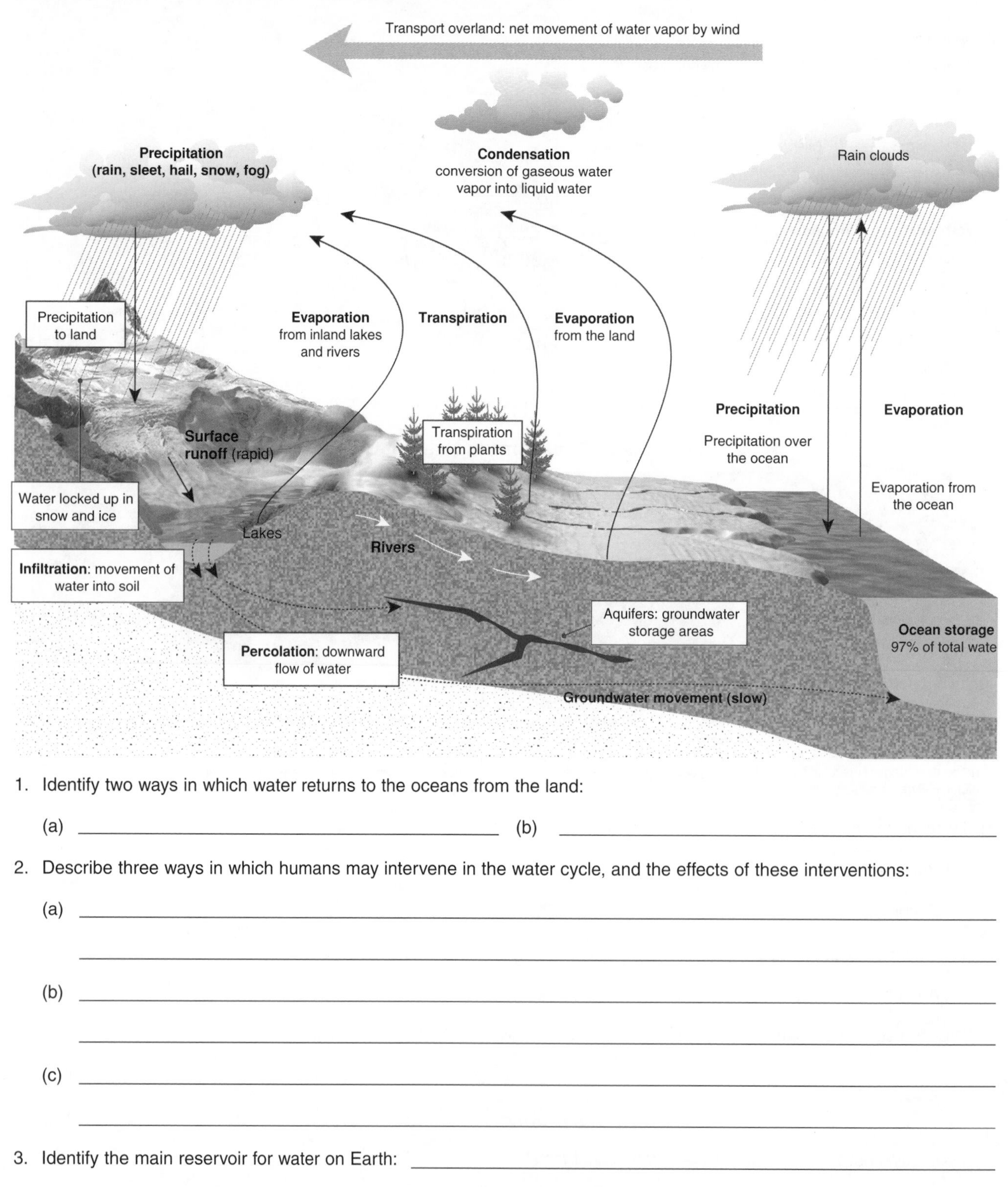

1. Identify two ways in which water returns to the oceans from the land:

   (a) ____________________ (b) ____________________

2. Describe three ways in which humans may intervene in the water cycle, and the effects of these interventions:

   (a) ____________________

   (b) ____________________

   (c) ____________________

3. Identify the main reservoir for water on Earth: ____________________

4. Identify the main reservoirs for **fresh water**: ____________________

5. Describe the important role of plants in the cycling of water through ecosystems: ____________________

**ISBN: 978-1-927173-12-1**

# The Carbon Cycle

Carbon is an essential element in living systems, providing the chemical framework to form the molecules that make up living organisms (e.g. proteins, carbohydrates, fats, and nucleic acids). Carbon also makes up approximately 0.03% of the atmosphere as the gas carbon dioxide ($CO_2$), and it is present in the ocean as carbonate and bicarbonate, and in rocks such as limestone. Carbon cycles between the living (biotic) and non-living (abiotic) environment: it is fixed in the process of photosynthesis and returned to the atmosphere in respiration. Carbon may remain locked up in biotic or abiotic systems for long periods of time as, for example, in the wood of trees or in fossil fuels such as coal or oil. Human activity has disturbed the balance of the carbon cycle (the global carbon budget) through activities such as combustion (e.g. the burning of wood and **fossil fuels**) and deforestation.

Respiration
Atmospheric carbon dioxide
Combustion
Dissolving
Carbon in plant and animal tissues
Photosynthesis
Diatoms A
Drilling for oil and gas
Death and excretion
Fossilization
Fossilization
B
Decomposers
Oil and gas
Coal
Limestone

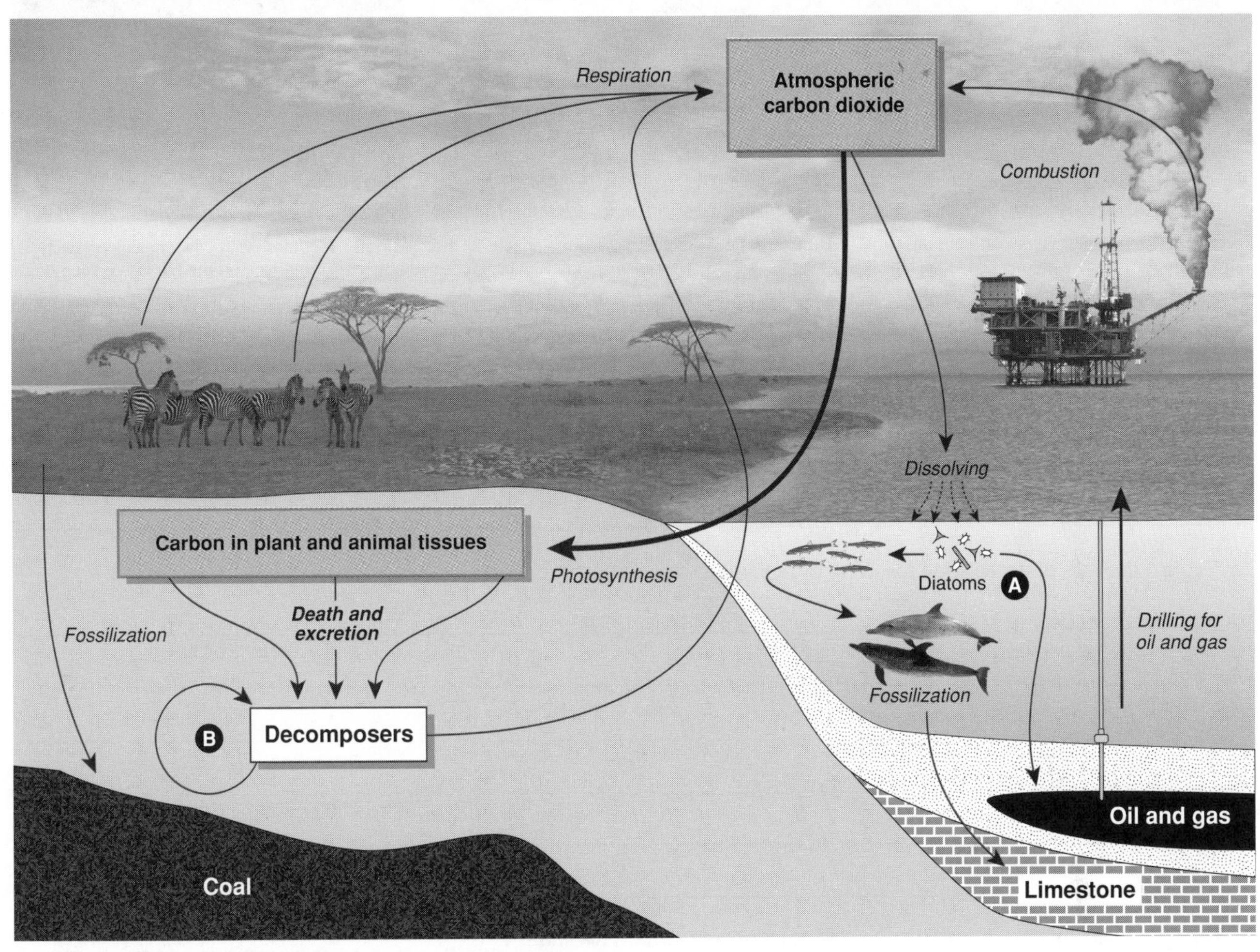

1. Add **arrows** and **labels** to the diagram above to show:
   (a) Dissolving of limestone by acid rain
   (b) Release of carbon from the marine food chain
   (c) Mining and burning of coal
   (d) Burning of plant material.

2. Describe the **biological origin** of the following geological deposits:
   (a) Coal: ____
   (b) Oil: ____
   (c) Limestone: ____

3. (a) What two processes release carbon into the atmosphere? ____
   ____
   (b) In what form is the carbon released? ____

4. Name the four geological reservoirs (sinks), in the diagram above, that can act as a source of carbon:
   (a) ____ (c) ____
   (b) ____ (d) ____

5. (a) Identify the process carried out by diatoms at point [**A**]: ____
   (b) Identify the process carried out by decomposers at [**B**]: ____



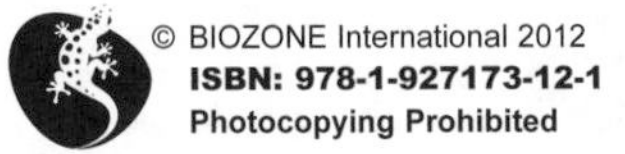

**ISBN: 978-1-927173-12-1**




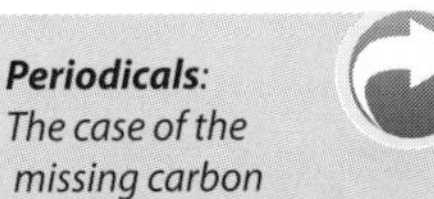

Termite mound in rainforest

**Termites**: These insects play an important role in nutrient recycling. With the aid of symbiotic protozoans and bacteria in their guts, they can digest the tough cellulose of woody tissues in trees. Termites fulfill a vital function in breaking down the endless rain of debris in tropical rainforests.

DEQ

Dung beetle on cow pat

**Dung beetles:** Beetles play a major role in the decomposition of animal dung. Some beetles merely eat the dung, but true dung beetles, such as the scarabs and *Geotrupes*, bury the dung and lay their eggs in it to provide food for the beetle grubs during their development..

Bracket fungus on tree trunk

**Fungi:** Together with decomposing bacteria, fungi perform an important role in breaking down dead plant matter in the leaf litter of forests. Some mycorrhizal fungi have been found to link up to the root systems of trees where an exchange of nutrients occurs (a mutualistic relationship).

6. What would be the effect on carbon cycling if there were **no decomposers** present in an ecosystem?

7. Explain the role of each of the following organisms in the carbon cycle:

   (a) Dung beetles:

   (b) Termites:

   (c) Fungi:

8. Using specific examples, explain the role of insects in carbon cycling:

9. In natural circumstances, accumulated reserves of carbon such as peat, coal and oil represent a sink or natural diversion from the cycle. Eventually, the carbon in these sinks returns to the cycle through the action of geological processes which return deposits to the surface for oxidation.

   (a) What is the effect of human activity on the amount of carbon stored in sinks?

   (b) Describe two **global effects** resulting from this activity:

   (c) What could be done to prevent or alleviate these effects?


**ISBN: 978-1-927173-12-1**

# The Nitrogen Cycle

Nitrogen is a crucial element for all living things, forming an essential part of the structure of proteins and nucleic acids. The Earth's atmosphere is about 80% nitrogen gas ($N_2$), but molecular nitrogen is so stable that it is only rarely available directly to organisms and is often in short supply in biological systems. Bacteria play an important role in transferring nitrogen between the biotic and abiotic environments. Some bacteria are able to fix atmospheric nitrogen, while others convert ammonia to nitrate and thus make it available for incorporation into plant and animal tissues. Nitrogen-fixing bacteria are found living freely in the soil *(Azotobacter)* and living symbiotically with some plants in root nodules *(Rhizobium)*. Lightning discharges also cause the oxidation of nitrogen gas to nitrate which ends up in the soil. Denitrifying bacteria reverse this activity and return fixed nitrogen to the atmosphere. Humans intervene in the nitrogen cycle by producing, and applying to the land, large amounts of nitrogen fertilizer. Some applied fertilizer is from organic sources (e.g. green crops and manures) but much is inorganic, produced from atmospheric nitrogen using an energy-expensive industrial process. Overuse of nitrogen fertilizers may lead to pollution of water supplies, particularly where land clearance increases the amount of leaching and runoff into ground and surface waters.

Atmospheric nitrogen ($N_2$ gas)
Lightning
Atmospheric nitrogen fixation
Nitrogen in plant and animal tissues
Excretion (urea)
Dead organisms
Decomposers
Ammonification
Ammonia $NH_3$
Nitrogen fixing bacteria
*Azotobacter*
*Rhizobium*
Denitrification
Denitrifying bacteria (anaerobic)
Nitrogen fixing cyanobacteria
Assimilation
Nitrate available to plants ($NO_3^-$)
Ammonia $NH_3$
Nitrifying bacteria *Nitrosomonas* spp.
Nitrite ($NO_2^-$)
Nitrification
Nitrifying bacteria *Nitrobacter* spp.
Nitrate ($NO_3^-$)
Nitrification
Via nitrification processes

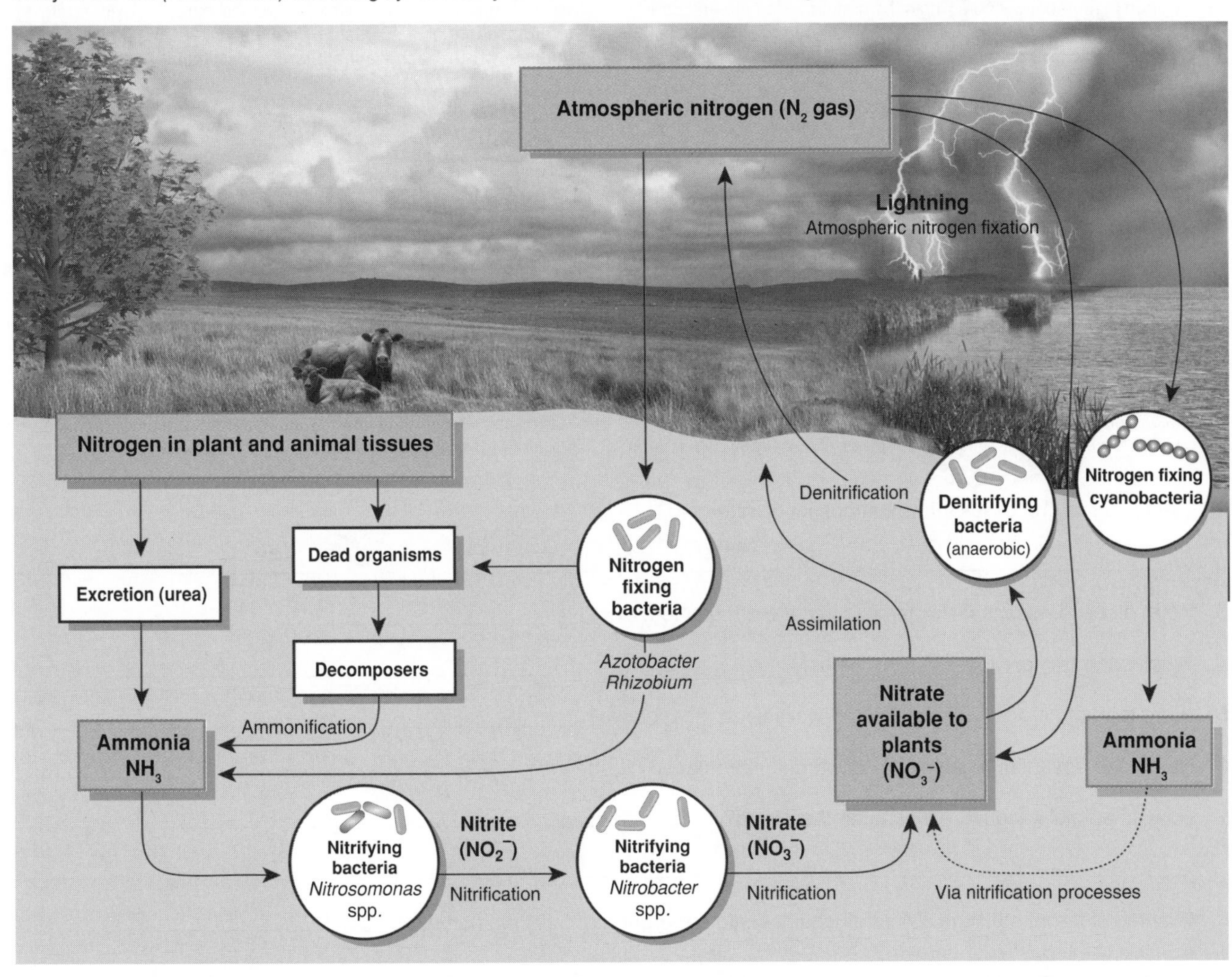

1. Describe five instances in the nitrogen cycle where **bacterial** action is important. Include the name of each of the processes and the changes to the form of nitrogen involved:

(a) ______________________________

______________________________

(b) ______________________________

______________________________

(c) ______________________________

______________________________

(d) ______________________________

______________________________

(e) ______________________________

______________________________

ISBN: 978-1-927173-12-1

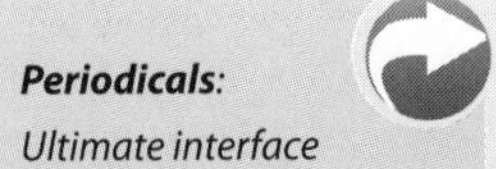

## Nitrogen Fixation in Root Nodules

**Root nodules** are a root **symbiosis** between a higher plant and a bacterium. The bacteria fix atmospheric nitrogen and are extremely important to the nutrition of many plants, including the economically important legume family. Root nodules are extensions of the root tissue caused by entry of a bacterium. In legumes, this bacterium is *Rhizobium*. Other bacterial genera are involved in the root nodule symbioses in non-legumes.

The bacteria in these symbioses live in the nodule where they fix atmospheric nitrogen and provide the plant with most, or all, of its nitrogen requirements. In return, they have access to a rich supply of carbohydrate. The fixation of atmospheric nitrogen to ammonia occurs within the nodule, using the enzyme **nitrogenase**. Nitrogenase is inhibited by oxygen and the nodule provides a low $O_2$ environment in which fixation can occur.

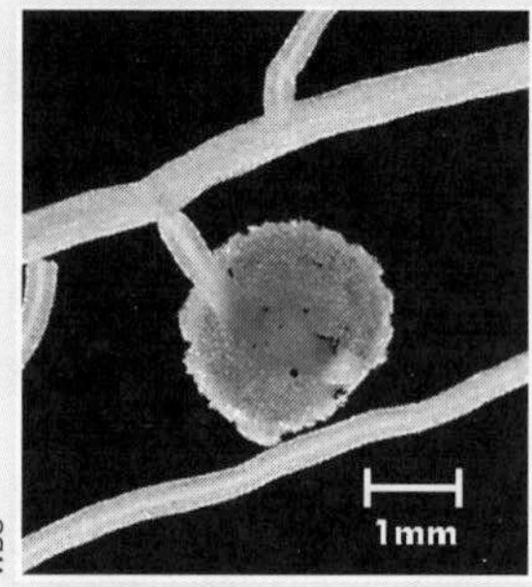


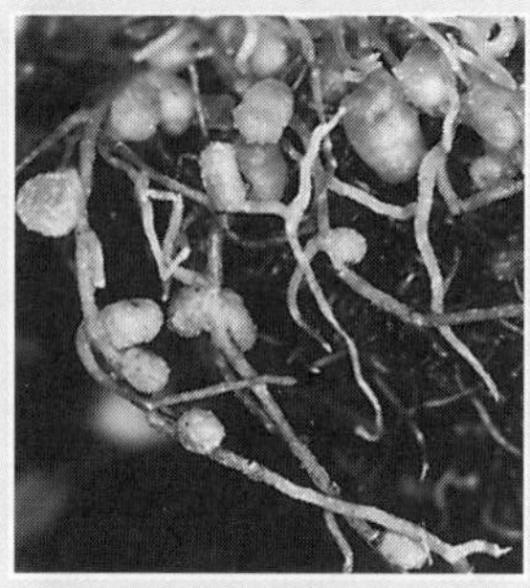

Two examples of legume nodules caused by *Rhizobium*. The photographs above show the size of a single nodule (left), and the nodules forming clusters around the roots of *Acacia* (right).

## Human intervention in the Nitrogen Cycle

Until about sixty years ago, microbial nitrogen fixation (left) was the only mechanism by which nitrogen could be made available to plants. However, during WW II, Fritz Haber developed the **Haber process** whereby nitrogen and hydrogen gas combine to form gaseous ammonia. The ammonia is converted into ammonium salts and sold as inorganic fertilizer. Its application has revolutionized agriculture by increasing crop yields.

As well as adding nitrogen fertilizers to the land, humans use anaerobic bacteria to break down livestock wastes and release $NH_3$ into the soil. They also intervene in the nitrogen cycle by discharging **effluent** into waterways. Nitrogen is removed from the land through burning, which releases nitrogen oxides into the atmosphere. It is also lost by mining, harvesting crops, and irrigation, which leaches nitrate ions from the soil.

Two examples of human intervention in the nitrogen cycle. The photographs above show the aerial application of a commercial fertilizer (left), and the harvesting of an agricultural crop (right).

2. Identify three processes that **fix** atmospheric nitrogen:

   (a) ______________________ (b) ______________________ (c) ______________________

3. What process releases nitrogen gas into the atmosphere? ______________________

4. What is the primary reservoir for nitrogen? ______________________

5. What form of nitrogen is most readily available to most plants? ______________________

6. Name one essential organic compound that plants need nitrogen for: ______________________

7. How do animals acquire the nitrogen they need? ______________________

8. Why might farmers plough a crop of legumes into the ground rather than harvest it? ______________________

9. Describe five ways in which humans may intervene in the nitrogen cycle and the effects of these interventions:

   (a) ______________________

   (b) ______________________

   (c) ______________________

   (d) ______________________

   (e) ______________________


**ISBN: 978-1-927173-12-1**

# Nitrogen Pollution

The effect of excess nitrogen compounds on the environment is varied. Depending on the compound formed, nitrogen can cause smog in cities or algal blooms in lakes and seas. Nitrogen gas makes up almost 80% of the atmosphere but is unreactive at normal pressure and temperature. At the high pressures and temperatures reached in factories and combustion engines nitrogen gas forms nitric oxide along with other nitrogen oxides, most of which contribute to atmospheric pollution. Nitrates in fertilizers are washed into ground water by rain and slowly make their way to lakes and rivers and eventually out to sea. This process can take time to become noticeable as ground water can take many decades to reach a waterway. In many places where nitrate effects are only just becoming apparent, the immediate cessation of their use could take a long time to have any effect as it might take many years before the last of the ground water carrying the nitrates reaches a waterway.

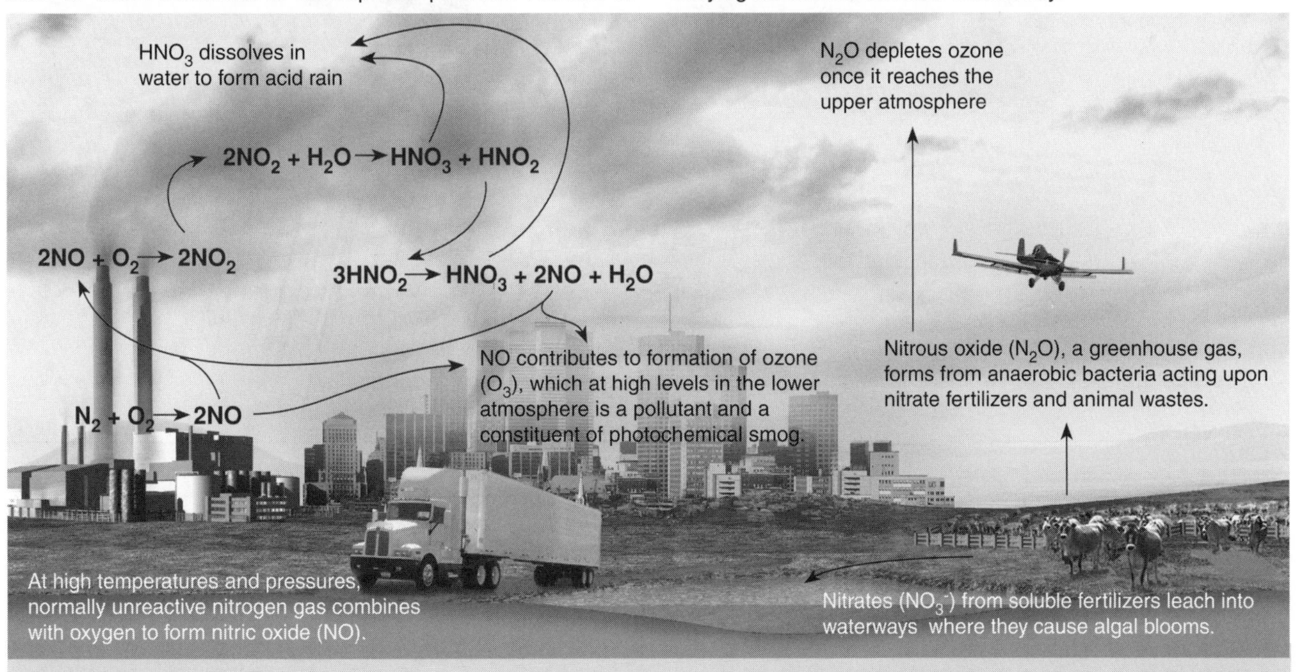


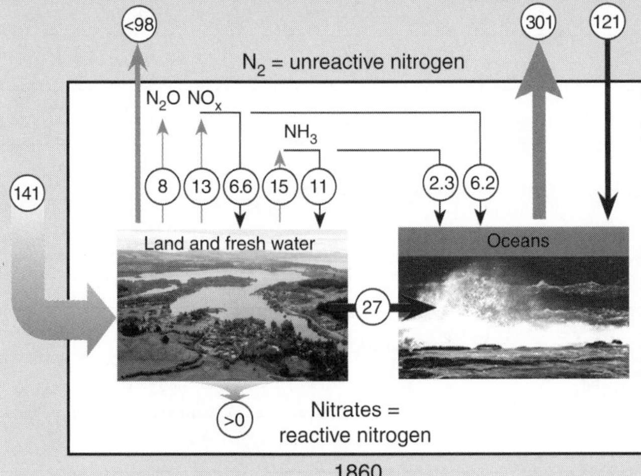


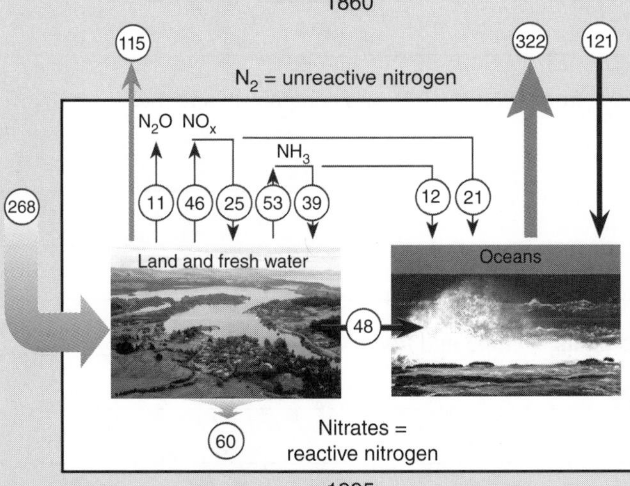


Changes in nitrogen inputs and outputs between 1860 and 1995 in million Tonne (modified from Galloway *et al* 2004)

Early last century, the Haber-Bosch process made nitrate fertilizers readily available for the first time. Since then, the use of nitrogen fertilizers has increased at an almost exponential rate. Importantly, this has led to an increase in the levels of nitrogen in land and water by up to 60 times those of 100 years ago. This extra nitrogen load is one of the causes of accelerated enrichment (**eutrophication**) of lakes and coastal waters. An increase in algal production also results in higher decomposer activity and, consequently, oxygen depletion, fish deaths, and depletion of aquatic biodiversity. Many aquatic microorganisms also produce toxins, which may accumulate in the water, fish, and shellfish. The diagrams (left) show the increase in nitrates in water sources from 1860 to 1995. The rate at which nitrates are added has increased faster than the rate at which nitrates are returned to the atmosphere as unreactive $N_2$ gas. This has led to the widespread accumulation of nitrogen.

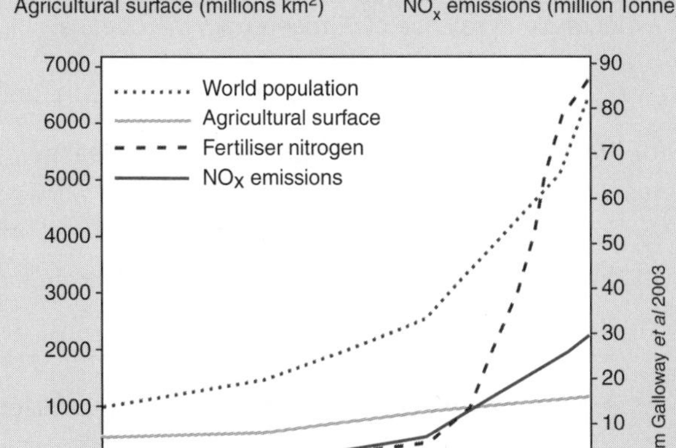


Energy Flow & Nutrient Cycles

ISBN: 978-1-927173-12-1


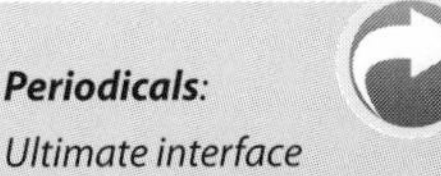

***Periodicals:***
*Ultimate interface*

***Related activities:*** *The Nitrogen Cycle*

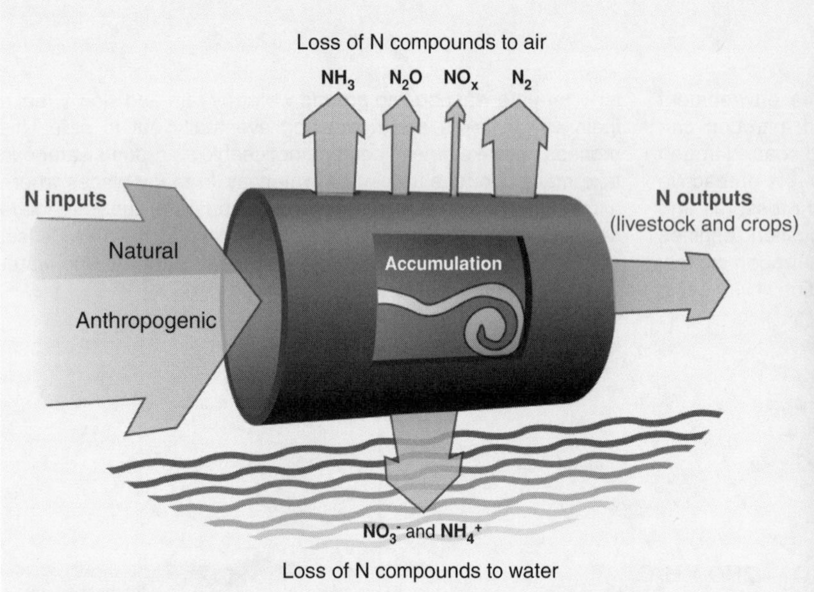


From O. Oenema *et al* 2007

The "hole in the pipe" model (left) demonstrates inefficiencies in nitrogen fertilizer use. Nitrogen that is added to the soil and not immediately taken up by plants is washed into waterways or released into the air by bacterial action. These losses can be minimized to an extent by using slow release fertilizers during periods of wet weather and by careful irrigation practices.

Satellite photo of algal blooms around Florida. Excessive nitrogen contributes to algal blooms in both coastal and inlands waters. *Image: NASA*

1. Describe the effect each of the following nitrogen compounds have on air and water quality:

   (a) NO: ________________

   (b) $N_2O$: ________________

   (c) $NO_2$: ________________

   (d) $NO_3^-$: ________________

2. Explain why the formation of NO can cause large scale and long term environmental problems: ________________

3. Why would an immediate halt in the use of nitrogen fertilizers not cause an immediate stop in their effects?

4. (a) Calculate the increase in nitrogen deposition in the oceans from 1860 to 1995 and compare this to the increase in release of nitrogen from the oceans.

   (b) What is the effect of this increase on the oceans? ________________

5. (a) Why do nitrogen inputs tend to be so much more than outputs in crops and from livestock? ________________

   (b) Suggest how the nitrogen losses could be minimized: ________________


ISBN: 978-1-927173-12-1

# KEY TERMS: Mix and Match

*INSTRUCTIONS: Test your vocabulary by matching each term to its definition, as identified by its preceding letter code.*

autotroph ..........

carbon cycle ..........

chemoautotroph ..........

consumer ..........

decomposer ..........

detritivore ..........

ecological pyramid ..........

food chain ..........

food web ..........

gross primary production ..........

heterotroph ..........

hydrologic cycle ..........

net primary production ..........

nitrogen cycle ..........

nutrient cycle ..........

photoautotroph ..........

primary consumer ..........

producer ..........

primary productivity ..........

saprotroph ..........

ten percent rule ..........

trophic efficiency ..........

trophic level ..........

**A** Biogeochemical cycle by which carbon is exchanged among the biotic and abiotic components of the Earth.

**B** An organism capable of manufacturing its own food.

**C** The total amount of energy fixed by producers per unit area, less the costs of respiration. It is effectively the amount of biomass available to consumer levels.

**D** A sequence of steps describing how an organism derives energy from the ones before it.

**E** An organism that manufactures its own food from simple inorganic substances.

**F** An organism that uses inorganic energy sources, such as hydrogen sulfide or elemental sulfur to synthesize its own organic compounds.

**G** An organism that obtains energy by ingesting dead material mixed with inorganic material, e.g. earthworm.

**H** This cycle describes the continuous movement of water on, above, and below the surface of the Earth.

**I** An organism that obtains its energy and carbon from the extracellular digestion and absorption of dead or decaying matter.

**J** The efficiency of energy transfer between trophic levels, e.g. the ratio of secondary production to primary production consumed.

**K** The collective biological and non-biological processes by which nitrogen is converted between its various chemical forms.

**L** An organism that uses the free energy in sunlight to synthesize its own organic compounds.

**M** An organism that obtains its carbon and energy from other organisms.

**N** Organisms that obtain their energy from other living organisms or their dead remains.

**O** The position an organism occupies on the food chain.

**P** The rate of production (effectively biomass produced) of producers (may be gross or net).

**Q** A complex series of interactions showing the feeding relationships between organisms in an ecosystem.

**R** An organism that obtains energy from dead material by extracellular digestion.

**S** Cycle in which inorganic nutrients move from the soil through the environment and back.

**T** The rule of thumb governing the maximum amount of energy that typically passes from one trophic level to another.

**U** An organism at the second trophic level. A herbivore.

**V** The total amount of energy fixed by producers per unit area.

**W** A graphical representation of the numbers, energy, or biomass at each trophic level in an ecosystem. Pyramidal in shape, but sometimes inversely so.



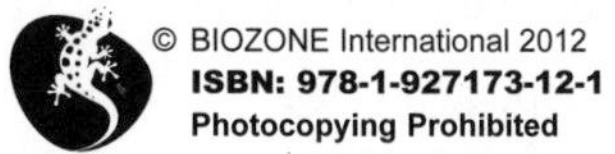


**ISBN: 978-1-927173-12-1**

Enduring Understanding

**4.A**
**4.B**

# Populations

## Key concepts

- Population specific characteristics include age structure, natality, and mortality.
- We can quantify the attributes of populations.
- Population growth is regulated by density dependent and density independent factors.
- Demographic data related to age and fecundity can be used to study human populations.
- Species interact in ways that influence survival and population growth.

## Key terms

abundance
age structure
biotic potential (*r*)
carrying capacity (K)
community
demography
density
density dependent factor
density independent factor
distribution
environmental resistance
exponential growth
interspecific competition
intraspecific competition
K selection
life tables
limiting factor
logistic growth
mark and recapture
migration
mortality
mutualism
natality
parasite / parasitism
population
population growth
population sampling
population size
predator / predation
quadrat
*r* selection
survivorship curve
symbiosis

## Essential Knowledge

☐ 1. Use the **KEY TERMS** to compile a glossary for this topic.

**Population Dynamics** *(4.A.5: c, 4.B.3)* pages 319-334

☐ 2. Distinguish between a **population** and a **community**.

☐ 3. Understand that populations are dynamic and exhibit attributes not shown by the individuals themselves. Describe population-specific characteristics including **age structure**, **natality**, **mortality**, and **distribution**. Distinguish population **density** and population **size**.

☐ 4. Describe how the attributes of populations can be assessed quantitatively using population sampling methods (e.g. **quadrats** and **mark and recapture**).

☐ 5. Explain how population size can be affected by births, deaths, and migration and express the relationship in an equation.

☐ 6. Explain how **life tables** provide information about patterns of population natality and mortality. Explain the role of **survivorship curves** in analyzing populations. Describe and explain the features of type I, II, and III survivorship curves.

☐ 7. Describe and explain the typical features of *r* and K selection.

☐ 8. Describe how the trends in population change can be shown in a population growth curve of population numbers (Y axis) against time (X axis).

☐ 9. Describe factors affecting final population size. Include reference to **carrying capacity** (K), **environmental resistance**, **density dependent factors**, and **density independent factors**.

☐ 10. Describe **exponential** and **logistic growth**. Explain patterns of population growth in colonizing, stable, and declining populations.

☐ 11. Describe and explain population cycles in interacting species (e.g. predator and prey) in which the fluctuations in population growth in one species lag behind those of the other, usually in a predictable way that can be modelled.

☐ 12. Describe and explain demographic trends in human populations.

**Population Interactions** *(4.A.5: b, 4.A.6: e, 4.B.3)* pages 334-344

☐ 13. Describe and explain **interspecific** interactions in communities, including **competition**, **mutualism**, and **exploitation** (parasitism, predation, herbivory).

☐ 14. Explain the role of **interspecific competition** in constraining niche breadth in species. Describe possible consequences of interspecific competition, e.g. **niche differentiation** or **competitive exclusion**.

☐ 15. Describe **intraspecific competition** in populations with and without a social structure. Recognize **territoriality** and hierarchical systems as strategies for managing the impact of intraspecific competition on population survival.

☐ 16. Describe and explain the role of competition in limiting population size.

*Periodicals:*
*Listings for this chapter are on page 376*

*Weblinks:*
*www.thebiozone.com/weblink/AP2-3121.html*

# Features of Populations

Populations have a number of attributes that may be of interest. Usually, biologists wish to determine **population size** (the total number of organisms in the population). It is also useful to know the **population density** (the number of organisms per unit area). The density of a population is often a reflection of the **carrying capacity** of the environment, i.e. how many organisms a particular environment can support. Populations also have structure; particular ratios of different ages and sexes. These data enable us to determine whether the population is declining or increasing in size. We can also look at the **distribution** of organisms within their environment and so determine what particular aspects of the habitat are favored over others. One way to retrieve information from populations is to **sample** them. Sampling involves collecting data about features of the population from samples of that population (since populations are usually too large to examine in total). Sampling can be done directly through a number of sampling methods or indirectly (e.g. monitoring calls, looking for droppings or other signs). Some of the population attributes that we can measure or calculate are illustrated on the diagram below.

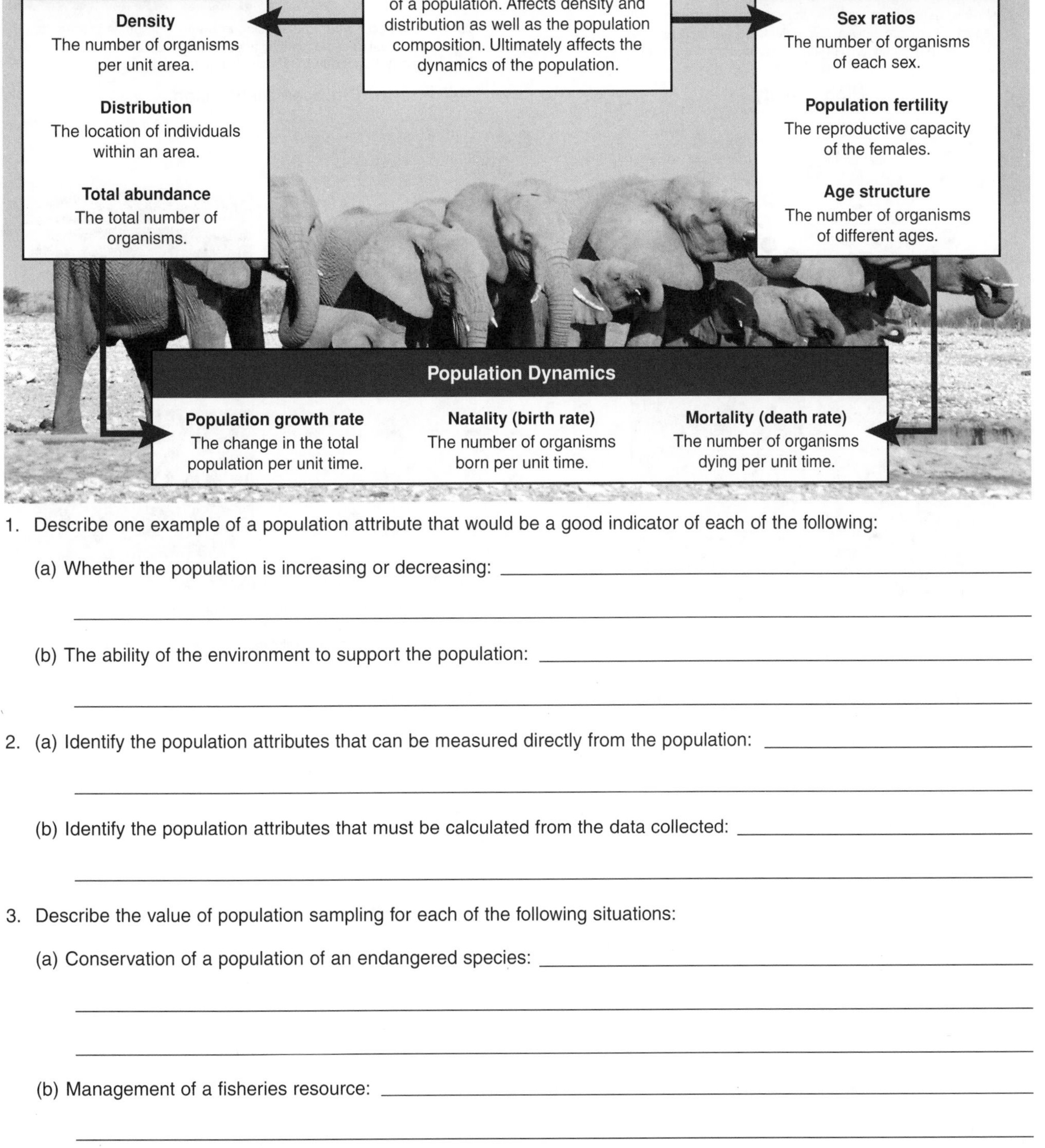


1. Describe one example of a population attribute that would be a good indicator of each of the following:

   (a) Whether the population is increasing or decreasing: ______

   (b) The ability of the environment to support the population: ______

2. (a) Identify the population attributes that can be measured directly from the population: ______

   (b) Identify the population attributes that must be calculated from the data collected: ______

3. Describe the value of population sampling for each of the following situations:

   (a) Conservation of a population of an endangered species: ______

   (b) Management of a fisheries resource: ______


ISBN: 978-1-927173-12-1


***Related activities:*** *Density and Distribution, Population Age Structure*

# Density and Distribution

Distribution and density are two interrelated properties of populations. Population density is the number of individuals per unit area (for land organisms) or volume (for aquatic organisms). Careful observation and precise mapping can determine the distribution patterns for a species. The three basic distribution patterns are: random, clumped and uniform. In the diagram below, the circles represent individuals of the same species. It can also represent populations of different species.

### Low Density

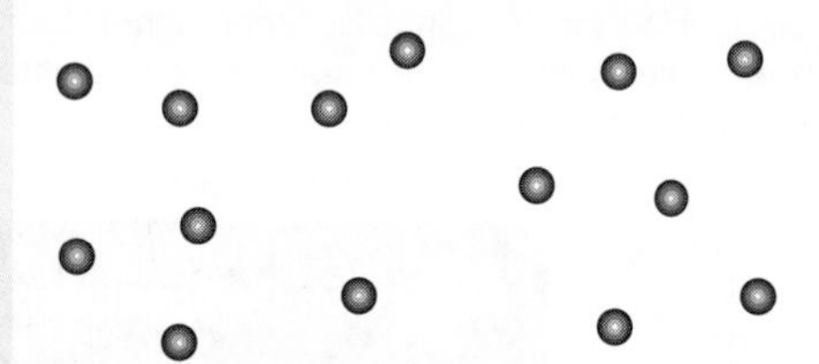

In low density populations, individuals are spaced well apart. There are only a few individuals per unit area or volume (e.g. highly territorial, solitary mammal species).

### High Density

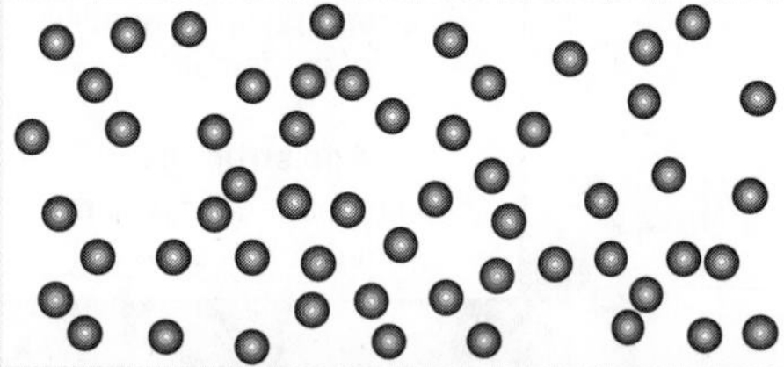

In high density populations, individuals are crowded together. There are many individuals per unit area or volume (e.g. colonial organisms, such as many corals).

Tigers are solitary animals, found at low densities.

Termites form well organized, high density colonies.

### Random Distribution

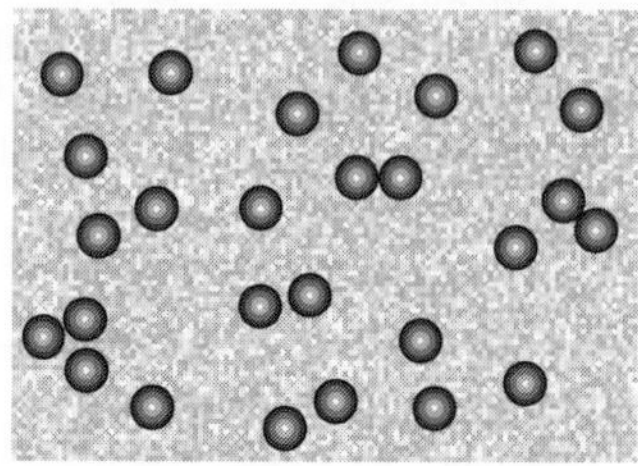

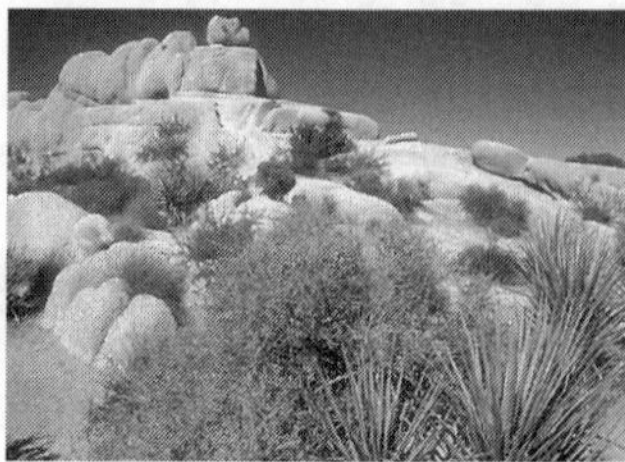

Random distributions occur when the spacing between individuals is irregular. The presence of one individual does not directly affect the location of any other individual. Random distributions are uncommon in animals but are often seen in plants.

### Clumped Distribution

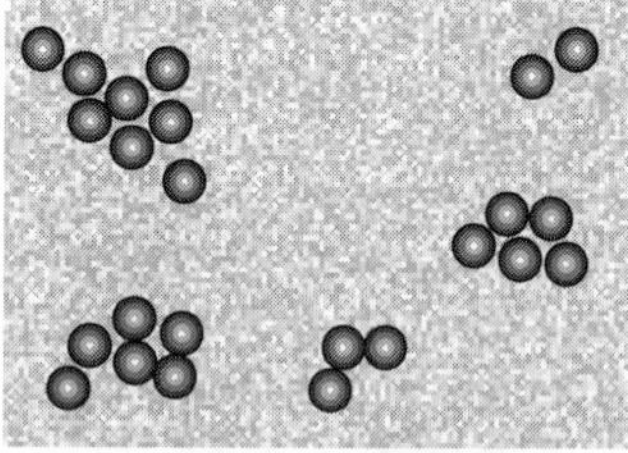

Clumped distributions occur when individuals are grouped in patches (sometimes around a resource). The presence of one individual increases the probability of finding another close by. Such distributions occur in herding and highly social species.

### Uniform Distribution

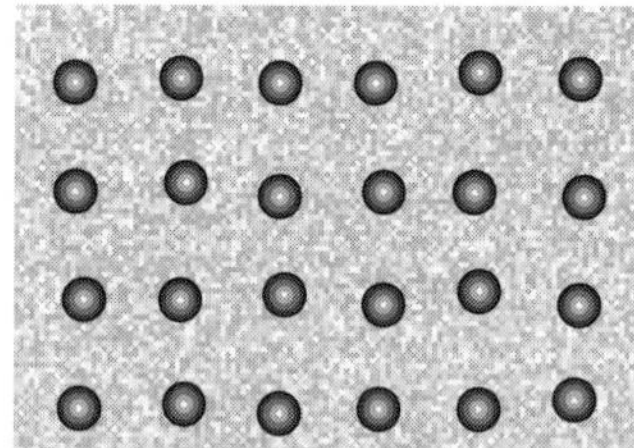

Regular distribution patterns occur when individuals are evenly spaced within the area. The presence of one individual decreases the probability of finding another individual very close by. The penguins illustrated above are also at a high density.

1. Describe why some organisms may exhibit a clumped distribution pattern because of:

   (a) Resources in the environment: ______________________

   (b) A group social behavior: ______________________

2. Describe a social behavior found in some animals that may encourage a uniform distribution: ______________________

3. Describe the type of environment that would encourage uniform distribution: ______________________

4. Describe an example of each of the following types of distribution pattern:

   (a) Clumped: ______________________

   (b) Random (more or less): ______________________

   (c) Uniform (more or less): ______________________



**Related activities:** *Features of Populations*


**ISBN: 978-1-927173-12-1**

# Population Age Structure

The **age structure** of a population refers to the relative proportion of individuals in each age group in the population. The age structure of populations can be categorized according to specific age categories (such as years or months), but also by other measures such as life stage (egg, larvae, pupae, instars), of size class (height or diameter in plants). Population growth is strongly influenced by age structure; a population with a high proportion of reproductive and prereproductive aged individuals has a much greater potential for population growth than one that is dominated by older individuals. The ratio of young to adults in a relatively stable population of most mammals and birds is approximately 2:1 (below, left). Growing populations in general are characterized by a large and increasing number of young, whereas a population in decline typically has a decreasing number of young. Population age structures are commonly represented as pyramids, in which the proportions of individuals in each age/size class are plotted with the youngest individuals at the pyramid's base. The number of individuals moving from one age class to the next influences the age structure of the population from year to year. The loss of an age class (e.g. through overharvesting) can profoundly influence a population's viability and can even lead to population collapse.

## Age Structures in Animal Populations

These theoretical age pyramids, which are especially applicable to birds and mammals, show how growing populations are characterized by a high ratio of young (white bar) to adult age classes (blue bars). Ageing populations with poor production are typically dominated by older individuals.

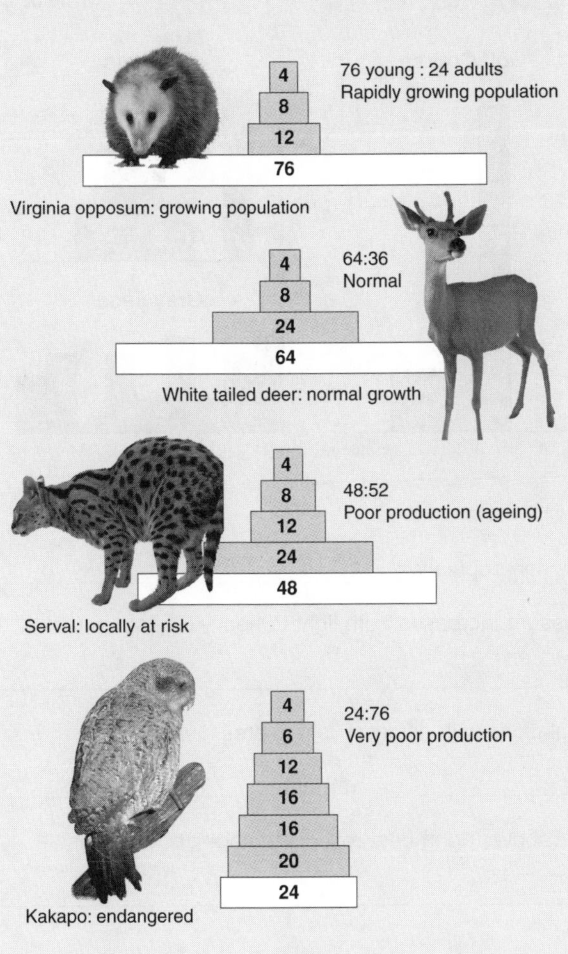


## Age Structures in Human Populations

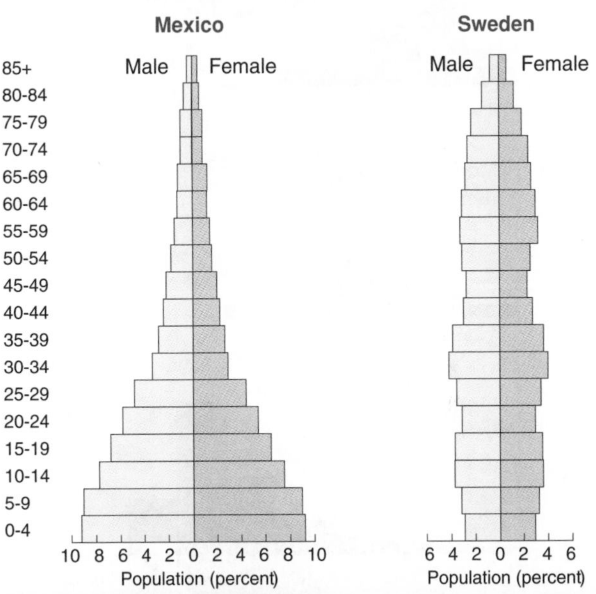

Most of the growth in human populations in recent years has occurred in the developing countries in Africa, Asia, and Central and South America. This is reflected in their age structure; a large proportion of the population comprises individuals younger than 15 years (age pyramid above, left). Even if each has fewer children, the population will continue to increase for many years. The stable age structure of Sweden is shown for comparison.

1. For the theoretical age pyramids above left:

   (a) State the approximate ratio of young to adults in a rapidly increasing population: ________________

   (b) Suggest why changes in population age structure alone are not necessarily a reliable predictor of population trends:

2. Explain why the population of Mexico is likely to continue to increase rapidly even if the rate of population growth slows:


ISBN: 978-1-927173-12-1

Analysis of the age structure of populations can assist in their management because it can indicate where most population mortality occurs and whether or not reproductive individuals are being replaced. The age structure of plant and animal populations can be examined; a common method is through an analysis of size which is often related to age in a predictable way.

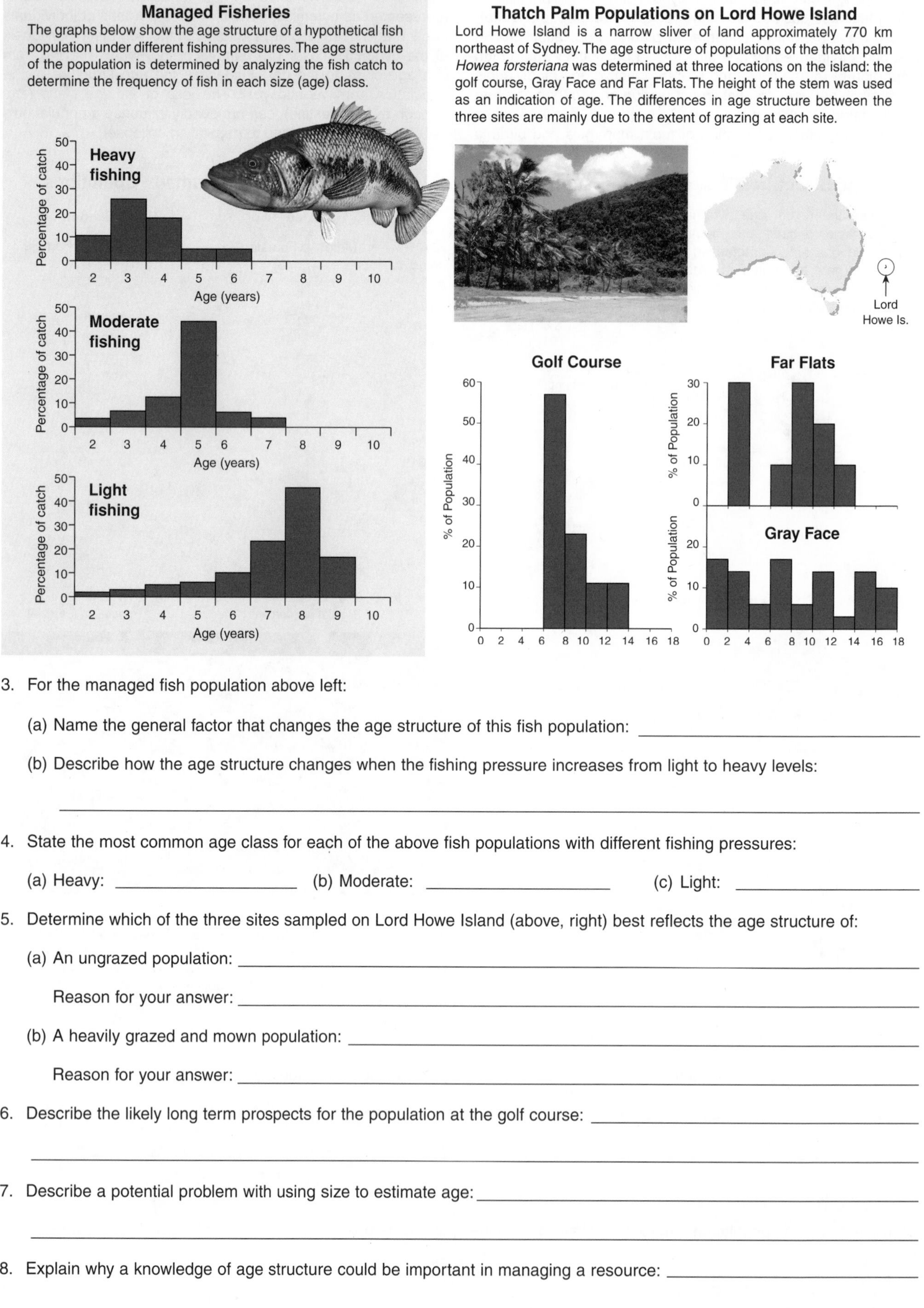


3. For the managed fish population above left:

   (a) Name the general factor that changes the age structure of this fish population: ______________________

   (b) Describe how the age structure changes when the fishing pressure increases from light to heavy levels:

   ______________________________________________

4. State the most common age class for each of the above fish populations with different fishing pressures:

   (a) Heavy: ____________ (b) Moderate: ____________ (c) Light: ____________

5. Determine which of the three sites sampled on Lord Howe Island (above, right) best reflects the age structure of:

   (a) An ungrazed population: ______________________

   Reason for your answer: ______________________

   (b) A heavily grazed and mown population: ______________________

   Reason for your answer: ______________________

6. Describe the likely long term prospects for the population at the golf course: ______________________

   ______________________________________________

7. Describe a potential problem with using size to estimate age: ______________________

   ______________________________________________

8. Explain why a knowledge of age structure could be important in managing a resource: ______________________

   ______________________________________________


ISBN: 978-1-927173-12-1

# Life Tables and Survivorship

**Life tables**, such as those shown below, provide a summary of mortality for a population (usually for a group of individuals of the same age or **cohort**). The basic data are just the number of individuals remaining alive at successive sampling times (the **survivorship** or **$l_x$**). Life tables are an important tool when analyzing changes in populations over time. They can tell us the ages at which most mortality occurs in a population and can also provide information about life span and population age structure. Biologists use the $l_x$ column of a basic life table to derive a survivorship curve. Survivorship curves are standardized as the number of survivors per 1000 individuals so that populations of different types can be easily compared.

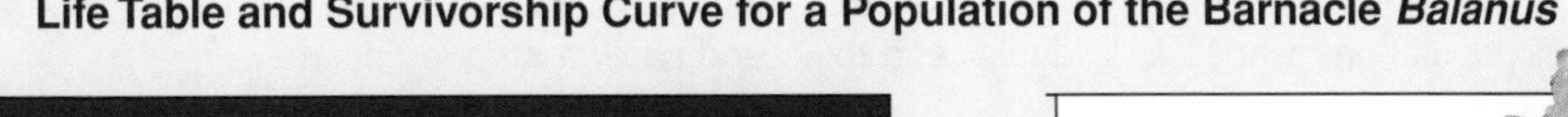

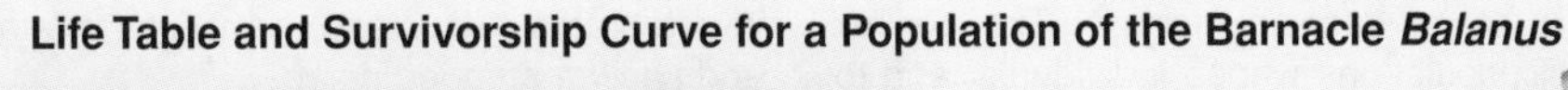

### Life Table and Survivorship Curve for a Population of the Barnacle *Balanus*

| Age in years (*x*) | No. alive each year ($N_x$) | Proportion surviving at the start of age *x* ($l_x$) | Proportion dying between *x* and *x* +1 ($d_x$) | Mortality ($q_x$) |
|---|---|---|---|---|
| 0 | 142 | 1.000 | 0.563 | 0.563 |
| 1 | 62 | 0.437 | 0.198 | 0.452 |
| 2 | 34 | 0.239 | 0.098 | 0.412 |
| 3 | 20 | 0.141 | 0.035 | 0.250 |
| 4 | 15 | 0.106 | 0.028 | 0.267 |
| 5 | 11 | 0.078 | 0.036 | 0.454 |
| 6 | 6 | 0.042 | 0.028 | 0.667 |
| 7 | 2 | 0.014 | 0.0 | 0.000 |
| 8 | 2 | 0.014 | 0.014 | 1.000 |
| 9 | 0 | 0.0 | 0.0 | – |

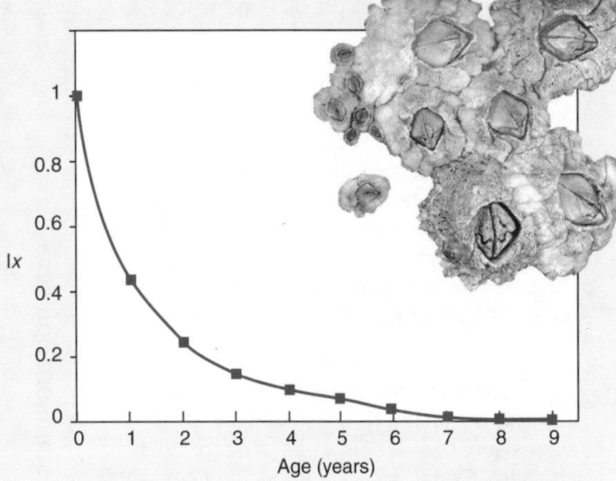


### Life Table for Female Elk, Northern Yellowstone

| *x* | $l_x$ | $d_x$ | $q_x$ |
|---|---|---|---|
| 0 | 1000 | 323 | 0.323 |
| 1 | 677 | 13 | 0.019 |
| 2 | 664 | 2 | 0.003 |
| 3 | 662 | 2 | 0.003 |
| 4 | 660 | 4 | 0.006 |
| 5 | 656 | 4 | 0.006 |
| 6 | 652 | 9 | 0.014 |
| 7 | 643 | 3 | 0.005 |
| 8 | 640 | 3 | 0.005 |
| 9 | 637 | 9 | 0.014 |
| 10 | 628 | 7 | 0.001 |
| 11 | 621 | 12 | 0.019 |
| 12 | 609 | 13 | 0.021 |
| 13 | 596 | 41 | 0.069 |
| 14 | 555 | 34 | 0.061 |
| 15 | 521 | 20 | 0.038 |
| 16 | 501 | 59 | 0.118 |
| 17 | 442 | 75 | 0.170 |
| 18 | 367 | 93 | 0.253 |
| 19 | 274 | 82 | 0.299 |
| 20 | 192 | 57 | 0.297 |
| 21+ | 135 | 135 | 1.000 |

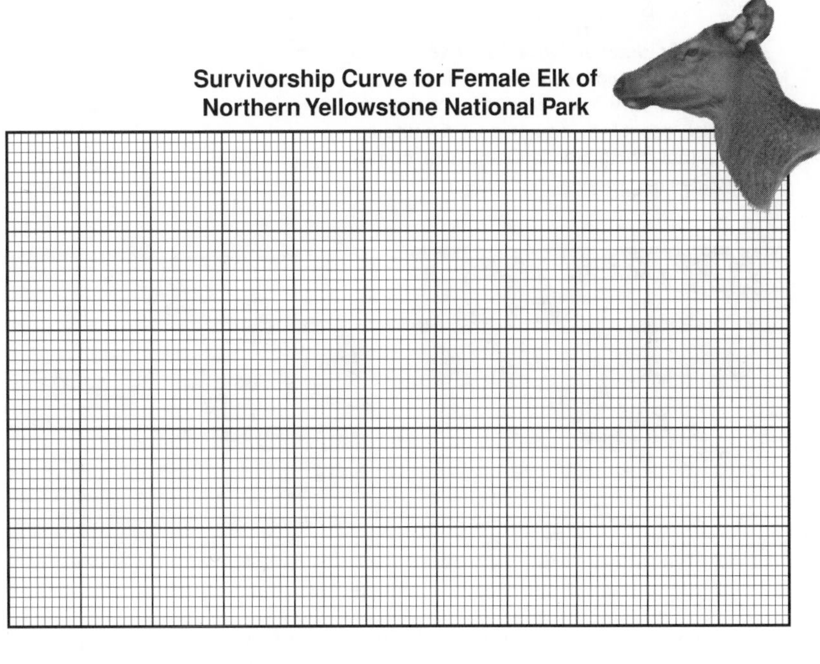

**Survivorship Curve for Female Elk of Northern Yellowstone National Park**

1. (a) In the example of the barnacle *Balanus* above, state when most of the group die: ____________

   (b) Identify the type of survivorship curve is represented by these data (see opposite): ____________

2. (a) Using the grid, plot a survivorship curve for elk hinds (above) based on the life table data provided:

   (b) Describe the survivorship curve for these large mammals: ____________

3. Explain how a biologist might use life table data to manage an endangered population: ____________


**ISBN: 978-1-927173-12-1**

# Survivorship Curves

The survivorship curve depicts age-specific mortality. It is obtained by plotting the number of individuals of a particular cohort against time. Survivorship curves are standardized to start at 1000 and, as the population ages, the number of survivors progressively declines. The shape of a survivorship curve thus shows graphically at which life stages the highest mortality occurs. Survivorship curves in many populations fall into one of three hypothetical patterns (below). Wherever the curve becomes steep, there is an increase in mortality. The convex Type I curve is typical of populations whose individuals tend to live out their physiological life span. Such populations usually produce fewer young and show some degree of parental care. Organisms that suffer high losses of the early life stages (a Type III curve) compensate by producing vast numbers of offspring. These curves are conceptual models only, against which real life curves can be compared. Many species exhibit a mix of two of the three basic types. Some birds have a high chick mortality (Type III) but adult mortality is fairly constant (Type II). Some invertebrates (e.g. crabs) have high mortality only when molting and show a stepped curve.

## Hypothetical Survivorship Curves

### Type I

**Late loss survivorship curve**

Mortality (death rate) is very low in the infant and juvenile years, and throughout most of adult life. Mortality increases rapidly in old age. **Examples**: Humans (in developed countries) and many other large mammals (e.g. big cats, elephants).

### Type II

**Constant loss survivorship curve**

Mortality is relatively constant through all life stages (no one age is more susceptible than another). **Examples**: Some invertebrates such as *Hydra*, some birds, some annual plants, some lizards, and many rodents.

### Type III

**Early loss survivorship curve**

Mortality is very high during early life stages, followed by a very low death rate for the few individuals reaching adulthood. **Examples**: Many fish (not mouth brooders) and most marine invertebrates (e.g. oysters, barnacles).

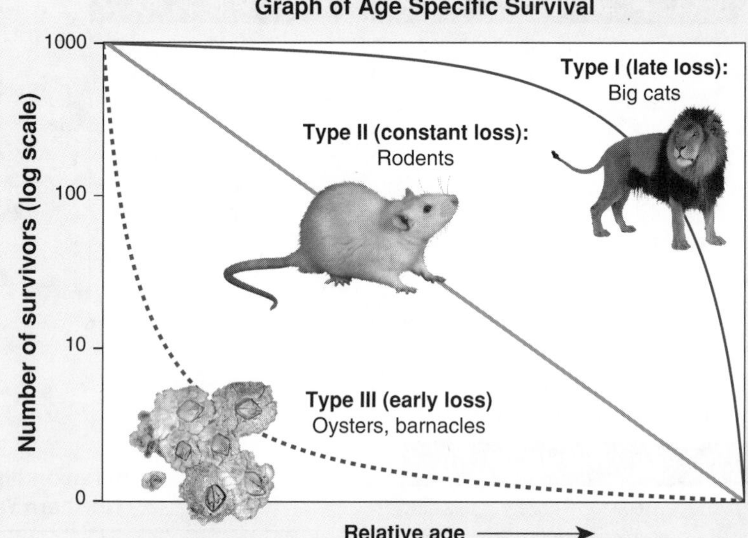


Three basic types of survivorship curves and representative organisms for each type. The vertical axis may be scaled arithmetically or logarithmically.

Elephants have a close matriarchal society and a long period of parental care. Elephants are long-lived and females usually produce just one calf.

Rodents are well known for their large litters and prolific breeding capacity. Individuals are lost from the population at a more or less constant rate.

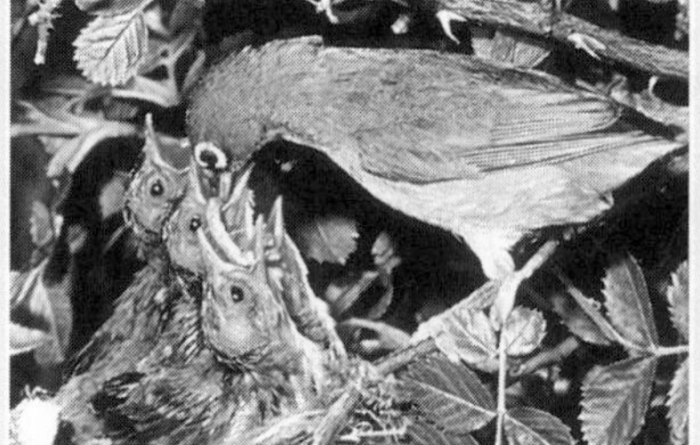


Despite vigilant parental care, many birds suffer high juvenile losses (Type III). For those surviving to adulthood, deaths occur at a constant rate.

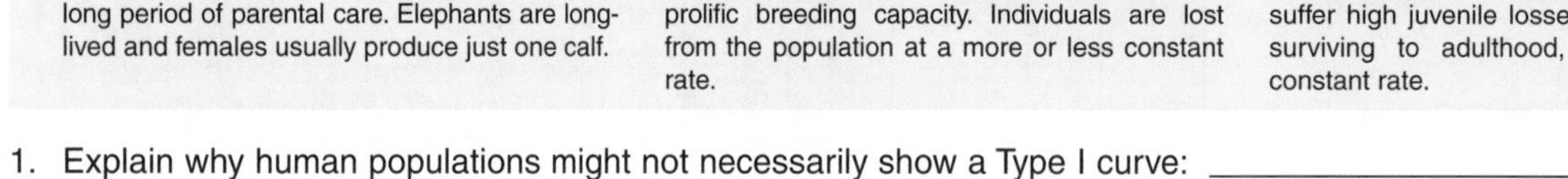

1. Explain why human populations might not necessarily show a Type I curve: ______

2. Explain how organisms with a Type III survivorship compensate for the high mortality during early life stages:

3. Describe the features of a species with a Type I survivorship that aid in high juvenile survival: ______

4. Discuss the following statement: "There is no standard survivorship curve for a given species; the curve depicts the nature of a population at a particular time and place and under certain environmental conditions.":



**Related activities:** *Life Tables and Survivorship, Survivorship and Life Expectancy*


**ISBN: 978-1-927173-12-1**

# Survivorship and Life Expectancy

**Life expectancy** is the average number of years of life remaining when at any given age. Life expectancy changes as an individual ages. For example in the US, at birth a human has a life expectancy of around 78 years. However a 65 year old has a life expectancy of around 18 years meaning they should live to the age of 83. In human societies, life expectancy is heavily dependent on aspects of the socio-economic structure such as public health facilities, presence and treatment of endemic disease, and poverty. Countries where war, famine, or disease are common have low life expectancies. Life expectancy is also affected by gender; In the US, life expectancy for males is 75 compared with 80 for females.

### Interpreting Survivorship Curves

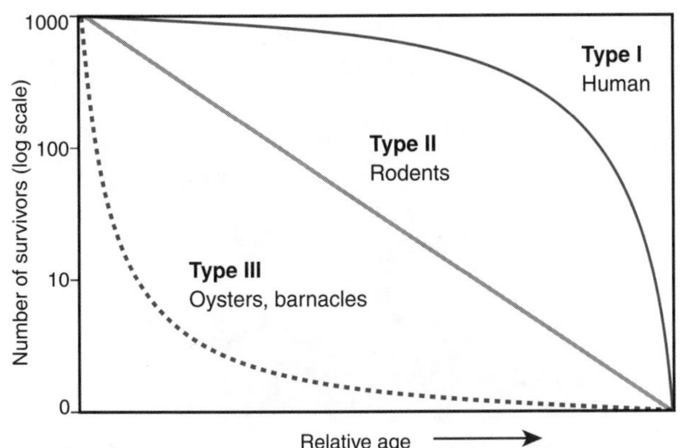


Survivorship curves give a graphical representation of the surviving members of a population per age group. Species with type I (late loss) curves have low mortality when young, which increases with age. Species with type II (constant loss) curves have a relatively constant mortality throughout all stages of life. Species with type III (early loss) curves have high mortality when young, which decreases with age.

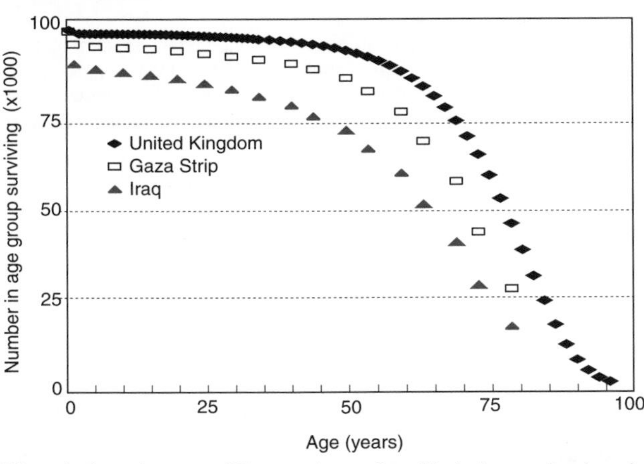


Although there is some difference is survivorship between developed and developing countries, human populations all still follow the type I curve showing the effort humans put into a small number of offspring. The average life expectancy can be estimated from a survivorship curve being the age at which 50% of the people in the sample are still alive.

### Life Expectancy, Mortality and Wealth

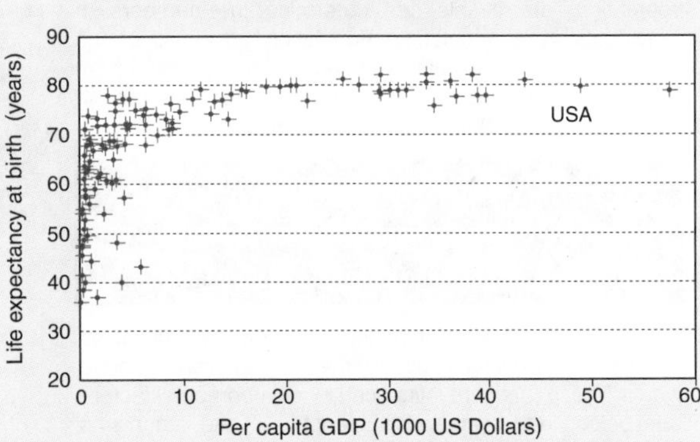


Life expectancy is linked to many factors but in human society there is a close correlation between life expectancy and a country's per capita gross domestic product (GDP). Those countries with high GDP can be expected to have citizens with long life expectancies.

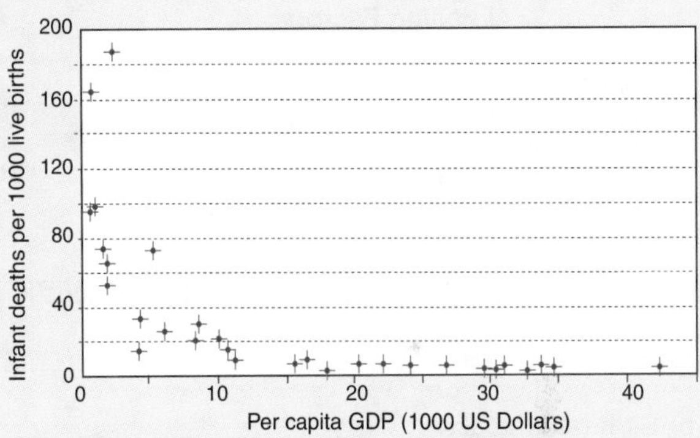


In developed nations, the **infant mortality rate** (IMR) is low because of better standards of living and advanced medical technology. People in developing countries often lack access to quality medical care and the IMR can be high, especially where there are high rates of endemic disease. However, populations in developing countries often have high birth rates so that population growth rates remain high despite a high IMR.

1. Explain why life expectancy changes as one ages: __________

2. Describe the relationship between a country's per capita GDP and life expectancy of its citizens: __________

3: Explain why IMR might be linked to a nation's wealth: __________

4. Explain why human survivorship curves might vary from country to country: __________

5. Estimate average life expectancy for: (a) United Kingdom: ______ (b) Gaza Strip: ______ (c) Iraq: ______

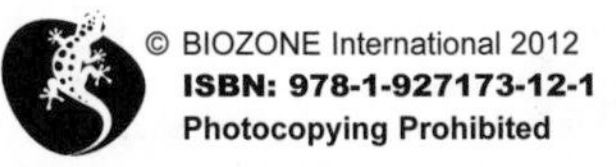


**ISBN: 978-1-927173-12-1**


***Related activities:*** *Survivorship Curves, World Population Growth*

# Population Growth

Organisms do not generally live alone. A **population** is a group of organisms of the same species living together in one geographical area. This area may be difficult to define as populations may comprise widely dispersed individuals that come together only infrequently (e.g. for mating). The number of individuals comprising a population may also fluctuate considerably over time. These changes make populations dynamic: populations gain individuals through births or immigration, and lose individuals through deaths and emigration. For a population in **equilibrium**, these factors balance out and there is no net change in the population abundance. When losses exceed gains, the population declines.

Births, deaths, immigrations (movements into the population) and emigrations (movements out of the population) are events that determine the numbers of individuals in a population. Population growth depends on the number of individuals added to the population from births and immigration, minus the number lost through deaths and emigration. This is expressed as:

**Population growth =**

**Births – Deaths + Immigration – Emigration**
**(B) (D) (I) (E)**

The difference between immigration and emigration gives net migration. Ecologists usually measure the **rate** of these events. These rates are influenced by environmental factors (see below) and by the characteristics of the organisms themselves. Rates in population studies are commonly expressed in one of two ways:

- Numbers per unit time, e.g. 20,150 live births per year.
- Per capita rate (number per head of population), e.g. 122 live births per 1000 individuals per year (12.2%).

### Limiting Factors

Population size is also affected by limiting factors; factors or resources that control a process such as organism growth, or population growth or distribution. Examples include availability of food, predation pressure, or available habitat.

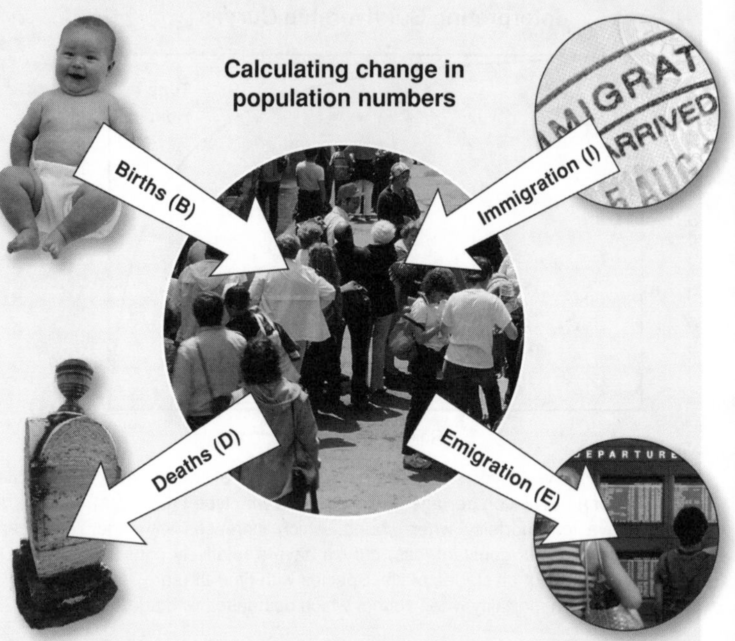


Human populations often appear exempt from limiting factors as technology and efficiency solve many food and shelter problems. However, as the last arable land is used and agriculture reaches its limits of efficiency, it is estimated that the human population may peak at around 10 billion by 2050.

1. Define the following terms used to describe changes in population numbers:
   (a) Death rate (mortality): ______
   (b) Birth rate (natality): ______
   (c) Net migration rate: ______
2. Explain how the concept of limiting factors applies to population biology: ______
3. Using the terms, B, D, I, and E (above), construct equations to express the following (the first is completed for you):
   (a) A population in equilibrium: B + I = D + E
   (b) A declining population: ______
   (c) An increasing population: ______
4. A population started with a total number of 100 individuals. Over the following year, population data were collected. Calculate birth rates, death rates, net migration rate, and rate of population change for the data below (as percentages):
   (a) Births = 14: Birth rate = ______
   (b) Net migration = +2: Net migration rate = ______
   (c) Deaths = 20: Death rate = ______
   (d) Rate of population change = ______
   (e) State whether the population is increasing or declining: ______
5. The human population is around 6.7 billion. Describe and explain two limiting factors for population growth in humans: ______




**ISBN: 978-1-927173-12-1**

# Patterns of Population Growth

Populations becoming established in a new area for the first time are often termed **colonizing populations** (below, left). They may undergo a rapid **exponential** (logarithmic) increase in numbers as there are plenty of resources to allow a high birth rate, while the death rate is often low. Exponential growth produces a J-shaped growth curve that rises steeply as more and more individuals contribute to the population increase. If the resources of the new habitat were endless (inexhaustible) then the population would continue to increase at an **exponential** rate. However, this rarely happens in natural populations. Initially, growth may be exponential (or nearly so), but as the population grows, its increase will slow and it will stabilize at a level that can be supported by the environment (called the carrying capacity or K). This type of growth is called sigmoidal and produces the **logistic growth curve** (below, right). **Established populations** will fluctuate about K, often in a regular way (blue area on the graph below, right). Some species will have populations that vary little from this stable condition, while others may oscillate wildly.

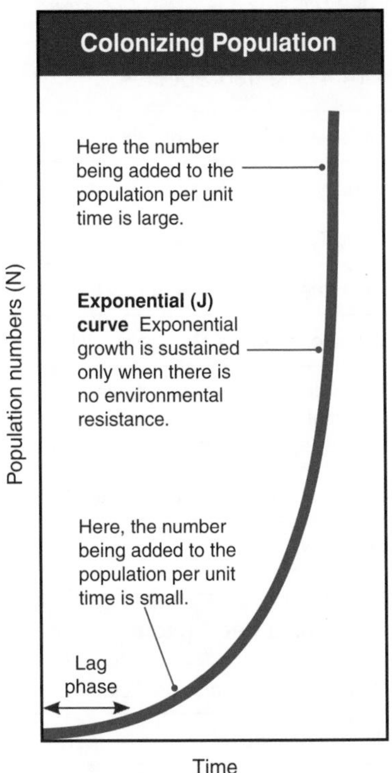


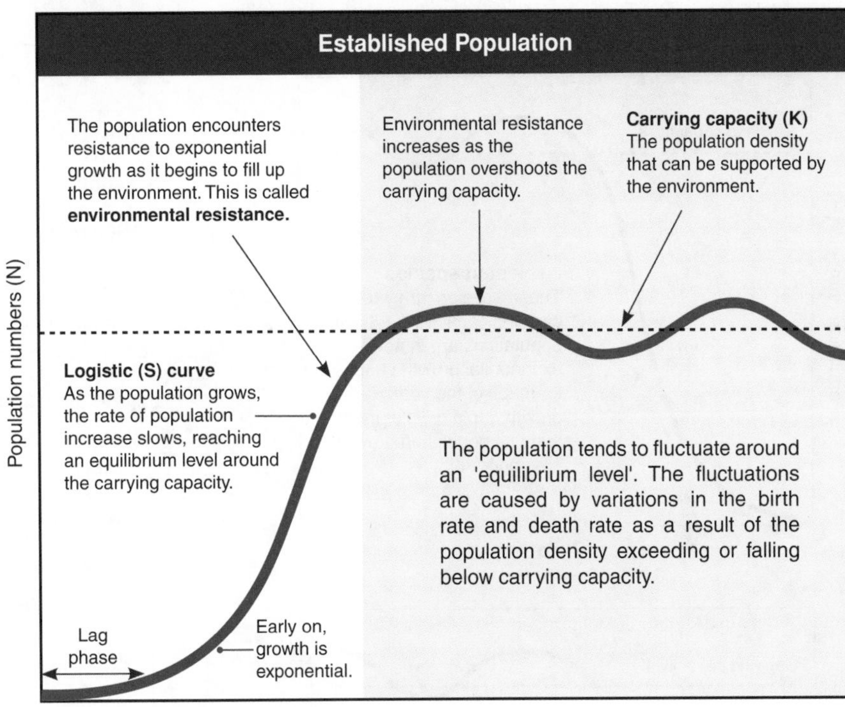


1. Explain why populations tend not to continue to increase exponentially in an environment:

2. Explain what is meant by environmental resistance:

3. (a) Explain what is meant by carrying capacity:

   (b) Explain the importance of **carrying capacity** to the growth and maintenance of population numbers:

4. Species that expand into a new area, such as rabbits did in areas of Australia, typically show a period of rapid population growth followed by a slowing of population growth as density dependent factors become more important and the population settles around a level that can be supported by the carrying capacity of the environment.

   (a) Explain why a newly introduced consumer (e.g. rabbit) would initially exhibit a period of exponential population growth:

   (b) Describe a likely outcome for a rabbit population after the initial rapid increase had slowed:

5. Describe the effect that introduced grazing species might have on the carrying capacity of the environment:


ISBN: 978-1-927173-12-1

# *r* and K Selection

The capacity of a species to increase in numbers is called its **biotic potential**. It is a measure of reproductive capacity and is assigned a set value (denoted by the letter ***r***) that is specific to the organism involved. Species with a high biotic potential are called ***r*-selected species**. They include algae, bacteria, rodents, many insects, and most annual plants. These species show life history features associated with rapid growth in disturbed environments. To survive, they must continually invade new areas to compensate for being replaced by more competitive species. The population growth of species with lower biotic potential tends to depend on the carrying capacity of the environment (**K**).These species, which include most large mammals, birds of prey, and large, long-lived plants, exist near the **carrying capacity** of their environments and are forced, through their interactions with other species, to use resources more efficiently. These species have fewer offspring and longer lives, and put their energy into nurturing their young to reproductive age. Most organisms have reproductive patterns between these two extremes.

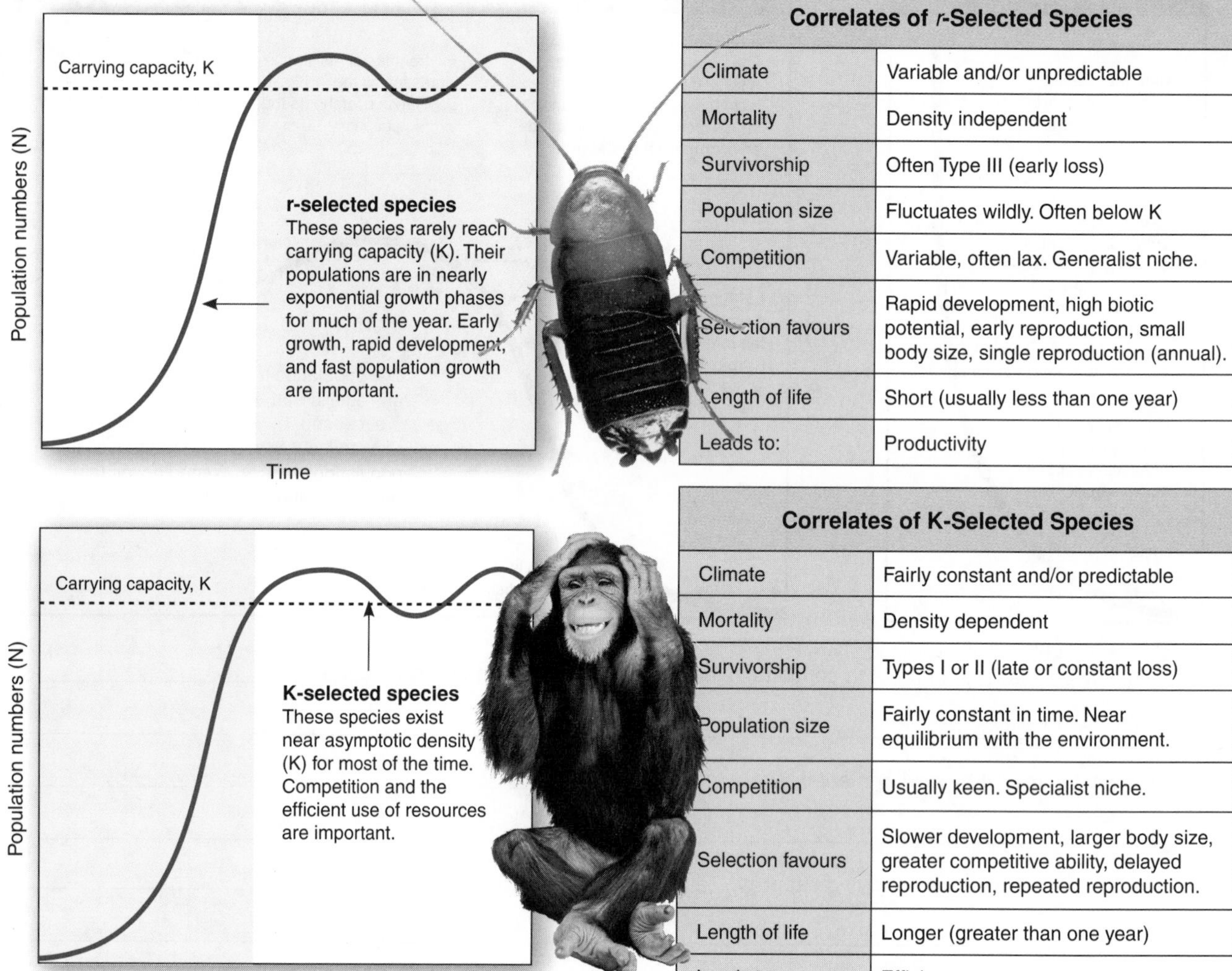


| Correlates of *r*-Selected Species | |
|---|---|
| Climate | Variable and/or unpredictable |
| Mortality | Density independent |
| Survivorship | Often Type III (early loss) |
| Population size | Fluctuates wildly. Often below K |
| Competition | Variable, often lax. Generalist niche. |
| Selection favours | Rapid development, high biotic potential, early reproduction, small body size, single reproduction (annual). |
| Length of life | Short (usually less than one year) |
| Leads to: | Productivity |

| Correlates of K-Selected Species | |
|---|---|
| Climate | Fairly constant and/or predictable |
| Mortality | Density dependent |
| Survivorship | Types I or II (late or constant loss) |
| Population size | Fairly constant in time. Near equilibrium with the environment. |
| Competition | Usually keen. Specialist niche. |
| Selection favours | Slower development, larger body size, greater competitive ability, delayed reproduction, repeated reproduction. |
| Length of life | Longer (greater than one year) |
| Leads to: | Efficiency |

1. Explain why *r*-selected species tend to predominate in unstable, disturbed, or early successional communities:

2. Explain why many K-selected species tend to predominate in stable, climax communities:

3. Describe factors that might cause a change in the predominance of K-selected species in a climax community:




**ISBN: 978-1-927173-12-1**

# Human Demography

Human populations through time have undergone demographic shifts related to societal changes and economic development. The demographic transition model (DTM) was developed in 1929 to explain the transformation of countries from high birth rates and high death rates to low birth rates and low death rates as part of their economic development from a pre-industrial to an industrialized economy. The transition involves four stages, or possibly five (with some nations recognized as moving beyond stage four). Each stage of the transition reflects the changes in birth and death rates observed in human societies over the last 200 years. Most developed countries are beyond stage three of the model; the majority of developing countries are in stage two or stage three. The model was based on the changes seen in Europe, so these countries follow the DTM relatively well. Many developing countries have moved into stage three. The exceptions include some poor countries, mainly in sub-Saharan Africa and some Middle Eastern countries, which are poor or affected by government policy or civil strife.

Stage one: Birth and death rates balanced but high as a result of starvation and disease.

Stage two: Improvement in food supplies and public health result in reduced death rates.

Stage three moves the population towards stability through a decline in the birth rate.

Stage four: Birth and death rates are both low and the total population is high and stable.

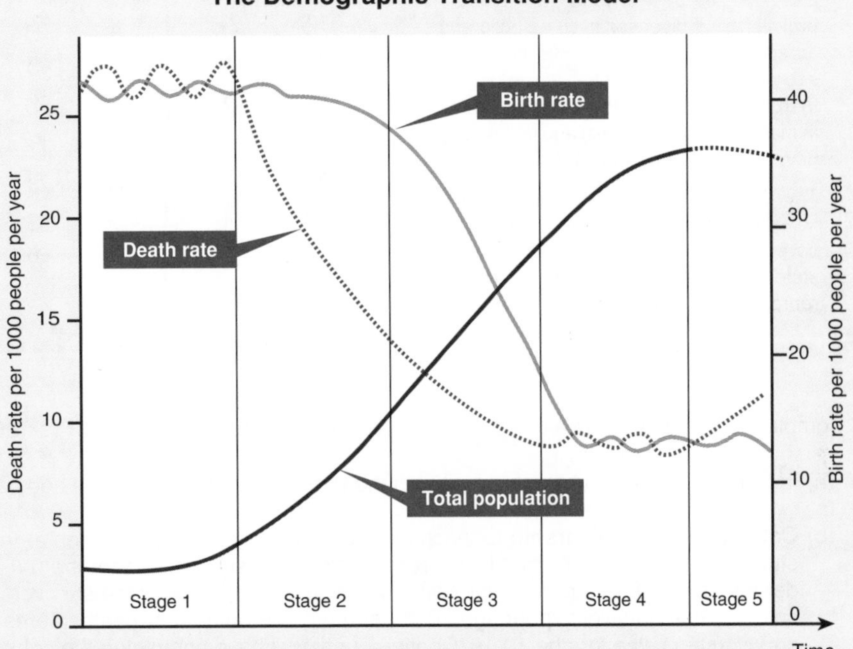


**Stage one (pre-modern)**: A balance between birth and death rates as was true of all populations until the late 18th Century. Children are important contributors to the household economy. Losses as a result of starvation and disease are high. Stage one is sometimes called the "High Stationary Stage" of population growth (high birth and death rates and stationary total population numbers).

**Stage two (early expanding)**: Rapid population expansion as a result of a decline in death rates. The changes leading to this stage in Europe were initiated in the Agricultural Revolution of the 18th century but have been more rapid in developing countries since then. Stage two is associated with more reliable food supplies and improvements in public health.

**Stage three (late expanding)**: The population moves towards stability through a decline in the birth rate. This stage is associated with increasing urbanization and a decreased reliance on children as a source of family wealth. Family planning in nations such as Malaysia (photo left) has been instrumental in their move to stage three.

**Stage four (post-industrial)**: Birth and death rates are both low and the total population is high and stable. The population ages and in some cases the fertility rate falls below replacement.

**Stage five (declining)**: Proposed by some theorists as representing countries that have undergone the economic transition from manufacturing based industries into service and information based industries and the population reproduces well below replacement levels. Countries in stage five include the United Kingdom (the earliest nation recognized as reaching Stage Five) and Germany.

1. Each of the first four stages of the DTM is associated with a particular age structure. Identify which of the diagrams (right) corresponds to stage one of the DTM and explain your choice:

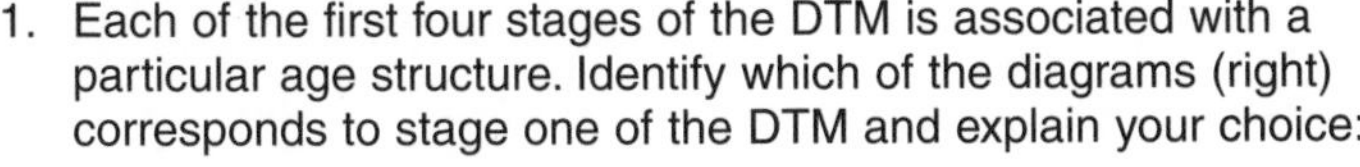

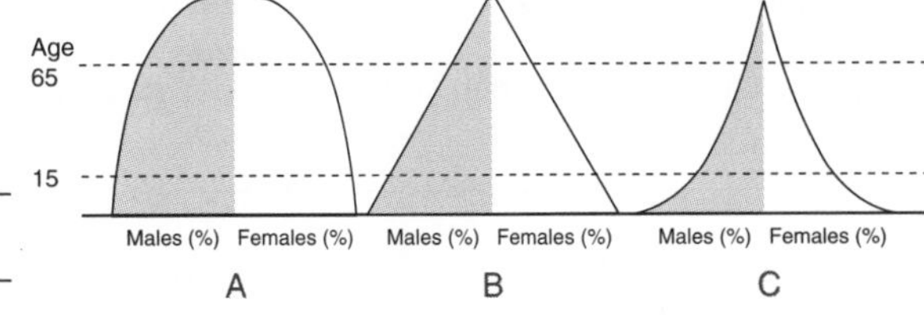


2. Suggest why it might become less important to have a large number of children in more economically developed nations:


ISBN: 978-1-927173-12-1

Cemeteries are an excellent place to study changes in human demographics. Data collected from headstones can be used to calculate death rates and produce survivorship curves. It is also possible to compare survivorship curves over different periods and see how certain factors (e.g. war, medical advances) have altered survivorship.

The data (right) represents age of death data for males and females collected over two different time periods; pre-1950 and post 1950. The pre-1950s was characterized by two world wars, and the prevalence of diseases such as polio and tuberculosis. The post 1950s have also seen global conflict, but to a lesser degree than the pre-1950 period. Many advances in medicine (e.g. vaccines) and technology have been made during this time.

The data used in this exercise has been collected from the online records of several cemeteries across five different states in the United States to provide representative data.

| Pre-1950 | | | | Post 1950 | | | |
|---|---|---|---|---|---|---|---|
| Males age at death | | Females age at death | | Males age at death | | Females age at death | |
| 81 | 5 | 9 | 4 | 80 | 31 | 92 | 76 |
| 40 | 0 | 76 | 18 | 81 | 78 | 46 | 92 |
| 54 | 24 | 0 | 71 | 79 | 56 | 44 | 96 |
| 70 | 70 | 78 | 2 | 81 | 86 | 70 | 65 |
| 75 | 39 | 69 | 63 | 8 | 80 | 80 | 54 |
| 64 | 71 | 6 | 1 | 30 | 64 | 71 | 87 |
| 45 | 27 | 46 | 84 | 88 | 41 | 88 | 82 |
| 22 | 64 | 60 | 68 | 90 | 76 | 65 | 80 |
| 71 | 0 | 84 | 58 | 84 | 17 | 51 | 90 |
| 62 | 41 | 75 | 19 | 64 | 40 | 80 | 85 |
| 89 | 77 | 43 | 24 | 60 | 79 | 87 | 63 |
| 31 | 21 | 64 | 62 | 71 | 74 | 76 | 58 |
| 10 | 1 | 67 | 52 | 62 | 46 | 63 | 89 |
| 42 | 75 | 42 | 29 | 63 | 71 | 33 | 56 |
| 1 | 50 | 39 | 8 | 83 | 90 | 99 | 86 |

Data source: http://www.interment.net/us/index.htm

3. Complete the following table using the cemetery data provided. The males Pre-1950 data have been completed for you.

(a) In the number of deaths column, record the number of deaths for each age category.

(b) **Calculate the survivorship** for each age category. For each column, enter the total number of individuals in the study (30) in the 0-9 age survivorship cell. This is the survivorship for the 0-9 age group. Subtract the number of deaths at age 0-9 from the survivorship value at age 0-9. This is the survivorship at the 10-19 age category. To calculate the survivorship for age 20-29, subtract the number of deaths at the age 10-19 age category from the survivorship value for age 10-19. Continue until you have completed the column.

| | Males Pre-1950 | | Females Pre-1950 | | Males Post 1950 | | Females Post 1950 | |
|---|---|---|---|---|---|---|---|---|
| **Age** | **No. of deaths** | **Survivorship** | **No. of deaths** | **Survivorship** | **No. of deaths** | **Survivorship** | **No. of deaths** | **Survivorship** |
| 0-9 | 5 | 30 | | | | | | |
| 10-19 | 1 | 25 | | | | | | |
| 20-29 | 4 | 24 | | | | | | |
| 30-39 | 2 | 20 | | | | | | |
| 40-49 | 4 | 18 | | | | | | |
| 50-59 | 2 | 14 | | | | | | |
| 60-69 | 3 | 12 | | | | | | |
| 70-79 | 7 | 9 | | | | | | |
| 80-89 | 2 | 2 | | | | | | |
| 90-99 | 0 | 0 | | | | | | |
| Total | 30 | | | | | | | |

4. (a) On a separate piece of graph paper, construct a graph to compare the survivorship curves for each category. Staple the graph into this workbook once you have completed the activity.

(b) What conclusions can you make about survivorship before 1950 and after 1950? ____________________

____________________

____________________

(c) What factors might cause these differences? ____________________

____________________

____________________

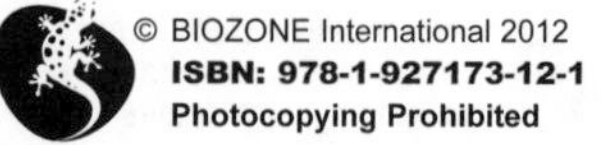


ISBN: 978-1-927173-12-1

# World Population Growth

For most of human history, humans have not been very numerous compared to other species. It took all of human history to reach a population of 1 billion in 1804, but little more than 150 years to reach 3 billion in 1960. The world's population, now at 6.8 billion, is growing at the rate of about 80 million per year. This growth is slower than predicted but the world's population is still expected to increase substantially before stabilizing (see Figure 1). World population increase carries important environmental consequences, particularly when it is associated with increasing **urbanization**. Although the world as a whole still has an average fertility rate of 2.8, growth rates are now lower than at any time since World War II. Continued declines may give human populations time to address some to the major problems posed by the increasing the scope and intensity of human activities.

## World Population Density

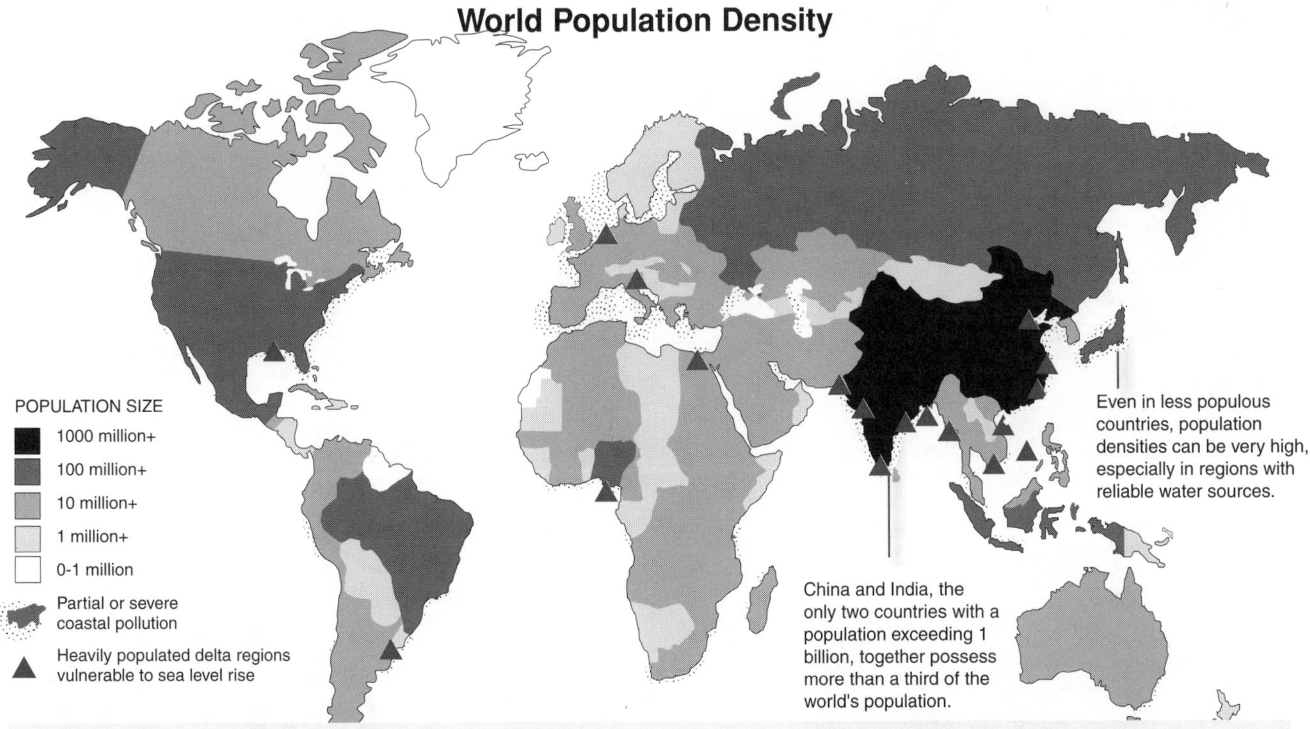


**Figure 1: Progress towards population stabilization**

Stabilization ratio (births/deaths)
South & Central America
Africa
Asia
Developing
Developed
(1 = stable population)
Projected

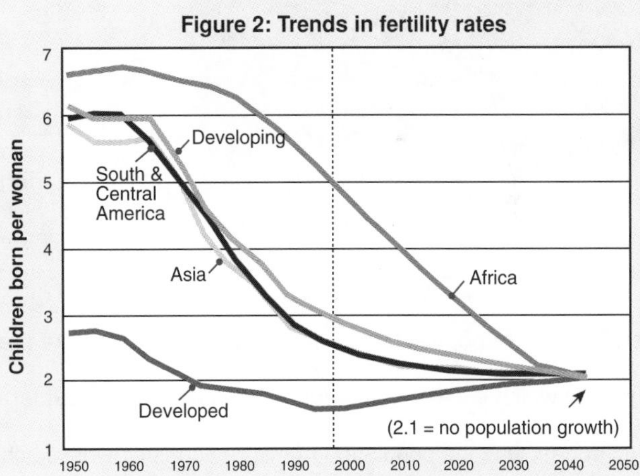


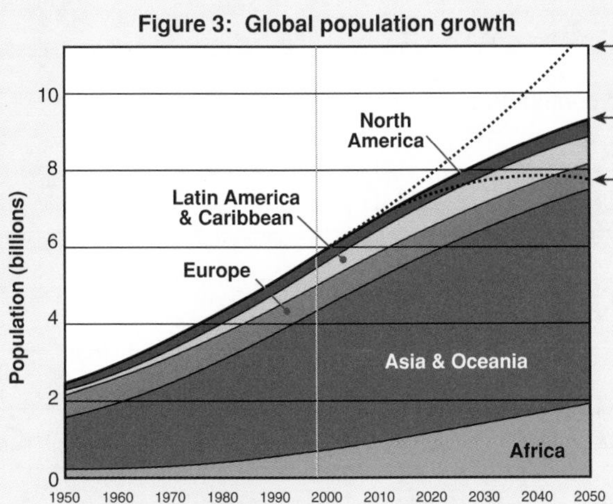


**High** fertility rate: 11.2 billion

**Medium** fertility rate: 9.4 billion

**Low** fertility rate: 7.7 billion

Source for graphs: United Nations Population Division, *World Population Prospects 1950-2050 (The 1996 Revision)*, U.N., New York, 1996.

The graphs on this page attempt to show the likely future growth of the human population. Estimates are highly uncertain and projections made in 1996 for 2050 ranged from a low of 7.7 billion to a high of 11.2 billion. Since the world's population reached 7 billion in October 2011, only 12 years after reaching the 6 billion mark, the lower projection was clearly incorrect. The latest (2014) "medium variant" U.N. projection of 9.6 billion is close to that predicted in 1996 and depends on a number of important assumptions: fertility rates will continue to decline and life expectancy will continue to increase (as in the industrialized countries) and that developing countries will broadly follow these demographic trends.


**ISBN: 978-1-927173-12-1**


***Periodicals:*** *Human population grows up*

***Related activities:*** *Population Growth, Population Age Structure*
***Weblinks:*** *Human Impact: Overpopulation*

## The Shift to Urban Living

The traditional villages characteristic of the rural populations of less economically developed countries have a close association with the land. The households depend directly on agriculture or harvesting natural resources for their livelihood and are linked through family ties, culture, and economics.

Cities are differentiated communities, where the majority of the population does not depend directly on natural resource-based occupations. While cities are centers of commerce, education, and communication, they are also centers of crowding, pollution, and disease.

Cities, especially those that are growing rapidly, face a range of problems associated with providing residents with adequate water, food, sanitation, housing, jobs, and basic services, such as health care. Slums or squatter settlements are found in most large cities in developing countries as more poor people migrate from rural to urban areas.

The redistribution of people from rural to urban environments, or **urbanization**, has been an important characteristic of human societies. Almost half of the people in the world already live in urban areas and by the end of the 21st century, this figure is predicted to increase to 80-90%. Urban populations can grow through natural increase (i.e. more births than deaths) or by **immigration**. Immigration is driven both by **push factors** that encourage people to leave their rural environment and **pull factors** that draw them to the cities.

### Immigration push factors

- Rural overpopulation
- Lack of work or food
- Changing agricultural practices
- Desire for better education
- Racial or religious conflict
- Political instability

### Immigration pull factors

- Opportunity for better jobs
- Chance of better housing
- More reliable food supply
- Opportunity for greater wealth
- Freedom from village traditions
- Government policy

### Rural and Urban Populations in the USA

— % of US population in urban areas
····· % of US population in rural areas
Total urban population
Total rural population

Population (millions): 0, 20, 40, 60, 80, 100, 120, 140, 160, 180, 200
Percentage of population: 0, 10, 20, 30, 40, 50, 60, 70, 80
1900 1910 1920 1930 1940 1950 1960 1970 1980 1990

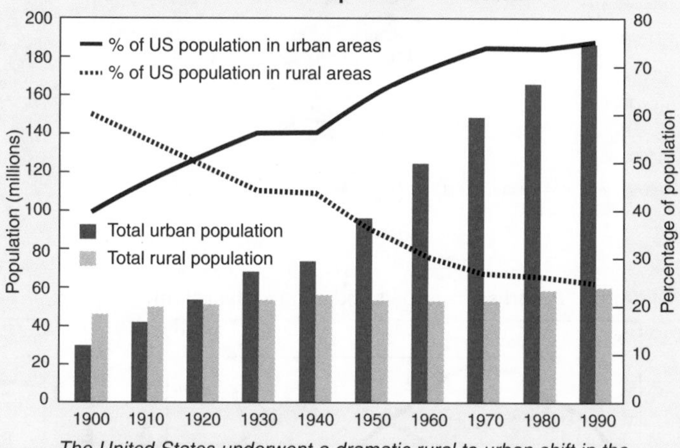

*The United States underwent a dramatic rural to urban shift in the 19th and early 20th centuries. Many developing countries are now experiencing similar shifts.* Graph compiled from UN data

1. Fertility rates of populations for all geographic regions are predicted to decline over the next 50 years.

   (a) State which continent is predicted to have the highest fertility rate at the beginning of next century: ______________

   (b) Suggest why the population of this region is slower to achieve a low fertility rate than other regions: ______________

2. Describe the kinds of changes in agricultural practices that could contribute to urbanization: ______________

3. (a) Describe some of the positive effects of urbanization: ______________

   (b) Describe some negative effects of urbanization: ______________


ISBN: 978-1-927173-12-1

# The Rise and Fall of Human Populations

Human populations are subject to rises and collapses in the same way as natural animal populations. Throughout history there have been a number of peaks of human civilization followed by collapse. These collapses have been triggered by various events but can generally be attributed either to the spread of disease or to the collapse of a food source (normally agriculture). Examples can be traced right back to the origins of humans.

Mitochondrial DNA analyses show that the human population may have been on the brink of extinction with only around 10,000 individuals alive 150,000 years ago. The population remained low for virtually the whole of human prehistory. When the first towns and cities were being built, around 10,000 years ago, the human population had reached barely 5 million. By around 700 AD, the human population had reached 150 million, and the first very large cities were developing. One such city was the Mayan city of Tikal.

**TIKAL**: At its peak around 800 AD, Tikal and the surrounding area, was inhabited by over 400,000 people. Extensive fields were used to cultivate crops and the total area of the city and its satellite towns and fields may have reached over 250 km$^2$. Eventually the carrying capacity of the tropical, nutrient-poor land was overextended and people began to starve. By 900 AD the city had been deserted and the surrounding area abandoned.

**EASTER ISLAND**: Similar events happened elsewhere. Easter Island is located 3,000 km from South America and 2,000 km from the nearest occupied land (the tiny, isolated Pitcairn Island). Easter Island has a mild climate and fertile volcanic soil, but when Europeans discovered it in the 1700s, it was covered in dry grassland, lacking trees or any vegetation above 1m high. Around 2,000 people survived on the island by subsistence farming, yet all around stood huge stone statues, some 30 m tall and weighing over 200 tonnes. Clearly a much larger more advanced society had been living on the island at some time in the past. Archaeological studies have found that populations reached 20,000 people prior to 1500AD. Exhaustion of the island's resources by the population was followed by war and civil unrest and the population fell to the subsistence levels found in the 1700s.

**EUROPE**: Despite isolated events, the world population continued to grow so that by 1350 AD it had reached around 450 million. As a result of the continued rise of urban populations, often living in squalid conditions, disease spread rapidly. The bubonic plague, which swept through Europe at this time, reduced its population almost by half, and reduced the world's population to 350 million. Despite further outbreaks of plague and the huge death tolls of various wars, the human population had reached 2.5 billion by 1950. By 1990 it was 5 billion and today it is around 6.5 billion. In slightly less than 60 years the human population has grown almost twice as much as it did in the whole of human history up until 1950. Much of this growth can be attributed to major advances in agriculture and medicine. However, signs are appearing that the human population are approaching maximum sustainable levels. Annual crop yields have ceased increasing and many common illnesses are becoming more difficult to treat. The rapid spread of modern pandemics, such as H1N1 swine flu, illustrates the vulnerability of modern human populations. Could it be, perhaps, that another great reduction in the human population is imminent?

1. Describe the general trend of human population growth over the last 100,000 years: ______

2. Explain why the human population has grown at such a increased rate in the last 60 years: ______

3. Discuss similarities between the events at Tikal and on the Easter Islands and how they can help us plan for the future:


**ISBN: 978-1-927173-12-1**

# Population Regulation

Very few species show continued exponential growth. Population size is regulated by factors that limit population growth. The diagram below illustrates how population size can be regulated by environmental factors. **Density independent factors** may affect all individuals in a population equally. Some, however, may be better able to adjust to them. **Density dependent factors** have a greater affect when the population density is higher. They become less important when the population density is low.



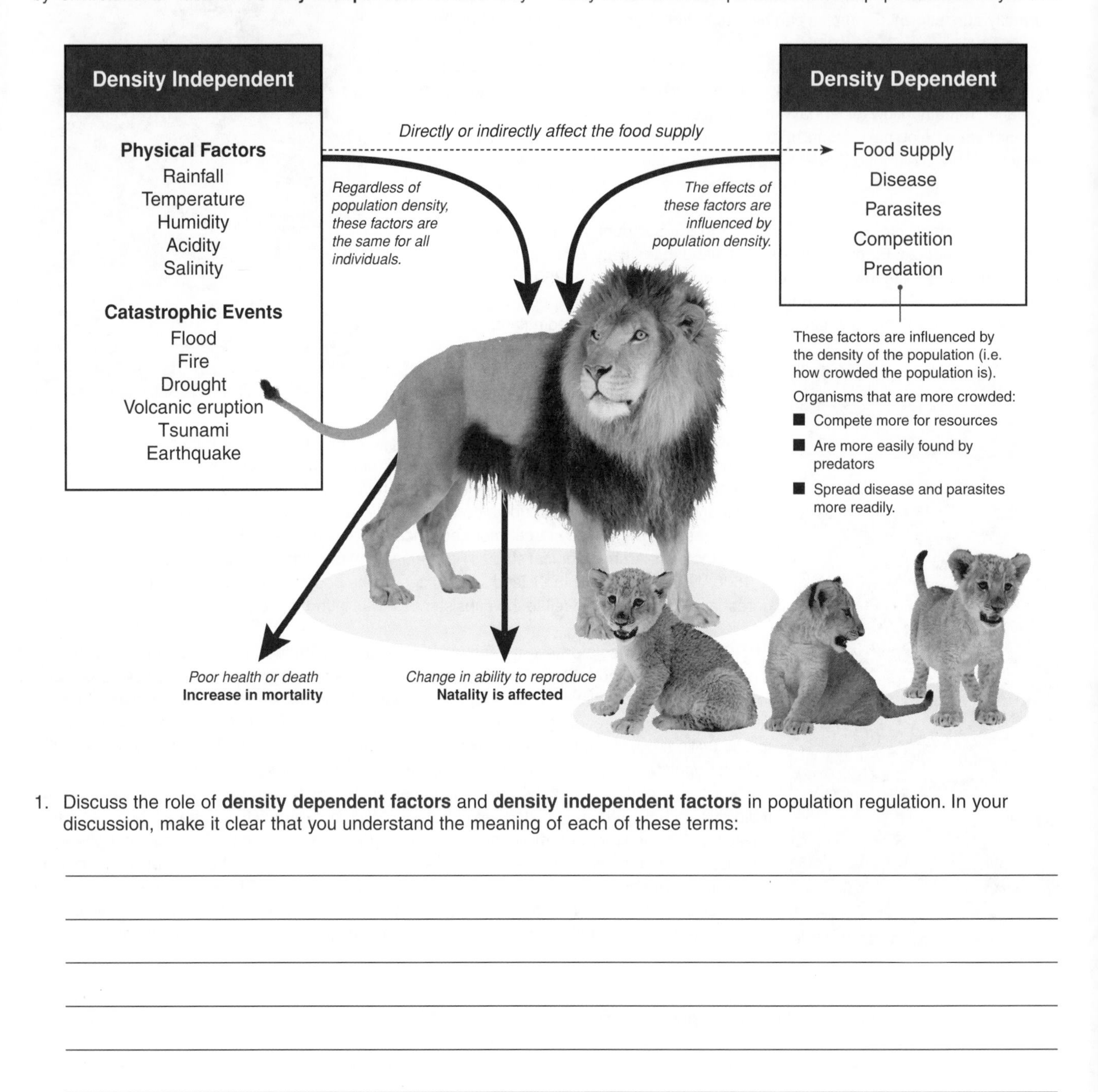

1. Discuss the role of **density dependent factors** and **density independent factors** in population regulation. In your discussion, make it clear that you understand the meaning of each of these terms:

2. Explain how an increase in population density allows disease to have a greater influence in regulating population size:

3. In cooler climates, aphids go through a huge population increase during the summer months. In autumn, population numbers decline steeply. Describe a density dependent and a density independent factor regulating the population:

   (a) Density dependent:

   (b) Density independent:



***Related activities:*** *Density and Distribution, r and K Selection*

**ISBN: 978-1-927173-12-1**

# Species Interactions

The particular characteristics of ecosystems and their communities arise as a result of the physical environment, the species that live there, and the complex interactions occurring between those species. Species interactions may involve only occasional or indirect contact (**predation** or **competition**) or they may involve close association or **symbiosis**. Symbiosis is a term that encompasses a variety of interactions involving close species contact. There are three types of symbiosis: **parasitism** (a form of exploitation), **mutualism**, and **commensalism**.

## Species Interactions on the African Savannah

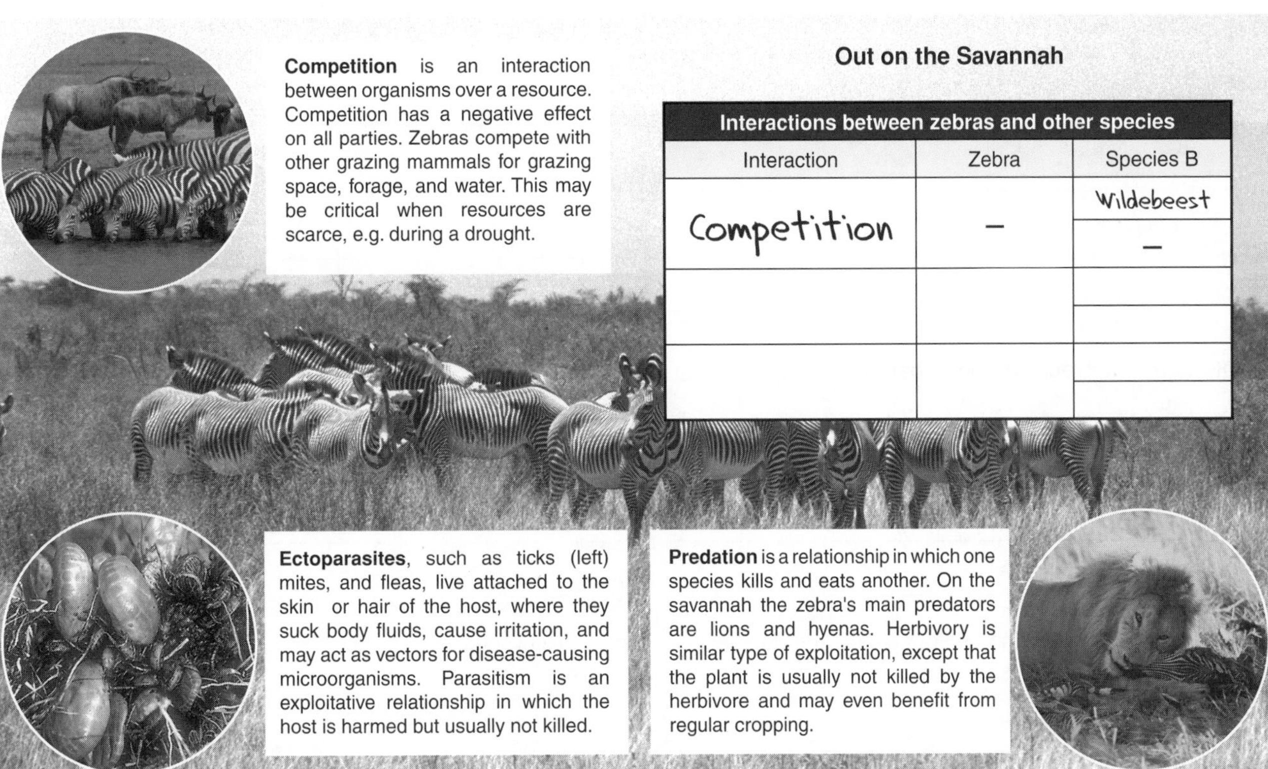

**Competition** is an interaction between organisms over a resource. Competition has a negative effect on all parties. Zebras compete with other grazing mammals for grazing space, forage, and water. This may be critical when resources are scarce, e.g. during a drought.

### Out on the Savannah

| Interactions between zebras and other species | | |
|---|---|---|
| Interaction | Zebra | Species B |
| Competition | – | Wildebeest |
| | | – |
| | | |
| | | |
| | | |
| | | |

**Ectoparasites**, such as ticks (left) mites, and fleas, live attached to the skin or hair of the host, where they suck body fluids, cause irritation, and may act as vectors for disease-causing microorganisms. Parasitism is an exploitative relationship in which the host is harmed but usually not killed.

**Predation** is a relationship in which one species kills and eats another. On the savannah the zebra's main predators are lions and hyenas. Herbivory is similar type of exploitation, except that the plant is usually not killed by the herbivore and may even benefit from regular cropping.

## Species Interactions in Redwood Forests

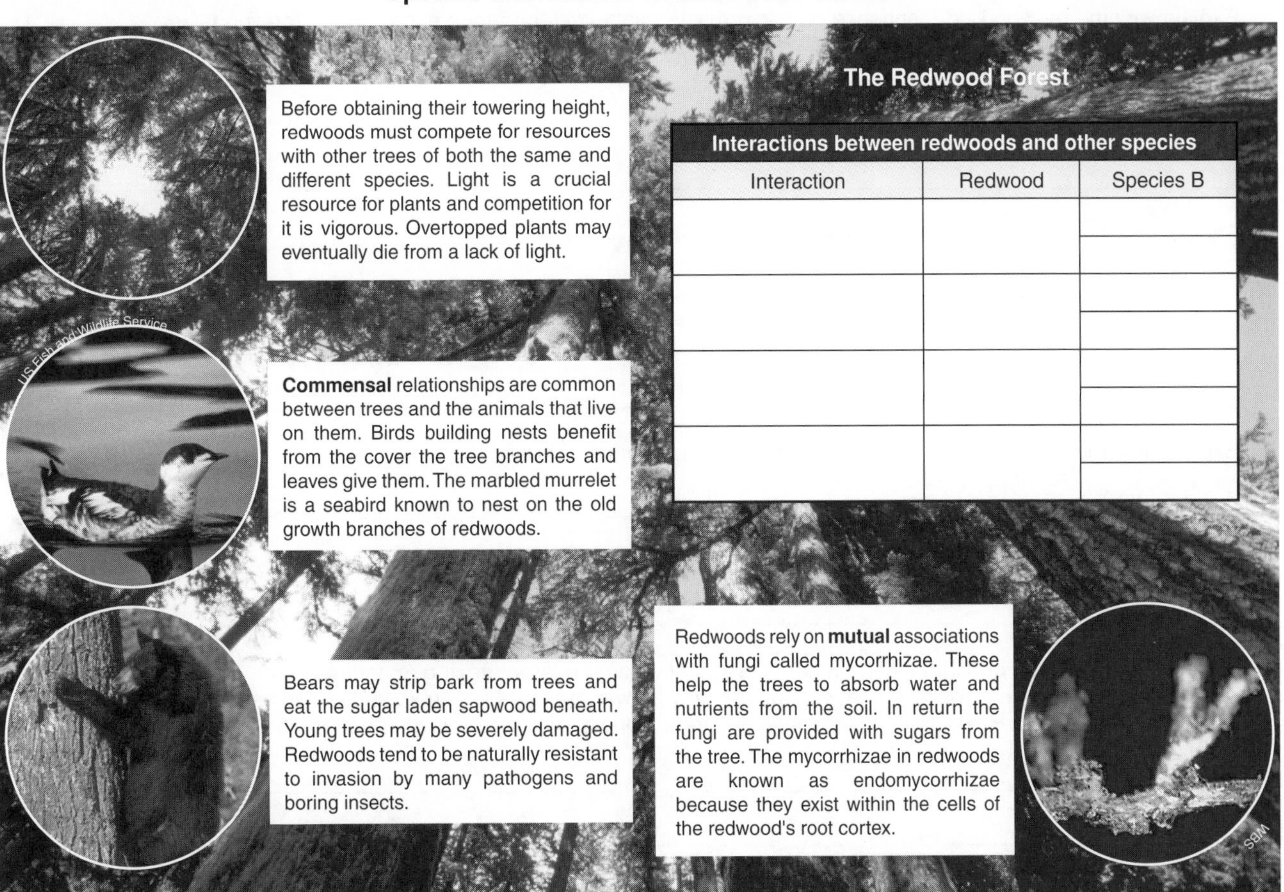

Before obtaining their towering height, redwoods must compete for resources with other trees of both the same and different species. Light is a crucial resource for plants and competition for it is vigorous. Overtopped plants may eventually die from a lack of light.

### The Redwood Forest

| Interactions between redwoods and other species | | |
|---|---|---|
| Interaction | Redwood | Species B |
| | | |
| | | |
| | | |
| | | |
| | | |
| | | |
| | | |
| | | |

US Fish and Wildlife Service

**Commensal** relationships are common between trees and the animals that live on them. Birds building nests benefit from the cover the tree branches and leaves give them. The marbled murrelet is a seabird known to nest on the old growth branches of redwoods.

Bears may strip bark from trees and eat the sugar laden sapwood beneath. Young trees may be severely damaged. Redwoods tend to be naturally resistant to invasion by many pathogens and boring insects.

Redwoods rely on **mutual** associations with fungi called mycorrhizae. These help the trees to absorb water and nutrients from the soil. In return the fungi are provided with sugars from the tree. The mycorrhizae in redwoods are known as endomycorrhizae because they exist within the cells of the redwood's root cortex.

SBM

**ISBN: 978-1-927173-12-1**

***Periodicals:***
*Inside story*

***Related activities:*** *Predator-Prey Strategies*
***Weblinks:*** *Ecological Interactions from EcoLibrary*

1. Complete the tables of relationships on the proceeding page for each of the examples given, filling in the type of relationship the effect (+, –, 0) and the species involved. The first one has been done for you.

2. Summarize your knowledge of species interactions by completing the following, entering a (+), (–), or (0) for species B, and writing a brief description of each term. Codes: (+): species benefits, (–): species is harmed, (0): species is unaffected.

| Interaction | Species | | Description of relationship |
|---|---|---|---|
| | A | B | |
| (a) Mutualism | + | | |
| (b) Commensalism | + | | |
| (c) Parasitism | – | | |
| (d) Predation | – | | |
| (e) Competition | – | | |

3. Distinguish a **predator** from a **parasite**:

4. Explain why **competition** for a resource has negative effects on all parties:

5. Explain how competition for light affects the population density of redwoods:

6. Explain why the redwood's dependence on mycorrhizae might limit the range expansion of redwoods:

7. Explain how the marbled murrelet is affected by the density of the redwoods:

8. Explain how redwood density may affect the feeding preference of bears:

9. Oxpeckers are African birds that feed on parasites that live on the large grazing mammals of the African savannah. This is of benefit to the large grazers. However, they are now known to actively keep wounds open caused by ectoparasites and feed off the blood. How would you classify the interactions between oxpeckers and their hosts? Explain your answer:


ISBN: 978-1-927173-12-1

# Interspecific Competition

In naturally occurring populations, direct competition between different species (**interspecific competition**) is usually less intense than intraspecific competition. This is because coexisting species have evolved slight differences in their realized niches, even though their fundamental niches may overlap (a phenomenon termed **niche differentiation**). However, when two species with very similar niche requirements are brought into direct competition through the introduction of a foreign species, one usually benefits at the expense of the other. The inability of two species with the same described niche to coexist is referred to as the **competitive exclusion principle**. In Britain, introduction of the larger, more aggressive, gray squirrel in 1876 has contributed to a contraction in range of the native red squirrel (below), and on the Scottish coast, this phenomenon has been well documented in barnacle species. The introduction of ecologically aggressive species is often implicated in the displacement or decline of native species, although there may be more than one contributing factor. Displacement of native species by introduced ones is more likely if the introduced competitor is also adaptable and hardy. It can be difficult to provide evidence of decline in a species as a direct result of competition, but it is often inferred if the range of the native species contracts and that of the introduced competitor shows a corresponding increase.

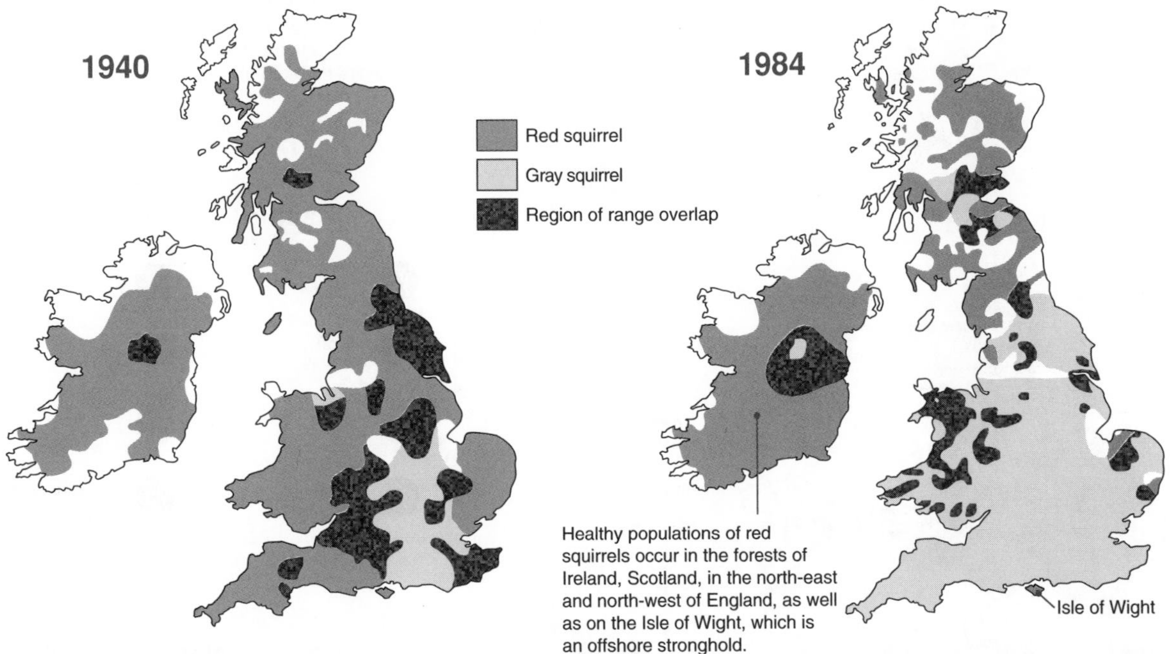

The **European red squirrel**, *Sciurus vulgaris*, was the only squirrel species in Britain until the introduction of the **American gray squirrel**, *Sciurus carolinesis*, in 1876. In 44 years since the 1940 distribution survey (map left), the adaptable grey squirrel has displaced populations of the native reds over much of the British Isles, particularly in the south (with the exception of the Isle of Wight). Whereas the red squirrels once occupied both coniferous and broad leafed woodland, they are now often restricted to coniferous forest and are absent from much of their former range.

1. Outline the evidence to support the view that the red-gray squirrel distributions in Britain are an example of the competitive exclusion principle:

2. Some biologists believe that competition with grey squirrels is only one of the factors contributing to the decline in the red squirrels in Britain. Explain the evidence from the 1984 distribution map that might support this view:


ISBN: 978-1-927173-12-1


***Related activities:*** *The Ecological Niche*
***Weblinks:*** *Red Squirrel, Gray Squirrel, Black Squirrel...*

RA 2

# Niche Differentiation

Competition is most intense between members of the same species because their habitat and resource requirements are identical. In naturally occurring populations, **interspecific competition** (between different species) is usually less intense than intraspecific competition because coexisting species have developed (through evolution) slight differences in their realized niches. In fact, when the niches of naturally coexisting species are described, there is seldom much overlap. Species with similar ecological requirements may reduce competition by exploiting microhabitats within the ecosystem. In the eucalypt forest below, different bird species exploit tree trunks, leaf litter, different levels within the canopy, and air space. Competition may also be reduced by exploiting the same resources at a different time of the day or year.

Canopy
Subcanopy
Understorey
Ground
Sw St Lf Wt Bt Rf Ys Gt

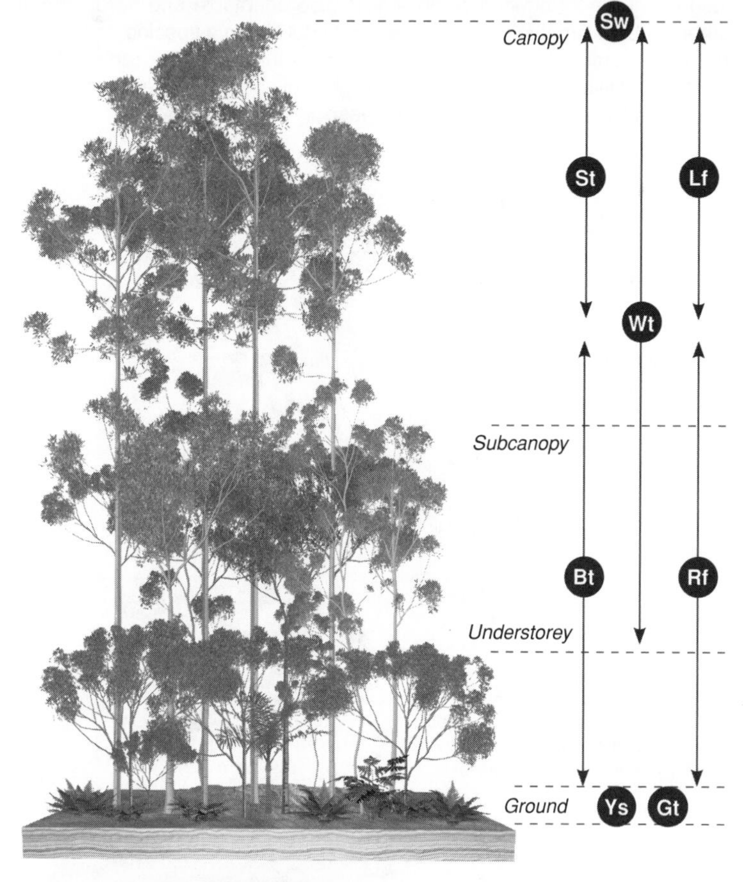

## Reducing competition in a eucalypt forest

The diagram on the left shows the foraging heights of birds in an eastern Australian eucalypt forest. A wide variety of food resources are offered by the structure of the forest. Different layers of the forest allow birds to specialize in foraging at different heights. The ground-dwelling yellow-throated scrubwren and ground thrush have robust legs and feet, while the white-throated treecreeper has long toes and large curved claws and the swifts are extremely agile fliers capable of catching insects on the wing.

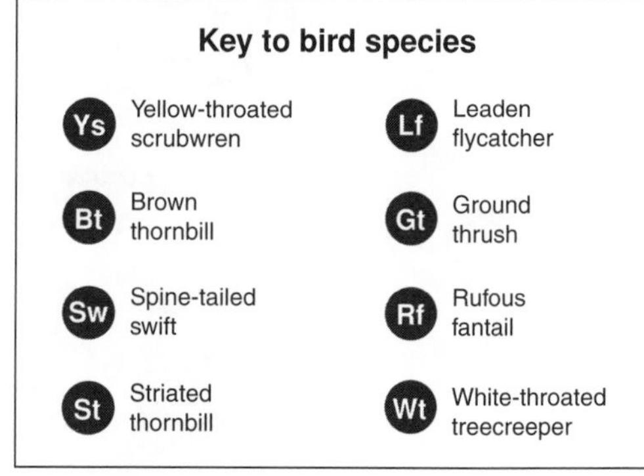


Adapted from: Recher et al., 1986. A Natural Legacy: Ecology in Australia. Maxwell Macmillan Publishing Australia.

## Distribution of ecologically similar fish

The diagram below shows the distribution of ecologically similar damselfish over a coral reef at Heron Island, Queensland, Australia. The habitat and resource requirements of these species overlap considerably.

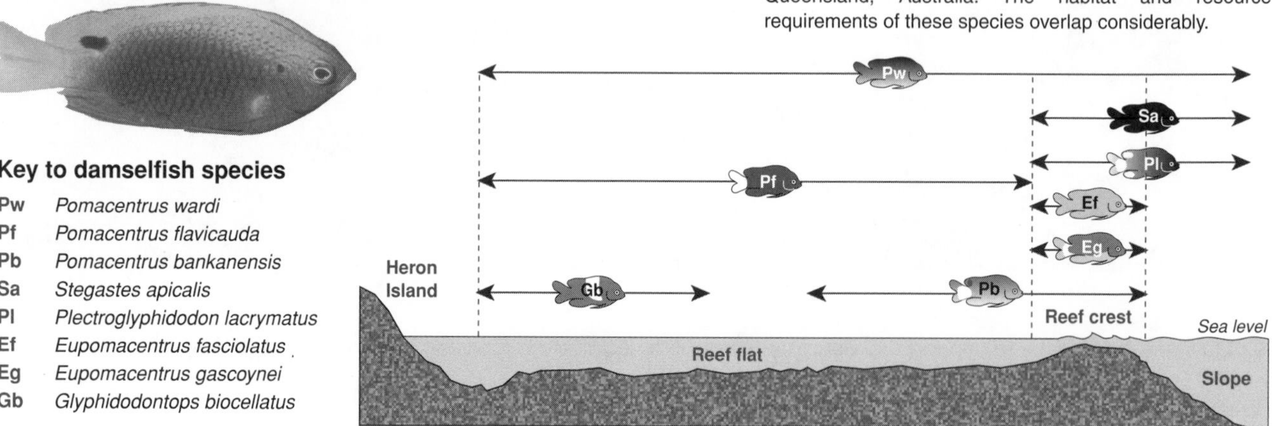


1. Describe two ways in which species can avoid directly competing for the same resources in their habitat:

   (a) ______

   (b) ______

2. Explain why **intraspecific** competition is more intense than **interspecific** competition: ______

3. Suggest how the damsel fish on the reef at Heron Island (above) might reduce competition: ______



*Related activities:* *The Ecological Niche, Interspecific Competition, Intraspecific Competition*


**ISBN: 978-1-927173-12-1**

# Intraspecific Competition

Some of the most intense competition occurs between individuals of the same species (**intraspecific competition**). Most populations have the capacity to grow rapidly, but their numbers cannot increase indefinitely because environmental resources are finite. Every ecosystem has a **carrying capacity** (K), defined as the number of individuals in a population that the environment can support. Intraspecific competition for resources increases with increasing population size and, at carrying capacity, it reduces the per capita growth rate to zero. When the demand for a particular resource (e.g. food, water, nesting sites, nutrients, or light) exceeds supply, that resource becomes a **limiting factor**. Populations respond to resource limitation by reducing their population growth rate (e.g. through lower birth rates or higher mortality). The response of individuals to limited resources varies depending on the organism. In many invertebrates and some vertebrates such as frogs, individuals reduce their growth rate and mature at a smaller size. In some vertebrates, territoriality spaces individuals apart so that only those with adequate resources can breed. When resources are very limited, the number of available territories will decline.

## Intraspecific Competition

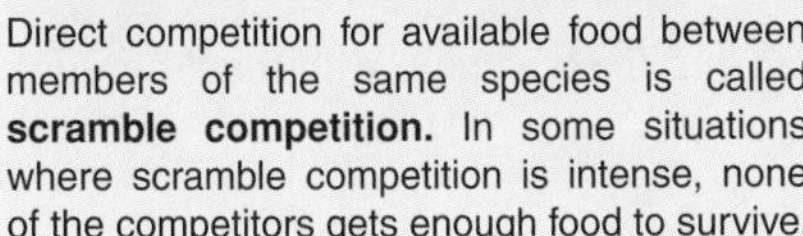

Scramble competition in caterpillars

Direct competition for available food between members of the same species is called **scramble competition.** In some situations where scramble competition is intense, none of the competitors gets enough food to survive.

Contest competition in wolves

In some cases, competition is limited by hierarchies existing within a social group. Dominant individuals receive adequate food, but individuals low in the hierarchy must **contest** the remaining resources and may miss out.

Display of a male anole

Intraspecific competition may be for mates or breeding sites, or for food. In anole lizards (above), males have a bright red throat pouch and use much of their energy displaying to compete with other males for available mates.

## Competition Between Tadpoles of *Rana tigrina*

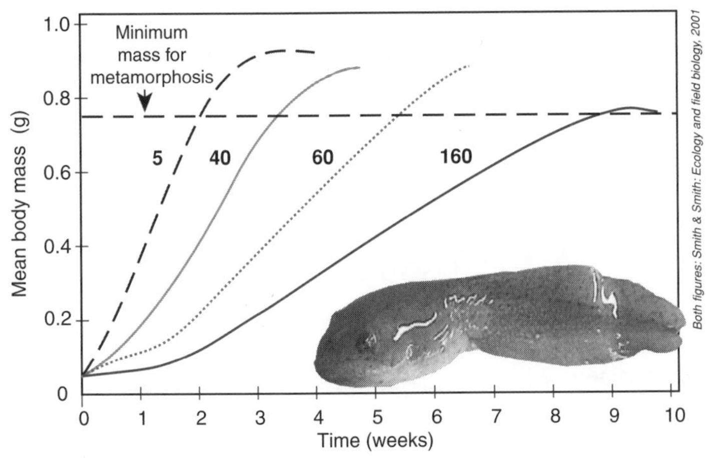


Food shortage reduces both individual growth rate and survival, and population growth. In some organisms, where there is a metamorphosis or a series of moults before adulthood (e.g. frogs, crustacean zooplankton, and butterflies), individuals may die before they mature.

The graph (left) shows how the growth rate of tadpoles (*Rana tigrina*) declines as the density increases from 5 to 160 individuals (in the same sized space).

- At high densities, tadpoles grow more slowly, take longer to reach the minimum size for metamorphosis (0.75 g), and have less chance of metamorphosing into frogs.
- Tadpoles held at lower densities grow faster to a larger size, metamorphosing at an average size of 0.889 g.
- In some species, such as frogs and butterflies, the adults and juveniles reduce the intensity of intraspecific competition by exploiting different food resources.

1. Using an example, predict the likely effects of **intraspecific competition** on each of the following:

   (a) Individual growth rate: ______________________

   (b) Population growth rate: ______________________

   (c) Final population size: ______________________

ISBN: 978-1-927173-12-1


***Related activities:*** *Population Growth Curves*
***Weblinks:*** *Intraspecific Relations*

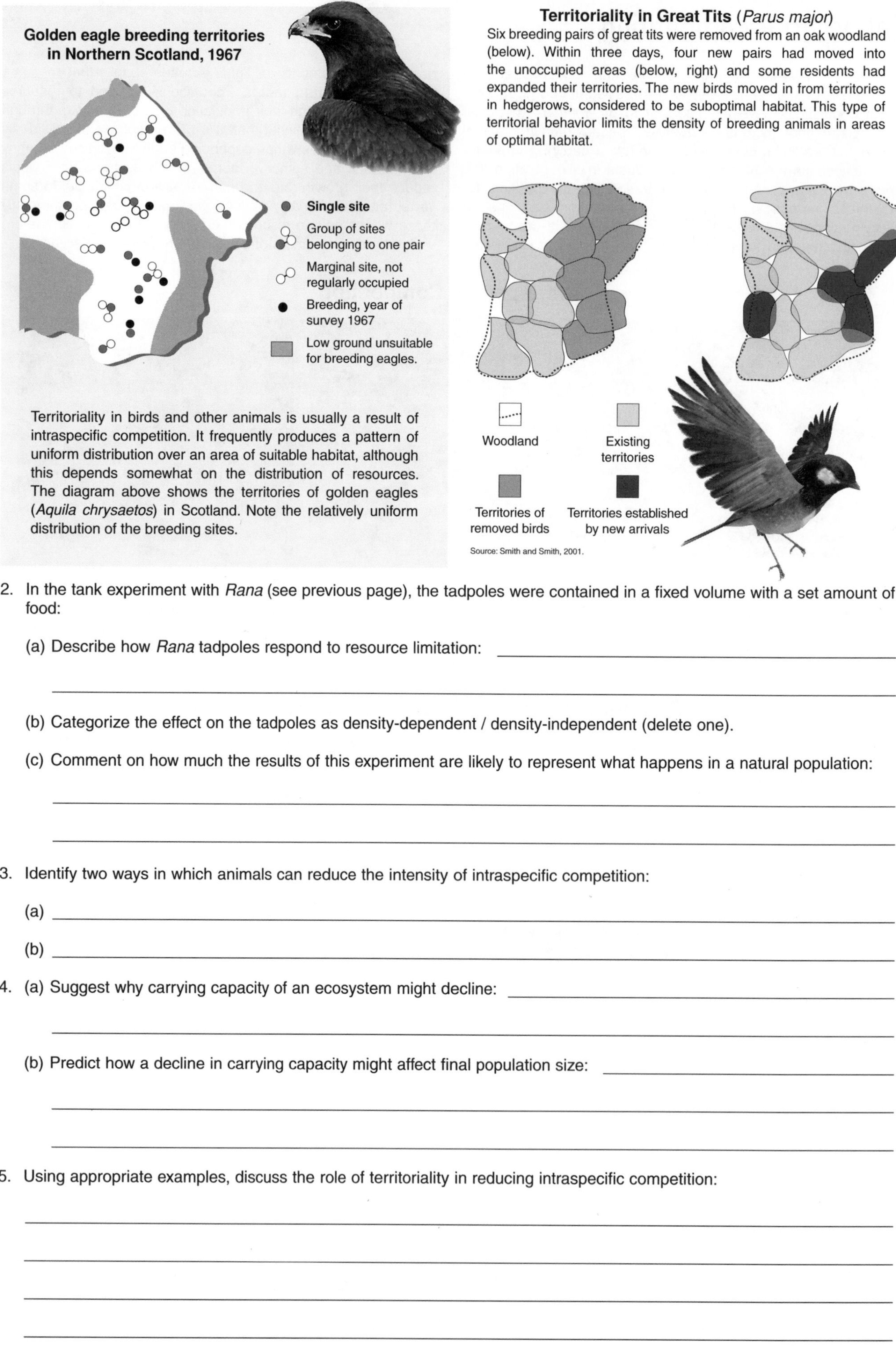
Golden eagle breeding territories in Northern Scotland, 1967
Single site
Group of sites belonging to one pair
Marginal site, not regularly occupied
Breeding, year of survey 1967
Low ground unsuitable for breeding eagles.
Territoriality in Great Tits (Parus major)
Six breeding pairs of great tits were removed from an oak woodland (below). Within three days, four new pairs had moved into the unoccupied areas (below, right) and some residents had expanded their territories. The new birds moved in from territories in hedgerows, considered to be suboptimal habitat. This type of territorial behavior limits the density of breeding animals in areas of optimal habitat.
Woodland
Existing territories
Territories of removed birds
Territories established by new arrivals
Source: Smith and Smith, 2001.

Territoriality in birds and other animals is usually a result of intraspecific competition. It frequently produces a pattern of uniform distribution over an area of suitable habitat, although this depends somewhat on the distribution of resources. The diagram above shows the territories of golden eagles (*Aquila chrysaetos*) in Scotland. Note the relatively uniform distribution of the breeding sites.

2. In the tank experiment with *Rana* (see previous page), the tadpoles were contained in a fixed volume with a set amount of food:

   (a) Describe how *Rana* tadpoles respond to resource limitation: ______________________

   ______________________

   (b) Categorize the effect on the tadpoles as density-dependent / density-independent (delete one).

   (c) Comment on how much the results of this experiment are likely to represent what happens in a natural population:

   ______________________

   ______________________

3. Identify two ways in which animals can reduce the intensity of intraspecific competition:

   (a) ______________________

   (b) ______________________

4. (a) Suggest why carrying capacity of an ecosystem might decline: ______________________

   ______________________

   (b) Predict how a decline in carrying capacity might affect final population size: ______________________

   ______________________

   ______________________

5. Using appropriate examples, discuss the role of territoriality in reducing intraspecific competition:

______________________

______________________

______________________

______________________

______________________

______________________

ISBN: 978-1-927173-12-1

# Mark and Recapture Sampling

The mark and recapture method of estimating population size is used in the study of animal populations where individuals are highly mobile. It is of no value where animals do not move or move very little. The number of animals caught in each sample must be large enough to be valid. The technique is outlined in the diagram below.

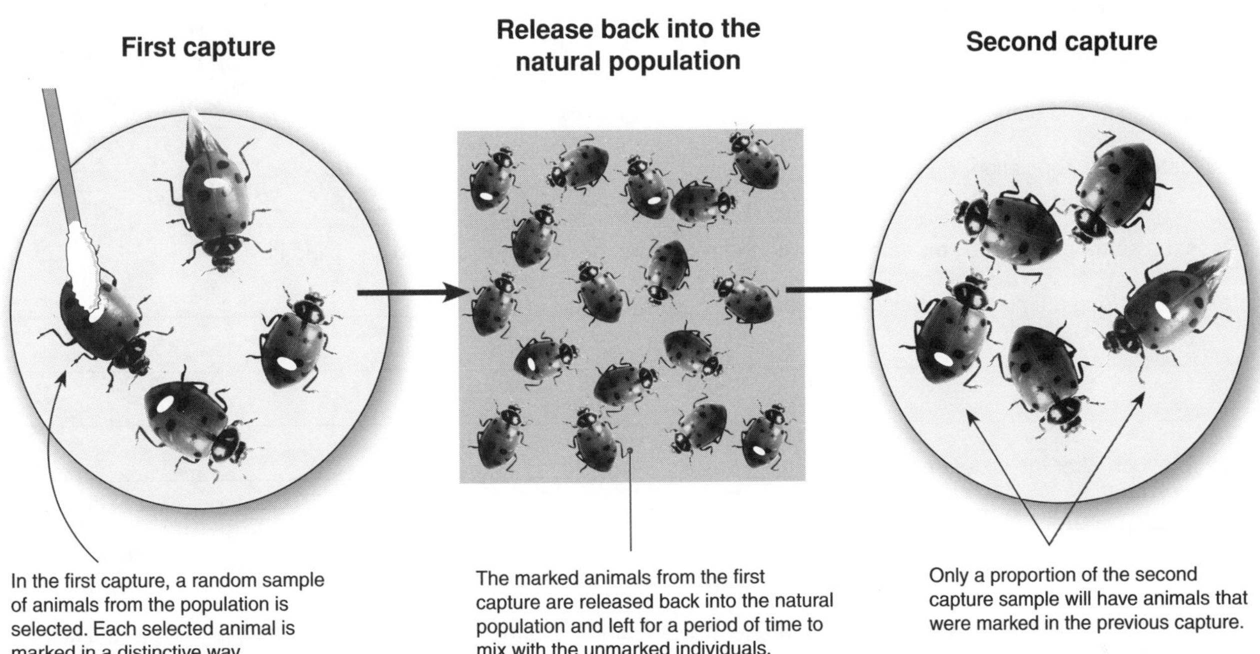


In the first capture, a random sample of animals from the population is selected. Each selected animal is marked in a distinctive way.

The marked animals from the first capture are released back into the natural population and left for a period of time to mix with the unmarked individuals.

Only a proportion of the second capture sample will have animals that were marked in the previous capture.

## The Lincoln Index

$$\text{Total population} = \frac{\text{No. of animals in 1st sample (all marked)} \times \text{Total no. of animals in 2nd sample}}{\text{Number of marked animals in the second sample (recaptured)}}$$

The mark and recapture technique comprises a number of simple steps:

1. The population is sampled by capturing as many of the individuals as possible and practical.
2. Each animal is marked in a way to distinguish it from unmarked animals (unique mark for each individual not required).
3. Return the animals to their habitat and leave them for a long enough period for complete mixing with the rest of the population to take place
4. Take another sample of the population (this does not need to be the same sample size as the first sample, but it does have to be large enough to be valid).
5. Determine the numbers of marked to unmarked animals in this second sample. Use the equation above to estimate the size of the overall population.

1. For this exercise you will need several boxes of matches and a pen. Work in a group of 2-3 students to 'sample' the population of matches in the full box by using the mark and recapture method. Each match will represent one animal.

   (a) Take out 10 matches from the box and mark them on 4 sides with a pen so that you will be able to recognize them from the other unmarked matches later.
   (b) Return the marked matches to the box and shake the box to mix the matches.
   (c) Take a sample of 20 matches from the same box and record the number of marked matches and unmarked matches.
   (d) Determine the total population size by using the equation above.
   (e) Repeat the sampling 4 more times (steps b-d above) and record your results:

| | Sample 1 | Sample 2 | Sample 3 | Sample 4 | Sample 5 |
|---|---|---|---|---|---|
| Estimated Population | | | | | |

   (f) Count the actual number of matches in the matchbox : ______________________

   (g) Compare the actual number to your estimates and state by how much it differs: ______________________

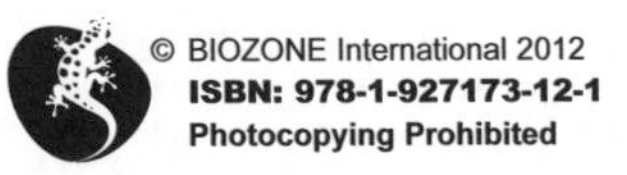


ISBN: 978-1-927173-12-1


*Periodicals*: *Classroom mark-recapture with crickets*

***Related activities***: *Sampling Communities*

2. In 1919 a researcher by the name of Dahl wanted to estimate the number of trout in a Norwegian lake. The trout were subject to fishing so it was important to know how big the population was in order to manage the fish stock. He captured and marked 109 trout in his first sample. A few days later, he caught 177 trout in his second sample, of which 57 were marked. Use the **Lincoln index** (on the previous page) to estimate the total population size:

   Size of 1st sample: ______________________

   Size of 2nd sample: ______________________

   No. marked in 2nd sample: ______________________

   Estimated total population: ______________________

3. Describe some of the problems with the mark and recapture method if the second sampling is:

   (a) Left too long a time before being repeated: ______________________

   (b) Too soon after the first sampling: ______________________

4. Describe two important assumptions being made in this method of sampling, that would cause the method to fail if they were not true:

   (a) ______________________

   (b) ______________________

5. Some types of animal would be unsuitable for this method of population estimation (i.e. the method would not work).

   (a) Name an animal for which this method of sampling would not be effective: ______________________

   (b) Explain your answer above: ______________________

6. Describe three methods for marking animals for mark and recapture sampling. Take into account the possibility of animals shedding their skin, or being difficult to get close to again:

   (a) ______________________

   (b) ______________________

   (c) ______________________

7. Scientists in the UK and Canada have, at various times since the 1950s, been involved in computerized tagging programs for Northern cod (a species once abundant in Northern Hemisphere waters but now severely depleted). Describe the type of information that could be obtained through such tagging programs:

   ______________________

**ISBN: 978-1-927173-12-1**

# Predator-Prey Interactions

Most predators have more than one prey species, although one may be preferred. Typically, when one prey species becomes scarce, predation on other species increases (prey switching), so the proportion of each prey species in the predator's diet fluctuates. Predators, especially vertebrate predators, may not always limit prey populations; prey species are more likely to be regulated by other factors such as food availability and climate. In contrast, predator populations may be greatly affected by the availability of prey, especially when there is little opportunity for prey switching.

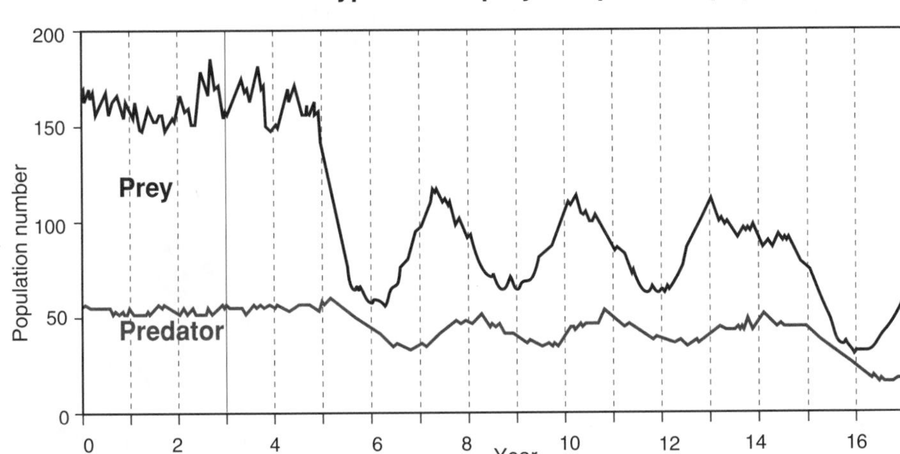

Predators capture and eat their prey, but do they control their population numbers?

### Population cycles in crown-of-thorns starfish

The crown-of-thorns starfish (left) is a voracious predator of coral on the Great Barrier Reef (right). A single starfish can destroy 5 $m^2$ of coral in a year. There have been two major starfish population explosions in recent times. One lasted from the early 1960s until the mid 1970s. The other began in the late 1970s and receded in the early 1990s. The reason for these cyclic outbreaks is not yet known. A favorite explanation is that shell collecting has led to a decline in the numbers of triton shells; one of the crown-of-thorn's few predators. However, tritons have a varied diet and feed on other organisms and they have never been common.

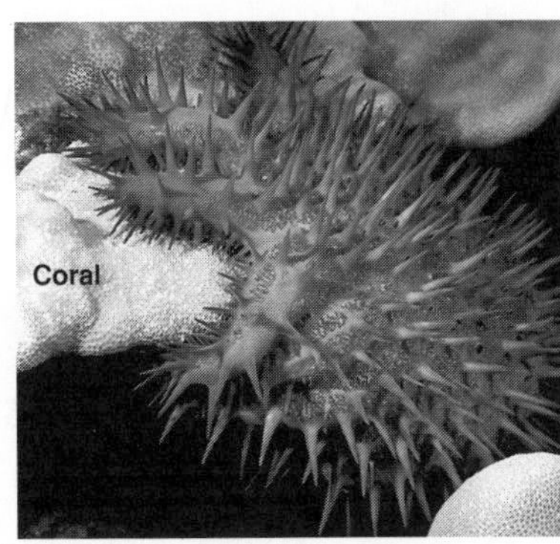


1. The graph above shows cycles in hypothetical predator and prey populations over 17 years. Answer the following:
   (a) State which population shows the greatest variation in population numbers: ______
   (b) Give the year at which a change occurred in the stability of the two population levels: ______
   (c) Describe the nature of the pattern of population numbers after the change: ______
   (d) Determine the period (duration) of the cycle exhibited by the two populations: ______
   (e) Determine how much the peaks of the two populations are out of phase (time lag between peaks): ______

2. The **crown-of-thorns starfish** on the Great Barrier Reef have undergone huge population increases in recent times.
   (a) Describe a favorite suggested cause for this population increase: ______
   (b) Why is this unlikely to be the major reason for the increase in starfish numbers? ______

3. At various times, a number of exotic species (such as wild boar and European starlings) were released into the United States' natural ecosystems. Why did many of these introduced species become uncontrollable pests?


ISBN: 978-1-927173-12-1

*Periodicals:* *Worm charmers*

***Related activities:*** *Patterns of Population Growth*
***Weblinks:*** *History of Crown-of-Thorns Starfish Outbreaks*

# Population Cycles

Some mammals, particularly in highly seasonal environments, exhibit regular cycles in their population numbers. Snowshoe hares in Canada exhibit such a cycle of population fluctuation that has a periodicity of 9–11 years. Populations of lynx in the area show a similar periodicity. Contrary to early suggestions that the lynx controlled the size of the hare population, it is now known that the fluctuations in the hare population are governed by other factors, probably the availability of palatable grasses. The fluctuations in the lynx numbers however, do appear to be the result of fluctuations in the numbers of hares (their principal food item). This is true of most **vertebrate** predator-prey systems: predators do not usually control prey populations, which tend to be regulated by other factors such as food availability and climatic factors. Most predators have more than one prey species, although one species may be preferred. Characteristically, when one prey species becomes scarce, a predator will "switch" to another available prey item. Where one prey species is the principal food item and there is limited opportunity for prey switching, fluctuations in the prey population may closely govern predator cycles.

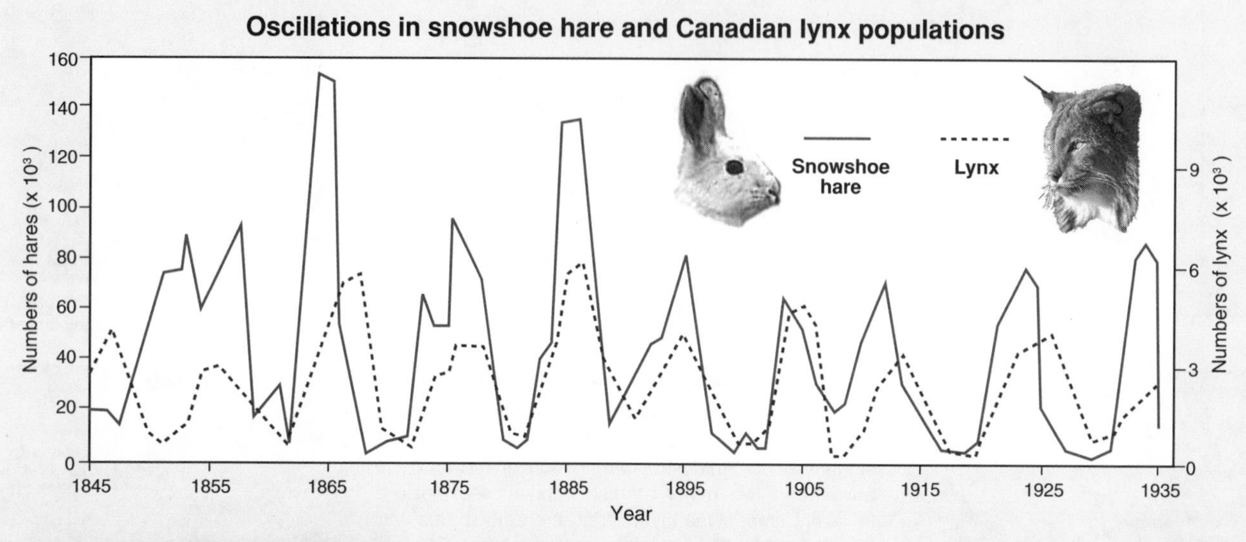

**Canadian lynx and snowshoe hare**

Regular trapping records of Canadian lynx (left) over a 90 year period revealed a cycle of population increase and decrease that was repeated every 10 years or so. The oscillations in lynx numbers closely matched those of the snowshoe hare (right), their principal prey item. There is little opportunity for prey switching in this system and the lynx are very dependent on the hares for food. Consequently, the oscillations in the two populations have a similar periodicity, with the lynx numbers lagging slightly behind those of the hare.

1. (a) From the graph above, determine the lag time between the population peaks of the hares and the lynx:

   (b) Explain why there is this time lag between the increase in the hare population and the response of the lynx:

2. Suggest why the lynx populations appear to be so dependent on the fluctuations on the hare:

3. (a) In terms of birth and death rates, explain how the availability of palatable food might regulate the numbers of hares:

   (b) Explain how a decline in available palatable food might affect their ability to withstand predation pressure:



*Related activities:* Predator-Prey Strategies, Population Growth


ISBN: 978-1-927173-12-1

# KEY TERMS: Flash Card Game

The cards below have a keyword or term printed on one side and its definition printed on the opposite side. The aim is to win as many cards as possible from the table. To play the game.....

1) Cut out the cards and lay them definition side down on the desk. You will need one set of cards between two students.
2) Taking turns, choose a card and, BEFORE you pick it up, state your own best definition of the keyword to your opponent.
3) Check the definition on the opposite side of the card. If both you and your opponent agree that your stated definition matches, then keep the card. If your definition does not match then return the card to the desk.
4) Once your turn is over, your opponent may choose a card.

| | | |
|---|---|---|
| Carrying capacity | Limiting factor | Interspecific competition |
| Demography | Density independent factor | Symbiosis |
| *r*-selection | Natality | Distribution |
| Density dependent factor | Intraspecific competition | Population |
| Survivorship curve | K-selection | Mutualism |
| Exponential growth | Mortality | Migration |


ISBN: 978-1-927173-12-1

**When you've finished the game keep these cutouts and use them as flash cards!**

| | | |
|---|---|---|
| Competitive interactions that occur between different species. | A factor affecting the maximum number a population can reach. | The maximum number of a specific organism that the environment can provide for. |
| A close and often long term interaction between different species. | A factor that affects population size but acts in the same way regardless of population density. | The statistical study of human populations. |
| The arrangement or spread of individuals in a population in an area. | The birth rate of a population, i.e. the number of live births in a population per unit time (usually expressed as per 1000 per year). | Selection favoring rapid rates of population increase especially prominent in species that colonize transient environments. |
| The total number of individuals of a species within a set habitat or area. | Competitive interactions that occur between members of the same species. | A factor that regulates the size of a population in proportion to the density of the population. |
| A biological interaction in which each party in the relationship benefits. | Selection that occurs in an environment at or near carrying capacity, favouring the production of a few, highly competitive offspring. | A curve showing the age specific mortality of a population. |
| The movement of individuals into or out of a population. | The death rate of a population, i.e. the number of deaths in a population per unit time (usually expressed as per 1000 per year). | Growth that occurs in multiples based on earlier populations. |

**ISBN: 978-1-927173-12-1**

Enduring Understanding

**2.D**
**4.B**
**4.C**

# The Diversity & Stability of **Ecosystems**

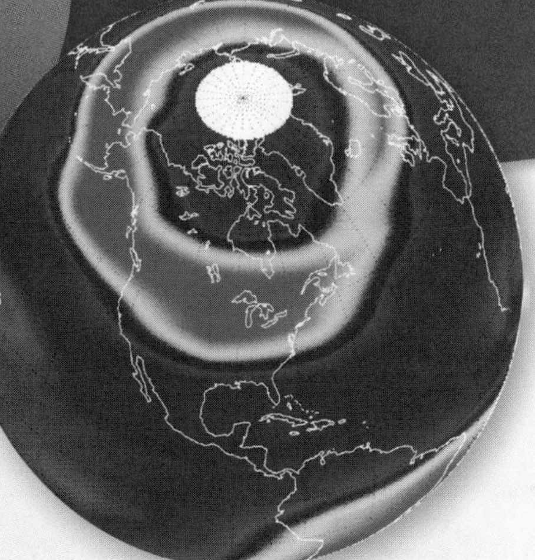

## Key concepts

- Successional changes in ecosystems can occur naturally as the result of interactions between the biotic and abiotic environments.
- Biotic interactions in diverse systems contribute to ecosystem stability and resilience.
- The ability of a population to respond to environmental change is influenced by its genetic diversity.
- Human activity can alter species distribution and abundance and accelerate ecosystem change.

## Key terms

abundance
biodiversity
climate change
conservation
distribution
ecological succession
ecosystem stability
endangered species
genetic diversity
global warming
greenhouse effect
greenhouse gas
introduced species
keystone (=key) species
pollution
resilience
species diversity
threatened species
urbanization

## Essential Knowledge

☐ 1. Use the **KEY TERMS** to compile a glossary for this topic.

### Stability and Disturbance in Ecosystems *(2.D.1: c, 4.C.4)* pages 348-354

☐ 2. Recognize **ecological succession** as a change in community structure and function over time as a result of interactions between the biotic and abiotic environment. Compare primary and secondary succession. Explain that successional changes rarely follow a standard progression (*cross ref. with 4.B.4:b*).

☐ 3. Explain how biotic interactions contribute to **ecosystem stability** and discuss the relationship between stability and biodiversity. Using examples, describe and explain the role of **keystone species** in ecosystem structure and function. Explain the consequences of loss of keystone species (*cross ref. with 4.B.3:c*).

☐ 4. Recognize **resilience** as a component of ecosystem stability and discuss its significance, especially with respect to human induced or disturbance.

### Biodiversity *(4.C.3)* pages 355-356, 370-372

☐ 5. Recognize the components of biodiversity, including **species diversity** and **genetic diversity**. Explain how a population's ability to respond to environmental change is influenced by its genetic diversity. Using examples, describe and explain the consequences of low genetic diversity to population survival and viability. Comment on the implications of this to **conservation** efforts.

☐ 6. Use examples to illustrate how genetic diversity within a population provides resilience because of the differential responses of individuals to change. Examples include resistance to disease and the behavioral responses of herds.

☐ 7. Describe factors influencing the ability of populations to survive environmental change. Recognize that species unable to adapt face **extinction**. Identify attributes in **threatened species** that make their extinction more likely.

☐ 8. Recall that the genetic diversity of populations, as measured by allele frequencies, can be modeled by the Hardy-Weinberg equation.

### Human Impact on Populations *(4.B.3-4.B.4)* pages 355-371

☐ 9. Using examples, describe and explain the impact of human activity on the **distribution** and **abundance** of populations. Examples could include shifts in species distribution in response to climate change, the loss of keystone species, or the impact of **introduced species** (e.g. kudzu in the Southeastern US).

☐ 10. Using examples, explain how human activity can accelerate the pace of change in ecosystems, usually with detrimental effects. Examples could include tropical deforestation (slash-and-burn agriculture and logging), the impact of **urbanization** (including the development of infrastructure such as dams and roading), **global warming** and **climate change**, and intensive agriculture.

☐ 11. Using examples, describe the impact of introduced species on ecosystems (e.g. fire ants in the USA). Explain why introduced species (including introduced diseases) usually have a detrimental effect on the established community.

***Periodicals:***
*Listings for this chapter are on page 376*

***Weblinks:***
*www.thebiozone.com/weblink/AP2-3121.html*

*BIOZONE APP: Student Review Series Pollution & Global Change*

# The Modern Atlantis?

Computer models of accelerated global warming show that the mean sea level may rise by between 100 and 900 mm over the next century. This is mainly due to the thermal expansion of the oceans as they increase in temperature, but also includes the melting of large ice sheets. This rise in sea level will have a significant effect on low lying islands and coral atolls as many are presently only a few metres above sea level. The following news article focuses on the island nation of Kiribati in the Pacific.

## Vanishing Lands

*The newspaper, and author name of the following article are fictitious, but the text is based on real events and information.*

The Tribune
By Michael Anton: Saturday 7 June 2008

After many years of unanswered appeals for action on climate change, the tiny South Pacific nation of Kiribati has concluded that it is doomed. On Thursday its President, Anote Tong, used World Environment Day to request international help to evacuate his country before it disappears.

Kiribati consists of just 33 coral atolls scattered across 5 million square kilometers of the Pacific Ocean and it has limited scope for coping with impending global climate changes with most of the land being barely 2 metres above sea level.

Speaking from New Zealand, Mr Tong said his fellow countrymen, i-Kiribati, as they are known, had no alternative but to leave. "We may be beyond redemption," he said. "We may be at the point of no return, where the emissions in the atmosphere will carry on contributing to climate change, to produce a sea level change. So in time our small, low-lying islands will be submerged."

A London economics graduate, Mr Tong said the emigration of his people needed to start immediately. "We don't want to believe this, and our people don't want to believe this. It gives us a deep sense of frustration. What do we do?"

Kiribati is home to 97,000 people most of them living on the main atoll of Tarawa, a ring of islets surrounding a central lagoon. It is regarded as one of the places most vulnerable to climate change along with Vanauta, Tuvalu and the Marshall Islands.

Currently, the most serious problem Kiribati faces is erosion caused by flooding and storms. "We have to find the next highest spot," said Mr Tong. "At the moment there's only the coconut trees." But even the coconut trees are dying, caused by drought and a rising level of salt in ground water, which is also not being replaced due to the fact there has little rain at all for the last three years.

Mr Tong was in New Zealand – which, after committing itself to becoming carbon neutral, was chosen to host the UN's World Environment Day – for talks with Prime Minister, Helen Clark, whom he hopes to persuade to help resettle his people. But he also appealed to other countries for help relocating the i-Kiribati.

However, New Zealand, already with a large population of Pacific Islanders, would have immense trouble absorbing the 97,000 immigrants which would strain its generosity to the limit and total almost 2.5% of its current national population.

And that is just Kiribati. Talks have not yet begun with many of the other island nations which soon may also be submerged. In 2006, the Australian government issued a warning of a flood of environmental refugees across the Asian-Pacific region.

President Tong said he had heard many national leaders argue that measures to combat climate change could negatively affect their economic development, but pointed out that for the i-Kiribati it was not a matter of economics, it was a matter of survival. He said that while international scientists argue about the causes of climate change, the effects were already beginning to show on his nation. "I am not a scientist, but what I know is that things are happening we did not experience in the past... Every second week, when we get the high tides, there are always reports of erosion."

Villages, after occupying the same site for centuries, are having to be relocated due to the encroaching water. "We're doing it now... it's that urgent," he said. "Where they have been living over the past few decades is no longer there. It is being eroded."

Worse case scenarios suggest the i-Kiribati could be uninhabitable within 60 years, Mr Tong said. "I've appealed to the international community that we need to address this challenge. It's a challenge for the whole global community."

Leading industrialized nations last month pledged to cut their carbon emissions in half by 2050. But they stopped short of setting firm targets for 2020, which many scientists argue is crucial if the planet is to be saved. But for Kiribati, its saviour may have come too late.

1. From the information provided in the article, explain why Kiribati is vulnerable to the effects of global warming:

2. The article states that global warming is already adversely affecting the island nation of Kiribati. What are the physical effects it refers to, and what will be their impact on the inhabitants of Kiribati in the short term?

3. The inhabitants of Kiribati may be forced to relocate to other countries. What could this mean for the identity of the i-Kiribati and their culture and language?

4. Using other research tools, such as the library or internet, find and read other articles relating to sea level change in the Pacific. Do they all agree that there will be a rise in sea level if carbon emissions are not reduced? Explain why you think some articles take different viewpoints:




**ISBN: 978-1-927173-12-1**

# Primary Succession

Ecological succession is the process by which communities change over time. Succession occurs as a result of the interactions between biotic and abiotic factors. Early communities modify the physical environment. This in turn alters the biotic community, which further alters the physical environment and so on. Each successive community makes the environment more favorable for the establishment of new species. An idealized succession proceeds in stages to a **climax community**, which is generally stable until further disturbance. Early successional communities have low species diversity and a simple structure. Climax communities are more complex, with a greater number of species interactions and higher species diversity.

Gregg M. Erickson CC 3.0

**Primary succession** is the biological colonization and development of a region with no preexisting vegetation or soil. A primary succession may take hundreds of years, but the time scale depends strongly on factors such as distance from vegetated areas and availability of pollinating species such as birds and insects.

Examples include the emergence of new volcanic islands, new coral atolls, or islands where the previous community has been destroyed by a volcanic eruption.

Many texts show a sequence of colonization beginning with lichens, mosses, and liverworts, then progressing to ferns, grasses, shrubs, and finally a **climax community** of mature forest. In reality, this sequence is very rare. The sequence below shows the recolonization of Mount St Helens (left) after it erupted in 1980.

Pumice fields → Lichen → Nitrogen fixing plant: e.g. *Lupinus lepidus* → Grasses, shrubs, small saplings

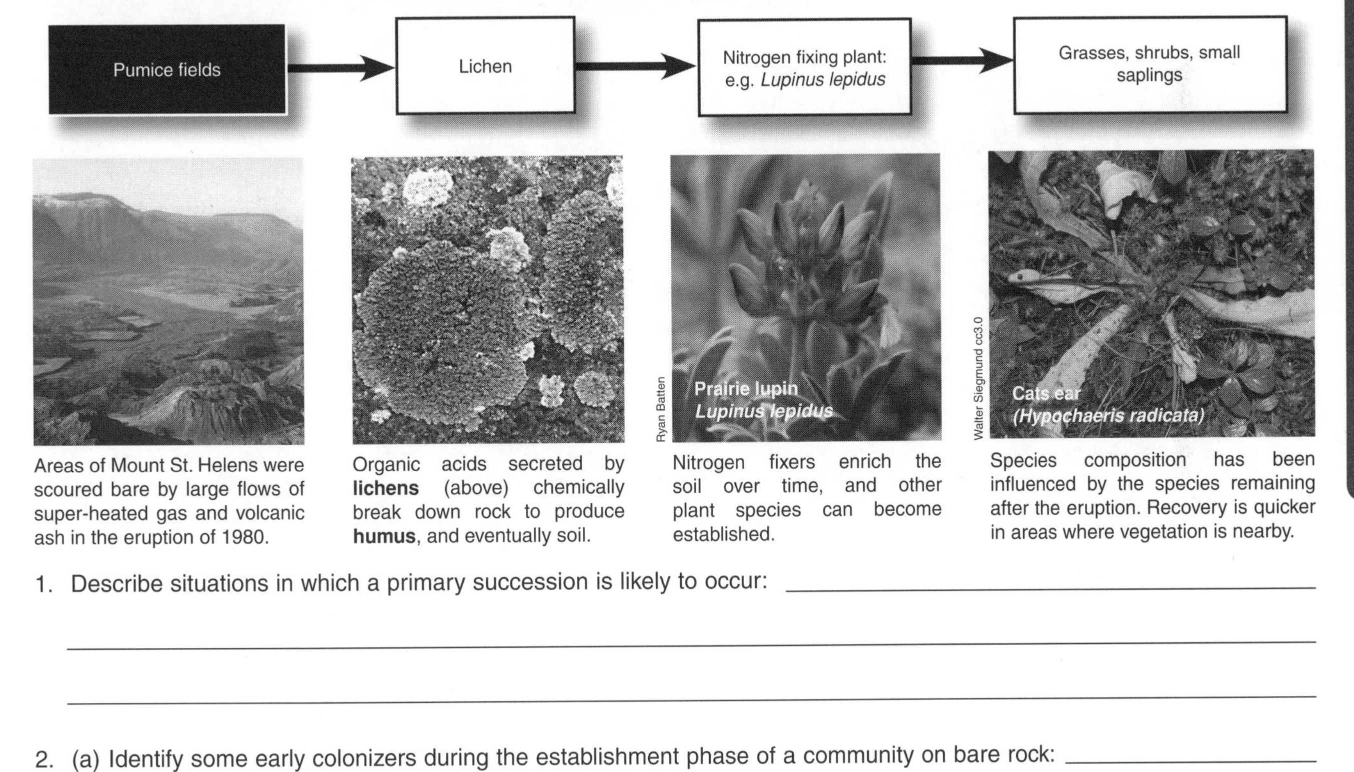

Areas of Mount St. Helens were scoured bare by large flows of super-heated gas and volcanic ash in the eruption of 1980.

Organic acids secreted by **lichens** (above) chemically break down rock to produce **humus**, and eventually soil.

Ryan Batten

Prairie lupin *Lupinus lepidus*

Nitrogen fixers enrich the soil over time, and other plant species can become established.

Walter Siegmund cc3.0

Cats ear *(Hypochaeris radicata)*

Species composition has been influenced by the species remaining after the eruption. Recovery is quicker in areas where vegetation is nearby.

1. Describe situations in which a primary succession is likely to occur: ____________________

2. (a) Identify some early colonizers during the establishment phase of a community on bare rock: ____________________

   (b) Describe two important roles of the species that are early colonizers of bare slopes: ____________________

3. Explain why climax communities are more stable and resistant to disturbance than early successional communities:


ISBN: 978-1-927173-12-1


*Periodicals:* Mountain transformed

*Related activities:* Succession on Surtsey Island *Weblinks:* Mount St Helens, Forest Succession Animation, Primary Succession

# Succession on Surtsey Island

Surtsey Island is a volcanic island lying 33 km off the southern coast of Iceland. The island was formed over four years from 1963 to 1967 when a submarine volcano 130 m below the ocean surface built up an island that initially reached 174 m above sea level and covered 2.7 km$^2$. Erosion has since reduced the island to around 150 m above sea level and 1.4 km$^2$. As an entirely new island, Surtsey was able to provide researchers with an ideal environment to study primary succession in detail. The colonization of the island by plants and animals has been recorded since the island's formation. The first vascular plant there (sea rocket) was discovered in 1965, two years before the eruptions on the island ended. Since then, 69 plant species have colonized the island and there are a number of established seabird colonies.

Sea rocket

*H. peploides*

Plant colonization of Surtsey can be divided into four stages. The first stage (1965-1974) was dominated by shore plants colonizing the northern shores. The most successful of these was *Honckenya peploides*, which established on tephra sand and gravel flats. It set seed in 1971 and subsequently spread across the island. Carbon and nitrogen levels in the soil were recorded as being very low during this time. This initial colonization by shore plants was followed by a lag phase (1975-1984). Shore plants continued to establish but there were few new colonizers. This slowed the rate of succession.

*P. annua*

*S. phylicifolia*

A number of new plant species arrived after a gull colony became established at the southern end of the island, (1985-1994). Populations of plants within or near the colony expanded rapidly to about 3 ha, while populations outside the colony remained low but stable. Grasses such as *Poa annua* formed extensive patches of vegetation. After this rapid increase in plant species, the arrival of new colonizers again slowed (1995-2008). A second wave of colonizers began to establish following this slower phase and soil organic matter increased markedly. The first bushy plants established in 1998, with the arrival of willow *Salix phylicifolia*. The area of vegetation cover near the gull colony expanded to about 10 ha.

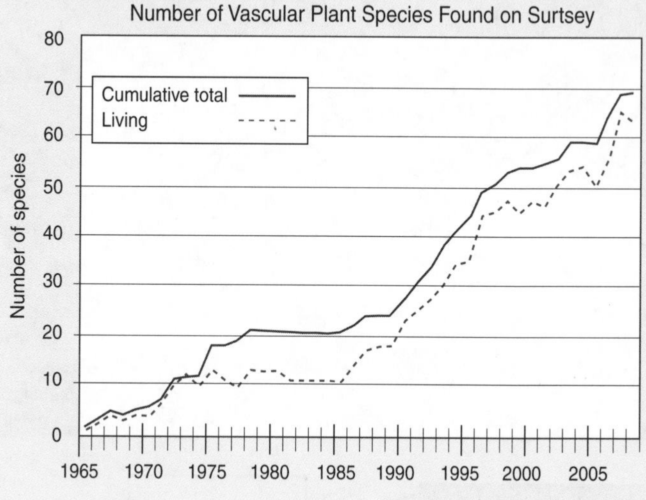


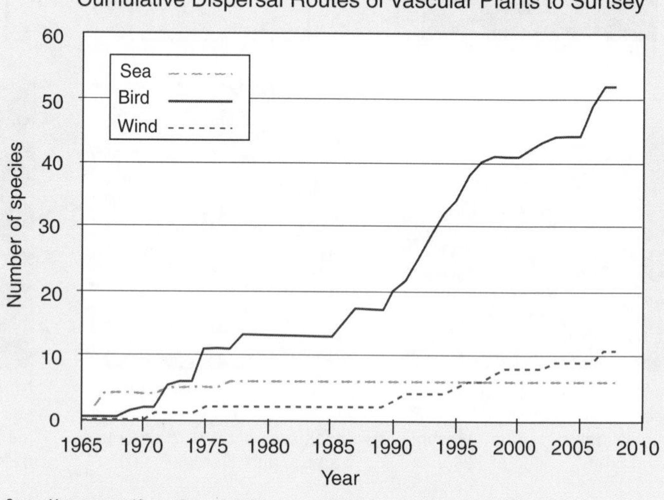


1. Explain why Surtsey provided ideal conditions for studying primary succession: ______

2. Explain why the first colonizing plants established in the north of the island, while later colonizers established in the south.

3. Use the graphs to identify the following:

   (a) The year the gull colony established: ______

   (b) The most common method for new plant species to arrive on the island: ______

   (c) The year of the arrival of the second wave of plant colonizers. Suggest a reason for this second wave of colonizers:




**ISBN: 978-1-927173-12-1**

# Secondary Succession

**Secondary succession** occurs when land is cleared of vegetation (e.g. after a fire or landslide). These events do not involve the loss of soil or seed, and root stocks are often undamaged. As a result, secondary succession tends to be more rapid than primary succession, although the time scale depends on the species involved, soil composition, and climate. Secondary succession may occur over a wide area (as after a forest fire), or in smaller areas where single trees have fallen or abandoned farmland has been left to regenerate. Tree falls result in a type of secondary succession called **gap regeneration** (below).

Redwoods (*Sequoia* and *Sequoiadendron*) are very long-lived species (500-1000 years on average). They are a keystone species in the ecosystem, supporting many different species, some of which are endangered.

DUK cc3.0

The light open spaces in the canopy allow smaller understory plants to become established as a brushy growth. Examples include tan oak, huckleberry (left), blackberry, and rhododendrons.

BH

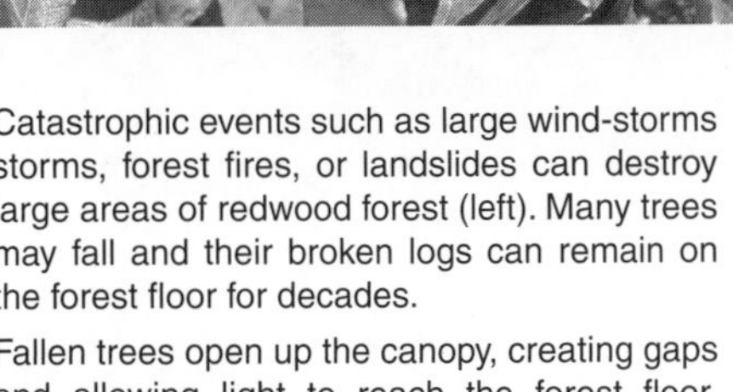

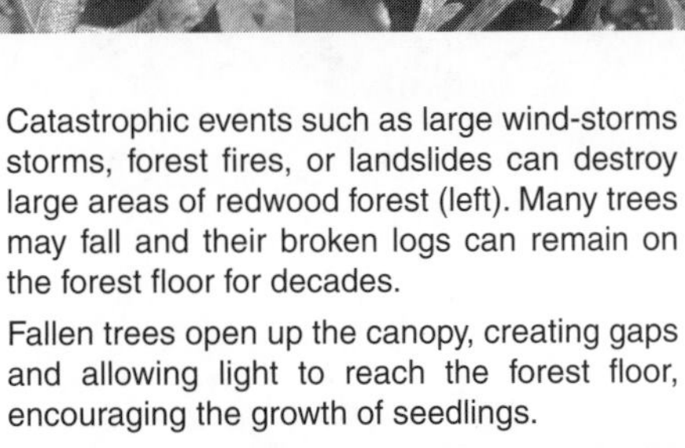

Catastrophic events such as large wind-storms storms, forest fires, or landslides can destroy large areas of redwood forest (left). Many trees may fall and their broken logs can remain on the forest floor for decades.

Fallen trees open up the canopy, creating gaps and allowing light to reach the forest floor, encouraging the growth of seedlings.

Fallen trees play an important role in the ecosystem diversity of a forest. They support a variety of invertebrates, bacteria, and fungi.

Victorgrigas cc3.0

Redwood seedlings begin to grow through the understory. As they grow taller, they begin to out-compete the smaller plants. While the saplings grow up, the crowns of the remaining canopy trees will close some of the gap. This may take decades.

1. Describe how **secondary succession** differs from primary succession: ______________________________

2. Explain why secondary succession usually takes place more rapidly than primary succession: ______________________________

3. Using an example, explain how the outcome of a secondary succession may depend on an interplay of unpredictable biotic and abiotic factors:

______________________________


ISBN: 978-1-927173-12-1

# Wetland Succession

Wetland areas present a special case of ecological succession. Wetlands are constantly changing as plant invasion of open water leads to siltation and infilling. This process is accelerated by **eutrophication**. In well drained areas, pasture or **heath** may develop as a result of succession from freshwater to dry land. When the soil conditions remain non-acid and poorly drained, a swamp will eventually develop into a seasonally dry **fen**. In special circumstances (see below) an acid **peat bog** may develop. The domes of peat that develop produce a hummocky landscape with a unique biota. Wetland peat ecosystems may take more than 5000 years to form but are easily destroyed by excavation and lowering of the water table.

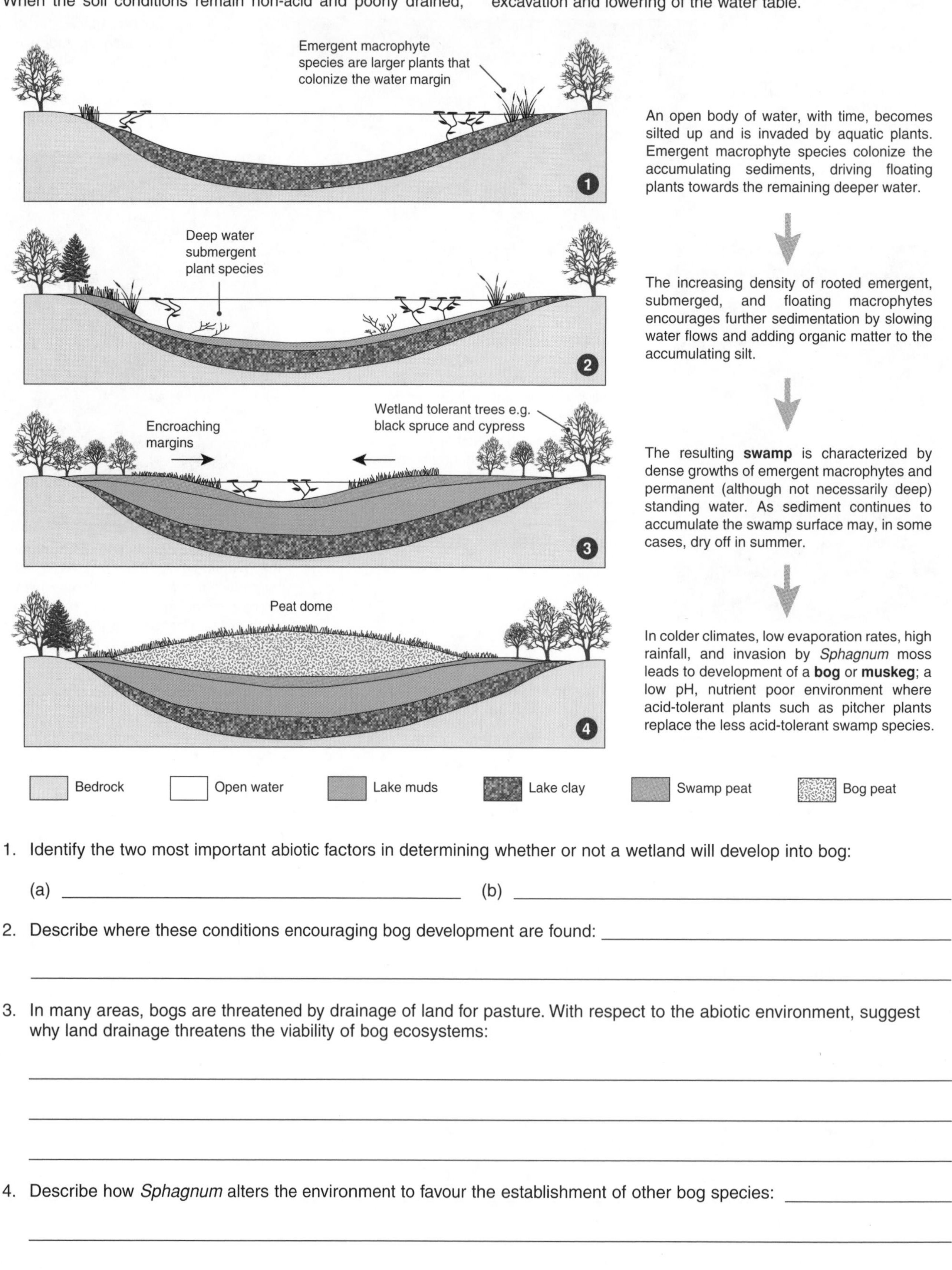


1. Identify the two most important abiotic factors in determining whether or not a wetland will develop into bog:

   (a) ______________________ (b) ______________________

2. Describe where these conditions encouraging bog development are found: ______________________

3. In many areas, bogs are threatened by drainage of land for pasture. With respect to the abiotic environment, suggest why land drainage threatens the viability of bog ecosystems:

4. Describe how *Sphagnum* alters the environment to favour the establishment of other bog species: ______________________



***Related activities:*** *Secondary Succession*
***Weblinks:*** *Wetland Succession, A Hydrosere*


**ISBN: 978-1-927173-12-1**

# Disturbance and Community Structure

Ecological theory suggests that all species in an ecosystem contribute in some way to ecosystem function. Therefore, species loss past a certain point is likely to have a detrimental effect on the functioning of the ecosystem and on its ability to resist change over time (its **stability**). Although many species still await discovery, we do know that the rate of species extinction is increasing. Scientists estimate that human destruction of natural habitats is implicated in the extinction of up to 100,000 species every year. This substantial loss of biodiversity has serious implications for the long term stability of many ecosystems.

## The Concept of Ecosystem Stability

Ecosystem stability has various components, including **inertia** (the ability to resist disturbance) and **resilience** (ability to recover from external disturbances). Ecosystem stability is closely linked to the biodiversity of the system, although it is difficult to predict which factors will stress an ecosystem beyond its range of tolerance. It was once thought that the most stable ecosystems were those with the most species, because they had the greatest number of biotic interactions operating to buffer them against change. This assumption is supported by experimental evidence but there is uncertainty over what level of biodiversity provides an insurance against catastrophe.

Monoculture

Natural grassland

©Digital Image 2000. T. Worden

Rainforest

Deforestation

Single species crops (monocultures), such as the soy bean crop (above, left), represent low diversity systems that can be vulnerable to disease, pests, and disturbance. In contrast, natural grasslands (above, right) may appear homogeneous, but contain many species which vary in their predominance seasonally. Although they may be easily disturbed (e.g. by burning) they are very resilient and usually recover quickly.

Tropical rainforests (above, left) represent the highest diversity systems on Earth. Whilst these ecosystems are generally resistant to disturbance, once degraded, (above, right) they have little ability to recover. The biodiversity of ecosystems at low latitudes is generally higher than that at high latitudes, where climates are harsher, niches are broader, and systems may be dependent on a small number of key species.

### Community Response to Environmental Change

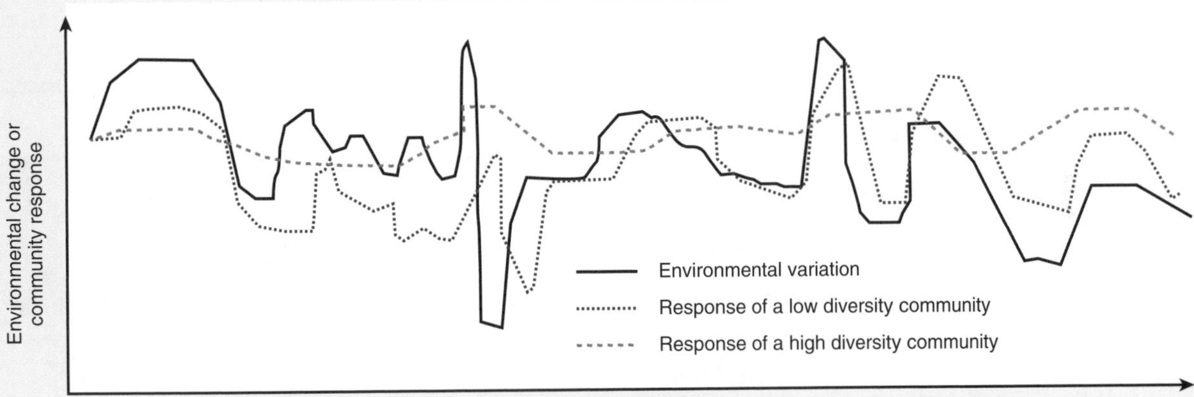


In models of ecosystem function, higher species diversity increases the stability of ecosystem functions such as productivity and nutrient cycling. In the graph above, note how the low diversity system varies more consistently with the environmental variation, whereas the high diversity system is buffered against major fluctuations. In any one ecosystem, some species may be more influential than others in the stability of the system. Such **keystone (key) species** have a disproportionate effect on ecosystem function due to their pivotal role in some ecosystem function such as nutrient recycling or production of plant biomass.

Elephants can change the entire vegetation structure of areas into which they migrate. Their pattern of grazing on taller plant species promotes a predominance of lower growing grasses with small leaves.

Termites are amongst the few larger soil organisms able to break down plant cellulose. They shift large quantities of soil and plant matter and have a profound effect on the rates of nutrient processing in tropical environments.

The starfish *Pisaster* is found along the coasts of North America where it feeds on mussels. If it is removed, the mussels dominate, crowding out most algae and leading to a decrease in the number of herbivore species.




ISBN: 978-1-927173-12-1


**Related activities:** *Threats to Biodiversity*

## Keystone Species in North America

Gray wolf

Beaver, *Castor canadensis*

Sea otter, *Enhydra lutris*

Quaking aspen

**Gray** or **timber wolves** (*Canis lupus*) are a keystone predator and were once widespread in North American ecosystems. Historically, wolves were eliminated from Yellowstone National Park because of their perceived threat to humans and livestock. As a result, elk populations increased to the point that they adversely affected other flora and fauna. Wolves have since been reintroduced to the park and balance is returning to the ecosystem.

Two smaller mammals are also important keystone species in North America. **Beavers** (top) play a crucial role in biodiversity and many species, including 43% of North America's endangered species, depend partly or entirely on beaver ponds. **Sea otters** are also critical to ecosystem function. When their numbers were decimated by the fur trade, sea urchin populations exploded and the kelp forests, on which many species depend, were destroyed.

**Quaking aspen** (*Populus tremuloides*) is one of the most widely distributed tree species in North America, and aspen communities are among the most biologically diverse in the region, with a rich understorey flora supporting an abundance of wildlife. Moose, elk, deer, black bear, and snowshoe hare browse its bark, and aspen groves support up to 34 species of birds, including ruffed grouse, which depends heavily on aspen for its winter survival.

1. Suggest one probable reason why high biodiversity promotes greater ecosystem stability: ________________

2. Explain why **keystone species** are so important to ecosystem function: ________________

3. For each of the following species, discuss features of their biology that contribute to their position as keystone species:

   (a) Sea otter: ________________

   (b) Beaver: ________________

   (c) Gray wolf: ________________

   (d) Quaking aspen: ________________

4. Giving examples, explain how the actions of humans to remove a keystone species might result in ecosystem change:


**ISBN: 978-1-927173-12-1**

# Threats to Biodiversity

The simplest definition of biodiversity is as the sum of all biotic variation from the level of genes to ecosystems, but often the components of total biodiversity are distinguished. Background (natural) extinction rates for all organisms (including bacteria and fungi) are estimated to be 10-100 species a year. The actual extinction rate is estimated to be 100-1000 times higher, mainly due to the effects of human activity. Over 41,000 species are now on the International Union for Conservation's (IUCN) red list, and 16,000 are threatened with extinction. Loss of biodiversity reduces the stability and resilience of natural ecosystems and decreases the ability of their communities to adapt to changing environmental conditions. Humans rely heavily on the biodiversity in nature and a loss of species richness has a deleterious effect on us all.

Insects make up 80% of all known animal species. There are an estimated 6-10 million insect species on Earth, but only 900,000 have been identified. Some 44,000 species may have become extinct over the last 600 years. The Duke of Burgundy butterfly (*Hamearis lucina*), right, is an endangered British species.

Just over 5% of the 8225 reptile species are at risk. These include the two tuatara species (right) from New Zealand, which are the only living members of the order Sphenodontia, and the critically endangered blue iguana. Only about 200 blue iguanas remain, all in the Grand Caymans.

| | Total number of species* | Number of IUCN listed species |
|---|---|---|
| Plants | 310,000 - 422,000 | 8474 |
| Insects | 6 -10 million | 622 |
| Fish | 28,000 | 126 |
| Amphibians | 5743 | 1809 |
| Reptiles | 8225 | 423 |
| Birds | 10,000 | 1133 |
| Mammals | 5400 | 1027 |

* Estimated numbers

The giant panda (above), is one of many critically endangered terrestrial mammals, with fewer than 2000 surviving in the wild. Amongst the 120 species of marine mammals, approximately 25% (including the humpback whale and Hector's dolphin) are on the ICUN's red list.

Prior to the impact of human activity on the environment, one bird species became extinct every 100 years. Today, the rate is one every year, and may increase to 10 species every year by the end of the century. Some at risk birds, such as the Hawaiian crow (right), are now found only in captivity.

Current estimates suggest as many as 47% of plant species may be endangered. Some, such as the South African cycad *Encephalartos woodii* (above), is one of the rarest plants in the world. It is extinct in the wild and all remaining specimens are clones.

## Threats to Biodiversity

Rainforests in some of the most species-rich regions of the world are being destroyed at an alarming rate as world demand for tropical hardwoods increases and land is cleared for the establishment of agriculture.

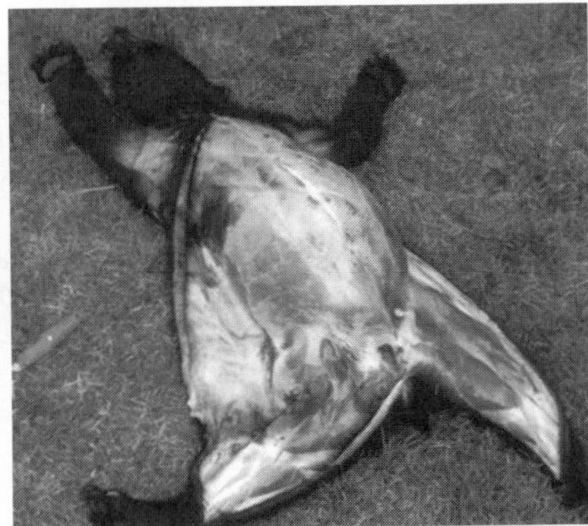

Illegal trade in species (for food, body parts, or for the exotic pet trade) is pushing some species to the brink of extinction. Despite international bans on trade, illegal trade in primates, parrots, reptiles, and big cats (among others) continues.

Pollution and the pressure of human populations on natural habitats threatens biodiversity in many regions. Environmental pollutants may accumulate through food chains or cause harm directly, as with this bird trapped in oil.

1. Discuss, in general terms, the effects of loss of biodiversity on an ecosystem: ______




ISBN: 978-1-927173-12-1


*Periodicals:* Earth's nine lives

***Related activities:*** *Tropical Deforestation*
***Weblinks:*** *Biodiversity Hotspots*

## Global Biodiversity

Spider monkeys

White-headed duck

Burmese river turtle

Hawaiian silversword

Arrow poison frogs

Giant anteater

Lemurs

Orangutan

Numbat

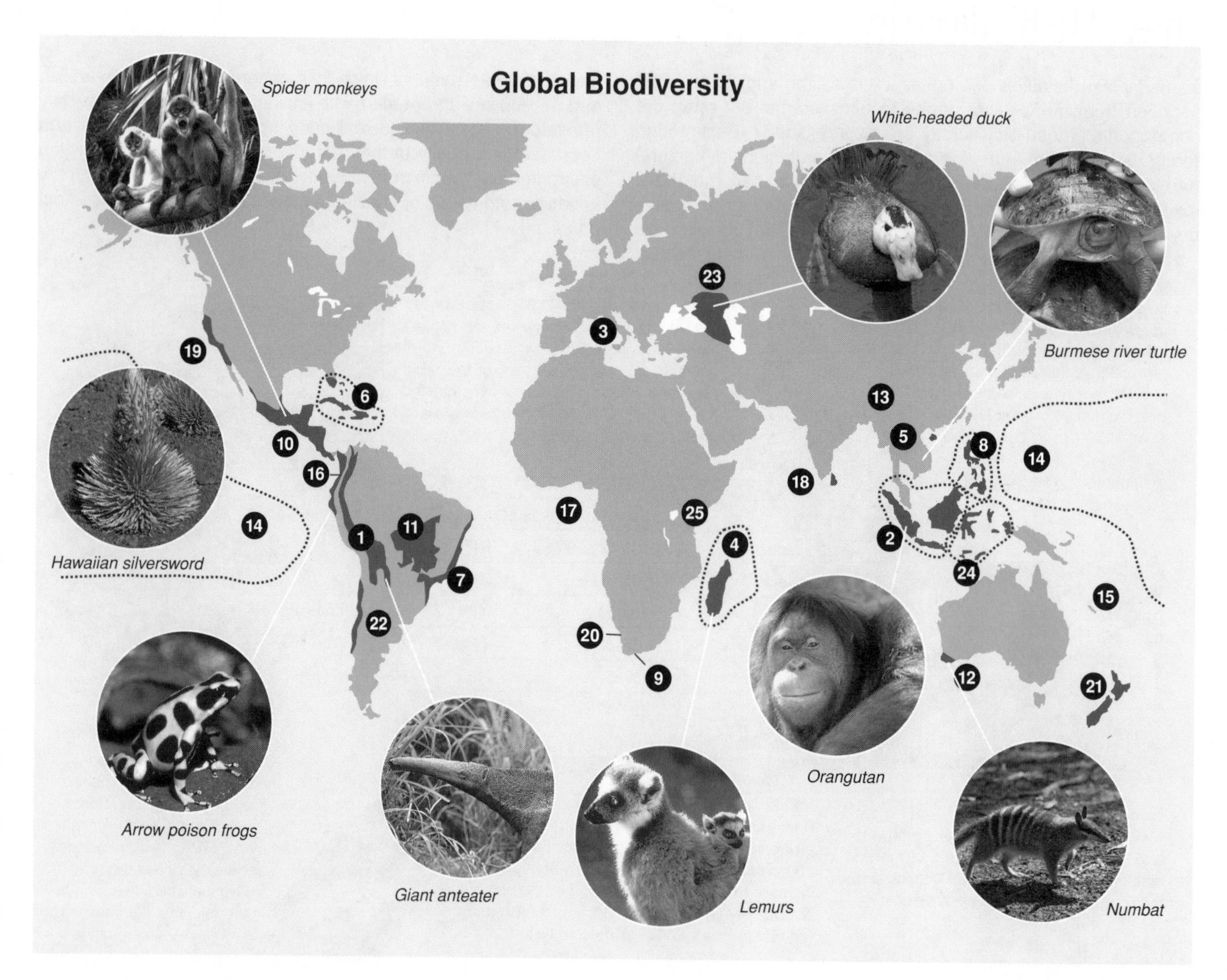

Biodiversity is not distributed evenly on Earth, being consistently richer in the tropics and concentrated more in some areas than in others. The simplest definition of biodiversity is as the sum of all biotic variation from the level of genes to ecosystems, but often the components of total biodiversity are distinguished. **Species diversity** describes the number of different species in an area (**species richness**), **genetic diversity** is the diversity of genes within a species, and **ecosystem diversity** refers to the diversity at the higher ecosystem level of organization. **Habitat diversity** is also sometimes described and is essentially a subset of ecosystem diversity expressed per given unit area. Total biological diversity is often threatened because of the loss of just one of these components. More than a third of the planet's known terrestrial plant and animal species are found in these 25 regions, which cover only 1.4% of the Earth's land area. Unfortunately, biodiversity hotspots often occur near areas of dense human habitation and rapid human population growth. Most are located in the tropics and most are forests. Loss of biodiversity reduces the stability and resilience of natural ecosystems and decreases the ability of their communities to adapt to changing environmental conditions. With increasing pressure on natural areas from urbanization, roading, and other human encroachment, maintaining species diversity is paramount and should concern us all today.

2. Distinguish between species diversity and genetic diversity and explain the importance of both of these to our definition of total biological diversity:

3. Identify the threat to biodiversity that you perceive to be the most important and explain your choice:

4. Use your research tools (including the *Weblinks* identified on the previous page) to identify each of the 25 biodiversity hotspots illustrated in the diagram above. For each region, summarize the characteristics that have resulted in it being identified as a biodiversity hotspot. Present your summary as a short report and attach it to this page of your workbook.

**ISBN: 978-1-927173-12-1**

# Global Warming

The Earth's atmosphere comprises a mix of gases including nitrogen, oxygen, and water vapor. Small quantities of carbon dioxide ($CO_2$), methane, and a number of other trace gases are also present. The term 'greenhouse effect' describes the natural process by which heat is retained within the atmosphere by these greenhouse gases, which act as a thermal blanket around the Earth, letting in sunlight, but trapping the heat that would normally radiate back into space. The greenhouse effect results in the Earth having a mean surface temperature of about 15°C, 33°C warmer than it would have without an atmosphere. About 75% of the natural greenhouse effect is due to water vapor. The next most significant agent is $CO_2$. Fluctuations in the Earth's surface temperature as a result of climate shifts are normal, and the current period of warming climate is partly explained by the recovery after the most recent ice age that finished 10,000 years ago. However since the mid 20th century, the Earth's surface temperature has been increasing. This phenomenon is called global warming and the majority of researchers attribute it to the increase in atmospheric levels of $CO_2$ and other greenhouse gases emitted into the atmosphere as a result of human activity (i.e. it is anthropogenic). Nine of the ten warmest years on record were in the 2000s (1998 being the third warmest on record). Global surface temperatures in 2005 set a new record but are now tied with 2010 as being the hottest years on record.

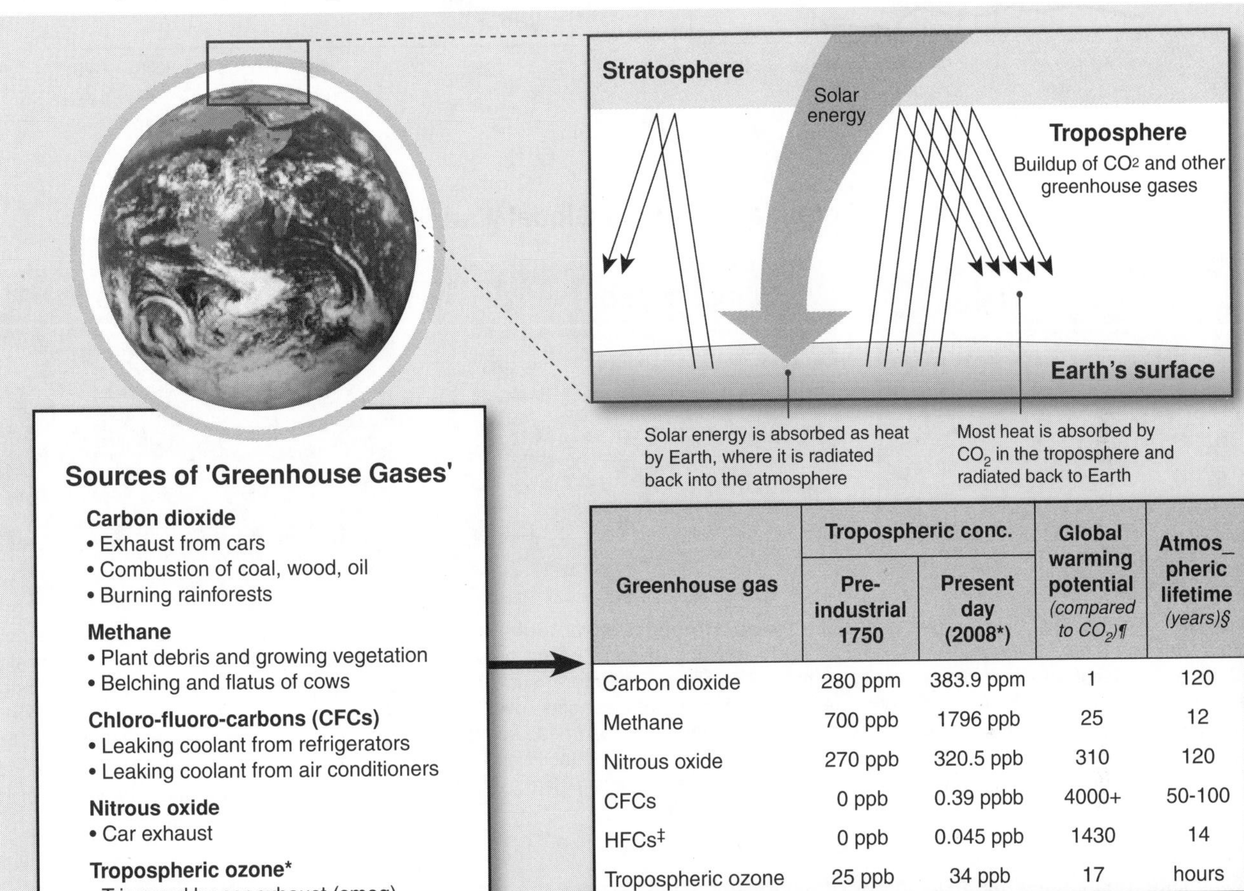


## Sources of 'Greenhouse Gases'

**Carbon dioxide**
- Exhaust from cars
- Combustion of coal, wood, oil
- Burning rainforests

**Methane**
- Plant debris and growing vegetation
- Belching and flatus of cows

**Chloro-fluoro-carbons (CFCs)**
- Leaking coolant from refrigerators
- Leaking coolant from air conditioners

**Nitrous oxide**
- Car exhaust

**Tropospheric ozone***
- Triggered by car exhaust (smog)

*Tropospheric ozone is found in the lower atmosphere (not to be confused with ozone in the stratosphere)

| Greenhouse gas | Tropospheric conc. | | Global warming potential *(compared to $CO_2$)¶* | Atmos_pheric lifetime *(years)§* |
|---|---|---|---|---|
| | Pre-industrial 1750 | Present day (2008*) | | |
| Carbon dioxide | 280 ppm | 383.9 ppm | 1 | 120 |
| Methane | 700 ppb | 1796 ppb | 25 | 12 |
| Nitrous oxide | 270 ppb | 320.5 ppb | 310 | 120 |
| CFCs | 0 ppb | 0.39 ppbb | 4000+ | 50-100 |
| HFCs‡ | 0 ppb | 0.045 ppb | 1430 | 14 |
| Tropospheric ozone | 25 ppb | 34 ppb | 17 | hours |

**ppm** = parts per million; **ppb** = parts per billion; ‡Hydrofluorcarbons were introduced in the last decade to replace CFCs as refrigerants; * Data from July 2007-June 2008. ¶ Figures contrast the radiative effect of different greenhouse gases relative to $CO_2$ over 100 years, e.g. over 100 years, methane is 25 times more potent as a greenhouse gas than $CO_2$ § How long the gas persists in the atmosphere. *Source: $CO_2$ Information Analysis Centre, Oak Ridge National Laboratory, USA.*

This graph shows how the mean temperature for each year from 1860-2010 (bars) compares with the average temperature between 1961 and 1990. The blue line represents the fitted curve and shows the general trend indicated by the annual data. Most anomalies since 1977 have been above normal; warmer than the long term mean, indicating that global temperatures are tracking upwards. The decade 2001-2010 has been the warmest on record.

*Source: Hadley Center for Prediction and Research*

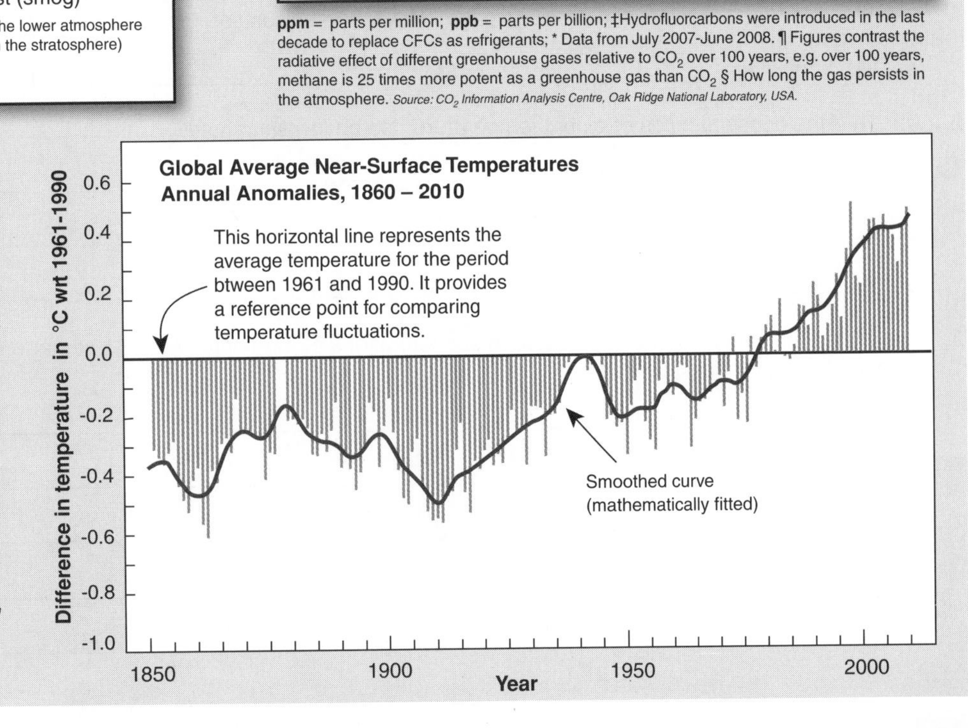

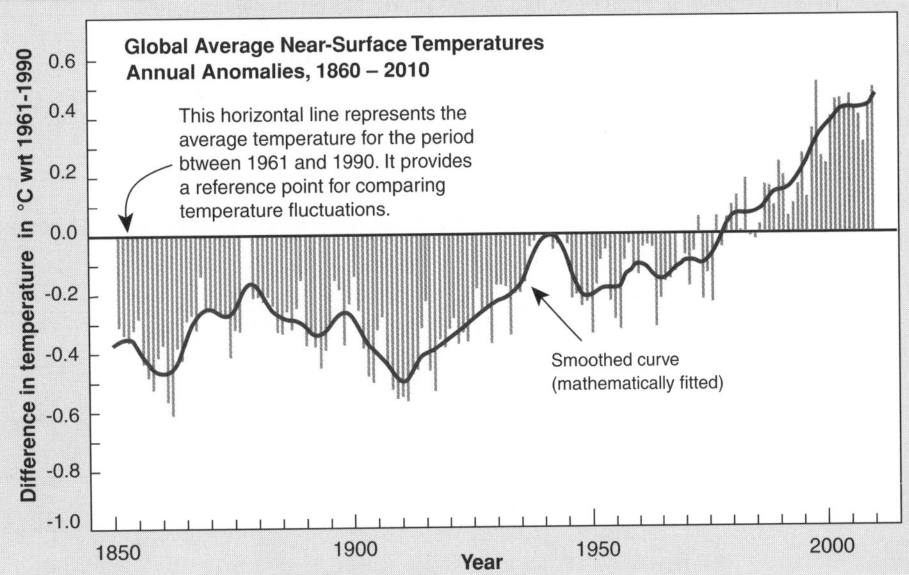

## Changes in Atmospheric $CO_2$

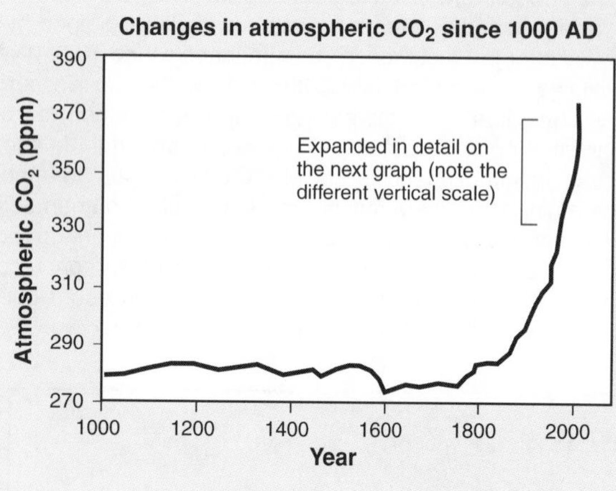


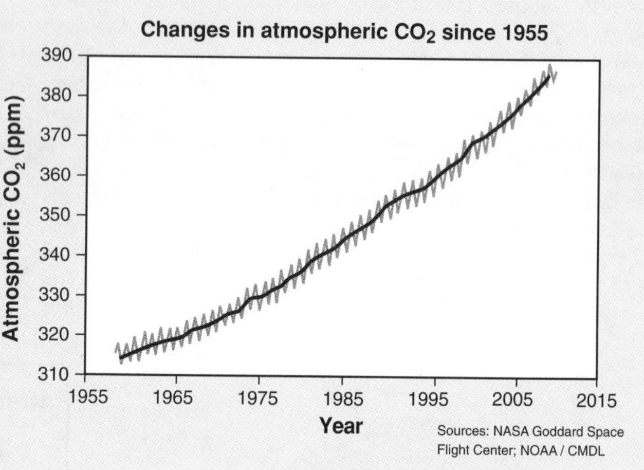


## Potential Effects of Global Warming

**Sea levels** are expected to rise by 30-50 cm by the year 2100. This is the result of the thermal expansion of ocean water and melting of glaciers and ice shelves. Many of North America's largest cities are near the coast. The predicted rises in sea levels could result in inundation of these cities and entry of salt water into agricultural lands.

**The ice-albedo effect** refers to the ability of ice to reflect sunlight. Cooling tends to increase ice cover, so more sunlight is reflected. Warming reduces ice cover, and more solar energy is absorbed, so more warming occurs. Ice has a stabilizing effect on global climate, reflecting nearly all the sun's energy that hits it.

Global warming may cause regional changes in **weather patterns**, affecting the intensity and frequency of storms. High intensity hurricanes now occur more frequently, driven by higher ocean surface temperatures. The devastating effects of disasters, such as hurricane Katrina, illustrate the vulnerability of low lying cities to sea level rises.

1. Calculate the increase (as a %) in the 'greenhouse gases' between the pre-industrial era and the 2008 measurements (use the data from the table, see previous page). **HINT**: The calculation for carbon dioxide is: $(383.9 - 280) \div 280 \times 100 =$

   (a) Carbon dioxide: ______________ (b) Methane: ______________ (c) Nitrous oxide: ______________

2. What are the consequences of global temperature rise on low lying land? ______________

3. Explain the relationship between the rise in concentrations of atmospheric $CO_2$, methane, and oxides of nitrogen, and the enhanced greenhouse effect:


**ISBN: 978-1-927173-12-1**

# Models of Climate Change

There is much debate over how much global warming is due to anthropogenic (human) causes and what actions need to be taken to remedy it. Many of the points of debate are raised by parties with stakes in the ultimate outcome of the global warming issue. For some a rise in global temperatures is seen as desirable, producing longer growing seasons or an opportunity to grow new economically valuable crops. For others the effects of global warming could be disastrous. A country may find much of its coastline inundated, or climate shifts may result in the loss of established crops or the introduction of new pests. In addition, there is debate over responsibility for global warming and who should pay to fix the problem. Should wealthy industrialized nations pay because they began the industrial revolution, or should developing countries pay because they produce $CO_2$ at a much higher rate that other countries and consume large amounts of energy?

## Climate Modeling

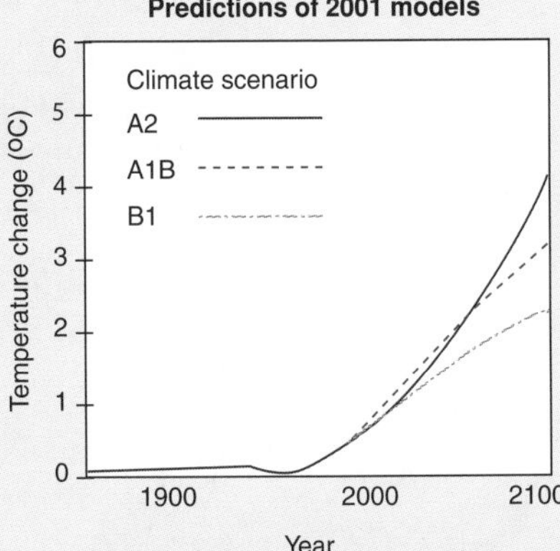


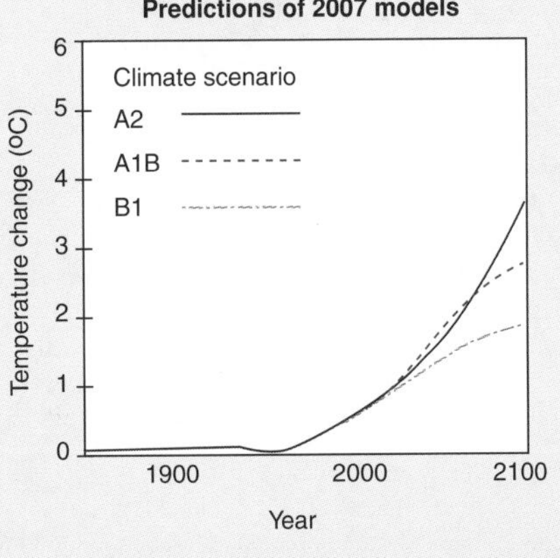


**Factors affecting climate models:**

- Complex natural systems.
- A lack of long term accurate data .
- Uncertainty about the way human and natural activity influence climate.
- Climate models often need to be simplified in order to get them to work.

Data: IPCC assessment reports 2001/2007

Computer modeling can be a valuable tool for predicting climate change. However, predicting the outcomes can be difficult because of the complexity and number of factors involved (top right). The result is often complex and sometimes models produce conflicting data. Scientists often produce a series of climate models based upon different scenarios and update them when new information is available. The new and old models can then be compared (above).



## Confusing the Debate

### Media coverage

Global warming is a complex issue, and most people obtain their information from the popular media. However, some media sources provide unbalanced information which may dismiss or ignore one side of the debate. In order to make an informed decision, people must read or listen to a wide range of media, or read scientific documents and make their own decision. Many people have become overwhelmed and disengaged with the issue, believing it receives too much media coverage.

### Lobby Groups

Lobby groups with specific interests are constantly trying to influence policy makers on a range of topics, including global warming. For example, reducing $CO_2$ emissions by restricting coal and oil use will help reduce global warming. However, fossil fuel consumption generates billions of dollars of revenue for the coal and oil companies, so they will lobby against legislation that penalizes fossil fuel consumption. If successful, lobbying could result in less effective climate change policies.

### Controversy

All scientific bodies of international standing agree that human activity has contributed disproportionately to global warming. However, there are still some in the political, scientific, and commercial community who claim that global warming is not occurring. Often these people provide highly visible, highly emotive arguments which attract media attention. In addition, a highly skeptical public believe that many scientists falsify research data to support their own theories on global warming.

1. Explain why climate modellers can never be certain of their predictions: ________________________________

ISBN: 978-1-927173-12-1


*Periodicals:*
*Behind the predictions*

***Related activities:*** *Global Warming*

# Ice Sheet Melting

The surface temperature of the Earth is in part regulated by the amount of ice on its surface, which reflects a large amount of heat into space. However, the area and thickness of the polar sea-ice is rapidly decreasing. From 1980 to 2008 the Arctic summer sea-ice minimum almost halved, decreasing by more than 3 million square kilometers. This melting of sea-ice can trigger a cycle where less heat is reflected into space during summer, warming seawater and reducing the area and thickness of ice forming in the winter. At the current rate of reduction, it is estimated that there may be no summer sea-ice left in the Arctic by 2050.

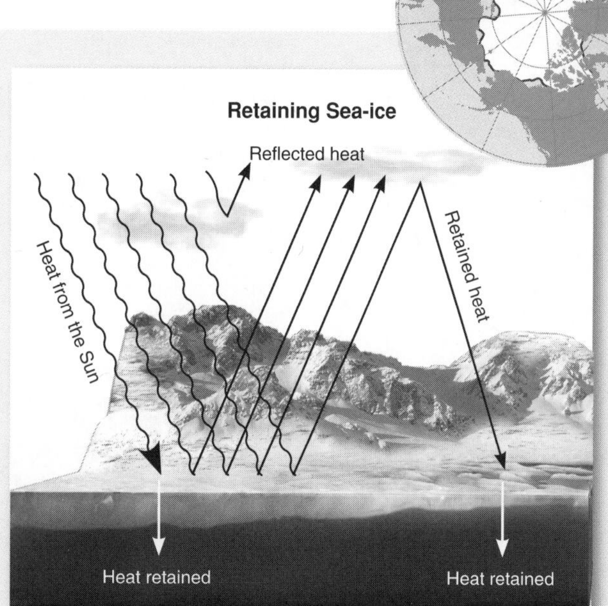


The **albedo** (reflectivity of sea-ice) helps to maintain its presence. Thin sea-ice has a lower albedo than thick sea-ice. More heat is reflected when sea-ice is thick and covers a greater area. This helps to regulate the temperature of the sea, keeping it cool.

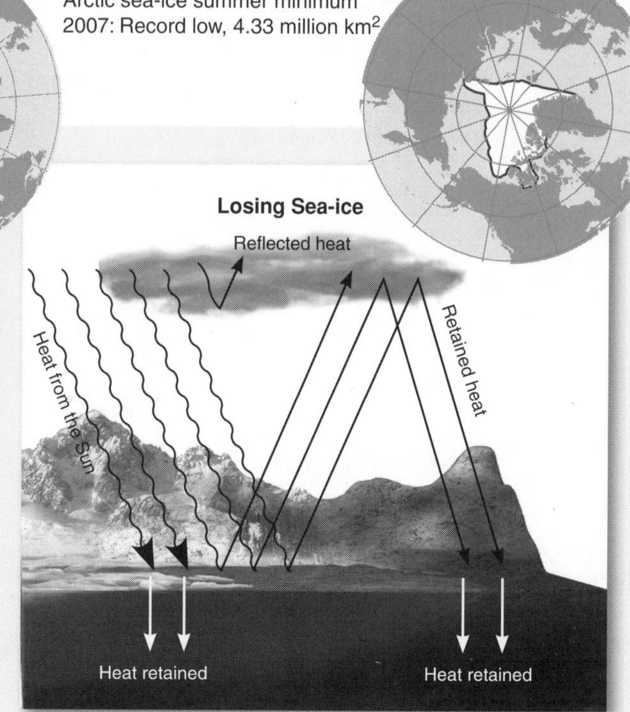


As sea-ice retreats, more non-reflective surface is exposed. Heat is retained instead of being reflected, warming both the air and water and causing sea-ice to form later in the autumn than usual. Thinner and less reflective ice forms and perpetuates the cycle.

The temperature in the Arctic has been above average every year since 1988. Coupled with the reduction in summer sea-ice, this is having dire effects on Arctic wildlife such as polar bears, which hunt out on the ice. The reduction in sea-ice reduces their hunting range and forces them to swim longer distances to firm ice. Studies have already shown an increase in drowning deaths of polar bears.

**Average* Arctic Air Temperature Fluctuations**

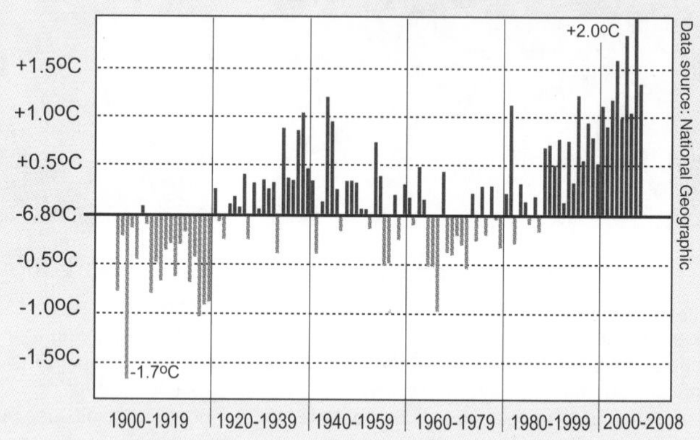


*Figure shows deviation from the average annual surface air temperature over land. Average calculated on the years 1961-2000.

1. Explain how low sea-ice albedo and volume affects the next year's sea-ice cover:

2. Discuss the effects of decreasing summer sea-ice on polar wildlife:



Related activities: Global Warming

Periodicals: Greenland poised on a knife edge


ISBN: 978-1-927173-12-1

# Global Warming and Agriculture

The impacts of climate change on agriculture and horticulture in North America will vary because of the size and range of its geography. In some regions, temperature changes will increase the growing season for existing crops, or enable a wider variety of crops to be grown. Changes in temperature or precipitation patterns may benefit some crops, but have negative effects on others. Increasing atmospheric $CO_2$ levels will enhance the growth of some crops (e.g. wheat, rice, and soybeans).

## Effects of increases in temperature on crop yields

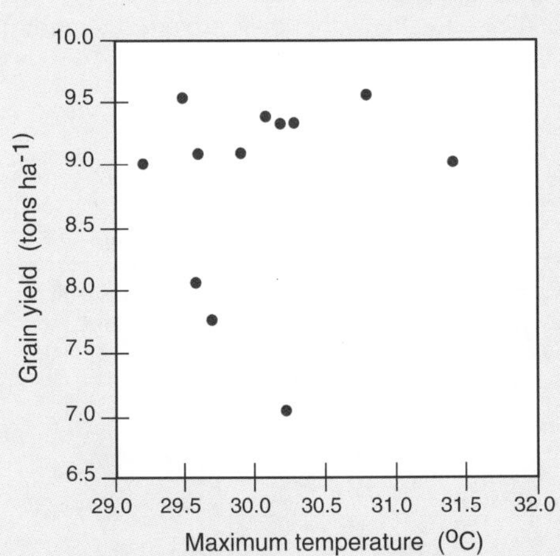


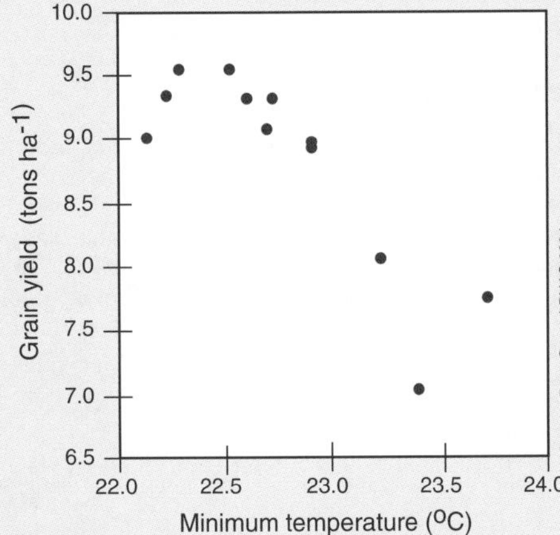


Studies on the grain production of rice have shown that maximum daytime temperatures have little effect on crop yield. However minimum night time temperatures lower crop yield by as much as 5% for every 0.5°C increase in temperature.

## Possible effects of increases in temperature on crop damage

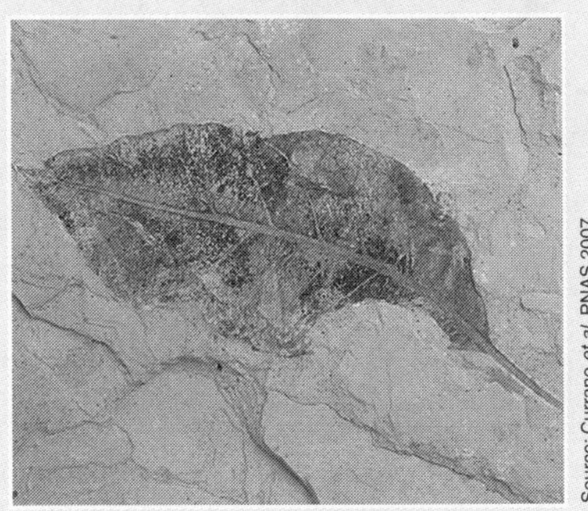
Source: Currano et.al. PNAS 2007

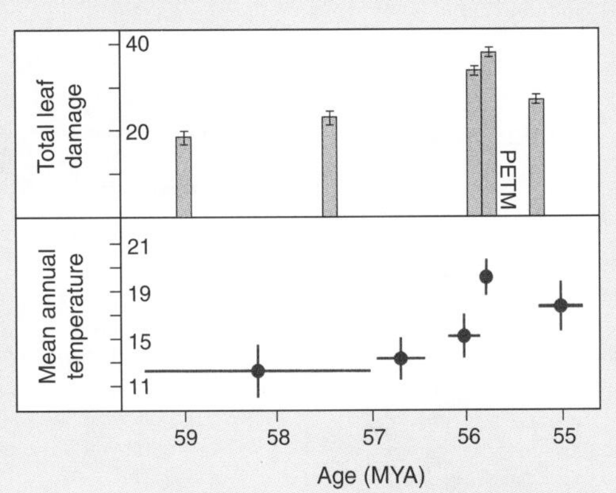


The fossil record shows that global temperatures rose sharply around 56 million years ago. Studies of fossil leaves with insect browse damage indicate that leaf damage peaked at the same time as the Paleocene Eocene Thermal Maximum (PETM). This gives some historical evidence that as temperatures increase, plant damage caused by insects also rises. This could have implications for agricultural crops.

Ellen Levy Finch cc3.0

Citrus production will shift slightly north with reduced yields in Texas and Florida.

Grain crops (such as wheat, above)are at higher risk of crop failures if precipitation decreases.

Brocken Inaglory cc3.0

Crops grown near to their climate threshold may suffer reductions in yield, quality, or both.

USDA

Milder winters and longer growing seasons may see the distribution of agricultural pest species spread.

1. Why will global warming benefit some agricultural crops, while disadvantaging others?

2. Explain how global warming can influence the distribution of pest species, and in turn affect agriculture:


ISBN: 978-1-927173-12-1

# Temperature and Enzyme Activity

If global warming increases temperature by 1°C to 3°C, the metabolic rates and enzyme activity of many organisms may be affected. All enzymes, including those involved in metabolic processes, have a limited range of conditions over which they can function (e.g. pH, temperature range). Each enzyme has a set of optimal conditions in which they work most effectively. Outside these conditions, enzyme activity rapidly declines. As global temperatures rise, poikilothermic organisms (those whose body temperature varies with the environmental temperature) may find that higher environmental temperatures prevent their enzymes working optimally. Mobile organisms may be forced to migrate to more suitable climates, or they may be restricted to narrower ranges. Poikilothermic organisms with limited or no mobility could become extinct if they are unable to adapt to the rising temperature. Homeotherms will have fewer problems as internal body temperature is regulated but they still may have to shift to other locations to stay within their preferred temperature zone.

## Enzymes and Temperature

Temperature affects the rate of enzyme activity. All enzymes have a temperature range they function within. At temperatures below this range, enzyme activity is very slow. As the temperature increases, so too does the enzyme activity, until the temperature is high enough to damage the enzyme's functional structure. At this point, the enzyme is **denatured** and ceases to function.

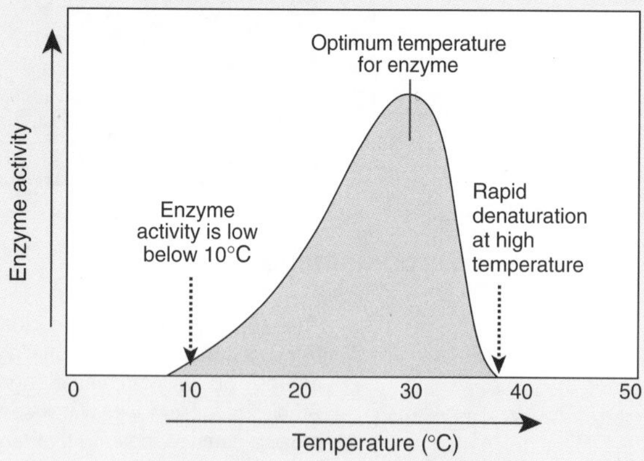


## Coral Reef Bleaching

An increase in sea temperatures could mean the death of coral reefs. Healthy coral reefs depend upon the symbiotic relationship between a coral polyp that builds the reef, and photosynthetic protozoa called zooxanthellae. Zooxanthellae live within the polyp tissues and provides it with most of its energy. A 1-2°C temperature increase is sufficient to disrupt the photosynthetic enzymes. The zooxanthellae either dies, or is expelled from the coral due to stress. The result is coral bleaching

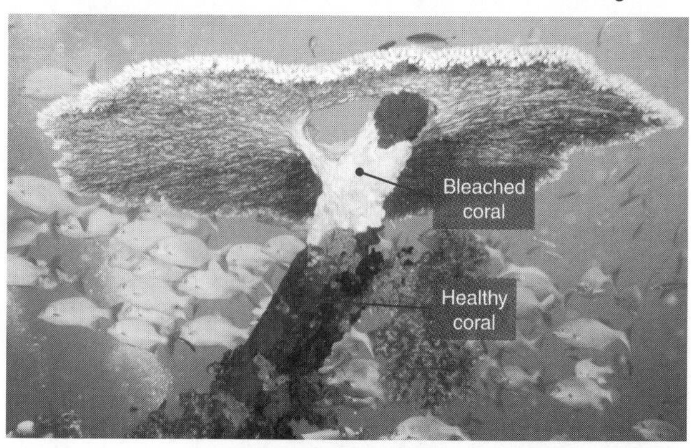


## Enzyme Activity in Rice

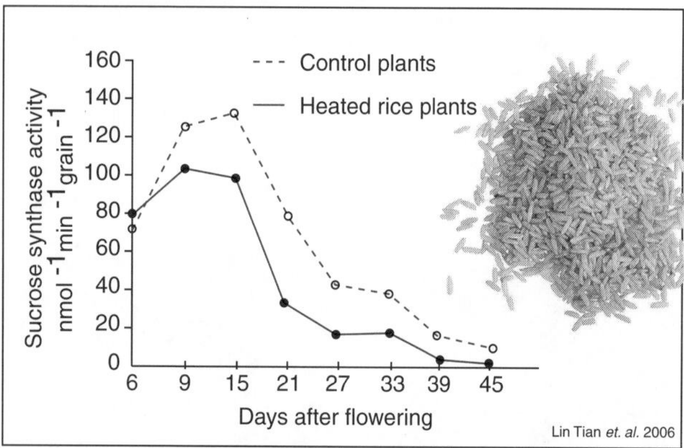


Rice contains starch and it is one of the world's most important food plants. To make starch, sugars (including sucrose) are first hydrolyzed into monomers and then the components are linked together. The degradation of sucrose by the enzyme sucrose synthase is the first step in the synthesis of starch. Recent studies (above) have shown that a temperature increase of 0.3 - 2°C reduces sucrose synthase activity. This would inhibit or slow the formation of starch within a rice grain and change its nutritional value.

## Photosynthesis

The photosynthetic enzyme **RuBisCO** (below) is activated by a companion enzyme, **rubisco activase**. Rubisco activase is inhibited at temperatures above 35°C and also by high $CO_2$ levels. In many countries, global warming could push the temperature above 35°C. Rubisco activase would become less active, reducing the activity of the RuBisCo enzyme. The result would be a reduction in net photosynthesis and productivity.

The RuBisCO enzyme (above) is the most abundant protein found in plant leaves. It catalyzes the first major step in the carbon fixation process.

1. Describe why enzymes could be affected by an increase in global temperature: ______________________________

2. Explain why it can be difficult to predict the effect of global warming on specific enzymes: ______________________________



**Related activities:** *Extinction or Evolution?*


**ISBN: 978-1-927173-12-1**

# Temperature and the Distribution of Species

Global warming will have important implications for the development, distribution, and survival of many species. For example, many similar species avoid competition by breeding at different times and under different conditions. Temperature shifts may force similar species into competition within overlapping range. Many habitats will change in a warmer climate; climatic zones will shift and some species will need to relocate or adapt to the new conditions in order to survive.

## Distribution and Breeding of Leopard Frogs in North America

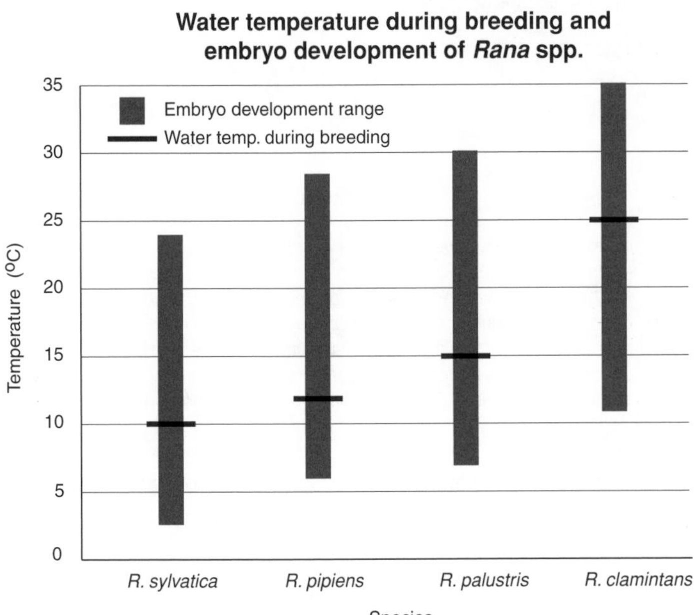


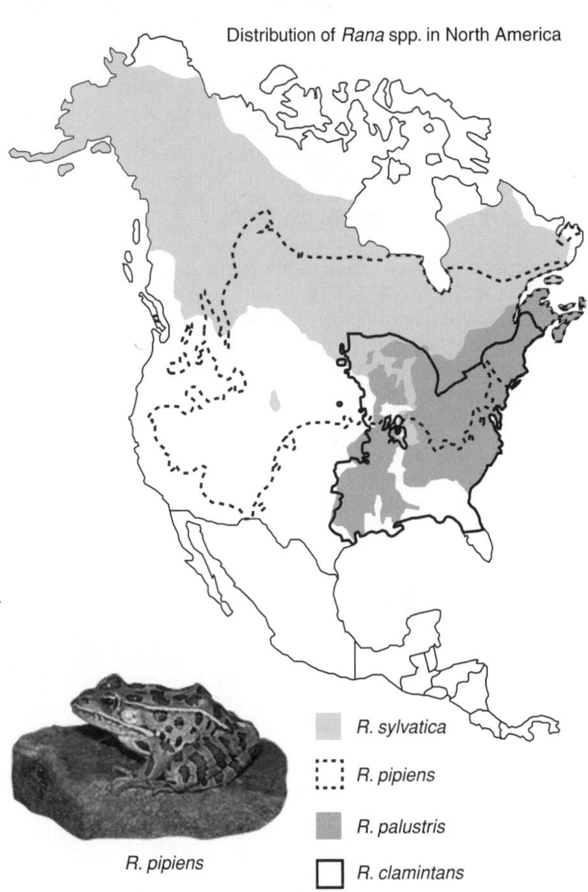


The frog genus *Rana* is relatively common and widely distributed in North America (top right). The graph above shows the preferred water temperature for breeding in four common species of *Rana*, as well as the temperature tolerance range for embryonic development. Outside these ranges, embryonic development rate decreases or the embryos die. Increases in temperature could reduce the available breeding habitat for some species (e.g. *Rana sylvatica* requires low temperature to breed). Another likely outcome would be a change to the timing of breeding or a shift in the distribution patterns of populations.

## Effects of Temperature on Life History Parameters in *Ceriodaphnia*

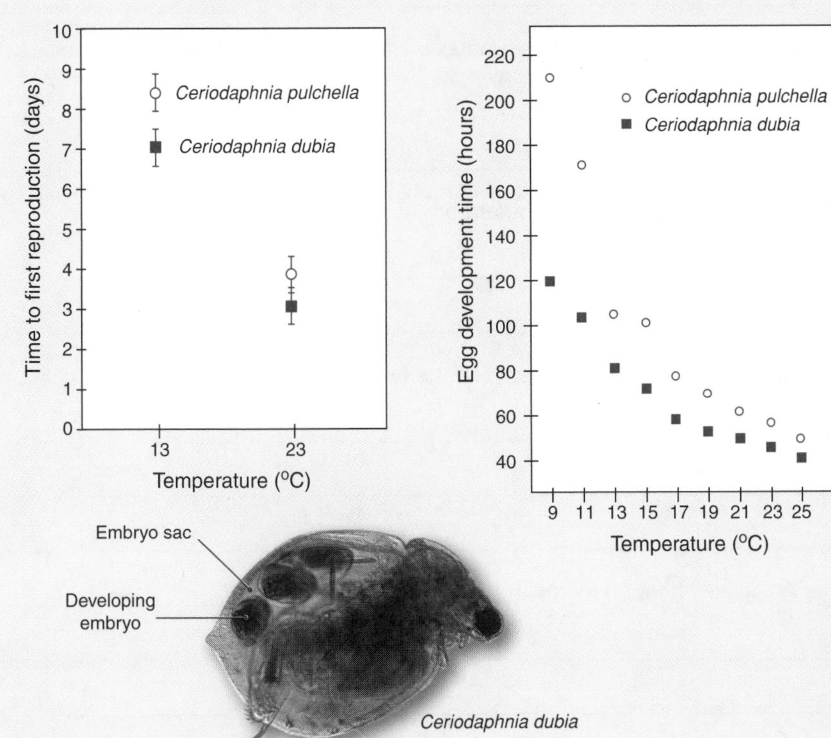


Image and data: Tracey Greenwood

Studies of lake ecosystems in New Zealand highlight the significance of temperature on the generation times and competitive outcome between organisms.

Two small species of daphnid (*Ceriodaphnia pulchella* and *C. dubia*) show marked temperature-dependent differences in important aspects of their life histories, such as **egg development time** and time to maturity. At lower temperatures, *C. dubia* has a significant breeding advantage over *C. pulchella* because its eggs develop faster and it matures earlier. This means that fall populations can establish quickly without competition from *C. pulchella,* which matures one and a half days later. At higher temperatures *C. dubia* loses this competitive advantage (far left).

The competitive advantage enjoyed by *C. dubia* during cooler seasons could be lost if lake temperatures were to rise by even 1 or 2°C. The warming could provide a subtle but measurable change in selection pressures, shift the species balance, and so influence the ecology of the lake.


ISBN: 978-1-927173-12-1


*Periodicals:*
*A world 4°C warmer,*
*Life at the edge*

***Related activities:*** *Extinction or Evolution?*

## Disappearing Sea Ice

**Arctic summer sea ice 1980**

**Arctic summer sea ice 2007**

Temperatures have risen in the Arctic every year for the last 27 years. At the same time, the average summer sea ice minimum has decreased by almost half (above). Polar bears mainly hunt seals which surface at breathing holes in the ice. The reduced sea ice levels have changed the distribution patterns of seals, and many polar bears have been forced to swim for miles to hunt. Between 1979 and 1991, 87% of bears observed were on sea ice during summer. However in the period between 1992 to 2004, this reduced to just 33%. In addition, the thinner sea ice cannot hold the bear's weight, and they are forced to the mainland before they have built up their winter fat stores. The loss of condition is affecting reproductive rates, and juvenile survival rates are lower as a consequence. Pregnant females must also swim for longer distances to reach their dens, and so lose more condition in the process.

1. Discuss the potential effects of a rise in global temperature on the North American distribution of the frog genus *Rana*:

2. (a) Describe the general effect of temperature on egg development rates of *C. pulchella* and *C. dubia*:

   (b) Describe the implication of this to competition between the two species:

3. (a) Describe the effect of temperature on differences in time to first reproduction between *C. pulchella* and *C. dubia*:

   (b) Describe the effect of increasing temperature on egg development for both *C. pulchella* and *C. dubia:*

   (c) Predict the effect of an increase in temperature on competition between the two species:

4. Discuss the impact of the reduction of the Arctic ice sheet on the polar bear population:


**ISBN: 978-1-927173-12-1**

# Biodiversity and Global Warming

Since the last significant period of climate change at the end of the ice age 10,000 years ago, plants and animals have adapted to survive in their current habitats. Accelerated global warming is again changing the habitats that plants and animals live in and this could have significant effects on the biodiversity of specific regions as well as on the planet overall. As temperatures rise, organisms will be forced to move to new areas where temperatures are similar to their current level. Those that cannot move face extinction, as temperatures move outside their limits of tolerance. Changes in precipitation as a result of climate change also affect where organisms can live. Long term changes in climate could see the contraction of many organisms' habitats while at the same time the expansion of others. Habitat migration, the movement of a habitat from its current region into another, will also become more frequent. Already there are a number of cases showing the effects of climate change on a range of organisms.

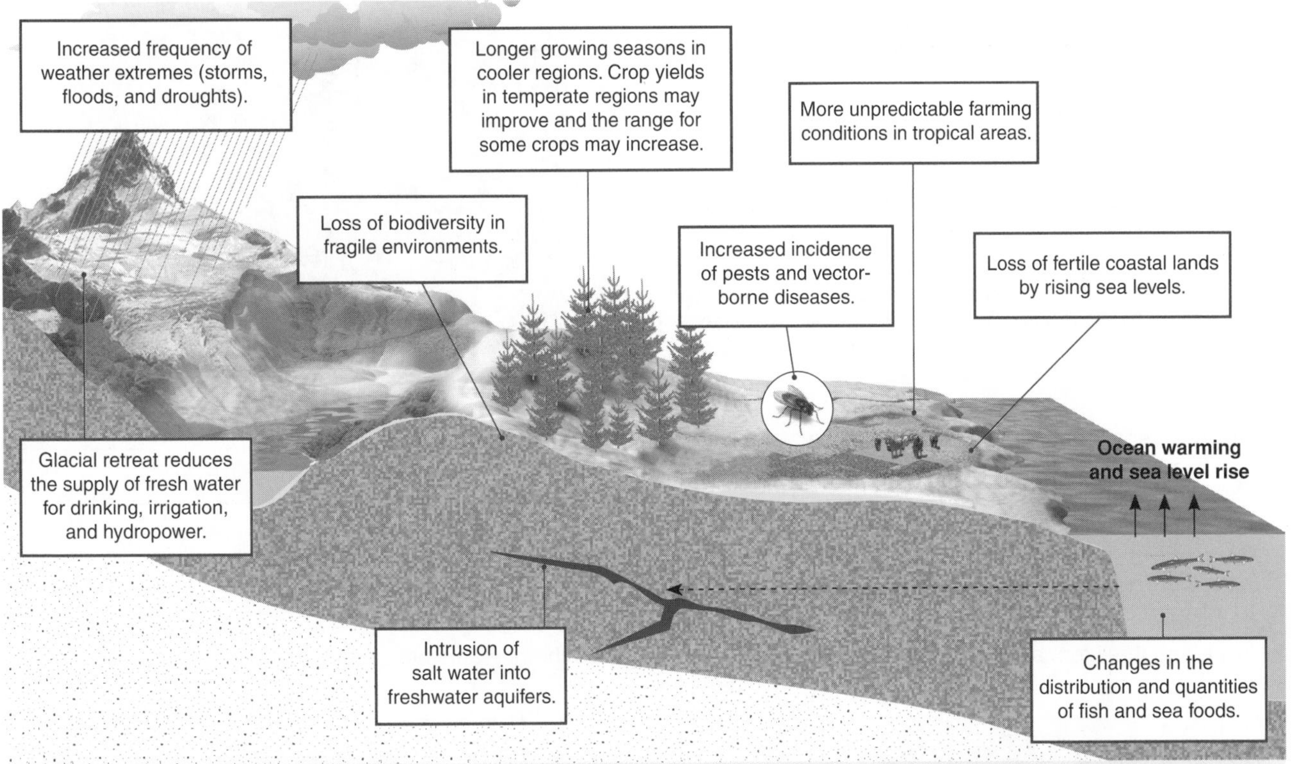

Studies of forests in the United States have shown that although there will be increases and decreases in the distribution ranges of various tree species, overall there will be an 11% decrease in forest cover, with an increase in savanna and arid woodland. Communities of oak/pine and oak/hickory are predicted to increase in range while spruce/fir and maple/beech/birch communities will decrease.

Studies of the distributions of butterfly species in many countries show their populations are shifting. Surveys of Edith's checkerspot butterfly (*Euphydryas editha*) in western North America have shown it to be moving north and to higher altitudes.

Studies of sea life along the Californian coast have shown that between 1931 and 1996, shoreline ocean temperatures increased by 0.79°C and populations of invertebrates including sea stars, limpets and snails moved northward in their distributions.

An Australian study in 2004 found the centre of distribution for the AdhS gene in *Drosophila*, which helps survival in hot and dry conditions, had shifted 400 kilometers south in the last twenty years.

A 2009 study of 200 million year old plant fossils from Greenland has provided evidence of a sudden collapse in biodiversity that is correlated with, and appears to be caused by, a very slight rise in $CO_2$ levels.

**ISBN: 978-1-927173-12-1**


***Related activities**: Global Warming*

***Weblinks**: Climate Change, NRDC: Global Warming*

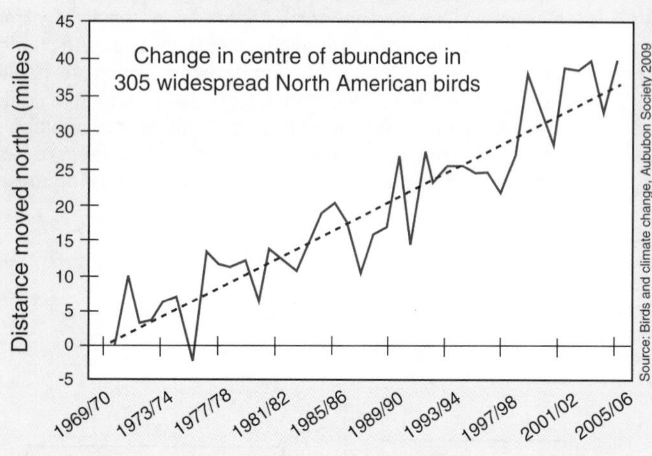


A number of studies indicate that animals are beginning to be affected by increases in global temperatures. Data sets from around the world show that birds are migrating up to two weeks earlier to summer feeding grounds and are often not migrating as far south in winter.

Animals living at altitude are also affected by warming climates and are being forced to shift their normal range. As temperatures increase, the snow line increases in altitude pushing alpine animals to higher altitudes. In some areas of North America this has resulting the local extinction of the North American pika (*Ochotona princeps*).

1. Describe some of the likely effects of global warming on physical aspects of the environment: ____________________

____________________

____________________

____________________

2. (a) Using the information on this and the previous activity, discuss the probable effects of global warming on plant crops:

____________________

____________________

(b) Suggest how farmers might be able to adjust to these changes: ____________________

____________________

____________________

____________________

3. Discuss the evidence that insect populations are affected by global temperature: ____________________

____________________

____________________

4. (a) Describe how increases in global temperatures have affected some migratory birds: ____________________

____________________

____________________

(b) Explain how these changes in migratory patterns might affect food availability for these populations: ____________________

____________________

____________________

5. Explain how global warming could lead to the local extinction of some alpine species: ____________________

____________________

____________________

____________________


ISBN: 978-1-927173-12-1

# Ocean Acidification

The oceans act as a **carbon sink,** absorbing much of the $CO_2$ produced by burning fossil fuels. When $CO_2$ reacts with water it forms carbonic acid, which decreases the pH of the oceans. This could have major effects on marine life, especially shell making organisms. Ocean acidification is relative term, referring to the oceans becoming less basic as the pH decreases.

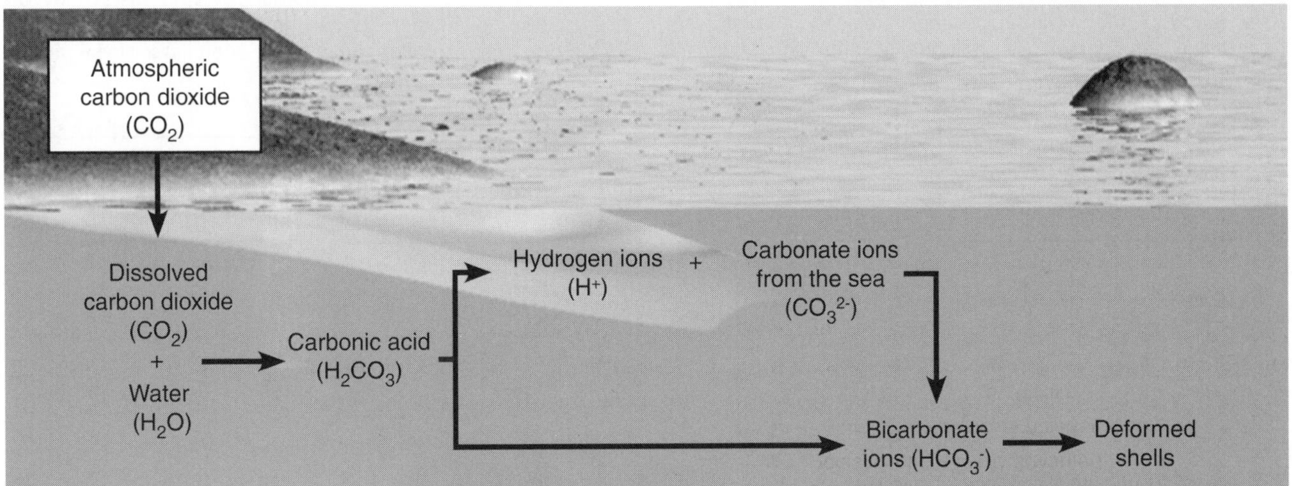

Atmospheric carbon dioxide ($CO_2$)

Dissolved carbon dioxide ($CO_2$) + Water ($H_2O$) → Carbonic acid ($H_2CO_3$)

Hydrogen ions ($H^+$) + Carbonate ions from the sea ($CO_3^{2-}$) → Bicarbonate ions ($HCO_3^-$) → Deformed shells

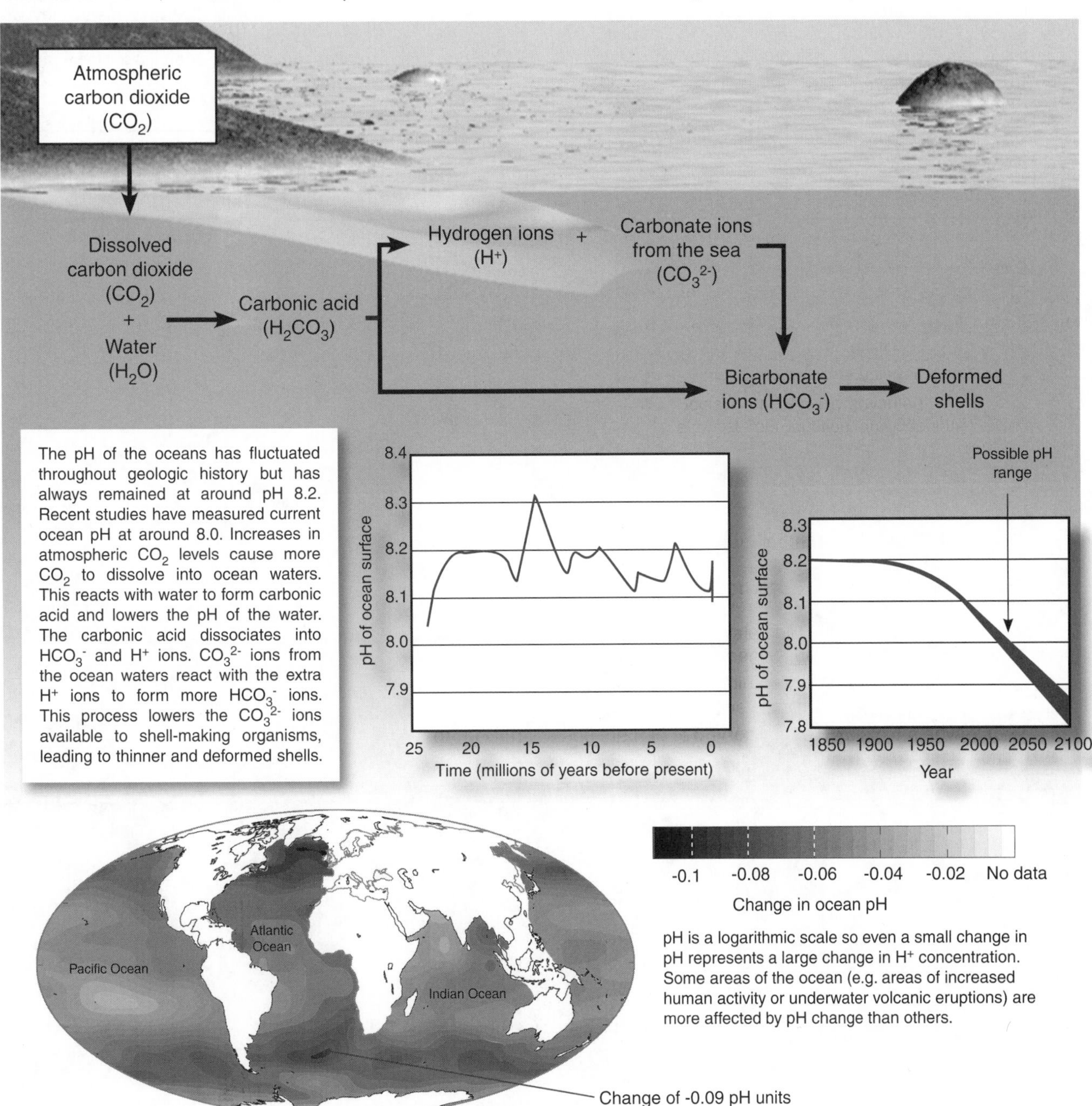

The pH of the oceans has fluctuated throughout geologic history but has always remained at around pH 8.2. Recent studies have measured current ocean pH at around 8.0. Increases in atmospheric $CO_2$ levels cause more $CO_2$ to dissolve into ocean waters. This reacts with water to form carbonic acid and lowers the pH of the water. The carbonic acid dissociates into $HCO_3^-$ and $H^+$ ions. $CO_3^{2-}$ ions from the ocean waters react with the extra $H^+$ ions to form more $HCO_3^-$ ions. This process lowers the $CO_3^{2-}$ ions available to shell-making organisms, leading to thinner and deformed shells.

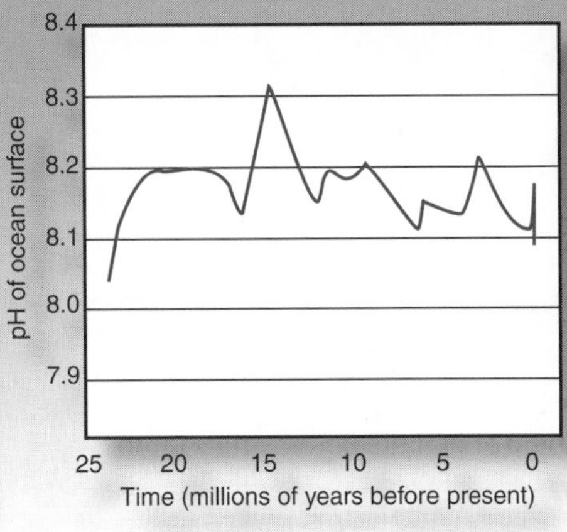

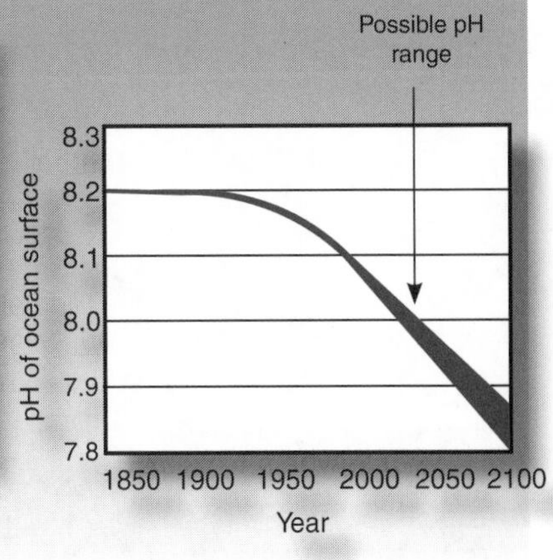

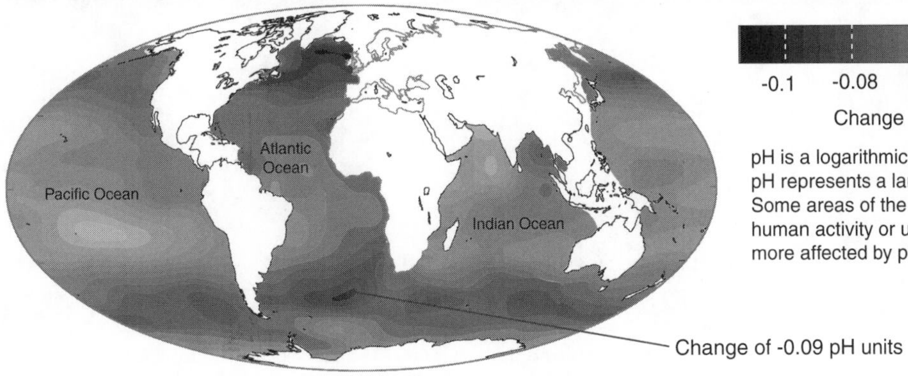

pH of ocean surface (8.4, 8.3, 8.2, 8.1, 8.0, 7.9)
Time (millions of years before present) (25, 20, 15, 10, 5, 0)

Possible pH range
pH of ocean surface (8.3, 8.2, 8.1, 8.0, 7.9, 7.8)
Year (1850, 1900, 1950, 2000, 2050, 2100)

Pacific Ocean
Atlantic Ocean
Indian Ocean
Change of -0.09 pH units

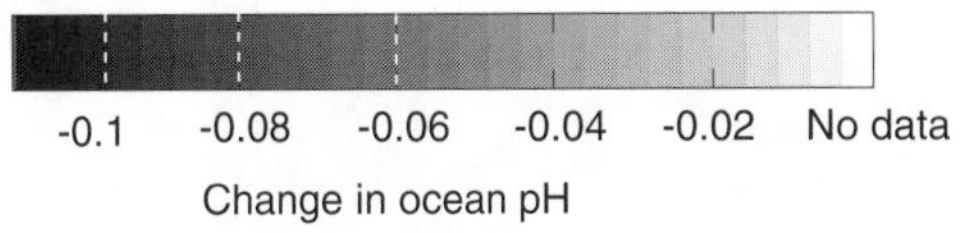

-0.1, -0.08, -0.06, -0.04, -0.02, No data
Change in ocean pH

pH is a logarithmic scale so even a small change in pH represents a large change in $H^+$ concentration. Some areas of the ocean (e.g. areas of increased human activity or underwater volcanic eruptions) are more affected by pH change than others.

For questions 1 to 4, circle the letter with the correct answer:

1. Ocean acidification is the process of:
   A. The pH of the oceans rising
   B. The oceans absorbing $CO_2$
   C. Decreasing concentrations of $HCO_3^-$ ions
   D. The oceans becoming less alkaline

2. Ocean acidification has the effect of:
   A. Increasing the $CO_2$ absorbed by the oceans
   B. Dissolving seashells
   C. Decreasing the $CO_3^{2-}$ ions available to shell making organisms
   D. Increasing the pH of the oceans

3. The oceanic area most affected by ocean acidification is:
   A. The Indian Ocean, near Australia
   B. The North Atlantic Ocean, near Greenland
   C. The North Pacific near Japan
   D. The Southern Ocean, near Antarctica

4. Even allowing for error, ocean pH in 2100 is predicted to be:
   A. 7.8 Below
   B. 7.9
   C. The same as presently
   D. About the same as 22 mya

5. Describe the relationship between ocean acidity and ocean pH: ______________________

______________________________________________

# Tropical Deforestation

Tropical rainforests prevail in places where the climate is very moist throughout the year (200 to 450 cm of rainfall per year). Almost half of the world's rainforests are in just three countries: **Brazil** in South America, **Zaire** in Africa, and **Indonesia** in Southeast Asia. Much of the world's biodiversity resides in rainforests. Destruction of the forests will contribute towards global warming through a large reduction in photosynthesis. In the Amazon, 75% of deforestation has occurred within 50 km of Brazil's roads. Many potential drugs could still be discovered in rainforest plants, and loss of species through deforestation may mean they will never be found. Rainforests can provide economically sustainable crops (rubber, coffee, nuts, fruits, and oils) for local people.

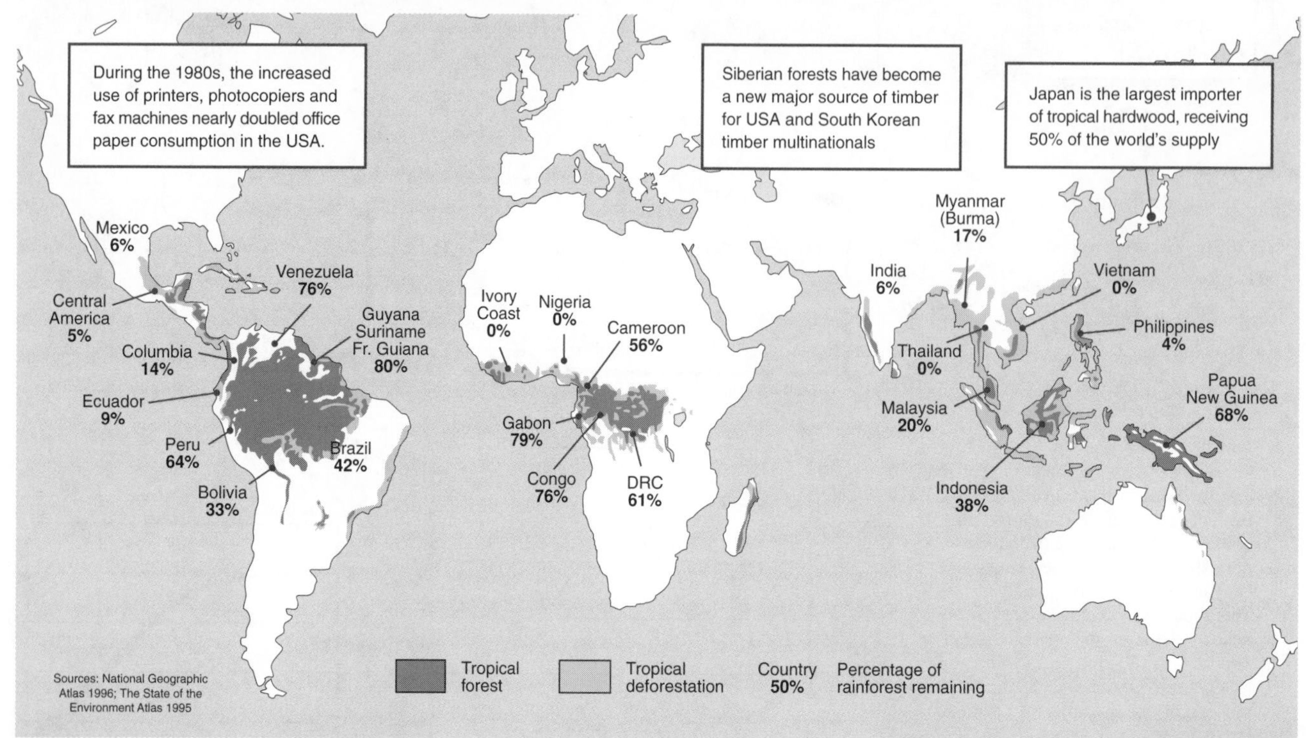

The felling of rainforest trees is taking place at an alarming rate as world demand for tropical hardwoods increases and land is cleared for the establishment of agriculture. The resulting farms and plantations often have shortlived productivity.

Huge forest fires have devastated large amounts of tropical rainforest in Indonesia and Brazil in 1997/98. The fires in Indonesia were started by people attempting to clear the forest areas for farming in a year of particularly low rainfall.

The building of new road networks into regions with tropical rainforests causes considerable environmental damage. In areas with very high rainfall there is an increased risk of erosion and loss of topsoil.

1. Describe three reasons why tropical rainforests should be conserved:

   (a) __________

   (b) __________

   (c) __________

2. Identify the three main human activities that cause tropical deforestation and briefly describe their detrimental effects:

   (a) __________

   (b) __________

   (c) __________



***Related activities**: Secondary Succession, Global Warming*

***Periodicals**:*
*A forest in tatters…, Last of the Amazon*


**ISBN: 978-1-927173-12-1**

# The Impact of Introduced Species

**Introduced species** are those that have evolved at one place in the world and have been transported by humans, either intentionally or in advertently, to another region. Some of these introductions are beneficial, e.g. introduced agricultural plants and animals, and Japanese clams and oysters (the mainstays of global shellfish industries). **Invasive species** are those alien species that have a detrimental effect on the ecosystems into which they have been imported. They number in their hundreds with varying degrees of undesirability to humans. Humans have brought many exotic species into new environments for use as pets, food, ornamental specimens, or decoration, while others have hitched a ride with cargo shipments or in the ballast water of ships. Some have been deliberately introduced to control another pest species and have themselves become a problem. Some of the most destructive of all alien species are aggressive plants, e.g. mile-a-minute weed, a perennial vine from Central and South America, miconia, a South American tree invading Hawaii and Tahiti, and *Caulerpa* seaweed, the aquarium strain now found in the Mediterranean. Two introductions, one unintentional and the other deliberate, are described below.

## Kudzu

### A deliberate introduction

Kudzu (*Pueraria lobata*) is a climbing vine native to south-east Asia. It spreads aggressively by vegetative reproduction and is a serious invasive pest in the southern US, where it has been spreading at a rate of 61,000 ha per annum. Kudzu was first introduced to the US in the 1800s as an ornamental plant for shade porches, and was subsequently widely distributed as a high-protein cattle fodder and as a cover plant to prevent soil erosion. It grew virtually unchecked in the climate of the Southeastern US and was finally listed as a weed in 1970, more than a decade after it was removed from a list of suggested cover plants. Today, kudzu is estimated to cover 3 million ha of land in the southeastern US.

## Red Imported Fire Ant

### An accidental invasion

Red fire ants (*Solenopsis invicta*) were accidentally introduced into the United States from South America in the 1920s and have spread north each year from their foothold in the Southeast. Red fire ants are now resident in 14 US states where they displace populations of native insects and ground-nesting wildlife. They also damage crops and are very aggressive, inflicting a nasty sting. The USDA estimates damage and control costs for red fire ants at more than $6 billion a year. Red fire ants lack natural control agents in North America and thrive in disturbed habitats such as agricultural lands, where they feed on cereal crops and build large mounded nests.

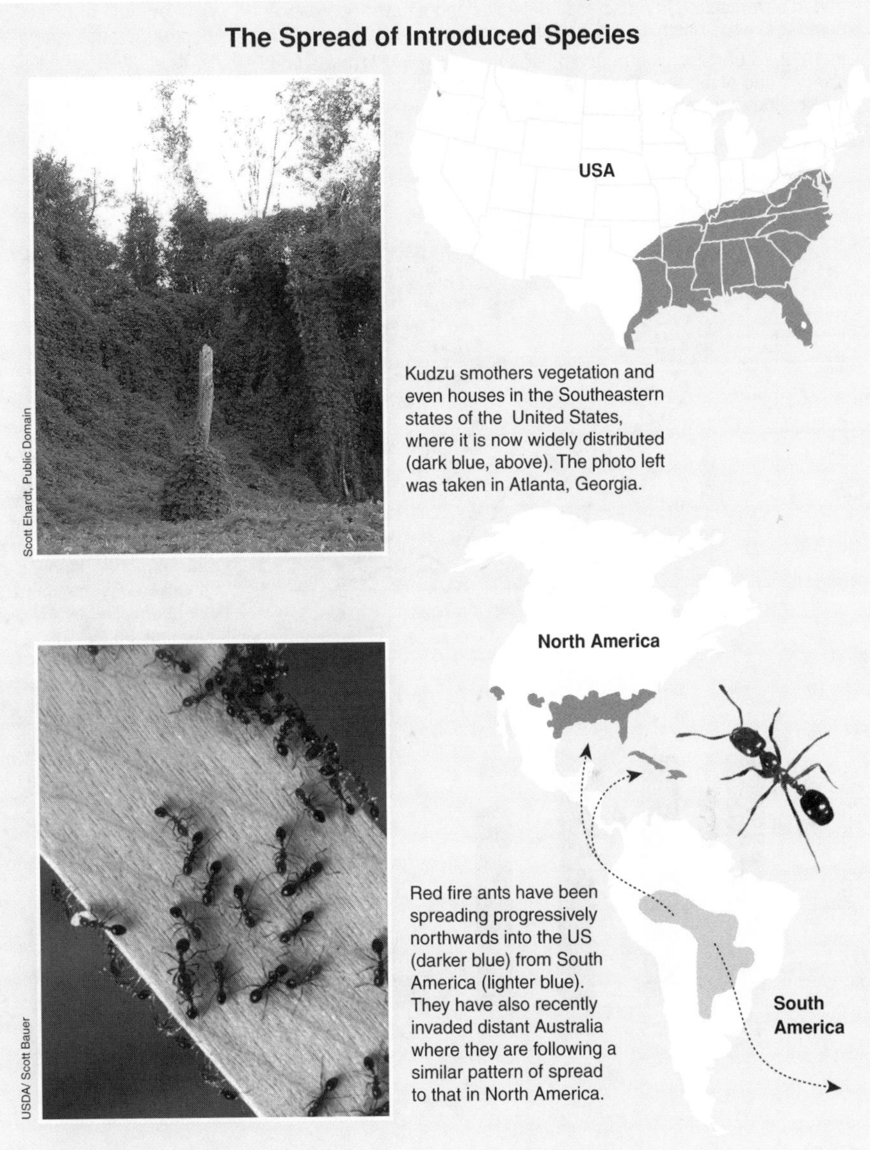


Kudzu smothers vegetation and even houses in the Southeastern states of the United States, where it is now widely distributed (dark blue, above). The photo left was taken in Atlanta, Georgia.

Red fire ants have been spreading progressively northwards into the US (darker blue) from South America (lighter blue). They have also recently invaded distant Australia where they are following a similar pattern of spread to that in North America.

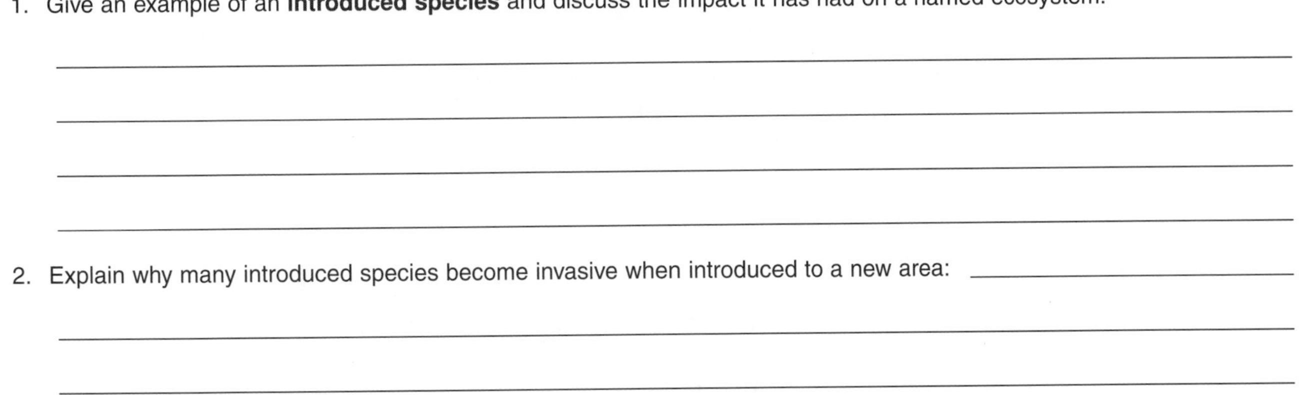

1. Give an example of an **introduced species** and discuss the impact it has had on a named ecosystem:

2. Explain why many introduced species become invasive when introduced to a new area:

**ISBN: 978-1-927173-12-1**

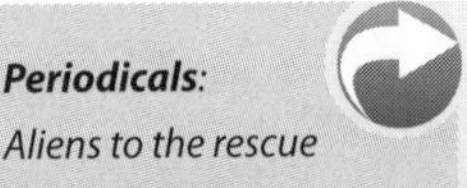

***Periodicals:*** *Aliens to the rescue*

***Related activities:*** *Interspecific Competition*
***Weblinks:*** *Fire Ants Invade and Evolve*

# Responses to Environmental Change

**Genetic diversity** refers to the variety of alleles and genotypes present in a population. Genetic diversity is important to the survival and adaptability of a species. Populations with low genetic diversity may not be able to respond to environmental change, and are at risk of becoming extinct. In contrast, species with more genetic diversity have an improved ability to be able to adapt and respond to environmental change. This increases their chance of surviving.

## The Effects of Low Genetic Diversity

**Illinois prairie chicken**: Until 1992, the Illinois prairie chicken was virtually destined for extinction. The population had fallen from millions before European arrival to 25,000 in 1933 and then to 50 in 1992. The dramatic decline in the population in such a short time resulted in a huge loss of genetic diversity, which led to inbreeding and in turn resulted in a decrease in fertility and an ever-decreasing number of eggs hatching successfully.

In 1992, a translocation program began, bringing in 271 birds from Kansas and Nebraska. There was a rapid population response, as fertility and egg viability increased. The population is now recovering.

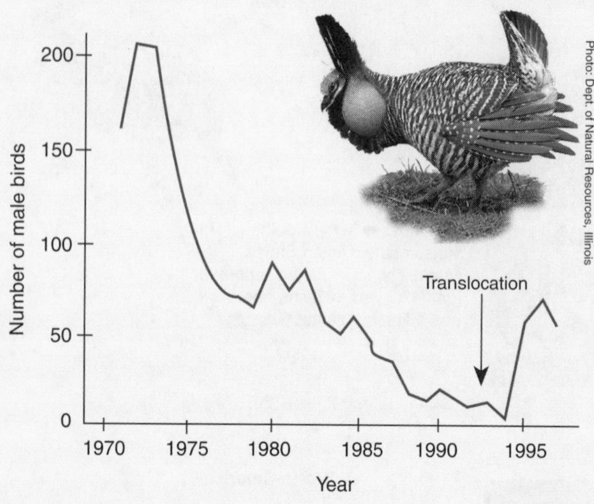


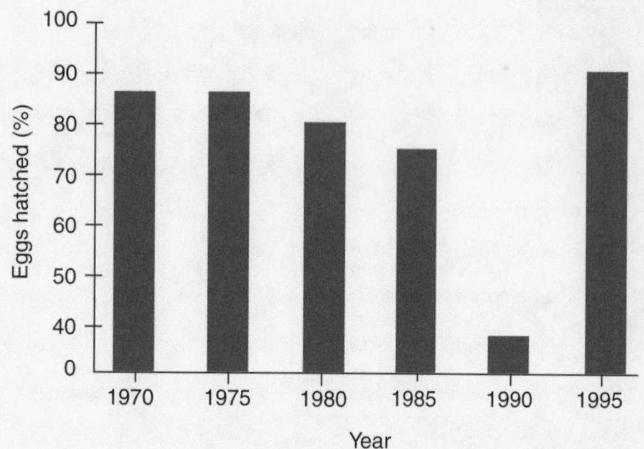


**The great potato famine**: Potatoes were the staple food for most rural Irish families during the 17th century. At the time, farmers favored one particular variety, and because potatoes are cultivated by vegetative propagation, genetic diversity between potato crops was very low. In the mid 1840s a fungal disease, called late blight, swept through the country, infecting all non-resistant potato plants. Affected crops rotted during storage, creating a great famine in which over one million people died. If the farmers had planted a wider variety of potato crops (with greater genetic diversity), the chances of some of them being resistant to the disease would have been higher, and the effects of the famine might not have been so great.

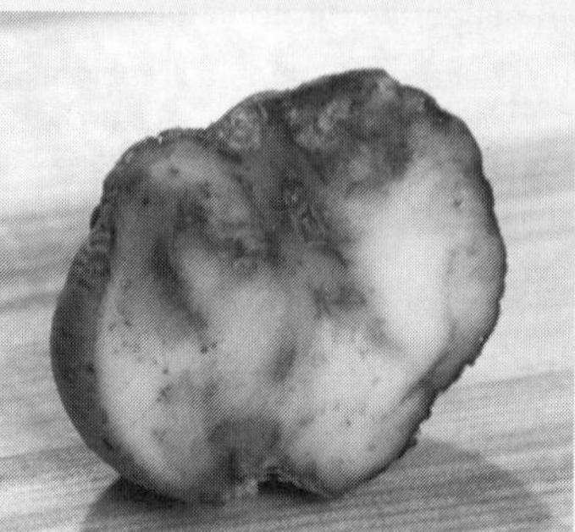
Infected potatoes are shrunken and rotted.

**Tasmanian devil**: The Tasmanian devil is a carnivorous marsupial found only in Australia. Genetic diversity amongst the species is very low, particularly in a region of the genome that codes for the major histocompatibility complex (MHC) of the immune response. This low genetic diversity has made the Tasmanian devil vulnerable to an infectious facial cancer that is spread between individuals by biting. The MHC does not recognize the cancer cells as foreign, so no immune response is launched, leaving the cancer to spread unchecked. Researchers estimate that 60% of animals in the wild have been killed by the disease since 1996, but in some regions the figure may be as high at 90%. Tasmanian devils may become extinct within 25 years if the disease can not be halted.

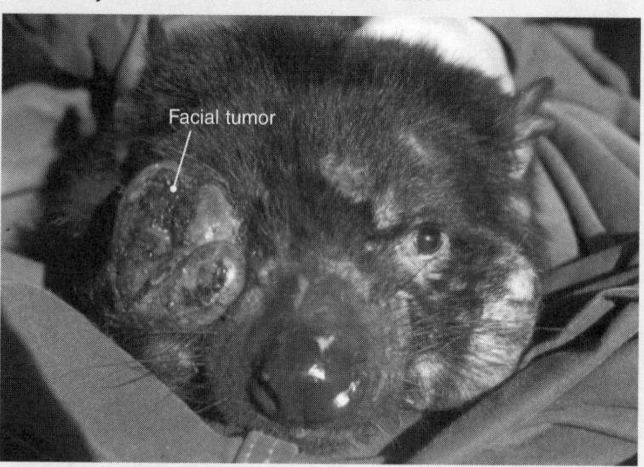


1. Explain what is meant by **genetic diversity**: ____________________

2. (a) The decline of the Illinois prairie chicken before 1992 is sometimes referred to as an extinction vortex. Explain the causes and effects of this extinction vortex in relation to the Illinois prairie chicken:

(b) Why did the translocation of 271 birds from outside Illinois into the Illinois population halt the population decline?



**Related activities**: *Threats to Biodiversity, Extinction or Evolution*

**Periodicals**: *Tasmanian devils.., Catching Condors...*


**ISBN: 978-1-927173-12-1**

## Genetic Diversity and HIV-1 Resistance

Before the advent of large scale and global travel (effectively only 150 years old) human populations were largely isolated, and populations developed regional variations in alleles. Some of these variations (diversity) affects our immune system and causes some of the variation in immunity that is seen in humans. This variation may allow some people to be naturally resistant to some diseases.

A modern example of genetic diversity leading to differential responses to selective pressures is the HIV/AIDS epidemic. HIV/AIDS has killed more than 25 million people globally, and infected another 33 million since it was first recognized in 1981.

In the mid 1990s, it was found that the HIV-1 virus entered T-cells of the immune system by docking with the receptor encoded by the CCR5 gene. Soon after this, it was discovered that a mutation in the gene (called CCR5Δ32) caused resistance to HIV-1.

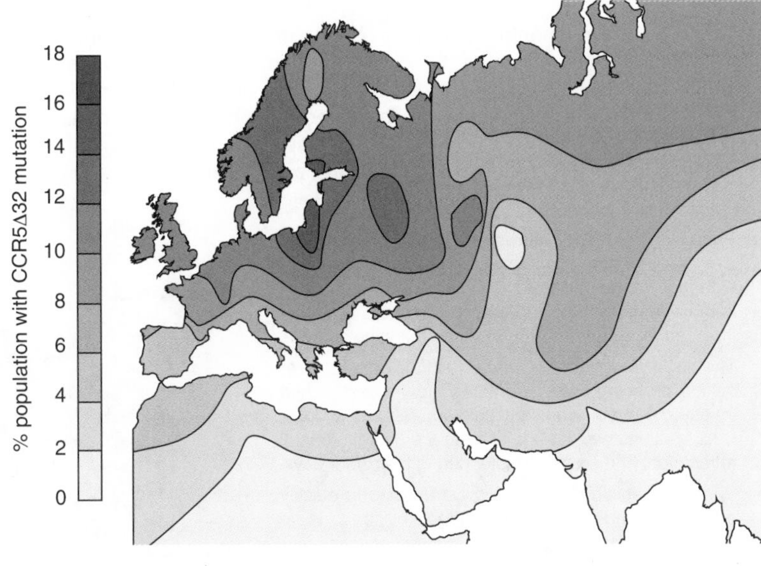


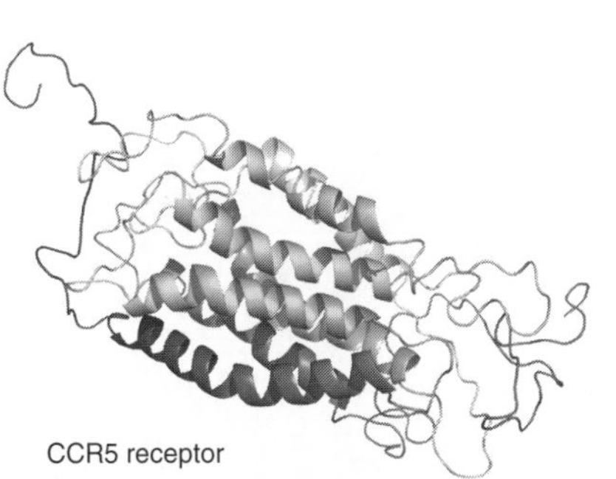
CCR5 receptor

Geographical studies have found that the CCR5Δ32 mutation is found in Caucasian populations in some areas of northern Europe where it is carried by up to 18% of population. The mutation is virtually absent in Asian, Middle Eastern, and American Indian populations.

The CCR5Δ32 mutation produces a premature stop codon in the mRNA. People with this mutation in one allele produce T-cells with a reduced number of CCR5 receptors. HIV-1 infects these cell only slowly, taking 2-3 years longer than normal to progress to AIDS. People with mutations in both alleles produce T-cells with no CCR5 receptors. HIV-1 is effectively unable to enter these cells.

3. (a) Explain how low genetic diversity contributed to the great potato famine: ______

   (b) Explain why the Tasmanian devil is so susceptible to an infectious facial cancer: ______

4. (a) Explain the effect of the CCR5Δ32 mutation on the entry of HIV-1 into T-cells: ______

   (b) In which populations in this mutation found? ______

   (c) Explain how this genetic diversity could naturally halt the spread of HIV-1: ______

5. Explain the significance of **genetic diversity** to the resilience of a population to environmental change: ______

The Diversity & Stability of Ecosystems


**ISBN: 978-1-927173-12-1**

# Extinction or Evolution?

Changes in environmental conditions (e.g. climate change) can put stress on organisms. If they can not adapt to the new environmental conditions, populations may decline or become extinct. Many recent studies have investigated the phenotypic response of organisms to climate change. Responses include changes to life history patterns (e.g. changing breeding times), and morphological changes (e.g. becoming smaller). However, most of these variations are a result of **phenotypic plasticity** rather than true adaptive change through evolution. Phenotypic plasticity involves changes to the phenotype only; the genotype is unchanged. To date, very few studies have been able to show any genetic response to climate change.

Phenotypic plasticity allows an organism to shift its phenotype within certain, genetically constrained, extremes in response to environmental changes. It is more important for immobile organisms because if they cannot adapt to the new environment they will die.

The buckeye butterfly (*Junonia coenia*) changes color as the breeding season progresses. Those hatched in early summer are light colored, (almost yellow). Buckeyes hatched in early autumn are darker, so they attract heat and warm up faster during the cooler days.

Such phenotypic changes allow a species to quickly adapt to changes in its environment, but there is no overall change to the gene pool.

Early season Buckeye

Late season Buckeye

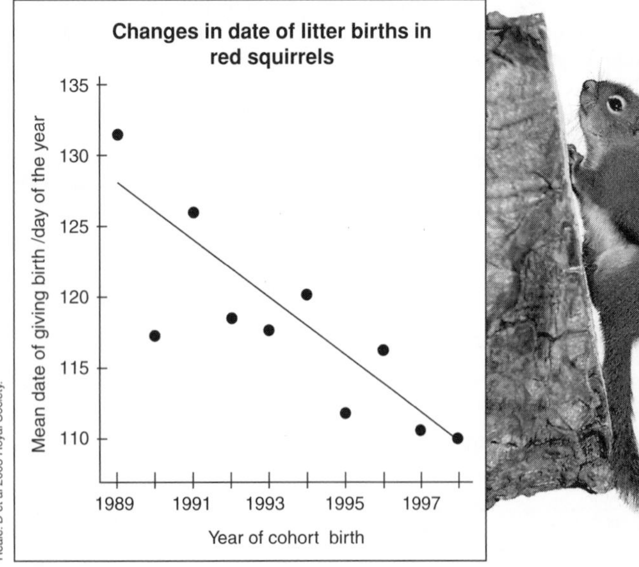


North American red squirrels (*Tamiasciurus hudsonicus*) in Canada have adapted to a 2°C increase in spring temperature by breeding earlier in the year. Records were kept of cohorts of female squirrels to determine the day of the year they gave birth. Over a period of ten years, the squirrel breeding time shifted to be earlier in the year (by 18 days). The change was linked to an increase in the abundance of spruce cones, an important food source for the red squirrels.

Some organisms have limited phenotypic plasticity and will be unable to adapt to long term environmental changes. For these organisms, relocation may be their only chance for survival. However, the habitat of some organisms is very narrow (e.g. polar bears, above), and for them there may no place to go, resulting in their extinction.

1. a) Explain the difference between phenotypic plasticity and adaptation through evolution: ____________

   b) Explain why adapting to the environment takes longer than a physiological adjustment to it: ____________

2. Explain why some organisms are more likely to become extinct as a result of climate change than others:



Related activities: *Biodiversity and Global Warming, Ice Sheet Melting*


ISBN: 978-1-927173-12-1

# KEY TERMS: Word Find

Use the clues below to find the relevant key terms in the WORD FIND grid

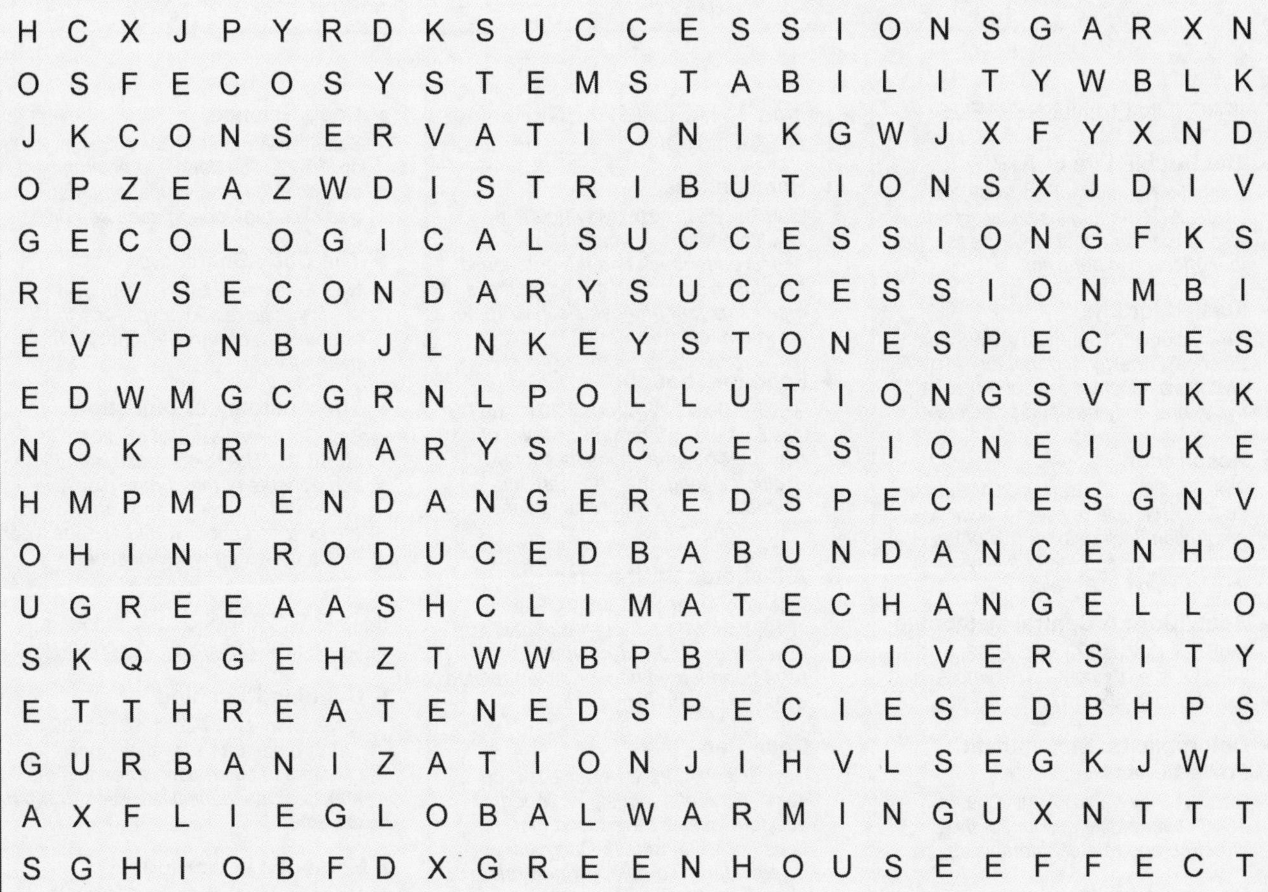


The sum of all biotic variation from the level of genes to ecosystems

The term describing the active management of natural populations in order to rebuild numbers and ensure species survival.

The term used to describe the ability of an ecosystem to resist change over time.

A species present in such small numbers that it is at risk of extinction.

The accelerated increase in global temperature.

Warming of the troposphere caused by the absorption of the Earth's heat by greenhouse gases.

The general term for any gas in the atmosphere that causes the retention of heat in the Earth's atmosphere.

Any undesirable change in the quality of air, water or soil as a result of contaminants.

A succession sequence that occurs on land that has not had plants or soil in the past is termed this.

The terms used for a succession sequence that takes place after a land clearance event and does not involve the loss of seeds and root stock.

Change in an ecological community over time.

The migration of people into urban environments and the growth of urban regions is called this.

The description of how a species is spread throughout an area and is interrelated with the density of a population.

The progression from initial colonization of a newly cleared area to a climax community.

A species that is vulnerable to endangerment or extinction in the future.

The relative representation of a species within an ecosystem.

A species that is influential in the stability of an ecosystem and has a disproportionate effect due to their pivotal role.

Name for an organism that is brought into an area it did not previously inhabit (non-native).

A significant and lasting change in the statistical distribution of weather patterns over periods ranging from decades to millions of years.




ISBN: 978-1-927173-12-1

# Appendix

## PERIODICAL REFERENCES

### ENERGY IN LIVING SYSTEMS

- **The Double Life of ATP**
Scientific American, Dec. 2009, pp. 60-67. *ATP the fuel inside living cells, also serves as a molecular messenger that affects cell behavior.*

- **Fuelled for Life**
New Scientist, 13 January 1996 (Inside Science). *Energy and metabolism: ATP, glycolysis, electron transport, Krebs cycle, and enzymes and cofactors.*

- **Respiration**
Biol. Sci. Rev., 23(2) Nov. 2010, pp. 20-21. *An article revising cellular respiration. Includes the role of the mitochondria, and the pathways involved, and ATP yield.*

- **AcetylCoA: A Central Metabolite**
Biol. Sci. Rev., 20(4) April 2008, pp.38-40. *The role of acetyl coenzyme A in metabolizing fat and carbohydrate.*

- **Chloroplasts: Biosynthetic Powerhouses**
Biol. Sci. Rev., 21(4) April 2009, pp. 25-27. *Informative account of the structure and role of chloroplasts.*

- **Photosynthesis**
Biol. Sci. Rev., 23(2) Sept. 2010, pp. 20-21. *Photosynthetic processes including the absorption spectrum and chloroplast structure and function.*

- **Rubisco: The World's Most Important Enzyme**
Biol. Sci. Rev., 23(4) April 2011, pp. 2-5. *The role of Rubisco in the Calvin cycle, its efficiency as a catalyst, and the significance of photorespiration.*

### ENZYMES AND METABOLISM

- **Making the Rate: Enzyme Dynamics Using Pop-it Beads**
The American Biology Teacher, 66(9), November 2004, pp. 621-626. *Describes an enzyme exercise to help student's visualize chemical reactions at the molecular level.*

- **Enzymes**
Biol. Sci. Rev., 23(3) Feb. 2011, pp. 20-21. *Informative article revising enzymes. Graphs showing factors affecting the rate of enzyme reactions. Cofactors, inhibitors, synoptic possibilities.*

- **Battling Biofilms**
Scientific American, 285(1), july 2001, pp. 60-67. *The nature of biofilms, how they form, their properties, and their role in infection and disease.*

### HOMEOSTASIS AND ENERGY ALLOCATION

- **Homeostasis**
Biol. Sci. Rev., 12(5) May 2000, pp. 2-5. *Homeostasis: what it is, the role of negative feedback and the autonomic nervous system, and the adaptations of organisms for homeostasis in extreme environments (excellent).*

- **Food for Thought**
Biol. Sci. Rev., 22(4), April 2010, pp. 22-25. *A clear, thorough account of how the body maintains its supply of glucose long after the nutrients absorbed from a meal have been exhausted.*

- **A Diabetes Cliffhanger**
Scientific American, 306(2), Feb. 2012, , pp. 17-19. *Type 1 diabetes is increasing world wide. Hypotheses have been formulated for gluten related diets, fungi, and hygienic living.*

- **Code Red**
New Scientist, 22 October 2011, pp. 48-51. *25% of seriously wounded accident victims have blood that does not clot normally. Coagulation is regulated by a complex network of blood-born proteins and a process involving the 'thrombin switch' can lead to a wash of anticoagulant in the body.*

### PLANT STRUCTURE AND ADAPTATION

- **Plants, Water and Climate**
New Scientist, 25 Feb. 1989, (Inside Science). *Aspects of plant transport: osmosis and turgor, transport from the root to the leaf, transpiration, and stomatal control.*

- **Does Water Climb Trees?**
Biol. Sci. Rev., 24(2), Nov. 2011, pp. 27-29. *The cohesion-tension theory, which is dependent on the properties of water, explains the movement of water up even very tall trees.*

- **Cacti**
Biol. Sci. Rev., 20(1), Sept. 2007, pp. 26-30. *The growth forms and structural and physiological adaptations of cacti.*

### OBTAINING NUTRIENTS AND ELIMINATING WASTES

- **Getting in and Out**
Biol. Sci. Rev., 20(3), Feb. 2008, pp. 14-16. *Diffusion: some adaptations and some common misunderstandings*

- **Breathless**
New Scientist, 8 March 2003, pp. 46-49. *Adaptations for gas exchange in shark species able to withstand anoxia.*

- **Deep Science**
Scientific American, 204(3), Sept. 2003, pp. 78-93. *The deep sea environment, including the challenges to humans of gas exchange during deep sea dives.*

- **The Body Snatchers**
New Scientist, 24 July 1999, pp.42-46. *Parasitic relationships, and the adaptations of parasites that promote their survival.*

- **The Anatomy of Digestion**
Biol. Sci. Rev., 23 (3) Feb. 2010, pp. 18-21. *The role of each of the components of the human digestive system is described, and explanations are provided about what happens when things go wrong with digestion.*

### INTERACTIONS IN PHYSIOLOGICAL SYSTEMS

- **Cunning Plumbing**
New Scientist, 6 Feb. 1999, pp. 32-37. *The arteries can actively respond to changes in blood flow, spreading the effects of mechanical stresses to avoid extremes.*

- **Metabolic Powerhouse**
New Scientist, 11 Nov. 2000 (Inside Science). *The myriad roles of the liver in metabolism, including discussion of amino acid and glucose metabolism.*

- **Breaking Out of the Box**
The American Biology Teacher, 63(2), February 2001, pp. 101-115. *Investigating cardiovascular activity: a web-based activity on the cardiac cycle.*

- **Humans with Altitude**
New Scientist, 2 Nov. 2002, pp. 36-39. *The short term adjustments and long term adaptations to life at altitude.*

### DEFENSE MECHANISMS

- **Red Alert!**
New Scientist, 28 Sept. 2003, pp. 40-44. *Plant defences and the benefits of red pigmentation in leaves.*

- **Misery for all Seasons**
National Geographic, 209(5) May 2006, pp. 116-135. *The causes, effects, and prevention of common allergies.*

- **Hard to Swallow**
New Scientist, 26 Jan. 2008, pp. 37-39. *Many people fear that vaccines are unsafe and cause health problems. Particular reference to the polio and measles vaccines.*

- **Immunology**
Biol. Sci. Rev., 22(4) April 2010, pp. 20-21. *A pictorial but information-packed review of the basic of internal defense functions.*

# Appendix

## PERIODICAL REFERENCES

▶ **Boosting Vaccine Power**
Scientific American, October 2009, pp. 56-59. *This article looks at vaccines and immunization, and how researchers are boosting the effectiveness of vaccines to be even more specific.*

▶ **Fast Track to Vaccines**
Scientific American, 304(5), may 2011, pp. 50-55. *New strategies for the development of an HIV/AIDs vaccine. By measuring and comparing changes in genetic activity, protein levels, and cellular activity, researchers can quickly determine the success of the vaccine.*

### TIMING, COORDINATION AND SOCIAL BEHAVIOR

▶ **Don't Talk, Reproduce**
Scientific American, May 2009, pp. 11-12. *Bacteria can monitor their population density, and moderate their behavior accordingly.*

▶ **Times of our Lives**
Scientific American, Sept. 2002, pp. 40-47. *Biological clocks and their neurological basis; how the brain synchronises bodily functions.*

▶ **Keeping in Time**
New Scientist, 16 March 1996 (Inside Science). *Biological clocks: the time keepers for circadian rhythms. The activities of plants and animals may seem like randomly occurring events but the lives of organisms are controlled by their internal clocks.*

▶ **The Hunger, The Horror?**
New Scientist, 30 May 2008, pp. 42-45. *The adaptive dispersal behaviour of locusts when they swarm.*

▶ **Flight of the Navigators**
New Scientist, 26 July 2008, pp. 36-39. *Better tracking technologies are solving the mysteries of bird migrations.*

▶ **High Flyers**
New Scientist, 5 June 2010, pp. 32-34. *To complete their long migration, moths and butterflies have some spectacularly elaborate navigation strategies.*

▶ **The Compass Within**
Scientific American, January 2012, pp. 36-41. *Explores the role the Earth's magnetic field plays in animal migration.*

▶ **All for One**
New Scientist, 13 June 1998, pp. 32-35. *Social insects such as ants and termites all show cooperative behaviour: workers are altruistic and the colony functions as a superorganism.*

▶ **Kiss and Make Up**
New Scientist, 7 May 2005, pp. 35-37. *Species where the individuals get along together have an evolutionary advantage over those that don't, such as lower stress levels and stable social groups..*

▶ **Curious Liaisons**
New Scientist, 3 July 2010, pp. 36-39. *Discusses sex, reproduction, mating rituals, and courtship in a variety of organisms.*

▶ **Animal Attraction**
National Geographic, July 2003, pp. 28-55. *An engaging and expansive account of mating in the animal world.*

▶ **Show me the Honey**
New Scientist, 19 September 2009, pp. 40-41. *The waggle dance may not be as important to bees for finding food as originally thought.*

▶ **Relative Distance**
Scientific American, Jan. 2008, pp. 15-16. *Cooperative breeding behaviour in hyaenas: spatial groups of striped hyenas comprise just one female defended by up to three males, which means some males don't breed.*

### NERVOUS SYSTEMS & RESPONSES

▶ **Sense and Sense Ability**
New Scientist, 20 August 2011, pp. 32-37. *Examines the sensory abilities of many different animals, and discusses the components of sensory organs.*

▶ **Infinite Sensation**
New Scientist, 11 August, 2001, pp. 24-28. *The nature of sensory perception. The article has links to web sites where students explore their own responses to sensory inputs.*

▶ **Through the Mind's Eye**
New Scientist, 6 May, 2006, pp. 32-36. *fMRI can allow researchers to examine active areas of the brain, and in be able to be used to discover the images the brain is creating or remembering.*

▶ **Alzheimer's: Forestalling the Darkness**
Scientific American, June 2010, pp. 32-39. *Testing for Alzheimer's before symptoms arise can allow drug treatments to be used early. This may increase the chances of the treatment being effective.*

▶ **Just Can't Get Enough**
New Scientist, 26 August, 2006, pp. 30-35. *The role of dopamine release in the brain's reward circuits may be fuelling many addictions including gambling, shopping, and exercise.*

▶ **Circuit Training**
New Scientist, 9 April, 2011, pp. 35-39. *Discusses methods for treating illness such as depression and Parkinson's disease, and looks if the treatments can be used to enhance everyday performance and feelings.*

▶ **Disco Inferno**
New Scientist, 1 September, 2001, pp. 19. *A half page article describing how ecstasy alters metabolic rate and prevents the body from cooling down.*

▶ **Generation Specs**
New Scientist, 7 November, 2009, pp. 45-51. *Article covers the human eye and vision, and some conditions such as myopia that can affect the eye.*

▶ **From Genes to Color Vision**
Biol. Sci. Rev., 22(4) April 2010, pp. 2-5. *How do we distinguish color? This article describes the physiological basis of color vision and examines the case for why and how it may have evolved.*

▶ **Exquisite Sense**
New Scientist, 17 September, 2011, pp. 44-47. *Humans have around 400 olfactory receptors, allowing us to distinguish subtle difference between molecules. Smell can influence human mood and behavior.*

### THE NATURE OF ECOSYSTEMS

▶ **Grasslands**
Biol. Sci. Rev., 22(2), Nov. 2009, 2-5. *Grasslands are the dominant vegetation over vast regions of the world. The distribution of this immense biome is governed by rainfall and temperature patterns across the globe.*

▶ **Getting to Grips with Ecology**
Biol. Sci. Rev., 22(3), Feb. 2010, pp. 14-16. *A good overview on ecological principles using ladybirds and aphids as example organisms.*

▶ **Fatal Attraction**
National Geographic, 217(3), March 2010, pp. 80-95. *The diversity of insectivorous plants is explored. Beautiful photographs illustrate modifications.*

### ENERGY FLOW & NUTRIENT CYCLES

▶ **All Life is Here**
New Scientist, 24 April 2010, pp. 31-35. *A look at the variation of biodiversity from the tropics to the poles. Biodiversity hotspots are also covered.*

▶ **Meltdown!**
New Scientist, 2 November 2002, pp. 44-48. *How human activity and global warming affects the water cycle.*

▶ **The Case of the Missing Carbon**
National Geographic, 205(2), Feb. 2004, pp. 88-117. *The role of carbon sinks in the Earth's carbon cycling.*

▶ **Ultimate Interface**
New Scientist, 14 November 1998 (Inside Science). *The cycling of the Earth's elements and the influence of human activity on nutrient cycling.*

# Appendix

PERIODICAL REFERENCES

## POPULATIONS

▶ **Population Bombshell**
New Scientist, 11 July 1998 (Inside Science). *Current and predicted growth rates in human populations, including analyses of population age distributions.*

▶ **The Shock of the Old**
New Scientist, 10 April 2010, pp. 26-27. *Human demographics and, in particular, how the human population is aging.*

▶ **Human Population Grows Up**
Scientific American, September 2005, pp. 26-33. *Human demographics, including population movement, age structures and relative wealth.*

▶ **Inside Story**
New Scientist, 29 April 2000, pp. 36-39. *Ecological interactions between fungi and plants and animals: what are the benefits?*

▶ **Classroom: Mark-Recapture with Crickets**
The Am. Biology Teacher, 69(5), May. 2007, pp. 292-297. *A hands-on exercise where students sample crickets to understand about ecological principles and sampling methods.*

▶ **Worm Charmers**
Scientific American, March 2010, pp. 56-59. *The behavioral adaptations earthworms use to escape from hungry moles. It looks at the predator-prey relationship of earthworms and moles.*

## THE DIVERSITY & STABILITY OF ECOSYSTEMS

▶ **Mountain Transformed**
National Geographic, 217(5), Life has returned to Mount St. Helens thirty years after its catastrophic eruption.

▶ **Earth's Nine Lives**
New Scientist, 27 February 2010, pp. 31-35. *Looks at how human activity is threatening the planetary life-support systems that keep Earth viable.*

▶ **Global Warming**
Time, special issue, 9 April 2007. *A special issue on global warming: the causes, perils, solutions, and actions. Comprehensive and engaging.*

▶ **Think or Swim**
New Scientist, 18 September 2010, pp. 40-43. *Sea levels are predicted to rise as a result of global warming. This article looks at the implications of sea level rise on human populations.*

▶ **Behind the Predictions**
New Scientist, 15 January 2011, pp. 38-41. *The Earth is a complex system, so modelling climate change predictions can be very difficult, but the models are getting better.*

▶ **Greenland Poised on a Knife Edge**
New Scientist, 8 January 2011, pp. 8-9. *The glaciers and ice sheets of Greenland have begun to melt because of global warming. They have the potential to raise sea level by 7 meters.*

▶ **A World 4°C Warmer**
New Scientist, 3 Oct 2009, pp. 14-15. *An article outlining some of the potential outcomes if the Earth's temperature was to increase by 4°C.*

▶ **Life At the Edge**
National Geographic, 211(6), Jun. 2007, pp. 32-55. *Rising temperatures are shrinking the Arctic ice sheet. This could impact on the survival of the polar bear.*

▶ **Life At the Edge**
National Geographic, 219(4), April 2011, pp. 100-121. *A look at how the acidification of the oceans is affecting coral and pteropods.*

▶ **Threatening Ocean Life from the Inside Out**
Scientific American, August 2010, pp. 52-59. *Oceans are becoming more acidified as they absorb $CO_2$. This article looks at how sea life is coping with the change, and how the food chains may be affected.*

▶ **Last of the Amazon**
National Geographic, 212(1), January 2007, pp. 40-71. *Deforestation by human activity is rapidly altering the Amazon rain forest. This has implications for the people who live there.*

▶ **A Forest in Tatters and on the Brink**
New Scientist, 7 May 2009, pp. 12. *The Atlantic forest in Brazil is a biodiversity hotspot and home to thousands of different species. It is becoming endangered by deforestation activities.*

▶ **Aliens to the Rescue**
New Scientist, 15 January 2011, pp. 34-37. *Some alien species may actually be good for their new environments. The pros and cons of introduced species are discussed.*

▶ **Tasmanian Devils were Sitting Ducks for Cancer**
New Scientist, 2 July 2011, pp. 10-11. *Low genetic diversity of Tasmanian devils has made them susceptible to catching facial cancer. A conservation programme is underway to save them.*

▶ **Catching Condors in Grand Canyon Country**
New Scientist, 17 December 2011, pp. 14. *Arizona condors have suffered a genetic bottleneck because they were all breed from only 22 birds. Their survival is at risk because of this.*

# Index

# Index